THE CITY CULTURES READER
Second edition

Cities are both products of culture, and sites where culture is made and received. By presenting the very best of classic and contemporary writing on the culture of cities, *The City Cultures Reader* provides an accessible overview of the diverse material on the interface between cities and culture.

The extensively revised and updated second edition of *The City Cultures Reader* now features fifty generous writings (of which thirty-eight are new) organised into ten parts which explore themes such as: what is a city?; what is culture?; symbolic economies; the culture industry; cultures and technologies; everyday lives; contesting identity; boundaries and transgressions; utopias and dystopias; and possible urban futures.

Designed to aid student understanding, this new edition now features extensive introductory sections that define both the city and culture. Part introductions outline the major themes, whilst introductions to the individual writings explain their interest and significance to wider debates. Annotated further reading is also provided at the end of each part.

Malcolm Miles is Reader in Cultural Theory at the University of Plymouth. **Tim Hall** is Senior Lecturer in Human Geography at the Geography and Environmental Management Research Unit, University of Gloucestershire. **Iain Borden** is Professor of Architecture and Urban Culture, and Director of the Bartlett School of Architecture, University College London.

THE ROUTLEDGE URBAN READER SERIES

Series editors

Richard T. LeGates
Professor of Urban Studies, San Francisco State University

Frederic Stout
Lecturer in Urban Studies, Stanford University

The Routledge Urban Reader Series responds to the need for comprehensive coverage of the classic and essential texts that form the basis of intellectual work in the various academic disciplines and professional fields concerned with cities.

The readers focus on the key topics encountered by undergraduates, graduates and scholars in urban studies and allied fields. They discuss the contributions of major theoreticians and practitioners and other individuals, groups, and organizations that study the city or practise in a field that directly affects the city.

As well as drawing together the best of classic and contemporary writings on the city, each reader features extensive general, section and selection introductions prepared by the volume editors to place the selections in context, illustrate relations among topics, provide information on the author and point readers towards additional related bibliographic material.

Each reader will contain:

- Approximately thirty-six *selections* divided into approximately six sections. Almost all of the selections will be previously published works that have appeared as journal articles or portions of books.
- A *general introduction* describing the nature and purpose of the reader.
- Two- to three-page *section introductions* for each section of the reader to place the readings in context.
- A one-page *selection introduction* for each selection describing the author, the intellectual background of the selection, competing views of the subject matter of the selection and bibliographic references to other readings by the same author and other readings related to the topic.
- A plate section with twelve to fifteen plates and illustrations at the beginning of each section.
- An index.

The types of readers and forthcoming titles are as follows:

THE CITY READER

The City Reader: third edition – an interdisciplinary urban reader aimed at urban studies, urban planning, urban geography and urban sociology courses – will be the *anchor urban reader*. Routledge published a first edition of *The City Reader* in 1996 and a second edition in 2000. *The City Reader* has become one of the most widely used anthologies in urban studies, urban geography, urban sociology and urban planning courses in the world.

URBAN DISCIPLINARY READERS

The series will contain *urban disciplinary readers* organized around social science disciplines. The urban disciplinary readers will include both classic writings and recent, cutting-edge contributions to the respective disciplines. They will be lively, high-quality, competitively priced readers which faculty can adopt as course texts and which will also appeal to a wider audience.

TOPICAL URBAN ANTHOLOGIES

The urban series will also include *topical urban readers* intended both as primary and supplemental course texts and for the trade and professional market.

INTERDISCIPLINARY ANCHOR TITLE

The City Reader: third edition
Richard T. LeGates and Frederic Stout (eds)

URBAN DISCIPLINARY READERS

The Urban Geography Reader
Nick Fyfe and Judith Kenny (eds)

The Urban Sociology Reader
Jan Lin and Christopher Mele (eds)

The Urban Politics Reader
Elizabeth Strom and John Mollenkopf (eds)

The Urban and Regional Planning Reader
Eugenie Birch (ed.)

TOPICAL URBAN READERS

The City Cultures Reader: second edition
Malcolm Miles and Tim Hall, with Iain Borden (eds)

The Cybercities Reader
Stephen Graham (ed.)

The Sustainable Urban Development Reader
Stephen M. Wheeler and Timothy Beatley (eds)

The Global Cities Reader
Neil Brenner and Roger Keil (eds)

For further information on The Routledge Urban Reader Series
please contact:

Andrew Mould
Routledge
11 New Fetter Lane
London EC4P 4EE
England
andrew.mould@routledge.co.uk

Richard T. LeGates
Urban Studies Program
San Francisco State University
1600 Holloway Avenue
San Francisco, California 94132
(415) 338-2875
dlegates@sfsu.edu

Frederic Stout
Urban Studies Program
Stanford University
Stanford, California 94305-6050
(650) 725-6321
fstout@stanford.edu

The City Cultures Reader
Second edition

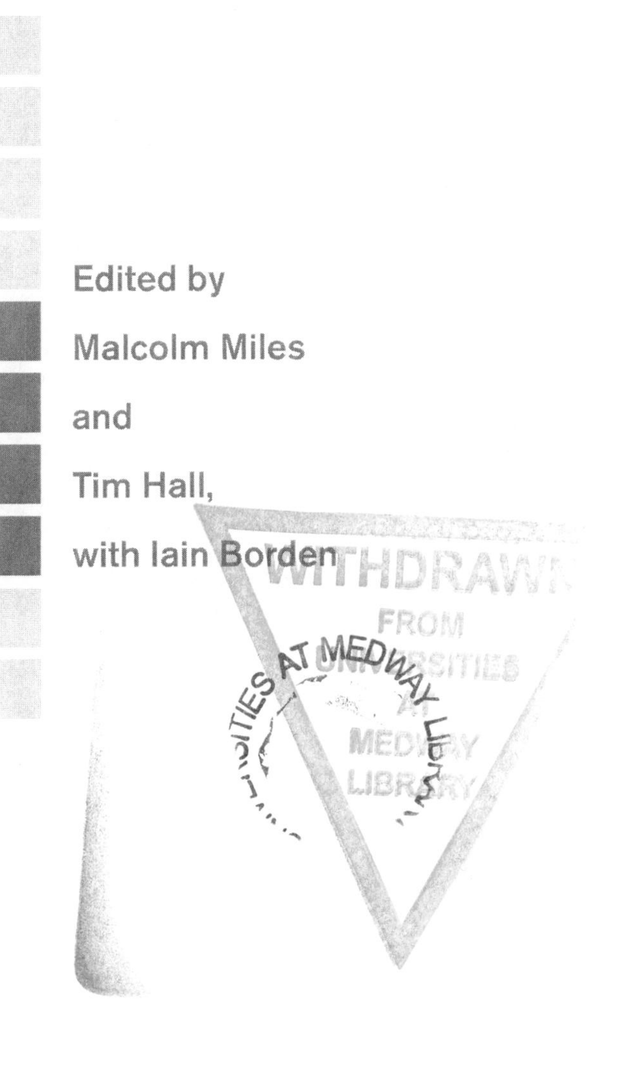

Edited by

Malcolm Miles

and

Tim Hall,

with Iain Borden

LONDON AND NEW YORK

First published 2000
by Routledge
2 Park Square, Milton Park, Abingdon, Oxon OX14 4RN

Simultaneously published in the USA and Canada
by Routledge
270 Madison Ave, New York, NY 10016

Second edition 2004

Transferred to Digital Printing 2008

Routledge is an imprint of the Taylor & Francis Group, an informa business

Designed and typeset in Amasis and Akzidenz Grotesk by
Keystroke, Jacaranda Lodge, Wolverhampton
Printed and bound in Great Britain by
TJI Digital, Padstow, Cornwall

British Library Cataloguing in Publication Data
A catalogue record for this book is available from the British Library

Library of Congress Cataloging in Publication Data
The city cultures reader / [edited by] Malcolm Miles and Tim Hall;
with Iain Borden. – 2nd ed.
p. cm.—(The Routledge urban reader series; no. 3)
Includes bibliographical references and index.
1. Cities and towns. 2. City and town life. 3. Culture. I. Miles, Malcolm.
II. Hall, Tim. III. Borden, Iain. IV. Series.

HT151.C5822 2003
307.76—dc21 2003007419

ISBN 10: 0-415-30244-7 (hbk)
ISBN 10: 0-415-30245-5 (pbk)

ISBN 13: 978-0-415-30244-9 (hbk)
ISBN 13: 978-0-415-30245-6 (pbk)

Contents

Plates

TOUCH SANITATION BY MIERLE LADERMAN UKELES (BETWEEN PAGES 346 AND 347)

RING DER ERINNERUNG BY HERMAN PRIGANN (BETWEEN PAGES 446 AND 447)

Acknowledgements

The Editors would like to express gratitude to rights holders who have granted permission to reprint the texts included in this book, as listed under copyright information (pp. 497–500); and to the artists and galleries who have enabled us to reproduce visual material: John Goto, Marjetica Potrč and the Max Protetch Gallery, Herman Prigann, and Mierle Laderman Ukeles and Ronald Feldman Fine Arts.

We would like to thank Iain Borden, with whom we worked as co-editors on the first edition, for providing us with constructive and helpful feedback on the selection of material and new introductions written for the second edition, and for providing a text for inclusion. We are grateful to Nicola Kirkham for efficiently and patiently carrying out the work of gaining permissions to reprint the texts. And we are grateful to Andrew Mould and his colleagues at Routledge for their encouragement and support in developing a revised, second edition.

INTRODUCTION TO THE SECOND EDITION

This general introduction states the aims and scope of the second edition of *The City Cultures Reader*, says how it differs from the first edition, and explains its arrangement in ten parts. Each part then has a specific introduction establishing its terrain and giving suggestions for further reading. Each text has a short introduction situating it in relation to the author.

AIMS AND SCOPE

The aim of *The City Cultures Reader* is to acquaint students, professionals, and interested lay readers with material from diverse sources at the interface between cities and cultures. The book is organised in a thematic rather than discipline-based structure, and includes the work of four visual artists whose work constitutes a critical commentary on contemporary urban, cultural and social realities. The resulting juxtapositions enable readers working in one field to access material from others, and may extend the scope of critical discussion. The aim is also to balance topical interests, in an area which continues to develop rapidly and in several competing directions, with awareness of some of the histories and theories by which such interests are contextualised. The book's point of departure, however, is the present. Where texts from the early or mid-twentieth century have been included it is because they pose questions which the Editors see as retaining validity now.

The scope of the anthology reflects contemporary discourses of cities and cultures. Some of the texts will be familiar, others obscure. But which texts are familiar and which obscure will vary for readers in different fields. Because the terms cities and cultures have multiple meanings and derive particular nuances from the contexts in which they are used, and because the literatures relevant to each term are extensive, the book has an unusually wide potential scope, while the limit to how many texts can be reprinted is finite (in this case fifty). Within an anglophone set of literatures, then, a selection has been made on the following basis:

- a text should be of sufficient intrinsic interest and clarity that it is worth reprinting;
- it is likely to be of interest in more than one field;
- it creates a synergy with other texts;
- it is likely to retain interest for several years.

Other criteria are more specific: a need to give a rounded picture of a wide-ranging terrain; and to focus on recent and contemporary debates. The book reprints no text dating from prior to the twentieth century, and includes mainly discursive rather than descriptive writing. The ease with which a text could be found elsewhere – in a widely available paperback edition, for instance – was a factor in a few cases. Pragmatic decisions were made, too, about the comparative value for money of publishers' fees for permission to reprint a text. A few texts which might have been expected are thus not here.

Having said the book is intended for readers working at an interface between fields, it is recognised that most are likely to be located in a conventionally defined discipline or profession. The growth of trans-disciplinary research has had little impact so far on the structures of University departments. The book is seen, then, as relevant to readers in architecture, art and design, cultural and urban geographies, cultural planning and policy, cultural studies, urban sociology, and urban studies. A majority of the texts will be useful in undergraduate courses, and a key aspect of the book's scope is in providing a teaching aid at this level. Many of the texts will also be of interest to graduate students working in trans-disciplinary fields. For professionals in an area such as cultural planning, it is hoped the book will offer critical tools for opening fresh enquiry.

DIFFERENCES BETWEEN THE TWO EDITIONS

The second edition of *The City Cultures Reader* differs from the first in several respects. First, it offers a different body of texts. Of the fifty pieces of writing included in this edition only twelve were in the first edition, though several others included there are cited in introductory material here. These changes reflect the large amount of writing on cities and urban cultures which has been published since the list of contents for the first edition was compiled in 1998–9. Fifteen texts published since then are included in this edition. The changes also reflect a wish to give the volume a more rounded feeling by including more material from an earlier period; and to construct a clearer focus on fundamental questions as to what constitutes a city or a culture. A few texts are included in this book which appear also in *The City Reader*, to which the book is a companion volume, when they are helpful in creating a synergy within a group of texts. Second, a number of visual sections have been included in the second edition. These are, like the texts, critical commentaries on contemporary urban conditions but in a visual rather than verbal language. In a book on cities and cultures, in a world where so much communication is visual, no excuse is required for the inclusion of this material. Third, while some very short extracts were given in the first edition we have where possible revised this to include longer texts in the second. Some of the shorter extracts included in the first edition have been omitted to make the necessary space. Fourth, individual introductions have been added for every text, an omission in retrospect in the first edition, and a change bringing the format of this anthology closer to that of *The City Reader*. Fifth, the categories have been revised so that the questions 'what is a city?' and 'what is culture?' are posed at the outset, with longer section introductions than for the remaining sections. Sixth, suggestions for further reading are now given more selectively than in the first edition, after each part introduction. The Editors hope these modifications will make the second edition a book of more durable and scholarly attraction, which will continue to provoke argument as to how cities and cultures shape each other, and what kinds of shaping not only does take place but might do so in futures yet to be determined.

There is one notable absence in the list of contents of this volume (and of the first edition, in that case because permission to reprint a text was refused by the owner of the rights): the work of Walter Benjamin, whose *Arcades Project* is perhaps the most extensive critical commentary to date on an urban cultural phenomenon. Since other representatives of the Frankfurt School are included in the book a word of explanation is necessary. There are two reasons for the decision: first, that it is difficult to adequately extract passages from Benjamin's work without doing violence to the intricacies of his arguments; and second, because there is already available a wide range of paperback editions of his work in English translation, as well as a substantial and still growing critical literature. The extent of this critical literature attests, too, to the need not so much for inclusion of his writing in anthologies as much as for detailed explanation of its arguments and background, if it is to be useful in undergraduate study.

ARRANGEMENT

The book is arranged in ten parts. The first two are keynote, dealing with the questions of what constitutes a city or a culture. The remaining eight articulate a range of broad and to an extent overlapping themes, working loosely in pairs around a set of plates. The arrangement is not chronological, but it seemed logical to put a part on future possibilities at the end, and to link it with another on utopian thought.

The first part asks, then, what is a city. This does not imply that any satisfactory answer to the question is yet available or likely to be. All that can be attempted is to offer a sample of key texts from different disciplines, from comparison of which new insights may arise. The question entails particular kinds of complexity: cities are seen today not simply as accumulations of buildings and spaces but as sites of occupation where processes of identity and cultural formation occur. Those processes lend meaning to built forms and social environments while being at the same time conditioned by them. It is a two-way street, and the part tries to avoid neat definitions of a concept called 'the city'. The second part revolves around different meanings of culture: anthropologically as a collective system of value; as a term for the refinement of the intellect and taste; and more immediately as the production, mediation, and reception of cultural goods in the arts and mass media. Confusions arise when these meanings are used casually, as if they are interchangeable. The part includes texts which assume different meanings for culture, but in such a way as to draw out a relation between them in the context of the growth of cities. Between the first and second parts is a set of photographs of the waterfront in Barcelona. The city has a history of liberal planning from the Cerdà plan of 1859, which continued in the provision of new public spaces in residential neighbourhoods in the late 1980s and early 1990s. The new waterfront indicates a shift towards a more market-based approach as Barcelona seeks to be a world city.

The third and fourth parts deal with symbolic economies and the culture industry. Cultural reception (but not always production) tends to figure centrally in symbolic economies, as cities compete for investment and tourism revenues by re-presenting themselves as vibrant cultural centres. This raises the issue of whose culture is being marketed to whom for what purpose or benefit. While the part on symbolic economies looks at these issues with reference to socio-economic factors, a focus on production and distribution is given in the fourth part, on the culture industry. This term is used by Adorno to refer to so-called popular culture and the mass media. It has its baggage, but so, equally and no less ideologically, does the term cultural and creative industries in vogue today in discussion of culturally led urban redevelopment. Adorno's term is used here because it embodies a critical position. The texts cover both local and global cultural production and reception, and ask in differing ways whose is the culture produced or marketed in today's cultural resurgences. Selected images from John Goto's *Capital Arcade* are reproduced between parts Three and Four. This is one element of a larger work, *Tales of the Twentieth Century*. Originally in full-colour, *Capital Arcade* takes its compositions from art history but reworks them through digital techniques. Similarly, the figures are contemporary types from images taken by Goto. The result is a postmodern equivalent of the European tradition, and in a way of the panorama and the nineteenth-century arcade as well, but in which the only certainty now is consumption.

Part Five looks at cultures and technologies in urban settings and change. It ranges in coverage from modern technologies of mobility, and postmodern technologies in high-tech subcultures, to appropriate or intermediate technologies of communication in a non-affluent context. The part has an underpinning question as to what use different technologies of movement and communication might be for different interests in the shaping of cities and settlements. If information is power, who has access to its dissemination? Part Six extends this concern for the non-spectacular by moving to the area of everyday life, or everyday lives, to emphasise the human rather than conceptual dimension of urban cultures. This draws on an emerging critical literature, together with English translations of sources in French cultural and social theory from the 1970s and 1980s, seen now as marking out a field between architecture, cultural studies and the social sciences. The part includes both kinds of text, and references activities from tango to skateboarding. Between parts Five and Six is a series of works by Slovenian artist Marjetica Potrč. They include plates capturing sightings of bears and a coyote in urban environments in which they have become increasingly if strangely familiar; and installation views of objects which construct a human survival culture, such as a pepper spray for use on bears, and a container of emergency water as issued to unauthorised immigrants crossing the Mexican–US border. Each group in its way lends an uncanny sense to human culture.

Part Seven moves from the setting to the process of identity formation, in terms of groups and individuals within the dominant society. This section has a background in questions of class, gender and ethnicity, but these categories are overtaken today by those of global and local, majority and minority, or centre and margin, while the atomistic self of liberal humanism is increasingly replaced now by awareness of group (and multi-valent) identities. Part Eight follows this by looking at the permeability, or not, of boundaries. The same contested categories of sociation and culture appear, but in this section in terms of practices, or tactics, rather than of broad conceptual frameworks. Where boundaries occur, so do transgressions; or the prevention of transgression through new technologies of surveillance. What emerges is the extent of self-policing on the part of disciplined publics. Between parts Seven and Eight is a series of documentary images of participatory artwork by Mierle Laderman Ukeles. Ukeles adopted the unfunded role of artist-in-residence in the New York Sanitation Department in 1979, and now, officially recognised, has an office in their building in lower Manhattan. Her current work concerns the reclamation of the Fresh Kills land-fill site on Statten Island, but one of her first interventions – shown here – was to walk the city's five boroughs shaking the hands of garbage collectors, touching the hands of those who touch filth. This is more than a gesture. It questions how those who produce garbage perceive the workers who keep the city moving by collecting it, but also interrogates the culture which requires the process of collection and removal, and the collectors, to be more or less invisible.

Finally, parts Nine and Ten look to the duality of utopia–dystopia, and to possible urban futures. There is a substantial literature over five centuries on the ideal of utopia, which tends to take the form of a description of a city. The doom scenario is also well-rehearsed today in disaster movies and fantasies of planetary-scale destruction, and in some writing on cities such as Los Angeles or New York. Elements of utopian literature and philosophy are cited within recent critical writing included here. The part also covers current perceptions of urban or social dysfunctionality and how the actualities which give rise to them can be approached, through perspectives from political theory and planning as well as culture. In the final part, the accent is on alternative approaches to settlement in affluent and non-affluent places. In one case, a specific architectural practice articulates a departure from architectural and planning conventions which, though perhaps idealistic, is a possible future encountered in the present. In another case resistant action is proposed to shape the process of determining whose city is produced. Between the ninth and tenth parts is a sequence of photographs of a work by Herman Prigann, *Ring der Erinnerung* (*Ring of Remembrance*) sited in the Harz mountains, Germany, on the ex-border between the Democratic and Federal Republics. The work consists of a circle of tree trunks piled with earth, broken by four entrances, which will in time be taken over by natural growth. In time, too, memories will mutate – but remain marked by experiences which took place in the border zone.

PART ONE

What Is a City?

INTRODUCTION TO PART ONE

The question 'what is a city?' is fundamental to almost all responses to and experiences of the urban. The engineer sees the city as a problem of circulation, the planner sees order and disorder, the novelist as an accumulation of interconnected stories, the clubber as terrain animated by and through hedonistic experiences, the criminal as opportunities, the pensioner, stranded in decaying public housing in the inner city, as a jungle, but also as the repository of rich, meaningful memories. The art lover sees Paris, Florence and now Bilbao, the stockmarket trader London, New York, Tokyo, the football fan Manchester, Barcelona and Rio de Janeiro, the film buff New York, Los Angeles and perhaps Mumbai.

This part does not seek to answer the question 'what is a city?' This would be impossible because there is no correct answer. Rather, it seeks to open the question up by presenting five extracts from some of the twentieth-century's most important urban thinkers, all of whom consider this question in various ways. There has been no attempt at achieving representativeness in arriving at this sample. Again, this would be impossible, or at least would represent only one perspective. Instead, these extracts touch on some of the fundamental conceptualisations of what a city is, that have been explored by these and other writers on the city. In each of these five extracts the writers conceive of the city in different and often sharply contrasting ways. Before moving to these texts, a little more can be said, however, on the broad conceptualisation of the city in modern times and in the postmodern present.

The archetypal modern city was, perhaps, that imagined in the plans and manifestos of a number of highly influential modern architects, among them Charles Edouard Generet, better known as Le Corbusier – the name he adopted after moving from Switzerland to Paris. Le Corbusier laid a blueprint for the imagination and planning of the modern city in his book *The City of Tomorrow and its Planning* (New York, The Architectural Press, 1947), first published in 1929. Le Corbusier is associated with a certain kind of rationality in town planning, but this can equally be seen as a utopian (or dystopian, according to viewpoint) projection. Critiques of Le Corbusier can be found in John Gold's *The Experience of Modernism* (London, Routledge, 1998), and James Holston's account of Brasilia, *The Modernist City: An Anthropological Critique of Brasilia* (Chicago, University of Chicago Press, 1989). Le Corbusier's approach underpinned much in postwar planning, together with ideas from the Chicago School and to a lesser extent the Bauhaus. There is an idealism in this move to, as it were, engineer a new society by good design, which in one way extends nineteenth-century utopianism, but in another gives very little scope for matters outside design, such as the occupation of spaces by dwellers. The outcome, all too well rehearsed, is the de-humanising environment of the social housing project, which has nothing in common with another key idea about the city, that it is a place of engagement.

That sense of engagement is found in Canon Barnett's vision of a city of interactions, written in 1893–4. Based on his experiences as a clergyman in London and Bristol, Barnet sees in high-density living, despite the realities this tended to produce – of overcrowding and ill health, for instances – in the nineteenth century, a possibility for the optimum condition of human dwelling:

> They forget that the highest possible life for men [*sic*] may be a city life; and that the prophets foresaw, not a paradise or a garden, but a city with its streets and its markets, its manifold interests

and its hum of life . . . We have our neighbours in a city, not the trees and the beasts but fellow human beings. We can from them learn greater lessons, and with them do greater deeds. We can become more human. (Canon S. A. Barnett, *The Ideal City*, ed. H. E. Meller, Leicester, Leicester University Press, [1893–4], 1975, p. 55)

The actual living conditions of the industrial working class were less conducive to it becoming more human through living among others, though by the end of the nineteenth century, after a period of liberal reformism which saw major engineering projects to provide clean water and to separate it from waste products, conditions were not as bad as those described in Manchester by Friedrich Engels in 1844:

under the railway bridge there stands a court, the filth and horrors of which surpass all the others by far, just because it was hitherto so cut off, so secluded that the way to it could not be found without a good deal of trouble, I should never have discovered it myself, without the breaks made by the railway, though I thought I knew this whole region thoroughly. Passing along a rough bank, among stakes and washing lines, one penetrates into this chaos of small one-storied, one-roomed huts, in most of which there is no artificial floor; kitchen, living and sleeping-room all in one. In such a hole, scarcely five feet long by six feet broad, I found two beds – and such bedsteads and beds! – which, with a staircase and chimney-place, exactly filled the room. In several others I found absolutely nothing, while the door stood open, and the inhabitants leaned against it. Everywhere before the doors refuse and offal; that any sort of pavement lay underneath could not be seen but only felt, here and there, with the feet. (from Friedrich Engels, *The Condition of the Working Class in England, 1844*, London, Allen and Unwin, [1845] 1892, cited in James Donald, *Imagining the Urban*, London, Athlone, 1999, p. 35.)

The modernist vision failed to deliver what would have been in many ways an antithesis of the conditions described by Engels. A key voice for a re-integration of city life at street level was Jane Jacobs, a writer more than a professional planner whose *The Death and Life of Great American Cities* (New York, Random House, 1961) is frequently cited. Based on her feeling for the informality of city living in New York's Greenwich Village and Boston's North End, this book argues that city form and spaces should provide more than narrowly defined optimal solutions to certain problems such as the distribution of people, traffic and amenities. Her ideal city, in as much as there is or can be one, is, primarily, a populated city. A similar feeling is found in Elizabeth Wilson's *The Sphinx in the City* (Berkeley, University of California Press, 1993). Wilson writes:

It is essential to acknowledge that city centres have become increasingly places of paradox: playgrounds for the rich, but dustbins for the very poor. Nor do I have a solution for the complex problems facing city dwellers and city governments. I am arguing that we will never solve the problems of cities unless we *like* the urban-ness of urban life. Cities aren't villages; they aren't machines; they aren't works of art; and they aren't telecommunications stations. They are spaces for face to face contact of amazing variety and richness. They are spectacle – and what's wrong with that? (*The Sphinx in the City*, p. 158)

Within this complexity are many layers, and several different but overlapping forms of spatial organisation. These do not form neat zones, as if lines on a map corresponded to the intricacies of human sociation, but produce a palimpsest of constantly over-written and changing patterns. Peter Marcuse describes such a multi-layered city of quarters which are social and cultural rather than geographical in 'The Layered City':

The quartering of the city runs along a number of dimensions . . . for a city like New York, the key ones that may be experienced run according to lines of race, of class, of occupation, of ethnicity. While these are correlated among themselves, they are also overlapping in their spatial configuration. The result is not only a quartered city but a layered city, in which one line of division overlaps another, sometimes creating congruent quarters, sometimes not. (Peter Marcuse, 'The Layered City', in *The Urban Lifeworld*, edited by Pete Madsen and Richard Plunz, London, Routledge, 2002, pp. 94–114)

Perhaps this is the most accurate description of the postmodern city, which is increasingly the post-industrial city. It is a long way from Le Corbusier's sweeping gestures of urban cleansing.

The study of urban form, and especially the creation of models of urban structure, has been a major preoccupation of sociologists and subsequently geographers for over one hundred years, from the work of the Chicago School sociologists in the early twentieth century to more recent mappings of the postmodern city by Edward Soja and Witold Rybczynski. Without doubt the writer who most influenced this strand of urban analysis was Ernest W. Burgess. Many since have returned to and explored the notion that there is an underlying logic to the apparent chaos of the city, even if they did not agree with Burgess's arguments about what processes constituted this logic. One of Burgess's most famous pieces of writing 'The Growth of the City: An Introduction to a Research Project' is included here. It was in this extract that Burgess outlined his ecological model of urban land use and social segregation, based on fieldwork in Chicago in the 1920s. This is more famously known as the 'concentric zone model', an ideal-type representation of the city.

This model embodied Burgess's two key ideas about the city. These were that the city is a dynamic organism and that social processes and physical form are interrelated. Burgess's thinking was based on the idea that the city expanded through a process of ecological succession. Namely, as immigrant groups became established in the city they made their way from low rent districts surrounding the city centre to better, more expensive neighbourhoods further out. As this outward movement occurred more immigrants would enter the lowest rent district to take up vacated spaces there. This process of expansion, therefore, caused the city to grow outwards in a series of largely homogeneous concentric zones. The model was based on the notion of ecological competition for land use between different groups of residents with differential resources at their disposal. The model was used by Burgess, others in the Sociology department at the University of Chicago, most notably Robert Park, and their students to explain the existence of social problems within certain districts in the city.

One of the major influences on the Chicago School sociologists was Georg Simmel. Simmel was concerned with the question of whether there is a distinctive urban culture. This, like the physical forms of cities, is a question that has been at the centre of much urban enquiry. Another hugely influential essay concerned with the definition of urban culture, that can be contrasted with Simmel's, is Louis Wirth's 'Urbanism As a Way of Life' written in 1938 and included in Richard T. LeGates and Frederic Stout *The City Reader* (third edition, London, Routledge, 2004). Wirth, a latter member of the Department of Sociology at the University of Chicago, attempted a spatial definition of culture, contrasting urban culture with rural culture by identifying three 'independent variables' – size, density and heterogeneity – which he argued caused a distinctive urban culture. While Wirth failed to sustain his thesis, Simmel, by contrast, proposed a temporal definition of urban culture, differentiating it from earlier cultural formations.

Simmel's most famous and influential essay was the classic 'Metropolis and Mental Life' published in 1903, an extract from which is included here. In this essay it was not cities *per se* that Simmel argued created a distinctive culture but rather the dominance of money in modern society. Money, Simmel argued, is inherently instrumental and consequently modern society is intellectual, instrumental, blasé and reserved. The equation in Simmel's work between urban culture and the culture of modernity derived from his belief that the dominance of money was most well developed within cities and hence cities display these traits to a greater degree than rural areas and smaller settlements. Simmel certainly did not see rural areas as immune from the traits of modern culture, which he saw penetrating beyond the boundaries of the city. Unlike Wirth, Simmel was concerned with highlighting contrasts between modern society and earlier more traditional, non-money societies. Simmel argued that social ties in modern society were weak and people lonely and alienated. Their everyday exchanges were like the exchanges of money commodities – impersonal, distant and cold – while other aspects of city life (notably time and punctuality) are rendered into matters of quantity rather than quality. He saw this as deriving from the separation of subject and object, a retreat into intellectuality as a defensive response against the sensory overload characteristic of the modern urban experience. This has been something of a key theme amongst many writers on the sociology of modernity. Simmel's ideas in this essay are more fully worked out in what he regarded as his most important work, *The Philosophy of Money*, published in 1900.

Of the influential sociologists of modernity Siegfried Kracauer is one of the less well known. However, despite this, he occupied something of a central position within the evolution of sociologies of modernity and urban culture. He was taught by Georg Simmel and was heavily influenced by Walter Benjamin and was himself a teacher of Theodor Adorno. In the extract 'The Hotel Lobby' from *The Mass Ornament*, originally published in 1927, Kracauer draws on the influence of Simmel and uses the spatial type of the hotel lobby not so much to explain this kind of internalised architecture in itself but as a means to understand society at large. Hence the lobby appears as a place of silence, avoided eye contact, distant relations, unreality, anonymity and insubstantial aesthetics – all aspects paralleled in Kracauer's view of the meaningless existence of modern life, governed by a dominant 'ratio' or rationality.

Lewis Mumford's concern in his considerable output on the city was with the relationship between cities, human culture and personality. Mumford is one of the most widely read and famous writers on the city to be published during the twentieth century. His writing career spanned sixty years from his first book in 1922. Mumford's perspective was humanist and he saw both the potentially negative consequences of urban development, he was a lifelong opponent of mega developments, most notably those of Robert Moses in New York, and the liberating and uplifting potentials of cities and urban life. The extract here, 'What is a City?' (1937), contains many of Mumford's central concerns. In it he outlines his fundamental principles for the planning of cities if the potential of urban life is to be achieved. His account prefigures later accounts by people such as Jane Jacobs and Don Appleyard, who recognised the importance of the theatricality of diversity encouraged by sympathetic urban design and planning. Mumford frequently provided an answer to the question 'what is a city?' in his writings where he talked of it as a theatre. The monolithic, uniform mega redevelopment of cities, characteristic of the postwar American highway city, was anathema to Mumford.

Of course 'what a city is' depends on who you are. In the most recent extract in this section Elizabeth Wilson reflects current concerns with inequality and division within the city by exploring the dimension of gender. In a chapter from her book *The Sphinx in the City* (Berkeley, University of California Press, 1991), she considers cities of Latin America and Africa. Wilson notes that the severe infrastructural and architectural deficiencies of, for example, shanty towns, may also be countered by easy social contacts and supportive friendships and familial relations. In terms of gender, and arguing that the everyday life of cities provides both dangers and exhilaration for women, Wilson seeks in places opportunities to break out of constricting family stereotypes, work patterns, community activities and architectural design processes. Wilson's book was an attempt to map the city from a variety of women's viewpoints and acted as a powerful critique of various forms of utopian planning and the ways that they acted to constrain women, ethnic minorities and the working class. As well as being a critique, the power of Wilson's influential thesis lay in its imagination of possible alternative futures. As she concludes, the 'gulf between what is and what might be may appear to widen; on the other hand, the city both raises aspirations and gives more chance of their realisation'.

Each of the writers in this part reflects, at least implicitly, on the question 'what is a city?' They explore it as a dynamic organism, a stage, a form of consciousness, a realm of experience, a physical entity, as both rational and informal and as both liberating and constraining. There are many more answers to the question. While some of the writers presented in this part represent some of the most significant writers on the city of the twentieth century they do not provide a definite answer; rather, their writing opens up the possibilities that the question offers.

SUGGESTED FURTHER READING

James Donald, *Imagining the Modern City*, London, Athlone, 1999
Arie Graafland and Deborah Hauptmann, eds, *Cities in Transition*, Rotterdam, 010 Publishers, 2001

Peter Marcuse, 'The Layered City', in *The Urban Lifeworld: Formation, Perception, Representation*, edited by Peter Madsen and Richard Plunz, London, Routledge, 2002, pp. 94–114

Michael Paccione, ed., *Britain's Cities: Geographies of Division in Urban Britain*, London, Routledge, 1997

Heinz Paetzold, ed., *City Life: Essays on Urban Culture*, Maastricht, Jan van Eyck Akademie, 1997

Steve Pile, 'The Un(known) City . . . or, an Urban Geography of What Lies Buried Below the Surface'; and William Menking, 'From Tribeca to Triburbia: A New Concept of the City', both in *The Unknown City: Contesting Architecture and Social Space*, edited by Iain Borden *et al.* Cambridge, MA, MIT, 2001, pp. 262–79

'The Metropolis and Mental Life'

from *The Sociology of Georg Simmel* (1950) [1903]

Georg Simmel

Editors' Introduction

Georg Simmel (1858–1918) was a major figure in German philosophy and sociology who did much to establish urban culture as a legitimate object of academic study. His influence was significant both during his own lifetime and in the years since his death. Simmel's work was a major influence on the Chicago School, especially Robert Park, as well as on the evolution of Marxist philosophy through figures such as Ernst Bloch. His work and its legacy is still widely read and debated and he is recognised as one of the most significant urban theorists of the twentieth century. Despite a prolific writing career, Simmel is best known for his work on urban culture and particularly his essay 'Metropolis and mental life' an extract from which follows.

Born in Berlin, Simmel was clearly influenced by the modern urban culture unfolding around him in the latter half of the nineteenth century. Simmel's expertise spanned a vast field from history and philosophy to the social sciences. This catholicism was apparent by the time he received his doctorate in philosophy from the University of Berlin in 1881. His thesis was entitled 'The Nature of Matter According to Kant's Physical Monadology'. While remaining a popular *Privatdozent* – effectively an unpaid lecturer whose income derived from student fees and who was unable to take part in many of the affairs of the academic community – at the University of Berlin for fifteen years, Simmel was something of an outsider academically. The position of *Privatdozent* marked him as only marginal to the life and affairs of the academic community. Despite being internationally famous, widely published and highly regarded by numerous leading scholars, Simmel failed to secure a senior position at any German university until he was appointed a professor at the University of Strasbourg in 1914. One theory for this is that many within the academic community felt threatened by the ground-breaking brilliance of Simmel's work.

Simmel was a hugely prolific writer who addressed a variety of audiences on numerous topics. Over 200 of his articles appeared in academic journals, newspapers and magazines and many more were published after his death. In addition to this, he produced twenty-one books covering sociology, philosophy and cultural criticism. His two most significant works were *The Philosophy of Money* (1900) which expanded many of the themes in 'Metropolis and mental life' and *Sociology: Investigations on the Forms of Sociation* (1908).

One of Simmel's most perceptive advocates is David Frisby. He discusses Simmel's work in *Fragments of Modernity: Theories of Modernity in the Work of Simmel, Kracauer and Benjamin* (Cambridge, Polity, 1985) and situates it amongst a range of other responses to and representations of the modern metropolis in *Cityscapes of Modernity*, (Oxford, Blackwell, 2001). In addition Wirth's and Simmel's work are discussed together in Mike Savage, Alan Warde and Kevin Ward's *Urban Sociology, Capitalism and Modernity* (London, Palgrave, 2002).

The deepest problems of modern life derive from the claim of the individual to preserve the autonomy and individuality of his existence in the face of overwhelming social forces, of historical heritage, of external culture, and of the technique of life. The fight with nature which primitive man has to wage for his *bodily* existence attains in this modern form its latest transformation. The eighteenth century called upon man to free himself of all the historical bonds in the state and in religion, in morals and in economics. Man's nature, originally good and common to all, should develop unhampered. In addition to more liberty, the nineteenth century demanded the functional specialization of man and his work; this specialization makes one individual incomparable to another, and each of them indispensable to the highest possible extent. However, this specialization makes each man the more directly dependent upon the supplementary activities of all others. Nietzsche sees the full development of the individual conditioned by the most ruthless struggle of individuals; socialism believes in the suppression of all competition for the same reason. Be that as it may, in all these positions the same basic motive is at work: the person resists being levelled down and worn out by a social-technological mechanism. An enquiry into the inner meaning of specifically modern life and its products, into the soul of the cultural body, so to speak, must seek to solve the equation which structures like the metropolis set up between the individual and the supraindividual contents of life. Such an enquiry must answer the question of how the personality accommodates itself in the adjustments to external forces. This will be my task today.

The psychological basis of the metropolitan type of individuality consists in the *intensification of nervous stimulation* which results from the swift and uninterrupted change of outer and inner stimuli. Man is a differentiating creature. His mind is stimulated by the difference between a momentary impression and the one which preceded it. Lasting impressions, impressions which differ only slightly from one another, impressions which take a regular and habitual course and show regular and habitual contrasts – all these use up, so to speak, less consciousness than does the rapid crowding of changing images, the sharp discontinuity in the grasp of a single glance, and the unexpectedness of onrushing impressions. These are the psychological conditions which the metropolis creates. With each crossing of the street, with the tempo and multiplicity of economic, occupational and social life, the city sets up a deep contrast with small town and rural life with reference to the sensory foundations of psychic life. The metropolis exacts from man as a discriminating creature a different amount of consciousness than does rural life. Here the rhythm of life and sensory mental imagery flows more slowly, more habitually, and more evenly. Precisely in this connection the sophisticated character of metropolitan psychic life becomes understandable – as over against small town life which rests more upon deeply felt and emotional relationships. These latter are rooted in the more unconscious layers of the psyche and grow most readily in the steady rhythm of uninterrupted habituations. The intellect, however, has its locus in the transparent, conscious, higher layers of the psyche; it is the most adaptable of our inner forces. In order to accommodate to change and to the contrast of phenomena, the intellect does not require any shocks and inner upheavals; it is only through such upheavals that the more conservative mind could accommodate to the metropolitan rhythm of events. Thus the metropolitan type of man – which, of course, exists in a thousand individual variants – develops an organ protecting him against the threatening currents and discrepancies of his external environment which would uproot him. He reacts with his head instead of his heart. In this an increased awareness assumes the psychic prerogative. Metropolitan life, thus, underlies a heightened awareness and a predominance of intelligence in metropolitan man. The reaction to metropolitan phenomena is shifted to that organ which is least sensitive and quite remote from the depth of the personality. Intellectuality is thus seen to preserve subjective life against the overwhelming power of metropolitan life, and intellectuality branches out in many directions and is integrated with numerous discrete phenomena.

The metropolis has always been the seat of the money economy. Here the multiplicity and concentration of economic exchange gives an importance to the means of exchange which the scantiness of rural commerce would not have allowed. Money economy and the dominance of the intellect are intrinsically connected. They share a matter-of-fact attitude in dealing with men and with things; and, in this attitude, a formal justice is often coupled with an inconsiderate hardness. The intellectually sophisticated person is indifferent to all genuine individuality, because relationships and reactions result from it which cannot be exhausted with logical operations. In the same manner, the individuality of phenomena is not commensurate with the pecuniary principle. Money is concerned only with what is common to all: it asks for the exchange value, it reduces all quality and individuality to the question: How much? All intimate emotional relations between persons are founded in their individuality, whereas in rational relations man is reckoned with like a number, like an element which is in itself indifferent. Only the objective measurable achievement is of interest. Thus metropolitan man reckons with his merchants and customers, his domestic servants and often even with persons with whom he is obliged to have social intercourse. These features of intellectuality contrast with the nature of the small circle in which the inevitable knowledge of individuality as inevitably produces a warmer tone of behaviour, a behaviour which is beyond a mere objective balancing of service and return. In the sphere of the economic psychology of the small group it is of importance that under primitive conditions production serves the customer who orders the good, so that the producer and the consumer are acquainted. The modern metropolis, however, is supplied almost entirely by production for the market, that is, for entirely unknown purchasers who never personally enter the producer's actual field of vision. Through this anonymity the interests of each party acquire an unmerciful matter-of-factness; and the intellectually calculating economic egoisms of both parties need not fear any deflection because of the imponderables of personal relationships. The money economy dominates the metropolis; it has displaced the last survivals of domestic production and the direct barter of goods; it minimizes, from day to day, the amount of work ordered by customers. The matter-of-fact attitude is obviously so intimately interrelated with the money economy, which is dominant in the metropolis, that nobody can say whether the intellectualistic mentality first promoted the money economy or whether the latter determined the former. The metropolitan way of life is certainly the most fertile soil for this reciprocity, a point which I shall document merely by citing the dictum of the most eminent English constitutional historian: throughout the whole course of English history, London has never acted as England's heart but often as England's intellect and always as her moneybag!

In certain seemingly insignificant traits, which lie upon the surface of life, the same psychic currents characteristically unite. Modern mind has become more and more calculating. The calculative exactness of practical life which the money economy has brought about corresponds to the ideal of natural science: to transform the world into an arithmetic problem, to fix every part of the world by mathematical formulas. Only money economy has filled the days of so many people with weighing, calculating, with numerical determinations, with a reduction of qualitative values to quantitative ones. Through the calculative nature of money a new precision, a certainty in the definition of identities and differences, an unambiguousness in agreements and arrangements has been brought about in the relations of life elements – just as externally this precision has been effected by the universal diffusion of pocket watches. However, the conditions of metropolitan life are at once cause and effect of this trait. The relationships and affairs of the typical metropolitan usually are so varied and complex that without the strictest punctuality in promises and services the whole structure would break down into an inextricable chaos. Above all, this necessity is brought about by the aggregation of so many people with such differentiated interests, who must integrate their relations and activities into a highly complex organism. If all clocks and watches in Berlin would suddenly go wrong in different ways, even if only by one hour, all economic life and communication of the city would be disrupted for a long time. In addition, an apparently mere external factor – long distances – would make all waiting and broken appointments result in an ill-afforded waste of time. Thus, the technique of metropolitan life is unimaginable without the most punctual integration of all activities and mutual relations into a stable and impersonal time schedule. Here again the general conclusions of this entire task of reflection become obvious, namely, that from each point on the surface of existence – however closely attached to the surface alone – one may drop a sounding into the depth of the psyche so that all the most banal externalities of life finally are connected with the ultimate decisions concerning the meaning and style of life. Punctuality, calculability, exactness are forced upon life by the complexity and extension of metropolitan existence and are not only most intimately connected with its money economy and intellectualistic character. These traits must also colour the contents of life and favour the exclusion of those irrational, instinctive, sovereign traits and impulses which aim at determining the mode of life from within, instead of receiving the general and precisely schematized form of life from without. Even though sovereign types of personality, characterized by irrational impulses, are by no means impossible in the city, they are, nevertheless, opposed to typical city life. The passionate hatred of men like Ruskin and Nietzsche for the metropolis is understandable in these terms. Their natures discovered the value of life alone in the unschematized existence which cannot be defined with precision for all alike. From the same source of this hatred of the metropolis surged their hatred of money economy and of the intellectualism of modern existence.

The same factors which have thus coalesced into the exactness and minute precision of the form of life have coalesced into a structure of the highest impersonality; on the other hand, they have promoted a highly personal subjectivity. There is perhaps no psychic phenomenon which has been so unconditionally reserved to the metropolis as has the blasé attitude. The blasé attitude results first from the rapidly changing and closely compressed contrasting stimulations of the nerves. From this, the enhancement of metropolitan intellectuality, also, seems originally to stem. Therefore, stupid people who are not intellectually alive in the first place usually are not exactly blasé. A life in boundless pursuit of pleasure makes one blasé because it agitates the nerves to their strongest reactivity for such a long time that they finally cease to react at all. In the same way, through the rapidity and contradictoriness of their changes, more harmless impressions force such violent responses, tearing the nerves so brutally hither and thither that their last reserves of strength are spent; and if one remains in the same milieu they have no time to gather new strength. An incapacity thus emerges to react to new sensations with the appropriate energy. This constitutes that blasé attitude which, in fact, every metropolitan child shows when compared with children of quieter and less changeable milieus.

This physiological source of the metropolitan blasé attitude is joined by another source which flows from the money economy. The essence of the blasé attitude consists in the blunting of discrimination. This does not mean that the objects are not perceived, as is the case with the half-wit, but rather that the meaning and differing values of things, and thereby the things themselves, are experienced as insubstantial. They appear to the blasé person in an evenly flat and gray tone; no one object deserves preference over any other. This mood is the faithful subjective reflection of the completely internalized money economy. By being the equivalent to all the manifold things in one and the same way, money becomes the most frightful leveller. For money expresses all qualitative differences of things in terms of 'how much?' Money, with all its colourlessness and indifference, becomes the common denominator of all values; irreparably it hollows out the core of things, their individuality, their specific value, and their incomparability. All things float with equal specific gravity in the constantly moving stream of money. All things lie on the same level and differ from one another only in the size of the area which they cover. In the individual case this coloration, or rather discoloration, of things through their money equivalence may be unnoticeably minute. However, through the relations of the rich to the objects to be had for money, perhaps even through the total character which the mentality of the contemporary public everywhere imparts to these objects, the exclusively pecuniary evaluation of objects has become quite considerable. The large cities, the main seats of the money exchange, bring the purchasability of things to the fore much more impressively than do smaller localities. That is why cities are also the genuine locale of the blasé attitude. In the blasé attitude the concentration of men and things stimulate the nervous system of the individual to its highest achievement so that it attains its peak. Through the mere quantitative intensification of the same conditioning factors this achievement is transformed into its opposite and appears in the peculiar adjustment of the blasé attitude. In this phenomenon the nerves find in the refusal to react to their stimulation the last possibility of accommodating to the contents and forms of metropolitan life. The self-preservation of certain personalities is bought at the price of devaluating the whole objective world, a devaluation which in the end unavoidably drags one's own personality down into a feeling of the same worthlessness.

Whereas the subject of this form of existence has to come to terms with it entirely for himself, his self-preservation in the face of the large city demands from him a no less negative behaviour of a social nature. This mental attitude of metropolitans toward one another we may designate, from a formal point of view, as reserve. If so many inner reactions were responses to the continuous external contacts with innumerable people as are those in the small town, where one knows almost everybody one meets and where one has a positive relation to almost everyone, one would be completely atomized internally and come to an unimaginable psychic state. Partly this psychological fact, partly the right to distrust which men have in the face of the touch-and-go elements of metropolitan life, necessitates our reserve. As a result of this reserve we frequently do not even know by sight those who have been our neighbours for years. And it is this reserve which in the eyes of the small-town people makes us appear to be cold and heartless. Indeed, if I do not deceive myself, the inner aspect of this outer reserve is not only indifference but, more often than we are aware, it is a slight aversion, a mutual strangeness and repulsion, which will break into hatred and fight at the moment of a closer contact, however caused. The whole inner organization of such an extensive communicative life rests upon an extremely varied hierarchy of sympathies, indifferences, and aversions of the briefest as well as of the most permanent nature. The sphere of indifference in this hierarchy is not as large as might appear on the surface. Our psychic activity still responds to almost every impression of somebody else with a somewhat distinct feeling. The unconscious, fluid and changing character of this impression seems to result in a state of indifference. Actually this indifference would be just as unnatural as the diffusion of indiscriminate mutual suggestion would be unbearable. From both these typical dangers of the metropolis, indifference and indiscriminate suggestibility, antipathy protects us. A latent antipathy and the preparatory stage of practical antagonism effect the distances and aversions without which this mode of life could not at all be led. The extent and the mixture of this style of life, the rhythm of its emergence and disappearance, the forms in which it is satisfied – all these, with the unifying motives in the narrower sense, form the inseparable whole of the metropolitan style of life. What appears in the metropolitan style of life directly as dissociation is in reality only one of its elemental forms of socialization.

This reserve with its overtone of hidden aversion appears in turn as the form or the cloak of a more general mental phenomenon of the metropolis: it grants to the individual a kind and an amount of personal freedom which has no analogy whatsoever under other conditions. The metropolis goes back to one of the large developmental tendencies of social life as such, to one of the few tendencies for which an approximately universal formula can be discovered. The earliest phase of social formations found in historical as well as in contemporary social structures is this: a relatively small circle firmly closed against neighbouring, strange, or in some way antagonistic circles. However, this circle is closely coherent and allows its individual members only a narrow field for the development of unique qualities and free, self-responsible movements. Political and kinship groups, parties and religious associations begin in this way. The self-preservation of very young associations requires the establishment of strict boundaries and a centripetal unity. Therefore they cannot allow the individual freedom and unique inner and outer development. From this stage, social development proceeds at once in two different, yet corresponding, directions. To the extent to which the group grows – numerically, spatially, in significance and in content of life – to the same degree the group's direct, inner unity loosens, and the rigidity of the original demarcation against others is softened through mutual relations and connections. At the same time, the individual gains freedom of movement, far beyond the first jealous delimitation. The individual also gains a specific individuality to which the division of labour in the enlarged group gives both occasion and necessity. The state and Christianity, guilds and political parties, and innumerable other groups have developed according to this formula, however much, of course, the special conditions and forces of the respective groups have modified the general scheme. This scheme seems to me distinctly recognizable also in the evolution of individuality within urban life. The small-town life in antiquity and in the Middle Ages set barriers against movement and relations of the individual toward the outside, and it set up barriers against individual independence and differentiation within the individual self. These barriers were such that under them modern man could not have breathed. Even today a metropolitan man who is placed in a small town feels a restriction similar, at least, in kind. The smaller the circle which forms our milieu is, and the more restricted those relations to others are which dissolve the boundaries of the individual, the more anxiously the circle guards the achievements, the conduct of life, and the outlook of the individual, and the more readily a quantitative and qualitative specialization would break up the framework of the whole little circle.

The ancient *polis* in this respect seems to have had the very character of a small town. The constant threat to its existence at the hands of enemies from near and afar effected strict coherence in political and military respects, a supervision of the citizen by the citizen, a jealousy of the whole against the individual whose particular life was suppressed to such a degree that he could compensate only by acting as a despot in his own household. The tremendous agitation and excitement, the unique colourfulness of Athenian life, can perhaps be understood in terms of the fact that a people of incomparably individualized personalities struggled against the constant inner and outer pressure of a de-individualizing small town. This produced a tense atmosphere in which the weaker individuals were suppressed and those of stronger natures were incited to prove themselves in the most passionate manner. This is precisely why it was that there blossomed in Athens what must be called, without defining it exactly, 'the general human character' in the intellectual development of our species. For we maintain factual as well as historical validity for the following connection: the most extensive and the most general contents and forms of life are most intimately connected with the most individual ones. They have a preparatory stage in common, that is, they find their enemy in narrow formations and groupings, the maintenance of which places both of them into a state of defence, against expanse and generality lying without and the freely moving individuality within. Just as in the feudal age, the 'free' man was the one who stood under the law of the land, that is, under the law of the largest social orbit, and the unfree man was the one who derived his right merely from the narrow circle of a feudal association and was excluded from the larger social orbit – so today metropolitan man is 'free' in a spiritualized and refined sense, in contrast to the pettiness and prejudices which hem in the small-town man. For the reciprocal reserve and indifference and the intellectual life conditions of large circles are never felt more strongly by the individual in their impact upon his independence than in the thickest crowd of the big city. This is because the bodily proximity and narrowness of space makes the mental distance only the more visible. It is obviously only the obverse of this freedom if, under certain circumstances, one nowhere feels as lonely and lost as in the metropolitan crowd. For here as elsewhere it is by no means necessary that the freedom of man be reflected in his emotional life as comfort.

It is not only the immediate size of the area and the number of persons which because of the universal historical correlation between the enlargement of the circle and the personal inner and outer freedom, has made the metropolis the locale of freedom. It is rather in transcending this visible expanse that any given city becomes the seat of cosmopolitanism. The horizon of the city expands in a manner comparable to the way in which wealth develops; a certain amount of property increases in a quasi-automatical way in ever more rapid progression. As soon as a certain limit has been passed, the economic, personal, and intellectual relations of the citizenry, the sphere of intellectual predominance of the city over its hinterland, grow as in geometrical progression. Every gain in dynamic extension becomes a step, not for an equal, but for a new and larger extension. From every thread spinning out of the city, ever new threads grow as if by themselves, just as within the city the unearned increment of ground rent, through the mere increase in communication, brings the owner automatically increasing profits. At this point, the quantitative aspect of life is transformed directly into qualitative traits of character. The sphere of life of the small town is, in the main, self-contained and autarchic. For it is the decisive nature of the metropolis that its inner life overflows by waves into a far-flung national or international area. Weimar is not an example to the contrary, since its significance was hinged upon individual personalities and died with them; whereas the metropolis is indeed characterized by its essential independence even from the most eminent individual personalities. This is the counterpart to the independence, and it is the price the individual pays for the independence, which he enjoys in the metropolis. The most significant characteristic of the metropolis is this functional extension beyond its physical boundaries. And this efficacy resets in turn and gives weight, importance, and responsibility to metropolitan life. Man does not end with the limits of his body or the area comprising his immediate activity. Rather is the range of the person constituted by the sum of effects emanating from him temporally and spatially. In the same way, a city consists of its total effects which extend beyond its immediate confines. Only this range is the city's actual extent in which its existence is expressed. This fact makes it obvious that individual freedom, the logical and historical complement of such extension, is not to be understood only in the negative sense of mere freedom of mobility and elimination of prejudices and petty philistinism. The essential point is that the particularity and incomparability, which ultimately every human being possesses, be somehow expressed in the working out of a way of life. That we follow the laws of our own nature – and this after all is freedom – becomes obvious and convincing to ourselves and to others only if the expressions of this nature differ from the expressions of others. Only our unmistakability proves that our way of life has not been superimposed by others.

Cities are, first of all, seats of the highest economic division of labour. They produce thereby such extreme phenomena as in Paris the renumerative occupation of the *quatorzième*. They are persons who identify themselves by signs on their residences and who are ready at the dinner hour in correct attire, so that they can be quickly called upon if a dinner party should consist of thirteen persons. In the measure of its expansion, the city offers more and more the decisive conditions of the division of labour. It offers a circle which through its size can absorb a highly diverse variety of services. At the same time, the concentration of individuals and their struggle for customers compel the individual to specialize in a function from which he cannot be readily displaced by another. It is decisive that city life has transformed the struggle with nature for livelihood into an inter-human struggle for gain, which here is not granted by nature but by other men. For specialization does not flow only from the competition for gain but also from the underlying fact that the seller must always seek to call forth new and differentiated needs of the lured customer. In order to find a source of income which is not yet exhausted, and to find a function which cannot readily be displaced, it is necessary to specialize in one's services. This process promotes differentiation, refinement, and the enrichment of the public's needs, which obviously must lead to growing personal differences within this public.

All this forms the transition to the individualization of mental and psychic traits which the city occasions in proportion to its size. There is a whole series of obvious causes underlying this process. First, one must meet the difficulty of asserting his own personality within the dimensions of metropolitan life. Where the quantitative increase in importance and the expense of energy reach their limits, one seizes upon qualitative differentiation in order somehow to attract the attention of the social circle by playing upon its sensitivity for differences. Finally, man is tempted to adopt the most tendentious peculiarities, that is, the specifically metropolitan extravagances of mannerism, caprice, and preciousness. Now, the meaning of these extravagances does not at all lie in the contents of such behaviour, but rather in its form of 'being different', of standing out in a striking manner and thereby attracting attention. For many character types, ultimately the only means of saving for themselves some modicum of self-esteem and the sense of filling a position is indirect, through the awareness of others. In the same sense a seemingly insignificant factor is operating, the cumulative effects of which are, however, still noticeable. I refer to the brevity and scarcity of the interhuman contacts granted to the metropolitan man, as compared with social intercourse in the small town. The temptation to appear 'to the point', to appear concentrated and strikingly characteristic, lies much closer to the individual in brief metropolitan contacts than in an atmosphere in which frequent and prolonged association assures the personality of an unambiguous image of himself in the eyes of the other.

The most profound reason, however, why the metropolis conduces to the urge for the most individual personal existence – no matter whether justified and successful – appears to me to be the following: the development of modern culture is characterized by the preponderance of what one may call the 'objective spirit' over the 'subjective spirit' This is to say, in language as well as in law, in the technique of production as well as in art, in science as well as in the objects of the domestic environment, there is embodied a sum of spirit. The individual in his intellectual development follows the growth of this spirit very imperfectly and at an ever increasing distance. If, for instance, we view the immense culture which for the last hundred years has been embodied in things and in knowledge, in institutions and in comforts, and if we compare all this with the cultural progress of the individual during the same period – at least in high status groups – a frightful disproportion in growth between the two becomes evident. Indeed, at some points we notice a retrogression in the culture of the individual with reference to spirituality, delicacy, and idealism. This discrepancy results essentially from the growing division of labour. For the division of labour demands from the individual an ever more one-sided accomplishment, and the greatest advance in a one-sided pursuit only too frequently means dearth to the personality of the individual. In any case, he can cope less and less with the overgrowth of objective culture. The individual is reduced to a negligible quantity, perhaps less in his consciousness than in his practice and in the totality of his obscure emotional states that are derived from this practice. The individual has become a mere cog in an enormous organization of things and powers which tear from his hands all progress, spirituality, and value in order to transform them from their subjective form into the form of a purely objective life. It needs merely to be pointed out that the metropolis is the genuine arena of this culture which outgrows all personal life. Here in buildings and educational institutions, in the wonders and comforts of space-conquering technology, in the formations of community life, and in the visible institutions of the state, is offered such an overwhelming fullness of crystallized and impersonalized spirit that the personality, so to speak, cannot maintain itself under its impact. On the one hand, life is made infinitely easy for the personality in that stimulations, interests, uses of time and consciousness are offered to it from all sides. They carry the person as if in a stream, and one needs hardly to swim for oneself. On the other hand, however, life is composed more and more of these impersonal contents and offerings which tend to displace the genuine personal colorations and incomparabilities. This results in the individual's summoning the utmost in uniqueness and particularization, in order to preserve his most personal core. He has to exaggerate this personal element in order to remain audible even to himself. The atrophy of individual culture through the hypertrophy of objective culture is one reason for the bitter hatred which the preachers of the most extreme individualism, above all Nietzsche, harbour against the metropolis. But it is, indeed, also a reason why these preachers are so passionately loved in the metropolis and why they appear to the metropolitan man as the prophets and saviours of his most unsatisfied yearnings.

If one asks for the historical position of these two forms of individualism which are nourished by the quantitative relation of the metropolis, namely, individual independence and the elaboration of individuality itself, then the metropolis assumes an entirely new rank order in the world history of the spirit. The eighteenth century found the individual in oppressive bonds which had become meaningless – bonds of a political, agrarian, guild, and religious character. They were restraints which, so to speak, forced upon man an unnatural form and outmoded, unjust inequalities. In this situation the cry for liberty and equality arose, the belief in the individual's full freedom of movement in all social and intellectual relationships. Freedom would at once permit the noble substance common to all to come to the fore, a substance which nature had deposited in every man and which society and history had only deformed. Besides this eighteenth-century ideal of liberalism, in the nineteenth century, through Goethe and Romanticism, on the one hand, and through the economic division of labour, on the other hand, another ideal arose: individuals liberated from historical bonds now wished to distinguish themselves from one another. The carrier of man's values is no longer the 'general human being' in every individual, but rather man's qualitative uniqueness and irreplaceability. The external and internal history of our time takes its course within the struggle and in the changing entanglements of these two ways of defining the individual's role in the whole of society. It is the function of the metropolis to provide the arena for this struggle and its reconciliation. For the metropolis presents the peculiar conditions which are revealed to us as the opportunities and the stimuli for the development of both these ways of allocating roles to men. Therewith these conditions gain a unique place, pregnant with inestimable meanings for the development of psychic existence. The metropolis reveals itself as one of those great historical formations in which opposing streams which enclose life unfold, as well as join one another with equal right. However, in this process the currents of life, whether their individual phenomena touch us sympathetically or antipathetically, entirely transcend the sphere for which the judge's attitude is appropriate. Since such forces of life have grown into the roots and into the crown of the whole of the historical life in which we, in our fleeting existence, as a cell, belong only as a part, it is not our task either to accuse or to pardon, but only to understand.

'The Growth of the City: An Introduction to a Research Project'

from Robert E. Park, Ernest W. Burgess and Roderick McKenzie (eds), *The City* (1925)

Ernest W. Burgess

Editors' Introduction

Ernest W. Burgess (1886–1966) was a key member of the 'Chicago School' who reinvented the study of urban sociology by using their local urban environment as a 'living laboratory' for their work. The Department of Sociology at the University of Chicago was founded in 1892 and became home to the highly influential journal, *American Journal of Sociology*. While best known for his 'concentric zone' model of the city, detailed in the extract contained here, Burgess enjoyed a long and varied academic career which included publications covering issues such as the problems facing the elderly, family and marriage.

Burgess's work, and that of other members of the Chicago School, is available now in a series of collections published by the University of Chicago Press. These include: Ernest W. Burgess and Donald J. Bogue (eds), *Contributions to Urban Sociology* (1964), Ernest W. Burgess and Donald J. Bogue (eds.), *Urban Sociology* (1967), Robert Park and Ernest W. Burgess, *Introduction to the Science of Sociology* (1921) and Robert Park, Ernest W. Burgess and Roderick D. McKenzie, *The City* (1925). The story of the Chicago School and an assessment of its legacy is told in Martin Bulmer's *The Chicago School of Sociology: Institutionalization, Diversity, and the Rise of Sociological Research* (Chicago, University of Chicago Press, 1984). Burgess's enduring contribution to the disciplines of geography and urban sociology will remain his urban modelling. An interesting extended discussion of the significance and importance of urban models can be found in the following three articles by Michael Pacione published in 2001: 'Models of Urban Land Use Structure in Cities of the Developed World' *Geography*, 86, 2: 97–120; 'Internal Structure of Cities in the Third World' *Geography*, 86, 3: 189–210 and 'The Future of the City: Cities of the Future' *Geography*, 86, 4: 277–86.

The outstanding fact of modern society is the growth of great cities. Nowhere else have the enormous changes which the machine industry has made in our social life registered themselves with such obviousness as in the cities. In the United States the transition from a rural to an urban civilization, though beginning later than in Europe, has taken place, if not more rapidly and completely, at any rate more logically in its most characteristic forms.

All the manifestations of modern life which are peculiarly urban – the skyscraper, the subway, the department store, the daily newspaper, and social work

– are characteristically American. The more subtle changes in our social life, which in their cruder manifestations are termed "social problems," problems that alarm and bewilder us, such as divorce, delinquency, and social unrest, are to be found in their most acute forms in our largest American cities. The profound and "subversive" forces which have wrought these changes are measured in the physical growth and expansion of cities. That is the significance of the comparative statistics of Weber, Bucher, and other students.

These statistical studies, although dealing mainly with the effects of urban growth, brought out into clear relief certain distinctive characteristics of urban as compared with rural populations. The larger proportion of women to men in the cities than in the open country, the greater percentage of youth and middle-aged, the higher ratio of the foreign-born, the increased heterogeneity of occupation increase with the growth of the city and profoundly alter its social structure. These variations in the composition of population are indicative of all the changes going on in the social organization of the community. In fact, these changes are a part of the growth of the city and suggest the nature of the processes of growth.

The only aspect of growth adequately described by Bucher and Weber was the rather obvious process of the aggregation of urban population. Almost as overt a process, that of expansion, has been investigated from a different and very practical point of view by groups interested in city planning, zoning, and regional surveys. Even more significant than the increasing density of urban population is its correlative tendency to overflow, and so to extend over wider areas, and to incorporate these areas into a larger communal life. This paper, therefore, will treat first of the expansion of the city, and then of the less-known processes of urban metabolism and mobility which are closely related to expansion.

EXPANSION AS PHYSICAL GROWTH

The expansion of the city from the standpoint of the city plan, zoning, and regional surveys is thought of almost wholly in terms of its physical growth. Traction studies have dealt with the development of transportation in its relation to the distribution of population throughout the city. The surveys made by the Bell Telephone Company and other public utilities have attempted to forecast the direction and the rate of growth of the city in order to anticipate the future demands for the extension of their services. In the city plan the location of parks and boulevards, the widening of traffic streets, the provision for a civic center, are all in the interest of the future control of the physical development of the city.

This expansion in areas of our largest cities is now being brought forcibly to our attention by the Plan for the Study of New York and Its Environs, and by the formation of the Chicago Regional Planning Association, which extends the metropolitan district of the city to a radius of 50 miles, embracing 4,000 square miles of territory. Both are attempting to measure expansion in order to deal with the changes that accompany city growth. In England, where more than one-half of the inhabitants live in cities having a population of 100,000 and over, the lively appreciation of the bearing of urban expansion on social organization is thus expressed by C. B. Fawcett:

> One of the most important and striking developments in the growth of the urban populations of the more advanced peoples of the world during the last few decades has been the appearance of a number of vast urban aggregates, or conurbations, far larger and more numerous than the great cities of any preceding age. These have usually been formed by the simultaneous expansion of a number of neighboring towns, which have grown out toward each other until they have reached a practical coalescence in one continuous urban area. Each such conurbation still has within it many nuclei of denser town growth, most of which represent the central areas of the various towns from which it has grown, and these nuclear patches are connected by the less densely urbanized areas which began as suburbs of these towns. The latter are still usually rather less continuously occupied by buildings, and often have many open spaces.
>
> These great aggregates of town dwellers are a new feature in the distribution of man over the earth. At the present day there are from thirty to forty of them, each containing more than a million people, whereas only a hundred years ago there were, outside the great centers of population on the waterways of China, not more than two or three. Such aggregations of people are phenomena of great geographical and social importance; they give rise to new problems in the organization of the life and well-being of their inhabitants and in their

varied activities. Few of them have yet developed a social consciousness at all proportionate to their magnitude, or fully realized themselves as definite groupings of people with many common interests, emotions and thoughts.

In Europe and America the tendency of the great city to expand has been recognized in the term "the metropolitan area of the city," which far overruns its political limits, and, in the case of New York and Chicago, even state lines. The metropolitan area may be taken to include urban territory that is physically contiguous, but it is coming to be defined by that facility of transportation that enables a business man to live in a suburb of Chicago and to work in the loop, and his wife to shop at Marshall Field's and attend grand opera in the Auditorium.

EXPANSION AS A PROCESS

No study of expansion as a process has yet been made, although the materials for such a study and intimations of different aspects of the process are contained in city planning, zoning, and regional surveys. The typical processes of the expansion of the city can best be illustrated, perhaps, by a series of concentric circles, which may be numbered to designate both the successive zones of urban extension and the types of areas differentiated in the process of expansion [Figure 1].

[Figure 1] represent an ideal construction of the tendencies of any town or city to expand radially from its central business district – on the map "the Loop" (I). Encircling the downtown area there is normally an area in transition, which is being invaded by business and light manufacture (II). A third area (III) is inhabited by the workers in industries who have escaped from the area of deterioration (II) but who desire to live within easy access of their work. Beyond this zone is the "residential area" (IV) of high-class apartment buildings or of exclusive "restricted" districts of single family dwellings. Still farther out, beyond the city limits, is the commuters' zone: suburban areas, or satellite cities, within a thirty- to sixty-minute ride of the central business district.

This [figure] brings out clearly the main fact of expansion, namely, the tendency of each inner zone to extend its area by the invasion of the next outer zone. This aspect of expansion may be called *succession*, a process which has been studied in detail in plant ecology. If this [figure] is applied to Chicago, all four of these zones were in its early history included in the circumference of the inner zone, the present business district. The present boundaries of the area of deterioration were not many years ago those of the zone now inhabited by independent wage-earners, and within the memories of thousands of Chicagoans contained the residences of the "best families." It hardly needs to be added that neither Chicago nor any other city fits perfectly into this ideal scheme. Complications are introduced by the lake front, the Chicago River, railroad lines, historical factors in the location of industry, the relative degree of the resistance of communities to invasion, etc.

Besides extension and succession, the general process of expansion in urban growth involves the antagonistic and yet complementary processes of concentration and decentralization. In all cities there is the natural tendency for local and outside transportation to converge in the central business district. In the downtown section of every large city we expect to find the department stores, the skyscraper office buildings, the railroad stations, the great hotels, the theaters, the art museum, and the city hall. Quite naturally, almost inevitably, the economic, cultural, and political life centers here. The relation of centralization to the other processes of city life may be roughly gauged by the fact that over half a million people daily enter and leave Chicago's "loop." More recently sub-business centers have grown up in outlying zones. These "satellite loops" do not, it seems, represent the "hoped for" revival of the neighborhood, but rather a telescoping of several local communities into a larger economic unity. The Chicago of yesterday, an agglomeration of country towns and immigrant colonies, is undergoing a process of reorganization into a centralized decentralized system of local communities coalescing into sub-business areas visibly or invisibly dominated by the central business district. The actual processes of what may be called centralized decentralization are now being studied in the development of the chain store, which is only one illustration of the change in the basis of the urban organization.

Expansion, as we have seen, deals with the physical growth of the city, and with the extension of the technical services that have made city life not only livable, but comfortable, even luxurious. Certain of these basic necessities of urban life are possible only

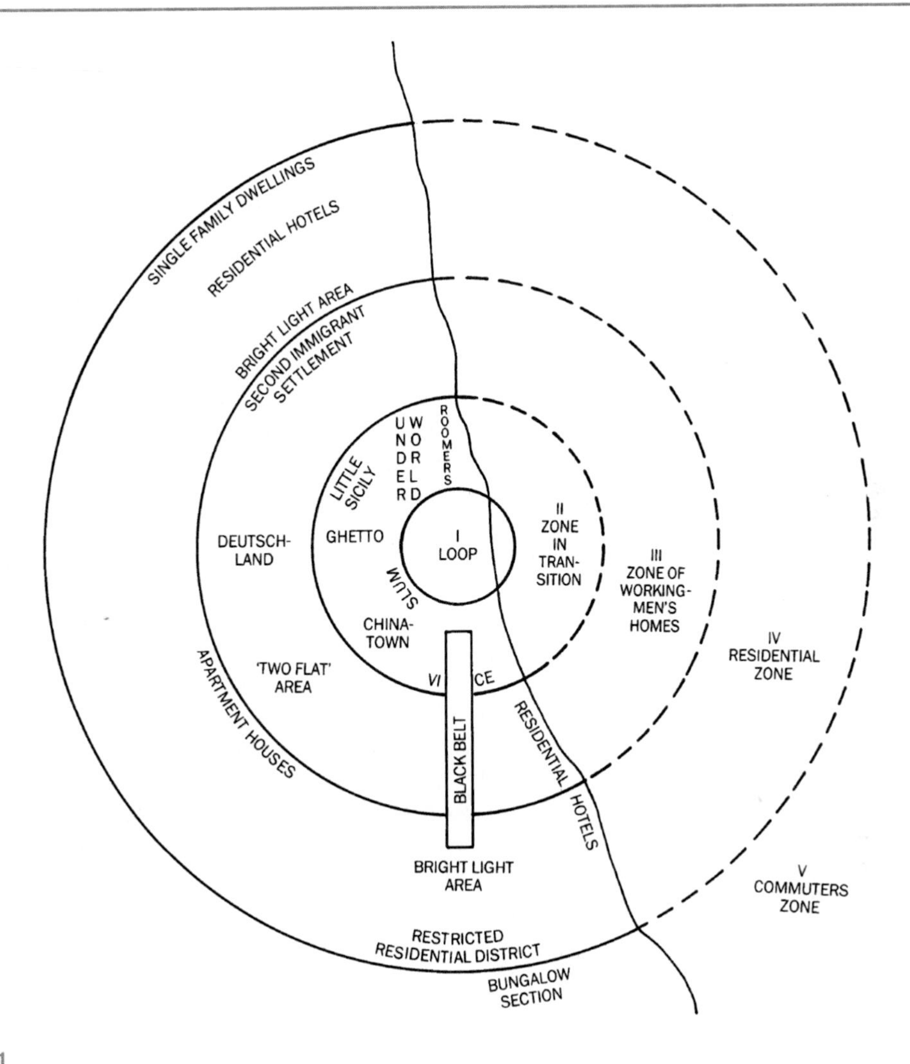

Figure 1

through tremendous development of communal existence. Three millions of people in Chicago are dependent upon one unified water system, one giant gas company, and one huge electric light plant. Yet, like most of the other aspects of our communal urban life, this economic co-operation is an example of co-operation without a shred of what the "spirit of co-operation" is commonly thought to signify. The great public utilities are a part of the mechanization of life in great cities, and have little or no other meaning for social organization.

Yet the processes of expansion, and especially the rate of expansion, may be studied not only in the physical growth and business development, but also in the consequent changes in the social organization and in personality types. How far is the growth of the city, in its physical and technical aspects, matched by a natural but adequate readjustment in the social organization? What, for a city, is a normal rate of expansion, a rate of expansion with which controlled changes in the social organization might successfully keep pace?

SOCIAL ORGANIZATION AND DISORGANIZATION AS PROCESSES OF METABOLISM

These questions may best be answered, perhaps, by thinking of urban growth as a resultant of organization and disorganization analogous to the anabolic and katabolic processes of metabolism in the body. In what way are individuals incorporated into the life of a city? By what process does a person become an organic part of his society? The natural process of acquiring culture is by birth. A person is born into a family already adjusted to a social environment – in this case the modern city. The natural rate of increase of population most favorable for assimilation may then be taken as the excess of the birth-rate over the death-rate, but is this the normal rate of city growth? Certainly, modern cities have increased and are increasing in population at a far higher rate. However, the natural rate of growth may be used to measure the disturbances of metabolism caused by any excessive increase, as those which followed the great influx of southern Negroes into northern cities since the war. In a similar way all cities show deviations in composition by age and sex from a standard population such as that of Sweden, unaffected in recent years by any great emigration or immigration. Here again, marked variations, as any great excess of males over females, or of females over males, or in the proportion of children, or of grown men or women, are symptomatic of abnormalities in social metabolism.

Normally the processes of disorganization and organization may be thought of as in reciprocal relationship to each other, and as co-operating in a moving equilibrium of social order toward an end vaguely or definitely regarded as progressive. So far as disorganization points to reorganization and makes for more efficient adjustment, disorganization must be conceived not as pathological, but as normal. Disorganization as preliminary to reorganization of attitudes and conduct is almost invariably the lot of the newcomer to the city, and the discarding of the habitual, and often of what has been to him the moral, is not infrequently accompanied by sharp mental conflict and sense of personal loss. Oftener, perhaps, the change gives sooner or later a feeling of emancipation and an urge toward new goals.

In the expansion of the city a process of distribution takes place which sifts and sorts and relocates individuals and groups by residence and occupation. The resulting differentiation of the cosmopolitan American city into areas is typically all from one pattern, with only interesting minor modifications. Within the central business district or on an adjoining street is the "main stem" of "hobohemia," the teeming Rialto of the homeless migratory man of the Middle West. In the zone of deterioration encircling the central business section are always to be found the so-called "slums" and "bad lands," with their submerged regions of poverty, degradation, and disease, and their underworlds of crime and vice. Within a deteriorating area are rooming-house districts, the purgatory of "lost souls." Nearby is the Latin Quarter, where creative and rebellious spirits resort. The slums are also crowded to over flowing with immigrant colonies – the Ghetto, Little Sicily, Greek town, Chinatown – fascinatingly combining old world heritages and American adaptations. Wedging out from here is the Black Belt with its free and disorderly life. The area of deterioration, while essentially one of decay, of stationary or declining population, is also one of regeneration, as witness the mission, the settlement, the artists' colony, radical centers – all obsessed with the vision of a new and better world.

The next zone is also inhabited predominatingly by factory and shop workers, but skilled and thrifty. This is an area of second immigrant settlement, generally of the second generation. It is the region of escape from the slum, the *Deutschland* of the aspiring Ghetto family. For *Deutschland* (literally "Germany") is the name given, half in envy, half in derision, to that region beyond the Ghetto where successful neighbors appear to be imitating German Jewish standards of living. But the inhabitant of this area in turn looks to the "Promised Land" beyond, to its residential hotels, its apartment-house region, its "satellite loops," and its "bright light" areas.

This differentiation into natural economic and cultural groupings gives form and character to the city. For segregation offers the group, and thereby the individuals who compose the group, a place and a role in the total organization of city life. Segregation limits development in certain directions, but releases it in others. These areas tend to accentuate certain traits, to attract and develop their kind of individuals, and so to become further differentiated.

The division of labor in the city likewise illustrates disorganization, reorganization and increasing differentiation. The immigrant from rural communities in Europe and America seldom brings with him economic skill of any great value in our industrial, commercial, or professional life. Yet interesting occupational selection has taken place by nationality, explainable more by racial temperament or circumstance than by old-world economic background as Irish policemen, Greek ice-cream parlors, Chinese laundries, Negro porters, Belgian janitors, etc.

The facts that in Chicago one million (996,589)

individuals gainfully employed reported 509 occupations, and that over 1,000 men and women in *Who's Who* gave 116 different vocations give some notion of how in the city the minute differentiation of occupation "analyzes and sifts the population, separating and classifying the diverse elements." These figures also afford some intimation of the complexity and complication of the modern industrial mechanism and the intricate segregation and isolation of divergent economic groups. Interrelated with this economic division of labor is a corresponding division into social classes and into cultural and recreational groups. From this multiplicity of groups, with their different patterns of life, the person finds his congenial social world and – what is not feasible in the narrow confines of a village – may move and live in widely separated, and perchance conflicting, worlds. Personal disorganization may be but the failure to harmonize the canons of conduct of two divergent groups.

If the phenomena of expansion and metabolism indicate that a moderate degree of disorganization may and does facilitate social organization, they indicate as well that rapid urban expansion is accompanied by excessive increases in disease, crime, disorder, vice, insanity and suicide, rough indexes of social disorganization. But what are the indexes of the causes, rather than of the effects, of the disordered social metabolism of the city? The excess of the actual over the natural increase of population has already been suggested as a criterion. The significance of this increase consists in the immigration into a metropolitan city like New York and Chicago of tens of thousands of persons annually. Their invasion of the city has the effect of a tidal wave inundating first the immigrant colonies, the ports of first entry, dislodging thousands of inhabitants who overflow into the next zone, and so on and on until the momentum of the wave has spent its force on the last urban zone. The whole effect is to speed up expansion, to speed up industry, to speed up the "junking" process in the area of deterioration (II). These internal movements of the population become the more significant for study. What movement is going on in the city, and how may this movement be measured? It is easier, of course, to classify movement within the city than to measure it. There is the movement from residence to residence, change of occupation, labor turnover, movement to and from work, movement for recreation and adventure. This leads to the question: what is the significant aspect of movement for the study of the changes in city life? The answer to this question leads directly to the important distinction between movement and mobility.

MOBILITY AS THE PULSE OF THE COMMUNITY

Movement, per se, is not an evidence of change or of growth. In fact, movement may be a fixed and unchanging order of motion, designed to control a constant situation, as in routine movement. Movement that is significant for growth implies a change of movement in response to a new stimulus or situation. Change of movement of this type is called *mobility*. Movement of the nature of routine finds its typical expression in work. Change of movement, or mobility, is characteristically expressed in adventure. The great city, with its "bright lights," its emporiums of novelties and bargains, its palaces of amusement, its underworld of vice and crime, its risks of life and property from accident, robbery, and homicide, has become the region of the most intense degree of adventure and danger, excitement and thrill.

Mobility, it is evident, involves change, new experience, stimulation. Stimulation induces a response of the person to those objects in his environment which afford expression for his wishes. For the person, as for the physical organism, stimulation is essential to growth. Response to stimulation is wholesome so long as it is a correlated integral reaction of the entire personality. When the reaction is segmental, that is, detached from, and uncontrolled by, the organization of personality, it tends to become disorganizing or pathological. That is why stimulation for the sake of stimulation, as in the restless pursuit of pleasure, partakes of the nature of vice.

The mobility of city life, with its increase in the number and intensity of stimulations, tends inevitably to confuse and to demoralize the person. For an essential element in the mores and in personal morality is consistency, consistency of the type that is natural in the social control of the primary group. Where mobility is the greatest, and where in consequence primary controls break down completely, as in the zone of deterioration in the modern city, there develop areas of demoralization, of promiscuity, and of vice.

In our studies of the city it is found that areas of mobility are also the regions in which are found juvenile delinquency, boys' gangs, crime, poverty, wife desertion, divorce, abandoned infants, vice.

These concrete situations show why mobility is perhaps the best index of the state of metabolism of the city. Mobility may be thought of, in more than a fanciful sense, as the "pulse of the community." Like the pulse of the human body, it is a process which reflects and is indicative of all the changes that are taking place in the community, and which is susceptible of analysis into elements which may be stated numerically.

The elements entering into mobility may be classified under two main heads: (1) the state of mutability of the person, and (2) the number and kind of contacts or stimulations in his environment. The mutability of city populations varies with sex and age composition, and the degree of detachment of the person from the family and from other groups. All these factors may be expressed numerically. The new stimulations to which a population responds can be measured in terms of change of movement or of increasing contacts. Statistics on the movement of urban population may only measure routine, but an increase at a higher ratio than the increase of population measures mobility. In 1860 the horsecar lines of New York City carried about 50,000,000 passengers; in 1890 the trolley cars (and a few surviving horse-cars) transported about 500,000,000; in 1921, the elevated, subway, surface, and electric and steam suburban lines carried a total of more than 2,500,000,000 passengers. In Chicago the total annual rides per capita on the surface and elevated lines were 164 in 1890; 215 in 1900; 320 in 1910; and 338 in 1921. In addition, the rides per capita on steam and electric suburban lines almost doubled between 1916 (23) and 1921 (41), and the increasing use of the automobile must not be overlooked. For example, the number of automobiles in Illinois increased from 131,140 in 1915 to 833,920 in 1923.

Mobility may be measured not only by these changes of movement, but also by increase of contacts. While the increase of population of Chicago in 1912–22 was less than 25 percent (23.6 percent), the increase of letters delivered to Chicagoans was double that (49.6 percent) – from 693,048,196 to 1,038,007,854. In 1912 New York had 8.8 telephones; in 1922, 16.9 per 100 inhabitants. Boston had, in 1912, 10.1 telephones; ten years later, 19.5 telephones per 100 inhabitants. In the same decade the figures for Chicago increased from 12.3 to 21.6 per 100 population. But increase of the use of the telephone is probably more significant than increase in the number of telephones. The number of telephone calls in Chicago increased from 606,131,928 in 1914 to 944,010,586 in 1922, an increase of 55.7 percent, while the population increased only 13.4 percent.

Land values, since they reflect movement, afford one of the most sensitive indexes of mobility. The highest land values in Chicago are at the point of greatest mobility in the city, at the corner of State and Madison streets, in the Loop. A traffic count showed that at the rush period 31,000 people an hour, or 210,000 men and women in sixteen and one-half hours, passed the southwest corner. For over ten years land values in the Loop have been stationary but in the same time they have doubled, quadrupled and even sextupled in the strategic corners of the "satellite loops," an accurate index of the changes which have occurred. Our investigations so far seem to indicate that variations in land values, especially where correlated with differences in rents, offer perhaps the best single measure of mobility, and so of all the changes taking place in the expansion and growth of the city.

In general outline, I have attempted to present the point of view and methods of investigation which the department of sociology is employing in its studies in the growth of the city, namely, to describe urban expansion in terms of extension, succession, and concentration; to determine how expansion disturbs metabolism when disorganization is in excess of organization; and, finally, to define mobility and to propose it as a measure both of expansion and metabolism, susceptible to precise quantitative formulation, so that it may be regarded almost literally as the pulse of the community. In a way, this statement might serve as an introduction to any one of five or six research projects under way in the department. The project, however, in which I am directly engaged is an attempt to apply these methods of investigation to a cross-section of the city – to put this area, as it were, under the microscope, and so to study in more detail and with greater control and precision the processes which have been described here in the large. For this purpose the West Side Jewish community has been selected. This community includes the so-called "Ghetto," or area of first settlement, and Lawndale, the so-called "Deutschland," or area of second settlement. This area has certain obvious advantages for this study, from the standpoint of expansion, metabolism, and mobility. It exemplifies the tendency to expansion radially from the business center of the city. It is now relatively a homogeneous cultural group. Lawndale is itself an area in flux, with the tide of migrants still flowing in from the Ghetto and a constant egress to more desirable regions of the residential zone. In this area, too, it is also possible to study how the expected outcome of this high rate of mobility in social and personal disorganization is counteracted in large measure by the efficient communal organization of the Jewish community.

'What is a City?'

Architectural Record (1937)

Lewis Mumford

Editors' Introduction

Lewis Mumford (1895–1990) is a huge figure in the history of urban writing and thinking. Born in Flushing, New York, he became an urban historian, sociologist, an architectural critic and an urban planner. During his lifetime he published twenty-five highly influential volumes. His first book was *The Study of Utopia* in 1922 and his last, his autobiography, *Sketches from Life*, in 1982. Between these he published some of the landmark texts on the city including *The Culture of Cities* (1938), *The Urban Prospect* (1968) and *The City in History* (1961). The latter work is probably the most widely read and highly regarded texts on urban history published in the twentieth century. The book attempts to trace the evolution of towns and cities, beginning with cave dwellings, through every period of urban history. His interests in *The City in History*, go far beyond the physical form of cities and include architecture, art, technology, culture and politics. It is a book that is staggering in both its ambition and its achievements. It is testimony to the scope and range of Mumford's career that it is not overshadowed by this vast tome. Among Mumford's many other achievements include co-founding the Regional Planning Association of America in 1923, regular contributions to *The New Yorker*, acting as an advisor on the postwar replanning of British cities and acting as a fierce opponent to Robert Moses's destructive redevelopment of New York City.

Mumford was heavily influenced by Patrick Geddes and despite the diversity, scope and ambition of Mumford's work a number of themes run throughout. Mumford was particularly concerned with the relationship between cities and nature, he wrote an introduction for Ian McHarg's *Design with Nature* (1969), for example, and the relationship between cities and human culture and personality. A powerful strand of his analysis of cities was his attempt to trace the links between urban form, architectural style and social and cultural values, something he examined for every urban epoch. This is an approach that has influenced a number of sociologists and others, such as Richard Sennett, since.

Most of our housing and city planning has been handicapped because those who have undertaken the work have had no clear notion of the social functions of the city. They sought to derive these functions from a cursory survey of the activities and interests of the contemporary urban scene. And they did not, apparently, suspect that there might be gross deficiencies, misdirected efforts, mistaken expenditures here that would not be set straight by merely building sanitary tenements or straightening out and widening irregular streets.

The city as a purely physical fact has been subject to numerous investigations. But what is the city as a social institution? The earlier answers to these questions, in Aristotle, Plato, and the Utopian writers from Sir Thomas More to Robert Owen, have been on the whole more satisfactory than those of the more systematic sociologists: most contemporary treatises on "urban sociology" in America throw no important light upon the problem. One of the soundest definitions of the city was that framed by John Stow, an honest observer of Elizabethan London, who said:

> Men are congregated into cities and commonwealths for honesty and utility's sake, these shortly be the commodities that do come by cities, commonalties and corporations. First, men by this nearness of conversation are withdrawn from barbarous fixity and force, to certain mildness of manners, and to humanity and justice . . . Good behavior is yet called urbanitas because it is rather found in cities than elsewhere. In sum, by often hearing, men be better persuaded in religion, and for that they live in the eyes of others, they be by example the more easily trained to justice, and by shamefastness restrained from injury.
>
> And whereas commonwealths and kingdoms cannot have, next after God, any surer foundation than the love and good will of one man towards another, that also is closely bred and maintained in cities, where men by mutual society and companying together, do grow to alliances, commonalties, and corporations.

It is with no hope of adding much to the essential insight of this description of the urban process that I would sum up the sociological concept of the city in the following terms:

The city is a related collection of primary groups and purposive associations: the first, like family and neighborhood, are common to all communities, while the second are especially characteristic of city life. These varied groups support themselves through economic organizations that are likewise of a more or less corporate, or at least publicly regulated, character; and they are all housed in permanent structures, within a relatively limited area. The essential physical means of a city's existence are the fixed site, the durable shelter, the permanent facilities for assembly, interchange, and storage; the essential social means are the social division of labor, which serves not merely the economic life but the cultural processes. The city in its complete sense, then, is a geographic plexus, an economic organization, an institutional process, a theater of social action, and an aesthetic symbol of collective unity. The city fosters art and is art; the city creates the theater and is the theater. It is in the city, the city as theater, that man's more purposive activities are focused, and work out, through conflicting and cooperating personalities, events, groups, into more significant culminations.

Without the social drama that comes into existence through the focusing and intensification of group activity there is not a single function performed in the city that could not be performed – and has not in fact been performed – in the open country. The physical organization of the city may deflate this drama or make it frustrate; or it may, through the deliberate efforts of art, politics, and education, make the drama more richly significant, as a stage-set, well-designed, intensifies and underlines the gestures of the actors and the action of the play. It is not for nothing that men have dwelt so often on the beauty or the ugliness of cities: these attributes qualify men's social activities. And if there is a deep reluctance on the part of the true city dweller to leave his cramped quarters for the physically more benign environment of a suburb – even a model garden suburb! – his instincts are usually justified: in its various and many-sided life, in its very opportunities for social disharmony and conflict, the city creates drama; the suburb lacks it.

One may describe the city, in its social aspect, as a special framework directed toward the creation of differentiated opportunities for a common life and a significant collective drama. As indirect forms of association, with the aid of signs and symbols and specialized organizations, supplement direct face-to-face intercourse, the personalities of the citizens themselves become many-faceted: they reflect their specialized interests, their more intensively trained aptitudes, their finer discriminations and selections: the personality no longer presents a more or less unbroken traditional face to reality as a whole. Here lies the possibility of personal disintegration; and here lies the need for reintegration through wider participation in a concrete and visible collective whole. What men cannot imagine as a vague formless society, they can live through and experience as citizens in a city. Their unified plans and buildings become a symbol of their social relatedness; and when the physical environment itself becomes disordered and incoherent, the social functions that it harbors become more difficult to express.

One further conclusion follows from this concept of the city: social facts are primary, and the physical organization of a city, its industries and its markets, its lines of communication and traffic, must be subservient to its social needs. Whereas in the development of the city during the last century we expanded the physical plant recklessly and treated the essential social nucleus, the organs of government and education and social service, as mere afterthought, today we must treat the social nucleus as the essential element in every valid city plan: the spotting and inter-relationship of schools, libraries, theaters, community centers is the first task in defining the urban neighborhood and laying down the outlines of an integrated city.

In giving this sociological answer to the question: What is a City? one has likewise provided the clue to a number of important other questions. Above all, one has the criterion for a clear decision as to what is the desirable size of a city – or may a city perhaps continue to grow until a single continuous urban area might cover half the American continent, with the rest of the world tributary to this mass? From the standpoint of the purely physical organization of urban utilities – which is almost the only matter upon which metropolitan planners in the past have concentrated – this latter process might indeed go on indefinitely. But if the city is a theater of social activity, and if its needs are defined by the opportunities it offers to differentiated social groups, acting through a specific nucleus of civic institutes and associations, definite limitations on size follow from this fact.

In one of Le Corbusier's early schemes for an ideal city, he chose three million as the number to be accommodated: the number was roughly the size of the urban aggregate of Paris, but that hardly explains why it should have been taken as a norm for a more rational type of city development. If the size of an urban unit, however, is a function of its productive organization and its opportunities for active social intercourse and culture, certain definite facts emerge as to adequate ratio of population to the process to be served. Thus, at the present level of culture in America, a million people are needed to support a university. Many factors may enter which will change the size of both the university and the population base; nevertheless one can say provisionally that if a million people are needed to provide a sufficient number of students for a university, then two million people should have two universities. One can also say that, other things being equal, five million people will not provide a more effective university than one million people would. The alternative to recognizing these ratios is to keep on overcrowding and overbuilding a few existing institutions, thereby limiting, rather than expanding, their genuine educational facilities.

What is important is not an absolute figure as to population or area: although in certain aspects of life, such as the size of city that is capable of reproducing itself through natural fertility, one can already lay down such figures. What is more important is to express size *always as a function of the social relationships to be served* . . . There is an optimum numerical size, beyond which each further increment of inhabitants creates difficulties out of all proportion to the benefits. There is also an optimum area of expansion, beyond which further urban growth tends to paralyze rather than to further important social relationships. Rapid means of transportation have given a regional area with a radius of from forty to a hundred miles, the unity that London and Hampstead had before the coming of the underground railroad. But the activities of small children are still bounded by a walking distance of about a quarter of a mile; and for men to congregate freely and frequently in neighborhoods the maximum distance means nothing, although it may properly define the area served for a selective minority by a university, a central reference library, or a completely equipped hospital. The area of potential urban settlement has been vastly increased by the motor car and the airplane; but the necessity for solid contiguous growth, for the purposes of intercourse, has in turn been lessened by the telephone and the radio. In the Middle Ages a distance of less than a half a mile from the city's center usually defined its utmost limits. The block-by-block accretion of the big city, along its corridor avenues, is in all important respects a denial of the vastly improved type of urban grouping that our fresh inventions have brought in. For all occasional types of intercourse, the region is the unit of social life but the region cannot function effectively, as a well-knit unit, if the entire area is densely filled with people – since their very presence will clog its arteries of traffic and congest its social facilities.

Limitations on size, density, and area are absolutely necessary to effective social intercourse; and they are therefore the most important instruments of rational economic and civic planning. The unwillingness in the past to establish such limits has been due mainly to two facts: the assumption that all upward changes in magnitude were signs of progress and automatically "good for business," and the belief that such limitations were essentially arbitrary, in that they proposed to "decrease economic opportunity" – that is, opportunity for profiting by congestion – and to halt the inevitable course of change. Both these objections are superstitious.

Limitations on height are now common in American cities; drastic limitations on density are the rule in all municipal housing estates in England: that which could not be done has been done. Such limitations do not obviously limit the population itself: they merely give the planner and administrator the opportunity to multiply the number of centers in which the population is housed, instead of permitting a few existing centers to aggrandize themselves on a monopolistic pattern. These limitations are necessary to break up the functionless, hypertrophied urban masses of the past. Under this mode of planning, the planner proposes to replace the "mononucleated city," as Professor Warren Thompson has called it, with a new type of "polynucleated city," in which a cluster of communities, adequately spaced and bounded, shall do duty for the badly organized mass city. Twenty such cities, in a region whose environment and whose resources were adequately planned, would have all the benefits of a metropolis that held a million people, without its ponderous disabilities: its capital frozen into unprofitable utilities, and its land values congealed at levels that stand in the way of effective adaptation to new needs.

Mark the change that is in process today. The emerging sources of power, transport, and communication do not follow the old highway network at all. Giant power strides over the hills, ignoring the limitations of wheeled vehicles; the airplane, even more liberated, flies over swamps and mountains, and terminates its journey, not on an avenue, but in a field. Even the highway for fast motor transportation abandons the pattern of the horse-and-buggy era. The new highways, like those of New Jersey and Westchester, to mention only examples drawn locally, are based more or less on a system definitively formulated by Benton MacKaye in his various papers on the Townless Highway. The most complete plans form an independent highway network, isolated both from the adjacent countryside and the towns that they bypass: as free from communal encroachments as the railroad system. In such a network no single center will, like the metropolis of old, become the focal point of all regional advantages: on the contrary, the "whole region" becomes open for settlement.

Even without intelligent public control, the likelihood is that within the next generation this dissociation and decentralization of urban facilities will go even farther. The Townless Highway begets the Highwayless Town in which the needs of close and continuous human association on all levels will be uppermost. This is just the opposite of the earlier mechanocentric picture of Roadtown, as pictured by Edgar Chambless and the Spanish projectors of the Linear City. For the highwayless town is based upon the notion of effective zoning of functions through initial public design, rather than by blind legal ordinances. It is a town in which the various functional parts of the structure are isolated topographically as urban islands, appropriately designed for their specific use with no attempt to provide a uniform plan of the same general pattern for the industrial, the commercial, the domestic, and the civic parts.

The first systematic sketch of this type of town was made by Messrs. Wright and Stein in their design for Radburn in 1929; a new type of plan that was repeated on a limited scale – and apparently in complete independence – by planners in Köln and Hamburg at about the same time. Because of restrictions on design that favored a conventional type of suburban house and stale architectural forms, the implications of this new type of planning were not carried very far in Radburn. But in outline the main relationships are clear: the differentiation of foot traffic from wheeled traffic in independent systems, the insulation of residence quarters from through roads; the discontinuous street pattern; the polarization of social life in specially spotted civic nuclei, beginning in the neighborhood with the school and the playground and the swimming pool. This type of planning was carried to a logical conclusion in perhaps the most functional and most socially intelligent of all Le Corbusier's many urban plans: that for Nemours in North Africa, in 1934.

Through these convergent efforts, the principles of the polynucleated city have been well established. Such plans must result in a fuller opportunity for the primary group, with all its habits of frequent direct meeting and face-to-face intercourse: they must also result in a more complicated pattern and a more comprehensive life for the region, for this geographic area can only now, for the first time, be treated as an instantaneous whole for all the functions of social existence. Instead of trusting to the mere massing of population to produce the necessary social concentration and social drama, we must now seek these results through deliberate local nucleation and a finer regional articulation. The words are jargon; but the importance of their meaning should nor be missed. To embody these new possibilities in city life, which come to us not merely through better technical organization but through acuter sociological understanding, and to dramatize the activities themselves in appropriate individual and urban structures, forms the task of the coming generation.

'The Hotel Lobby'

from *The Mass Ornament: Weimar Essays* (1922–25)

Siegfried Kracauer

Editors' Introduction

Siegfried Kracauer (1889–1966) was a cultural theorist who is known now primarily for his interest in mass culture. This interest is demonstrated in the extract 'The hotel lobby' from *The Mass Ornament* (originally published in 1927). Kracauer was born in Frankfurt, Germany. After fleeing the Nazis he settled in the United States and eventually died in New York. His early work, from which this extract is drawn, explored a phenomenology of modernity and mass culture. His themes in *The Mass Ornament* include the relationship of the individual to the mass, the urban culture of the masses, alienation and everyday life. The range of, seemingly unconnected, topics explored by Kracauer in *The Mass Ornament* is staggering and includes architectural settings, in addition to the hotel lobby, such as the shopping arcade and the cinema, amusements and activities such as dance and bestsellers, as well as other topics including photography, Kafka and the Bible. Kracauer saw the ubiquity of the surface in modern culture as revelatory. He was one of the first major thinkers to realise and demonstrate the power of studying everyday life and its apparently mundane, unimportant and ephemeral aspects.

His later works included an autobiography of Jacques Offenbach (1937), which situated its subject's works within their social and cultural context. In avoiding the idealism that characterised studies of individual artists Kracauer's concern for social and cultural history shared much with the work of influential art historians such as Erwin Panofsky. He also wrote on film, most notably in *The Nature of Film* (1960) and *From Caligari to Hitler* (1947), the latter subtitled 'A Pyschological History of German Film'. Again, here his concern was as much with context as text.

Notable amongst various accounts of Kracauer's life and work is *Siegfried Kracauer: An Introduction* by Gertrud Koch (Princeton University Press, 2000). Kracauer's work has enjoyed something of a revival in recent years and he is now recognised as one of the major cultural critics of the twentieth century who was both ahead of his time in his choice of subjects and approaches and whose enduring influence has begun to be recognised.

The community of the higher realms that is fixated upon God is secure in the knowledge that as an oriented community – both in time and for all eternity – it lives within the law and beyond the law, occupying the perpetually untenable middle ground between the natural and the supernatural. It not only presents itself in this paradoxical situation but also experiences and names it as well. In spheres of lesser reality, consciousness of existence and of the authentic conditions dwindles away in the existential stream, and clouded sense becomes lost in the labyrinth of distorted events whose distortion it no longer perceives.

The *aesthetic* rendering of such a life bereft of reality, a life that has lost the power of self-observation, may be able to restore to it a sort of language; for even if the artist does not force all that has become mute and illusory directly up into reality, he does express his directed self by giving form to this life. The more life is submerged, the more it needs the artwork, which unseals its withdrawnness and puts its pieces back in place in such a way that these, which were lying strewn about, become organized in a meaningful way. The unity of the aesthetic construct, the manner in which it distributes the emphases and consolidates the event, gives a voice to the inexpressive world, gives meaning to the themes broached within it. Just what these themes mean, however, must still be brought out through translation and depends to no small extent on the level of reality evinced by their creator. Thus, while in the higher spheres the artist confirms a reality that grasps itself, in the lower regions his work becomes a harbinger of a manifold that utterly lacks any revelatory word. His tasks multiply in proportion to the world's loss of reality, and the cocoon-like spirit [*Geist*] that lacks access to reality ultimately imposes upon him the role of educator, of the observer who not only sees but also prophetically foresees and makes connections. Although this overloading of the aesthetic may well accord the artist a mistaken position, it is understandable, because the life that remains untouched by authentic things recognizes that it has been captured in the mirror of the artistic construct, and thereby gains consciousness, albeit negative, of its distance from reality and of its illusory status. For no matter how insignificant the existential power that gives rise to the artistic formation may be, it always infuses the muddled material with intentions that help it become transparent.

Without being an artwork, the *detective novel* still shows civilized society its own face in a purer way than society is usually accustomed to seeing it. In the detective novel, proponents of that society and their functions give an account of themselves and divulge their hidden significance. But the detective novel can coerce the self-shrouding world into revealing itself in this manner only because it is created by a consciousness that is not circumscribed by that world. Sustained by this consciousness, the detective novel really thinks through to the end the society dominated by autonomous *Ratio* – a society that exists only as a concept – and develops the initial moments it proposes in such a way that the idea is fully realized in actions and figures. Once the stylization of the one-dimensional unreality has been completed, the detective novel integrates the individual elements – now adequate to the constitutive presuppositions – into a self-contained coherence of meaning, an integration it effects through the power of its existentiality, the latter transformed not into critique and exigency but into principles of aesthetic composition. It is only this entwinement into a unity that really makes possible the interpretation of the presented findings. For, like the philosophical system, the aesthetic organism aims at a totality that remains veiled to the proponents of civilized society, a totality that in some way disfigures the entirety of experienced reality and thereby enables one to see it afresh. Thus, the true meaning of these findings can be found only in the way in which they combine into an aesthetic totality. This is the minimum achievement of the artistic entity: to construct a whole out of the blindly scattered elements of a disintegrated world – a whole that, even if it seems only to mirror this world, nevertheless does capture it in its wholeness and thereby allows for the projection of its elements onto real conditions. The fact that the structure of the life presented in the detective novel is so typical indicates that the consciousness producing it is not an individual, coincidental one; at the same time, it shows that what has been singled out are the seemingly metaphysical characteristics. Just as the detective discovers the secret that people have concealed, the detective novel discloses in the aesthetic medium the secret of a society bereft of reality, as well as the secret of its insubstantial marionettes. The composition of the detective novel transforms an ungraspable life into a translatable analogue of actual reality.

In the *house of God*, which presupposes an already extant community, the congregation accomplishes the task of making connections. Once the members of the congregation have abandoned the relation on which the place is founded, the house of God retains only a decorative significance. Even if it sinks into oblivion, civilized society at the height of its development still maintains privileged sites that testify to its own nonexistence, just as the house of God testifies to the existence of the community united in reality. Admittedly society is unaware of this, for it cannot see beyond its own sphere; only the aesthetic construct, whose form renders the manifold as a projection, makes it possible to demonstrate this correspondence. The typical characteristics of the *hotel lobby*, which appears repeatedly in detective novels, indicate that it is conceived as the inverted image of the house of God. It is a negative church, and can be transformed into a church so long as one observes the conditions that govern the different spheres.

In both places people appear there as *guests*. But

whereas the house of God is dedicated to the service of the one whom people have gone there to encounter, the hotel lobby accommodates all who go there to meet no one. It is the setting for those who neither seek nor find the one who is always sought, and who are therefore guests in space as such – a space that encompasses them and has no function other than to encompass them. The impersonal nothing represented by the hotel manager here occupies the position of the unknown one in whose name the church congregation gathers. And whereas the congregation invokes the name and dedicates itself to the service in order to fulfill the relation, the people dispersed in the lobby accept their host's incognito without question. Lacking any and all relation, they drip down into the vacuum with the same necessity that compels those striving in and for reality to lift themselves out of the nowhere toward their destination.

The congregation, which gathers in the house of God for prayer and worship, outgrows the imperfection of communal life in order not to overcome it but to bear it in mind and to reinsert it constantly into the tension. Its gathering is a *collectedness* and a unification of this directed life of the community, which belongs to two realms: the realm covered by law and the realm beyond law. At the site of the church – but of course not only here – these separate currents encounter each other; the law is broached here without being breached, and the paradoxical split is accorded legitimacy by the sporadic suspension of its languid continuity. Through the edification of the congregation, the community is always reconstructing itself, and this elevation above the everyday prevents the everyday itself from going under. The fact that such a returning of the community to its point of origin must submit to spatial and temporal limitations, that it steers away from worldly community, and that it is brought about through special celebrations – this is only a sign of man's dubious position between above and below, one that constantly forces him to establish on his own what is given or what has been conquered in the tension.

Since the determining characteristic of the lower region is its lack of tension, the togetherness in the hotel lobby has no meaning. While here, too, people certainly do become detached from everyday life, this detachment does not lead the community to assure itself of its existence as a congregation. Instead it merely displaces people from the unreality of the daily hustle and bustle to a place where they would encounter the void only if they were more than just reference points. The lobby,[1] in which people find themselves *vis-à-vis de rien*,[2] is a mere gap that does not even serve a purpose dictated by *Ratio* (like the conference room of a corporation), a purpose which at the very least could mask the directive that had been perceived in the relation. But if a sojourn in a hotel offers neither a perspective on nor an escape from the everyday, it does provide a groundless distance from it which can be exploited, if at all, *aesthetically* – the aesthetic being understood here as a category of the nonexistent type of person, the residue of that positive aesthetic which makes it possible to put this nonexistence into relief in the detective novel. The person sitting around idly is overcome by a disinterested satisfaction in the contemplation of a world creating itself, whose purposiveness is felt without being associated with any representation of a purpose. The Kantian definition of the beautiful is instantiated here in a way that takes seriously its isolation of the aesthetic and its lack of content. For in the emptied-out individuals of the detective novel – who, as rationally constructed complexes, are comparable to the transcendental subject – the aesthetic faculty is indeed detached from the existential stream of the total person. It is reduced to an unreal, purely formal relation that manifests the same indifference to the self as it does to matter. Kant himself was able to overlook this horrible last-minute sprint of the transcendental subject, since he still believed there was a seamless transition from the transcendental to the preformed subject-object world. The fact that he does not completely give up the total person even in the aesthetic realm is confirmed by his definition of the "sublime," which takes the ethical into account and thereby attempts to reassemble the remaining pieces of the fractured whole. In the hotel lobby, admittedly, the aesthetic – lacking all qualities of sublimity – is presented without any regard for these upward-striving intentions, and the formula "purposiveness without purpose"[3] also exhausts its content. Just as the lobby is the space that does not refer beyond itself, the aesthetic condition corresponding to it constitutes itself as its own limit. It is forbidden to go beyond this limit, so long as the tension that would propel the breakthrough is repressed and the marionettes of *Ratio* – who are not human beings – isolate themselves from their bustling activity. But the aesthetic that has become an end in itself pulls up its own roots; it obscures the higher level toward which it should refer and signifies only its own emptiness, which, according to the literal meaning of the Kantian definition, is a mere relation of faculties. It rises above a meaningless formal harmony only when it is in the service of something, when instead of making claims to autonomy it inserts itself into the tension that does not concern it in particular. If human beings orient themselves beyond the form, then a kind of beauty may also mature that is a fulfilled beauty, because it is the consequence and not the aim – but where beauty is chosen as an aim without further consequences, all that remains is its empty shell. Both the hotel lobby and the house of God respond to the aesthetic sense that articulates its legitimate demands in them. But whereas in the latter the beautiful employs a language with which it also testifies against itself, in the former it is involuted in its muteness, incapable of finding the other. In tasteful lounge chairs a civilization intent on rationalization comes to an end, whereas the decorations of the church pews are born from the tension that accords them a revelatory meaning. As a result, the chorales that are the expression of the divine service turn into medleys whose strains encourage pure triviality, and devotion congeals into erotic desire that roams about without an object.

The *equality* of those who pray is likewise reflected in distorted form in the hotel lobby. When a congregation forms, the differences between people disappear, because these beings all have one and the same destiny, and because, in the encounter with the spirit that determines this destiny, anything that does not determine that spirit simply ceases to exist – namely, the limit of necessity, posited by man, and the separation, which is the work of nature. The provisional status of communal life is experienced as such in the house of God, and so the sinner enters into the "we" in the same way as does the upright person whose assurance is here disturbed. This – the fact that everything human is oriented toward its own contingency – is what creates the equality of the contingent. The great pales next to the small, and good and evil remain suspended when the congregation relates itself to that which no scale can measure. Such a relativization of qualities does not lead to their confusion but instead elevates them to the status of reality, since the relation to the last things demands that the penultimate things be convulsed without being destroyed. This equality is positive and essential, not a reduction and foreground; it is the fulfillment of what has been differentiated, which must renounce its independent singular existence in order to save what is most singular. This singularity is awaited and sought in the house of God. Relegated to the shadows so long as merely human limits are imposed, it throws its own shadow over those distinctions when man approaches the absolute limit.

In the hotel lobby, equality is based not on a relation to God but on a relation to the nothing. Here, in the space of unrelatedness, the change of environments does not leave purposive activity behind, but brackets it for the sake of a freedom that can refer only to itself and therefore sinks into relaxation and indifference. In the house of God, human differences diminish in the face of their provisionality, exposed by a seriousness that dissipates the certainty of all that is definitive. By contrast, an aimless lounging, to which no call is addressed, leads to the mere play that elevates the unserious everyday to the level of the serious. Simmel's definition of society as a "play form of sociation"[4] is entirely legitimate, but does not get beyond mere description. What is presented in the hotel lobby is the formal similarity of the figures, an equivalence that signifies not fulfillment but evacuation. Removed from the hustle and bustle, one does gain some distance from the distinctions of "actual" life, but without being subjected to a new determination that would circumscribe from above the sphere of validity for these determinations. And it is in this way that a person can vanish into an undetermined void, helplessly reduced to a "member of society as such" who stands superfluously off to the side and, when playing, intoxicates himself. This invalidation of togetherness, itself already unreal, thus does not lead up toward reality but is more of a sliding down into the doubly unreal mixture of the undifferentiated atoms from which the world of appearance is constructed. Whereas in the house of God a creature emerges which sees itself as a supporter of the community, in the hotel lobby what emerges is the inessential foundation at the basis of rational socialization. It approaches the nothing and takes shape by analogy with the abstract and formal *universal concepts* through which thinking that has escaped from the tension believes it can grasp the world. These abstractions are inverted images of the universal concepts conceived within the relation; they rob the ungraspable given of its possible content, instead of raising it to the level of reality by relating it to the higher determinations. They are irrelevant to the oriented and total person who, the world in hand, meets them halfway; rather, they are posited by the transcendental subject, which allows them to become part of the powerlessness into which that transcendental subject degenerates as a result of its claim to be creator of the world. Even if free-floating *Ratio* – dimly aware of its limitation – does acknowledge the concepts of God, freedom, and immortality, what it discovers are not the homonymic existential concepts, and the categorical imperative is surely no substitute for a commandment that arises out of an ethical resolution. Nevertheless, the weaving of these concepts into a system confirms that people do not want to abandon the reality that has been lost; yet, of course, they will not get hold of it precisely because they are seeking it by means of a kind of thinking which has repudiated all attachment to that reality. The desolation of *Ratio* is complete only when it removes its mask and hurls itself into the void of random abstractions that no longer mimic higher determinations, and when it renounces seductive consonances and desires itself even as a concept. The only immediacy it then retains is the now openly acknowledged nothing, in which, grasping upward from below, it tries to ground the reality to which it no longer has access. Just as God becomes, for the person situated in the tension, the beginning and end of all creation, so too does the intellect that has become totally self-absorbed create the appearance of a plenitude of figures from zero. It thinks it can wrench the world from this meaningless universal, which is situated closest to that zero and distinguishes itself from it only to the extent necessary in order to deduct a something. But the world is world only when it is interpreted by a universal that has been really experienced. The intellect reduces the relations that permeate the manifold to the common denominator of the concept of energy, which is separated merely by a thin layer from the zero. Or it robs historical events of their paradoxical nature and, having leveled them out, grasps them as progress in one-dimensional time. Or, seemingly betraying itself, it elevates irrational "life" to the dignified status of an entity in order to recover itself, in its delimitation, from the now liberated residue of the totality of human being, and in order to traverse the realms across their entire expanse. If one takes as one's basis these extreme reductions of the real, then (as Simmel's philosophy of life confirms) one can obtain a distorted image of the discoveries made in the upper spheres – an image that is no less comprehensive than the one provided by the insistence of the words "God" and "spirit." But even less ambiguously than the abusive employment of categories that have become incomprehensible, it is the deployment of empty abstractions that announces the actual position of a thinking that has slipped out of the tension. The visitors in the hotel lobby, who allow the individual to disappear behind the peripheral equality of social masks, correspond to the exhausted terms that coerce differences out of the uniformity of the zero. Here, the visitors suspend the undetermined special being – which, in the house of God, gives way to that invisible equality of beings standing before God (out of which it both renews and determines itself) – by devolving into tuxedos. And the triviality of their conversation, haphazardly aimed at utterly insignificant objects so that one might encounter oneself in their exteriority, is only the obverse of prayer, directing downward what they idly circumvent.

The observance of *silence*, no less obligatory in the hotel lobby than in the house of God, indicates that in both places people consider themselves essentially as equals. In "Death in Venice" Thomas Mann formulates this as follows: "A solemn stillness reigned in the room, of the sort that is the pride of all large hotels. The attentive waiters moved about on noiseless feet. A rattling of the tea service, a half-whispered word was all that one could hear."[5] The contentless solemnity of this conventionally imposed silence does not arise out of mutual courtesy, of the sort one encounters everywhere, but rather serves to eliminate differences. It is a silence that abstracts from the differentiating word and compels one downward into the equality of the encounter with the nothing, an equality that a voice resounding through space would disturb. In the house of God, by contrast, silence signifies the individual collecting himself as firmly directed self, and the word addressed to human beings is effaced solely in order to release another word, which, whether uttered or not, sits in judgment over human beings.

Since what counts here is not the dialogue of those who speak, the members of the congregation are anonymous. They outgrow their names because the very empirical being which these names designate disappears in prayer; thus, they do not know one another as particular beings whose multiply determined existences enmesh them in the world. If the proper name reveals its bearer, it also separates him from those whose names have been called; it simultaneously discloses and obscures, and it is with good reason that lovers want to destroy it, as if it were the final wall separating them. It is only the relinquishing of the name – which abolishes the semisolidarity of the intermediate spheres – that allows for the extensive solidarity of those who step out of the bright obscurity of reciprocal contact and into the night and the light of the higher mystery. Now that they do not know who the person closest to them is, their neighbor becomes the closest, for out of his disintegrating appearance arises a creation whose traits are also theirs. It is true that only those who stand before God are sufficiently estranged from one another to discover they are brothers; only they are exposed to such an extent that they can love one another without knowing one another and without using names. At the limit of the human they rid themselves of their naming, so that the word might be bestowed upon them – a word that strikes them more directly than any human law. And in the seclusion to which such a relativization of form generally pushes them, they inquire about their form. Having been initiated into the mystery that provides the name, and having become transparent to one another in their relation to God, they enter into the "we" signifying a commonality of creatures that suspends and grounds all those distinctions and associations adhering to the proper name.

This limit case "we" of those who have dispossessed themselves of themselves – a "we" that is realized vicariously in the house of God due to human limitations – is transformed in the hotel lobby into the isolation of anonymous atoms. Here profession is detached from the person and the name gets lost in the space, since only the still unnamed crowd can serve *Ratio* as a point of attack. It reduces to the level of the nothing – out of which it wants to produce the world – even those pseudo-individuals it has deprived of individuality, since their anonymity no longer serves any purpose other than meaningless movement along the paths of convention. But if the meaning of this anonymity becomes nothing more than the representation of the insignificance of this beginning, the depiction of formal regularities, then it does not foster the solidarity of those liberated from the constraints of the name; instead, it deprives those encountering one another of the possibility of association that the name could have offered them. Remnants of individuals slip into the nirvana of relaxation, faces disappear behind newspapers, and the artificial continuous light illuminates nothing but mannequins. It is the coming and going of unfamiliar people who have become empty forms because they have lost their password, and who now file by as ungraspable flat ghosts. If they possessed an interior, it would have no windows at all, and they would perish aware of their endless abandonment, instead of knowing of their homeland as the congregation does. But as pure exterior, they escape themselves and express their nonbeing through the false aesthetic affirmation of the estrangement that has been installed between them. The presentation of the surface strikes them as an attraction; the tinge of exoticism gives them a pleasurable shudder. Indeed, in order to confirm the distance whose definitive character attracts them, they allow themselves to be bounced off a proximity that they themselves have conjured up: their monological fantasy attaches designations to the masks, designations that use the person facing them as a toy. And the fleeting exchange of glances which creates the possibility of exchange is acknowledged only because the illusion of that possibility confirms the reality of the distance. Just as in the house of God, here too namelessness unveils the meaning of naming; but whereas in the house of God it is an awaiting within the tension that reveals the preliminariness of names, in the hotel lobby it is a retreat into the unquestioned groundlessness that the intellect transforms into the names' site of origin. But where the call that unifies into the "we" is not heard, those that have fled the form are irrevocably isolated.

In the congregation the entire community comes into being, for the immediate relation to the supralegal *mystery* inaugurates the paradox of the law that can be suspended in the actuality of the relation to God. That law is a penultimate term that withdraws when the connection occurs that humbles the self-assured and comforts those in danger. The tensionless people in the hotel lobby also represent the entire society, but not because transcendence here raises them up to its level; rather, this is because the hustle and bustle of immanence is still hidden. Instead of guiding people beyond themselves, the mystery slips between the masks; instead of penetrating the shells of the human, it is the veil that surrounds everything human; instead of confronting man with the question of the provisional, it paralyzes the questioning that gives access to the realm of provisionality. In his all-too contemplative detective novel *Der Tod kehrt im Hotel ein* (Death Enters the Hotel), Sven Elvestad writes: "Once again it is confirmed that a large hotel is a world unto itself and that this world is like the rest of the large world. The guests here roam about in their light-hearted, careless summer existence without suspecting anything of the strange mysteries circulating among them."[6] "Strange mysteries": the phrase is ironically ambiguous. On the one hand, it refers quite generally to the disguised quality of lived existence as such; on the other, it refers to the higher mystery that finds distorted expression in the illegal activities that threaten safety. The clandestine character of all legal and illegal activities – to which the expression initially and immediately refers – indicates that in the hotel lobby the pseudo-life that is unfolding in pure immanence is being pushed back toward its undifferentiated origin. Were the mystery to come out of its shell, mere possibility would disappear in the fact: by detaching the illegal from the nothing, the Something would have appeared. The hotel management therefore thoughtfully conceals from its guests the real events which could put an end to the false aesthetic situation shrouding that nothing. Just as the formerly experienced higher mystery pushes those oriented toward it across the midpoint, whose limit is defined by the law, so does the mystery – which is the distortion of the higher ground and as such the utmost abstraction of the dangers that disrupt immanent life – relegate one to the lapsed neutrality of the meaningless beginning from which the pseudo-middle arises. It hinders the outbreak of differentiations in the service of emancipated *Ratio*, which strengthens its victory over the Something in the hotel lobby by helping the conventions take the upper hand. These are so worn out that the activity taking place in their name is at the same time an activity of dissimulation – an activity that serves as protection for legal life just as much as for illegal life, because as the empty form of all possible societies it is not oriented toward any particular thing but remains content with itself in its insignificance.

NOTES

1 Kracauer here and elsewhere uses the anglicism *Hall* instead of the German term *Halle* ("lobby"), which is derived from it.
2 In French in the original. "Face-to-face with nothing."
3 This hallmark phrase from Kant's *Critique of Judgment* is put in quotation marks in the later republication of this essay. See *Schriften* 1, 130.
4 "Soziologie der Geselligkeit," in *Grundfragen der Soziologie: Individuum und Gesellschaft* (Berlin: Göschen, 1917; 4th ed. Berlin: de Gruyter, 1984), 53; translated by Everett C. Hughes as "Sociability," in Donald N. Levine, ed., *Georg Simmel: On Individuality and Social Forms* (Chicago: University of Chicago Press, 1971), 130. Strictly speaking, Simmel's phrase is here used to define not society but sociability.
5 Thomas Mann, "Tod in Venedig," *Sämtliche Erzählungen* (Frankfurt: Fischer Verlag, 1963), 376. Compare the translation by H. T. Lowe-Porter, in *Death in Venice and Seven Other Stories* (New York: Vintage, 1955), 28.
6 Sven Elvestad, *Doden tar ind paa hotellet* (1921), in German as *Der Tod kehrt im Hotel ein: Roman*, trans. F. Koppel (Munich: Georg Müller, 1923).

'World Cities'

from *The Sphinx in the City* (1993)

Elizabeth Wilson

Editors' Introduction

Elizabeth Wilson's *The Sphinx in the City* was, when first published by Virago Press in 1991 and then by the University of California (the edition used here), a seminal contribution to a gendered approach to urbanism. It introduced personal experiences and a first-person narrative in a critical view of metropolitan cities, in a way which can be compared with Doreen Massey's writing (see Part Seven). Wilson argues that a revival of cities depends on looking at them in fresh ways, while dominant ways of seeing the city in the past have produced cities of division and exclusion. A key aspect of Wilson's writing is her ability to highlight what makes cities work, and what makes them vibrant. Wilson remains committed to such environments, seeing them as engaging and exciting, offering opportunities for carnival for all, and is committed also to the need for regulation, though on a more equitable basis than in planning's largely patriarchal (and at times, as in the garden city movement, anti-urban) history. The chapter reprinted below deals with 'World Cities', not in the contemporary use of the term for cities seen as having a world role, or being world-class cities, but in the sense of cities throughout the world, particularly in colonial countries and what was once called the 'third world'. Her remarks on informal settlements can be read in conjunction with Jeremy Seabrook's text in Part Ten; and on gender issues in such cities in conjunction with Farha Ghannam's essay in Part Seven.

Until recently Professor of Cultural Studies at North London University (now merged as London Metropolitan University), Elizabeth Wilson currently teaches at the Architectural Association, London. A recent publication is *The Contradictions of Cuture: Cities, Culture, Women* (London, Sage, 2000), part of which revisits the material of *The Sphinx in the City* a decade after it was written. She has also produced a book on bohemianism from the nineteenth century to the present, *Bohemians: The Glamorous Outcasts* (London, I. B. Tauris, 2000).

> The train swept in . . . From the third class coaches there emerged first the experienced Nairobi wives, hefty women with calf-length skirts and aggressively set sleeves, passing tin and wooden suitcases through windows, bunches of green bananas, squawky hens and passive children. Next . . . came the men . . . After them the rearguard, the mothers-in-law and the young brides, not very pushing . . .
>
> At last Paulina came into sight, clutching a triangular bundle . . . She was wearing a faded blue cotton dress and a white headscarf. Her rubber shoes were scuffed and brown . . . He could see that . . . the journey had frightened her . . .
>
> The front of the station was full of taxis and cars meeting trains. People thronged together. Ahead of them lay a street of tall buildings and rushing traffic. She supposed it was normal for big cities to be like this, but . . . she wanted to leap away from the kerb each time a car came close and felt, being new and strange, that she must be the one to give way whenever she came face to face with someone hurrying in the opposite direction.
>
> (Marjorie Oludhe Macgoye, *Coming to Birth* (1964))

When in the nineteenth and early twentieth centuries writers and planners viewed urban growth with anxiety, feeling that these vast conurbations had run out of control, the city they discussed was the western city. Theoretical writings usually either defined the city in such a way as to exclude all but western cities, or simply did not take non-western cities into account. The assumptions of the town-planning movement, which grew out of this anxiety, were subsequently applied to the fast-growing cities of the 'third world', to their great detriment. (The term 'third world' is used as a familiar, although unsatisfactory, shorthand.)

Since the Second World War, new forms of colonialism and capitalism have contributed to the creation of huge population centres in the non-western world. Between 1950 and 1970 the population of São Paulo grew from just under two and a half million to eight and a half million; that of Delhi from 1,737,000 to 3,100,000; while that of Lagos grew from 267,000 in 1952 to between two and a half and three million by 1980.[1] Thus the experience of urbanisation in nineteenth-century Europe and America is being repeated in Latin America, South-East Asia and Africa.

At least two-thirds of this growth has been due to the migration of rural peasants to the city. Sometimes migrants travel from one country to another. Workers from Mozambique and Malawi find work in South Africa, for example. As in the early years of the Industrial Revolution in Europe, rural migrants are often attracted to the cities by the promise of work. Numbers in the present-day world city are also swelled by those who have been pushed off the land by changes in farming patterns and land development, as was also to some extent the case in pre-Victorian Britain. Patterns vary from one country and one continent to another; in tropical Africa, the majority of town dwellers retain strong links with their rural family of origin and often continue to farm in the countryside.

It has been usual to take the view that the cities of the third world are becoming, at least superficially, more like western cities, but that in some more fundamental way urbanisation in the third world is actually different from that both of the west and of the Soviet Union and eastern Europe. Such a view is itself ethnocentric and oversimplified. For one thing it ignores the differences among various third-world countries and their cities. In Africa and South Asia, for example, the level of urbanisation is much lower than in Latin America – although Africans are now migrating to the cities at a faster rate than are the populations of Latin America. It is also ahistorical to lump all third-world cities together, given that most Latin American cities were founded in the sixteenth century and are therefore much older than the majority of African, or indeed North American cities, while many cities of the Middle East are older still.[2]

Another difference is that in many Latin American and Middle Eastern countries power in the cities lies with long-established local elite families, part of a rigid class structure. By contrast, the African urban elite have maintained close links with rural-based extended families, many members of which remain very poor. In African cities there is usually no one dominant language, and this too makes for a difference from Latin America, where Portuguese or Spanish is universal.[3]

In Latin America urbanisation has, as in the west, gone in tandem with industrialisation. In Africa 'urbanisation without industrialisation' has been more usual – although of course there is some industrialisation as well.

At issue in the whole debate about urbanisation in the third world is an underlying presumption that urbanisation must always follow the same path – that taken historically by western cities in the industrial period. This may not be so for Africa; a form of urbanisation based on manufacturing industry may be a pattern that will no longer apply by the time full industrialisation reaches Africa – if it does.[4]

Despite the changing form of colonialism, the west has not lost interest in former dependencies and new 'spheres of influence'. On the contrary, research on a huge scale has been undertaken into the problems of the new world mega-cities. Sometimes this research is financed by the countries concerned, but often by United Nations, American, or other global interests. As a result, there is today a mass of information about non-western cities, but this knowledge is itself used as an instrument of power and domination. Researchers and planners have extended their former concern for the moral, physical and eugenic well-being of the inhabitants of their own cities, and have often treated third-world population centres as caricatures of the 'parent' cities of the west – as reproducing their worst problems in grotesquely exaggerated form. 'Detached observers' have attempted to investigate and describe the life of the 'teeming' cities, not in order to enter into the experience of urban life – as the *flâneur* of nineteenth-century Paris or twentieth-century São Paulo had done – but in order to change and reform. Plans for third-world cities have been stamped, like those of the nineteenth century, with the utopianism of those who aim to do away with poverty and crime while preserving capitalist interests.

Yet the history of urbanism and planning in Europe and the United States has continued largely to overlook the way in which the economies of countries that were being colonised in the nineteenth and twentieth centuries contributed to the urbanisation of the west. The histories of Bournville and Port Sunlight do not refer to the source of the cocoa that supported Cadbury's profits or the coconut oil that went into the making of Lever's soaps.

As part of the same process of colonialism, the British (and other colonising nations) sent more than cheap cotton or other manufactured goods to their colonies. They also exported the emergent town planning of the early twentieth century. 'We want not only England but all parts of the Empire to be covered with Garden Cities', wrote one of Ebenezer Howard's disciples.[5] In 1912, Captain Swinton, Chairman of the London County Council and a member of the New Delhi Planning Committee, enthused: 'I hope that in the new Delhi we shall be able to show how those ideas which Mr [Ebenezer] Howard put forward . . . can be brought in to assist this first Capital created in our time. The fact is that no new city or town should be permissible in these days to which the word "Garden" cannot be rightly applied.'[6]

In the 1950s, *Town and Country Planning*, the vehicle of the Garden City movement, was still praising the benign influence that British town planning had had in the 1930s on the countries of the British Empire. The 1932 Town and Country Planning Act had been used first in Trinidad, later in Uganda, Fiji, Aden, Sarawak, Mauritius and in Sierra Leone and East Africa: 'thus, the 1932 Act has left its mark in all corners of the world . . . Modern developments in British planning procedure are followed closely by colonial planning officers who are always ready to profit by the experience of the mother country.'[7] The earliest colonial settlements were often explicitly military towns. The later developments, built in order to promote the garden-city obsessions of physical health, sanitation and social orderliness, constituted a form of 'cultural colonialism'.[8]

The influence of British colonialism was not confined to developing countries; it equally determined the shape of cities in Canada and Australia. Australia, which is economically part of the 'west', is an interesting example of a highly urbanised society (85 per cent of its population was living in cities in 1971), which urbanised faster than either the United States or Canada. Its suburbs and urban transport systems anticipated similar developments in Europe and the United States; in the late nineteenth century Melbourne and Sydney rivalled London as centres of culture and commerce – Augustus Sala referred to 'marvellous Melbourne', and Melbourne was also compared to Chicago and to Paris.[9] Yet the Australian heritage has been largely defined in terms of the outback and the 'bush ethos', – just as Americans continued to be obsessed with the Frontier. The intensity and savagery of the Australian landscape, such a stark contrast to a caricatured suburban banality, has been perceived as a source of truth and inspiration. 'Depressed urban workers were led to believe that the bush was a place of healing away from the diseased life of the city. Furthermore the bush was regarded as masculine in gender, a place to escape from the ladylike refinements of the city and women's challenge to [male] supremacy.'[10] In the loneliness of the outback, the solitary male could pit himself against raw nature, although in practice, of course, he was usually supported by the unstinting and unrecognised labour of women, a situation subtly explored by Henry Handel Richardson in her great novel of Australian pioneering, *The Fortunes of Richard Mahony*. Recent attempts to atone for the great wrongs done to the Aboriginal peoples, however well intentioned, have served to reinforce the contrast between inauthentic city and the spiritual qualities of the Bush.

In India, the British imposed their own form of urban planning on a civilisation that already boasted many cities, although they paid little attention to the traditional forms of Indian urban life. New Delhi, built by the English architect Edwin Lutyens, was an administrative and bureaucratic centre, not a modern industrial city. In this it was typical of the way in which urbanisation in dependent imperial territories, such as India then was, took place without the industrialisation which had been key to urbanism in the West.

In Africa the old urban centres such as Kano in Northern Nigeria, Ibadan in Southern Nigeria and Benin were likewise disregarded. West Africa was the site of long-established pre-industrial indigenous cities, especially amongst the Yoruba. The Yoruba civilisation was based on an agricultural economy, but the farmers lived in the cities. The farms were set in belts of agricultural land that surrounded the towns to a distance of about fifteen miles. Within the cities craft specialisation and guilds on the one hand and market trading on the other were important activities. The markets were large, involving entrepreneurs and a money economy based on cowrie shells. The traditional extended family, or lineage system, remained important in the cities, partly because of the value of the land. Women played a subordinate role in the lineage family system, but occupied a central place in the market system, retailing being largely in their hands.[11]

In fact the lineage family remained important in urban life all over west Africa. A variety of urban housing forms developed to house these extended families. In the towns settled by the Mende tribe, the household typically consisted of a group of houses built round a compound, with separate one-room apartments for the male head of the family, for young wives and their children, for a head wife and for the male head of household's mother and for younger men and other dependants. The Ga, living in older districts of Accra, in the Gold Coast, had a residential unit consisting of two adjoining establishments, one occupied by men and the other by their wives. By the 1950s, though, educated Ga, working in clerical occupations in Accra, had adopted a much smaller nuclear family type of unit.[12]

New cities along more colonial lines also grew up in twentieth-century west Africa. These were not 'the place of the white man' in the way that the copperbelt towns of southern Africa, or Cape Town, or Johannesburg were. The colonial administration adapted in some degree to them. Control over housing and housing conditions was less direct, and the cities offered many different forms of employments, the whole way of life was diversified, and the inhabitants were more independent. The commercial nature of these cities, based on the export of cash crops, favoured the expansion of an educated indigenous professional and business middle class.[13]

These cities brought changes to the lives of women. They represented choice. The lives of women in the traditional lineage family outside the city, living in a compound which a head of family shared with his wives and mother and where a wife would come under the authority of these women and the wives of her husband's brothers had been lives of submission to the family, who would be primarily interested in the economic advantages, in the shape of bridewealth, to be obtained from an impending marriage. Increasingly in the twentieth century girls could escape an unwelcome marriage by leaving for the city; and young couples could also break free to some extent of family restrictions by the same route.[14] This is not to imply that women always or even often wished for a different kind of life; simply to indicate the way in which the very existence of cities made it clear that other ways of life were possible.

In any case, life for the west African woman in the towns of mid-twentieth-century west Africa was not easy. Freedom from the restrictions of traditional ways of life brought insecurity as well as opportunities. Traditional marriage might be replaced by a statutory form of legal marriage, made possible by the introduction of western legal systems. Formal polygamy declined, but men were rarely monogamous. Nor did a husband have an individual obligation to support a deserted wife; that was still the responsibility of the lineage. Many men found it hard to earn enough money to support a family single-handed. Yoruba women were expected to wear as much new clothing and jewellery as possible on the many social occasions and ceremonial events in which they participated, and the employment of servants often also fell to the wife. Therefore, the majority of women in west African towns worked, and set great store by work. From their traditional sphere of marketing they expanded into nursing and other professional vocational training. They also set up voluntary associations, some for both sexes, which played a social and professional role, and some of which were women's associations. Through these associations they were enabled to raise capital to expand their trading opportunities. For this reason the associations presented to some extent an alternative to the family.[15]

In east Africa new urban forms were imposed upon a rural population. Lusaka, for example, in what is now Zambia, was planned and built as the capital of British Northern Rhodesia. Like New Delhi it was to be a governmental centre, and here – as in South Africa today – the colonial authorities attempted to restrict the participation of the African population in urban life. They were to be migrant labourers whose period of residence in an urban area was regarded as strictly temporary. It was assumed that African women would remain permanently in the rural areas.

The garden-city ideal inspired Lusaka, which was planned by Stanley D. Adshead, an enthusiastic disciple of Ebenezer Howard. The colonial version of the Letchworth/Welwyn ideal lacked, however, even the moderate reformism of those experiments. It is easy to dissent from the reformism, or laugh at the quaint utopian socialist lifestyle of the more eccentric

Letchworth residents, but at least they were attempting to create a less hierarchical and non-oppressive community; the garden city in its English incarnation had aimed to foster neighbourhoods and communities. In Africa, by contrast, reading rooms, libraries and lecture halls were replaced by the races, the club and the golf course. In Lusaka and Delhi, an outmoded British way of life was artificially preserved. Middle-class bureaucrats and professionals were able to aspire to an aristocratic way of life with retinues of servants and an aura of ruling-class grandeur.[16]

In the building of colonial cities the environment created was a visible expression of the ideas upon which colonial rule was based. Lusaka, for example, was built as a city for Europeans. The Africans were to remain in the rural hinterland, with the exception of able-bodied men who would be needed to undertake manual work, and of men and women needed for domestic service.

Housing for these workers was minimal, since they were expected to make periodic migrations from the countryside, returning there when not needed. Sometimes workers built their own hutments. Sometimes employers provided very basic housing free. They did not mind doing this, because they were then able to pay labourers very low wages on the grounds that they had no housing costs.

In the original plans of the nineteenth- and early-twentieth-century utopias of the western world the intention had been to zone women safely into areas of domesticity and consumption. They were to be excluded altogether from the colonial city. Like the British working class, the indigenous population, particularly the female population, was totally written out of the colonial city utopia.

After the Second World War local authorities began to provide hostel accommodation, but as the African population expanded and changed, 'unauthorised' dwellings (squatters' areas or shanty towns) provided an important source of housing for Africans in Lusaka. By 1963 the idea that the African population consisted of a circulating labour force that frequently returned to a rural family was far from the reality; the census that year revealed that nearly as many African women as men were living in the city, and that over 50 per cent of its black inhabitants were under the age of twenty-one.[17]

In Latin America cities grew up in a number of different ways. There, the strong planning tradition of Spanish and Portuguese colonialism determined the form of the older cities. São Paulo, Brazil, for example, was founded in 1554 by the Jesuits as a bridgehead to the hinterland to further their 'domestication' of indigenous Indians. After independence in 1822 it became a provincial administrative and political centre. By the mid nineteenth century coffee-growing was substantial; this, and the related growth of a railroad network, transformed São Paulo by 1900 into a city of a quarter of a million inhabitants (as against 65,000 in 1890) and an important industrial centre.[18]

An infrastructure of sewage, roads, water supplies and lighting always lagged behind the needs of the raggedly growing city. By the 1930s, when Claude Lévi-Strauss was living there, the city was dealing with its housing problem by a variety of hand-to-mouth solutions, and 'in 1935 the citizens of São Paulo boasted that, on an average, one house per hour was built in their town . . . The town is developing so fast that it is impossible to obtain a map of it; a new edition would be required every week.' Everything was chaotic, and 'vast roadworks' were being built adjacent to old crafts quarters like Syrian bazaars. A new public avenue cut through a once exclusive neighbourhood, 'where painted wooden villas were falling to pieces in gardens full of eucalyptus and mango trees'; then came a working-class neighbourhood, 'along which lay a red light district, consisting of hovels with raised entresols, from the windows of which the prostitutes hailed their clients. Finally, on the outskirts of the town, the lower-middle-class residential areas . . . were making headway.'[19]

Today, many city dwellers live in shanty towns, and these have become the most telling and guilt-inducing image of 'third-world poverty', inviting the voyeuristic horror of the westerner. In the 1960s about one quarter of the population of cities such as Manila and Djakarta, one third of the population of Mexico City and half that of Lima lived in shanty towns. In the late 1970s it was estimated that by 1990 three-quarters of Lima's poor would be living in such conditions.[20]

The literature on shanty towns usually emphasises their squalor and poverty. Dwellings are constructed of wattle, cardboard or corrugated iron, and placed in close proximity, with narrow lanes running between them which also serve as open drains. Seven or eight members of a household sleep together in one room, and animals often share the human living space. Whether located in Venezuela or Calcutta, Cairo or Djakarta, the descriptions of the '*barrios*' are extraordinarily reminiscent of the Victorians' lurid depictions of nineteenth-century slums. Like the London or Paris poor of that time, women of the *barrios* queue for water, which is supplied – often intermittently – from standpipes. There are no proper sewage systems in the 'typical' shanty town, which in general lacks all basic amenities.

It is often assumed that the people living in such conditions must somehow themselves be inadequate, just as the Victorian reformers assumed for the most part that the poor were locked into the slums because they were lacking in moral fibre. For example, an American academic, M.H. Ross, described the Nairobi shanty towns as follows:

> Downtown Nairobi is beautiful, with its tall buildings, modern architecture and flowering trees . . . Four miles [away] . . . live some 10,000 to 20,000 urban squatters . . . The houses, crammed together in an apparently haphazard fashion dictated by the uneven terrain of the valley's walls, are built of mud and wattle and have roofs made of cardboard, flattened-out tin cans, or even sheet metal. A visitor entering the area is struck by the lack of social services; the roads are makeshift, garbage is piled high in open areas, and children play in the dust.[21]

He dismissed the inhabitants of Mathare as 'generally urban misfits and rural outcasts . . . [who] lack the skills necessary to find jobs in the modern economy' – yet went on to reveal that the community organised nursery schools, a co-op and social events, and that 'the most striking aspect of Mathare is that it is highly organised and politically integrated . . . There is a clearly identifiable group of community leaders.'

In the 1950s and 1960s governments in many countries responded to spontaneous settlements with harassment and evictions. Since the 1970s a different strategy has been more often used: a 'site and services' policy. This recognises the impossibility of providing better housing, and also that the squatters will not, indeed cannot, return to their former rural homes. The trend has therefore been to legalise the occupation of the shanty towns. Where possible sewage, electricity and water may be provided. The residents are encouraged to build their own homes (which they were already doing). National government and international agencies recognised that this self-help alternative could be a low-cost housing policy, and since 1974 self-help has been officially endorsed by the World Bank.[22]

Some researchers and planners have given these policies enthusiastic support. Critics, on the other hand, have argued that they simply let governments off the hook and act as a justification for low wages and frightful living conditions.[23]

Running through many of these debates, which for the most part originate in western institutions, has been the continuing theme of the planners' contempt for the poor. Individual studies of African or other non-western cities have testified to their ebullience, diversity and variety. There is friendliness, and social contacts are established with ease. Most of those who come to the towns do not come as strangers but already have members of the family established there. This family will cushion the shock of the transition to urban life, and provide much-needed help and support in the initial months. The new arrival relies precisely on ties of kinship to see her or him through the bewildering early days of life in the town.

Yet many who have written of third-world city life have emphasised poverty, the breakdown of family life, prostitution, crime and psychological maladjustment, and have blamed the impersonal nature of city life with its alienation and anomie. The work of Oscar Lewis, who described an alleged 'culture of poverty', was extremely influential in the 1950s and 1960s in the development of this view,[24] which, consciously or not, built on the assumptions of Georg Simmel and Louis Wirth, who had emphasised the impersonal factors in urban life.

More recently, there has been a greater recognition of the persistence of family obligations in the city. In the cities of the third world as in nineteenth-century New York or Chicago, or for that matter in the new industrial towns of nineteenth-century Lancashire, family connections play an important role in the process by which immigrants and rural workers are transformed into urban dwellers.

Yet the negative view of the effects of city life on the individual and the family has persisted. In addition, writers, planners and officials have then imposed the weight of the theories and assumptions of postwar western planning with its emphasis on zoning, segregation and surveillance. The result has been that in third-world cities, as in the cities of postwar Britain and the United States, redevelopment, zoning, skyscraper business districts and dormitory suburbs are to be the answers to the 'chaos' and 'moral breakdown' of the unplanned industrial city. The same exaggerated faith in the ability of neutral, scientific planning to solve the urban crisis has led to a reproduction of the same mistakes as in western cities, but with far more extreme results.

Western assumptions about the normality and universality of the nuclear family have been extended to the shanty towns and urban populations of the third world. The view that the slums of Latin America fostered unmarried motherhood, delinquency, prostitution and a psychology of apathy and living for the moment has ceded to that of writers who were more likely to emphasise the 'normality' and stability in western terms of family life and social existence in the shanty towns. 'Normality', however, has meant the stereotype of family life in which the husband works and the woman remains at home, engaged in full-time domestic work and child rearing. This is not the actual experience even of the majority of western families. It has been used to marginalise women in the paid workforce, and has also perpetuated inequalities generally between men and women inside and outside the family.[25]

It is even more inappropriate as a model of the family in non-western urban settlements. Its basic assumptions do not take into account the situation and crucial importance of women in and to these housing settlements. National and international investment in self-help housing projects has failed to answer the needs of women and, on the contrary, has excluded them. Governmental and financial policies tend to be based on the assumption that the nuclear family with a male breadwinner is the usual and indeed natural family form, whereas in fact different household forms coexist. Recent figures have suggested that one in three of the world's households are now headed by women. In urban areas, however, approximately 50 per cent of households are headed by women, and in the refugee camps of Central America the figure is closer to 90 per cent.[26]

Women-headed households are placed in a kind of double jeopardy. Their very existence is denied, minimised or not taken account of. Then, in addition, it is often assumed that only male-headed households are sufficiently stable to merit inclusion in housing schemes. In one project in São Paulo, funded by the Brazilian Housing Bank Profilurb programme, women who headed households were excluded by the criteria of eligibility. Yet some observers have argued that families headed by women may be more stable, show more responsibility in paying back loans, in paying rent and in improving properties.[27]

Governments have also failed to recognise that women play an especially important role in 'community management'. Because women are more directly concerned than men with the welfare of the household, and with 'community' issues such as water supply and safety, they are normally more aware than men of the needs of a housing project, and more committed to its success.[28]

A neglect of the needs of women has often militated against the success of self-help housing projects. For example, a failure to recognise that women as well as men engage in paid employment leads to a male bias when 'squatters' are relocated, often miles from the centre of the city. This bias arises either when rehousing is near factories or other sites of male employment, or simply because it is assumed that men can travel long distances to work. Planners pay little heed to the fact that women both need work, and need to be near their work if, as they invariably have to, they are successfully to combine it with domestic duties. For example, many women take in laundry or have jobs as maids, and for this they need to live near to their employers. Even if it is acknowledged that the relocation has created transport difficulties, these may be interpreted as difficulties for men. In Belo Horizonte, Brazil, for example, public transport was laid on at peak hours, so that men could travel to work, but was withdrawn during the day when women needed to use the service, either to take children to school, to go shopping, or get to their own part-time jobs.[29]

Another way in which women are ignored and excluded is that they are not consulted on issues such as the design of houses. For example, the object of the Tanzanian 'Better Housing Campaign' was to persuade people to build more durable houses, built of imported materials, to replace traditional ones built with local materials. One unintended effect of the higher costs of building with imported materials was that it impeded the tradition of building separate accommodation for male and female members of families. As a result, women were redefined as dependants, and their traditional autonomy was undermined. In two housing projects in Tunis, houses were designed with a much smaller courtyard than in traditional design. Because Muslim women, spending most of their time in the home, needed this internal space, its absence caused depression and even suicide.[30]

In spite of all the obstacles and prejudice, women have managed to involve themselves in many local housing projects. They have organised locally against the fearful hardships of the majority of the *barrios*. In one settlement in Guayaquil, capital of Ecuador, initial conditions were terrible, as the settlement consisted of swampland. Settlers bought plots on which they then had to build their own houses. To begin with, the settlement had no services, not even water. It was approached by perilous catwalks above the swamps. These were so dangerous that two children drowned, while two women died as a result of wading through the swamp, and a man was electrocuted while trying to fix a light connection.

Women, given that they less frequently went out to work than did their male partners, were particularly isolated. Many had only just left their parents' home, others had given up paid work, and for those who tried to carry on with their laundry work the shortage of water was a disaster. They became more economically dependent than hitherto on their husbands or partners. So, while to settle on the swamp had economic advantages – they no longer had to pay rent and therefore had more money for education and consumer goods – the disadvantages of the area bore particularly upon women. Soon, however, they established mutual-aid networks, and this led to the formation of organised committees to agitate for change. To take a leading role in these did involve a challenge to traditional views of how a woman should behave (submissively, with the emphasis on her role as mother and homemaker): on the other hand, the traditional role itself validated membership of groups seeking to improve domestic living conditions.[31]

In Nicaragua, the Sandinista Revolution in 1979 resulted in a political commitment to tackle problems such as these, but changes were brought about only with great difficulty. In one project in Managua, men objected to women sharing in the actual work of house construction, but, after considerable discord, some women did manage to learn the basic building skills. The women in the project were more radical than the men in that they wanted the finished houses allocated according to need, while the men wanted them to be allocated according to the amount of work put in (advantageous to them, since on the whole the men had fewer alternative calls on their time). During a dispute over the allocation of one particular house, the issues of *machismo* (also referred to as *somocismo*, i.e. behaviour worthy of the Somoza regime, overthrown by the Sandinistas) was raised. This was possible because the revolution had initiated a reassessment of the roles of men and women, and was committed in principle to the equality of women.[32] Increasingly throughout the 1980s, however, the Nicaraguan experience was distorted by the incursions of the Contras and American destabilisation, so that towards the end of the decade, the mounting economic crisis was testing the survival strategies of urban women to the limit.[33]

Throughout Latin America, the position of single mothers is ambiguous. Confined to low-paid work, they, with their children, are among the poorest in the community, and still tend to be stigmatised, blamed for their plight, although many have been irresponsibly deserted by husband or lover. Some, however, have made the choice to bring up the children on their own, preferring the hardships of this way of life to the domineering behaviour of their menfolk, many of whom refuse to allow their women freedom of movement, abuse them physically, and spend the family income on heavy drinking. An Ecuador study showed that participation in the organisation of the *barrio* had also enabled some of the cohabiting women involved to achieve economic independence so that they were in a much more powerful position in relation to their husbands/partners.[34]

In countries such as Brazil, where feminism influenced radical middle-class women in the 1970s, the general turbulence of the political situation made possible, for a time at least, alliances between the feminists and women's groups from the working class and the poorest sections of society, including women in the shanty towns who were fighting for the provision of basic needs. In this way, the impact of feminist ideas on an educated middle class affected a much broader group of women, although in different ways. Female domestic servants, for example, became a militant and well-organised force – a far cry from the usual picture of maids as unorganised, and, indeed, impossible to organise.[35] For these women, life in the *barrios* and settlements was not a life of apathy and despair. Most of those living there neither came from nor wished to return to a rural life.

Life in the city provides the preconditions for continuing struggle, since in the city the poor, although 'excluded from the comforts of the city, are exposed to its modernity'.[36] The existence of the benefits of urban life – even though they are excluded from them – justifies their demands for *inclusion*. The gulf between what is and what might be may appear to widen; on the other hand, the city both raises aspirations and gives more chance of their realisation.[37]

Although cities in the non-western world differ both from one another and from the cities of Europe and North America, it has become customary to refer to the 'third-world city' or the 'world city' as a separate and recognisable entity. A global capitalism must surely be creating a global city: this seems to be the assumption upon which such a generalisation is based, although it is also recognised that global capital may act to differentiate one city, and one economy, from another. A fear remains that the world, and its cities, are becoming homogenised: difference is ironed out and everything is the same. At the same time, within every city a growing distance between rich and poor makes for another kind of unreality, and a gulf in experience that cannot be bridged. We move, then, from world city to postmodern city.

NOTES

1. Moser C and Peake L, eds (1987) *Women: Human Settlements and Housing*, London, Tavistock, pp4–5
2. O'Connor A (1983) *The African City*, London, Hutchinson, p315
3. Ibid
4. Ibid (p316)
5. King A D (1980) 'Exporting Planning: the Colonial and neo-Colonial Experience', in Cherry G, ed. (1980) *Shaping an Urban World*, London, Mansell
6. Ibid, quoting Swinton G (1912) 'Planning an Imperial Capital', *Garden Cities and Town Planning*, #2, p4
7. Ibid, quoting Steven P H M (1955) 'Planning Legislation in the Colonies', *Town and Country Planning*, March
8. Ibid
9. Briggs A (1968) *Victorian Cities*, Harmondsworth, Penguin, p278
10. Butel E (1985) *Margaret Preston: The Art of Constant Rearrangement*, Ringwood, Australia, Penguin p3; Sandercock L (1976) *Cities for Sale: Property Politics and Urban Planning in Australia*, London, Heinemann, p7
11. Bascom W (1955) 'Urbanisation Among the Yoruba', *American Journal of Sociology*, vol. 60, 5, p239
12. Little K (1959) 'Some Urban Patterns of Marriage and Domesticity in West Africa', *Sociological Review*, vol. 7, 1
13. O'Connor (1983) Ch. 2
14. Baker T and Bird M (1959) 'Urbanization and the Position of Women', *Sociological Review*, vol 7, 1
15. Ibid
16. Collins J (1980) 'Lusaka: Urban Planning in a British Colony 1931–64;, in Cherry (1980) op cit
17. See O'Connor op cit
18. Cunningham S (1980) 'Brazilian Cities Old and New: Growth and Planning Experiences', in Cherry (1980) op cit
19. Lévi-Strauss C (1976) *Tristes Tropiques*, Harmondsworth, Penguin, pp120–23
20. Lloyd P (1979) *Slums of Hope: Shanty Towns of the Third World*, Harmondsworth, Penguin, p20
21. Lloyd, op cit, pp5–8 citing Ross M H (1973) *Grass Roots in an African City*, Cambridge (Mass), MIT
22. Moser and Peake p5
    p95
23. Ibid
24. Lewis O (1961) *The Children of Sanchez*, New York, Random House
25. Moser and Peake, op cit
26. Moser and Peake, p14
27. See Machado L (1987) 'The Problems for Women-Headed Households in a Low Income Housing Programme in Brazil', in Moser and Peake, op cit
28. Chant S (1987) 'Domestic Labour, Decision Making and Dwelling Construction' in Moser and Peake, op cit
29. Moser and Peake, p21; see Machado op cit
30. Moser and Peake, p19
31. Moser and Peake, pp169–78
32. Vance I (1987) 'More than Bricks and Mortar: Women's Participation in Self-Help Housing in Managua' in Moser and Peake, op cit
33. Collinson H (1990) *Women and Revolution in Nicaragua*, London, Zed Books
34. Moser and Peake, op cit
35. Sarti C (1989) 'The Panorama of Feminism in Brazil', *New Left Review*, 173
36. Ibid p75
37. Ibid

The images which follow were taken in 2001–2 on the rapidly developing waterfront of Barcelona. The city's recent cultural policy, based on cultural tourism, is described in Dianne Dodd's essay in Part Four, but these images show Barcelona's new real estate as it moves from a Mediterranean to a world city. This contrasts with the progressive humanism of the city's planning history. The images move from the World Trade Centre at Port Vel, past the towers of a hotel built for the Olympics to new development in the furthest, north-eastern part of the city's shoreline. Signs of global culture appear in fast food outlets, though the beaches are used by a diversity of publics. Inland, a new knowledge-economy zone takes shape in the old working-class quarter of Poble Nou, and at the end a vast mall and apartment complex has been developed by a US-based company, with a landscape park designed by Enric Miralles. Further away are the three chimneys of the power station at St Adria de Besòs, one of the last sites of resistance in the war against fascism. For further reading on the Cerdà plan of 1859, see Arturo Soria y Puig, editor, *Cerda: The Five Bases of the General Theory of Urbanization*, Madrid, Electa, 1999; and François Choay, *The Rule and the Model*, Cambridge, MA, MIT, 1997, pp. 233–255.

Plate 1 **Claes Oldenburg and Coosje van Bruggen, *Book matches*, 1991, Parc de la Vall d'Hebron** (M. Miles, 2002).

Plate 2 **Museum of Contemporary Art, Barcelona (MACBA)** (M. Miles, 2002).

Plate 3 **Old and new apartment blocks in el Raval: the absence of balconies in the new facades** (M. Miles, 2002).

Plate 4 **World Trade Centre, Port Vel** (M. Miles, 2002).

Plate 5 **Hotel des Arts and casino, 1992 waterfront development** (M. Miles, 2002).

Plate 6 **'Beach Club'** (M. Miles, 2002).

Plate 7 **Parc de Poblenou, designed by X. Vendrell and M. Ruisànchez** (M. Miles, 2002).

Plate 8 **Preserved nineteenth-century brick chimney seen from mall development site** (M. Miles, 2002).

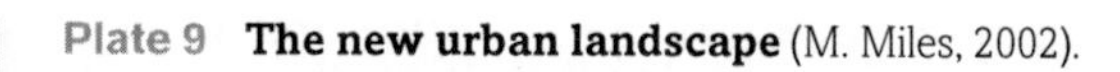

Plate 9 **The new urban landscape** (M. Miles, 2002).

Plate 10 **Beyond the construction site: the three chimneys of the power station at 'red Besòs'** (M. Miles, 2002).

PART TWO

What Is Culture?

INTRODUCTION TO PART TWO

To ask 'What is culture?' leads to no more easy or single an answer than to ask 'What is a city?'. In *Keywords* (London, Fontana, 1976), Raymond Williams writes: 'Culture is one of the two or three most complicated words in the English language' on account of its 'intricate historical development, in several European languages' (p. 76). The complications are as evident in dictionary definitions as in colloquial speech. *The Shorter Oxford English Dictionary* (2 volumes, 1983 edition, Oxford, Oxford University Press), for instance, gives several meanings from the word's long history. It begins with worship, then cultivation in the sense of tilling the land, and the growth of microscopic organisms in a laboratory. Only after that does it move to a more modern usage denoting an improvement of the mind by education, and a refinement of tastes and manners. The word's root – the latin *cultura* – has meanings appropriate to all these, and at the level of metaphor the growing of crops is related to the growing of consciousness in education. Yet none of the meanings given above are quite what is needed in a discussion of city cultures in the twenty-first century. Ordinary language mirrors the dictionary's definitions when we speak of religious cults, agriculture and horticulture, the culture which is the active ingredient in yoghurt, and of a cultured person as possessing the refinements of taste and behaviour produced in a certain kind of education. Today, we might add a knowledge of (or participation in) the arts, but all these definitions are less than satisfactory. So, what *is* culture?

The dictionary rightly avoids privileging any single meaning as *the* meaning, but even if the religious sense (from Caxton in 1483) is discarded as outdated, and the sense of agrarian and garden pursuits are set aside along with the laboratory meaning as having gained their own specialist senses, what we have left is a process of education and a refinement of taste. To arrive at a contemporary meaning, or two contrasting meanings, we need to extend the eighteenth-century idea of taste in the high arts to include a diversity of cultural production in the arts and media – the kinds of activities for which a Ministry of Culture makes policies – and to take into account the anthropological meaning, initially from German thought in the nineteenth century, of culture as a way of life. These more recent meanings both involve collectivities: the collective name for forms of cultural production, dissemination, and reception; and the collective values of a social group as expressed in the habits and expressions of everyday lives. The two senses overlap, directly when immersion in cultural activity *is* a way of life for an artist; and indirectly when cultural objects denote values produced in specific historical conditions. Cultural consumption in museums, or museum shops and cafés, for instance, may be part of the way a privileged social group displays its capacity to spend time non-productively and to buy objects which are in effect markers of social status. The latter process, which is mainly linked to metropolitan cities, is called conspicuous consumption by Thorsten Veblen in *The Theory of the Leisure Class* (New York, New America Library, 1899).

But, though the several uses of the term culture are linked, confusion follows when they are used as if interchangeable. For example, although flagship cultural institutions are not infrequently used to anchor urban redevelopment in post-industrial zones, it would be confusing to say their presence, as successors to the museums of the nineteenth century, had a regenerating effect on local cultures (as ways of life). The divergence of the terms becomes clear when the anthropological meaning is juxtaposed with that relating to the arts, not least because the concept of the arts is particular to western, industrialised societies. This does not mean, as anthropologists have been at pains to show for a century or more, that non-western, non-industrial social groups lack cultures of their own. Obviously they do not, but they may not have a direct equivalent for the concept of culture as the arts, or refinement of tastes which the arts manifest in their production and engender in their appreciation.

There seems a need to be clear which of the available meanings of culture is being used in any discussion. Williams offers a helpful summary of the term's development in *Keywords*, on which the above already draws. He notes the term's origin in a verb associated with processes of growth and nurture, and its shift into nouns for the products of such processes. He notes, too, the metaphorical link between a growth of mental sensibilities and a growth of plants, so that human subjects can be cultivated under particular conditions. The three definitions he summarises are:

> i) the independent and abstract noun which describes a general process of intellectual, spiritual and aesthetic development . . .
> ii) the independent noun, whether used generally or specifically, which indicates a particular way of life, whether of a people, a period or a group . . .
> iii) the independent and abstract noun which describes the works and practices of intellectual and especially artistic activity (p. 80).

The first begins to be used in the eighteenth century, and leads into the third as the products of an aesthetic sensibility. But it is probably, as indicated above, the second and third definitions which are most commonly used now. Again as already indicated, the second meaning – a way of life – is derived from the social sciences. It is from this rather than the third meaning – as cultural objects – that the term sub-culture is derived for the culture of a divergent group within a society (see *Subculture: The Meaning of Style*, D. Hebdige, London, Methuen, 1979). This tends to consolidate the link between culture as the arts and an elite when, in contrast, the practices – as in fashion, body decoration, and music – of sub-groups within a society are excluded from discussion alongside the museum objects and concert performances which are still seen as embodiments of a national culture. The separation enshrined in language is problematic, and part of Williams's aim was to construct a critical method which could encompass both dominant and dominated cultural forms. This project is extended today when anthropologists and ethnographers look, not to the cultures of far-away societies, but to those of their own.

Keywords was begun as an appendix to Williams's earlier *Culture and Society, 1780–1950* (London, Chatto and Windus, 1958), in which the concepts culture, class, art, industry and democracy (all included in *Keywords*) are taken as focal points in a history of socio-cultural change through the period of the English industrial revolution. He writes that it was the variability of the term culture which first attracted him: 'The very fact that it was important in two areas often thought of as separate – *art* and *society* – posed new questions and suggested new kinds of connection' (p. 12). Williams, whose background is in literary criticism, sees the modern separation of art from society and social issues, its displacement to an aesthetic dimension, as reflecting a new relation between writers and their middle-class readers in the nineteenth century. A parallel is found in the new relation of middle-class publics to annual art exhibitions at London's Royal Academy or the Paris Salon, and dependence of artists on an art market after the demise of aristocratic patronage. Williams argues specifically in *Culture and Society* that the separation of art from social reality results in the nineteenth century from a fear of the urban working class. The conditions of that class were reported by Friedrich Engels in 1845 (cited in the Introduction to Part One; see also *Imagining the Modern City* by James Donald, London, Athlone, 1999, pp. 33–7, and *The City Reader*, edited by Richard T. LeGates and Frederick Stout, London, Routledge, second edition, 2000, pp. 47–55). Subjected to appalling conditions and insecurities, the urban-industrial poor were perceived as likely to rise up and destroy bourgeois society, the cultural production of the period being an antidote to this fear.

Herbert Marcuse takes this argument in a more philosophical direction in the text reprinted below, first published in 1937, going back to classical Greek thought as well as more explicitly into the twentieth-century crisis of the rise of fascism in a nation which regarded itself as being highly cultured. Marcuse uses the term affirmative culture to describe this (non-)relation of cultural production to social and political conditions. Williams, in his essay on Modernism reprinted below, moves in another direction not anticipated by Marcuse in the 1930s to emphasise the extension of cultural production and reception through new technologies of sound and image, and new forms such as film and the mass media. He then sees the use of the term modern to describe the arts in the late twentieth century as anachronistic. Similarly, he sees new cultural technologies as creating new spatialities to replace those, more finite (if still complex), of the metropolitan spaces in which modern culture was for the most part produced. Yet, as he also argues, it was the mobility of the metropolis, and the new social arrangements such conditions produce, which were the ground for that newer sense of a dispersed spatiality.

To summarise the positions set out so far: the term culture as used today has distinct meanings referring to, on one hand, the arts (which may or not extend to the mass media – a question discussed further below) and, on the other, a way of life or collective set of values. The uses have different origins, and although they overlap they are not interchangeable.

But if, in western, industrial society, and as a reaction to the social consequences of industrialisation under capitalism, the arts have been for the most part displaced to an independent realm of aesthetic value where they cease to impinge on social or political struggles, this state of affairs is not inevitable. The arts may, from their eighteenth-century association with the taste of a leisured class, be elitist, and elitism may be maintained by the rituals of cultural consumption discussed by Carol Duncan in her chapter on the art museum reprinted below (1995), but elitism is not their defining condition any more than the substitution of culture for the sacred which conventional museum architecture implies. Both elitism and culture's sacred function are questioned by Williams in his contribution, with Richard Hoggart and Stuart Hall (see *Culture/Metaculture* by Francis Mulhern, London, Routledge, 2000, p. 98), to the establishment of the discipline of cultural studies. There, as in anthropology, the question 'What is culture?' becomes 'What is studied as culture?', giving rise to an open situation in which any cultural object, including vernacular forms and the ephemeral forms of sub-cultures, can be a legitimate object of analysis. This is more than a matter of a confrontation between high and low culture as representing opposed class interests. The categories of gender and race added, with a focus on intermediate (neither high nor low) art forms such as jazz and film. A frequently cited text is Walter Benjamin's essay 'The Work of Art in the Age of its Technical Reproducibility' (see *Walter Benjamin: Overpowering Conformism*, Esther Leslie, London, Pluto, 2000, p. 132; and the standard translation in *Illuminations*, edited by Hannah Arendt, London, Fontana, 1975, pp. 219–53). Cultural studies, then, established that cultural forms such as a comic strip or a newspaper cartoon are as much reflections of social trends as are works of art. In the 1980s, design history established a similar possibility for readings of everyday, mass-produced objects such as cars, toys, or cereal packets. This is not to say that forms in advertising and the mass media, or in industrial or graphic design, have the same aesthetic qualities as a painting by Leonardo or a drama by Racine – because that notion of aesthetic value is more limited in application – but that any item produced in a culture (as a way of life or state of social relations) is a sedimentation (to borrow a term from Adorno) of those relations and the value structure informing them.

This open-endedness parallels that of culture's anthropological meaning as a set of collective practices and values through which one society is differentiated from another. Though a rationalisation of those practices, culture is thus all-encompassing. In *A Scientific Theory of Culture* (Chapel Hill, University of North Carolina Press, 1944), Bronislaw Malinowski defines culture as 'the widest context of human behaviour . . . as important to the psychologist as to the social student, to the historian as to the linguist' (p. 5). For Malinowski, it is a set of processes through which people adapt to their material and psychological environments, giving culture an active, not merely reflective, sense. Among the categories of adaptation Malinowski lists are kinship, shelter, and hygiene – that is, a structure of human relations, the building of material structures, a set of habits required for membership of a social group. But while conventional anthropology tended to study structures such as kinship as if they were ossified, fixed in a traditional society's reproduction of the conditions of its origins (through creation stories, for instance), cultural studies drew attention to the mutability of these socially produced institutions. For Williams, there was also a necessity to balance received understandings of culture as the product of elites with other insights drawn from areas of social life hitherto regarded as uncultured. Near the end of *Culture and Society*, Williams writes, for instance, that:

> The working class, because of its position, has not . . . produced a culture in the narrower sense. The culture which it has produced . . . is the collective democratic institution, whether in the trades unions, the co-operative movement, or a political party (p. 315, cited in *Raymond Williams: Writing Culture, Politics*, Alan O'Connor, Oxford, Blackwell, 1989, p. 65).

The introduction of an idea of solidarity *as* culture renders culture an active force. It bears an imprint of the conditions of its production, but creates a sense of common cause. This may enhance opposition, and make intervention more likely; and brings culture directly into a social formation. This links it to the Marxist concept of a philosophy of practice in which, as Marx put it in his 'Theses on Feuerbach', written in Brussels in 1845 (and reprinted in many editions of his work), the point is not to explain the world in various ways but to change it. This implies that alongside intervention at the barricade, then, there is a possibility for intervention in the ways in which society is perceived, and from that for new perceptions of how it might be organised.

The idea of solidarity suggests a link between cultural studies and sociology. Solidarity was the topic of Emile Durkheim's first lecture course, and his book *The Division of Labour in Society* (Basingstoke, Macmillan, 1984). Asking what holds a social group together, Durkheim drew a distinction between mechanical-industrial and organic-community forms of cohesion, seeing the former as an imposition through social institutions, and the latter, perhaps romantically, as self-organised. This dualism, like similarly sweeping contrasts between rural and metropolitan forms of sociation and settlement in early twentieth-century sociology, and Durkheim's lack of attention to gender and race, is frequently questioned. Yet his point remains valid that forms of sociation are not fixed, but vary according to the conditions in which people and their interests are bound together. Graham Crow summarises:

> the form which social solidarity takes has to be seen in its appropriate social structural context and classified in this light . . . [it is] open to being expressed in several forms, depending on whether social structure functioned by promoting people's similarities or their differences (Graham Crow, *Social Solidarities: Theories, Identities and Social Change*, Buckingham, Open University Press, 2002, p. 21).

Williams sees solidarity in working-class movements, which tend to be self-organised. It might also be found in everyday practices including, to give two cases, growing vegetables on an allotment, where the production of food is sometimes a secondary outcome of a primary act of occupation and sociation, or the temporary but renewable associations of caravan users meeting on vacation or at club events.

In cultural studies and in the longer-established disciplines of anthropology and sociology, culture assumes an inclusive definition which contrasts with the exclusivity of its narrow definition as the arts. In *Keywords*, Williams cites a range of meanings for the latin verb *colere* – to inhabit, to protect, and to honour, as well as to cultivate – which is the root of such active derivations. He explains (as noted above) that, from the sixteenth century, people were seen metaphorically as objects of cultivation to be improved by education or training, and continues that this leads to the idea of culture as a class of aesthetic objects. But there is a further sense of culture, beginning like the anthropological meaning in the nineteenth century and related to it, as a set of characteristics which differentiates one nation from another. These characteristics may be the properties of national literatures or national schools of painting and sculpture, or of habits of consumption and the design of habitations. In this way we can speak of American culture, or French culture, though such shorthand definitions conceal differences within nations and usually reflect only the dominant cultural production of a state at a given time. The meaning which Williams has in mind is in any case more specific than that, and is expressed in the German word *Kultur*: as enshrining national traditions and values which cannot be articulated in technical progress but are expressions of the human spirit under particular conditions.

In German thought this idea of culture is seen in opposition to the idea of civilisation, the latter taken as industrial society's development through mechanical invention.

The meaning of culture as an enduring expression of spirit is informed by German Idealism, and rejects the Materialism associated with industrial societies. Williams notes that in Germany, *Kultur* at first meant the process of civilisation which was a society-wide equivalent of the cultivation of the individual mind – a national refinement, as it were – but goes on to say that, following Johann Herder's argument that culture is not universal but specific to the groups who produce it, it comes to mean the distinctive qualities of *different* civilisations or societies, or of different groups or classes within a society. While, then, anthropology established the validity of cultures in colonial countries, in the home states of the colonisers culture became a means not only to differentiate European cultures from non-European cultures, but also to assert a privileged status among European societies. Williams sees both radical and reactionary aspects to the identification of culture with non-material, or spiritual, qualities in human development. Culture's use as an antidote to the industrial revolution again comes to mind – *Kultur* expresses an equivalent sentiment to the adoption of a pseudo-sacred role for culture in England in the nineteenth century, for example by Matthew Arnold (*Keywords*, p. 80). That attitude persists in the twentieth century in T. S. Eliot's *Notes Towards the Definition of Culture* (London, Faber and Faber, 1948), in F. R. Leavis's *New Bearings in English Poetry* (Harmondsworth, Penguin, 1963), and in another way in Adorno's cultural critique which retains for different reasons an aversion for low culture in favour of a necessary difficulty of interpretation in high culture. Given the continuing association of cultural life and elitism, it is worth looking further at this.

Leavis asserts that in the modern world, poetry does not matter. If he is right, then poets are representatives of a discarded world. This is not the same as being outsiders or bohemians – Eliot worked first in a bank and then in a publishing house, cultivating his high-church English accent to be an insider. But it is to yearn for a world which does not presently exist, the danger of which is that it may lead to projections of a world which never existed (except in the poet's fantasy or, possibly, childhood memories – the safety of the nursery, and so forth) onto reality. That, however, is not Leavis's concern. On Eliot, he writes of an 'incessant rapid change that characterizes the Machine Age' and breaches social continuity to produce an 'uprooting of life' (pp. 11, 71). Eliot's poetry reflects the poet's privileged insight into this unreality. Leavis also cites Arnold's 'strange disease of modern life' (p. 18), and, despite (or because of) his concern with a specifically modern poetry, concludes that the only viable way forwards is to occupy an autonomous aesthetic realm of refined expression. It seems that the two main nineteenth-century meanings of culture outlined above are in direct confrontation: in anthropology and sociology culture is inclusive of all aspects of daily living; in the narrower sense of cultural criticism (*Kulturkritik*), and of the objects of high culture, it is exclusive of such diversities. There is an irony in that both cultural criticism as developed by the Frankfurt School, and British cultural studies, are vehicles for social criticism from a left-leaning position, while taking more or less opposite routes to that end. Williams was not drawn to the Frankfurt School, but perhaps Adorno would have found the tension between the two approaches interesting.

For social critics, there is a possible attraction in cultural criticism in its refusal of market forces. This is an aspect noted by Francis Mulhern in a discussion of *Kulturkritik* in *Culture/Metaculture* (London, Routledge, 2000). He cites Leavis, together with Thomas Mann, Julien Benda, Karl Mannheim, and José Ortega y Gasset as standing for a defence against modernisation. But just as culture holds to older values which do not include commercialisation, so it refuses, too, the fluctuations of political allegiance shown by electorates in a democracy. This has a dark side, but Mannheim interprets it as a case for an intellectual elite (or intelligentsia) safeguarding a view of the social whole against partisan interests. This depends on access to an education system, because the elite is not defined by class, while the detachment of intellect from specific interests is not unlike the autonomy of modern art from natural appearances. Mulhern summarises Mannheim's sociology:

> all knowledge, and especially political knowledge, was 'interest-bound' . . . the major social classes – workers and bourgeoisie alike – 'have their outlooks and activities directly and exclusively

determined by their specific social situations' (Karl Mannheim, *Ideology and Utopia*, trans. Wirth, L. and Shills, E., London, Routledge and Kegan Paul, 1936, p. 144, in Mulhern, p. 10).

Culture is antidote to class as well as industrialisation. Yet the main outcome of this position is not an art without class associations, but the model of rational planning developed in the 1920s in the Chicago School of Sociology by Robert Park, Ernest W. Burgess, and Louis Wirth, in which a professional elite make politically neutral judgements on urban growth.

But to return to Mannheim: while both workers and the bourgeoisie are conditioned in their attitudes to society and each other by their circumstances, intellectuals constitute a hybrid class drawn from lower and upper strata, but which is not itself a class. Mulhern describes Mannheim's image of the education system in which the intelligentsia are trained as 'an everyday constituent assembly of the mind'; there, a synthesis of competing knowledges is produced for a common good which can be identified only in that shared space (Mulhern, pp. 10–11). This is more progressive than the nostalgia of Ortega y Gasset and Leavis, or the Symbolist-inspired vision expressed by W. B. Yeats in 'Poetry and Tradition':

> Three types of men [*sic*] have made all beautiful things, Aristocracies have made beautiful manners, because their place in the world puts them above the fear of life, and the countrymen have made beautiful stories and beliefs, because they have nothing to lose and so do not fear, and the artists have made all the rest, because Providence has filled them with recklessness (W. B. Yeats, 'Poetry and Tradition', in *Selected Criticism and Prose*, London, Pan, [1907] 1980, p. 160).

Yeats later adopted a dubious political stance, but his views owe much to the Pre-Raphaelites and Baudelaire. His early poetry is perhaps the most succinct articulation of a culture of aesthetic retreat. Whether this can become, through a more critical and incisive edge, a tool for the imagination of a freedom not-yet present but potentially realisable is another question. It preoccupies Ernst Bloch in his massive *The Principle of Hope* (completed in 1959), and concerns Adorno and Marcuse in their later work on the aesthetic dimension.

For Adorno (*The Culture Industry: Selected Essays on Mass Culture*, edited and introduced by J. M. Bernstein, London, Routledge, 1991) mass culture is a means of mass deception. Although it is stated in a quite different way, Guy Debord's *Society of the Spectacle* (Detroit, Black Rose, 1967), part of which is reprinted in this part, makes a similar case. Debord defines the spectacle as mass representation: 'all of life presents itself as an immense accumulation of *spectacles*. Everything that was directly lived has moved away into a representation' (paragraph 1). The spectacle, as in the visual universe of the mass media in which unifying myths of social relations are played out, makes no pretence to a refinement of taste. Its distancing is less aesthetic than ideological (though aesthetic distancing is ideological, too). In *Critical and Cultural Theory*, Dani Cavallaro writes:

> Ideology can use all sorts of strategies . . . not always explicitly political. In fact, some of ideology's most successful ploys rely on concepts that may at first seem to have little to do with politics. One such concept . . . is the aesthetic. The aesthetic has time and again been harnessed to ideological imperatives (Dani Cavallaro, *Critical and Cultural Theory*, London, Athlone, 2001, pp. 76–7).

Drawing on Terry Eagleton's *The Ideology of the Aesthetic* (Oxford, Blackwell, 1990) Cavallaro alludes to the dissemination of stereotypical images in television programmes and advertising. Culture, then, through its ideological content, may re-locate social development beyond intervention.

In a period of globalisation, the ideological content of western mass culture has spread around the globe. Its aim is to ensure an ever-expanding (western) economy by extending the consumption patterns of the affluent society to all other places. Through Hollywood movies, Disney's fantasy parks, and satellite television, a culture based on consumption now establishes a global hegemony. This is the issue approached by John Tomlinson in his essay 'Globalised Culture: The Triumph of the West?', from *Culture and Global Change* (edited by Tracey Skelton and Tim Allen, London, Routledge, 1999). Tomlinson differentiates global culture from utopian notions of universality, and argues that it has a detrimental impact on local cultures. Culture, in other words, is domination. Or it may be, as Mike Featherstone suggests in *Undoing Culture: Globalization, Postmodernism and Identity* (London, Sage, 1995, pp. 86–101), that the connections between local and global cultures are complex, and although most of the traffic is one-way there are cross-currents as well. Discussion of that issue could usefully reference work on the sociology of consumption which balances consumers' victim status in some accounts from the 1970s and 1980s with a more recent emphasis on their abilities at playing with the market on their own terms, or resisting it – see M. Oh and J. Arditi, 'Shopping and Postmodernism: Consumption, Production, Identity, and the Internet', in *New Forms of Consumption*, edited by Mark Gottdiener (Lanham, Rowman & Littlefield, 2000). But at this point, culture shades into identity-formation, the construction of a subject as actor and acted-upon in social development, which is the theme of Part Seven.

To end this introduction, it can be said that culture is, as Williams says, a complex term the meaning of which does not stay the same. It begins as a verb, to tend or cultivate, from which are derived meanings of worship (from the tending of a temple) and refinement (the tending of a mind). Its roots remain visible in terms such as agriculture and horticulture, but a shift has taken place from a verb to a noun, and this refers to abstract ideas such as a refinement of sensibility, or taste. From that, culture is used to refer to the arts and, in a late twentieth-century context, the mass media as well. At the same time, culture means a way of life, a set of habits and practices in ordinary lives which embody a collective set of values or assumptions about the world, which may in turn be responses to specific conditions of living and the needs of survival. Added to these meanings is the more specialist sense of culture in cultural criticism (*Kulturkritik*), set against the technical progress denoted by the term civilisation (in that duality it, too, has complex meanings). Both cultural criticism and some areas of cultural production and reception exhibit a retreat from material and social realities to an aesthetic realm which may yet offer critical possibilities, or be a realm of ideology concealing rather than exposing contradictions in the prevailing value structure and social system.

SUGGESTED FURTHER READING

Dani Cavallaro, *Critical and Cultural Theory*, London, Athlone, 2001
Ian Chambers, *Culture after Humanism: History, Culture, Subjectivity*, London, Routledge, 2001
Mike Featherstone, *Undoing Culture: Globalization, Postmodernism and Identity*, London, Sage, 1995
Francis Mulhern, *Culture/Metaculture*, London, Routledge, 2000
Alan O'Connor, *Raymond Williams: Writing, Culture, Politics*, Oxford, Blackwell, 1989
Marcia Pointon, ed., *Art Apart: Art Institutions and Ideology across England and North America*, Manchester, Manchester University Press, 1994
Raymond Williams, *Keywords*, London, Fontana, 1976

'Metropolitan Perceptions and the Emergence of Modernism'

from *The Politics of Modernism* (1989)

Raymond Williams

Editors' Introduction

This is one of eleven essays in *The Politics of Modernism*, a collection of lectures and essays by Williams on Modernist culture (London, Verso, 1989). Others cover topics including the avant-garde, political theatre, cinema and socialism, and the future of cultural studies. The essay reprinted here was first published as a keynote essay in *Unreal City: Urban Experience in Modern European Literature and Art*, edited by Edward Timms and David Kelley (Manchester, Manchester University Press, 1985). The collection was published after Williams' death, but compiled from notes left by him to bring together such a book from previous writing.

This essay has been selected for inclusion here because it deals specifically with metropolitan culture, or the metropolis as standing for the set of conditions in which most cultural production in modernity has taken place. The metropolis is itself a cultural construct, more a matter of certain kinds of experiences rationalised as patterns of sociation than of data – see Georg Simmel's 'The Metropolis and Mental Life' (1903) in Part One. Williams sees metropolitan life as the ground for avant-garde art, though he notes that what is called 'modern' was already, when he wrote in 1985, quite old. In another essay in the collection, 'The Politics of the Avant-Garde', he describes Modernism as beginning in the late nineteenth century with a tendency for artists to assume increasing control over the means of production and dissemination of their work, and the avant-garde, as a movement within it which reacts against art's institutions in the early twentieth century. A case of the former would be the Secessions in Vienna, Berlin and Munich in the 1890s, or the Independent Salon in Paris in the 1880s; of the latter, Futurism in the early 1910s, or Dada a few years later. The metropolis which Williams takes as his terrain here is, then, of roughly the same period as Simmel's.

In the course of the essay, Williams gives several quotations from English literature, mainly from the nineteenth century, which emphasise the industrial condition of the metropolis. As discussed in the Introduction to this part, one direction in culture at this time was a reaction against the industrial revolution. Williams draws out two further aspects: a sense of anonymity possible only in large cities where traditional relations between people break down, and are replaced by more contingent patterns of sociation; and, extending it, a sense of loneliness within a crowd, not in isolation but amidst teeming masses. Williams sees the alienation produced by metropolitan conditions, together with (he implies) that of exchange mechanisms, as evident in avant-garde art. He concludes that 'the metropolis . . . moved into a quite new cultural dimension. It was now much more than the very large city . . . It was the place where new social and economic and cultural relations . . . were beginning to be formed'. (p. 44).

This text can be compared with Walter Benjamin's writing in the 1930s on Paris in the nineteenth century, in *Charles Baudelaire* (London, Verso, 1997), which includes both 'The Paris of the Second Empire in Baudelaire', and 'Paris: The Capital of the Nineteenth Century' (Benjamin's draft preface to his *Arcades Project*); or at greater length in *The Arcades Project*, translated by Howard Eiland and Kevin McLaughlin (Cambridge, MA, Harvard, 1999). While Williams gives a critical summary of certain tendencies, Benjamin, in his researches on the Paris

Arcades of Baudelaire's time gives a vast array of documentation and incidental observations derived from his immersion in his subject. A different kind of comparison would be with James Donald's *Imagining the Modern City* (London, Athlone, 1999), which includes a section on Engels' research on the living conditions of the working class in Manchester in the late nineteenth century.

It is now clear that there are decisive links between the practices and ideas of the avant-garde movements of the twentieth century and the specific conditions and relationships of the twentieth-century metropolis. The evidence has been there all along, and is indeed in many cases obvious. Yet until recently it has been difficult to disengage this specific historical and cultural relationship from a less specific but widely celebrated (and execrated) sense of 'the modern'.

In the late twentieth century it has become increasingly necessary to notice how relatively far back the most important period of 'modern art' now appears to be. The conditions and relationships of the early-twentieth-century metropolis have in many respects both intensified and been widely extended. In the simplest sense, great metropolitan aggregations, continuing the development of cities into vast conurbations, are still historically increasing (at an even more explosive rate in the Third World). In the old industrial countries, a new kind of division between the crowded and often derelict 'inner city' and the expanding suburbs and commuter developments has been marked. Moreover, within the older kinds of metropolis, and for many of the same reasons, various kinds of avant-garde movement still persist and even flourish. Yet at a deeper level the cultural conditions of the metropolis have decisively changed.

The most influential technologies and institutions of art, though they are still centred in this or that metropolis, extend and indeed are directed beyond it, to whole diverse cultural areas, not by slow influence but by immediate transmission. There could hardly be a greater cultural contrast than that between the technologies and institutions of what is still mainly called 'modern art' – writing, painting, sculpture, drama, in minority presses and magazines, small galleries and exhibitions, city-centre theatres – and the effective output of the late-twentieth-century metropolis, in film, television, radio and recorded music. Conservative analysts still reserve the categories 'art' or 'the arts' to the earlier technologies and institutions, with continued attachment to the metropolis as the centre in which an enclave can be found for them or in which they can, often as a 'national' achievement, be displayed. But this is hardly compatible with a continued intellectual emphasis on their 'modernity', when the actual modern media are of so different a kind. Secondly, the metropolis has taken on a much wider meaning, in the extension of an organized global market in the new cultural technologies. It is not every vast urban aggregation, or even great capital city, which has this cultural metropolitan character. The effective metropolis – as is shown in the borrowing of the word to indicate relations between nations, in the neocolonial world – is now the modern transmitting metropolis of the technically advanced and dominant economies.

Thus the retention of such categories as 'modern' and 'Modernism' to describe aspects of the art and thought of an undifferentiated twentieth-century world is now at best anachronistic, at worst archaic. What accounts for the persistence is a matter for complex analysis, but three elements can be emphasized. First, there is a factual persistence, in the old technologies and forms but with selected extensions to some of the new, of the specific relations between minority arts and metropolitan privileges and opportunities. Secondly, there is a persistent intellectual hegemony of the metropolis, in its command of the most serious publishing houses, newspapers and magazines, and intellectual institutions, formal and especially informal. Ironically, in a majority of cases, these formations are in some important respects residual: the intellectual and artistic forms in which they have their main roots are for social reasons – especially in their supporting formulations of 'minority' and 'mass', 'quality' and 'popular' – of that older, early-twentieth-century period, which for them is the perennially 'modern'. Thirdly and most fundamentally, the central product of that earlier period, for reasons which must be explored, was a new set of 'universals', aesthetic, intellectual and psychological, which can be sharply contrasted with the older 'universals' of specific cultures, periods and faiths, but which in just that quality resist all further specificities, of historical change or of cultural and social diversity: in the conviction of what is beyond question and for all effective time the 'modern absolute', the defined universality of a human condition which is effectively permanent.

There are several possible ways out of this intellectual deadlock, which now has so much power over a whole range of philosophical, aesthetic and political thinking. The most effective involve contemporary analysis in a still rapidly changing world. But it is also useful, when faced by this curious condition of cultural stasis – curious because it is a stasis which is continually defined in dynamic and experientially precarious terms – to identify some of the processes of its formation: seeing a present beyond 'the modern' by seeing how, in the past, that specifically absolute 'modern' was formed. For this identification, the facts of the development of the city into the metropolis are basic. We can see how certain themes in art and thought developed as specific responses to the new and expanding kinds of nineteenth-century city and then, as the central point of analysis, see how these went through a variety of actual artistic transformations, supported by newly offered (and competitive) aesthetic universals, in certain metropolitan conditions of the early twentieth century: the moment of 'modern art'.

It is important to emphasize how relatively old some of these apparently modern themes are. For that is the inherent history of themes at first contained within 'pre-modern' forms of art which then in certain conditions led to actual and radical changes of form. It is the largely hidden history of the conditions of these profound internal changes which we have to explore, often against the clamour of the 'universals' themselves.

For convenience I will take examples of the themes from English literature, which is particularly rich in them. Britain went through the first stages of industrial and metropolitan development very early, and almost at once certain persistent themes were arrived at. Thus the effect of the modern city as a crowd of strangers was identified, in a way that was to last, by Wordsworth:

> O Friend! one feeling was there which belonged
> To this great city, by exclusive right;
> How often, in the overflowing streets,
> Have I gone forward with the crowd and said
> Unto myself, 'The face of every one
> That passes by me is a mystery!'
>
> Thus have I looked, nor ceased to look, oppressed
> By thoughts of what and whither, when and how,
> Until the shapes before my eyes became
> A second-sight procession, such as glides
> Over still mountains, or appears in dreams.
> And all the ballast of familiar life,
> The present, and the past; hope, fear; all stays,
> All laws of acting, thinking, speaking man
> Went from me, neither knowing me, nor known.[1]

What is evident here is the rapid transition from the mundane fact that the people in the crowded street are unknown to the observer – though we now forget what a novel experience that must in any case have been to people used to customary small settlements – to the now characteristic interpretation of strangeness as 'mystery'. Ordinary modes of perceiving others are seen as overborne by the collapse of normal relationships and their laws: a loss of 'the ballast of familiar life'. Other people are then seen as if in 'second sight' or, crucially, as in dreams: a major point of reference for many subsequent modern artistic techniques.

Closely related to this first theme of the crowd of strangers is a second major theme, of an individual lonely and isolated within the crowd. We can note some continuity in each theme from more general Romantic motifs: the general apprehension of mystery and of extreme and precarious forms of consciousness; the intensity of a paradoxical self-realization in isolation. But what has happened, in each case, is that an apparently objective milieu, for each of these conditions, has been identified in the newly expanding and overcrowded modern city. There are a hundred cases, from James Thomson to George Gissing and beyond, of the relatively simple transition from earlier forms of isolation and alienation to their specific location in the city. Thomson's poem 'The Doom of a City' (1857) addresses the theme explicitly, as 'Solitude in the midst of a great City':

> The cords of sympathy which should have
>   bound me
> In sweet communication with earth's brotherhood
> I drew in tight and tighter still around me,
> Strangling my lost existence for a mood.[2]

Again, in the better-known 'City of Dreadful Night' (1870), a direct relationship is proposed between the city and a form of agonized consciousness:

> The City is of Night, but not of Sleep;
> There sweet sleep is not for the weary brain;
> The pitiless hours like years and ages creep,
> A night seems termless hell. This dreadful strain

Of thought and consciousness which never ceases,
Or which some moment's stupor but increases,
This, worse than woe, makes wretches there
insane.[3]

There is direct influence from Thomson in Eliot's early city poems. But more generally important is the extension of the association between isolation and the city to alienation in its most subjective sense: a range from dream or nightmare (the formal vector of 'Doom of a City'), through the distortions of opium or alcohol, to actual insanity. These states are being given a persuasive and ultimately conventional social location.

On the other hand, alienation in the city could be given a social rather than a psychological emphasis. This is evident in Elizabeth Gaskell's interpretation of the streets of Manchester in *Mary Barton*, in much of Dickens, especially in *Dombey and Son*, and (though here with more emphasis on the isolated and crushed observer) in Gissing's *Demos* and *The Nether World*. It is an emphasis drawn out and formally argued by Engels:

> They crowd by one another as though they had nothing in common, nothing to do with one another. . . . The brutal indifference, the unfeeling isolation of each in his private interest becomes the more repellent and offensive, the more these individuals are crowded together, within a limited space. And, however much one may be aware that this isolation of the individual, this narrow self-seeking is the fundamental principle of our society everywhere, it is nowhere so shamelessly barefaced, so self-conscious as just here in the crowding of the great city. The dissolution of mankind into monads . . . is here carried out to its utmost extremes.[4]

These alternative emphases of alienation, primarily subjective or social, are often fused or confused within the general development of the theme. In a way their double location within the modern city has helped to override what is otherwise a sharp difference of emphasis. Yet both the alternatives and their fusion or confusion point ahead to observable tendencies in twentieth-century avant-garde art, with its at times fused, at times dividing, orientations towards extreme subjectivity (including subjectivity as redemption or survival) and social or social/cultural revolution.

There is also a third theme, offering a very different interpretation of the strangeness and crowding and thus the 'impenetrability' of the city. Already in 1751 Fielding had observed:

> Whoever considers the Cities of London and Westminster, with the late vast increases of their suburbs, the great irregularity of their buildings, the immense numbers of lanes, alleys, courts and bye-places, must think that had they been intended for the very purpose of concealment they could not have been better contrived.[5]

This was a direct concern with the facts of urban crime, and the emphasis persisted. The 'dark London' of the late nineteenth century, and particularly the East End, were often seen as warrens of crime, and one important literary response to this was the new figure of the urban detective. In Conan Doyle's *Sherlock Holmes* stories there is a recurrent image of the penetration by an isolated rational intelligence of a dark area of crime which is to be found in the otherwise (for specific physical reasons, as in the London fogs, but also for social reasons, in that teeming, mazelike, often alien area) impenetrable city. This figure has persisted in the urban 'private eye' (as it happens, an exact idiom for the basic position in consciousness) in cities without the fogs.

On the other hand, the idea of 'darkest London' could be given a social emphasis. It is already significant that the use of statistics to understand an otherwise too complex and too numerous society had been pioneered in Manchester from the 1830s. Booth in the 1880s applied statistical survey techniques to London's East End. There is some relation between these forms of exploration and the generalizing panoramic perspectives of some twentieth-century novels (Dos Passos, Tressell). There were naturalistic accounts from within the urban environment, again with an emphasis on crime, in several novels of the 1890s, for example, Morrison's *Tales of Mean Streets* (1894). But in general it was as late as the 1930s, and then in majority in realist modes, before any of the actual inhabitants of these dark areas wrote their own perspectives, which included the poverty and the squalor but also, in sharp contradiction to the earlier accounts, the neighbourliness and community which were actual working-class responses.

A fourth general theme can, however, be connected with this explicit late response. Wordsworth, interestingly, saw not only the alienated city but new possibilities of unity:

among the multitudes
Of that huge city, oftentimes was seen
Affectingly set forth, more than elsewhere
Is possible, the unity of men.[6]

What could be seen, as often in Dickens, as a deadening uniformity, could be seen also, in Dickens and indeed, crucially, in Engels, as the site of new kinds of human solidarity. The ambiguity had been there from the beginning, in the interpretation of the urban crowd as 'mass' or 'masses', a significant change from the earlier 'mob'. The masses could indeed be seen, as in one of Wordsworth's emphases, as:

slaves unrespited of low pursuits,
Living amid the same perpetual flow
Of trivial objects, melted and reduced
To one identity.[7]

But 'mass' and 'masses' were also to become the heroic, organizing words of working-class and revolutionary solidarity. The factual development of new kinds of radical organization within both capital and industrial cities sustained this positive urban emphasis.

A fifth theme goes beyond this, but in the same positive direction. Dickens's London can be dark, and his Coketown darker. But although, as also later in H.G. Wells, there is a conventional theme of escape to a more peaceful and innocent rural spot, there is a specific and unmistakable emphasis on the vitality, the variety, the liberating diversity and mobility of the city. As the physical conditions of the cities were improved, this sense came through more and more strongly. The idea of the pre-industrial and pre-metropolitan city as a place of light and learning, as well as of power and magnificence, was resumed with a special emphasis on physical light: the new illuminations of the city. This is evident in very simple form in Le Gallienne in the 1890s:

London, London, our delight,
Great flower that opens but at night,
Great city of the midnight sun,
Whose day begins when day is done.

Lamp after lamp against the sky
Opens a sudden beaming eye,
Leaping a light on either hand
The iron lilies of the Strand.[8]

It is not only the continuity, it is also the diversity of these themes, composing as they do so much of the repertory of modern art, which should now be emphasized. Although Modernism can be clearly identified as a distinctive movement, in its deliberate distance from and challenge to more traditional forms of art and thought, it is also strongly characterized by its internal diversity of methods and emphases: a restless and often directly competitive sequence of innovations and experiments, always more immediately recognized by what they are breaking from than by what, in any simple way, they are breaking towards. Even the range of basic cultural positions within Modernism stretches from an eager embrace of modernity, either in its new technical and mechanical forms or in the equally significant attachments to ideas of social and political revolution, to conscious options for past or exotic cultures as sources or at least as fragments *against* the modern world, from the Futurist affirmation of the city to Eliot's pessimistic recoil.

Many elements of this diversity have to be related to the specific cultures and situations within which different kinds of work and position were to be developed, though within the simpler ideology of modernism this is often resisted: the innovations being directly related only to themselves (as the related critical procedures of formalism and structuralism came to insist). But the diversity of position and method has another kind of significance. The themes, in their variety, including as we have seen diametrically opposite as well as diverse attitudes to the city and its modernity, had formerly been included within relatively traditional forms of art. What then stands out as new, and is in this defining sense 'modern', is the series (including the competitive sequence) of breaks in form. Yet if we say only this we are carried back inside the ideology, ignoring the continuity of themes from the nineteenth century and isolating the breaks of form, or worse, as often in subsequent pseudohistories, relating the formal breaks to the themes as if both were comparably innovative. For it is not the general themes of response to the city and its modernity which compose anything that can be properly called Modernism. It is rather the new and specific location of the artists and intellectuals of this movement within the changing cultural milieu of the metropolis.

For a number of social and historical reasons the metropolis of the second half of the nineteenth century and of the first half of the twentieth century moved into a quite new cultural dimension. It was now much more than the very large city, or even the capital city of an important nation. It was the place where new social and economic and cultural relations, beyond both city and nation in their older senses, were beginning to be formed: a distinct historical phase which was in fact to be extended, in the second half of the twentieth century, at least potentially, to the whole world.

In the earliest phases this development had much to do with imperialism: with the magnetic concentration of wealth and power in imperial capitals and the simultaneous cosmopolitan access to a wide variety of subordinate cultures. But it was always more than the orthodox colonial system. Within Europe itself there was a very marked unevenness of development, both within particular countries, where the distances between capitals and provinces widened, socially and culturally, in the uneven developments of industry and agriculture, and of a monetary economy and simple subsistence or market forms. Even more crucial differences emerged between individual countries, which came to compose a new kind of hierarchy, not simply, as in the old terms, of military power, but in terms of development and thence of perceived enlightenment and modernity.

Moreover, both within many capital cities, and especially within the major metropolises, there was at once a complexity and a sophistication of social relations, supplemented in the most important cases – Paris, above all – by exceptional liberties of expression. This complex and open milieu contrasted very sharply with the persistence of traditional social, cultural and intellectual forms in the provinces and in the less developed countries. Again, in what was not only the complexity but the miscellaneity of the metropolis, so different in these respects from traditional cultures and societies beyond it, the whole range of cultural activity could be accommodated.

The metropolis housed the great traditional academies and museums and their orthodoxies; their very proximity and powers of control were both a standard and a challenge. But also, within the new kind of open, complex and mobile society, small groups in any form of divergence or dissent could find some kind of foothold, in ways that would not have been possible if the artists and thinkers composing them had been scattered in more traditional, closed societies. Moreover, within both the miscellaneity of the metropolis – which in the course of capitalist and imperialist development had characteristically attracted a very mixed population, from a variety of social and cultural origins – and its concentration of wealth and thus opportunities of patronage, such groups could hope to attract, indeed to form, new kinds of audience. In the early stages the foothold was usually precarious. There is a radical contrast between these often struggling (and quarrelling and competitive) groups, who between them made what is now generally referred to as 'modern art', and the funded and trading institutions, academic and commercial, which were eventually to generalize and deal in them. The continuity is one of underlying ideology, but there is still a radical difference between the two generations: the struggling innovators and the modernist establishment which consolidated their achievement.

Thus the key cultural factor of the modernist shift is the character of the metropolis: in these general conditions, but then, even more decisively, in its direct effects on form. The most important general element of the innovations in form is the fact of immigration to the metropolis, and it cannot too often be emphasized how many of the major innovators were, in this precise sense, immigrants. At the level of theme, this underlies, in an obvious way, the elements of strangeness and distance, indeed of alienation, which so regularly form part of the repertory. But the decisive aesthetic effect is at a deeper level. Liberated or breaking from their national or provincial cultures, placed in quite new relations to those other native languages or native visual traditions, encountering meanwhile a novel and dynamic common environment from which many of the older forms were obviously distant, the artists and writers and thinkers of this phase found the only community available to them: a community of the medium; of their own practices.

Thus language was perceived quite differently. It was no longer, in the old sense, customary and naturalized, but in many ways arbitrary and conventional. To the immigrants especially, with their new second common language, language was more evident as a medium – a medium that could be shaped and reshaped – than as a social custom. Even within a native language, the new relationships of the metropolis, and the inescapable new uses in newspapers and advertising attuned to it, forced certain productive kinds of strangeness and distance: a new consciousness of conventions and thus of changeable, because now open, conventions. There had long been pressures towards the work of art as artefact and commodity, but these now greatly intensified, and their combined pressures were very complex indeed. The preoccupying visual images and styles of particular cultures did not disappear, any more than the native languages, native tales, the native styles of music and dance, but all were now passed through this crucible of the metropolis, which was in the important cases no mere melting pot but an intense and visually and linguistically exciting process in its own right, from which remarkable new forms emerged.

At the same time, within the very openness and complexity of the metropolis, there was no formed and settled society to which the new kinds of work could be related. The relationships were to the open and complex and dynamic social process itself, and the only accessible form of this practice was an emphasis on the medium: the medium as that which, in an unprecedented way, defined art. Over a wide and diverse range of practice, this emphasis on the medium, and on what can be done in the medium, became dominant. Moreover, alongside the practice, theoretical positions of the same kind, most notably the new linguistics, but also the new aesthetics of significant form and structure, rose to direct, to support, to reinforce and to recommend. So nearly complete was this vast cultural reformation that, at the levels directly concerned – the succeeding metropolitan formations of learning and practice – what had once been defiantly marginal and oppositional became, in its turn, orthodox, although the distance of both from other cultures and peoples remained wide. The key to this persistence is again the social form of the metropolis, for the facts of increasing mobility and social diversity, passing through a continuing dominance of certain metropolitan centres and a related unevenness of all other social and cultural development, led to a major expansion of metropolitan forms of perception, both internal and imposed. Many of the direct forms and media processes of the minority phase of modern art thus became what could be seen as the common currency of majority communication, especially in films (an art form created, in all important respects, by these perceptions) and in advertising.

It is then necessary to explore, in all its complexity of detail, the many variations in this decisive phase of modern practice and theory. But it is also time to explore it with something of its own sense of strangeness and distance, rather than with the comfortable and now internally accommodated forms of its incorporation and naturalization. This means, above all, seeing the imperial and capitalist metropolis as a specific historical form, at different stages: Paris, London, Berlin, New York. It involves looking, from time to time, from outside the metropolis: from the deprived hinterlands, where different forces are moving, and from the poor world which has always been peripheral to the metropolitan systems. This need involve no reduction of the importance of the major artistic and literary works which were shaped within metropolitan perceptions. But one level has certainly to be challenged: the metropolitan interpretation of its own processes as universals.

The power of metropolitan development is not to be denied. The excitements and challenges of its intricate processes of liberation and alienation, contact and strangeness, stimulation and standardization, are still powerfully available. But it should no longer be possible to present these specific and traceable processes as if they were universals, not only in history but as it were above and beyond it. The formulation of the modernist universals is in every case a productive but imperfect and in the end fallacious response to particular conditions of closure, breakdown, failure and frustration. From the necessary negations of these conditions, and from the stimulating strangeness of a new and (as it seemed) unbonded social form, the creative leap to the only available universality – of raw material, of medium, of process – was impressively and influentially made.

At this level as at others – 'modernization' for example – the supposed universals belong to a phase of history which was both creatively preceded and creatively succeeded. While the universals are still accepted as standard intellectual procedures, the answers come out as impressively as the questions determine. But then it is characteristic of any major cultural phase that it takes its local and traceable positions as universal. This, which Modernism saw so clearly in the past which it was rejecting, remains true for itself. What is succeeding it is still uncertain and precarious, as in its own initial phases. But it can be foreseen that the period in which social strangeness and exposure isolated art as only a medium is due to end, even within the metropolis, leaving from its most active phases the new cultural monuments and their academies which in their turn are being challenged.

NOTES

1 William Wordsworth, *The Prelude*, VII, in *Poetical Works*, edited by De Selincourt and Darbishire, eds, London 1949, p. 261.
2 A. Ridler, ed., *Poems and Some Letters of James Thomson*, London 1963, p. 25.
3 *Ibid.*, p. 180.
4 Friedrich Engels, *The Condition of the Working Class in England in 1844*, translated by F.K. Wischnewetzky, London, 1934, p. 24.
5 Henry Fielding, *Inquiry into the Cause of the Late Increase in Robbers*, 1751, p. 76.
6 Wordsworth, p. 286.
7 *Ibid.*, p. 292.
8 C. Trent, *Greater London*, London 1965, p. 200.

'The Affirmative Character of Culture'

from *Negations* (1968)

Herbert Marcuse

Editors' Introduction

The extract from this text reprinted below sets out the basis of Marcuse's theory of bourgeois culture as a displacement mechanism, which contains desires for a better world in an aesthetic dimension separate from ordinary life and political intervention. The full text is quite long, and the style can be assertive, in the way of German academic writing of the time. Before evaluating the thesis put forward, it could be recalled that Marcuse wrote the essay in the late 1930s, after the flight of the Frankfurt School's members to other countries following Hitler's rise to power in Germany in 1933. Not only were the academics at the Frankfurt Institute for Social Research mainly Jewish, they were also Marxists seeking to recover the ground lost when revolution failed in Germany in 1918–19. The founding of the German Communist Party by Rosa Luxemburg and Karl Liebknecht, a brief period of socialism immediately following the defeat of 1918, and the disastrous uprising in Berlin of 1919 constitute an overlooked history, overshadowed by the October Revolution in Russia in 1917. But these were events through which Marcuse lived, before embarking on his doctoral research. Indeed, he was an active member of an independent socialist group led by the poet Kurt Eisner. A good summary of this period of his development is given by Barry Katz in *Herbert Marcuse and the Art of Liberation* (London, Verso, 1982).

The essay below was written, then, in the shadow of the failure of revolution in industrialised Germany followed by the rise of fascism and anti-semitism. One of the points Marcuse makes in it is that affirmative culture has some responsibility for that situation, aestheticising hope rather than translating it into practice. Towards the end he writes, 'That individuals freed for over four hundred years march with so little trouble in the communal columns of the authoritarian state is due in no small part to affirmative culture' (p. 125). This may seem sweeping, and in his later writing – notably *The Aesthetic Dimension* (Boston, Beacon Press, 1978) – Marcuse sees aesthetic distancing as the space of the only viable criticality remaining in a grim situation in which political change is as remote as ever. But in the 1930s, when the most urgent need was to resist fascism, it seemed appropriate.

The extract reprinted is from an earlier part of the essay and explains the theory and its relation to a separation of knowledges of the beautiful and the useful in Greek classical thought. This gives rise to problems later investigated by Marcuse and Adorno in more depth, and a tension between art's social and aesthetic dimensions which may be impossible to resolve, though for Adorno that is not a defeat but maintenance of an openness of argument (see *Aesthetic Theory*, translated by Robert Hullot-Kentor, London, Athlone, 1997).

THE AFFIRMATIVE CHARACTER OF CULTURE[1]

The doctrine that all human knowledge is oriented toward practice belonged to the nucleus of ancient philosophy. It was Aristotle's view that the truths arrived at through knowledge should direct practice in daily life as in the arts and sciences. In their struggle for existence, men need the effort of knowledge, the search for truth, because what is good, beneficial, and right for them is not immediately evident. Artisan and merchant, captain and physician, general and statesman – each must have correct knowledge in his field in order to be capable of acting as the changing situation demands.

While Aristotle maintained the practical character of every instance of knowledge, he made a significant distinction between forms of knowledge. He ordered them, as it were, in a hierarchy of value whose nadir is functional acquaintance with the necessities of everyday life and whose zenith is philosophical knowledge. The latter has no purpose outside itself. Rather, it occurs only for its own sake and to afford men felicity. Within this hierarchy there is a fundamental break between the necessary and useful on the one hand and the "beautiful" on the other. "The whole of life is further divided into two parts, business and leisure, war and peace, and of actions some aim at what is necessary and useful, and some at what is beautiful [τὰ καλὰ]."[2] Since this division is not itself questioned, and since, together with other regions of the "beautiful," "pure" theory congeals into an independent activity alongside and above other activities, philosophy's original demand disintegrates: the demand that practice be guided by known truths. Separating the useful and necessary from the beautiful and from enjoyment initiated a development that abandons the field to the materialism of bourgeois practice on the one hand and to the appeasement of happiness and the mind within the preserve of "culture" on the other.

One theme continually recurs in the reasons given for the relegation of the highest form of knowledge and of pleasure to pure, purposeless theory: the world of necessity, of everyday provision for life, is inconstant, insecure, unfree – not merely in fact, but in essence. Disposal over material goods is never entirely the work of human industry and wisdom, for it is subject to the rule of contingency. The individual who places his highest goal, happiness, in these goods makes himself the slave of men and things. He surrenders his freedom. Wealth and well-being do not come or persist due to his autonomous decision but rather through the changeable fortune of opaque circumstances. Man thus subjects his existence to a purpose situated outside him. Of itself, such an external purpose can vitiate and enslave men only if the material conditions of life are poorly ordered, that is, if their reproduction is regulated through the anarchy of opposing social interests. In this order the preservation of the common existence is incompatible with individual happiness and freedom. Insofar as philosophy is concerned with man's happiness – and the theory of classical antiquity held it to be the highest good – it cannot find it in the established material organization of life. That is why it must transcend this order's facticity.

Along with metaphysics, epistemology, and ethics, this transcendence also affects psychology. Like the extrapsychic[3] world, the human soul is divided into a lower and a higher region. The history of the soul transpires between the poles of sensuality[4] and reason. The devaluation of sensuality results from the same motives as that of the material world: because sensuality is a realm of anarchy, of inconstancy, and of unfreedom. Sensual pleasure is not in itself bad. It is bad because, like man's lower activities, it is fulfilled in a bad order. The "lower parts of the soul" drive man to covet gain and possessions, purchase and sale. He is led to "admire and value nothing but wealth and its possessors."[5] Accordingly the "appetitive" part of the soul, which is oriented toward sensual pleasure, is also termed by Plato the "money-loving" part, "because money is the principal means of satisfying desires of this kind."[6]

All the ontological classifications of ancient idealism express the badness of a social reality in which knowledge of the truth about human existence is no longer incorporated into practice. The world of the true, the good, and the beautiful is in fact an "ideal" world insofar as it lies beyond the existing conditions of life, beyond a form of existence in which the majority of men either work as slaves or spend their life in commerce, with only a small group having the opportunity of being concerned with anything more than the provision and preservation of the necessary. When the reproduction of material life takes place under the rule of the commodity form and continually renews the poverty of class society, then the good, beautiful, and true are transcendent to this life. And if everything requisite to preserving and securing material life is produced in this form, then whatever lies beyond it is certainly "superfluous." What is of authentic import to man, the highest truths, the highest goods, and the highest joys, is separated in significance from the necessary by an abyss. They are a "luxury." Aristotle did not conceal this state of affairs. "First philosophy," which includes the highest good and the highest pleasure, is a function of the leisure of the few, for whom all necessities of life are already adequately taken care of. "Pure theory" is appropriated as the profession of an elite and cordoned off with iron chains from the majority of mankind. Aristotle did not assert that the good, the beautiful, and the true are universally valid and obligatory values which should also permeate and transfigure "from above" the realm of necessity, of the material provision for life. Only when this claim is raised are we in the presence of the concept of culture that became central to bourgeois practice and its corresponding weltanschauung. The ancient theory of the higher value of truths above the realm of necessity includes as well the "higher" level of society. For these truths are supposed to have their abode in the ruling social strata, whose dominant status is in turn confirmed by the theory insofar as concern with the highest truths is supposed to be their profession.

In Aristotelian philosophy, ancient theory is precisely at the point where idealism retreats in the face of social contradictions and expresses them as ontological conditions. Platonic philosophy still contended with the social order of commercial Athens. Plato's idealism is interlaced with motifs of social criticism. What appears as facticity from the standpoint of the Ideas is the material world in which men and things encounter one another as commodities. The just order of the soul is destroyed by

> the passion for wealth which leaves a man not a moment of leisure to attend to anything beyond his personal fortunes. So long as a citizen's whole soul is wrapped up in these, he cannot give a thought to anything but the day's takings.[7]

And the authentic, basic demand of idealism is that this material world be transformed and improved in accordance with the truths yielded by knowledge of the Ideas. Plato's answer to this demand is his program for a reorganization of society. This program reveals what Plato sees as the root of evil. He demands, for the ruling strata, the abolition of private property (even in women and children) and the prohibition of trade. This same program, however, tries to root the contradictions of class society in the depths of human nature, thereby perpetuating them. While the majority of the members of the state are engaged for their entire lives in the cheerless business of providing for the necessities of life, enjoyment of the true, the good, and the beautiful is reserved for a small elite. Although Aristotle still lets ethics terminate in politics, for him the reorganization of society no longer occupies a central role in philosophy. To the extent to which he is more "realistic" than Plato, his idealism is more resigned in the face of the historical tasks of mankind. The true philosopher is for him no longer essentially the true statesman. The distance between facticity and Idea has increased precisely because they are conceived of as in closer relationship. The purport of idealism, viz, the realization of the Idea, dissipates. The history of idealism is also the history of its coming to terms with the established order.

Behind the ontological and epistemological separation of the realm of the senses and the realm of Ideas, of sensuousness and reason, of necessity and beauty, stands not only the rejection of a bad historical form of existence, but also its exoneration. The material world (i.e. the manifold forms of the respective "lower" member of this relation) is in itself mere matter, mere potentiality, akin more to Non-Being than to Being. It becomes real only insofar as it partakes of the "higher" world. In all these forms the material world remains bare matter or stuff for something outside it which alone gives it value. All and any truth, goodness, and beauty can accrue to it only "from above" by the grace of the Idea. All activity relating to the material provision of life remains in its essence untrue, bad, and ugly. Even with these characteristics, however, such activity is as necessary as matter is for the Idea. The misery of slave labor, the degradation of men and things to commodities, the joylessness and lowliness in which the totality of the material conditions of existence continuously reproduces itself, all these do not fall within the sphere of interest of idealist philosophy, for they are not yet the actual reality that constitutes the object of this philosophy. Due to its irrevocably material quality, material practice is exonerated from responsibility for the true, good, and beautiful, which is instead taken care of by the pursuit of theory. The ontological cleavage of ideal from material values tranquillizes idealism in all that regards the material processes of life. In idealism, a specific historical form of the division of labor and of social stratification takes on the eternal, metaphysical form of the relationship of necessity and beauty, of matter and Idea.

In the bourgeois epoch the theory of the relationship between necessity and beauty, labor and enjoyment, underwent decisive changes. First, the view that concern with the highest values is appropriated as a profession by particular social strata disappears. In its place emerges the thesis of the universality and universal validity of "culture." With good conscience, the theory of antiquity had expressed the fact that most men had to spend their lives providing for necessities while a small number devoted themselves to enjoyment and truth. Although the fact has not changed, the good conscience has disappeared. Free competition places individuals in the relation of buyers and sellers of labor power. The pure abstractness to which men are reduced in their social relations extends as well to intercourse with ideas. It is no longer supposed to be the case that some are born to and suited to labor and others to leisure, some to necessity and others to beauty. Just as each individual's relation to the market is immediate (without his personal qualities and needs being relevant except as commodities), so his relations to God, to beauty, to goodness, and to truth are relations of immediacy. As abstract beings, all men are supposed to participate equally in these values. As in material practice the product separates itself from the producers and becomes independent as the universal reified form of the "commodity," so in cultural practice a work and its content congeal into universally valid "values." By their very nature the truth of a philosophical judgment, the goodness of a moral action, and the beauty of a work of art should appeal to everyone, relate to everyone, be binding upon everyone. Without distinction of sex or birth, regardless of their position in the process of production, individuals must subordinate themselves to cultural values. They must absorb them into their lives and let their existence be permeated and transfigured by them. "Civilization" is animated and inspired by "culture."

This is not the place to discuss the various attempts to define culture. There is a concept of culture that can serve as an important instrument of social research because it expresses the implication of the mind in the historical process of society. It signifies the totality of social life in a given situation, insofar as both the areas of ideational reproduction (culture in the narrower sense, the "spiritual world") and of material reproduction ("civilization") form a historically distinguishable and comprehensible unity.[8] There is, however, another fairly widespread usage of the concept of culture, in which the spiritual world is lifted out of its social context, making culture a (false) collective noun and attributing (false) universality to it. This second concept of culture (clearly seen in such expressions as "national culture," "Germanic culture," or "Roman culture") plays off the spiritual world against the material world by holding up culture as the realm of authentic values and self-contained ends in opposition to the world of social utility and means. Through the use of this concept, culture is distinguished from civilization and sociologically and valuationally removed from the social process.[9] This concept itself has developed on the basis of a specific historical form of culture, which is termed "affirmative culture" in what follows. By affirmative culture is meant that culture of the bourgeois epoch which led in the course of its own development to the segregation from civilization of the mental and spiritual world as an independent realm of value that is also considered superior to civilization. Its decisive characteristic is the assertion of a universally obligatory, eternally better and more valuable world that must be unconditionally affirmed: a world essentially different from the factual world of the daily struggle for existence, yet realizable by every individual for himself "from within," without any transformation of the state of fact. It is only in this culture that cultural activities and objects gain that value which elevates them above the everyday sphere. Their reception becomes an act of celebration and exaltation.

Although the distinction between civilization and culture may have joined only recently the mental equipment of the social and cultural sciences, the state of affairs that it expresses has long been characteristic of the conduct of life and the weltanschauung of the bourgeois era. "Civilization and culture" is not simply a translation of the ancient relation of purposeful and purposeless, necessary and beautiful. As the purposeless and beautiful were internalized and, along with the qualities of binding universal validity and sublime beauty, made into the cultural values of the bourgeoisie, a realm of apparent unity and apparent freedom was constructed within culture in which the antagonistic relations of existence were supposed to be stabilized and pacified. Culture affirms and conceals the new conditions of social life.

In antiquity, the world of the beautiful beyond necessity was essentially a world of happiness and enjoyment. The ancient theory had never doubted that men's concern was ultimately their worldly gratification, their happiness. Ultimately, not immediately; for man's first concern is the struggle for the preservation and protection of mere existence. In view of the meager development of the productive forces in the ancient economy, it never occurred to philosophy that material practice could ever be fashioned in such a way that it would itself contain the space and time for happiness. Anxiety stands at the source of all idealistic doctrines that look for the highest felicity in ideational practice: anxiety about the uncertainty of all the conditions of life, about the contingency of loss, of dependence, and of poverty, but anxiety also about satiation, ennui, and envy of men and the gods. Nonetheless, anxiety about happiness, which drove philosophy to separate beauty and necessity, preserves the demand for happiness even within the separated sphere. Happiness becomes a preserve, in order for it to be able to be present at all. What man is to find in the philosophical knowledge of the true, the good, and the beautiful is ultimate pleasure, which has all the opposite characteristics of material facticity: permanence in change, purity amidst impurity, freedom amidst unfreedom.

The abstract individual who emerges as the subject of practice at the beginning of the bourgeois epoch also becomes the bearer of a new claim to happiness, merely on the basis of the new constellation of social forces. No longer acting as the representative or delegate of higher social bodies, each separate individual is supposed to take the provision of his needs and the fulfillment of his wants into his own hands and be in immediate relation to his "vocation," to his purpose and goals, without the social, ecclesiastical, and political mediations of feudalism. In this situation the individual was allotted more room for individual requirements and satisfactions: room which developing capitalist production began to fill with more and more objects of possible satisfaction in the form of commodities. To this extent, the bourgeois liberation of the individual made possible a new happiness.

But the universality of this happiness is immediately canceled, since the abstract equality of men realizes itself in capitalist production as concrete inequality. Only a small number of men dispose of the purchasing power required for the quantity of goods necessary in order to secure happiness. Equality does not extend to the conditions for attaining the means. For the strata of the rural and urban proletariat, on whom the bourgeoisie depended in their struggle against the feudal powers, abstract equality could have meaning only as real equality. For the bourgeoisie, when it came to power, abstract equality sufficed for the flourishing of real individual freedom and real individual happiness, since it already disposed of the material conditions that could bring about such satisfaction. Indeed, stopping at the stage of abstract freedom belonged to the conditions of bourgeois rule, which would have been endangered by a transition from abstract to concrete universality. On the other hand, the bourgeoisie could not give up the general character of its demand (that equality be extended to all men) without denouncing itself and openly proclaiming to the ruled strata that, for the majority, everything was still the same with regard to the improvement of the conditions of life. Such a concession became even less likely as growing social wealth made the real fulfillment of this general demand possible while there was in contrast the relatively increasing poverty of the poor in city and country. Thus the demand became a postulate, and its object a mere idea. The vocation of man, to whom general fulfillment is denied in the material world, is hypostatized as an ideal.

The rising bourgeois groups had based their demand for a new social freedom on the universality of human reason. Against the belief in the divinely instituted eternity of a restrictive order they maintained their belief in progress, in a better future. But reason and freedom did not extend beyond these groups' interest, which came into increasing opposition to the interest of the majority. To accusing questions the bourgeoisie gave a decisive answer: affirmative culture. The latter is fundamentally idealist. To the need of the isolated individual it responds with general humanity, to bodily misery with the beauty of the soul, to external bondage with internal freedom, to brutal egoism with the duty of the realm of virtue. Whereas during the period of the militant rise of the new society all of these ideas had a progressive character by pointing beyond the attained organization of existence, they entered increasingly into the service of the suppression of the discontented masses and of mere self-justifying exaltation, once bourgeois rule began to be stabilized. They concealed the physical and psychic vitiation of the individual.

NOTES

1 This essay was prompted by Max Horkheimer's remarks about "affirmative culture" and the "false idealism" of modern culture. Cf. *Zeitschrift für Sozialforschung*, V (1936), p. 219.

2 Aristotle *Politics* 1333 a 30ff., trans. by Benjamin Jowett in *The Basic Works of Aristotle*, Richard McKeon, ed. (New York: Random House. 1941), p. 1298 (with change in translation).

3 *Translator's note*: While "*Seele*" has an adjectival form, "*seelisch*," its English counterpart "soul" does not. I have used "psychic" or "spiritual," depending on the context. Accordingly, although the word "*geistig*" means both "spiritual" and "mental," in the present essay I have rendered it as "mental," and "spiritual" refers to a quality of "soul," not of "mind."

4 *Translator's note*: "*Sinnlich*" means simultaneously "sensual," which stresses its appetitive aspect, and "sensuous," which stresses its aesthetic aspect. I have translated it in each case according to the emphasis of the context, but both meanings are always implied. For further discussion, see Herbert Marcuse, *Eros and Civilization* (Boston: Beacon Press, 1955), pp. 166–167.

5 Plato *Republ.* 553 in *The Republic of Plato*, trans. by Francis M. Cornford (New York: Oxford, 1945), p. 277. Cf. *Republ.* 525.

6 *Ibid.*, pp. 306–307.

7 Plato *Leges* 831, trans. by A. E. Taylor in *The Collected Dialogues of Plato*, Edith Hamilton and Huntington Cairns, eds. (New York: Bollingen Foundation-Pantheon Books 1964), p. 1397. Cf. J. Brake, *Wirtschaften und Charakter in der antiken Bildung* (Frankfurt am Main, 1935), pp. 124ff.

8 See *Studien über Autorität und Familie* ("Schriften des Instituts für Sozialforschung," V [Paris, 1936]), pp. 7ff.

9 Spengler interprets the relationship of culture and civilization not as simultaneity, but as "necessary organic succession." Civilization is the inevitable fate and end of every culture. See *Der Untergang des Abendlandes*, 23d to 32d editions (Munich, 1920), I, pp. 43–44. Such reformulation does not modify the abovementioned traditional evaluation of culture and civilization.

'The Art Museum as Ritual'

from *Civilizing Rituals: Inside Public Art Museums* (1995)

Carol Duncan

Editors' Introduction

The two previous texts have looked at general conditions of culture. This one looks at the specific case of the modern institution of the art museum. Although much intellectual and academic effort goes into defining culture in an abstract way, it can be seen, too, that institutional structures define it in practical ways which are highly influential on those who experience them. Because public museums, many founded in the nineteenth century in a spirit of liberal reformism, had an educational role, they inculcated particular histories and values into the minds of their publics. This involved two distinct but related kinds of interpretation: of taste, in universal terms as refinement of sensibilities; and of national schools and international styles. The museum buildings themselves tended, either through classical allusion or a in new, Modernist architecture to house modern art, to instill certain values – culture as high culture, as public good, as civilising.

Duncan is also author of *The Aesthetics of Power* (Cambridge, Cambridge University Press, 1993), an enquiry into patriarchal relations in art, among other concerns. In the chapter reprinted below from *Civilising Rituals* she focuses on the power of monuments, seeing art museums as equivalents to the monuments which inscribe dominant meanings on public spaces. Going to an art museum is thus a social ritual which reproduces given values in much the same way, if by different means, as observing public ceremonies. This consolidates the argument that culture is a secular equivalent of religious observance – see the Introduction to this part – though Duncan sees a dichotomy between the sacred and the secular in the terms of Enlightenment rationality. It is not, however, a matter of museums reproducing older or other forms of ritual: 'Museums resemble older ritual sites not so much because of their specific architectural references but because they, too, are settings for rituals' (p. 10).

This text can be read in conjunction with those included in *Art Apart: Art Institutions and Ideology Across England and North America*, edited by Marcia Pointon (Manchester, Manchester University Press, 1994), particularly Brandon Taylor's 'From Penitentiary to Temple of Art: Early Metaphors of Improvement at the Millbank Tate' (pp. 9–32) and Christoph Grunenberg's 'The Politics of Presentation: The Museum of Modern Art, New York' (pp. 192–211). In relation to ethnographic museums, see Annie Coombes essay in this collection, 'Blinded by Science: Ethnography at the British Museum' (pp. 102–19). For a series of papers on museums and social relations, see *Museums and Communities: The Politics of Public Culture*, edited by Ivan Karp, Christine M. Kreamer, and Steven D. Lavine (Washington, Smithsonian, 1992).

This chapter sets forth the basic organizing idea of this study, namely, the idea of the art museum as a ritual site. Unlike the chapters that follow, where the focus is on specific museums and the particular circumstances that shaped them, this chapter generalizes more broadly about both art museums and ritual. Besides introducing the concept of ritual that informs the book as a whole, it argues that the ritual character of art museums has, in effect, been recognized for as long as public art museums have existed and has often been seen as the very fulfillment of the art museum's purpose.

Art museums have always been compared to older ceremonial monuments such as palaces or temples. Indeed, from the eighteenth through the mid-twentieth centuries, they were deliberately designed to resemble them. One might object that this borrowing from the architectural past can have only metaphoric meaning and should not be taken for more, since ours is a secular society and museums are secular inventions. If museum facades have imitated temples or palaces, is it not simply that modern taste has tried to emulate the formal balance and dignity of those structures, or that it has wished to associate the power of bygone faiths with the present cult of art? Whatever the motives of their builders (so the objection goes), in the context of our society, the Greek temples and Renaissance palaces that house public art collections can signify only secular values, not religious beliefs. Their portals can lead to only rational pastimes, not sacred rites. We are, after all, a post-Enlightenment culture, one in which the secular and the religious are opposing categories.

It is certainly the case that our culture classifies religious buildings such as churches, temples, and mosques as different in kind from secular sites such as museums, court houses, or state capitals. Each kind of site is associated with an opposite kind of truth and assigned to one or the other side of the religious/secular dichotomy. That dichotomy, which structures so much of the modern public world and now seems so natural, has its own history. It provided the ideological foundation for the Enlightenment's project of breaking the power and influence of the church. By the late eighteenth century, that undertaking had successfully undermined the authority of religious doctrine – at least in western political and philosophical theory if not always in practice. Eventually, the separation of church and state would become law. Everyone knows the outcome: secular truth became authoritative truth; religion, although guaranteed as a matter of personal freedom and choice, kept its authority only for voluntary believers. It is secular truth – truth that is rational and verifiable – that has the status of "objective" knowledge. It is this truest of truths that helps bind a community into a civic body by providing it a universal base of knowledge and validating its highest values and most cherished memories. Art museums belong decisively to this realm of secular knowledge, not only because of the scientific and humanistic disciplines practiced in them – conservation, art history, archaeology – but also because of their status as preservers of the community's official cultural memory.

Again, in the secular/religious terms of our culture, "ritual" and "museums" are antithetical. Ritual is associated with religious practices – with the realm of belief, magic, real or symbolic sacrifices, miraculous transformations, or overpowering changes of consciousness. Such goings-on bear little resemblance to the contemplation and learning that art museums are supposed to foster. But in fact, in traditional societies, rituals may be quite unspectacular and informal-looking moments of contemplation or recognition. At the same time, as anthropologists argue, our supposedly secular, even anti-ritual, culture is full of ritual situations and events – very few of which (as Mary Douglas has noted) take place in religious contexts.[1] That is, like other cultures, we, too, build sites that publicly represent beliefs about the order of the world, its past and present, and the individual's place within it.[2] Museums of all kinds are excellent examples of such microcosms; art museums in particular – the most prestigious and costly of these sites[3] – are especially rich in this kind of symbolism and, almost always, even equip visitors with maps to guide them through the universe they construct. Once we question our Enlightenment assumptions about the sharp separation between religious and secular experience – that the one is rooted in belief while the other is based in lucid and objective rationality – we may begin to glimpse the hidden – perhaps the better word is disguised – ritual content of secular ceremonies.

We can also appreciate the ideological force of a cultural experience that claims for its truths the status of objective knowledge. To control a museum means precisely to control the representation of a community and its highest values and truths. It is also the power to define the relative standing of individuals within that community. Those who are best prepared to perform its ritual – those who are most able to respond to its various cues – are also those whose identities (social, sexual, racial, etc.) the museum ritual most fully confirms, It is precisely for this reason that museums and museum practices can become objects of fierce struggle and impassioned debate. What we see and do not see in art museums – and on what terms and by whose authority we do or do not see it – is closely linked to larger questions about who constitutes the community and who defines its identity.

I have already referred to the long-standing practice of museums borrowing architectural forms from monumental ceremonial structures of the past. Certainly when Munich, Berlin, London, Washington, and other western capitals built museums whose facades looked like Greek or Roman temples, no one mistook them for their ancient prototypes. On the contrary, temple facades – for 200 years the most popular source for public art museums[4] – were completely assimilated to a secular discourse about architectural beauty, decorum, and rational form. Moreover, as coded reminders of a pre-Christian civic realm, classical porticos, rotundas, and other features of Greco-Roman architecture could signal a firm adherence to Enlightenment values. These same monumental forms, however, also brought with them the spaces of public rituals – corridors scaled for processions, halls implying large, communal gatherings, and interior sanctuaries designed for awesome and potent effigies.

Museums resemble older ritual sites not so much because of their specific architectural references but because they, too, are settings for rituals. (I make no argument here for historical continuity, only for the existence of comparable ritual functions.) Like most ritual space, museum space is carefully marked off and culturally designated as reserved for a special quality of attention – in this case, for contemplation and learning. One is also expected to behave with a certain decorum. In the Hirshhorn Museum, a sign spells out rather fully the dos and don'ts of ritual activity and comportment. Museums are normally set apart from other structures by their monumental architecture and clearly defined precincts. They are approached by impressive flights of stairs, guarded by pairs of monumental marble lions, entered through grand doorways. They are frequently set back from the street and occupy parkland, ground consecrated to public use. (Modern museums are equally imposing architecturally and similarly set apart by sculptural markers. In the United States, Rodin's *Balzac* is one of the more popular signifiers of museum precincts, its priapic character making it especially appropriate for modern collections.[5])

By the nineteenth century, such features were seen as necessary prologues to the space of the art museum itself:

> Do you not think that in a splendid gallery . . . all the adjacent and circumjacent parts of that building should . . . have a regard for the arts, . . . with fountains, statues, and other objects of interest calculated to prepare [visitors'] minds before entering the building, and lead them the better to appreciate the works of art which they would afterwards see?

The nineteenth-century British politician asking this question[6] clearly understood the ceremonial nature of museum space and the need to differentiate it (and the time one spends in it) from day-to-day time and space outside. Again, such framing is common in ritual practices everywhere. Mary Douglas writes:

> A ritual provides a frame. The marked off time or place alerts a special kind of expectancy, just as the oft-repeated 'Once upon a time' creates a mood receptive to fantastic tales.[7]

"Liminality," a term associated with ritual, can also be applied to the kind of attention we bring to art museums. Used by the Belgian folklorist Arnold van Gennep,[8] the term was taken up and developed in the anthropological writings of Victor Turner to indicate a mode of consciousness outside of or "betwixt-and-between the normal, day-to-day cultural and social states and processes of getting and spending."[9] As Turner himself realized, his category of liminal experience had strong affinities to modern western notions of the aesthetic experience – that mode of receptivity thought to be most appropriate before works of art. Turner recognized aspects of liminality in such modern activities as attending the theatre, seeing a film, or visiting an art exhibition. Like folk rituals that temporarily suspend the constraining rules of normal social behavior (in that sense, they "turn the world upside-down"), so these cultural situations, Turner argued, could open a space in which individuals can step back from the practical concerns and social relations of everyday life and look at themselves and their world – or at some aspect of it – with different thoughts and feelings. Turner's idea of liminality, developed as it is out of anthropological categories and based on data gathered mostly in non-western cultures, probably cannot be neatly superimposed onto western concepts of art experience. Nevertheless, his work remains useful in that it offers a sophisticated general concept of ritual that enables us to think about art museums and what is supposed to happen in them from a fresh perspective.[10]

It should also be said, however, that Turner's insight about art museums is not singular. Without benefit of the term, observers have long recognized the liminality of their space. The Louvre curator Germain Bazin, for example, wrote that an art museum is "a temple where Time seems suspended"; the visitor enters it in the hope of finding one of "those momentary cultural epiphanies" that give him "the illusion of knowing intuitively his essence and his strengths."[11] Likewise, the Swedish writer Goran Schildt has noted that museums are settings in which we seek a state of "detached, timeless and exalted" contemplation that "grants us a kind of release from life's struggle and . . . captivity in our own ego." Referring to nineteenth-century attitudes to art, Schildt observes "a religious element, a substitute for religion."[12] As we shall see, others, too, have described art museums as sites which enable individuals to achieve liminal experience – to move beyond the psychic constraints of mundane existence, step out of time, and attain new, larger perspectives.

Thus far, I have argued the ritual character of the museum experience in terms of the kind of attention one brings to it and the special quality of its time and space. Ritual also involves an element of performance. A ritual site of any kind is a place programmed for the enactment of something. It is a place designed for some kind of performance. It has this structure whether or not visitors can read its cues. In traditional rituals, participants often perform or witness a drama – enacting a real or symbolic sacrifice. But a ritual performance need not be a formal spectacle. It may be something an individual enacts alone by following a prescribed route, by repeating a prayer, by recalling a narrative, or by engaging in some other *structured* experience that relates to the history or meaning of the site (or to some object or objects on the site). Some individuals may use a ritual site more knowledgeably than others – they may be more educationally prepared to respond to its symbolic cues. The term "ritual" can also mean habitual or routinized behavior that lacks meaningful subjective context. This sense of ritual as an "empty" routine or performance is not the sense in which I use the term.

In art museums, it is the visitors who enact the ritual.[13] The museum's sequenced spaces and arrangements of objects, its lighting and architectural details provide both the stage set and the script – although not all museums do this with equal effectiveness. The situation resembles in some respects certain medieval cathedrals where pilgrims followed a structured narrative route through the interior, stopping at prescribed points for prayer or contemplation. An ambulatory adorned with representations of the life of Christ could thus prompt pilgrims to imaginatively re-live the sacred story. Similarly, museums offer well-developed ritual scenarios, most often in the form of art-historical narratives that unfold through a sequence of spaces. Even when visitors enter museums to see only selected works, the museum's larger narrative structure stands as a frame and gives meaning to individual works.

Like the concept of liminality, this notion of the art museum as a performance field has also been discovered independently by museum professionals. Philip Rhys Adams, for example, once director of the Cincinnati Art Museum, compared art museums to theatre sets (although in his formulation, objects rather than people are the main performers):

> The museum is really an impresario, or more strictly a *régisseur*, neither actor nor audience, but the controlling intermediary who sets the scene, induces a receptive mood in the spectator, then bids the actors take the stage and be their best artistic selves. And the art objects do have their exits and their entrances; motion – the movement of the visitor as he enters a museum and as he goes or is led from object to object – is a present element in any installation.[14]

The museum setting is not only itself a structure; is also constructs its *dramatis personae*. These are, ideally, individuals who are perfectly predisposed socially, psychologically, and culturally to enact the museum ritual. Of course, no real visitor ever perfectly corresponds to these ideals. In reality, people continually "misread" or scramble or resist the museum's cues to some extent; or they actively invent, consciously or unconsciously, their own programs according to all the historical and psychological accidents of who they are. But then, the same is true of any situation in which a cultural product is performed or interpreted.[15]

Finally, a ritual experience is thought to have a purpose, an end. It is seen as transformative: it confers or renews identity or purifies or restores order in the self or to the world through sacrifice, ordeal, or enlightenment. The beneficial outcome that museum rituals are supposed to produce can sound very like claims made for traditional, religious rituals. According to their advocates, museum visitors come away with a sense of enlightenment, or a feeling of having been spiritually nourished or restored. In the words of one well-known expert,

> The only reason for bringing together works of art in a public place is that . . . they produce in us a kind of exalted happiness. For a moment there is a clearing in the jungle: we pass on refreshed, with our capacity for life increased and with some memory of the sky.[16]

One cannot ask for a more ritual-like description of the museum experience. Nor can one ask it from a more renowned authority. The author of this statement is the British art historian Sir Kenneth Clark, a distinguished scholar and famous as the host of a popular BBC television series of the 1970s, "Civilization." Clark's concept of the art museum as a place for spiritual transformation and restoration is hardly unique. Although by no means uncontested, it is widely shared by art historians, curators, and critics everywhere. Nor, as we shall see below, is it uniquely modern.

We come, at last, to the question of art museum objects. Today, it is a commonplace to regard museums as the most appropriate places in which to view and keep works of art. The existence of such objects – things that are most properly used when contemplated as art – is taken as a given that is both prior to and the cause of art museums. These commonplaces, however, rest on relatively new ideas and practices. The European practice of placing objects in settings designed for contemplation emerged as part of a new and, historically speaking, relatively modern way of thinking. In the course of the eighteenth century, critics and philosophers, increasingly interested in visual experience, began to attribute to works of art the power to transform their viewers spiritually, morally, and emotionally. This newly discovered aspect of visual experience was extensively explored in a developing body of art criticism and philosophy. These investigations were not always directly concerned with the experience of art as such, but the importance they gave to questions of taste, the perception of beauty, and the cognitive roles of the senses and imagination helped open new philosophical ground on which art criticism would flourish. Significantly, the same era in which aesthetic theory burgeoned also saw a growing interest in galleries and public art museums. Indeed, the rise of the art museum is a corollary to the philosophical invention of the aesthetic and moral powers of art objects: if art objects are most properly used when contemplated as art, then the museum is the most proper setting for them, since it makes them useless for any other purpose.

In philosophy, Immanuel Kant's *Critique of Judgment* is one of the most monumental expressions of this new preoccupation with aesthetics. In it, Kant definitively isolated and defined the human capacity for aesthetic judgment and distinguished it from other faculties of the mind (practical reason and scientific understanding).[17] But before Kant, other European writers, for example, Hume, Burke, and Rousseau, also struggled to define taste as a special kind of psychological encounter with distinctive moral and philosophical import.[18] The eighteenth century's designation of art and aesthetic experience as major topics for critical and philosophical inquiry is itself part of a broad and general tendency so furnish the secular with new value. In this sense, the invention of aesthetics can be understood as a transference of spiritual values from the sacred realm into secular time and space. Put in other terms, aestheticians gave philosophical formulations to the condition of liminality, recognizing it as a state of withdrawal from the day-to-day world, a passage into a time or space in which the normal business of life is suspended. In philosophy, liminality became specified as the aesthetic experience, a moment of moral and rational disengagement that leads to or produces some kind of revelation or transformation. Meanwhile, the appearance of art galleries and museums gave the aesthetic cult its own ritual precinct.

Goethe was one of the earliest witnesses of this development. Like others who visited the newly created art museums of the eighteenth century, he was highly responsive to museum space and to the sacral feelings it aroused. In 1768, after his first visit to the Dresden Gallery, which housed a magnificent royal art collection,[19] he wrote about his impressions, emphasizing the powerful ritual effect of the total environment:

> The impatiently awaited hour of opening arrived and my admiration exceeded all my expectations. That *salon* turning in on itself, magnificent and so well-kept, the freshly gilded frames, the well-waxed parquetry, the profound silence that reigned, created a solemn and unique impression, akin to the

> emotion experienced upon entering a House of God, and it deepened as one looked at the ornaments on exhibition which, as much as the temple that housed them, were objects of adoration in that place consecrated to the holy ends of art.[20]

The historian of museums Niels von Holst has collected similar testimony from the writings of other eighteenth-century museum-goers. Wilhelm Wackenroder, for example, visiting an art gallery in 1797, declared that gazing at art removed one from the "vulgar flux of life" and produced an effect that was comparable to, but better than, religious ecstasy.[21] And here, in 1816, still within the age when art museums were novelties, is the English critic William Hazlitt, aglow over the Louvre:

> Art lifted up her head and was seated on her throne, and said, All eyes shall see me, and all knees shall bow to me. . . . There she had gathered together all her pomp, there was her shrine, and there her votaries came and worshipped as in a temple.[22]

A few years later, in 1824, Hazlitt visited the newly opened National Gallery in London, then installed in a house in Pall Mall. His description of his experience there and its ritual nature – his insistence on the difference between the quality of time and space in the gallery and the bustling world outside, and on the power of that place to feed the soul, to fulfill its highest purpose, to reveal, to uplift, to transform and to cure – all of this is stated with exceptional vividness. A visit so this "sanctuary," this "holy of holies," he wrote, "is like going on a pilgrimage – it is an act of devotion performed at the shrine of Art!"

> It is a cure (for the time at least) for low-thoughted cares and uneasy passions. We are abstracted to another sphere: we breathe empyrean air; we enter into the minds of Raphael, of Titian, of Poussin, of the Caracci, and look at nature with their eyes; we live in time past, and seem identified with the permanent forms of things. The business of the world at large, and even its pleasures, appear like a vanity and an impertinence. What signify the hubbub, the shifting scenery, the fantoccini figures, the folly, the idle fashions without, when compared with the solitude, the silence, the speaking looks, the unfading forms within? Here is the mind's true home. The contemplation of truth and beauty is the proper object for which we were created, which calls forth the most intense desires of the soul, and of which it never tires.[23]

This is not to suggest that the eighteenth century was unanimous about art museums. Right from the start, some observers were already concerned that the museum ambience could change the meanings of the objects it held, redefining them as works of art and narrowing their import simply by removing them from their original settings and obscuring their former uses. Although some, like Hazlitt and the artist Philip Otto Runge, welcomed this as a triumph of human genius, others were – or became – less sure. Goethe, for example, thirty years after his enthusiastic description of the art gallery at Dresden, was disturbed by Napoleon's systematic gathering of art treasures from other countries and their display in the Louvre as trophies of conquest. Goethe saw that the creation of this huge museum collection depended on the destruction of something else, and that it forcibly altered the conditions under which, until then, art had been made and understood. Along with others, he realized that the very capacity of the museum to frame objects as art and claim them for a new kind of ritual attention could entail the negation or obscuring of other, older meanings.[24]

In the late eighteenth and early nineteenth centuries, those who were most interested in art museums, whether they were for or against them, were but a minority of the educated – mostly poets and artists. In the course of the nineteenth century, the serious museum audience grew enormously; it also adopted an almost unconditional faith in the value of art museums. By the late nineteenth century, the idea of art galleries as sites of wondrous and transforming experience became commonplace among those with any pretensions to "culture" in both Europe and America.

Through most of the nineteenth century, an international museum culture remained firmly committed to the idea that the first responsibility of a public art museum is to enlighten and improve its visitors morally, socially, and politically. In the twentieth century, the principal rival to this ideal, the aesthetic museum, would come to dominate. In the United States, this new ideal was advocated most forcefully in the opening years of the century. Its main proponents, all wealthy, educated gentlemen, were connected to the Boston Museum of Fine Arts and would make the doctrine of the aesthetic museum the official creed of their institution.[25] The fullest and most influential statement of this doctrine is Benjamin Ives Gilman's *Museum Ideals of Purpose and Method*, published by the museum in 1918 but drawing on ideas developed in previous years. According to Gilman, works of art, once they are put in museums, exist for one purpose only: to be looked at as things of beauty. The first obligation of an art museum is to present works of art as just that, as objects of aesthetic contemplation and not as illustrative of historical or archaeological information. As he expounded it (sounding much like Hazlitt almost a century earlier), aesthetic contemplation is a profoundly transforming experience, an imaginative act of identification between viewer and artist. To achieve it, the viewer "must make himself over in the image of the artist, penetrate his intention, think with his thoughts, feel with his feelings."[26] The end result of this is an intense and joyous emotion, an overwhelming and "absolutely serious" pleasure that contains a profound spiritual revelation. Gilman compares it to the "sacred conversations" depicted in Italian Renaissance altarpieces – images in which saints who lived in different centuries miraculously gather in a single imaginary space and together contemplate the Madonna. With this metaphor, Gilman casts the modern aesthete as a devotee who achieves a kind of secular grace through communion with artistic geniuses of the past – spirits that offer a life-redeeming sustenance. "Art is the Gracious Message pure and simple," he wrote, "integral to the perfect life," its contemplation "one of the ends of existence."[27]

The museum ideal that so fascinated Gilman would have a compelling appeal to the twentieth century. Most of today's art museums are designed to induce in viewers precisely the kind of intense absorption that he saw as the museum's mission, and art museums of all kinds, both modern and historical, continue to affirm the goal of communion with immortal spirits of the past. Indeed, the longing for contact with an idealized past, or with things imbued by immortal spirits, is probably pervasive as a sustaining impetus not only of art museums but many other kinds of rituals as well. The anthropologist Edmond Leach noticed that every culture mounts some symbolic effort to contradict the irreversibility of time and its end result of death. He argued that themes of rebirth, rejuvenation, and the spiritual recycling or perpetuation of the past deny the fact of death by substituting for it symbolic structures in which past time returns.[28] As ritual sites in which visitors seek to re-live spiritually significant moments of the past, art museums make splendid examples of the kind of symbolic strategy Leach described.[29] Later sections of this book, will examine several museums in which a wish for immortality is given strong ritual expression.

Nowhere does the triumph of the aesthetic museum reveal itself more dramatically than in the history of art gallery design. Although fashions in wall colors, ceiling heights, lighting, and other details have over the years varied with changing museological trends, installation design has consistently and increasingly sought to isolate objects for the concentrated gaze of the aesthetic adept and to suppress as irrelevant other meanings the objects might have. The wish for ever closer encounters with art have gradually made galleries more intimate, increased the amount of empty wall space between works, brought works nearer to eye level, and caused each work to be lit individually.[30] Most art museums today keep their galleries uncluttered and, as much as possible, dispense educational information in anterooms or special kiosks at a tasteful remove from the art itself. Clearly, the more "aesthetic" the installations – the fewer the objects and the emptier the surrounding walls – the more sacralized the museum space. The sparse installations of the National Gallery in Washington, DC, take the aesthetic ideal to an extreme as do installations of modern art in many institutions. As the sociologist César Graña once suggested, modern installation practices have brought the museum-as-temple metaphor close to the fact. Even in art museums that attempt education, the practice of isolating important originals in "aesthetic chapels" or niches – but never hanging them to make an historical point – undercuts any educational effort.[31]

The isolation of objects for visual contemplation, something that Gilman and his colleagues in Boston ardently preached, has remained one of the outstanding features of the aesthetic museum and continues to inspire eloquent advocates. Here, for example, is the art historian Svetlana Alpers in 1988:

> Romanesque capitals or Renaissance altarpieces are appropriately looked at in museums (*pace* Malraux) even if not made for them. When objects like these are severed from the ritual site, the invitation to look attentively remains and in certain respects may even be enhanced.[32]

Of course, in Alpers' statement, only the original site has ritual meaning. In my terms, the attentive gazing she describes belongs to another, if different, ritual field, one which requires from the performer intense, undistracted visual contemplation.

In *The Museum Age*, Germain Bazin described with penetrating insight how modern installations help structure the museum as a ritual site, In his analysis, the isolation and illumination of objects induces visitors to fix their attention onto things that exist seemingly in some other realm. The installations thus take visitors on a kind of mental journey, a stepping out of the present into a universe of timeless values:

> Statues must be isolated in space, paintings hung far apart, a glittering jewel placed against a field of black velvet and spot-lighted; in principle, only one object at a time should appear in the field of vision. Iconographic meaning, overall harmony, aspects that attracted the nineteenth-century amateur, no longer interest the contemporary museum goer, who is obsessed with form and workmanship; the eye must be able to scan slowly the entire surface of a painting. The act of looking becomes a sort of trance uniting spectator and masterpiece.[33]

One could take the argument even farther: in the liminal space of the museum, everything – and sometimes anything – may become art, including fire-extinguishers, thermostats, and humidity gauges, which, when isolated on a wall and looked at through the aesthetizing lens of museum space, can appear, if only for a mistaken moment, every bit as interesting as some of the intended-as-art works on display, which, in any case, do not always look very different.

In this chapter, I have been concerned mainly with arguing the general ritual features of art museums. These are: first, the achievement of a marked-off, "liminal" zone of time and space in which visitors, removed from the concerns of their daily, practical lives, open themselves to a different quality of experience; and second, the organization of the museum setting as a kind of script or scenario which visitors perform. I also have argued that western concepts of the aesthetic experience, generally taken as the art museum's *raison d'être*, match up rather closely to the kind of rationales often given for traditional rituals (enlightenment, revelation, spiritual equilibrium or rejuvenation). In the chapters that follow, liminality will be assumed as a condition of art museum rituals, and attention will shift to the specific scenarios that structure the various museums discussed. As for the purposes of art museums – what they do and to whom or for whom they do it – this question, too, will be addressed, directly or indirectly, throughout much of what follows. Indeed, it is hardly possible to separate the purposes of art museums from their specific scenario structures. Each implies the other, and both imply a set of surrounding historical contingencies.

NOTES

1 Mary Douglas, *Purity and Danger*, London, Boston, and Henley, Routledge & Kegan Paul, 1966, p. 68.

On the subject of ritual in modern life, see Abner Cohen, *Two-Dimensional Man. An Essay on the Anthropology of Power and Symbolism in Complex Society*, Berkeley, Cal., University of California Press, 1974; Steven Lukes, "Political Ritual and Social integration," in *Essays in Social Theory*, New York and London, Columbia University Press, 1977, pp. 52–73. Sally F. Moore and Barbara Myerhoff, "Secular Ritual: Forms and Meanings," in Moore and Myerhoff (eds.), *Secular Ritual*, Assen/Amsterdam, Van Gorcum, 1977, pp. 3–24; Victor Turner, "Frame, Flow, and Reflection: Ritual and Drama as Public Liminality," in *Performance in Postmodern Culture*, Michel Benamou and Charles Caramello (eds.), Center for Twentieth Century Studies, University of Wisconsin–Milwaukee, 1977, pp. 33–55; and Turner, "Variations on a Theme of Liminality," in Moore and Myerhoff, op. cit., pp. 36–52. See also Masao Yamaguchi, "The Poetics of Exhibition in Japanese Culture," in I. Karp and S. Levine (eds.), *Exhibiting Cultures: The Poetics and Politics of Museum Display*, Washington and London, Smithsonian Institution, 1991, pp. 57–67. Yamaguchi discusses secular rituals and ritual sites in both Japanese and western culture, including modern exhibition space. The reference to our culture being anti-ritual comes from Mary Douglas, *Natural Symbols* (1973), New York, Pantheon Books, 1982, pp. 1–4, in a discussion of modern negative views of ritualism as the performance of empty gestures.

2 This is not to imply the kind of culturally or ideologically unified society that, according to many anthropological accounts, gives rituals a socially integrative function. This integrative function is much disputed, especially in modern society (see, for example, works cited in the preceding notes by Cohen, Lukes, and Moore and Myerhoff, and Edmond Leach, "Ritual," *International Encyclopedia of the Social Sciences*, vol. 13, David Sills (ed.), Macmillan Co. and The Free Press, 1968, pp. 521–6).

3 As Mary Douglas and Baron Isherwood have written, "the more costly the ritual trappings, the stronger we can assume the intention to fix the meanings to be" (*The World of Goods: Towards an Anthropology of Consumption* (1979), New York and London, W. W. Norton, 1982, p. 65).

4 See Nikolaus Pevsner, *A History of Building Types*, Princeton, NJ, Princeton University Press, 1976, pp. 118 ff.; Niels von Holst, *Creators, Collectors and Connoisseurs*, trans. B. Battershaw, New York, G. P. Putnam's Sons, 1967, pp. 228 ff.; Germain Bazin, *The Museum Age*, trans. J. van Nuis Cahill, New York, Universe Books, 1967, pp. 197–202; and William L. MacDonald, *The Parthenon: Design, Meaning, and Progeny*, Cambridge, Mass., Harvard University Press, 1976, pp. 125–32.

5 The phallic form of the *Balzac* often stands at or near the entrances to American museums, for example, the Los Angeles County Museum of Art or the Norton Simon Museum; or it presides over museum sculpture gardens, for example, the Museum of Modern Art in New York or the Hirshhorn Museum in Washington, DC.

6 William Ewart, MP, in *Report from the Select Committee on the National Gallery*, in House of Commons, Reports, vol. xxxv, 1853, p. 505.

7 *Purity and Danger*, op. cit. (note 1), p. 63.

8 Arnold van Gennep, *The Rites of Passage* (1908), trans. M. B. Vizedom and G. L. Caffee, Chicago, University of Chicago Press, 1960.

9 Turner, "Frame, Flow, and Reflection," op. cit. (note 1), p. 33. See also Turner's *Dramas, Fields, and Metaphors: Symbolic Action in Human Society*, Ithaca and London: Cornell University Press, 1974, especially pp. 13–15 and 231–2.

10 See Mary Jo Deegan, *American Ritual Dramas: Social Rules and Cultural Meanings*, New York, Westport, Conn., and London, Greenwood Press, 1988, pp. 7–12, for a thoughtful discussion of Turner's ideas and the limits of their applicability to modern art. For an opposing view of rituals and of the difference between traditional rituals and the modern experience of art, see Margaret Mead, "Art and Reality From the Standpoint of Cultural Anthropology," *College Art Journal*, 1943, vol. 2, no. 4, pp. 119–21. Mead argues that modern visitors in an art gallery can never achieve what primitive rituals provide, "the symbolic expression of the meaning of life."

11 Bazin, *The Museum Age*, op. cit. (note 4), p. 7.

12 Goran Schildt, "The Idea of the Museum," in L. Aagaard-Mogensen (ed.), *The Idea of the Museum: Philosophical, Artistic, and Political Questions*, Problems in Contemporary Philosophy, vol. 6, Lewiston, NY, and Quenstron, Ontario, Edwin Mellen Press, 1988, p. 89.

13 I would argue that this is the case even when they watch "performance artists" at work.

14 Philip Rhys Adams, "Towards a Strategy of Presentation," *Museum*, 1954, vol. 7, no. 1, p. 4.

15 For an unusual attempt to understand what museum visitors make of their experience, see Mary Beard, "Souvenirs of Culture: Deciphering (in) the Museum," *Art History*, 1992, vol. 15, pp. 505–32. Beard examines the purchase and use of postcards as evidence of how visitors interpret the museum ritual.

16 Kenneth Clark, "The Ideal Museum," *ArtNews*, January, 1954, vol. 52, p. 29.

17 Kant, *Critique of Judgment* (1790), trans. by J. H. Bernard, New York, Hafner Publishing, 1951.

18 Two classics in this area are: M. H. Abrams, *The Mirror and the Lamp*, New York, W. W. Norton, 1958, and Walter Jackson Bate, *From Classic to Romantic: Premises of Taste in Eighteenth-Century England*, New York, Harper & Bros., 1946. For a substantive summary of these developments, see Monroe C. Beardsley, *Aesthetics from Classical Greece to the Present: A Short History*, University, Alabama, University of Alabama Press, 1975, chs. 8 and 9.

19 For the Dresden Gallery, see von Holst, op.cit (note 4), pp. 121–3.

20 From Goethe's *Dichtung und Wahrheit*, quoted in Bazin, op. cit. (note 4), p. 160.

21 Von Holst, op. cit (note 4), p. 216.

22 William Hazlitt, "The Elgin Marbles" (1816), in P. P. Howe (ed.), *The Complete Works*, New York, AMS Press, 1967, vol. 18, p. 101. Thanks to Andrew Hemingway for the reference.

23 William Hazlitt, *Sketches of the Principal Picture-Galleries in England*, London, Taylor & Hessey, 1824, pp. 2–6.

24 See Goethe, cited in Elizabeth Gilmore Holt, *The Triumph of Art for the Public*, Garden City, New York, Anchor Press/Doubleday, 1979, p. 76. The Frenchman Quatremère de Quincy also saw art museums as destroyers of the historical meanings that gave value to art. See Daniel Sherman, "Quatremère/Benjamin/Marx: Museums, Aura, and Commodity Fetishism," in D. Sherman and I. Rogoff (eds.), *Museum Culture: Histories, Discourses, Spectacles*, Media and Society, vol. 6, Minneapolis and London, University of Minnesota Press, 1994, pp. 123–43. Thanks to the author for an advance copy of his paper.

25 See especially Paul Dimaggio, "Cultural Entrepreneurship in Nineteenth-Century Boston: The Creation of an Organized Base for High Culture in America," *Media, Culture and Society*, 1982, vol. 4, pp. 33–50 and 303–22;

and Walter Muir Whitehill, *Museum of Fine Arts, Boston: A Centennial History*, 2 vols., Cambridge, Mass., Harvard University Press, 1970.

In this chapter, I have quoted more from advocates of the aesthetic than the educational museum, because, by and large, they have valued and articulated more the liminal quality of museum space, while advocates of the educational museum tend to be suspicious of that quality and associate it with social elitism (see, for example, Dimaggio, op. cit.). But, the educational museum is no less a ceremonial structure than the aesthetic museum, as the following two chapters will show.

26 Benjamin Ives Gilman, *Museum Ideals of Purpose and Method*, Cambridge, Boston Museum of Fine Arts, 1918, p. 56.

27 Ibid., p. 108.

28 Leach, "Two Essays Concerning the Symbolic Representation of Time," in *Rethinking Anthropology*, London, University of London, Athlone Press, and New York, Humanities Press, Inc., 1961, pp. 124–36. Thanks to Michael Ames for the reference.

29 Recently, the art critic Donald Kuspit suggested that a quest for immortality is central to the meaning of art museums. The sacralized space of the art museum, he argues, by promoting an intense and intimate identification of visitor and artist, imparts to the visitor a feeling of contact with something immortal and, consequently, a sense of renewal. For Kuspit, the success of this transaction depends on whether or not the viewer's narcissistic needs are addressed by the art she or he is viewing ("The Magic Kingdom of the Museum," *Artforum*, April, 1992, pp. 58–63). Werner Muensterberger, in *Collecting: An Unruly Passion: Psychological Perspectives*, Princeton, NJ, Princeton University Press, 1994, brings to the subject of collecting the experience of a practicing psychoanalyst and explores in depth a variety of narcissistic motives for collecting, including a longing for immortality.

30 See, for example, Charles G. Loring, a Gilman follower, noting a current trend for "small rooms where the attention may be focused on two or three masterpieces" (in "A Trend in Museum Design," *Architectural Forum*, December 1927, vol. 47, p. 579).

31 César Graña, "The Private Lives of Public Museums," *Trans-Action*, 1967, vol. 4, no. 5, pp. 20–5.

32 Alpers, "The Museum as a Way of Seeing," in Karp and Levine, op. cit. (note 1), p. 27.

33 Bazin, op. cit. (note 4), p. 265.

'Separation Perfected'

from *Society of the Spectacle* (1983)

Guy Debord

Editors' Introduction

Guy Debord was a key figure in Situationism in Paris in the years up to 1968. Among his contributions to urbanism is the concept of the 'drift' (*dérive*), a form of playfully constructive behaviour in which participants walked parts of the city's terrain, not as tourists or incidental observers but as active makers of new meanings in their reading of its text, its traces of human occupation and institutional control. The drift was also a refusal of the productivity required in the dominant society, a kind of avant-garde vagrancy perhaps. Simon Sadler writes in *The Situationist City* (Cambridge, MA, MIT, 1999) that 'Wandering around the city, drifting without destination, neither going to work nor properly consuming, was a waste of time in the temporal economy' (p. 93). This resistant aspect of Situationism does not mean that Debord and other Situationists were not engaged in the urbanist discourses of the period. In paragraph 172 of *Society of the Spectacle*, Debord cites Lewis Mumford to the effect that urban sprawl and its attendant long-distance communication has become a means of social control. The Situationists had a closer link to Henri Lefebvre (see Part Six), though in the Paris uprising of 1968 their appropriation of Lefebvre's words as slogans in the street caused offence to Lefebvre while his publication of Situationist discussions caused offence to the Situationists.

Society of the Spectacle (*La société du spectacle*) was first published in Paris in 1967, and has appeared in several English editions since. Debord made a film of the same title – in which the camera pans across a vast office complex in Maine-Montparnasse as sign of a new Haussmannization of the city. The sections (or theses) reprinted here set out the concept of spectacle as a means of inducing passivity in the mass population. The argument is compatible with Adorno's that real life becomes more and more like the movies under the spell of the culture industry (see his text in Part Four). The spectacle is thus a product of a system of alienation which reproduces it: 'The spectacle within society corresponds to a concrete manufacture of alienation' (paragraph 32). In the following section – and the book deserves to be read in full – Debord continues on the theme of commodification and the spectacle, and commodity fetishism. This aspect of Debord's writing could be compared with Walter Benjamin's idea of the phantasmagoria presented by modern (that is nineteenth-century) commodity consumption in the Paris arcades (see *The Arcades Project*, cited under Williams above). For further reading on Situationism and the urban, see Simon Sadler's *The Situationist City* (noted above); and Andrew Hussey's essay 'The Map is Not the Territory: The Unfinished Journey of the Situationist International' in *Urban Visions*, edited by Steven Spier (Liverpool, Liverpool University Press, 2002, pp. 215–28). On Situationism and Lefebvre, see Rob Shields, *Lefebvre, Love and Struggle* (London, Routledge, 1999, pp. 89–92 and 103–7).

1

In societies where modern conditions of production prevail, all of life presents itself as an immense accumulation of spectacles. Everything that was directly lived has moved away into a representation.

2

The images detached from every aspect of life fuse in a common stream in which the unity of this life can no longer be reestablished. Reality considered *partially* unfolds, in its own general unity, as a pseudo-world *apart*, an object of mere contemplation. The specialization of images of the world is completed in the world of the autonomous image, where the liar has lied to himself. The spectacle in general, as the concrete inversion of life, is the autonomous movement of the non-living.

3

The spectacle presents itself simultaneously as all of society, as part of society, and as *instrument of unification*. As a part of society it is specifically the sector which concentrates all gazing and all consciousness. Due to the very fact that this sector is *separate*, it is the common ground of the deceived gaze and of false consciousness, and the unification it achieves is nothing but an official language of generalized separation.

4

The spectacle is not a collection of images, but a social relation among people, mediated by images.

5

The spectacle cannot be understood as an abuse of the world of vision, as a product of the techniques of mass dissemination of images. It is, rather, a *Weltanschauung* which has become actual, materially translated. It is a world vision which has become objectified.

6

The spectacle, grasped in its totality, is both the result and the project of the existing mode of production. It is not a supplement to the real world, an additional decoration. It is the heart of the unrealism of the real society. In all its specific forms, as information or propaganda, as advertisement or direct entertainment consumption, the spectacle is the present model of socially dominant life. It is the omnipresent affirmation of the choice *already* made in production and its corollary consumption. The spectacle's form and content are identically the total justification of the existing system's conditions and goals. The spectacle is also the *permanent presence* of this justification, since it occupies the main part of the time lived outside of modern production.

7

Separation is itself part of the unity of the world, of the global social praxis split up into reality and image. The social practice which the autonomous spectacle confronts is also the real totality which contains the spectacle. But the split within this totality mutilates it to the point of making the spectacle appear as its goal. The language of the spectacle consists of signs of the ruling production, which at the same time are the ultimate goal of this production.

8

One cannot abstractly contrast the spectacle to actual social activity: such a division is itself divided. The spectacle which inverts the real is in fact produced. Lived reality is materially invaded by the contemplation of the spectacle while simultaneously absorbing the spectacular order, giving it positive cohesiveness. Objective reality is present on both sides. Every notion fixed this way has no other basis than its passage into the opposite: reality rises up within the spectacle, and the spectacle is real. This reciprocal alienation is the essence and the support of the existing society.

9

In a world which *really is topsy-turvy*, the true is a moment of the false.

10

The concept of "spectacle" unifies and explains a great diversity of apparent phenomena. The diversity and the contrasts are appearances of a socially organized appearance, the general truth of which must itself be recognized. Considered in its own terms, the spectacle is *affirmation* of appearance and affirmation of all human life, namely social life, as mere appearance. But the critique which reaches the truth of the spectacle exposes it as the visible *negation* of life, as a negation of life which *has become visible*.

11

To describe the spectacle, its formation, its functions and the forces which tend to dissolve it, one must artificially distinguish certain inseparable elements. When *analyzing* the spectacle one speaks, to some extent, the language of the spectacular itself in the sense that one moves through the methodological terrain of the very society which expresses itself in the spectacle. But the spectacle is nothing other than the *sense* of the total practice of a social-economic formation, its *use of time*. It is the historical movement in which we are caught.

12

The spectacle presents itself as something enormously positive, indisputable and inaccessible. It says nothing more than "that which appears is good, that which is good appears." The attitude which it demands in principle is passive acceptance which in fact it already obtained by its manner of appearing without reply, by its monopoly of appearance.

13

The basically tautological character of the spectacle flows from the simple fact that its means are simultaneously its ends. It is the sun which never sets over the empire of modern passivity. It covers the entire surface of the world and bathes endlessly in its own glory.

14

The society which rests on modern industry is not accidentally or superficially spectacular, it is fundamentally *spectaclist*. In the spectacle, which is the image of the ruling economy, the goal is nothing, development everything. The spectacle aims at nothing other than itself.

15

As the indispensable decoration of the objects produced today, as the general exposé of the rationality of the system, as the advanced economic sector which directly shapes a growing multitude of image-objects, the spectacle is the *main production* of present-day society.

16

The spectacle subjugates living men to itself to the extent that the economy has totally subjugated them. It is no more than the economy developing for itself. It is the true reflection of the production of things, and the false objectification of the producers.

17

The first phase of the domination of the economy over social life brought into the definition of all human realization the obvious degradation of *being* into *having*. The present phase of total occupation of social life by the accumulated results of the economy leads to a generalized sliding of *having* into *appearing*, from which all actual "having" must draw its immediate prestige and its ultimate function. At the same time all individual reality has become social reality directly dependent on social power and shaped by it. It is allowed to appear only to the extent that it *is not*.

18

Where the real world changes into simple images, the simple images become real beings and effective motivations of hypnotic behavior. The spectacle, as a tendency to make one see the world by means of various specialized mediations (it can no longer be grasped directly), naturally finds vision to be the privileged human sense which the sense of touch was for other epochs; the most abstract, the most mystifiable sense corresponds to the generalized abstraction of present-day society. But the spectacle is not identifiable with mere gazing, even combined with hearing. It is that which escapes the activity of men, that which escapes reconsideration and correction by their work. It is the opposite of dialogue. Wherever there is independent *representation*, the spectacle reconstitutes itself.

19

The spectacle inherits all the weaknesses of the Western philosophical project which undertook to comprehend activity in terms of the categories of *seeing*; furthermore, it is based on the incessant spread of the precise technical rationality which grew out of this thought. The spectacle does not realize philosophy, it philosophizes reality. The concrete life of everyone has been degraded into a *speculative* universe.

20

Philosophy, the power of separate thought and the thought of separate power, could never by itself supersede theology. The spectacle is the material reconstruction of the religious illusion. Spectacular technology has not dispelled the religious clouds where men had placed their own powers detached from themselves; it has only tied them to an earthly base. The most earthly life thus becomes opaque and unbreathable. It no longer projects into the sky but shelters within itself its absolute denial, its fallacious paradise. The spectacle is the technical realization of the exile of human powers into a beyond; it is separation perfected within the interior of man.

21

To the extent that necessity is socially dreamed, the dream becomes necessary. The spectacle is the nightmare of imprisoned modern society which ultimately expresses nothing more than its desire to sleep. The spectacle is the guardian of sleep.

22

The fact that the practical power of modern society detached itself and built an independent empire in the spectacle can be explained only by the fact that this practical power continued to lack cohesion and remained in contradiction with itself.

23

The oldest social specialization, the specialization of power, is at the root of the spectacle. The spectacle is thus a specialized activity which speaks for all the others. It is the diplomatic representation of hierarchic society to itself, where all other expression is banned. Here the most modern is also the most archaic.

24

The spectacle is the existing order's uninterrupted discourse about itself, its laudatory monologue. It is the self-portrait of power in the epoch of its totalitarian management of the conditions of existence. The fetishistic, purely objective appearance of spectacular relations conceals the fact that they are relations among men and classes: a second nature with its fatal laws seems to dominate our environment. But the spectacle is not the necessary product of technical development seen as a *natural* development. The society of the spectacle is on the contrary the form which chooses its own technical content. If the spectacle, taken in the limited sense of "mass media" which are its most glaring superficial manifestation, seems to invade society as mere equipment, this equipment is in no way neutral but is the very means suited to its total self-movement. If the social needs of the epoch in which such techniques are developed can only be satisfied through their mediation, if the administration of this society and all contact among men can no longer take place except through the intermediary of this power of instantaneous communication, it is because this "communication" is essentially *unilateral*. The concentration of "communication" is thus an accumulation, in the hands of the existing system's administration, of the means which allow it to carry on this particular administration. The generalized cleavage of the spectacle is inseparable from the modern *State*, namely from the general form of cleavage within society, the product of the division of social labor and the organ of class domination.

25

Separation is the alpha and omega of the spectacle. The institutionalization of the social division of labor, the formation of classes, had given rise to a first sacred contemplation, the mythical order with which every power shrouds itself from the beginning. The sacred has justified the cosmic and ontological order which corresponded to the interests of the masters; it has explained and embellished that which society *could not do*. Thus all separate power has been spectacular, but the adherence of all to an immobile image only signified the common acceptance of an imaginary prolongation of the poverty of real social activity, still largely felt as a unitary condition. The modern spectacle, on the contrary, expresses what society *can do*, but in this expression the *permitted* is absolutely opposed to the *possible*. The spectacle is the preservation of unconsciousness within the practical change of the conditions of existence. It is its own product, and it has made its own rules: it is a pseudo-sacred entity. It shows what it is: separate power developing in itself, in the growth of productivity by means of the incessant refinement of the division of labor into a parcellization of gestures which are then dominated by the independent movement of machines; and working for an ever-expanding market. All community and all critical sense are dissolved during this movement in which the forces that could grow by separating are not yet *reunited*.

26

With the generalized separation of the worker and his products, every unitary view of accomplished activity and all direct personal communication among producers are lost. Accompanying the progress of accumulation of separate products and the concentration of the productive process, unity and communication become the exclusive attribute of the system's management. The success of the economic system of separation is the *proletarianization* of the world.

27

Due to the success of separate production as production of the separate, the fundamental experience which in primitive societies is attached to a central task is in the process of being displaced, at the crest of the system's development, by non-work, by inactivity. But this inactivity is in no way liberated from productive activity: it depends on productive activity and is an uneasy and admiring submission to the necessities and results of production; it is itself a product of its rationality. There can be no freedom outside of activity, and in the context of the spectacle all activity is negated, just as real activity has been captured in its entirety for the global construction of this result. Thus the present "liberation from labor," the increase of leisure, is in no way a liberation within labor, nor a liberation from the world shaped by this labor. None of the activity lost in labor can be regained in the submission to its result.

28

The economic system founded on isolation is a *circular production of isolation*. The technology is based on isolation, and the technical process isolates in turn. From the automobile to television, all the *goods selected* by the spectacular system are also its weapons for a constant reinforcement of the conditions of isolation of "lonely crowds." The spectacle constantly rediscovers its own assumptions more concretely.

29

The spectacle originates in the loss of the unity of the world, and the gigantic expansion of the modern spectacle expresses the totality of this loss: the abstraction of all specific labor and the general abstraction of the entirety of production are perfectly rendered in the spectacle, whose *mode of being concrete* is precisely abstraction. In the spectacle, one part of the world *represents itself* to the world and is superior to it. The spectacle is nothing more than the common language of this separation. What binds the spectators together is no more than an irreversible relation at the very center which maintains their isolation. The spectacle reunites the separate, but reunites it *as separate*.

30

The alienation of the spectator to the profit of the contemplated object (which is the result of his own unconscious activity) is expressed in the following way: the more he contemplates the less he lives; the more he accepts recognizing himself in the dominant images of need, the less he understands his own existence and his own desires. The externality of the spectacle in relation to the active man appears in the fact that his own gestures are no longer his but those of another who represents them to him. This is why the spectator feels at home nowhere, because the spectacle is everywhere.

31

The worker does not produce himself; he produces an independent power. The *success* of this production, its abundance, returns to the producer as an *abundance of dispossession*. All the time and space of his world become *foreign* to him with the accumulation of his alienated products. The spectacle is the map of this new world, a map which exactly covers its territory. The very powers which escaped us *show themselves* to us in all their force.

32

The spectacle within society corresponds to a concrete manufacture of alienation. Economic expansion is mainly the expansion of this specific industrial production. What grows with the economy in motion for itself can only be the very alienation which was at its origin.

33

Separated from his product, man himself produces all the details of his world with ever increasing power, and thus finds himself ever more separated from his world. The more his life is now his product, the more he is separated from his life.

34

The spectacle is *capital* to such a degree of accumulation that it becomes an image.

‘Globalised Culture: The Triumph of the West?’

from *Culture and Global Change* (1999)

John Tomlinson

Editors’ Introduction

Most of the material in the first two parts of this anthology (and most others, too) concerns the affluent world. It is that world which produces the dominant cultural products such as films and television programmes in the narrower sense of culture, and the attributes of a consumerist lifestyle such as certain kinds of drinks, cigarettes, clothes, and so forth in the broad sense of a way of life, which permeate local cultures globally. Disney’s currency, that is, is worldwide and still appreciating in value. But culture comes with its own ideological baggage, and this as well is exported globally. Just as Adorno sees mass culture as not produced by the mass public but for their control, so Tomlinson sees globalised culture as ‘not a culture that has arisen out of the mutual experiences and needs of all of humanity’ but one which ‘is the installation . . . of one particular culture born out of one particular, privileged historical experience’ (p. 23). The implications are that local diversity gives way to homogenisation, and that this figures in a wider script of control by the affluent world of the non-affluent world which it sees as providing its future growth markets as well as raw materials and cheap and often unregulated labour. In process, local cultures (and it could be added languages) become vulnerable as they cannot match the alluring spectacle of Western culture. But, as Tomlinson points out in a balanced argument, communications are increasingly global as well as cultural products, and can be used in more ways than those of the dominant society. Access remains uneven, yet new patterns of sociation and lattices of communication are nevertheless developed within and between marginalised groups and societies. The impact of this may be enormous in due course, and, though this is outside the scope of this text, has contributed to the organisation of mass protest against organisations such as the World Trade Organisation, and a feeling that perhaps resistant groups no longer need what was once called the oxygen of publicity of the news media.

John Tomlinson is Director of the Centre for Research in International Communication and Culture at Nottingham Trent University, and author of *Cultural Imperialism: A Critical Introduction* (London, Pinter, 1991). There is a rapidly increasing literature on development studies and work in the non-affluent world, of which *Liberation Ecologies: Environment, Development, Social Movements*, edited by Richard Peet and Michael Watts (London, Routledge, 1996) is particularly helpful. The essay which follows Tomlinson’s in *Culture and Global Change*, by Peter Worsley on ‘Culture and Development Theory’ is also suggested as further reading in this area.

GLOBAL AND GLOBALISED CULTURE

The idea of a single, unified culture encompassing the whole world has a long and, so far as I know, relatively undocumented history. An inventory of the various historical dreams, visions and speculations about a global culture would, I suppose, have to include at least those of: the imperial projects of the ancient 'world empires' such as China or Rome; the great proselytising 'world religions' and the communities of faith established around them – Christendom, the Ummah Islam etc.; the utopian global visions of early socialists such as Saint-Simon; the various movements dedicated to establishing world peace; the ideas, beginning in the nineteenth century, of enthusiasts for artificial 'international' languages such as Esperanto; and many more.[1] These ideas clearly differ from each other in all sorts of ways. For example, some (probably most) were simply naïve, unproblematised, often dogmatic, projections of a particular cultural outlook onto a 'universal' screen, while others were driven by the desire to reconcile cultural differences and to usher in a new, pacified ideal home for humankind. But two things unite all these visions. First, that they all approached the idea of a single global culture with enthusiasm, and second that none of them came anywhere near to seeing it achieved.

The ideas of a global culture in the air today – in the intellectual and critical discourses of the 1990s – are different. They are not, in the main, visionary or utopian ideas.[2] Rather they are speculations that arise in response to processes that we can actually see occurring around us. These processes, which are generally referred to collectively as 'globalisation', seem, whether we like it or not, to be tying us all – nations, communities, individuals – closer together. It is in the context of globalisation, then, that current discussions of an emergent global culture assume a different significance from earlier speculations. It is not only that the current social, economic and technological context makes a global culture in some senses more plausibly attainable – a concrete possibility rather than a mere dream. It is also that this very sense of imminence brings with it anxieties, uncertainties and suspicions.

Talk of a global culture today is just as likely, probably more likely, to focus on its dystopian aspect, to construct it as a threat rather than a promise. This, at any rate, is the sort of talk I want to consider here. To grasp its close relation to the processes of globalisation and to distinguish it from earlier traditions of thought, I shall refer in what follows to the idea of a *globalised* rather than global culture. A globalised culture refers here specifically to the way in which people, integrating the general signs of an increasing interdependence that characterises the globalisation process with other critical positions and assumptions, have constructed a pessimistic 'master scenario' (Hannerz 1991) of cultural domination. This is the speculative discourse that I want to criticise.

In order to develop this discussion in a relatively short piece, I shall have to leave on one side some pretty big and thorny related and contextual issues. Most of these relate to the way in which the notion of globalisation has been theorised. Though I shall offer later a brief description of what globalisation broadly means, it must be recognised that there are all sorts of unresolved controversies in globalisation theory which this discussion will necessarily rub up against from time to time without explicitly recognising. If readers recognise these, and develop the argument themselves, so much the better!

GLOBALISED CULTURE AS WESTERNISED CULTURE

The argument I want to focus on can be set out quite briefly in outline, though we shall see that it contains some crucial assumptions that will require unpacking presently. It goes like this. The globalised culture that is currently emerging is not a global culture in any utopian sense. It is not a culture that has arisen out of the mutual experiences and needs of all of humanity. It does not draw equally on the world's diverse cultural traditions. It is neither inclusive, balanced, nor, in the best sense, synthesising. Rather, globalised culture is the installation, world-wide, of one particular culture born out of one particular, privileged historical experience. It is, in short, simply the global extension of *Western* culture. The broad implications – and the causes of critical concern – are that: (a) this process is homogenising, that it threatens the obliteration of the world's rich cultural diversity; (b) that it visits the various cultural ills of the West on other cultures; (c) that this is a particular threat to the fragile and vulnerable cultures of peripheral, 'Third World' nations; and (d) that it is part and parcel of wider forms of domination – those involved in the ever-widening grip of transnational capitalism and those involved in the maintenance of post-colonial relations of (economic and cultural) dependency.

This is, of course, to view the globalisation process through a now familiar critical prism – that of the critique of Western 'cultural imperialism' (Friedman 1994; Hannerz 1991; and McQuail 1994). As I and others have argued elsewhere (Boyd-Barrett 1982; Schlesinger 1991; Sinclair 1992; Tomlinson 1991, 1995, 1997), this is a peculiarly vexed and often confused critical discourse which rolls a number of complex questions up together. In the space available here I shall have to take for granted most of this criticism. But before I come to my central argument it will be useful just to mention a couple of the most common objections to the general idea of Westernisation, so as to distinguish them from the specific, rather different line of criticism I want to follow later.

What do people mean when they talk about 'Westernisation'? A whole range of things: the consumer culture of Western capitalism with its now all-too-familiar icons (McDonald's, Coca-Cola, Levi Jeans), the spread of European languages (particularly English), styles of dress, eating habits, architecture and music, the adoption of an urban lifestyle based around industrial production, a pattern of cultural experience dominated by the mass media, a range of cultural values and attitudes – about personal liberty, gender and sexuality, human rights, the political process, religion, scientific and technological rationality and so on. Now, although all of these aspects of 'the West' can be found throughout the world today, they clearly do not exist as an indivisible package. To take but one example, an acceptance of the technological culture of the West and of aspects of its consumerism may well co-exist with a vigorous rejection of its sexual permissiveness and its generally secular outlook – as is common in many Islamic societies. A prime instance of this contradiction is the current attempt to regulate or even ban the use of satellite television receivers in countries like Iran, since they are seen by the authorities to be the source of various images of Western decadence. This sort of cultural-protectionist legislation is almost impossible to implement, partly because of the huge numbers of dishes involved (estimated at more than 500,000 in Tehran alone) but also because use of this technology is vital for education and scientific research. Restriction of access is thus resisted by these constituencies within the intelligentsia who might otherwise hold quite 'conventional' Muslim attitudes towards, for example, images of sexuality or nudity in Western television programmes (Haeri 1994).

So there is obviously a need to *discriminate* between various aspects of what is totalised as 'Westernisation', and such discrimination will reveal a much more complex picture: some cultural goods will be broader in their appeal than others, some values and attitudes easily adopted while others are actively resisted or found simply odd or irrelevant. All this will vary from society to society and between different groupings and divisions – class, age, gender, urban/rural, etc. – within societies. The first objection to the idea of Westernisation, then, is that it is too broad a generalisation. Its rhetorical force is bought at the price of glossing over a multitude of complexities, exceptional cases and contradictions. This criticism also connects with another one: that the Westernisation/homogenisation/cultural imperialism thesis itself, ironically, displays a sort of Western ethnocentrism (Hannerz 1991; Tomlinson 1991, 1995). Ulf Hannerz puts this point nicely:

> The global homogenisation scenario focuses on things that we, as observers and commentators from the centre, are very familiar with: our fast foods, our soft drinks, our sitcoms. The idea that they are or will be everywhere, and enduringly powerful everywhere, makes our culture even more important and worth arguing about, and relieves us of the real strains of having to engage with other living, complicated, puzzling cultures.
>
> (Hannerz 1991: 109)

A second set of objections concerns the way in which Westernisation suggests a rather crude model of the one-way flow of cultural influence. This criticism has – rightly – been the one most consistently applied to the whole cultural imperialism idea. Culture, it is argued, simply does not transfer in this way. Movement between cultural/geographical areas always involves translation, mutation and adaptation as the 'receiving culture' brings its own cultural resources to bear, in dialectical fashion, upon 'cultural imports' (Appadurai 1990; Garcia Canclini 1995; Lull 1995; Robins 1991; Tomlinson 1991). So, as Jesus Martin-Barbero describes the process of cultural influence in Latin America: 'The steady, predictable tempo of homogenising development [is] upset by the counter-tempo of profound differences and cultural discontinuities' (1993: 149). What follows from this argument is not simply the point that the Westernisation thesis underestimates the cultural resilience and dynamism of non-Western cultures, their capacity to 'indigenise' Western imports. It is also draws attention to the *counter-flow* of cultural influence – for instance in 'world music' (Abu-Lughod 1991) – from the periphery to the centre. Indeed this dialectical conception of culture can be further developed so as to undermine the sense of the West as a stable homogeneous cultural entity. As Pieterse puts it: 'It . . . implies an argument with Westernisation: the West itself may be viewed as a mixture and Western culture as Creole culture' (1995: 54). Of course, the ultimate implication is that whatever globalised culture is emerging, it will not bear the stamp of any particular cultural-geographical or national identity, but will be *essentially* a hybrid, *mestizaje*, 'cut-and-mix' culture (Pieterse 1995; Garcia Canclini 1995).

These sorts of criticism quickly take some of the wind out of the sails of the Westernisation argument, at least in its most dramatic, polemical formulations. However, they do not entirely resolve the issue of the contemporary cultural power of the West. For it could very reasonably be argued that, when all is said and done and all these criticisms met, Western cultural practices and institutions still remain firmly in the driving seat of global cultural development. No amount of attention to the processes of cultural reception and translation, no anthropological scruples about the complexities of particular local contexts, no dialectical theorising, can argue away the massive and everywhere manifest power of Western capitalism, both as a general cultural configuration (the commodification of everyday experience, consumer culture) and as a specific set of global cultural industries (CNN, Time-Warner, News International). What, it might be asked, is this, if not evidence of some sort of Western cultural hegemony? What ensues from this is a sort of critical stand-off. Both positions are convincing within their own terms, but seem somehow not precisely to engage.

To try to take the argument a little further I want, in what follows, to focus on one particular, largely implicit assumption that seems to be embedded in the idea of globalised culture as Westernised culture. This is the assumption that the process of globalisation is *continuous* with the long, steady, historical rise of Western cultural dominance. By this I mean that the sort of cultural power generally attributed to the West today is seen as of the same *order* of power that was manifest in, say, the great imperial expansion of European powers from the seventeenth century onwards. So, for example, this implicit understanding of globalised culture would see the massive and undeniable spread of Western cultural goods – 'Coca-colonisation' – as, at least broadly, part of the same process of domination as that which characterised the *actual* colonisation of much of the rest of the world by the West. I do not mean that no distinction is made between the obviously coercive and often bloody history of Western colonial expansion and the 'soft' cultural imperialism of McDonald's hamburgers, Michael Jackson videos or Hollywood movies. But I think it is often the case that these and many other instances of Western cultural power get lumped together – 'totalised' – by the term 'Westernisation' and that the result is an impression of the inexorable advance – even the 'triumph' of the West.

It is this particular totalising assumption that I think could benefit from being unpacked and critically examined. This is for two reasons: first because it mistakes the nature of the globalisation process and secondly because by conflating a number of different issues it overstates the general cultural power of the West. I do not want to deny that the West is in a certain sense 'culturally powerful', but I do want to suggest that this power, which is closely aligned with its technological, industrial and economic power, is not the whole story. It does not amount to the implicit claim that 'the way of life' of the West is now becoming installed, via globalisation, as the unchallengeable cultural model for all of humanity.

Indeed, as I shall now go on to argue, there are ways in which the globalisation process, properly conceived, can be shown to be actively problematic for the continuation of Western cultural dominance: to signal not the triumph of the West, but its imminent decline.

GLOBALISATION AS THE 'DECLINE OF THE WEST'

In this section I want to examine some observations by two important and influential British social theorists, Anthony Giddens and Zygmunt Bauman, which connect globalisation with the decline, rather than the triumph of the West. Giddens in particular has been highly influential in theorising the globalisation process and in relating this to the wider debate about the nature of social modernity. Neither of these particular arguments, however, is developed at any great length and there will not be space here to develop them much further. I present them simply as suggestive ways of thinking against the grain of the arguments we have so far reviewed. First, however, it will be useful to say a little more about the nature of the globalisation process itself.

Probably the most important thing to be clear about is that globalisation is not *itself* the emergence of a globalised culture. Rather, it refers to the complex pattern of interconnections and interdependencies that have arisen in the late-modern world. Globalisation is heavy with implications for all spheres of social existence – the economic, the political, the environmental, as well as the cultural. In all these spheres it has the effect of tying 'local' life to 'global' structures, processes and events. So, for example, the economic fate of local communities – levels of economic activity, employment prospects, standard of living – is increasingly tied to a capitalist production system and market that is global in scope and operation – to global 'market forces'. Similarly, our local environment (and consequently our health and physical quality of life) is subject to risks arising at a global level – global warming, ozone depletion, eco-disasters with global 'fall-out' such as Chernobyl. What these aspects of globalisation represent, then, is a rapidly growing context of global interdependence which already 'unites' us all, if only in the sense of making us all subject to certain common global influences, processes, opportunities and risks. But clearly this sort of 'structural unity' does not of itself imply the emergence either of a common 'global culture' (in the utopian sense) or of the globalised (Westernised) culture we have been discussing. Neither, it should be added, does this interdependence imply a levelling out of advantages and disadvantages globally. Globalisation is generally agreed to be an *uneven* process in which neither the opportunities nor the risks are evenly distributed (McGrew 1992; Massey 1994). But, again, this does not mean that it necessarily reproduces – or will reproduce – the precise historical patterns of inequality supposed in the dualism of the 'West versus the Rest'. More complex, contradictory patterns of winners and losers in the globalisation process may be emerging.

Another important aspect of globalisation is, of course, the increasing level of social-cultural *awareness* of global interdependence. As Robertson puts it, globalisation 'refers both to the compression of the world and the intensification of consciousness of the world as a whole' (1992: 8). Our sense of the rest of the world and of our connections with it are 'brought home to us' routinely via globalising media and communications technology. Now, of course, it can be argued that the *contents* of these images of a wider world are often highly selective and restricted ones. For instance, it has long been observed that the picture of developing countries portrayed on Western televisions tends to be restricted to 'the narrow agenda of conflicts and catastrophes' (Cleasby 1995: iii). Thus as Peter Adamson of UNICEF has argued, with 'no equivalent sense of the norms in poor countries to set against this constant reporting of the exceptional . . . the cumulative effect of the way in which the developing world is portrayed by the media is grossly misleading' (quoted, Cleasby 1995: iii).

However, the point I want to stress is the routine *availability* of distanciated imagery (however accurate) that globalising media technology provides. In the affluent Western world we take the experience provided by such technology pretty much for granted. We *expect* to have instant images of events happening in every remote corner of the world on our television screens. It is with no sense of wonder that we pick up the phone and speak to people on other continents. We just as quickly – some of us – become used to communicating globally on the Internet and the 'World Wide Web'. So globalisation seems also to involve, as Giddens puts it (1990: 187), the extension of our 'phenomenal worlds' from the local to the global. Of course, access to this technology is obviously not evenly distributed and so we must avoid the mistake of extrapolating from the Western experience. But, on the other hand, it would be equally misleading to treat such communications technology, and the experience it affords as, somehow, the exclusive property of the West. Again, we have to recognise the possibility that globalisation may be producing shifting global and local patterns in what has been called the 'information rich and the information poor'.

Perhaps the most widely recognised property of globalisation amongst those who have theorised it is its *ambiguous* nature – its mixture of risk and opportunity, its 'dialectical' counterposing of generalising and particularising tendencies, its confusing capacity both to enable and to disempower.

It is within this broad conceptualisation of globalisation that Anthony Giddens writes of '[t]he gradual decline in European or Western global hegemony, the other side of which is the increasing expansion of modern influences world-wide', of 'the declining grip of the West over the rest of the world' or of 'the evaporating of the privileged position of the West' (1990: 51–3). What can he mean by this?

Well, Giddens has written a good deal about the globalisation process and at a fairly high level of abstraction and really these comments need to be read in the context of his overall theorisation of the globalising nature of modernity (Giddens 1990, 1991, 1994a, 1994b). But to put it at its simplest, his argument is that, though the process of 'globalising modernity' may have *begun* in the extension of Western institutions, the very fact of the current global ubiquity of these institutions (capitalism, industrialism, the nation-state system and so on) – in a sense the West's 'success' in disseminating its institutional forms – also represents a decline in the differentials between it and the rest of the world, thus a loss of the West's (once unique) social/cultural 'edge'. As he puts the point in a more recent discussion:

> The first phase of globalisation was plainly governed, primarily, by the expansion of the West, and institutions which originated in the West. No other civilisation made anything like as pervasive an impact on the world, or shaped it so much in its own image . . . Although still dominated by Western power, globalisation today can no longer be spoken of only as a matter of one-way imperialism . . . increasingly there is no obvious 'direction' to globalisation at all and its ramifications are more or less ever present. The current phase of globalisation, then, should not be confused with the preceding one, whose structures it acts increasingly to subvert.
>
> (Giddens 1994b: 96)

There are various ways in which this 'loss of privilege' and even the 'subversion' of Western power may be understood. For example, it might be pointed out that certain parts of what we were used to calling the 'Third World' are now actually more advanced – technologically, industrially, economically – than some parts of the West. The comparison here might be, for example, between the socalled 'Asian Tiger' economies and some of the economically depressed heavy-industrial regions of Europe or the US. And there might be a complex causal relationship between the rise and decline of such regions connected by a globalised capitalist market (Giddens 1990: 65, 1994a: 65). Or, to put this slightly differently, it might be argued that capitalism has no 'loyalty' to its birthplace, and so provides no guarantees that the geographical patterns of dominance established in early modernity – the elective affinity between the interests of capitalism and of the West – will continue (Tomlinson 1997). There are signs of this, for example, in the increasingly uneasy relation between the capitalist money markets and the governments of Western nation-states – the periodic currency crises besetting the Western industrial nations. A rather spectacular instance of the capitalist system deserting the West could be seen in the débâcle of Britain's oldest merchant bank, Baring Brothers, in February 1995 as a result of its high risk globalising speculations carried out, appropriately enough, on the Singapore market.

On a more directly cultural level, the loss of privilege of the West can be seen in the shifting orientation and self-understanding of the discipline of anthropology, the academic discipline which, perhaps more than any other, displays the cultural assumptions on which the West has presumed to organise a discourse about other cultures. As Giddens points out (1994b: 97), anthropology in its formative stage was a prime example of the West's self-assured assumption of cultural superiority. Because of its 'evolutionary' assumptions, early taxonomising anthropology established itself as a practice to which the West had exclusive rights – the 'interrogation' of all other cultures. Other cultures were there, like the flora and fauna of the natural world, to be catalogued and observed, but there was no thought that they could ever *themselves* engage in the practice – they were categorised as 'if not inert, no more than a "subject" of enquiry' (1994b: 97).

Early anthropology was part of the cultural armoury of an imperialist West during 'early globalisation' precisely because it had not developed its inner logic. As this emerges, with the recognition of the integrity of traditions, the knowledgeability of all cultural agents and the growing sense of 'cultural relativism', so anthropology becomes simultaneously both a more modest and humble undertaking and, significantly, a globalised practice. Present day anthropologists have to approach their study in the role of the *ingenu* [*sic*] – the innocent abroad – rather than as the confident explorer and taxonomist. Without the assurance of a taken-for-granted superior cultural 'home-base', anthropological study becomes a matter of 'learning how to go on' rather than of detached, *de haut en bas* interrogation. Not only this, it becomes clear that in this later phase of globalisation, *all* cultures have a thoroughly reflexive anthropological sensibility: 'In British Columbia the present day Kwakiutl are busy reconstructing their traditional culture using [Franz] Boas' monographs as their guide, while Australian Aboriginals and other groups across the world are contesting land-rights on the basis of parallel anthropological studies' (Giddens 1994b: 100).

So, the trajectory of the development of anthropology, as Giddens puts it, 'leads to its effective dissolution today' (1994b: 97). This could also stand, more broadly, for the way in which current globalisation subverts and undermines the cultural power of the West from which it first emerged.

To conclude this section I want to comment briefly on an interesting distinction that Zygmunt Bauman makes between the 'global' and the 'universal':

> Modernity once deemed itself *universal*. Now it thinks of itself instead as *global*. Behind the change of term hides a watershed in the history of modern self-awareness and self-confidence. Universal was to be the rule of reason – the order of things that would replace the slavery to passions with the autonomy of rational beings, superstition and ignorance with truth, tribulations of the drifting plankton with self-made and thoroughly monitored history-by-design. 'Globality', in contrast, means merely that everyone everywhere may feed on McDonald's hamburgers and watch the latest made-for-TV docudrama. Universality was a proud project, a Herculean mission to perform. Globality in contrast, is a meek acquiescence to what is happening 'out there'.
>
> (Bauman 1995: 24)

Mapping this onto Giddens' distinction between early and late globalisation, the key difference becomes that between a Western culture with pretensions to universalism and one without.[3] The globalisation of the West's cultural practices is simply occurring without any real sense either that this is part of its collective project or 'mission', or that these practices are, indeed, the tokens of an ideal human civilisation. Early globalisation involves the self-conscious cultural project of universality, whilst late globalisation – globality – is mere ubiquity.

Now, whilst it may be argued that Bauman erects a rather contrived dualism here between the 'high cultural' project of enlightenment rationalism and some rather specific 'popular cultural' practices, I do think his stress on the cultural self-image and self-confidence of the West is an important one. The specific doubts he detects that now 'sap the ethical confidence and self-righteousness of the West' tend to be those of the intellectual. These are doubts about the capacity of the Enlightenment project ever to deliver full emancipation for all human beings, about 'whether the wedlock between the growth of rational control and the growth of social and personal autonomy, that crux of modern strategy, was not ill-conceived from the start . . .' (1995: 29).

However we can read the idea of the loss of Western self-confidence in more mundane ways. Bauman's description of globality as a meek acquiescence to what is happening 'out there' may be a little overstated, but it does grasp something of the spirit in which ordinary people in the West probably experience the global spread of their 'own' cultural practices. Indeed a lot probably hangs on the extent to which Westerners actually feel 'ownership' of the sorts of cultural practices that, typically, get globalised. Although this is an immensely complicated issue, my guess is that there is only a very low level of correspondence between people's routine interaction with the contemporary 'culture industry' and their sense of having a distinctive *Western* cultural identity, let alone feeling proud of it. It seems more likely to me that things like McDonald's restaurants are experienced as simply 'there' in our cultural environments: things we use and have become familiar and perhaps comfortable with, but which we do not – either literally or culturally – 'own'. In this sense the decline of Western cultural self-confidence may align with the structural properties of globalising modernity – the 'disembedding' (to use Giddens' term) of institutions from contexts of local to global control. In a world in which increasingly our mundane 'local' experience is governed by events and processes at a distance, it may become difficult to maintain a sharp sense of (at least 'mass') culture as distinctively 'the way we do things' in the West – to understand these practices as having any particular connection with our specific histories and traditions. Thus, far from grasping globalised culture in the complacent, proprietorial way that may have been associated with, say, the *Pax Britannica*, late-modern Westerners may experience it as a largely undifferentiated, 'placeless' modernity to which they relate effortlessly, but without much sense of either personal involvement or of control.

CONCLUSION

The arguments reviewed above are obviously not conclusive and leave many issues unresolved. In particular the complex phenomenology of cultural identity in a globalised world requires far more extensive and nuanced treatment than has been possible here. What I have tried to offer is simply a glimpse of alternative ways of thinking about the complex cultural issues forced upon us by the globalisation process. Nothing in this is meant to deny the continuing *economic* power of the West, nor even that particular, limited, sense of 'cultural' power that proceeds from this – the power of Western transnational capitalism to distribute its goods around the world. Nor, to be clear, do these arguments entail the idea of a simple 'turning of the tables'; the 'decline of the West' does not mean the inevitable 'rise' of any other particular hegemonic power (no matter how tempting it is to speculate about the 'Asian Tigers' and so forth). In the short term at least, much of the 'Third World' will probably continue to be marginalised by globalising technologies (Massey 1994). But, to look beyond this, these reflections do suggest that what is happening in globalisation is not a process firmly in the cultural grip of the West and that, therefore, the global future is much more radically open than the discourses of homogenisation and cultural imperialism suggest. We surely need to find new critical models to engage with the emerging 'power geometry' of globalisation, but we will not find these by rummaging through the theoretical box of tricks labelled 'Westernisation'.

REFERENCES

Abu-Lughod J (1991) 'Going Beyond Global Babble', in King A D, ed. (1991) *Culture, Globalization and the World System*, London, Macmillan

Appadurai A (1990) 'Disjuncture and Difference', in Featherstone M, ed. (1990) *Global Culture*, London, Sage

Bauman Z (1995) *Life in Fragments*, Oxford, Blackwell

Boyd-Barrett O (1982) 'Cultural Dependency and the mass media' in Gurevitch M, ed. (1982) *Culture, Society and the Media*, London, Methuen

Cleasby A (1995) *What in the World is Going On? British Television and Global Affair*, London, Third World and Environment Broadcasting Project

Friedman J (1994) *Cultural Identity and Global Process*, London, Sage

Garcia Clancini N (1995) *Hybrid Cultures*, Minneapolis, University of Minnesota Press

Giddens A (1990) *The Consequences of Modernity*, Cambridge, Polity

Giddens A (1991) *Modernity and Self-Identity: Self and Society in the Late Modern Age*, Cambridge, Polity

Giddens A (1994a) *Beyond Left and Right*, Cambridge, Polity

Giddens A (1994b) 'Living in a post-traditional society' in Beck U, Giddens A, and Lash S, eds (1994) *Reflexive Modernization*, Cambridge, Polity

Haeri S (1994) 'A fate wore than Saudi', *Index on Censorship* vol. 23, #4-5, pp49-51

Hannerz U (1991) 'Scenarios for Peripheral Culture', in King A D, ed. (1991) *Culture, Globalisation and the World System*, London, Macmillan

Lull J (1995) *Media, Communication, Culture: A Global Approach*, Cambridge, Polity

McGrew T (1992) 'A global society?' in Hall S, Held D, and McGrew T, eds (1992) *Modernity and its Futures*, Cambridge, Polity

McQuail D (1994) *Mass Communication Theory*, London, Sage

Massey D (1994) *Space, Place and Gender*, Cambridge, Polity

Pieterse J N (1995) 'Globalization a Hybridization', in Feathertone M, Lash S, and Robertson R, eds (1995) *Global Modernities*, London, Sage

Robins K (1991) 'Tradition and Translation: Natural Culture in its Global Context', in Corner J, ed. (1991) *Enterprise and Heritage*, London, Routledge

Schlesinger P (1991) *Media, state and Nation*, London, Sage

Sinclair J (1992) 'The decentering of globalization: Television and globalization', in Jacka E, ed. (1992), *Continental Shift: Globalization and Culture*, New York, Doubleday

Tomlinson J (1991) *Cultural Imperialism: A Critical Introduction*, London, Longman

Tomlinson J (1995) 'Homogenisation and globalisation' *History of European Ideas* vol. 20, #4-6, pp891-7

Tomlinson J (1997) 'Internationalism, globalization and cultural imperialism', in Thompson K, ed. (1997) *Cultural Change and Regulation: Policies and Controversies*, London, Sage and Open University

PART THREE

Symbolic Economies

INTRODUCTION TO PART THREE

One of the few consistent things about cities is that they are in a constant state of change. Studying, understanding and classifiying the processes of urban change, their outcomes and their impacts is a major preoccupation of geographers, planners, sociologists and others thinking about the city. Cities constantly demonstrate gradual, piecemeal change through processes of accretion, addition or demolition. This type of change can be regarded as largely cosmetic while the underlying processes of urbanisation and the overall structure of the city remain largely unaltered. However, at certain periods fundamentally different processes of urbanisation appear to have emerged; the result has been that the rate of urban change has accelerated and new, distinctly different, urban forms, spaces and types of city have emerged.

The city poses questions. Are observable changes part of the continual process of piecemeal change or are they are part of more fundamental processes of urban transformation? Just such a debate has been occupying geographers, sociologists and other social scientists since the 1980s. There has emerged a literature that has sought to map the postmodern city, its spaces, its meanings and its social and cultural geographies. Contributions to this debate have included David Harvey's *The Condition of Postmodernity* (Baltimore, Johns Hopkins University Press, 1989), Edward Soja's *Postmodern Geographies* (Oxford, Blackwell, 1989), and Sharon Zukin's *The Culture of Cities* (Basingstoke, Macmillan, 1996) among others.

Visually, the evidence to support the notion of a fundamental transformation in the processes of urbanisation, and consequently the cities they produce, appears compelling. The signs of this change are apparent in the urban landscapes of North America, the United Kingdom, Europe and many parts of the developing world. Similarly the language that academics (and property developers, local journalists and civic authorities) have used to represent cities since the late 1980s would suggest that something profoundly new has emerged in the urbanisation process. The language that academics have used to describe these apparent transformations, for example, has included: industrial–post-industrial, modern–postmodern, fordist–post-fordist.

Some of the most widely debated of these signs of change have been: the enhancement of city centres by extensive urban redevelopment, the redevelopment of derelict, formerly industrial areas such as factories and docks, the use of industrial and architectural heritage in new commercial and residential developments, the social, economic and environmental upgrading of inner city neighbourhoods by young, middle-class professionals, the emergence of brand-new 'city-like' settlements on the edges of existing urban areas and the emergence of large areas of poverty and degradation for example in old inner city areas and on council housing estates on the edges of numerous towns and cities. This part brings together critical social scientists who have taken a number of these new urban spaces as their subject.

Underpinning much discussion of new urban spaces is Pierre Bourdieu's notion of symbolic capital. The reinvention of city centre spaces since the 1980s has largely involved a pursuit of external sources of investment – jobs, companies, tourists and wealthy residents for example. For this to be successful cities have had to accumulate reserves of symbolic capital, for example, blue chip architecture, loft living spaces, public art, aesthetised heritage litter and other gilded spaces, to help create the appropriate 'aura' of distinction with which the providers of these sources of investment wish to attach themselves. David Harvey connects Bourdieu's notion of symbolic capital to Leon Krier's of symbolic richness to highlight the impulses underlying the apparent renaissance of so many cities since the 1980s:

> Conjoining the idea of symbolic capital with the search to market Krier's symbolic richness has much to tell us, therefore, about such urban phenomena as gentrification, the production of community (real, imagined, or simply packaged for sale by producers), the rehabilitation of urban landscapes, and the recuperation of history (again, real, imagined, or simply reproduced as pastiche). It also helps us to comprehend the present fascination with embellishment, ornamentation, and decoration as so many codes and symbols of social distinction (Harvey, 1989: 82)

Sociologist Sharon Zukin has recognised two distinct approaches to the study of the built environment: the political economy approach and the symbolic economy approach. The former is concerned with the material conditions of groups in urban society resulting from the process of local and global urban development. However, the latter is concerned with the relationship between dominant representations of the city, through architecture, urban design and advertising and what Rosalyn Deutsche (see Part Nine) calls the 'rights to the city', inclusion and exclusion of certain groups in urban society. As Zukin has argued, the symbolic economy approach is concerned with the relationship between culture and power. The two are strongly interconnected in the reshaping of cities through urban development.

> It is noticeable that as cities have developed service economies they have both propagated and been taken hostage by an aesthetic urge. On the one hand, there is a tendency to take a connoisseur's view of the past, 'reading' the legible practices of cultural discrimination through a reshaping of the city's collective memory. Historic preservation connects an ecology of urban buildings and streets with an ecology of images of the city's past. Historicist post-modern architecture instantly makes the present part of a classical age. On the other hand there is a desire to humanize the future, by viewing artists and art work as symbols of a postindustrial economy. Office buildings are not just monumentalised by height and façade; they are given another embodiment by video artists' screen installations and public concerts. Every well-designed down-town has a mixed-use shopping center and a nearby artists' quarter. The derelict factory district or waterfront has been converted into a marketplace for seasonal produce, cooking equipment, art galleries and an aquarium. Economic redevelopment plans have focused on museums, from Lowell, Massachusetts to downtown Los Angeles . . . Thus the symbolic economy of cultural meanings and representations implies real economic power (Zukin, 1995: 44).

This section opens with a discussion by Pierre Bourdieu of cultural goods and symbolic value in an extract entitled 'the production of belief: contribution to an economy of symbolic goods' from *The Field of Cultural Production*, (1993). Bourdieu takes as his subject Parisian theatre and the important cultural distinctions that circulate within and around it.

In the second extract Tim Hall takes one of the components of the postmodern city centre mentioned by Zukin, above, and considers its place in the gilding of urban space for external consumption. In 'Opening Up Public Art's Spaces: Art, Regeneration and Audience' (2003) Hall considers the roles of public art in the rejuvenation of urban space and marketing of the city. He goes on to note that critical writing on public art has largely been produced from the perspectives of the producers of space or from a privileged academic position. Little attempt has been made to understand the new urban spaces from the perspectives of the diverse publics who experience them. Hall notes that this reflects a failing of much writing on the postmodern city which see its spaces as carceral and without contest. As Ley and Mills argue that:

> Such models of hegemonic control present the consciousness of the masses as monolithic and unproblematic, passive and without the potential for resistance. The view of mass culture is distant and elitist. Soja's (1989) view of the surveillant state in Los Angeles, for example, is a view from on

> high; as the noose of total social control is drawn tightly around the city, we do not know if any member of the thousands of cultural worlds in that city has noticed, for no other voice, or values are admitted other than those of the author. As in so much of the literature on cultural hegemony, the social control of consciousness is alleged but never proven. When we look for the voices of the manipulated masses we encounter in the text a gasping silence (1993: 257–8).

Hall concludes by speculating on how other voices might be incorporated into the academic representation of the postmodern city. Part Six of this reader also includes texts which consider the hidden, everyday lives and geographies of the city.

The shopping mall has been one of the most debated urban forms since William Kowinski's *The Malling of America: Travels in the United States of Shopping* was published in 1985. Contributions have included, for example, semiotic deconstructions of the instrumental spaces of the mall, for example, in Mark Gottdiener's essay 'Recapturing the Centre: A Semiotic Analysis of Shopping Malls' in Gottdiener and Lagopoulos' edited collection *The City and the Sign* (1986), to explorations of the multiple meaning of the shopping arcade and the practice of shopping such as Taylor, Evans and Fraser's *A Tale of Two Cities* (1996) and Miller, Jackson, and Thrift's *Shopping, Place and Identity* (1998). Margaret Crawford takes the world's largest mall, the West Edmonton Mall in Canada, as her subject in 'The World in a Shopping Mall' (1992). She begins by outlining its mind-boggling attractions including an indoor amusement park, a water park, 800 shops, 11 department stores, over 100 restaurants, an ice skating rink, a hotel, a lake, 20 cinemas and 13 nightclubs. Crawford goes on to offer a deconstruction of the mall as a manufactured form and examines the simulacra constructed within the mall, its lures, the science that underpins the seemingly magical experience of consumption that it offers and its impacts on the local downtown and wider urban form. She observes that the mall is a symptom and a cause of the impoverished nature of the urban public realm and one that enjoys an expanded role as public space within the atomized, gated and privatised postmodern city.

Martha Rosler examines the view of the street as a marginal space, and considers the ways in which urban development has generated privatised, exclusive downtowns, and at the same time has produced a displaced population unable to afford access to such spaces. The article is a companion piece to her exhibition *If You Lived Here* in which she offers a critical commentary on the process of redevelopment in New York City, seeing such development as socially exclusive and creating conditions of abjection for economically marginalized populations within the city. The key to Rosler's argument is her representation of development as historical rather than natural. In revealing the consequences of development in material terms (access to housing and welfare), Rosler unsettles the dominant representations of the city (as in the adverts of property developers). Thus, she argues, a function of critical public art is to make visible the hidden histories and consequences of the production of space for exchange and the layering of symbolic value upon it, in doing so to play a role in the democratisation of space – in contrast to the public art discussed by Hall.

The question of social justice in the postmodern city is examined by Susan Fainstein in an extract from 'Justice, Politics and the Creation of Urban Space'. She considers how the issue of social justice has been treated by a variety of academic perspectives since the 1970s, the ways in which they differ in their conceptions of what constitutes a socially just city and how to achieve social transformations. She also considers policies through which social justice might be promoted, but remains critical of their limitations. She recognises the challenges that face policies that promote social justice in contexts where the majority political and public will is for economic development. A passage towards the end of the essay highlights this tension:

> A movement for social justice, if it is to mobilize large numbers of people, must focus less on the protection of the most deprived and more on broad benefits, less on the rights of the oppressed and more on security. Most people would prefer economic growth, if any of it trickles down to them, to redistribution, if redistribution does not produce an improvement in their standard to living. Strongly targeted redistributional policies may be the most efficient method of attaining broad

social benefits but they are suspect within the anti-tax sentiment of the present age to constitute an effective rallying point.

The tension between economic development and social justice is one that is at the heart of much debate about new urban spaces.

SUGGESTED FURTHER READING

David Harvey, *The Condition of Postmodernity: An Enquiry into the Origins of Cultural Change*, Oxford, Blackwell, 1989

D. Ley and C. Mills, 'Can There Be a Postmodernism of Resistance in the Urban Landscape?' In P. Knox, ed., *The Restless Urban Landscape*, Englewood Cliffs, NJ, Prentice-Hall, 1993

Edward Soja, *Postmodern Geographies: The Reassertion of Space in Critical Social Theory*, London, Verso, 1989

Sharon Zukin, *The Culture of Cities*, Oxford, Blackwell, 1995

‘The Production of Belief: Contribution to an Economy of Symbolic Goods’

from *The Field of Cultural Production* (1993)

Pierre Bourdieu

Editors’ Introduction

Pierre Bourdieu, born in 1930, was a French sociologists and anthropologist who, after studying philosophy with Louis Althusser and working initially in Algeria, went on to become possibly France’s leading intellectual of the 1980s and 1990s. A participant in the early developing of structuralism, Bourdieu’s early work was primarily concerned with producing anthropological accounts of traditional agriculture and ways of life in Algeria. However, this was no mere academic exercise but an attempt by Bourdieu to understand the clashes between the Algerian people and the French colonialists by reconstructing the social and economic conditions of people such as the Kabyles and the Berbers. The resulting publication *The Algerians* (1962) immediately established Bourdieu as an intellectual and political force. Bourdieu went on to publish a large number of very influential books on a range of issues and was a passionate opponent of French and other neo-liberal European governments of the 1990s who sought global authority rather than social justice.

Bourdieu’s work examined the enduring debates about the balance between free will and determinism. Bourdieu argued against extreme positions within this debate, taken up, for example by existentialists such as Jean-Paul Sartre and structuralists such as Claude Lévi-Strauss. Amongst his many influential ideas, one of the most widely applied was that of habitus. In *Distinction: A Social Critique of the Judgment of Taste* (1979) he argued against Kantian aesthetics that claimed that taste is not determined by social influences. Bourdieu argued that attitudes towards art and beauty are a crucial element in the distinction of class structures but are not determined by them. Elsewhere he attacked the media, French intellectuals, for extending an intellectual hegemony that they were little qualified to claim, and the numerous manifestations of late twentieth century globalisation. Pierre Bourdieu died on 23 January 2002.

One of the best guides to the work of Bourdieu is *Key Sociologists: Pierre Bourdieu*, by R. Jenkind (London, Routledge, 1992).

THE ESTABLISHMENT AND THE CHALLENGERS

Because the fields of cultural production are universes of belief which can only function in so far as they succeed in simultaneously producing products and the need for those products through practices which are the denial of the ordinary practices of the ‘economy’, the struggles which take place within them are ultimate conflicts involving the whole relation to the ‘economy’.

The 'zealots', whose only capital is their belief in the principles of the bad-faith economy and who preach a return to the sources, the absolute and intransigent renunciation of the early days, condemn in the same breath the merchants in the temple who bring 'commercial' practices and interests into the area of the sacred, and the pharisees who derive temporal profits from their accumulated capital of consecration by means of an exemplary submission to the demands of the field. Thus the fundamental law of the field is constantly reasserted by 'newcomers', who have most interest in the disavowal of self-interest.

The opposition between the 'commercial' and the 'non-commercial' reappears everywhere. It is the generative principle of most of the judgements which, in the theatre, cinema, painting or literature, claim to establish the frontier between what is and what is not art, i.e. in practice, between 'bourgeois' art and 'intellectual' art, between 'traditional' and 'avant-garde' art, or, in Parisian terms, between the 'right bank' and the 'left bank'. While this opposition can change its substantive content and designate very different realities in different fields, it remains structurally invariant in different fields and in the same field at different moments. It is always an opposition between small-scale and large-scale ('commercial') production, i.e. between the primacy of production and the field of producers or even the sub-field of producers for producers, and the primacy of marketing, audience, sales and success measured quantitatively; between the deferred, lasting success of 'classics' and the immediate, temporary success of best-sellers; between a production based on denial of the 'economy' and of profit (sales targets, etc.) which ignores or challenges the expectations of the established audience and serves no other demand than the one it itself produces, but in the long term, and a production which secures success and the corresponding profits by adjusting to a pre-existing demand. The characteristics of the commercial enterprise and the characteristics of the cultural enterprise, understood as a more or less disavowed relation to the commercial enterprise, are inseparable. The differences in the relationship to 'economic' considerations and to the audience coincide with the differences officially recognized and identified by the taxonomies prevailing in the field. Thus the opposition between 'genuine' art and 'commercial' art corresponds to the opposition between ordinary entrepreneurs seeking immediate economic profit and cultural entrepreneurs struggling to accumulate specifically cultural capital, albeit at the cost of temporarily renouncing economic profit. As for the opposition which is made within the latter group between consecrated art and avant-garde art, or between orthodoxy and heresy, it distinguishes between, on the one hand, those who dominate the field of production and the market through the economic and symbolic capital they have been able to accumulate in earlier struggles by virtue of a particularly successful combination of the contradictory capacities specifically demanded by the law of the field, and, on the other hand, the newcomers, who have and want no other audience than their competitors – established producers whom their practice tends to discredit by imposing new products – or other newcomers with whom they vie in novelty.

Their position in the structure of simultaneously economic and symbolic power relations which defines the field of production, i.e. in the structure of the distribution of the specific capital (and of the corresponding economic capital), governs the characteristics and strategies of the agents or institutions, through the intermediary of a practical or conscious evaluation of the objective chances of profit. Those in dominant positions operate essentially defensive strategies, designed to perpetuate the status quo by maintaining themselves and the principles on which their dominance is based. The world is as it should be, since they are on top and clearly deserve to be there; excellence therefore consists in being what one is, with reserve and understatement, urbanely hinting at the immensity of one's means by the economy of one's means, refusing the assertive, attention-seeking strategies which expose the pretensions of the young pretenders. The dominant are drawn towards silence, discretion and secrecy, and their orthodox discourse, which is only ever wrung from them by the need to rectify the heresies of the newcomers, is never more than the explicit affirmation of self-evident principles which go without saying and would go better unsaid. 'Social problems' are social relations: they emerge from confrontation between two groups, two systems of antagonistic interests and theses. In the relationship which constitutes them, the choice of the moment and sites of battle is left to the initiative of the challengers, who break the silence of the *doxa* and call into question the unproblematic, taken-for-granted world of the dominant groups. The dominated producers, for their part, in order to gain a foothold in the market, have to resort to subversive strategies which will eventually bring them the disavowed profits only if they succeed in overturning the hierarchy of the field without disturbing the principles on which the field is based. Thus their revolutions are only ever partial ones, which displace the censorships and transgress the conventions but do so in the name of the same underlying principles. This is why the strategy *par excellence* is the 'return to the sources' which is the basis of all heretical subversion and all aesthetic revolutions, because it enables the insurgents to turn against the establishment the arms which they use to justify their domination, in particular asceticism, daring, ardour, rigour and disinterestedness. The strategy of beating the dominant groups at their own game by demanding that they respect the fundamental law of the field, a denial of the 'economy', can only work if it manifests exemplary sincerity in its own denial.

Because they are based on a relation to culture which is necessarily also a relation to the 'economy' and the market, institutions producing and marketing cultural goods, whether in painting, literature, theatre or cinema, tend to be organized into structurally and functionally homologous systems which also stand in a relation of structural homology with the field of the fractions of the dominant class (from which the greater part of their clientele is drawn). This homology is most evident in the case of the theatre. The opposition between 'bourgeois theatre' and 'avant-garde theatre', the equivalent of which can be found in painting and in literature, and which functions as a principle of division whereby authors, works, styles and subjects can be classified practically, is rooted in reality. It is found both in the social characteristics of the audiences of the different Paris theatres (age, occupation, place of residence, frequency of attendance, prices they are prepared to pay, etc.) and in the – perfectly congruent – characteristics of the authors performed (age, social origin, place of residence, lifestyle, etc.), the works and the theatrical businesses themselves.

'High-brow' theatre in fact contrasts with 'middle-brow' theatre (*théâtre de boulevard*) in all these respects at once. On one side, there are the big subsidized theatres (Odéon, Théâtre de l'Est parisien, Théâtre national populaire) and the few small left-bank theatres (Vieux Colombier, Montparnasse, Gaston Baty, etc.), which are risky undertakings both economically and culturally, always on the verge of bankruptcy, offering unconventional shows (as regards content and/or *mise en scène*) at relatively low prices to a young, 'intellectual' audience (students, intellectuals, teachers). On the other side, there are the 'bourgeois' theatres (in order of intensity of the pertinent properties: Gymnase, Théâtre de Paris, Antoine, Ambassadeurs, Ambigu, Michodière, Variétés), ordinary commercial businesses whose concern for economic profitability forces them into extremely prudent cultural strategies, which take no risks and create none for their audiences, and offer shows that have already succeeded (adaptations of British and American plays, revivals of middle-brow 'classics') or have been newly written in accordance with tried and tested formulae. Their audience tends to be older, more 'bourgeois' (executives, the professions, businesspeople), and is prepared to pay high prices for shows of pure entertainment whose conventions and staging correspond to an aesthetic that has not changed for a century. Between the 'poor theatre' which caters to the dominant-class fractions richest in cultural capital and poorest in economic capital, and the 'rich theatre', which caters to the fractions richest in economic capital and poorest (in relative terms) in cultural capital, stand the classic theatres (Comédie Française, Atelier), which are neutral ground, since they draw their audience more or less equally from all fractions of the dominant class and share parts of their constituency with all types of theatre. Their programmes too are neutral or eclectic: 'avant-garde boulevard' (as the drama critic of *La Croix* put it), represented by Anouilh, or the consecrated avant-garde.

GAMES WITH MIRRORS

This structure is no new phenomenon. When Françoise Dorin, in *Le Tournant*, one of the great boulevard successes, places an avant-garde author in typical vaudeville situations, she is simply rediscovering (and for the same reasons) the same strategies which Scribe used in *La Camaraderie*, against Delacroix, Hugo and Berlioz: in 1836, to reassure a worthy public alarmed by the outrages and excesses of the Romantics, Scribe gave them Oscar Rigaut, a poet famed for his funeral odes but exposed as a hedonist, in short, a man like others, ill-placed to call the bourgeois 'grocers'.

Françoise Dorin's play, which dramatizes a middle-brow playwright's attempts to convert himself into an avant-garde playwright, can be regarded as a sort of sociological test which demonstrates how the opposition which structures the whole space of cultural production operated simultaneously in people's minds, in the form of systems of classification and categories of perception, and in objective reality, through the mechanisms which produce the complementary oppositions between playwrights and their theatres, critics and their newspapers. The play itself offers the contrasting portraits of two theatres: on the one hand, technical clarity and skill, gaiety, lightness and frivolity, 'typically French' qualities; on the other, 'pretentiousness camouflaged under ostentatious starkness', 'a confidence-trick of presentation', humourlessness, portentousness and pretentiousness, gloomy speeches and decors ('a black curtain and a scaffold certainly help . . .'). In short, dramatists, plays, speeches, epigrams that are 'courageously light', joyous, lively, uncomplicated, true-to-life, as opposed to 'thinking', i.e. miserable, tedious, problematic and obscure. 'We had a bounce in our backsides. They think with theirs. There is no overcoming this opposition, because it separates 'intellectuals' and 'bourgeois' even in the interests they have most manifestly in common. All the contrasts which Françoise Dorin and the 'bourgeois' critics mobilize in their judgements on the theatre (in the form of oppositions between the 'black curtain' and the 'beautiful set', 'the wall well lit, well decorated', 'the actors well washed, well dressed'), and, indeed, in their whole world view, are summed up in the opposition between *la vie en noir and la vie en rose* – dark thoughts and rose-coloured spectacles – which, as we shall see, ultimately stems from two very different ways of *denying the social world*.

Faced with an object so clearly organized in accordance with the canonical opposition, the critics, themselves distributed within the space of the press in accordance with the structure which underlies the object classified and the classificatory system they apply to it, reproduce, in the space of the judgements whereby they classify it and themselves, the space within which they are themselves classified (a perfect circle from which there is no escape except by objectifying it). In other words, the different judgements expressed on *Le Tournant* vary, in their form and content, according to the publication in which they appear, i.e. from the greatest distance of the critic and his readership *vis-à-vis* the 'intellectual' world to the greatest distance *vis-à-vis* the play and its 'bourgeois' audience and the smallest distance *vis-à-vis* the 'intellectual' world.

WHAT THE PAPERS SAY: THE PLAY OF HOMOLOGY

The subtle shifts in meaning and style which, from *L'Aurore* to *Le Figaro* and from *Le Figaro* to *L'Express*, lead to the neutral discourse of *Le Monde* and thence to the (eloquent) silence of *Le Nouvel Observateur* can only be fully understood when one knows that they accompany a steady rise in the educational level of the readership (which, here as elsewhere, is a reliable indicator of the level of transmission or supply of the corresponding messages), and a rise in the proportion of those class fractions – public-sector executives and teachers – who not only read most in general but also differ from all other groups by a particularly high rate of readership of the papers with the highest level of transmission (*Le Monde* and *Le Nouvel Observateur*); and, conversely, a decline in the proportion of those fractions – big commercial and industrial employers – who not only read least in general but also differ from other groups by a particularly high rate of readership of the papers with the lowest level of transmission (*France-Soir, L'Aurore*). To put it more simply, the structured space of discourses reproduces, in its own terms, the structured space of the newspapers and of the readerships for whom they are produced, with, at one end of the field, big commercial and industrial employers, *France-Soir* and *L'Aurore*, and, at the other end, public-sector executives and teachers, *Le Monde* and *Le Nouvel Observateur*, the central positions being occupied by private-sector executives, engineers and the professions and, as regards the press, *Le Figaro* and especially *L'Express*, which is read more or less equally by all the dominant-class fractions (except the commercial employers) and constitutes the neutral point in this universe. Thus the space of judgements on the theatre is homologous with the space of the newspapers for which they are produced and which disseminate them and also with the space of the theatres and plays about which they are formulated, these homologies and all the games they allow being made possible by the homology between each of these spaces and the space of the dominant class.

Let us now run through the space of the judgements aroused by the experimental stimulus of Françoise Dorin's play, moving from 'right' to 'left' and from 'right-bank' to 'left-bank'. First, *L'Aurore*:

Cheeky Françoise Dorin is going to be in hot water with our *snooty, Marxist* intelligentsia (the two go together). The author of 'Un sale égoïste' shows no respect for the solemn boredom, profound emptiness and vertiginous nullity which characterize so many so-called 'avant-garde' theatrical productions. She dares to profane with sacrilegious laughter the notorious 'incommunicability of beings' which is the alpha and omega of the contemporary stage. And this perverse *reactionary*, who flatters the lowest appetites of consumer society, far from acknowledging the error of her ways and wearing her boulevard playwright's reputation with humility, has the impudence to prefer the jollity of Sacha Guitry, or Feydeau's bedroom farces, to the darkness visible of Marguerite Duras or Arrabal. This is a crime it will be difficult to forgive. Especially since she commits it with cheerfulness and gaiety, using all the dreadful devices which make lasting successes.

(Gilbert Guilleminaud, *L'Aurore*, 12 January 1973)

Situated at the fringe of the intellectual field, at a point where he almost has to speak as an outsider ('our intelligentsia'), the *L'Aurore* Critic does not mince his words (he calls a reactionary a reactionary) and does not hide his strategies. The rhetorical effect of putting words into the opponent's mouth, in conditions in which his discourse, functioning as an ironic antiphrasis, objectively says the opposite of what it means, presupposes and brings into play the very structure of the field of criticism and his relationship of immediate connivance with his public, based on homology of position.

From *L'Aurore* we move to *Le Figaro*. In perfect harmony with the author of *Le Tournant* – the harmony of orchestrated habitus – the *Figaro* critic cannot but experience absolute delight at a play which so perfectly corresponds to his categories of perception and appreciation, his view of the theatre and his view of the world:

How grateful we should be to Mme Françoise Dorin for being a *courageously light* author, which means to say that she is *wittily* dramatic, and *smilingly serious*, irreverent without fragility, pushing the comedy into outright vaudeville, but in the subtlest way imaginable; an author who wields satire *with elegance*, an author who at all times demonstrates astounding virtuosity . . . Françoise Dorin knows *more than any of us* about the *tricks of the dramatist's art, the springs of comedy*, the *potential of a situation*, the comic or biting force of the *mot juste* . . . Yes, what skill in taking things apart, what irony in the deliberate side-stepping, what mastery in the way she lets you see her pulling the strings! *Le Tournant* gives every sort of enjoyment without an ounce of self-indulgence or vulgarity. And without ever being facile either, since it is quite clear that right now, *conformism lies with the avant-garde*, absurdity lies in gravity and imposture in tedium. Mme Françoise Dorin will *relieve a well-balanced audience* by bringing them back into *balance* with healthy laughter . . . Hurry and see for yourselves and I think you will *laugh so heartily* that you will forget to think how anguishing it can be for a writer to wonder if she is still in tune with the times in which she lives . . . In the end it is a question everyone asks themselves and only humour and *incurable optimism* can free them from it!

(Jean-Jacques Gautier, *Le Figaro*, 12 January 1973)

From *Le Figaro* one moves naturally to *L'Express*, which remains poised between endorsement and distance, thereby attaining a distinctly higher degree of euphemization:

'It's bound to be a runaway success . . . A witty and amusing play. A character. An actor who puts the part on like a glove: Jean Piat. With an *unfailing virtuosity that is only occasionally drawn out too long*, with a *sly cunning, a perfect mastery of the tricks of the trade*, Françoise Dorin has written a play on the 'turning point' in the Boulevard which is, ironically, the most traditional of Boulevard plays. *Only morose pedants will probe too far into the contrast between two conceptions of political life and the underlying private life*. The brilliant dialogue, full of *witticisms* and *epigrams*, is often viciously sarcastic. But Romain is not a caricature, he is much less stupid than the run-of-the-mill avant-garde writer. Philippe has the *plum role*, because he is on his own ground. What the author of 'Comme au théâtre' gently wants to suggest is that the Boulevard is where people speak and behave 'as in real life', and this is true, but it is only a partial truth, and not just because it is a class truth.

(Robert Kanters, *L'Express*, 15–21 January 1973)

Here the approval, which is still total, already begins to be qualified by systematic use of formulations that are ambiguous even as regards the oppositions involved: 'It's likely to be a runaway success', 'a sly cunning, a perfect mastery of the tricks of the trade', 'Philippe has the plum role', all formulae which could also be taken pejoratively. And we even find, surfacing through its negation, a glimmer of the other truth ('Only morose pedants will probe too far . . .') or even of the truth *tout court*, but doubly neutralized, by ambiguity and negation ('and not just because it is a class truth').

Le Monde offers a perfect example of ostentatiously neutral discourse, even-handedly dismissing both sides, both the overtly political discourse of *L'Aurore* and the disdainful silence of *Le Nouvel Observateur*.

> The simple or simplistic argument is complicated by a very subtle two-tier structure, as if there were two plays overlapping. One by Françoise Dorin, a conventional author, the other invented by Philippe Roussel, who tries to take 'the turning' towards modern theatre. This game performs a circular movement, like a boomerang. Françoise Dorin deliberately exposes the Boulevard clichés which Philippe attacks and, through his voice, utters a violent denunciation of the bourgeoisie. On the second tier, she contrasts this language with that of a young author whom she assails with equal vigour. Finally, the trajectory brings the weapon back on to the Boulevard stage, and the futilities of the mechanism are unmasked by the devices of the traditional theatre, which have therefore lost nothing of their value. Philippe is able to declare himself a 'courageously light' playwright, inventing 'characters who talk like everybody'; he can claim that his art is 'without frontiers' and therefore non-political. However, the demonstration is entirely distorted by the model avantgarde author chosen by Françoise Dorin. Vankovitz is an epigone of Marguerite Duras, a belated existentialist with militant leanings. He is caricatural in the extreme, as is the theatre that is denounced here ('A black curtain and a scaffold certainly help!' or the title of a play: 'Do take a little infinity in your coffee, Mr Karsov'). The audience gloats at this derisive picture of the modern theatre; the denunciation of the bourgeoisie is an amusing provocation inasmuch as it rebounds onto a detested victim and finishes him off . . . To the extent that it reflects the state of bourgeois theatre and reveals its systems of defence, *Le Tournant* can be regarded as an *important work*. Few plays let through so much anxiety about an 'external' threat and *recuperate* it with so much unconscious fury.
>
> (Louis Dandrel, *Le Monde*, 13 January 1973)

The ambiguity which Robert Kanters was already cultivating here reaches its peaks. The argument is 'simple or simplistic', take your pick; the play is split in two, offering two works for the reader to choose, a 'violent' but 'recuperatory' critique of the 'bourgeoisie' and a defence of non-political art. For anyone naïve enough to ask whether the critic is 'for or against', whether he finds the play 'good or bad', there are two answers: first, an 'objective informant's' dutiful report that the avant-garde author portrayed is 'caricatural in the extreme' and that 'the audience gloats [*jubile*]' (but without our knowing where the critic stands in relation to this audience, and therefore what the significance of this gloating is); then, after a series of judgements that are kept ambiguous by many reservations, nuances and academic attenuations ('To the extent that . . .', 'can be regarded as . . .'), the assertion that *Le Tournant* is 'an important work', but, be it noted, as a document illustrating the crisis of modern civilization, as they would say at *Sciences Po*.

Although the silence of *Le Nouvel Observateur* no doubt signifies something in itself, we can form an approximate idea of what its position might have been by reading its review of Felicien Marceau's play *La Preuve par quatre* or the review of *Le Tournant* by Philippe Tesson, then editor of *Combat*, published in *Le Canard enchaîné:*

> Theatre seems to me the wrong term to apply to these *society gatherings of tradesmen and business-women* in the course of which a famous and much loved actor recites the laboriously witty text of an equally famous author in the middle of an elaborate stage set, even a revolving one described with Folon's measured humour . . . No 'ceremony' here, no 'catharsis' or 'revelation' either, still less improvisation. Just a warmed-up dish of plain cooking [*cuisine bourgeoise*] for stomachs that have seen it all before . . . The audience, like all boulevard audiences in Paris, bursts out laughing, at the right time, in the most conformist places, wherever this spirit of easy-going rationalism comes into play. The connivance is perfect and the actors are all in on it.

This play could have been written ten, twenty or thirty years ago.

(M. Pierret, *Le Nouvel Observateur*, 12 February 1964, reviewing Felicien Marceau's *La Preuve par quatre*)

Françoise Dorin *really knows a thing or two.* She's a first-rate *recuperator* and terribly *well-bred.* Her *Le Tournant* is an excellent Boulevard comedy, which works mainly on bad faith and demagogy. The lady wants to prove that avant-garde theatre is tripe. To do so, she takes a *big bag of tricks* and need I say that as soon as she pulls one out the *audience* rolls in the aisles and shouts for more. Our author, who was just waiting for that, does it again. She gives us a young lefty playwright called Vankovitz – get it? – and puts him in various ridiculous, uncomfortable and rather shady situations, to show that this young gentleman is no more disinterested, no less bourgeois, than you and I. What common sense, Mme Dorin, what lucidity and what honesty! You at least have the courage to stand by your opinions, and very healthy, French ones they are too.

(Philippe Tesson, *Le Canard Enchaîné*, 17 March 1973)

'Opening up Public Art's Spaces: Art, Regeneration and Audience'

from *Cultures and Settlements* (2003)

Tim Hall

Editors' Introduction

Tim Hall is an urban geographer based at the University of Gloucestershire. His interests include cultural representations of urban change and regeneration. He has written extensively on the employment of public art within urban regeneration. Working within a poststructural framework, informed by the work of 'new' cultural geographers his early work examined the incorporation of local and regional mythologies into the landscapes and iconographies of new urban landscapes. A critic of the overly semiotic and productionist bias of much writing on public art, more recently he has begun to consider the ways in which non-representational theory can be applied to the study of public art.

Tim Hall is widely published within urban geography and is author of *Urban Geography* (1998/2001) and co-editor of *The Entrepreneurial City: Geographies of Politics, Regime and Representation* (1998) and *Urban Futures: Critical Commentaries on Shaping the City* (2003).

It has been argued that the last twenty years or so have seen a 'renaissance' of public art in the cities of Europe, the USA and beyond (Moody 1990, 2). This has been characterised by a rise in public and private sector commissions, an expansion of arts policy and administrative structures and the integration of artists into the urban design process. A further characteristic of this supposed renaissance has been the basis upon which art for the public realm has been advocated. Traditionally, art has been placed in the public realm for reasons of aesthetic enhancement, memorialisation, or simply because introducing art into everyday life has been seen as an inherently good thing. However, since the early 1980s public art has been advocated as contributing to the alleviation of a range of environmental, social and economic problems locating it squarely within the process of urban regeneration. Since the 1980s public art has been increasingly implicated in processes such as the rejuvenation of decaying urban spaces, the development of flagship projects of urban regeneration, the stimulation of central city economies and the enhancement or transformation of urban images (Goodey 1994; Hall 1995a).

In this paper I want to review some critical writing on public art and examine the question of how this writing has helped us uncover the meanings of prominent examples of public art employed in fashioning new cities. This involves a discussion of where meaning lies in public art, a reflection on the theoretical perspectives and methodological approaches employed by writers on the subject and thoughts on the partiality of much work to date. This echoes a number of ongoing debates in cultural studies, urban geography, urban studies and architectural geography. It concludes by drawing together examples of work that offer methodological clues to an alternative approach to uncovering meanings in public art and in everyday engagements with open space, that have been largely missed in critical writing to date.

There is great variety both in the urban regeneration programmes that public art has been incorporated within and within public art practice itself. However, in terms of urban regeneration projects we can recognise a broad distinction between flagship or spectacular regeneration projects, which often contain prominent works of public art by internationally famous artists, and neighbourhood or community arts and regeneration projects, typically, but not exclusively, publicly funded, often away from central city locations and with a greater emphasis on community development and participatory arts (Matarasso 1997; Dwelly 2001). This chapter is primarily concerned with public art in the former context.

Virtually all public art works found within major city centre projects of urban regeneration are examples of what has been called 'institutional' public art. Namely, they are art works that endorse 'official' views of the city, those of local authorities and commercial developers, for example, and celebrate and enhance the spaces produced by these interests. Unsurprisingly, this very prominent renaissance of public art has not gone unnoticed by a whole range of writers, critics and researchers. There are now a number of critical literatures concerned with public art within the urban regeneration process written from a number of theoretical and discipline based perspectives.

Critical writing on public art and urban regeneration has emanated from a number of perspectives including those of artists, arts advocates, cultural theorists, and urban and cultural geographers. Artists and arts advocates have been predominantly concerned with examining the processes of public art production. They have been concerned, for example, with the influence of the contexts of public art production (commissions, briefs, site, consultation and various other local constraints) on the public art works produced (see Jones 1992). This literature, some positive, some more critical, reflects the concern of writers from this perspective for quality in the production of public art works. By contrast much critical research and theoretical literature has emanated from a cultural studies or cultural geography perspective. This writing has reflected the approaches and concerns dominant in these disciplines since the mid-1980s. Prominent have been a concern for the politics of representation and, associated with this, a variety of deconstructive approaches. Typical examples have seen geographers and others 'read', 'unpack' or 'deconstruct' the meanings of a variety of cultural texts such as landscape paintings, films, television programmes or maps, as well as architecture and the built environment (see for example, Cosgrove and Daniels 1988; Shields 1991; Barnes and Duncan 1992; Bender 1993; Duncan and Ley 1993; Clarke 1997). These concerns and approaches have also informed writing on public art which has stemmed from this perspective. Broadly, writers on public art from cultural geography and other critical social sciences have been critical of the involvement of public art in projects of urban regeneration (Miles 1997). They have tended to situate public art within the politics of urban change and have been concerned with the ways that on-going processes of urban change have impacted unequally on the lives of different groups across the social spectrum. Public art then, has been viewed as a component of broader processes of uneven urban development.

These studies have sought to relate the narratives and myths promoted through the symbolism of public art to social, economic and/or cultural changes occurring in their local and global urban contexts. A persuasive strand of urban studies literature, which emerged in the late 1980s and early 1990s, argued that culture was being implicated in the process of uneven urban development as a kind of 'carnival mask'. This thesis argued that culture was being deployed in a commodified and sanitised form in cities to create the impression of affluence, vibrancy, conviviality, change and regeneration, while at the same time being used to mask the increasingly fractured and polarised social and economic realities that characterised life for the majority of urban dwellers (Harvey 1987; 1989a; 1989b). This critique has been applied to the landscapes of regenerated central city spaces and the disciplines of architecture, planning, urban design and public art that have produced them (Knox 1993; Hubbard 1996; Miles 1997; 1998) as well as the representation of regenerated spaces and cities through promotional materials and the media (Holcomb 1993; 1994; Thomas 1994; Kenny 1995; Wilson 1996; Short and Kim 1998). These landscapes and their representations have been read as texts, into which have been written elite visions of the city.

It has been argued that public art works have presented selective versions of history, or myths of harmony, offering another layer in the composition of elite images of the regenerated city. This is of significance, not just because the image (how cities are represented by the minority) is out of step with reality (how cities are experienced by the majority) but because there is a tangible relationship between the former and the latter. Image and appearance are important parts of the way that cities are understood and acted upon and hence they are embedded in the material reproduction of urban space. This critique of public art sees it squarely as one of the elite images of the city and thus it deconstructs its meanings within the context of the reproduction of social justice and injustice in the post-modern city. It is commonly discussed along with other such images of the city, produced, for example, through promotional campaigns or the local media and its architecture and landscapes.

Working within the critical framework outlined above I have been concerned with the role of public art in unveiling images of industry in the post-industrial landscape of urban regeneration in one British city, Birmingham in the Midlands (Hall 1995a; 1995b; 1997a; 1997b). I want to briefly reflect on the approach adopted in this work and, in doing so, illustrate some of the insights stemming from it, but also some of its limitations. The main subject of this enquiry has been a detailed study of Raymond Mason's statue *Forward*. *Forward* was unveiled in 1991 as part of an extensive public art programme linked to the redevelopment of the Broad Street area adjacent to the city centre as a conference, entertainment and business tourist zone.

At the heart of the research process was an attempt to deconstruct the symbolic meaning of the industrial iconography of the statue. However, the various influences that had been brought to bear on the production of the statue were also felt to be important insofar as they might offer clues both to the symbolic meanings of the statue and the reasons for the presence of industrial imagery in the context of the construction of a post-industrial image and identity for the City of Birmingham in the 1980s and 1990s. In this case attention was paid to the career trajectory of the artist Raymond Mason, in particular his concerns for industrial and working-class subjects (Farrington and Silber 1989; Edwards 1994), the position of the statue within a wider programme of urban regeneration and place promotion and finally its position within a public art commissioning strategy for the city with the stated aims of being integral to the regeneration of the city while exploring and expressing facets of the city's multiple identity and diverse character (Lovell 1988, 1). Exploring these multiple sites of meaning meant that a variety of research methods were utilised within the research. These included, as well as a deconstruction of the statue's iconographies, semi-structured interviews with representatives from the local authority, the Public Art Commissions Agency and the International Convention Centre Birmingham, archival research investigating the background to both the urban regeneration strategies in the city and the public art commissioning strategy, this involved reviewing official reports, strategies, briefs and minutes of meetings, a review of Raymond Mason's career and previous commissions and a review of published interviews with him, a review of coverage of the public art programme in the local and national press and a reading of social histories and reports of industrial and economic change in the city.

It also became obvious that there were two overriding and contrasting contexts that the statue was situated within. These were, first, a programme of urban regeneration, and its associated landscapes and geographies of local urban change, and, second, a powerful notion of the character and identity of the City of Birmingham and its region that saw them as strongly rooted in histories and experiences of industry. It was the contrast between these two contexts that initially appeared to offer some clue to the presence of the industrial imagery in the city's new public art work and which were explored within the research. On the one hand the statue was clearly intended to be viewed as part of a major urban regeneration project, an attempt to refashion the image and identity of the city and influence its economic development over the coming years. On the other it fitted into a tradition of British civic commemoration that venerates local histories, traditions and experiences and national values. In this context it represented a prominent articulation of civic and regional character and history. These two contexts produced two audiences who were likely to view the statue in very different ways and according to two very different sets of expectations. The answer to the apparent incongruity in the statue's exploration of local industrial histories lay in the contrast between these two contexts. Rather than seeing industry as polluting, mechanical and redundant, as it overwhelmingly was during the 1980s and 1990s, Raymond Mason in the statue associated it squarely with a positive set of values revolving around notions of craftsmanship and individual endeavour, a set of values likely to appeal to both local and outside audiences. As Raymond Mason said in an interview published at the time of the statue's unveiling: "For one precise moment in history Birmingham was unique . . . It founded a tradition of fine craftsmanship and fine machinery. That shouldn't be forgotten . . . It would be a great pity to forget what was a great moment in the human saga of fine work' (in Weideger 1991, 14). While not wishing to dismiss such semiotic approaches in themselves, I want to suggest that their perspective is partial and has produced, in debates about meaning in public art, something of an impasse that has characterised much writing on the subject over the past fifteen years or so.

We can appreciate these limitations by adopting a framework which situates sites of meaning within visual (and other) texts. This framework, outlined for example by Gillian Rose in her book *Visual Methodologies* (2001), is revealing when used as a device to review critical writings on public art. Rose, in reviewing theoretical writing on the visual recognises three sites of meaning with regard to visual texts.

- *Site 1. Production*: This recognises the influence of technologies, genre conventions, social, cultural and economic contexts within which images are produced and the biographies of individual producers of images on the meanings of images (Rose, 2001: 17–23).
- *Site 2. Text*: While recognising many of the above modalities, approaches that situate meanings in texts themselves typically use some form of semiotic deconstruction to unravel the iconographies of texts (Rose, 2001: 23–24).
- *Site 3. Audience*: This recognises the importance of the consumption or construction of meaning by the audiences of visual or other texts. It problematicises approaches that focus on other sites of meaning in that it recognises that there is no necessary correspondence between intended meanings and those available to the consumers of a visual text.

Much critical writing on public art has focused on the first two of these three sites of meaning. Published writing by artists and advocates tends to focus on the importance of the processes and contexts of production, for example, by discussing the influence of briefs, commissioners, processes of consultation, site and local constraints on the outcome of projects. There have also been a small number of serious sociologies of the production of public art (see for example Martorella's discussion of corporate art, 1990). By contrast critical writing from a cultural studies or cultural geography perspective typically seeks to situate the specific iconographies of public art works in the multiple contexts of their production and reception. While offering very sophisticated methods of saying a great deal about the meanings inherent in public art works, and the influence of the contexts within which this art is produced, the prevailing critical approaches within the literatures of public art have said, and are only able to say, very little about the public. Neither the intentions of the producers of public art, nor its iconographies necessarily correspond to the meanings derived from the incorporation of public art into the experiences of the public's everyday lives. While our understanding of the growth, production and intended meanings of public art is undoubtedly very good, our understanding of its readings by its diverse audiences is much less well developed.

Much of the theoretical sophistication of critical writing on public art derives from its employment of a Lefebvrian framework of analysis. However, this seems guilty, as Savage and Warde have recognised of much writing in this vein, of failing to adequately link the realms of signification (the representation of space) and production to that of experience.

> Lefebvre's laudable project to find a bridge between experienced space, representations of space, and spaces of representation has proved too hard to put into operation empirically. The crucial link between the construction of place in representation and at the level of everyday experience has not been demonstrated (1993: 132).

Loretta Lees, in reviewing broader approaches to the study of meaning in architecture and the urban landscape, of which critical writing on public art is part, makes a similar point:

> As Bondi has warned, this (semiotic) interpretation 'strips the built environment of the meanings it is given by the people who live in it and of the transformations, however modest, that they make'. While there has been a great deal of recent work in geography on the cultural production of the built environment, much less attention has been given to

its consumption. Contemporary architectural geographies do not emphasize enough the fact that 'urban meaning is not immanent to architectural form and space, but changes according to the social interaction of city dwellers' (2001: 55).

We might ask, following the logic of such critiques, to what extent the sophisticated readings of public art offered by cultural theorists and cultural geographers correspond to those (clearly multiple) meanings constructed by its audiences. In reviewing the critical literatures of public art, it is apparent that the voices of the public are largely absent. It seems not unreasonable to suppose that the origins of an invigorated critical approach to uncovering meanings in public art might focus on the audience as a site of meaning, rather than the contexts of production and the works themselves, as has been the case in the past. The fragments of such an approach lie in the few public voices that are admitted. Building on these fragments is, I would argue, the task awaiting cultural theorists approaching public art. I want now to move towards a discussion of a few examples of work from beyond the study of public art that offers some possible theoretical and methodological directions.

There are a number of precedents in geographical, and other, studies of a variety of urban spaces that the methodological tool-kit exists, supported by a respected raft of critical theory, to uncover the multiple meanings constructed by public art's audiences. For example, Jacqueline Burgess *et al.* (1988) have employed intensive, in-depth, small-group interviews to uncover the meanings and values attached to urban open spaces by local residents in South London. The researchers complimented this approach by using photographs of different physical settings to prompt open-ended reflection from interviewees. The findings of this project revealed a hidden richness in the human responses to, often apparently mundane open spaces, and challenged taken-for-granted assumptions that previously underpinned the provision and management of urban open spaces in the metropolis. There is little to suggest that the aims of the project and the methods employed are not applicable to uncovering the meanings constructed by the audiences of public art.

> Our primary objective has been to develop a methodology which is sensitive to the language, concepts and beliefs of people whose views about open spaces are rarely heard. Through empirical, qualitative research we hoped to explore the ways in which individuals and groups *read* urban landscapes and interpret their symbolic meaning . . . We emphasise the language that people used to talk about their experiences and activities for we believe it to be the key to understanding the internalisation of social and cultural values for the urban green (Burgess *et al.* 1988: 457).

By contrast Loretta Lees has employed techniques of ethnographic observation in her study of the architectural geography of Vancouver's new Public Library Building (2001). Lees was similarly concerned with uncovering the meanings held by the users of its spaces. She argues that:

> Architecture is about more than just representation. Both as a practice and a product, it is performative, in the sense that it involves ongoing social practices through which space is continually shaped and inhabited. Indeed, as the urban historian Dolores Hayden argues, the use and occupancy of the built environment is as important as its form and figuration . . . attention to the embodied practices through which architecture is lived requires some new approaches, just as it opens up some new concerns for a critical geography of architecture (2001: 53).

Such a shift echoes Nigel Thrift's call for geographers and other social scientists to move away from a theoretical concern for interpretation of representation and towards what he calls 'non-representational theory' or 'theories of practice' (2000). An example of such an approach would be Thrift *et al*'s study of the multiple meanings attached to the acts of shopping and its environments (Miller *et al.*'s, 1998; see also Taylor, Evans and Fraser, 1996) which have overturned earlier semiotic readings of apparently instrumental shopping environments (Gottdiener, 1986). We might also draw parallels between the concerns of this body of work and Jonathon Hill's (1998) work on the 'occupation' of architecture and Iain Borden's (2001) on skateboarding as a spatial practice.

Lees's concern is to capture the ways that meanings are performative, embodied and are continually constructed and sedimented through daily, ongoing practices, a dimension absent to an extent on approaches that rely exclusively on interviews to uncover public meanings. In doing this Lees presents a series of ethnographic vignettes that reveal the diverse ways that the library's spaces are appropriated by its numerous publics for their own purposes. I would argue that ethnographic observations might profitably complement interview-based approaches.

Interviews and ethnography are well known and widely employed methods within the social sciences. However, other less extensively employed methods may also be appropriate to the project of uncovering public meanings of urban spaces and public art specifically. For example, visual methodologies have gained increased attention in recent years across a range of social science disciplines including geography, sociology, ethnography and cultural studies (see Pink, 2001; Rose, 2001). Application of visual methods in these disciplines has tended, on the whole, to reflect broader concerns for the efficacy of representational theory. Most applications of visual methodologies, especially where they have been used to examine the city, have again involved deconstructing professional, official or elite representations of urban space (promotional materials, web sites, films, maps or plans, for example). There has been little attention, to date, paid to 'lay' visual knowledges of the city, visual texts produced by the city's publics.

This lack of attention is surprising given the richness of lay visual knowledges of the city that abound since photographic and video recording became widely available and affordable and the centrality of these visual texts to certain, especially, but not exclusively, tourist encounters with the city. Collections of snapshots and video recordings offer extensive archives of public engagements with urban space that appear, at the moment, to have been virtually untapped by academics studying the city. This source is especially appropriate to the study of public art, given that it is frequently the subject, or at least a subject, of such photography.

It is also surprising given the significance that has been attached to such knowledges and visual texts by those who have employed research techniques such as autophotography, or who have used family, tourist, or other photographs as a research source (see for example, Albers and James, 1988; Aitken and Wingate, 1993). In applying such approaches to uncovering meaning in the urban landscape, what is sought is not any notion of scientific representativeness but rather reflexive approaches to what are clues to the geography of engagement between the city's public(s) and its spaces. Collections of popular, 'snap-shot' photography, for example, are largely untapped archives of this engagement.

I have argued that researchers examining questions of the meanings of public art, its roles and functions in the context of fashioning new cities, should shift the focus of their concerns from production and text towards audience. Methodologically and theoretically the tools to enable this already exist. At the moment there are a number of basic questions that we seem to know little about. These include:

- Who are the audience(s) of public art?
- How do they engage with works of public art?
- What meanings do they attach to works of public art?
- What are the natures of the experiences they derive from this engagement?

A combination of ethnographic observation, qualitative interviews and analysis of popular visual knowledges of new urban spaces and public art offer the possibilities of addressing some of these issues. Conventional readings of the landscapes of public art and of regenerated city spaces have tended to paint them largely as socially sterile and closed to anything other than the intended meanings inscribed into them by their producers. However, geographers and other social scientists have begun to move beyond the semiotic in their readings of the city. As public art spaces have become increasingly central to the definition and experiences of the post-modern city, it is time that they were subject to such critical interrogations. Such projects offer the possibilities of opening up alternative geographies, histories and sociologies of the spaces of the post-modern city that have only been partially revealed in academic representations of it to date.

REFERENCES

Aitken, S. and Wingate, J. (1993) 'A preliminary study of the self-directed photography of middle-class, homeless and mobility-impaired children' *Professional Geographer*, 45, 65–72

Albers, P. C. and James, W. R. (1988) 'Travel photography: a methodological approach' *Annals of Tourism Research*, 15, 134–158

Barnes, T. and Duncan, J. eds. (1992) *Writing Worlds:*

Discourse, Text and Metaphor in the Representation of Landscape, Routledge, London

Bender, B., ed. (1993) *Landscape, Politics and Perspectives*, Berg, Oxford

Borden, I. (2001) *Skateboarding, Space and the City*, Berg, Oxford

Burgess, J., Harrison, C. and Limb, M. (1988) 'People, parks and the urban green: a study of popular meanings and values for open spaces in the city' *Urban Studies*, 25, 455–473

Clarke, D., ed. (1997) *The Cinematic City*, Routledge, London

Cosgrove, D. and Daniels, S., eds. (1988) *The Iconography of Landscape: Essays on the Symbolic Representation, Design and Use of Past Environments*, Cambridge University Press, Cambridge

Duncan, J. and Ley, D., eds. (1993) *Place / Culture / Representation*, Routledge, London

Dwelly, T. (2001) *Creative Regeneration. Lessons from Ten Community Arts Projects*, Joseph Rowntree Foundation, York

Edwards, M. (1994) *Raymond Mason*, Thames and Hudson, London

Farrington, J. and Silber, E., eds. (1989) *Raymond Mason: Sculptures and Drawings* Lund Humpheries in association with Birmingham Museum and Art Gallery, London

Goodey, B. (1994) 'Art-ful places: public art to sell public spaces?' in Gold, J. and Ward, S., eds. *Place Promotion: The Use of Publicity and Marketing to Sell Towns and Regions*, John Wiley, Chichester, 153–179

Gottdiener, M. (1986) 'Recapturing the center: a semiotic analysis of shopping malls', in Gottdiener, M. and Lagopoulos, A. P., eds. *The City and the Sign: An Introduction to Urban Semiotics*, Columbia University Press, New York, 288–302

Hall, T. (1995a) 'Public art, urban image' *Town and Country Planning* 64, 4, 122–123.

Hall, T., (1995b) 'The second industrial revolution: cultural reconstructions of industrial regions' *Landscape Research* 20, 3, 112–123

Hall, T. (1997a) 'Images of industry in the post-industrial city: Raymond Mason and Birmingham' *Ecumene* 4, 1, 46–68

Hall, T. (1997b) '(Re)placing the city: cultural relocation and the city as centre' in Westwood, S. and Williams, J., eds. *Imagining Cities: Scripts, Signs and Memories*, Routledge, London, 202–218

Harvey, D. (1987) 'Flexible accumulation through urbanisation: reflections on 'PostModernism' in the American city' *Antipode*, 19, 260–286

Harvey, D. (1989a) *The Condition of Postmodernity: An Enquiry into the Origins of Cultural Change*, Blackwell, Oxford

Harvey, D. (1989b) 'From managerialism to entrepreneurialism: the transformation of urban governance in late capitalism' *Geografiska Annaler* 71 B, 1–17

Hill, J., ed. (1998) *Occupying Architecture. Between The Architect and The User*, Routledge, London

Holcomb, B. (1993) 'Revisioning place: de- and re-constructing the image of the industrial city' in Philo, C. and Kearns, G., eds. *Selling Places: The City as Cultural Capital, Past, Present and Future*, Pergamon Press, Oxford, 133–143

Holcomb, B. (1994) 'City make-overs: marketing the post-industrial city' in Gold, J. and Ward, S., eds. *Place Promotion: The Use of Publicity and Marketing to Sell Towns and Regions*, John Wiley, Chichester, 115–131

Hubbard, P. (1996) 'Re-imagining the city: the transformation of Birmingham's urban landscape' *Cities* 12, 243–251

Jones, S., ed. (1992) *Art in Public: What, Why and How*, AN Publications, Sunderland

Kenny, J. (1995) 'Making Milwaukee Famous: cultural capital, urban image and the politics of place' *Urban Geography* 16, 5, 440–458

Knox, P. L., ed. (1993) *The Restless Urban Landscape* Prentice Hall, Englewood Cliffs, New Jersey

Lees, L. (2001) 'Towards a critical geography of architecture: the case of an ersatz colosseum' *Ecumene* 8, 1, 51–86

Lovell, V. (1988) 'Report by the Public Arts Commissions Agency to the Arts Working Party' 12/1/88. Unpublished

Martorella, R. (1990) *Corporate Art*, Rutgers University Press, New Brunswick NJ

Matarasso, F. (1997) *Use or Ornament? The Social Impact of Participation in the Arts*, Comedia, Stroud

Miles, M. (1997) *Art, Space and the City* Routledge, London

Miles, M. (1998) 'A game of appearance: public art in urban development, complicity or sustainability?' in Hall, T. and Hubbard, P. eds. *The Entrepreneurial City: Geographies of Politics, Regime and Representation*, John Wiley, Chichester, 203–224.

Miller, D., Jackson, P. and Thrift, N. (1998) *Shopping, Place and Identity*, London, Routledge

Moody, E. (1990) 'Introduction' in Public Art Forum,

eds., *Public Art Report*, Public Art Forum, London, 2–3
Pink, S. (2001) *Doing Visual Ethnography: Images, Media and Representation in Research*, Sage, London
Rose, G. (2001) *Visual Methodologies*, Sage, London
Savage, M. and Ward, A. (1993) *Urban Sociology: Capitalism and Modernity*, Macmillan, London
Shields, R. (1991) *Places on the Margin. Alternative Geographies of Modernity*, Routledge, London
Short, J. R. and Kim, Y.-K. (1998) 'Urban crises/urban representations: selling the city in difficult times' in Hall, T. and Hubbard, P., eds. *The Entrepreneurial City: Geographies of Politics, Regime and Representation*, John Wiley, Chichester, 55–76
Taylor, I., Evans, K. and Fraser, P. (1996) *A Tale of Two Cities: Global Change, Local Feeling and Everyday Life in the North of England, A Study in Manchester and Sheffield*, Routledge, London
Thomas, H. (1994) 'The local press and urban renewal: a South Wales case study' *International Journal of Urban and Regional Studies* 18, 2, 315–333
Thrift, N. (2000) 'Non-representational theory' in Johnston, R. Gregory, D., Pratt, G. and Watts, M., eds, *The Dictionary of Human Geography*, fourth edition, Blackwell, Oxford, 556
Weideger, P. (1991) 'Larger than life tribute to Brum's golden age', *The Independent* 8/6/91, 14
Wilson, D. (1996) 'Metaphors, growth coalitions, and black poverty neighbourhoods in a U.S. city', *Antipode*, 28, 72–97

'Fragments of a Metropolitan Viewpoint'

from *If You Lived Here* (1991)

Martha Rosler

Editors' Introduction

Martha Rosler is an artist who has exhibited internationally and whose work has explored a range of social issues concerning everyday life, the domestic sphere, phenomenologies of travel, the media and the city. Her work has included video, photography and photo-text projects and installations as well as writing. Much of Rosler's work has involved exposing the historical, rather than natural, separation of the private and public spheres under capitalism. Her work exposes the 'construction' of social and cultural reality for economic ends and considers the oppression of certain groups, notably women and the homeless, that result from this. Rosler has been producing work examining these issues since the early 1970s.

Some of her most widely debated work has been the photographic exhibition, *If You Lived Here* (1991). The extract included here is taken from the accompanying book of the exhibition. Between 1989 and 1992 Rosler examined the exclusion of homeless people by processes of uneven urban development, principally in New York City. The work coincided with an interest in politically committed public art within a range of disciplines including urban geography, cultural studies and sociology. Rosler's work, and the commentaries upon it by people such as Rosyln Deutsche, remain key texts in these debates.

ARTISTS IN THE CITIES

What variety of means *is* available in the effort to persuade and convince? How can one represent a city's "buried" life, the lives in fact of most city residents? How can one show the conditions of tenants' struggles, homelessness, alternatives to city planning as currently practiced – the subjects of "If You Lived Here . . ."? These have been the central issues shaping this project. Its forums, of course, provided an opportunity for direct speech. The three shows, however, also featured varieties of "direct evidence" and argumentation about the grounding of urban life. Artists, community groups, and activists made their points through an array of materials, from videotapes, films, and photographic works to pamphlets and posters to paintings, montages, and installations.

Certainly the conventionalized picture of the postmodern city, with its fortresses and deeply impoverished ghettos, with its epidemics of drugs and AIDS, reinforces the imagery of the urban frontier and discourages even partial approaches to poverty and homelessness. For artists, the image of the city's mean streets may feed a certain romantic Bohemianism. Yet, because artists often share city spaces with the underhoused, they have been positioned as both perpetrators and victims in the processes of displacement and urban planning. They have come to be seen as a pivotal group, easing the return of the middle class to center cities. Ironically, however, artists themselves are often displaced by the same wealthy professionals – their clientele – who have followed them into now-chic neighborhoods.

The "percent for art" programs put in place in a number of U.S. cities have also brought artists into the urban-planning blueprint, at a time when even the idea of public art – like the notion of public space – is being severely attacked. This isn't the place for a broad consideration of public art, but what is worth mentioning is the current high-profile version of "beautification," an ambition to improve the "quality of life" often invoked by anxious city administrations in canceling both taxes and unsightly urban elements for the benefit of powerful corporations. This sort of public art project is exemplified by Battery Park City, a megaproject on New York's Lower West Side. Financed by international capital (in this case, Olympia & York, the corporate entity of Montreal's Reichmann Brothers), Battery Park City imagines itself to be a fantasy enclave of residences, offices, parks, and gardens – something like the ruling-class rooftops in Fritz Lang's film, *Metropolis.* What is of interest here is the regularized incorporation of art by the authority running it – precisely as though this exclusive preserve reinvented the public, on privatized but publicly subsidized turf. Although the art program has been touted as showing risky "socially conscious" art, such work seems severely compromised by its context.

Irrespective of such public or corporate commissions, artists have always been capable of organizing and mobilizing around elements of social life; the city is art's habitat. But how do artists address directly the issues of city life and homelessness in which they are implicated? Most directly, of course, many artists engage in activism, including working with homeless people in shelters and hotels, as do Nancy Linn and Rachael Romero; producing posters and street works on urban issues, as do Robbie Conal, Ed Eisenberg, Janet Koenig, and Greg Sholette; or engaging in other forms of political activism, as do Marilyn Nance, Mel Rosenthal, and Juan Sanchez. Krzysztof Wodiczko and the Mad Housers work with homeless people in projects whose stop-gap solutions to homelessness show up the absurdity of official responses. But there are many other approaches as well.

Postmodern life is characterized by the erasure of history and the loss of social memory. Social life includes multiple streams of contesting momentary images, which, detached from particular locales, join the company of other images. Images, in appearing to capture history, become the great levelers, the informational counterpart of money, replacing material distinctions with their own "depthless" (that is, ahistorical) logic. One of the social functions of art is to crystallize an image or a response to a blurred social picture, bringing its outlines into focus. Many artists and critics engage with these dislocating politics of the image through critiques of signification. Such critical practices temporarily check the flow of (what passes for) public discourse. But such critiques-in-general, crucial as they are to a reorientation of social understanding, don't exhaust the avenues to urban meaning.

Consider the city once again. It is more than a set of relationships and a congeries of buildings, it is even more than a geopolitical locale – it is a set of unfolding historical processes. In short, a city embodies and enacts a history. In representing the city, in producing counterrepresentations, the specificity of a locale and its histories becomes critical. Documentary, rethought and redeployed, provides an essential tool, though certainly not the only one.

The arguments for documentary apparently need to be made anew. Image politics and still-contested notions of difference have prompted serious philosophical critiques of the claims to transparency and univocality of news, documentary, and photography in general – critiques made in the context of the growing distance between imagery and social meaning in the culture at large. Even past documentary works, which have taken on new meanings in textbooks, art history books, and gallery sales, are a matter of perpetual reinterpretation.

The "problem" of documentary is compounded by the art-world distrust of populist forms (for various reasons, some of which are valid and others simply manifestations of professional snobbery). Who could possibly deride a healthy skepticism in regard to the propaganda of the obvious that characterizes the myths of documentary transparency? On the other hand, the agitational intentions of activist social documentary aren't sufficient in themselves to secure a conviction except in the court of formalist aestheticism.

It would be ironic if those of us seeking a more complex account of experience and meaning were enjoined by our own theoretical strictures from presenting evidence in support of social meaning and social justice. Documentary practices are social practices, producing meanings within specific contexts. Rejecting various entrenched documentary practices hardly amounts to a negation of documentary in toto. The critical minefield surrounding practices rebuked for empiricism calls for careful negotiation. Social activists, certainly, continue to recognize the importance of documentary evidence in arguing for social change. It is the necessity to acknowledge the place – and time – from which one speaks that is an absolute requirement for meaningful social documentary. This requirement allows for an unspecifiable range of inventive forms but doesn't dispose of the historically derived ones. Naturally, this shifts the terrain of argument from the art object – the photograph, the film, the videotape, the picture book or magazine – to the context, to the processes of signification, and to social process. An underlying strategy of the project "If You Lived Here . . . " (of which this book is a part) has therefore been to use and extend documentary strategies.

A documentary photograph of a member of a social group composed of undifferentiated stereotypes – the "homeless," say – today serves the same purposes as did similar images at the inception of social documentary as a public photographic practice: it "humanizes" by particularizing. It suggests the character of a person's existence, in which material circumstances contradict human worth, and the more dire the conditions, the more the photo may have to tell us. Sometimes the "condition" is invisible, a conceptual understanding laid over the image by the viewer. But the problem is that of projection, of imagining that the characteristics we "see" in the person or scene are those that are "there." For that reason, the more patent the image, the more it accords with "commonsensical" presuppositions, the less it may have to tell us. This is not a condition that should make us vacate the territory of image making, for it is precisely the role of the context – especially the verbal text (written or otherwise supplied) linked with *this* image-text – that establishes a meaning beyond a simple ground for projection.

Documentarians – unlike "street photographers," another sort of practitioner entirely – have hardly relied on images alone to tell the right story. The development of high-profile, commercial, professional photojournalism, and the art-world appropriation of all kinds of photography into its own procrustean canon, paved the way for a photographic practice passing for social documentary to shorten its circuit from the street to the gallery wall. Lost along the way were more than symbolic claims for agitational intentions. The dead hand of "universalism" has lain heavily on documentary's shoulder, for a documentary work alibied as revealing an underlying human sameness becomes simply an excuse for spectacle. That is the basis of one of the most telling critiques of documentary, particularly of the subgenre exotica – a form of anthropology that masquerades as humanism when the subject is the down-and-outer in advanced Western society or in its familiar margins (Mexico or Bensonhurst). One of the problems of representations of the city is to make an argument without betraying people.

In one of the exhibitions for "If You Lived Here . . .," a pair of texts placed side by side on the wall argued for and against photographing the homeless. The first text, an excerpt from an essay of mine on documentary photography, criticized "victim photography" for rarely serving the purpose which (presumably) its makers intended – namely, to gather public support, to generate outrage, and to mobilize people for change. Rather, I argued, documentary photography may inadvertently support the viewers' sense of superiority or social paranoia. Especially in the case of homelessness, the viewers and the people pictured are never the same people. The images merely reproduce the situation of "us looking at them."

In the other text, "On Photographing the Homeless," photographer Mel Rosenthal argued for photographing the homeless. Although, he wrote, he was troubled by photographing people in desperate straits – people who, even when they gave their consent, may not have had much idea of how their photos would be used – on balance he felt that images of real individuals can dispel the numbness many people feel. Context, however, still remains crucial, and Rosenthal acknowledges this. (I've remarked elsewhere that political photography is repressed in our culture by being hung in a gallery.) Rosenthal's projects are never geared toward the gallery-museum circuit. His South Bronx photographs, for example – made during a period when he worked at a health clinic in the Bathgate area where he grew up – were published in activist and grass-roots magazines. Rosenthal gave prints to the people photographed, who often had no other photos of themselves. In exhibition form, these photos of resiliency in a war zone are accompanied by an array of quoted remarks (some of which are reproduced in this book) providing the necessary – damning – information.

It would be reductive to insist that no levels of mediation can exist between those who experience a situation and those who view it. In a fragment of an interview with Alexander Kluge reprinted in this book, Kluge takes up precisely this question of participatory versus supportive mediations – by chance, in relation to the eviction of squatters in Germany. There has to be room for an interested art practice that does not simply merge itself into its object. Interestingly, though, Bienvenida Matias, in Loisaida (the Hispanic Lower East Side of New York), and Nettie Wild, in Vancouver, B.C., were each invited to live in the housing communities whose struggles they were documenting on film and videotape (Matias in *El Corazon de Loisaida*, or *The Heart of Loisaida*, and Wild in *The Right to Fight*). Both accepted.

Ultimately, there's no denying that no matter how the works in "If You Lived Here . . ." originally were woven into the social fabric, the venue of the exhibitions was an art gallery, even if partly "transformed." The idea of these shows wasn't simply to thicken the context for the reception of "photographs of the Other." It was, first, to allow for a consideration of an underreported, underdescribed, multidetermined set of conditions producing simple results: homelessness and sadly inadequate housing. Perhaps no less importantly, the project intended to suggest how art communities (might) take on such questions. Since the problem of homelessness, like all social problems, exists in a stream of conflicting representations, it is not possible to change social reality without challenging its simplifying overlaid images. That was a main task of "If You Lived Here. . . ."

"Home Front," the first of the project's three exhibitions, meant to establish an ambience quite different from that of the usual art gallery. Substantively, it was conceived as a set of representations of contested neighborhoods. The term "Home Front" suggests a war zone, after all – and one outcome of a loss on that front is homelessness. The show provided a look at contested housing, primarily urban housing; it also offered help to embattled tenants, directing them to militant neighborhood groups and advocacy organizations. Some of the battles on the home front are protracted, some skirmishes have been all too visibly lost, but both successes and failures need to be considered.

In "Home Front," also, the truculence of official responses to the housing crisis was indicated by the prominently painted remark attributed to New York mayor Ed Koch: "IF YOU CAN'T AFFORD TO LIVE HERE, MO-O-VE!" (See Allan Sekula and others' Long Beach poster for a longer articulation of the same idea.) Statistical graphs and charts were arrayed around the room, above eye level, in the gallery equivalent of "waste space." These graphs were interspersed with real estate ads touting luxurious living in all those Manhattan high-rises with pretentious names; the prose and the poetry of profit – and loss.

Although homelessness was at the center of "If You Lived Here . . .," it was the entire focus only in the second exhibition, "Homeless: The Street and Other Venues." But it was critical, in this exhibition, not to reproduce the dichotomies that inform most discussions of homelessness – "us and them." Here, the wall text was a quotation from urbanist Peter Marcuse: "Homelessness exists not because the system is not working but because this is the way it works."

The third exhibition, "City: Visions and Revisions," offered some movement toward solutions to urban problems: from new designs for urban infill housing, to housing for people with AIDS or for homeless women, to utopic visions of the cities. In this exhibition, the production of urban space itself was conceived of as a matter of economic and social decisions and as a complex "metasignification." Some of the city revisions weren't victories. The slogan on the wall was drawn from the French student uprising of May 1968: "Under the Cobblestones, the Beach." Its romanticism may perhaps be excused by its reminder that the built environment is just that, and that, furthermore, the question of the body, of pleasure, and therefore of liberation cannot be divorced from rational considerations of urban life.

Throughout the project there was an effort to blur "inside" and "outside," to abolish the distinction between the gallery space as a large, squarish room and as a world apart, a zone of aestheticism. Couches and rugs faced video monitors in various places in each exhibition, and billboards and other oversized works originally installed "in the street" were hung on the gallery wall. A reading room provided a wide variety of material, from flyers for demonstrations and protests to organizational brochures for tenants and homeless people, activists, and volunteers. There were also photo books and catalogues, historical studies, scholarly books and critiques, project descriptions.

The reading room was reconfigured and repainted for each show; in its original design, the walls were on wheels. In "Home Front" it was a solid little castle against one gallery wall and harbored a living room space. In "Homeless" it was a shelter of empty beds with a desk screwed to the external wall. In "City" it was a desk on the outside of a hut in the middle of the gallery. It held a black-shrouded installation about the eviction of Latin American workers from San Diego County's brushland as tract towns spring up nearby – the waste space of displacement under the suburban street.

In the exhibition "Homeless," in addition to the reading room resources, there was counseling provided by Homeward Bound, and lists of institutions from private and public shelters to soup kitchens and counseling and employment services were posted and available to be taken away.

Many works in the project employed the customary means of traditional documentary, namely, photography, film, and video. It is worth considering, therefore, how some of the makers positioned themselves in relation to the "documentary problem." Often the videotapes and films show little evidence of questioning; they simply get on with their business. In videotapes like Julia Keydel's *St. Francis Residence* and Arlyn Gajilan's *Not Just a Number*, for example, the interview format is well-adapted to allowing the unheard to speak about their lives. Other films and videotapes were directly activist. For embattled tenants, *Don't Move, Fight Back* (made in conjunction with Strycker's Bay tenants' group in upper Manhattan); *How to Pull a Rent Strike* and *Techos y Derechos* (both by Tami Gold and Steve Krinsky for East Orange, New Jersey's Shelterforce); and *Clinton Coalition of Concern* (made by Brian Connell, a videomaker who is also a member of the coalition) are rallying tapes, informing people about others who are fighting or have fought successfully to save or improve their homes and providing a set of steps to follow.

Even failures can be instructive. Lost struggles are represented in the films, *The Fall of the I-Hotel*, by Curtis Choy and Chonk Moonhunter (a hotel housing primarily long-term elderly Filipino residents is lost to gentrification in downtown San Francisco), George Corsetti's *Poletown Lives!* (a working-class, largely Polish neighborhood in Detroit falls to a proposed auto plant), and Pablo Frasconi and Nancy Salzer's *Survival of a Small City* (gentrification displaces poor and working-class nonwhite residents of a former mill town in Connecticut).

Perhaps questioning documentary's historic reliance on physiognomic evidence, Mark Berghash's photographs showed very large, tight closeups of people's faces. First-person texts or audiotapes of the subjects were included. Some were of people in terrible circumstances, such as homelessness, and others were of well-situated people, but we don't know who is whom. Bob McKeown employed traditional social-documentary strategies in photographing the formation of the Homeless Union in Wayne County, Michigan. But the Urban Center for Photography, of which McKeown is a member, collectively produced a different kind of work, in which very large photos of people and buildings were placed in downtown Detroit, along with the stenciled legend "Demolished by Neglect."

Some photographers completely reject "humanist" documentary, with its multiplicity of hidden texts, especially in relation to women. Rhonda Wilson used only staged images in producing her poster series on women and homelessness in England. Also in England, members of the Docklands Community Poster Project use photomontage and also layer historical material into their work. Directly interrogating the voyeurism of documentary photography, Greg Sholette incorporated Jacob Riis' photo *Police Station Lodgers . . . in the West 47th Street Police Station* into a sculptural relief whose conceit centers on the interpretation of the facial expression of a principal female figure. Coincidentally, this photo from the late 1890s was hung in the entryway to the complex tenement-kitchen installation by the Chinatown History Project. That work provided a detailed examination of the narratives of life, historical and contemporary, on the Lower East Side and Chinatown. To develop its argument, the group included wall texts, a handout for gallery-goers (reproduced here), and a slide-and-tape show on the area's different groups and on current tenant organizing.

In many works, perhaps especially in videotapes, the subjects speak about and in some cases produce works about their lives. I'm thinking now of *2371 Second Avenue* and *Life in the G*, videotapes made by teenage Hispanic New Yorkers in conjunction with the Educational Video Center; and the photos and documents produced by photographer Marilyn Nance of her city-owned building in Brooklyn, that provided part of the tenants' court case.

In an entirely other sort of instance of the self-production of meaning, the group Homeward Bound maintained an office in the gallery (and participated in the forums), as advocates for themselves and other homeless people. Their portraits, taken the preceding summer by photographer Alcina Horstman during their hundred-day encampment in front of City Hall – during which they registered passers-by to vote – hung in their office area. These images, using an artified documentary approach, meant something very different in that office space. Homeward Bound's organizing efforts include both substantive movements toward bettering their lives and advocacy with municipal agencies, along with attempts to reposition themselves in relation to the reigning images of homeless people. Most homeless people aren't in a position to take on these roles.

FOCUS ON NEW YORK

The largest body of work in "If You Lived Here . . ." centered on New York City, particularly Manhattan, and this book concentrates that focus even more. New York is the largest city in the United States and Canada. New York is (still) a renter's city, an immigrant city, a city of great populations of color (including the largest number of African Americans on the continent), a city with a strong history of unionism and progressive politics but also of the uglier face of class struggle, such as police brutality and patrician rule, race riots, efforts to divide and contain immigrant populations and to segregate the city by race and class.

New York is also the home of Wall Street, which services international finance capital. In the past decade-and-a-half, New York has become a city ruled largely by banking and real estate interests. New York is an international city, with exclusive midtown *pieds à terre* for the jet set and less enticing accommodations for its immigrant groups. And New York is a city of vast abandonments, of decayed tenement stock that was never quite fit for habitation, and of glitzy new high-rise palaces and recently gentrified neighborhoods. Although New York is a city with strong rent protections, these protections have been eroded over the past couple of decades, and market rents have soared to the highest in the nation.

Under the first great modern urban-planning despot, Robert Moses, New York provided a model for the rest of the nation, not only for grandscale refiguring of the urban environment, but for the deployment of egalitarian rhetoric to justify social engineering ultimately devoted to the segregation of classes. Just as Moses' projects provided shaping models, more contemporary projects and situations, such as Battery Park City and the Times Square redevelopment on the upscale side and the Lower East Side and the South Bronx on the down, are exemplary.

New York doesn't just mean Midtown Manhattan. Although the four other boroughs (and the rest of Manhattan) have their share of expensive housing, suburban tracts, and gentrified districts, in three of them – the Bronx, Brooklyn, and in pockets of Queens – the poor, the nonwhite, the underhoused, and the homeless are collected. Not surprisingly, then, discussions of New York generally take in only Manhattan, with the spectral Other world represented by the South Bronx (collapse) and occasionally Brooklyn (a borough of Others) added in. Thousands of artists (and other middle-class people, including many whites) have wound up in Brooklyn, but Brooklyn – which would be the fifth largest city in the United States had it retained its separate status – doesn't figure in most discussions of urbanism, let alone of the art world. (Willie Birch, Erik Lewis, Marilyn Nance, Juan Sanchez, Dan Wiley, some of the makers of *Metropolitan Avenue* and of *People's Firehouse #1* are Brooklyn residents; of this list, the majority are people of color.) That lack is repeated in this volume.

New York, then, is a good model of a modern-day metropolis, and the way its living conditions are addressed, or not addressed, can serve as building blocks in wider explanations that can collect more than local examples.

New York, the center of the U.S. art world and the home of finance capital, is an appropriate place to tackle the intersection of art and real property. These shows were held in a gallery in Soho, an art district that forms part of the largest concentration of art-world institutions in the world. Soho – the first municipally mandated artist district – is a site of hypergentrification in a central urban area that has undergone several transformations of use in its hundred-and-fifty-year history. (The enameled lamppost texts designating Soho as a historic area begin with cast iron architecture, then relate that artists moved into the district in the early 1970s, and end by describing it as "now a lively residential and shopping area.")

During "If You Lived Here . . . " some people asked, "Why are you holding this project in Soho?" The question was asked only by people involved in art. And there could be no answer for those who feel that Soho is a true enclave, the Vatican of art, physically located in, but otherwise exempt from, the rules of New York. For those not involved in art, the question of showing in Soho may seem an incomprehensible quibble. Still, the Soho question is important, and it relates not simply to the gallery world in the abstract but to the project sponsor, the Dia Art Foundation. Dia established itself in the 1970s as an *haute moderniste* private foundation devoted to individual (white) (male) modernist artists, providing them with work space and generous stipends. Dia purchased a number of buildings and sites in Manhattan and elsewhere (as in Marfa, Texas), especially in Soho and Tribeca, becoming part of the real estate/art institution nexus. (Although most of its holdings have been sold, a reincorporated Dia now owns, in addition to its five Lower Manhattan properties, a site in Quemado, New Mexico, and a couple in New England.) It seemed important, therefore, to take the opportunity to challenge the paradigm of art production and distribution that Dia in its earlier incarnation had presupposed and which still clings to its exhibition practices, in step with most of the art world.

Earlier I remarked that the project meant to depart from the art gallery pattern. It appeared necessary to effect a significant transformation of the Dia gallery. Its front, with frosted windows and gray paint, was so self-effacing that it was common for intended visitors, and even for me, to walk right on by. I put "Come On In – We're Home" in large red letters on the doors, and ACT UP (AIDS Coalition To Unleash Power) put up posters on AIDS and homelessness. For "Homeless," housing activist and artist Stuart Nicholson painted a text comparing shelters to refugee camps on the sidewalk in front of the door. In the interior, I got away from the emptied-out look by filling it up.

Many commentators mentioned the transgressive character of these crowded exhibitions, and some seemed to miss the pristine quality of the modernist space, feeling intimidated by the volume of work and the reading room. But *who* was feeling intimidated? For some art-world professionals the project seemed to represent an outright rejection of art. Although by and large the work in the show was authored, framed, and neatly hung, accompanied by white labels, the show's organizational principles depended on other issues as I have described them here. The shows' inclusiveness annoyed some writers well known for their systematic dismissal of modernism's presuppositions.

The static and unconscious presuppositions about the art audience that some critics brought to these shows surprised me. Despite twenty years of rethinking the art system, a spotty amnesia has broken out in this regard, and some have forgotten that the art-world audience isn't born but constantly constructed and reconstructed, laboriously, just like any constituency. Many people, including artists and art students, come to Soho; they came to the Dia gallery, and they saw the shows. In addition, the diverse groups and people who made up these shows and forums brought a significant portion of the audience: church workers, elected representatives, New York City schoolchildren, college students, architects, urban planners, activists, advocates, homeless people, volunteers, filmmakers and videomakers, painters, poets, muralists, sculptors, photojournalists, and art photographers. Each event in the project – shows, poetry readings, film screenings, workshops, forums – was separately advertised; each brought interested people. Some of the project fliers didn't mention the art connection. Articles in mainstream newspapers left art out of it. Heterogeneity engendered heterogeneity, and people brought their friends.

'The World in a Shopping Mall'

from *Variations on a Theme Park* (1992)

Margaret Crawford

Editors' Introduction

Margaret Crawford is Professor of Urban Design and Planning Theory at the Graduate School of Design at Harvard University and has researched issues around the meanings and uses of urban space. Her publications have examined aspects of the built environment including the relationship between cars and urban life in the co-edited collection *The Car and the City: The Automobile, the Built Environment and Daily Urban Life* (1991), the design and historical geography of industrial company towns in *Building the Workingman's Paradise: The Design of American Company Towns* (1995) and the everyday experiences of urban spaces in the collection *Everyday Urbanism*, co-edited by Crawford with John Chase and John Kaliski (New York, Monacelli Press, 1999). In this more recent text Crawford writes on the spatial expressions of informal economies, as in street vending and garage sales in US cities, seeing such manifestations of a spatiality of occupation as offering a counter-argument to those of the dominant spaces of malls, and stating the survival of spaces of public association and exchange despite the encroachment of privatised spaces on a more conventionally defined public realm. She writes that 'Woven into the patterns of everyday life, it is difficult even to discern these places as public space', but that, located between the private, commercial, and domestic realms, they '. . . contain multiple and constantly shifting meanings rather than clarity of function' (p. 28).

The text below, however, is the full version of her chapter in *Variations on a Theme Park* (1992), included in this collection because it gives a succinct description as well as critical review of what has arguably become one of the main cases of a new architectural form in postmodernity. Crawford's more recent writing in *Everyday Urbanism* could usefully be read as well, and both texts can be seen in the light of Henri Lefebvre's theory of spatiality (see Part Six).

Larger than a hundred football fields, the West Edmonton Mall is, according to the *Guinness Book of Records*, the largest shopping mall in the world. At 5.2 million square feet, the world's first megamall is nearly twice as large as the runner-up, the Del Amo Mall in Los Angeles, which covers only 3 million square feet. Other *Guinness* titles the mall holds are World's Largest Indoor Amusement Park, World's Largest Indoor Water Park, and World's Largest Parking Lot. Besides its more than 800 shops, 11 department stores, and 110 restaurants, the mall also contains a full-size ice-skating rink, a 360-room hotel, a lake, a nondenominational chapel, 20 movie theaters, and 13 nightclubs. These activities are situated along corridors of repeated storefronts and in wings that mimic nineteenth-century Parisian boulevards and New Orleans's Bourbon Street. From the upper stories of the mall's hotel, the glass towers of downtown Edmonton are just visible in the distance.

Seen from above, the mall resembles an ungainly pile of oversized boxes plunked down in the middle of an enormous asphalt sea, surrounded by an endless landscape of single-family houses. Inside, the mall presents a dizzying spectacle of attractions and diversions: a replica of Columbus's *Santa Maria* floats in an artificial lagoon, where real submarines move through an impossible seascape of imported coral and plastic seaweed inhabited by live penguins and electronically controlled rubber sharks; fiberglass columns crumble in simulated decay beneath a spanking new Victorian iron bridge; performing dolphins leap in front of Leather World and Kinney's Shoes; fake waves, real Siberian tigers, Ching-dynasty vases, and mechanical jazz bands are juxtaposed in an endless sequence of skylit courts. Mirrored columns and walls further fragment the scene, shattering the mall into a kaleidoscope of ultimately unreadable images. Confusion proliferates at every level; past and future collapse meaninglessly into the present; barriers between real and fake, near and far, dissolve as history, nature, technology, are indifferently processed by the mall's fantasy machine.

Yet this implausible, seemingly random, collection of images has been assembled with an explicit purpose: to support the mall's claim to contain the entire world within its walls. At the opening ceremony aboard the *Santa Maria*, one of the mall's developers, Nader Ghermezian, shouted in triumph, "What we have done means you don't have to go to New York or Paris or Disneyland or Hawaii. We have it all here for you in one place, in Edmonton, Alberta, Canada!"[1] Publicity for the Fantasyland Hotel asks, "What country do you want to sleep in tonight?"– offering theme rooms based not only on faraway places such as Polynesia and Hollywood, and distant times such as ancient Rome and Victorian England, but also on modes of transportation, from horse-drawn carriages to pickup trucks.

The developer's claims imply that the goods for sale inside the mall represent the world's abundance and variety and offer a choice of global proportions. In fact, though, the mall's mixture of American and Canadian chains, with a few local specialty stores, rigorously repeats the range of products offered at every other shopping mall. Internal duplication reduces choice even further, since many stores operate identical outlets at different points in the mall. Despite the less than worldwide selection, shoppers still come from all over the world (70 percent of the mall's visitors are from outside Alberta) and spend enough to generate profits of $300 per square foot – more than twice the return of most malls. The West Edmonton Mall (WEM) dominates the local commercial economy. If superimposed onto downtown Edmonton, the mall and its parking lot would span most of the central business district. Commercially overshadowed by the mall, long-established downtown stores now open branches in the mall. As a gesture of urban goodwill, the WEM's developers have agreed to build another mall downtown to replace some of the revenue and activity drained off by the megamall.[2]

The inclusion of more and more activities into the mall has extended its operating day to twenty-four hours: a chapel offers services before shops open, nightclubs draw customers after they close, and visitors spend the night at the mall's hotel. The mall is also a workplace, with more than fifteen thousand people employed in its shops, services, and offices, many of whom also eat and spend their free time there. In the suburbs of Minneapolis, the WEM's developers are now erecting an even larger complex, the Mall of America, complete with office towers, three hotels, and a convention center. Orange County's Knott's Berry Farm theme park will supply the mall's entertainment centerpiece, "Camp Snoopy."[3] The mayor of Bloomington, Minnesota, exults, "Now people can come here and watch a Vikings game and stay for the weekend. It's a different world when you have a megamall."[4]

The mall's encyclopedic agglomeration of activities requires only the addition of housing, already present in other urban mall megastructures, to become fully inhabitable, a world complete in itself. In a sense, the fragmented forms and functions of modern living are being brought together under the mall's skylighted dome. This suggests the possibility that the unified world of premodern times might be reconstituted through the medium of consumption, an ironic reversal of the redemptive design projects imagined by nineteenth-century utopians such as Fourier and Owen, who sought unity through collective productive activity and social reorganization. Although Fourier's Phalanstery merged the arcade and the palace into a prefigurative mall form, its glass-roofed corridors were intended to encourage social intercourse and foster communal emotions, rather than stimulate consumption.

THE SCIENCE OF MALLING

The WEM's nonstop proliferation of attractions, activities, and images proclaims its uniqueness; but, beneath its myriad distractions, the mall is easily recognizable as an elephantine version of a generic type – the regional shopping mall. Indeed, the WEM is only the latest incarnation of a self-adjusting system of merchandising and development that has conquered the world by deploying standardized units in an extensive network. And, as the state-of-the-art mall is continually redefined, the WEM's jumbled collection of images is already on the verge of becoming obsolete. More seamless alternative worlds are coming off the drawing boards. Disney "imagineers" have recently designed an entertainment center and shopping mall for Burbank inspired by the "lure and magic of the movies." The cinematic medium, inherently fragmented and unreal, structures a sophisticated fantasy world that will be both more complex and more coherent than the WEM.[5]

Although it is, for the moment, unrivaled in size and spectacle, the WEM is not exempt from the rules of finance and marketing that govern the 28,500 other shopping malls in North America.[6] These rules date from the golden years between 1960 and 1980, when the basic regional mall paradigm was perfected and systematically replicated. Developers methodically surveyed, divided, and appropriated suburban cornfields and orange groves to create a new landscape of consumption. If a map of their efforts were to be drawn, it would reveal a continent covered by a wildly uneven pattern of overlapping circles representing mall-catchment areas, each circle's size and location dictated by demographic surveys measuring income levels and purchasing power. In a strangely inverted version of central-place theory, developers identified areas where consumer demand was not being met and where malls could fill the commercial voids.[7] Dense agglomerations of malls would indicate the richest markets, and empty spots the pockets of poverty: West Virginia, for example, has the lowest shopping-mall square footage per inhabitant in the country.[8]

The size and scale of a mall, then, reflects "threshold demand" – the minimum number of potential customers living within the geographical range of a retail item to enable it to be sold at a profit. Thus, *neighborhood* centers serve a local market within a two-mile radius; *community* centers draw from three to five miles. The next tier of 2,500 *regional* malls (at least two department stores and a hundred shops) attracts customers from as far as twenty miles away, while an elite group of 300 *super-regional* malls (at least five department stores and up to three hundred shops) serve a larger, often multistate, area within a hundred-mile radius. At the peak of the pyramid sits the West Edmonton megamall, an international shopping attraction. The system as a whole dominates retail sales in the United States and Canada, accounting for more than 53 percent of all purchases in both countries.[9]

The malling of America in less than twenty years was accomplished by honing standard real-estate, financing, and marketing techniques into predictive formulas. Generated initially by risk-free investments demanded by pension funds and insurance companies (sources of the enormous amounts of capital necessary to finance malls) the malling process quickly became self-perpetuating, as developers duplicated successful strategies. Specialized consultants developed techniques of demographic and market research, refined their environmental and architectural analysis, and produced *econometric* and *locational* models. Mall architect Victor Gruen proposed an ideal matrix for mall-building that combined the expertise of real-estate brokers, financial and marketing analysts, economists, merchandising experts, architects, engineers, transportation planners, landscape architects, and interior designers – each drawing on the latest academic and commercial methodologies. Gruen's highly structured system was designed to minimize guesswork and to allow him to accurately predict the potential dollar-per-square-foot-yield of any projected mall, thus virtually guaranteeing profitability to the mall's developers.

In a game with such high stakes, competition became irrelevant. The technical expertise and financial resources required for mall-building restricted participation to a small circle of large developers. The pioneers – DeBartolo, Rouse, Hahn, Bohannon, and Taubman – established their own institutions: the International Council of Shopping Centers and trade journals such as *Shopping Center World* and *National Mall Monitor* insured rapid circulation of investment and marketing information; the Urban Land Institute worked to standardize mall-development procedures. The application of such standardized methods of determining locations, structuring selling space, and controlling customers produced consistent and immense profits. In their first twenty-five years, less than one percent of shopping malls failed; profits soared, making malls, according to DeBartolo, the best investment known to man."[10]

For the consumer, the visible result of this intensive research is the "mix" – each mall's unique blend of tenants and department-store "anchors." The mix is established and maintained by restrictive leases with clauses that control everything from decor to prices. Even within the limited formula that the mix establishes for each mall, minute variations in the selection and location of stores can be critical. Detailed equations are used to determine exactly how many jewelry or shoe stores should be put on each floor. Since branches of national chains are the most reliable money-makers, individually owned stores are admitted only with shorter leases and higher rents. Mall managers constantly adjust the mix, using rents and leases to adapt to the rapidly changing patterns of consumption. The system operates much like television programming, with each network presenting slightly different configurations of the same elements. Apparent diversity masks fundamental homogeneity.

The various predictable mixes are fine-tuned to the ethnic composition, income levels, and changing tastes of a particular shopping area. Indexes such as VALS (the Values and Life Styles program), produced by the Stanford Research Institute, correlate objective measures such as age, income, and family composition with subjective indicators such as value systems, leisure preferences, and cultural backgrounds to analyze trade areas. For instance, Brooks Brothers and Ann Taylor are usually solid bets for areas populated by outer-directed *achievers* ("hardworking, materialistic, highly educated traditional consumers; shopping leaders for luxury products") and *emulators* ("younger, status-conscious, conspicuous consumers"). But since climate, geography, and local identity also play a role in spending patterns, these stores may not succeed in areas like Orange County, California, where good weather allows more informal dress. *Sustainers* ("struggling poor; anger toward the American system") and *belongers* ("middle-class, conservative, conforming shoppers, low to moderate income"), on the other hand, tend to be "value-oriented," making K mart or J. C. Penney good anchors for malls where these groups predominate. Shoppers' perceptions of themselves and their environment furnish more accurate predictions of shopping habits than income. According to the Lifestyle Cluster system, an alternative index, even with identical incomes, the *black enterprise* and *pools and patios* groups will exhibit very different consumption patterns.[11]

Through a careful study of such spending patterns, mall-builders can generate a mix that makes the difference between a mere profit-maker and a "foolproof money-machine" such as Southdale, outside of Minneapolis, the most successful of Equitable Life Assurance's one hundred shopping malls. Southdale's managers are constantly adjusting its mix to reflect increasingly refined consumer profiles. They know, for example, that their average customer is a 40.3-year-old female with an annual income of over $33,000, who lives in a household of 1.7 people. She is willing to spend more than $125 for a coat and buys six pairs of shoes a year in sizes 5 to 7. Southdale's mix reflects this ideal consumer; women's clothing stores and upscale boutiques have now replaced Woolworth's and the video arcade. The mall's decor and promotions target her tastes through "psychographics" – the detailed marketing profiles which identify the customer's aspirations as well as her stated needs in order to chart "identity" as well as income.[12]

Such precision in locating and satisfying consumers has become increasingly important since 1980, when malls approached the saturation point. The system demonstrated a surprising adaptability: in spite of its history of rigidly programmed uniformity, new economic and locational opportunities prompted new prototypes. Specialty malls were built without department stores, allowing a more flexible use of space. To fit urban sites, malls adopted more compact and vertical forms with stacked floors of indoor parking, as at the Eaton Center in Toronto and the Beverly Center in Los Angeles. To insure financing in uncertain markets, developers formed partnerships with redevelopment agencies. The Grand Avenue in Milwaukee and the Gallery at Market East in Philadelphia are both joint ventures by HUD, municipal redevelopment agencies, and the Rouse Company. To survive in high-rent downtown locations, malls added hotels, condominiums, and offices to become omni-centers, such as Trump Tower on Fifth Avenue, or Water Tower Place and Chicago Place on North Michigan Avenue.

Existing malls renewed themselves by upgrading their decor and amenities. Future archaeologists will read Orange County's social history in South Coast Plaza's successive levels: the lowest floors featuring Sears and J. C. Penney's, recall the suburbs' original lower-middle-class roots; the elaborate upper levels, with stores such as Gucci and Cartier, reflect the area's more recent affluence. Open-air plazas, once thought obsolete, have been revived and a new generation of consumers now stroll uncovered walkways.[13] Virtually any large building or historic area is a candidate for reconfiguration into a mall. Americans regularly browse through renovated factories (the

Cannery and Ghirardelli Square in San Francisco), piers (North Pier in Chicago), and government buildings (the Old Post Office in Washington, D.C.). The imposing neoclassical space of McKim, Mead, and White's Union Station, which once solemnly celebrated entry into the nation's capital, now contains a shopping mall. The city of New York has even considered developing the Brooklyn Bridge as a historic shopping mall, with the brick arches of its Manhattan approach enclosing retail shops and a health spa.[14]

Although by 1980 the American landscape was crowded with these palaces of consumption, the rest of the world was still open for development. The form could be exported intact into third-world economies, with local developers providing enclosed shopping malls as exotic novelties for upper-class consumers in Caracas or Buenos Aires. Planners of new towns such as Milton Keynes, England, and Marne-la-Vallée, outside Paris, followed the example of Columbia, Maryland, to create state-sponsored social-democratic malls, combining government and community facilities with retail space to create new town centers. Asian versions in Hong Kong and Singapore adapted local marketplace traditions, filling vast malls with small, individually owned shops. The enormous new market opening up in Eastern Europe will surely place Warsaw and Budapest on *Shopping Center World*'s list of hot spots ripe for development. The variations are endless, but whatever form the system adopts, the message conveyed is the same – a repeated imperative to consume.

THE UTOPIA OF CONSUMPTION

The ethos of consumption has penetrated every sphere of our lives. As culture, leisure, sex, politics, and even death turn into commodities, consumption increasingly constructs the way we see the world. As William Leiss points out, the best measure of social consciousness is now the *Index of Consumer Sentiment*, which charts optimism about the state of the world in terms of willingness to spend. The decision to buy a washing machine or a fur coat depends less on finances than on subjective reactions to everything from congressional debates to crime and pollution.[15] Consumption hierarchies, in which commodities define life-styles, now furnish indications of status more visible than the economic relationships of class positions. Status is thus easy to read, since the necessary information has already been nationally distributed through advertising. Moreover, for many, the very construction of the self involves the acquisition of commodities. If the world is understood through commodities, then personal identity depends on one's ability to compose a coherent self-image through the selection of a distinct personal set of commodities.

As central institutions in the realm of consumption, shopping malls constantly restructure both products and behavior into new combinations that allow commodities to penetrate even further into daily life. Most directly, the mall, as its domination of retail sales indicates, functions as an extremely efficient agent for the circulation of large numbers of goods. However, the rigid financial and merchandising formulas that guarantee and maximize its profits restrict the range and variety of goods it can offer. Retailers and shoppers are equally subject to a commercial logic that forces both to constantly justify themselves by concretely realizing the abstract concept of consumption in money terms. These economic imperatives are clearly expressed in the inescapable measurement of mall success in terms of dollars per square foot.

Faced with such restrictions, the mall can realize its profits only by efficiently mediating between the shopper and the commodity. The process of shopping begins even before the shopper enters the mall, in the commercialized contemporary social environment that William Leiss has characterized as the "high-intensity market setting." Primed by a barrage of messages about what he or she "needs" (before the age of twenty, the average American has seen 350,000 television commercials), the shopper arrives at the mall with "a confused set of wants." Presented with a constantly increasing range of products, each promising specialized satisfaction, the shopper is forced to fragment needs into constantly smaller elements. These are not false needs, distinct from objectively determined "real" needs; rather they conflate material and symbolic aspects of "needing" in an ambiguous, unstable state. Because advertising has already identified particular emotional and social conditions with specific products, the continuous fracturing of emotions and artifacts forces consumers to engage in intensive efforts to bind together their identity and personal integrity. Consumption is the easiest way to accomplish this task and achieve at least temporary resolution.[16]

Similarly fragmented attributes make up the commodities themselves. These bundles of objective and imputed characteristics and signals are in constant flux, rendered even more unstable by the consumer's fluctuating desires. As Leiss observes, "the realm of needs becomes identical with the range of possible objects, while the nature of the object itself becomes largely a function of the psychological state of those who desire it."[17] The shopping mall prolongs this exchange by offering a plethora of possible purchases that continuously accelerate the creation of new bonds between object and consumer. By extending the period of "just looking," the imaginative prelude to buying, the mall encourages "cognitive acquisition" as shoppers mentally acquire commodities by familiarizing themselves with a commodity's actual and imagined qualities. Mentally "trying on" products teaches shoppers not only what they want and what they can buy, but also, more importantly, what they don't have, and what they therefore need. Armed with this knowledge, shoppers can not only realize what they are but also imagine what they might become. Identity is momentarily stabilized even while the image of a future identity begins to take shape, but the endless variation of objects means that satisfaction always remains just out of reach.[18]

The shopping-mall mix is calculated to organize the disorienting flux of attributes and needs into a recognizable hierarchy of shops defined by cost, status, and life-style images. These shops, in turn, reflect the specific consumption patterns of the mall's marketing area. Merchandise contextualized by price and image orients the shopper, allowing the speculative spiral of desire and deprivation to be interrupted by purchases. The necessity of this double action – stimulating nebulous desire and encouraging specific purchases – establishes the mall's fundamentally contradictory nature. To survive profitably, it must operate within the enormous disjuncture created between the objective economic logic necessary for the profitable circulation of goods and the unstable subjectivity of the messages exchanged between consumers and commodities, between the limited goods permitted by this logic and the unlimited desires released by this exchange.

The physical organization of the mall environment mirrors this disjuncture; this is one reason why conventional architectural criticism, a discourse based on visible demonstrations of order, has not been able to penetrate its system. All the familiar tricks of mall design – limited entrances, escalators placed only at the end of corridors, fountains and benches carefully positioned to entice shoppers into stores – control the flow of consumers through the numbingly repetitive corridors of shops. The orderly processions of goods along endless aisles continuously stimulates the desire to buy. At the same time, other architectural tricks seem to contradict commercial considerations. Dramatic atriums create huge floating spaces for contemplation, multiple levels provide infinite vistas from a variety of vantage points, and reflective surfaces bring near and far together. In the absence of sounds from outside, these artful visual effects are complemented by the "white noise" of Muzak and fountains echoing across enormous open courts. The resulting "weightless realm" receives substance only through the commodities it contains.[19]

These strategies are effective; almost every mall-goer has felt their power. For Joan Didion the mall is an addictive environmental drug, where "one moves for a while in an aqueous suspension, not only of light, but of judgment, not only of judgment, but of personality." In the film *Dawn of the Dead*, both zombies and their victims are drawn to the mall, strolling the aisles in numb fascination, with fixed stares that make it difficult to tell the shoppers from the living dead. William Kowinski identified *mal de mall* as a perceptual paradox brought on by simultaneous stimulation and sedation, characterized by disorientation, anxiety, and apathy. The jargon used by mall management demonstrates not only their awareness of these side-effects, but also their partial and imprecise attempts to capitalize on them. The Gruen Transfer (named after architect Victor Gruen) designates the moment when a "destination buyer," with a specific purchase in mind, is transformed into an impulse shopper, a crucial point immediately visible in the shift from a determined stride to an erratic and meandering gait. Yet shoppers do not perceive these effects as negative: the expansion of the typical mall visit from twenty minutes in 1960 to nearly three hours today testifies to their increasing desirability.[20]

RETAIL MAGIC

Malls have achieved their commercial success through a variety of strategies that all depend on "indirect commodification," a process by which nonsalable objects, activities, and images are purposely placed in the commodified world of the mall. The basic marketing principle is "adjacent attraction," where "the most dissimilar objects lend each other mutual support when they are placed next to each other."[21] Richard Sennett explains this effect as a temporary suspension of the use value of the object, its decontextualized state making it unexpected and therefore stimulating. Thus, placing an ordinary pot in a window display of a Moroccan harem transforms the pot into something exotic, mysterious, and desirable. This logic of association allows noncommodified values to enhance commodities, but it also imposes the reverse process – previously noncommodified entities become part of the marketplace. Once this exchange of attributes is absorbed into the already open-ended and indeterminate exchange between commodities and needs, associations can resonate infinitely.

At an early stage, malls began to introduce a variety of services, such as movies and restaurants, fast-food arcades, video-game rooms, and skating rinks, which, while still requiring expenditure, signaled the malls' expanded recreational role. As "mall time" has become an increasingly standard unit of measure, more and more promotional activities have appeared; first fashion shows and petting zoos, then symphony concerts (the Chicago Symphony performs regularly at Woodfield Mall), and even high-school proms. Hanging out at the mall has replaced cruising the strip; for teenagers, malls are now social centers, and many even find their first jobs there. Now malls have become social centers for adults as well. The Galleria in Houston has achieved a reputation as a safe and benevolent place for singles to meet, and "mall-walkers" – senior citizens and heart patients seeking a safe place to exercise – arrive at malls before the shops open, to walk a measured route around the corridors. Popular culture also attests to the incorporation of the mall into daily life. Recent films such as *Scenes from a Mall* and *Phantom of the Mall* suggest that virtually any cinematic genre can be successfully transposed to this familiar setting. *Beverly Center*, the first novel named for a shopping mall, recounts the torrid adventures of retail employees in a place "where everything is for sale and nothing comes cheap."[22] Proximity has established an inescapable behavioral link between human needs – for recreation, public life, and social interaction – and the commercial activities of the mall, between pleasure and profit in an enlarged version of "adjacent attraction." As developer Bill Dawson sums it up: "The more needs you fulfill, the longer people stay."[23]

Indirect commodification can also incorporate fantasy, juxtaposing shopping with an intense spectacle of accumulated images and themes that entertain and stimulate and in turn encourage more shopping. The themes of the spectacle owe much to Disneyland and television, the most familiar and effective commodifiers in American culture. Theme-park attractions are now commonplace in shopping malls; indeed, the two forms converge – malls routinely entertain, while theme parks function as disguised marketplaces. Both offer controlled and carefully packaged public spaces and pedestrian experiences to auto-dependent suburban families already primed for passive consumption by television – the other major cultural product of the fifties.

While enclosed shopping malls suspended space, time, and weather, Disneyland went one step further and suspended reality. Any geographic, cultural, or mythical location, whether supplied by fictional texts (Tom Sawyer's Island), historical locations (New Orleans Square), or futuristic projections (Space Mountain), could be reconfigured as a setting for entertainment. Shopping malls easily adapted this appropriation of "place" in the creation of a specialized theme environment. In Scottsdale, the Borgata, an open-air shopping mall set down in the flat Arizona desert, reinterprets the medieval Tuscan bill town of San Gimignano with piazza and scaled-down towers (made of real Italian bricks). In suburban Connecticut, Olde Mystick Village reproduces a New England Main Street, circa 1720, complete with shops in saltbox houses, a waterwheel, and a pond. Again, the implied connection between unexpected settings and familiar products reinvigorates the shopping experience.

The larger the mall, the more sophisticated the simulation. The West Edmonton Mall borrowed yet another design principle from Disneyland: the spatial compression of themes. To simultaneously view Main Street and an African jungle from Tomorrowland was a feat previously reserved for science fiction. By eliminating the unifying concept of "land" – Disneyland's main organizing principle – the WEM released a frenzy of free-floating images. If Disneyland's abrupt shifts of space and time suggested that to change realities could be as easy as changing channels on a television, the WEM, as one writer observed, was more like turning on all the channels at once.[24] Again, the principle of "adjacent attraction" ensures that these images will exchange attributes with the commodities in the mall. The barrage of diverse images, though, may heighten the unstable relationship of commodity and consumer needs to such a degree that the resulting disorientation leads to acute shopper paralysis. This discouraging prospect makes oases of relative calm, such as the water park and the hotel, necessary for recuperation. Even the all-inclusive mall must acknowledge perceptual limits.

The contrived packaging, obvious manipulation, and mass-market imagery of formula malls was not without critics, particularly among affluent and educated shoppers. To please this more demanding audience, developer James Rouse expanded the definition of "adjacent attraction" to incorporate genuinely historic and scenic places into the world of the mall. Rouse's successful packaging of "authenticity" made him a legend in development circles. "Festival marketplaces" such as Faneuil Hall in Boston, Harborplace in Baltimore, and South Street Seaport in Manhattan reject the architectural homogeneity of the generic mall in favor of the unique character of a single location enhanced through "individualized" design. These scenic and historic areas use cultural attractions such as museums and historic ships to enliven predictable shopping experiences. Festival marketplaces, then, reverse the strategy employed at the WEM – imagery is reduced and activities focus on a single theme rooted in a genuine context – but with comparable results, the creation of a profitable marketplace. Faneuil Hall attracts as many visitors each year as Disneyland, confirming Rouse's slogan: "Profit is the thing that hauls dreams into focus."[25]

PUBLIC LIFE IN A PLEASURE DOME

The shift from a market economy to a consumer culture based on intensified commodity circulation became apparent in the first mass-consumption environment, the Parisian department store, which, after 1850, radically transformed the city's commercial landscape. The enormous number of goods presented in a single location dazzled shoppers accustomed to small shops with limited stocks. By 1870, the largest of the *grands magasins*, the Bon Marché, offered a huge assortment of goods to ten thousand customers a day.[26] Moreover, the department store's fixed prices altered the social and psychological relations of the marketplace. The obligation to buy implied by the active exchange of bargaining was replaced by the invitation to look, turning the shopper into a passive spectator, an isolated individual, a face in the department-store crowd, silently contemplating merchandise. Richard Sennett observed that haggling had been "the most ordinary instance of everyday theater in the city," weaving the buyer and the seller together socially; but the fixed-price system "made passivity into a norm."[27]

Department stores gradually discovered the marketing strategies required by this new passivity and began to theatricalize the presentation of goods. Emile Zola modeled his *Au Bonheur des Dames* on the Bon Marché; it portrays the modern retail enterprise as hardheaded commercial planning aimed at inducing fascination and fantasy. Zola vividly describes the display practices that dazzled and intoxicated the mostly female customers: "Amidst a deep bed of velvet, all the velvets, black, white, colored, interwoven with silk or satin, formed with their shifting marks a motionless lake on which reflections of sky and landscape seemed to dance. Women, pale with desire, leaned over as if to see themselves." Another shopper is "seized by the passionate vitality animating the great nave that day. Mirrors everywhere extended the shop spaces, reflecting displays with corners of the public, faces the wrong way round, halves of shoulders and arms." Zola's retail pleasure dome alternates such disorienting perspectives with comfortable resting places, reading and writing rooms, and a free buffet, countering the escapist fantasy world with comfortable homelike spaces where shoppers could reacquire a sense of control.[28]

In fact, the shopper's dream world was always firmly anchored to highly structured economic relations. The constant and rapid turnover of goods demanded standardized methods of organization, subjecting employees to a factorylike order that extended beyond working hours into the carefully supervised dormitories and eating halls. A strict hierarchy separated the sales clerks, drilled in middle-class manners and housed in attic dormitories, from the proletariat that staffed the workshops and stables and slept wherever they could. Class boundaries also put limits on the "magic" of merchandising. For instance, stores like the Magasins Dufayel and Bazar de l'Hôtel de Ville, located closer to the proletarian northern and eastern suburbs, offered more straightforward selections of inexpensive goods to their working-class clientele.[29]

The possibilities of material abundance and mass consumption first suggested in Zola's department store also inspired a number of other nineteenth-century writers and thinkers. While the naturalist Zola called his novel "a poem of modern life," more speculative thinkers imagined ideal futures in which the problematic realm of production withered away completely, leaving consumption the dominant mode of experience. In America, Edward Bellamy's *Looking Backward* outlined a future in which reorganized production systems efficiently supplied necessities to the entire population, reducing the workday or eliminating the need for work altogether. In this labor-free world of material plenty, the idle masses could now devote themselves to the pursuit of self-realization and aesthetic pleasure as well as the idle rich. Other writers enlarged the miniature dream world of the department store into a full-scale Utopia. Inspired by temples of abundance such as Wanamaker's and Macy's, the novel *The World a Department Store*, written by Ohio department-store owner Bradford Peck, proposed an ideal state modeled after a department store that equitably supplied housing, food, and endless goods to its contented citizens.[30]

America after World War II seemed to promise the realization of many such dreams. The booming consumer economy offered a previously unimaginable prosperity, with full employment supplying consumers for the large-scale distribution of affordable goods, while advertising and planned obsolescence insured their continuous circulation. Standardized work weeks allowed free time for new leisure activities. Jobs and housing quickly migrated to the suburbs, propelled by Federal subsidies and guaranteed mortgage insurance, and highway programs initiated a cycle of growth by stimulating the automobile, oil, and construction industries. In the cities, even the poor had housing and money to spend. On the cities' edges, suburban growth produced an economic landscape of single-family tracts connected by superhighways and punctuated by shopping centers.

With suburbs and automobiles, downtown department stores were no longer relevant. Interstate highways and suburbs created the demand for commercial services in newly developed areas. City stores built suburban branches; roadside strips and strip centers (collections of stores with shared parking) grew up along major routes and at important intersections; and developers continued to construct tasteful shopping centers in upscale suburbs, following earlier prototypes such as the Roland Park Shop Center outside of Baltimore (1907) and Market Square in Lake Forest (1916). All of these forms provided convenient off-street parking.

J. C. Nichols, generally regarded as the father of the shopping center for his role in developing Country Club Plaza in Kansas City (1924), established many of the financial, management, and merchandising concepts that were fundamental to postwar shopping centers.[31] Nichols's 1945 Urban Land Institute publication, *Mistakes We Have Made in Developing Shopping Centers*, codified his experience into a list of 150 maxims, which covered everything from strategies to ensure local political support to adequate ceiling heights. Although Country Club Plaza's elaborate Mediterranean architecture – complete with tiled fountains and wrought-iron balconies – distinguished it from the bland exteriors of later centers, Nichols argued against any unnecessary expenditure on decor. The key to shopping-center success, he claimed, lay in providing abundant, even unlimited, parking. By 1950, as the varieties of neighborhood shopping centers merged into a single new form – the regional mall – Nichols's wisdom was confirmed.

After several false starts, the successful prototype of the classic dumbbell format finally emerged at Northgate in Seattle in 1947: two department stores anchoring the ends of an open-air pedestrian mall, set in the middle of acres of parking. Designed by John Graham, Jr., the innovative combination of easy automobile access and free parking with pedestrian shopping offered both suburban convenience and downtown selection. Graham's mall, a narrow pedestrian corridor modeled after a downtown street, efficiently funneled shoppers from one department store to the other, taking them past every store in the mall.[32] Similar multi-million-dollar malls multiplied, spurred on by the abundance of cornfield sites at agricultural-land prices and encouraged by Reilly's Law of Retail Gravitation, which posits that, all other factors being equal, shoppers will patronize the largest shopping center they can get to easily. This served as the rationale for ever-larger centers optimally located near the exits of new interstate highways.[33]

The consumers were ready, armed with postwar savings and the benefits of recent prosperity – vital necessities in the newly created world of the suburbs, where the new way of life depended on new ways of consuming. The ideal single-family home – inhabited by the ideal family, commuting father, housewife, and two children – demanded an enormous range of purchases: house, car, appliances, furniture, televisions, lawnmowers, and bicycles. The mass production of standardized products found a market of consumers primed by advertising television, and magazines. In a landscape of stratified subdivisions, status, family roles, and personal identity found further expression in consumption. Without familiar neighborhoods and extended-family networks to set social standards, suburban families used their possessions as a mark of belonging. The suburb itself was a product: nature and community packaged and sold.[34]

Initially, shopping-mall design reinforced the domestic values and physical order of suburbia. Like the suburban house, which rejected the sociability of front porches and sidewalks for private backyards, the malls looked inward, turning their back on the public street. Set in the middle of nowhere, these consumer landscapes reflected the profound distrust of the street as a public arena visible in the work of such dissimilar urbanists as Frank Lloyd Wright and Le Corbusier. Instead, streets, preferably high-speed highways, served exclusively as automobile connections between functionally differentiated zones and structures. Although mall apologists cited earlier marketplace types to establish the mall's legitimacy, they ignored their different consequences for urban life. While Islamic bazaars and Parisian arcades reinforced existing street patterns, malls – pedestrian islands in an asphalt sea – further ruptured an already fragmented urban landscape. As suburbs sprawled, so did their only public spaces; the low-rise, horizontal forms of suburban centers reversed the tightly vertical order of traditional urban space.[35] Informal open areas landscaped with brick flowerbeds and spindly trees echoed frontyard imagery. Malls, composed of rows of basic boxes enlivened with porchlike overhangs, shared the design logic of the suburban tract; economics rather than aesthetics prevailed.

In 1956, the first enclosed mall – Southdale, in Edina, a suburb of Minneapolis – changed all this. Although its central court surrounded by two levels of shopping floors was quickly surpassed by more extravagant developments, Southdale's breakthrough design firmly established Victor Gruen in the pantheon of mall pioneers. By enclosing the open spaces and controlling the temperature, Gruen created a completely introverted building type, which severed all perceptual connections with the mall's surroundings. Inside, the commercial potential of enormous spaces was realized in theatrical "sets" where "retail drama" could occur. Mall developers rediscovered the lesson of the Parisian department store and transformed focused indoor spaces into fantasy worlds of shopping. Southdale was covered for practical reasons; Minnesota weather allows for only 126 outdoor shopping days. The contrast between the freezing cold or blistering heat outdoors and the mall's constant 72 degrees was dramatized by the atrium centerpiece, the Garden Court of Perpetual Spring, filled with orchids, azaleas, magnolias, and palms. Exaggerating the differences between the world outside and the world inside established a basic mall trope: an inverted space whose forbidding exteriors hid paradisiacal interiors. This combination was compelling enough to ensure that enclosed malls soon flourished even in the most temperate climates.

Recreating a "second" nature was only the first step; the next was to reproduce the single element missing in suburbia – the city. The enclosed mall compressed and intensified space. Glass-enclosed elevators and zigzagging escalators added dynamic vertical and diagonal movement to the basic horizontal plan of the mall. Architects manipulated space and light to achieve the density and bustle of a city downtown – to create essentially a fantasy urbanism devoid of the city's negative aspects: weather, traffic, and poor people. The consolidation of space also altered the commercial identity of the mall. Originally built to provide convenient one-stop shopping, newly glamorized malls now replaced stores serving practical needs – supermarkets, drugstores, hardware stores – with specialty shops and fast-food arcades. Infinitely expandable suburban strips became the new loci for commercial functions expelled from the increasingly exclusive world of the shopping mall. Sealed off from the tasks of everyday life, shopping became a recreational activity and the mall an escapist cocoon.

As the mall incorporated more and more of the city inside its walls, the nascent conflict between private and public space became acute. Supreme Court decisions confirmed an Oregon mall's legal right to be defined as a private space, allowing bans on any activity the owners deemed detrimental to consumption. Justice Thurgood Marshall's dissenting opinion argued that since the mall had assumed the role of a traditional town square, as its sponsors continually boasted, it must also assume its public responsibilities: "For many Portland citizens, Lloyd Center will so completely satisfy their wants, that they will have no reason to go elsewhere for goods and services. If speech is to reach these people, it much reach them in Lloyd Center."[36] Many malls now clarify the extent of their public role by posting signs that read: "Areas in this mall used by the public are not public ways, but are for the use of the tenants and the public transacting business with them. Permission to use said areas may be revoked at any time," thus "protecting" their customers from potentially disturbing petitions or pickets. According to the manager of Greengate Mall in Pennsylvania, "We simply don't want anything to interfere with the shopper's freedom to not be bothered and have fun."[37]

Repackaging the city in a safe, clean, and controlled form gave the mall greater importance as a community and social center. The enclosed mall supplied spatial centrality, public focus, and human density – all the elements lacking in sprawling suburbs. The mall served as the hub of suburban public life, and provided a common consumer focus for the amorphous suburbs. In New Jersey – which had already spawned settlements such as Paramus, "the town Macy's built" – the importance of the Cherry Hill Mall as a focal point and an object of considerable local pride led the inhabitants of adjacent Delaware Township to change the name of their town to Cherry Hill. Reversing the centrifugal pattern of suburban growth, malls became magnets for concentrated development, attracting offices, high-rise apartments, and hospitals to their vicinity, thereby reproducing a central business district.

The financial success of the simulated downtown-in-the-suburbs also restimulated the actual downtowns, which had previously been weakened by regional malls. Newly placed urban malls brought their suburban "values" back into the city. In urban contexts the suburban mall's fortresslike structures literalized their meaning, privatizing and controlling functions and activities formerly enacted in public streets. Heavily patrolled malls now provide a safe urban space with a clientele as homogeneous as that of their suburban counterparts. In many cities, the construction of urban malls served to resegregate urban shopping areas. In Chicago, for example, white suburbanites coming into the city flocked to the new Marshall Field's branch inside the Water Tower Place mall on upper Michigan Avenue, effectively abandoning the original Marshall Field's department store in the downtown Loop to mostly black and Hispanic patrons.[38]

In more than one way, downtown malls cash in on the paradoxical prospect of a new order of urban experience, well protected from the dangerous and messy streets outside. Attempting a double simulation of New York, Herald Center, when it opened on 34th Street, offered thematized floors named for the city's familiar sites, such as Greenwich Village, Central Park, and Madison Avenue, which imitated their namesakes with businesses approximating their commercial character: sandal shops, sporting goods, and European boutiques. Not only were the actual places represented in name only, the "typical" goods for purchase reduced to caricature the rich mixtures of a real urban neighborhood. By reproducing the city inside its walls, the mall suggested that it was safer and cleaner to experience New York inside its climate-controlled spaces than on the real streets outside. This particular experiment failed, but did not discourage new efforts. On Times Square, a new mall project designed by Jon Jerde, Metropolis Times Square, tries to upstage the flash and dazzle of its setting with its own indoor light show, featuring hundreds of televisions, neon lights and laser projections. This hyper-real Times Square mall, sanitizing the sleaze and vulgarity outside, offers instead the tamer delights of shops, restaurants, and a cineplex open twenty-four hours a day.[39]

While the city began to incorporate suburban-style development, the suburbs became increasingly urban. Large numbers of jobs have moved to the suburbs, turning these areas into new metropolitan regions, "urban villages" or "suburban downtowns." Super-regional malls at freeway interchanges – such as the Galleria outside Houston, South Coast Plaza in Orange County, and Tyson's Corners near Washington, D.C. – became catalysts for new suburban minicities, attracting a constellation of typically urban functions. Their current importance represents the culmination of several decades of suburban growth. The evolution of the Galleria – Post Oaks suburb in Houston, for example, began in the late fifties with shopping centers built to serve affluent residential areas. The construction of the 610 Loop freeway encouraged retail expansion, notably the Galleria, one of the first spectacular multi-use malls, followed by office buildings, high-rise apartments and hotels, and finally corporate headquarters. White-collar and executive employees moved to nearby high-income residential neighborhoods, which generated the critical mass necessary to support restaurants, movie complexes, and cultural centers. The result now surpasses downtown Houston, containing the city's highest concentration of retail space, high-rise apartment units, and hotels as well as the state's third-highest concentration of office space. It is also Houston's most visited attraction.[40]

Although these businesses and residences are concentrated spatially, they maintain the low-density suburban building pattern of isolated single-function buildings. Parallel to the 610 Loop and along Post Oak Boulevard rise clusters of freestanding towers, including the sixty-four-story Transco Tower. Each building stands alone, though, insulated by landscaping, parking, and roads. Sidewalks are rare, making each structure an enclave, accessible only by automobile. In this atomized landscape, the Galleria, pulsing with human activity, has expanded its role as town center even further, providing not only food, shopping, and recreation, but also urban experience. For many suburban inhabitants, the Galleria is the desirable alternative to the socially and economically troubled urban downtowns they fled. President Bush, casting his vote in the presidential election at the Galleria, symbolically verified the mall's status as the heart of the new suburban downtown.[41]

HYPERCONSUMPTION: SPECIALIZATION AND PROLIFERATION

Throughout the period of shopping-mall expansion, economic and social changes were significantly altering the character of the consumer market. After 1970, it became evident that the postwar economic and social system of mass production and consumption was breaking down, fragmenting income, employment, and spending patterns into a much more complex mosaic. More flexible types of production appeared, emphasizing rapid cycles of products that quickly responded to the consumer market's constantly changing needs and tastes. Restructured industries and markets in turn produced a differentiated and fragmented labor force. The pyramid model of income distribution that supported the regional mall was being replaced by a configuration more like a bottom-heavy hourglass, with a small group of very high incomes at the top, and the middle disappearing into a much larger group of low incomes. This picture was further complicated by an increasingly uneven geographic distribution of economic development, which produced equally exaggerated differences between zones of prosperity and poverty.

In this unstable situation, the continued development of existing mall types was no longer assured. Heightened competition – between corporations, entrepreneurs, and even urban regions – forced a series of shakedowns in the industry. Although the system of regional malls continued to flourish, it was clear that the generic-formula mix no longer guaranteed profits.[42] (Industry experts agree that there are few regional holes left to fill, although the system can still absorb three or even four more megamalls.) Instead, malls expanded by multiplying and diversifying into as many different fragments as the market. An enormous range of more specialized and flexible mall types appeared, focused on specific niches in the newly dispersed market. Such specialization permitted more coherent matching of consumer desires and commodity attributes at a single location, making consumption more efficient, while greater diversity allowed a much greater collection of commodities to be merchandised than ever before.

Specialization occurs across a wide economic spectrum. In the richest markets, luxury malls, like Trump Tower on Fifth Avenue or the Rodeo Collection in Beverly Hills, offer expensive specialty goods in sumptuous settings, more like luxurious hotels than shopping malls. At the other end of the market, outlet malls sell slightly damaged or out-of-date goods at discount prices; since low cost is the major attraction, undecorated, low-rent buildings only enhance their utilitarian atmosphere. New smaller malls eliminate social and public functions to allow more efficient shopping. Strip malls, with parking in front, are the most flexible type: their false fronts can assume any identity, their format can be adjusted to any site, and they can contain any mix of products. Some strip malls focus on specific products or services – furniture, automotive supplies, printing and graphic design, or even contemporary art. In Los Angeles, more than three thousand minimalls (fewer than ten stores) supply the daily needs of busy consumers with convenience markets, dry cleaners, video stores, and fast-food outlets.[43]

In this overcrowded marketplace, imagery has become increasingly critical as a way of attracting particular shops and facilitating acts of consumption. Through a selective manipulation of images, malls express a broad variety of messages about the world both outside and inside the mall. Large, diverse cities like Los Angeles offer veritable encyclopedias of specialized mall types that cater to recent immigrant groups. Here the images retain a vestige of their cultural heritage: Korean malls have blue-tile temple roofs, Japanese malls combine Zen gardens with slick modernism to attract both local residents and touring

Japanese. Minimall developers in Los Angeles also style their malls according to location: postmodern on the affluent Westside, high-tech in dense urban areas, and Spanish in the rest of the city.

Such imagery treads a thin line between invitation and exclusion. But if mall decor and design are not explicit enough to tell young blacks or the homeless that they are not welcome, more literal warnings can be issued. Since statistics show that shopping-mall crimes, from shoplifting and purse-snatching to car theft and kidnapping, have measurably increased, the assurance of safety implied by the mall's sealed space is no longer adequate.[44] At the WEM, the mall's security headquarters, Central Dispatch, is prominently showcased. Behind a glass wall, a high-tech command post lined with banks of closed-circuit televisions and computers is constantly monitored by uniformed members of the mall's security force. This electronic Panopticon surveys every corner of the mall, making patrons aware of its omnipresence and theatricalizing routine security activities into a spectacle of reassurance and deterrence. But the ambiguous attractions of a lively street life, although excluded from the WEM by a strictly enforced code of behavior, are not wholly absent. Rather, they are vicariously acknowledged, at a nostalgic distance to be sure, by Bourbon Street's collection of mannequins, "depicting the street people of New Orleans." Frozen in permanent poses of abandon, drunks, prostitutes, and panhandlers act out transgressions forbidden in the mall's simulated city.

Malls have not only responded to changing market conditions, but have also become trump cards in the increasing competition between developing cities and regions. The enormous success of projects like Faneuil Hall and the WEM have brought in revenue and attracted jobs, residents, and visitors to the cities. In a large-scale version of adjacent attraction, malls lend glamour and success to their urban setting, suggesting that the city is important, exciting, and prosperous.[45] Even if the WEM weakened the commercial power of Edmonton's downtown, as a whole it added luster and money to the urban region overall. Recognizing these potential benefits, cities now court developers with a range of financial incentives, from tax breaks to significant investments, in order to attract major mall projects. Faneuil Hall's success in generating adjacent development, such as condominiums, shops, and offices, led cities from Toledo to Norfolk into private-public ventures with the Rouse Company to build waterfront centers as catalysts for urban revitalization. This strategy can also backfire: Horton Plaza, San Diego's spectacular, enormously profitable, and heavily subsidized "urban theme park" mall has remained a self-contained environment, a city in itself – with little effect on its seedy surroundings.[46]

In Europe, political participation has gone even further. Municipal governments with extensive planning powers have taken over the developer's role themselves, though state sponsorship has produced no changes in the mall form. Thus, the Greater London Council developed festival marketplaces in Covent Gardens and the St. Katharine Docks as the commercial beachheads for larger urban redevelopment schemes. Built over the opposition of residents who demanded local services, the Covent Garden project produces municipal revenue by duplicating commercial formulas that attract tourists and impulse buyers. In Paris, the lengthy political battle over what would replace the razed Les Halles market was resolved by the decision to build a multi-level shopping center clad in the slick architectural modernism that the French state has adopted as its distinctive design image. This mall was the first step in the reorganization of the entire district and now stands at the center of a regional transport network, connected to a sequence of public sport, leisure, and cultural facilities through underground corridors.[47]

THE WORLD AS A SHOPPING MALL

The spread of malls around the world has accustomed large numbers of people to behavior patterns that inextricably link shopping with diversion and pleasure. The transformation of shopping into an experience that can occur in any setting has led to the next stage in mall development: "spontaneous malling," a process by which urban spaces are transformed into malls without new buildings or developers. As early as 1946, architects Ketchum, Gina, and Sharp proposed restructuring Main Street in Rye, New York, as a pedestrian shopping mall; later Victor Gruen planned to turn downtown Fort Worth into an enclosed mall surrounded by sixty thousand parking spaces. More recently, a number of cities have reconstituted certain areas as malls simply by designating them as pedestrian zones, which allows the development of concentrated shopping. Self-regulating real-estate values allow these new marketplaces to create their own tenant mix, organized around a unifying theme; this, in turn, attracts supporting activities such as restaurants and cafes. In Los Angeles, even without removing automobiles, urban streets like Melrose Avenue and Rodeo Drive have spontaneously regenerated themselves as specialty malls, thematically based on new-wave and European chic.

Different stimuli can initiate this process. The construction of a regional mall in a rural area of DuPage county, outside Chicago, completely transformed commercial activities in the area. Afraid of losing shoppers to the mall, local merchants in Naperville, an old railroad town, moved to transform its main street into a gentrified shopping area of antique shops and upscale boutiques. By emphasizing Naperville's historical small-town character, providing off-street parking, and offering specialized shops not available in malls, Naperville developed a commercial identity that allowed it to coexist harmoniously with the mall.[48] When its historical center was inundated by tourists, Florence turned the Via Calzaioli between the Duomo and the Piazza Signoria into a pedestrian zone, which soon resembled an outdoor Renaissanceland mall with the two monuments serving as authentic cultural anchors. Shoe and leather shops, fast-food restaurants, and the inevitable Benetton outlets – offering merchandise available at malls all over the world – took over from older stores, as tourists outnumbered local residents. In France, state policies to ensure the preservation of historical centers awarded large subsidies to small cities like Rouen, Grenoble, and Strasbourg. This unintentionally redefined commerce: as pedestrian zones brought more shoppers into the center and greater profits attracted national chains of luxury shops, stores for everyday needs disappeared, replaced by boutiques selling designer clothing, jewelry, and gifts.[49]

Clearly, the mall has transcended its shopping-center origins. Today, hotels, office buildings, cultural centers, and museums virtually duplicate the layouts and formats of shopping malls. A walk through the new additions to the Metropolitan Museum in New York with their enormous internal spaces, scenographic presentation of art objects, and frequent opportunities for purchasing other objects connected to them, produces an experience very similar to that of strolling through a shopping mall. The East Wing of the National Gallery of Art in Washington, D.C., designed by I. M. Pei, is an even closer match. The huge skylighted atrium is surrounded by promenades connected by bridges and escalators; individual galleries open off this space, placed exactly where the shops would be in a mall. Potted plants, lavish use of marble and brass, and, in the neon-lit basement concourse, fountains, shops, and fastfood counters make the resemblance even more striking.[50]

Indeed, as one observer has suggested, the entire Capitol Mall has been malled. A hodgepodge of outdoor displays, a giant dinosaur, a working 1890s carousel, the gothic fantasy of Smithson's sandstone castle, and NASA rockets hint at the range of time and space explored in the surrounding museums. Here, old-fashioned methods of systematically ordering and identifying artifacts have given way to displays intended for immediate sensory impact. Giant collages include authentically historical objects like *The Spirit of St. Louis*, supported by simulated backgrounds and sounds that recall Lindbergh's famous flight. In the Air and Space Museum, airplanes, rockets, and space capsules are suspended inside a huge central court, slick graphics direct visitors to the omni-max theater, and gift shops offer smaller replicas of the artifacts on display.[51] The barrage of images, the dazed crowds, are all too familiar; the museum could easily be mistaken for the WEM. The Museum of Science and Industry in Chicago presents a similar spectacle. Mannequins in glass cases reenact significant moments in the history of science; visitors line up to tour the full-size coal mine; families sample ice cream in the nostalgic ambience of Yesterday's Main Street, complete with cobblestones and gaslights. In the museum shops, posters and T-shirts serve as consumable surrogates for artifacts that stimulate the appetite but cannot themselves be purchased.

If commodities no longer dominate, this is because the salable product no longer carries the same importance, since history, technology, and art, as presented in the museums, have now become commodified. The principle of adjacent attraction is now operating at a societal level, imposing an exchange of attributes between the museum and the shopping mall, between commerce and culture. Even the Association of Museum Trustees, by meeting at Disney World to discuss new research-and-development strategies, acknowledges this new reality. The world of the shopping mall – respecting no boundaries, no longer limited even by the imperative of consumption[52] has become the world.

NOTES

1 Gordon M. Henry, "Welcome to the Pleasure Dome," *Time*, Oct. 27, 1986, p. 60. Other descriptions of the WEM include William S. Kowinski, "Endless Summer at the World's Biggest Shopping Wonderland," *Smithsonian*, Dec. 1986, pp. 35–41; Ian Pearson, "Shop Till You Drop," *Saturday Night*, May 1986, pp. 48–56. A more scholarly approach is offered by R. Shields, "Social Spatialization and the Built Environment: The West Edmonton Mall," *Environment and Planning D: Society and Space*, vol. 7, 1989, pp. 147–64.

2 Leonard Zehr, "Shopping and Show Biz Blend in Giant Center at Edmonton, Alberta," *Wall Street Journal*, Oct. 7, 1985.

3 Mary Ann Galante, "Mixing Marts and Theme Parks," *Los Angeles Times*, June 14, 1989.

4 *Newsweek*, June 19, 1989, p. 36.

5 Margaret Crawford, "I've Seen the Future and It's Fake," *L.A. Architect*, Nov. 1988, pp. 6–7.

6 N. R. Kleinfeld, "Why Everyone Goes to the Mall," *New York Times*, Dec. 21, 1986.

7 Central-place theory, developed by geographer Walter Christaller and economist August Losch, provides a hierarchical structure of market areas according to scale economies, transport costs, and number of household units. See Walter Christaller, *Central Places in Southern Germany* (Englewood Cliffs, N.J.: Prentice-Hall, 1966).

8 John Dawson and J. Dennis Lord, *Shopping Centre Development: Policies and Prospects* (Beckenham, Kent: Croom Helm, 1983), p. 123.

9 Peter Muller, *Contemporary Suburban America* (Englewood Cliffs, N.J.: Prentice-Hall, 1981), pp. 123–30.

10 "Why Shopping Centers Rode Out the Storm," *Forbes*, June 1, 1976, p. 35.

11 Interview with Linda Congleton (president of Linda Congleton and Associates: Market Research for Real Estate, Irvine, Calif.), who uses both of these systems in market analysis for mall development.

12 Kay Miller, "Southdale's Perpetual Spring," *Minneapolis Star and Tribune Sunday Magazine*, Sept. 28. 1986; Kleinfeld, "Why Everyone Goes."

13 William S. Kowinski, *The Malling of America* (New York: William Morrow, 1985), p. 218.

14 "Metropolitan Roundup," *New York Times*, Jan. 29, 1984.

15 William Leiss, *The Limits to Satisfaction* (Toronto: University of Toronto Press, 1976), p. 4; Lewis Mandell et al., *Surveys of Consumers 1971–72* (Ann Arbor: Institute for Social Research, University of Michigan, 1973), pp. 253–62, 274–75.

16 Leiss, *Limits*, pp. 19, 61.

17 Ibid., p. 92.

18 Rachel Bowlby, *Just Looking* (New York: Methuen 1985), pp. 1–30.

19 T. J. Jackson Lears, in *No Place of Grace* (New York: Pantheon, 1981) has provided a detailed exploration of the characteristic "weightlessness" of Victorian culture, closely linked to the penetration of market values into the educated middle class.

20 Joan Didion, "On the Mall," *The White Notebook* (New York: Simon and Schuster, 1979). p. 183; Kowinski, *Malling*, pp. 339–42.

21 Richard Sennett, *The Fall of Public Man* (New York: Vintage, 1976), pp. 144–45.

22 Muller, *Contemporary*, p. 92; Ryan Woodward, *Beverly Center* (New York: Leisure, 1985).

23 Galante, "Mixing Marts."

24 Ian Brown in the *Toronto Globe and Mail*, quoted in Kowinski, "Endless Summer," p. 41.

25 Michael Demarest, "He Digs Downtowns," *Time*, Aug. 24, 1981, p. 46.

26 Richard Cobb, "The Great Bourgeois Bargain," *New York Review of Books*, July 16, 1981, pp. 35–40.

27 Sennett, *Fall of Public Man*, p. 142.

28 Emile Zola, *Au Bonheur des Dames* (Paris, 1897); Rosalind Williams, *Dream Worlds* (Berkeley: California, 1982); Bowlby, *Just Looking*.

29 Cobb, "Bourgeois Bargain," p. 38; Meredith Clausen, "The Department Store – Development of the Type," *Journal of Architectural Education*, vol. 39, no. 1. (Fall, 1985), pp. 20–27.

30 Edward Bellamy, *Looking Backward* (New York: Penguin, 1982; reprint of 1888 original); Bradford Peck, *The World a Department Store* (Boston, 1900).

31 Didion, "On the Mall," p. 34.

32 Meredith Clausen, "Northgate Regional Shopping Center – Paradigm from the Provinces," *Journal of the Society of Architectural Historians*, vol. 43, no. 2 (May 1984), p. 160.

33 Urban Land Institute, *Shopping Center Development Handbook* (Washington D.C.: Urban Land Institute, 1977), pp. 29–31.

34 David Harvey, *The Urbanization of Capital* (Baltimore: Johns Hopkins, 1985), p. 128.

35 Clausen, "Northgate," p. 157.

36 Kowinski, *Malling*, p. 356.

37 "Shopping Centers," *Dollars and Sense*, July–Aug. 1978, p. 9.

38 Muller, *Suburban America*, p. 128.

39 Paolo Riani, "Metropolis Times Square," *L'Arca*, vol. 29 (July 1989), pp. 43–49.

40 E. B. Wallace "Houston's Clusters and the Texas Urban Agenda," *Texas Architect*, Sept.–Oct. 1984, p. 4.

41 Graham Shane, "The Architecture of the Street" (unpublished manuscript), chapt. 5, p. 10.

42 Mark McCAin, "After the Boom, Vacant Stores and Slow Sales," *New York Times*, June 5, 1988.

43 Richard Nordwind, "Cornering L.A.'s Markets," *Los Angeles Herald-Examiner*, June 28, 1987.

44 Linda Weber, "Protect Yourself from Shopping Mall Crime," *Good Housekeeping*, Mar. 1988, pp. 191–92.

45 Harvey, *Urbanization*, p. 68.

46 David Meyers, "Horton Plaza's Sales Booming," *Los Angeles Times*, Oct. 4, 1987.

47 Michele Beher and Manuelle Salama, *New/Nouvelle Architecture* (Paris: Editions Regirex-France, 1988), p. 44.

48 Robert Brueggmann, Suburban Downtowns Tour, Society of Architectural Historians Annual Meeting, Chicago, April 1988.

49 Ivon Forest, "Les Impacts des Aménagements pour Piétons dans les Centres Anciens des Villes en France" (M.A. thesis, Laval University, Quebec, 1982), pp. 47–57.

50 I am indebted to the unknown author of the manuscript "The Malling of the Mall" for many insights into the relationship between the museum and the mall. After coming across this manuscript a number of years ago. I have attempted, without success, to track down its author, to whom I would like to express my gratitude for not only stimulating my own thinking on the subject but also for providing useful concepts and examples.

51 "The Malling of the Mall."

52 Jean Baudrillard, in *For a Critique of the Political Economy of the Sign* (St. Louis: Telos, 1981) and subsequent writings, argues that this separation from the economic base has, indeed, occurred.

'Justice, Politics, and the Creation of Urban Space'

from Andrew Merrifield and Erik Swyngedouw (eds), *The Urbanization of Injustice* (1996)

Susan Fainstein

Editors' Introduction

Susan Fainstein is Professor of Urban Planning and Policy Development at Rutgers University, New Jersey. She has published a number of influential books, chapters and papers in the areas of international urban public policy, urban social movements, urban redevelopment and planning theory.

Amongst her most significant publications have been *Urban Political Movements* (1974), *Restructuring the City* (1986), the co-edited volume *Divided Cities: New York and London in the Contemporary World* (1992) and *The City Builders: Property, Politics, and Planning in London and New York* (1994).

Fainstein's work has consistently been at the forefront of debates about issues such as urban restructuring, globalization, urban planning, inequality and social justice. Diverse as it is in its choice of subject, her work is typically characterised by a concern to bring and evaluate theoretical perspectives to issues that have previously been under-theorised or only poorly theorised. This was most apparent in her work on urban social movements and community groups and is demonstrated in the extract here from 'Justice, Politics and the Creation of Urban Space' from Andrew Merrifield and Erik Swyngedouw's edited collection *The Urbanization of Injustice* (1996).

The 1970s marked the beginning of a new epoch in discussions of urban space. Although the project of developing theory and empirical research has advanced enormously since the initial writings of that decade, the themes developed then persevere today. In this chapter I use the early works of David Harvey, Manuel Castells, Richard Sennett, and Herbert Gans as a scaffolding for analysing how the issue of social justice has been treated in subsequent years. For the purposes of my argument I classify their approaches into a tripartite typology comprising: (1) political economy; (2) post-structuralism; and (3) urban populism.[1] These three tendencies, while all disparaging of capitalist domination, incorporate understandings of spatial relations and the built environment that differ in their visions of a socially just city and in their proposed strategies for achieving social transformation. Although all three types are built on a vision of social justice, they are vague about their normative frameworks, taking a critical stance without specifying clearly the standards by which they are evaluating the objects of their analyses. Thus, in my description of their definitions of social justice, I am imputing to them positions that are not fully spelled out or defended.[2]

The three perspectives attack the status quo of the capitalist city and seemingly occupy positions on the political left. Nevertheless, assessing their political content is complicated. Only the first is straightforward in giving primary emphasis to the causal connection between economic form, urban development, and social injustice. Consequently it remains alone in taking a traditional left stance derived from the socialist conflation of injustice with economic exploitation and justice with economic equality. Post-structuralism and urban populism are harder to pinpoint; the former, in its celebration of difference, can fade into cultural chauvinism even though its foundation lies in egalitarianism, while the latter often veers into right-wing defence of homeowner privilege and communal exclusionism despite its espousal of the democratic ethos. Indeed, the attractiveness of right-wing versions of identity politics and participatory democracy, which easily become incorporated into post-structuralist and populist political agendas, has constituted a major strategic difficulty for the left ever since the extension of the suffrage. For while the left claims to speak for the mass of people, it lacks the symbolic and material mass appeals of cultural identification and possessive individualism available to the right.

This chapter examines the critique, logic, aims, and weaknesses of each of the three bodies of thought, relying on analyses of a few thinkers to characterize the elements of each type. It identifies the greatest difficulty for the analyses presented in the tension between norms of equality, diversity, and democracy. It then inquires whether a political-economic approach can both maintain a coherent concept of social justice and develop an urban programme with majoritarian appeal. The thought of Karl Mannheim provides the basis for contending that an abstract formulation of urban social justice is possible; Mannheim's argument concerning the possibility of simultaneously holding a categorical ethic and recognizing the partiality embedded in any historical situation is used to justify the endeavour. Finally, the chapter provides examples of types of policies that could advance such an ethic within the present moment, but also recognizes that these policies may sacrifice the interests of some of society's most vulnerable and thus not wholly conform to the ethic proposed.

POLITICAL ECONOMY ANALYSES

Urban political economy encompasses a broad spectrum of viewpoints, ranging from explicitly Marxist to market-oriented approaches. For the purposes of my discussion, I restrict the definition of political-economic analysis to efforts at understanding urban development that start their explanations with economic processes and which criticize capitalist outcomes primarily on the basis of their impacts on the welfare of relatively deprived groups.[3] Political economists vary considerably according to the weight they give to purely economic causes, the extent to which they attribute autonomy to the state and to culture, and the potential that they see for reform within capitalism. Their frameworks need not be, and increasingly are not, Marxist, but there is a strong Marxist influence. Differing from liberal-pluralist analysts, they are sceptical concerning the potential for overcoming serious economic deprivation under capitalism and do not envision the possibility of full human development without the initial satisfaction of material needs.

David Harvey's *Social Justice and the City*[4] and Manuel Castells' *The Urban Question*[5] comprise the foundational works in the political economy tradition of urban analysis. Whatever the differences between Castells and Harvey and the divisions among the many scholars who subsequently elaborated and disputed their initial theses, they and those that followed them in the political-economy vein begin with economic relations and evaluate urban life in terms of a transcendent goal of economic equality.[6] Harvey's approach is to examine the history of urban development, essentially to demonstrate how the inequalities of the capitalist labour process play themselves out in spatial terms and then how the ensuing social space itself exacerbates inequality and exploitation:

> An increase in the total quantity of social surplus produced has historically been associated with the activity of urbanization. . . . Urban centres have frequently been 'generative' but the need to accomplish primitive accumulation [defined by Harvey as the exploitation of a section of the population in order to gain a surplus product to invest in enlarged reproduction] militates against the process being naturally and reciprocally beneficial as both Adam Smith and Jane Jacobs envision it, for the processes of primitive accumulation are, in Marx's words, 'anything but idyllic'.[7]

Further, in his emphasis on the capitalist property market, Harvey shows how spatial arrangements themselves are used to enhance the profitability of property capital at the expense of urban residents. Echoing Engels, he argues that governmental efforts at urban improvement, whether embodied in Haussmann's reconstruction of Paris or contemporary urban renewal schemes, inevitably recapitulate the miseries of the earlier situation, moving impoverished people out of one slum into another. The crux of

Harvey's argument in this book is that, within a capitalist economic system, there is no escape from extremes of inequality and that the built environment must both contribute to and embody the capitalist dynamic, regardless of the programmes of even well-meaning policy-makers.[8]

Castells moves away from Harvey's preoccupation with production and circulation. His emphasis is on collective consumption and the rise of urban social movements that aim at capturing control of the local state and at redistributing the social wage. He roots urban crisis in the clash between the accumulation and legitimation needs of capital, and in a prescient prophecy concerning the future of US cities, predicts 'a new and sinister urban form: the Wild City':

> What could emerge from the current urban crisis is a simplified and heightened version of the exploitative metropolitan model with the addition of massive police repression and control exercised in a rapidly deteriorating economic setting. The suburbs will remain fragmented and isolated, the single-family homes closed off, the families keeping to themselves, the shopping centres more expensive, the highways less well-maintained but people forced to drive further to reach jobs and to obtain services, the central districts still crowded during the office hours but deserted and curfewed after 5 p.m., the city services increasingly crumbling, the public facilities less and less public, the surplus population more and more visible, the drug culture and individual violence necessarily expanding, gang society and high society ruling the bottom and the top in order to keep a 'top and bottom' social order intact, urban movements repressed and discouraged and the urban planners eventually attending more international conferences in the outer, safer world.[9]

Even before his later break with Marxist structuralism, Castells focuses on battles over distribution, on the factors that give rise to social movements, and on the potential for such movements both to encompass coalitions that straddle class divides and to achieve social progress. While alleging that urban conflict is a displacement of workplace-generated antagonisms, he nevertheless concerns himself less with the logic of capital accumulation and pays more attention to overt political conflict than Harvey.

Despite their structuralist analysis, the action agenda of the political economists, as applied to the urban milieu, largely falls well short of calls for full-fledged, class-based revolutionary activity. Defending what Lefebvre termed 'the right to the city',[10] political economists attack schemes that enhance capital accumulation to the detriment of ordinary residents. With varying hopefulness concerning the possibilities for meaningful social action, they endorse community movements that oppose large-scale economic development projects and gentrification. They support construction of housing for low-income people, especially in the form of co-ops or public ownership, criticize 'right to buy' schemes that withdraw housing from the low-income rental stock, favour rent control, and stand up for the rights of tenants, squatters, and homeless people to housing security, even when their interests come into conflict with those of working-class neighbours seeking to protect their own interests in neighbourhood stability.

Political economists focus on substance rather than process; they evaluate the outcomes of actions as they affect social groupings and judge the process that produces these results primarily in terms of its contribution to equality. Nonetheless, even though the form and extent of this allegiance are problematic, the political economy tradition reveals a general commitment to participatory democracy for two reasons. First, programmes generated by the mass of people seem to comprise the only legitimate alternative to elite-dominated decision-making. Second, an analysis that perceives of society as dominated by a small class of the economically dominant leads logically to a conclusion that rule by the relatively deprived majority would result in the defeat of economic privilege.

Political economists have thus rhetorically supported participatory democracy. Each conceivable mode of aggregating popular interests, however, raises difficult questions and has been criticized by them. For example, Katznelson points to the parochialism of neighborhood groups and the inability of community interests to affect the important decisions that determine social outcomes.[11] Castells, despite his dependence on urban social movements as the mechanisms for 'changing the urban meaning', comments that they primarily play a defensive role and are 'not agents of structural social change'.[12] His disgust with the Communist Party prevents him from regarding party structures as more effective vehicles for social transformation. Therefore, he is forced to rely on 'core' – i.e. non-territorial – political movements like environmentalism and feminism as the best alternative even in the face of their evident weaknesses as methods of addressing economic forces.

Harvey finds at most only a glimmer of hope in the varying forms of resistance to capitalist domination that have been stimulated by urbanization:

> Can a co-ordinated attack against the power of capital be mounted out of the individualism of money, the more radical conceptions of community, the progressive elements of new family structures and gender relations, and the contested but potentially fruitful legitimacy of state power, all in alliance with the class resentments that derive from the conditions of labour and the buying and selling of labour power? The analysis of the conditions that define the urbanization of consciousness suggests that it will take the power of some such alliance to mount a real challenge to the power of capital. *But there is no natural basis of such an alliance and much to divide the potential participants.*[13]

Thus, the abstract commitment to mass equality and democracy that animates the political economy mode of analysis finds no expression in an agreed-upon mode of popular representation within a divided capitalist society.

The most obvious deficiency of the political economy approach is also its greatest strength – its starting point in the economic base of cities. By identifying the economic logic of capitalist urbanization, political economy delineates – I think correctly – the limits of reform and the recurring processes that continuously generate uneven economic development, subordination, and insecurity. But this privileging of the economic in the chain of causal explanation leads to an often mechanical calculation of real interests, as well as a denial of the validity of the subjective perceptions that drive human behaviour.[14] The tendency among political economists, as among economists of other ideological hues, is to equate economic motives with the rational and all other impetuses with the irrational. Thus, somewhat analogously to the way in which psychologists interpret individual aggressive behaviour as projections of underlying weakness, political economists have tended to ignore oppressive behaviour by working-class strata toward other non-elite groups, apparently seeing them as simple displacements of economic insecurity. The assumption is that economic competition is the only contest that really matters, and if it were eliminated other means for asserting superiority would lose their hold.

There are three serious problems with the logic that economic inequality subsumes all forms of subordination. First, both theory and empirical evidence point to the contrary argument. Thus, as Simmel argues, even after the introduction of socialism, individuals would continue to express 'their utterly inevitable passions of greed and envy, of domination and feeling of oppression, on the slight differences in social position that have remained. . . .'[15] Group antagonisms would likely also endure. Recollections of persecution of one group by another or feelings of group superiority based on colour, nationality, or religion will not go away simply because of economic equality. Socialism as it really existed demonstrated that abolition of private property does not dissolve ethnic and gender antagonism and may even increase the importance of symbolic differences.

People's interests are not defined only by their economic position; moreover, other status determinants interact with economic interests, and causality goes in both directions. Max Weber refers to ideal interests, meaning interests in establishing one's own values. Group affiliations involve such ideal aims, but in addition status group membership and gender identification are irretrievably intertwined with real material interests. Networks of influence, based on ethnicity, lineage, gender, or some other 'traditional' relationship combine with the relations of production to generate structures of domination regardless of the mode of property ownership. Marx regarded clientelism and patronage as remnants of traditional societies that would be eradicated by capitalism, while most contemporary political economists simply ignore the question of the non-economic bases of economic power.[16] These forces, however, as well as ordinary corruption endure tenaciously within both socialist and capitalist societies and seem more dependent on culture and political process than on economic system. Their persistence corroborates the correctness of the common view that having access to people in power improves one's chances for material benefits.

A feminist perspective further points to the way in which factors separable from economic relations contribute to subordination both within and outside the economy. Feminists correctly observe that the oppression of women has existed under all economic systems, and the experience of the former Soviet bloc and Communist China shows that full participation by women in production may only partially improve their situation within the household and in a number of respects may cause it to deteriorate.[17] In fact, to the extent that women have improved their economic and social position, this amelioration has largely occurred under bourgeois capitalism.

A second problem of the political economy approach is its silence concerning the need to maintain social order under any economic system.[18] Even among those political economists who have abandoned Marxist structuralism, little mention is made of problems of social hostility except as a product of the capitalist economic system, and the socialist penchant for tracing all forms of domination to economic inequality remains. Whereas liberal political theory has sought ways by which people with differing interests or lifestyles can remain dissimilar and live peacefully together, socialist thought has typically aimed at dissolving differences and thus has not been concerned with the problems of governing antagonistic groups.[19] Liberal thinkers are always vulnerable to the argument that their institutions and procedures function to perpetuate and disguise inequality. But a demonstration that the adjudication systems of liberal democracies are biased against the poor and minorities does not mean that greater equity would result from their dissolution. Harvey does finally recognize this issue when he declares that 'a just planning and policy practice must seek out non-exclusionary and non-militarized forms of social control to contain the increasing levels of both personal and institutionalized violence without destroying capacities for empowerment and self-expression.'[20] Harvey's simple statement of the problem, however, does not suggest a solution.

Third, an economic programme with redistribution as its central goal lacks strong appeal to popular majorities in societies where the majority feels that pursuit of such a policy would cause it to lose many of its advantages or would threaten aggregate economic prosperity. In particular, it alienates substantial portions of the stable working class, who see their security in homeownership and in separating themselves from the social strata beneath them. Moreover, solidarities based on non-economic communities have more emotional pull and indeed may provide many initially lower-income individuals with greater material gains than structural economic change that dampens class privilege.

The flaws of political economic analysis do not lie primarily in its explanation of urban phenomena as the product of capitalist economic forces. Rather, they stem first from its assurance that the problems it identifies are those that matter (or should matter) most to people; and second, from its political strategy which favours strong democracy but simultaneously rejects the choices of conservative majorities. The question of whether it is possible to retain a political economy approach and accommodate the insights of other traditions will be discussed again later.

THE POST-STRUCTURALIST VIEWPOINT

Post-structuralism is a fuzzy term encompassing a variety of formulations that emphasize contingency over structure in explaining outcomes and that therefore eschew reductionist explanations.[21] The urban post-structuralists emulate the post-modern literary critics in their use of the techniques of cultural criticism, mapping the ways in which spatial relations represent modes of domination, searching out the 'silences' and exclusions in the practices of planners and developers. In their examinations of the symbolic statements expressed by the built environment, the post-structuralists seek to show how urban form functions to manipulate consciousness.

Foucault is frequently identified as the progenitor of this theoretical tendency, but his work evades simple categorization.[22] Contemporary urban critics writing within this tradition nevertheless do take important elements from Foucault's work, especially his methodology of scrutinizing spatial elements for their social content, his insights into the use of space as an instrument of repression, and his elevation of freedom to the top of the pantheon of values.[23] Like Foucault the urban post-structuralists identify the way in which space embodies power without necessarily locating its source in particular groups of people.

The early texts of post-structuralist urbanism are Jane Jacobs's *Death and Life of Great American Cities* and Richard Sennett's *Uses of Disorder*.[24] Within urban studies the importance of post-structuralism lies in its recognition of diverse bases of social affiliation and multiple roots of oppression. This understanding in turn leads to a veneration of diversity. Whereas political-economic thought both predicts and celebrates the disappearance of racial, religious, and ethnic divides, post-structuralism expects and welcomes their persistence. In the words of Richard Sennett:

> It is the mixing of . . . diverse elements that provides the materials for the 'otherness' of visibly different life styles in a city; these materials of otherness are exactly what men need to learn about in order to become adults. Unfortunately, now these diverse city groups are each drawn into themselves, nursing their anger against the others without forums of expression. By bringing them together, we will increase the conflicts expressed and decrease the possibility of an eventual explosion of violence.[25]

This theme of diversity runs consistently through the post-structuralist critique of capitalist urbanization. Thus, when identifying the blighting effects of capitalism on urban form, rather than stressing uneven development, as do the political economists, it highlights exclusivity and sterility. Post-structuralists condemn the contemporary city as the product of a white male capitalist elite imposing order on other groups with potentially unruly lifestyles.[26] Subordination is achieved through the mechanisms of city planning, which segregates uses through zoning and which isolates social groupings through the development of large projects and separated suburban communities.[27] At its extreme the drive toward exclusionism expresses itself in the gated communities chronicled by Mike Davis in his description of Los Angeles.[28] In these suburban bourgeois utopias, social homogeneity and property interests combine to create a vicious politics in which middle-class homeowner groups ally themselves with capitalist elites in a battle against progressive taxation and publicly provided social welfare benefits.

Post-structuralist urban critics such as Christine Boyer and Michael Sorkin have 'read' the lineaments of power in the form of the built environment.[29] Frequently referring back to a golden age when the city was more heterogeneous, they excoriate the artificiality and 'false diversity' of theme-part projects.[30] Emphasizing the importance of symbols in determining human consciousness, critics in this vein reject projects like London's Covent Garden and New York's South Street Seaport, which although modelled on the busy marketplaces of earlier times, produce only a simulacrum of urbanism – an 'analogous city'.[31] The purpose of such enterprises is to create an illusion of safety and to foster consumption. Instead of linking the visitor with an authentic past, these imitative schemes project him or her into a fantasy world wherein an ostensibly meaningful existence is available for purchase.[32]

Culture rather than economics becomes the root of political identity in post-structuralist thought. Individuals exist as members of socio-cultural groups from which they draw their identities, derive their welfare, and deploy strategies of resistance and purposeful action. Although post-structuralists posit difference as indissoluble, they regard the particular bases on which it is created and its expressions as socially constructed and therefore subject to change through time and place. This malleability means that social hierarchies within capitalist societies are construed as less fixed than within the political-economic analysis of capitalism. The political aim embedded within the post-structuralist tradition is the empowerment of the least powerful, which may coincide with economic betterment but is by no means limited to it. Indeed, changes in the relations among groups and the overcoming of group domination may not coincide with economic transformations.[33]

One of the strongest statements of the post-structuralist position on difference within the urban milieu is made by Iris Marion Young:

> An alternative to the ideal of community . . . [is] an ideal of city life as a vision of social relations affirming group difference. As a normative ideal, city life instantiates social relations of difference without exclusion. Different groups dwell in the city alongside one another, of necessity interacting in city spaces. If city politics is to be democratic and not dominated by the point of view of one group, it must be a politics that takes account of and provides voice for the different groups that dwell together in the city without forming a community.[34]

Young claims that emancipation lies in the rejection of the assimilationist model and the assertion of a 'positive sense of group difference', wherein the group defines itself rather than being defined from the outside.[35] The politics of difference, she contends, 'promotes a notion of group solidarity against the individualism of liberal humanism'.[36]

Post-structuralists thus aim at eradicating social subordination and creating a civil society that allows the free expression of group difference. Strategically such a project fits into the American pluralist framework where ethnicity and interest groups politics have always formed the template for political activity and freedom has been the dominant value. Historically it reflects the experience of black protest movements in American cities and the women's rights movement. It clashes, however, not only with individualistic liberal humanism and assimilationism but also with the concept of the class-based mass party that has been the vehicle for the expansion of the welfare state and the rights of labour in Europe. Indeed, in their effort to transcend Marxist economic reductionism, the post-structuralists seem to have abandoned economic analysis and a recognition of class interests altogether.

Virtually by definition a view of society as consisting of multiple, dissimilar cultural groupings produces a conception of politics as based on coalitions. This is, of course, the standard perception of liberal pluralism, and within that context is wholly desirable. From a left perspective, however, such an approach is problematic. Where exclusion and oppression are identified as prevailing social characteristics of capitalist democracies and hopes for social emancipation come to rest with a coalition of out-groups that share little but their antagonism to the extant social hierarchy, expectations for a coherent alternative political force are shaky. Even the identification of the components of such an alliance is difficult. In Harvey's attempt to devise a programme sensitive to post-structuralist arguments, he declares that *'just planning and policy practices must empower rather than deprive the oppressed.*'[37] But who decides who is the oppressed? Without a universalistic discourse, oppression is in the eyes of the beholder. Many members of the American middle class would accept Harvey's dictum but consider themselves oppressed by welfare cheats and high taxes, while their European counterparts construe immigration as representing a similar imposition. Identification of oppressed groups within a pluralistic framework is hardly a simple matter.

Post-structuralist thought hence gives rise to only a weak – and largely oppositional rather than positive – political expression. The socialist dream of a working class united in its common commitment to justice and equality becomes reduced to, at best, demands for tolerance and redistribution with no programme of fundamental economic reconstruction. At worst, post-structuralism leads not to the tolerance and reduction of surplus repression prescribed by its more enlightened philosophers, but to essentialism, unproductive conflict, and new forms of oppression. In some of its versions, it leaves little room for individual deviation from group norms,[38] relegitimating some of the rationales for feudal tyranny that liberal absolutism combatted in the eighteenth century.[39] Most discomforting is the tendency of post-structuralist thought in its feminist and ethnic-culturalist manifestations to take a critical stance toward other groups but to avoid self-criticism.[40] Thus, we have an acceptance of the privileged position of the oppressed and the incapacity to deal with oppression carried out by members of groups that are themselves oppressed. Examples abound, from defense of the abuse of women in communities of color to black anti-Semitism to extreme stereotyping of Western thinkers and traditions by critics of Orientalism.[41]

The goals of both democracy and diversity are particularly difficult to combine in the real world of politics, where popular sentiment for the latter is often lacking. Left critics frequently attack highly popular urban forms, from suburban housing developments to festive markets, on the grounds that they are exclusionary and sterile. But the condemnation of suburbia comprises an assault on a type of living environment for which large numbers of people have a genuine emotional attachment. And the critique of projects designed to titillate consumption smacks of cultural elitism and seems to castigate people for engaging in harmless acts that afford them enjoyment. More seriously, by defending strong group identifications and simultaneously opposing spatial segregation, post-structuralism endorses a situation in which antagonisms are openly expressed and may easily result not in increased understanding of the Other, but in cycles of hostile action and revenge. Sennett seeks to overcome this problem by dismissing the desire of groups to segregate themselves as resulting from lack of experience with a heterogeneous situation:

> If the permeability of cities' neighborhoods were increased, through zoning changes and the need to share power across comfortable ethnic lines, I believe that working-class families would become more comfortable with people unlike themselves.[42]

This argument implies that in order to get people to consent to the new arrangements, they already would have had to experience them. To force the experience, however, is to override democratic considerations and possibly to cause a ratcheting up of animosities.

Sennett uses the example of South Boston's Irish Catholic enclave to demonstrate the xenophobia that ensues from spatial isolation. After he wrote *The Uses of Disorder*, in which this discussion appears, the substance of his fears for South Boston, in fact, became manifest, as that area became the locus of extremely aggressive opposition to school desegregation. This more recent history both supports his argument concerning the effects of isolation and undermines his contention that people should be required to confront the other – forced to be free, as it were. For the court-mandated attempt to break down racial isolation proved largely counter-productive, stimulating massive white flight and leaving a sullen legacy that militated against the genuine integration of the Boston schools.[43] A democratic transition to the desirable end state of tolerance and diversity is extraordinarily difficult to achieve, even in American cities where the melting pot remains a potent ideal and immigration still commands considerable popular sentiment. In Europe, where national cultures are more defined and deeply rooted, the juxtaposition of differing cultural traditions is fraught with tension and the enduring irredentism of Northern Irish Catholics, Spanish Basques, and Bosnian Serbs dispels any illusion that proximity necessarily leads to understanding.

URBAN POPULISM

Urban populism starts with democracy as its central value. This basic orientation encompasses two intertwined but sometimes separate strands. The first consists of a thrust toward economic democracy, aimed in particular at bringing down plutocratic elites. In Todd Swanstrom's words, 'the political analysis offered by urban populism was essentially a streetwise version of elite theory: a small closed elite, stemming from the upper economic class, uses its control over wealth to manipulate government for its own selfish purposes'.[44] The urban populists share the same egalitarian aims as the political economists, but writers and activists in this tradition rarely engage in sophisticated analyses of economic structure and tend to see wealth arising from power rather than vice versa. As a social philosophy urban populism exists less in theory than in practice, where it comprises the agenda of urban political movements that oppose urban redevelopment schemes or call for public ownership of factories and utilities.

While both the political-economy and post-structuralist approaches have difficulty with majoritarian opposition to their aims and thus resort to various contentions of false or untutored consciousness, urban populism begins with popular preference. Thus Gans attacks the schemes of planners that deviate from mass taste: 'The planner has advocated policies that fit the predispositions of the upper middle class, but not those of the rest of the population.'[45] He disparages Jane Jacobs's glorification of diversity and her contention that spatial arrangements are significant determinants of that aim. In a critique that applies also to later writings by, among others, Sennett, Boyer, and Sorkin, he maintains:

> [Jacobs's] argument is built on three fundamental assumptions: that people desire diversity; that diversity is ultimately what makes cities live and that the lack of it makes them die; and that buildings, streets, and the planning principles on which they are based shape human behaviour. . . . Middle-class people, especially those raising children, do not want working-class – or even Bohemian – neighbourhoods. They do not want the visible vitality of a North End, but rather the quiet and the privacy obtainable in low-density neighbourhoods and elevator apartment houses.[46]

Gans's accusations thus are that Jacobs – and by extension later post-structuralists – succumbs to both an undemocratic desire to impose on others her predisposition to Bohemian colour and a fallacious belief in physical determinism.

Other writers in the urban populist tradition emphasize the elitism of planners and intellectuals in disregarding the traditional affiliations and desires of ordinary people. Harry Boyte, while committed to the aims of diversity and economic equality, attacks Marxists for failing to appreciate the contribution of religion, family, and ethnicity to people's security and well-being.[47] Peter Saunders, incensed by the failure of the British left to comprehend the desire of ordinary people for homeownership, defends Margaret Thatcher's 'Right to Buy' programme in the name of democracy. He asserts:

> The socialist case against owner occupation . . . boils down to little more than an ill thought-out commitment to the ultimate value of collectivism coupled with an implicit fear of individualism. . . . Indeed, it can plausibly be argued that such intensely personal forms of ownership perform important psychological functions for the individual, whether in socialist, capitalist or precapitalist societies.[48]

Saunders argues that widespread homeownership can be achieved without exploitation and indicates how private owners can be prevented from profiting unduly from their possession of property.[49] Gans, Boyte, and Saunders, while seemingly susceptible to charges that they support a majoritarian effort to exploit or suppress powerless minorities, explicitly argue that their form of populism respects minorities. Thus, Gans while demanding that planners respect popular taste, also states that 'democracy is not inviolably equivalent to majority rule' and proposes 'a more egalitarian democracy' that responds to the need of minorities.[50] Boyte declares that 'democratic revolt requires . . . an important measure of cultural freedom, meaning both insulation from dominant individualist, authoritarian patterns and also openness to experimentation and diversity'.[51] Saunders calls for the development of 'a coherent socialist theory of individual property ownership' that would abolish relations of exploitation.[52]

Despite their efforts to deny the democratic authoritarianism latent in their arguments, these authors nevertheless fail to confront the genuine inconsistency that afflicts democratic theory in its effort to preserve minority rights. These three writers extol neighbourhood homogeneity, citizen activism, religion, family, ethnic ties, and home in the name of democracy, and even of equality, but their arguments can easily lead to a strongly illiberal, exclusionary politics that does reinforce inequality and minority exclusion. If democratic participation is the principal value underlying the anti-elitist orientation of urban populism, how is it possible to repudiate the injustices perpetrated by homeowner movements in the United States and community-based anti-immigrant mobilizations in Europe? Social theorists who foresaw 'the revolt of the masses' as destructive of civilized values have had endless corroboration in history.[53]

The recourse of democrats who fear the intolerance that can result from majority rule is a theory of rights. As embodied in the first ten amendments of the American Constitution and the Universal Declaration of Human Rights of the United Nations, such a theory protects individuals by attributing to them inalienable rights. The vesting of rights in individuals derives intellectually from concepts of natural law and offers the only logical resolution to the problem of imbuing a democratic philosophy with a transcendent ethic that limits the actions of democratic majorities.

The vulnerability of a theory of rights is in its failure to locate the source of individual rights except in a seemingly natural intuition concerning what is ethical behaviour. Historically liberal democratic theory has particularly stressed property rights as fundamental, and it is, in fact, that emphasis that has cropped up again in American homeowner movements rather than concerns over the exercise of free speech or the protection of privacy. Thus, what is in many respects a fundamentally intolerant movement justifies itself in the name of both democracy and individual rights.

A CONSISTENT THEORY OF URBAN SOCIAL JUSTICE?

Reconciliation of the values of egalitarianism, diversity, and democracy within societies divided by class and communalism presents a supreme challenge to any left agenda. Harvey seeks to meet this challenge by both recognizing the authenticity of multiple publics and specifying a discourse that can unify a popular majority around a positive programme for social change:

> If we accept that fragmented discourses are the only authentic discourses and that no unified discourse is possible, then there is no way to challenge the overall qualities of a social system. To mount that more general challenge we need some kind of unified or unifying set of arguments. For this reason, I chose . . . to take a closer look at the particular question of social justice as a basic ideal that might have more universal appeal.[54]

In taking this position Harvey maintains that transient coalitions among subordinate groups on particular issues do not suffice as a basis for effective mobilization. Instead he is calling for what Castells would designate as a 'core social movement'. This stance is not based on class, social status (e.g. feminism), or a substantive belief (e.g. environmentalism) but rather on a generalized concept of social justice as defined from the perspective of oppressed groups. Thus, for Harvey social transformation emanates from a coalition of people with a shared moral position in opposition to various forms of oppression.

Harvey enumerates a generalizable set of precepts on which such a movement would be based. It is possible, although he does not do so, to find the roots of his moral propositions in Kant's formulation (his 'categorical imperative') concerning the ethical actions of the individual in society.

> Kant's *first ethical formula* ran as follows: 'Act so that the maxim of thy action may serve as a general rule'; his *second formula* is: 'So act as to treat humanity, whether in thine own person or in that of any other, in every case as an end withal, never as a means only'. . . . Both formulas presuppose that we actually feel ourselves to be members of a kingdom of personal beings.[55]

In other words one should follow a consistent, generalizable ethic regardless of the specific circumstances; that ethic requires always putting oneself in the place of the other; and its basis is a view of people as sharing a common humanity within a social network. John Rawls later extends Kant's categorical imperative by arguing that it would be freely chosen by individuals in the 'original position' and that its meaning boils down to fairness.[56] Kant's prescription is also elaborated by Georg Simmel, who provides a concept of justice that expands the definition of fairness to encompass the ideals of freedom and equality, although not within a historical context.[57] In Simmel's interpretation, Kant's categorical imperative establishes the ideal of equality, but crucially, equality as a *freely chosen* ideal state: 'Equality supplies freedom, which is the mainspring of all ethics, with its content.'[58]

Such a choice, however, is outside history, where no choices can be made free of material and cultural constraints, as Marx and Engels never failed to point out. Indeed Kant's imperative, even as explicated by Rawls and Simmel, is stated as a pure abstraction rather than as a mandate to transform a complex world of already existing social hierarchies. A programme for social justice devised from a political-economy perspective, however, requires that history be brought into the analysis.

Harvey's ethical propositions are directed at achieving social justice within a political-economic framework that takes into account existing conditions. He essentially accepts the Kantian/Rawlsian contention that fairness comprises the basis for a social ethic, and like Simmel he defines fairness as being comprised by equality.[59] Harvey's use of a Kantian concept of social justice, however, raises fundamental problems, since any attempt to posit a general moral position is suspect within the Marxist tradition in which he places himself. Marx, in his later work, rejected the idealist tendencies of his youth and adopted a stance that Cornel West labels radical historicism: 'The radical historicist approach calls into question the very possibility of an ethics . . . that claims to rest upon philosophic notions of rational necessity and/or universal obligations.'[60] Harvey, by not arguing that everyone, if thinking rationally, would adopt equality as his or her aim, does not precisely present an argument that his ethical precepts are either necessary or universal. But he does claim that they comprise 'a basic ideal that might have more universal appeal.'[61] Moreover, he does not specify the groups for whom his assertions are warranted. Crucially, do they include the middle mass of the wealthy nations? Nor does he explain how to overcome resistance to his programme without engaging in acts of repression.

Harvey is not unique among Marxist philosophers in foregoing Marx's strict limitation on foundationalism. As West points out, Marx's followers could not accept the moral relativism inherent in his view and instead sought to ground the value of equality in historical materialism. For Friedrich Engels this quest led to a teleological justification in which actions became moral if they corresponded with the direction of history; while for Georg Lukács they were legitimated by conformity with the interests of the proletariat, which were defined as universal.[62] From the vantage point of the late twentieth century, we can have little confidence in either the inevitable triumph of communism or the universality of proletarian aims – and especially not in the implementation of freedom and equality by the state in the name of the working class. Thus, the approaches of Engels and Lukács to overcoming the issue of moral relativism – an issue also raised by the work of the post-modernists – are clearly inadequate.

According to West, for Marx an individual's concept of justice derives only from his or her position within a moral community, which itself is wholly a historical product.[63] Still, even within this constraint the search for a concept of social justice has a function:

> For the radical historicist, the search for philosophic foundations or grounds for moral principles is but an edifying way of reminding (and possibly further committing) oneself and others to what particular (old or new) moral community or group of believers one belongs to.[64]

Harvey, who does not even go as far as Engels and Lukács in attempting to justify his ethical propositions, might simply be restating Simmel's formulation to the community of those who are already predisposed toward it. He seemingly, however, intends his reach to stretch further when he claims that:

> Justice and rationality take on different meanings across space and time and persons, yet the existence of everyday meanings to which people do attach importance and which to them appear unproblematic, gives the terms a political and mobilizing power that can never be neglected.[65]

FAIRNESS AND HISTORY

Engels, Lukács, and their followers read into history a meaning that is not only unverifiable but easily used to legitimate egregiously inhumane actions. In contrast, the defect of the Kantian approach is its omission of history altogether. Is it possible to develop an argument for fairness that recognizes the historical embeddedness of identity groups and their particularistic constructions of reality? Can such a formulation avoid the pitfalls of blame and demands for retribution that run at times through everybody's conceptions of fairness and to which individuals who identify strongly with social groupings are particularly prone? Can such a framework be deployed to criticize and transform the capitalist city?

A good place to begin is with Mannheim's 'Sociology of Knowledge'.[66] Mannheim argues for the 'acquisition of perspective' so as to overcome the 'talking past one another' of groups in conflict.[67] He calls for both deconstruction and reconstruction through examining the social origins of ideas and the relationship of modes of interpreting the world to social structure. He names this process 'relationism' and asserts that:

> Relationism does not signify that there are no criteria of rightness and wrongness in a discussion. It does insist, however, that it lies in the nature of certain assertions that they cannot be formulated absolutely, but only in terms of the perspective of a given situation.[68]

Mannheim is arguing that concepts cannot be abstracted from their historical situatedness even while nor forgoing the possibility of a transcendent ethic. At the same time he is calling for demystification and empathy, thereby assuming that this ethic will be reshaped and reinterpreted within differing historical realities. Consciousness derives from group identification *and also* from rational formulations. He assumes that when history is taken into account by all participants, thinking individuals can find a common ground. His is essentially a Hegelian exercise in its acceptance of the concept of the rational subject who is capable of making comparisons and learning. Unlike Hegel, however, Mannheim does not contend that history is reason revealed, and he takes material as well as mental forces into account.

Within Mannheim's conceptualization we can continue to hold up fairness as the key to social justice while developing its content differently depending on our standpoint and historical location. Such an activity requires both self-criticism and criticism of the other. In particular, there is no 'other', whether high or low on the social hierarchy, so privileged as to be immunized from outside criticism.

If the concept of social justice acquires a socio-historical content, the meaning of freedom incorporated within it switches from the liberal formulation of 'freedom from', as well as from the Foucaultian notion of resistance, to a more complex idea incorporating the concept of self-development. Freedom then includes the acceptance of obligations for which, in Nancy Hirschmann's words, 'consent is not only unavailable but of questionable relevance.'[69] Equality likewise shifts its connotations and is measured not by sameness of condition but by reciprocity, communication, and the mutual acceptance of obligation. Indeed some inequalities of power or benefits, stemming from rewards to merit, response to need, or the provision of general welfare, can be interpreted as just.[70]

SOCIAL JUSTICE, ACTION, AND THE CITY

It is easier to recommend an abstract concept of social justice on which members of one's own moral community can agree than to find an expression for it that will attract a mass following. Fairness in the city requires the devising of programmes that do not offend most people's concepts of just rewards, that respond to individual aspirations as well as to a formulation of the social good, and that at a minimum do no harm. How to achieve this without pandering to the forces that wall off homogeneous groups in exclusionary communities, direct unconscionable profits to developers, and reinforce the exploitation of workers is no small task.

The three bodies of thought discussed here each give priority to a value that appears inherent in a concept of social justice. But the tensions among the values of equality, diversity, and democracy endure and cannot be as easily glossed over as the contemporary left, which supports all three, would like. In particular, a commitment to democratic process, in the present historical situation, requires a willingness to accept the outcomes of that process even if they do not favour the most disadvantaged groups.

A movement for social justice, if it is to mobilize large numbers of people, must focus less on the protection of the most deprived and more on broad benefits, less on the rights of the oppressed and more on security. Most people would prefer economic growth, if any of it trickles down to them, to redistribution, if redistribution does not produce an improvement in their standard of living. Strongly targeted redistributional policies may be the most efficient method of attaining broad social benefits, but they are too suspect within the anti-tax sentiment of the present age to constitute an effective rallying point.[71]

A progressive strategy with mass appeal requires supporting urban programmes that help most people. In particular, giving housing preferences to working and lower middle class households, improving access to excellent education for talented lower-income children, and providing venture capital for small business are programmatic options that do not fit well with strongly egalitarian goals but which are popular and would make life better for a considerable sector of the urban population. Combined with an effort to direct investment away from central business districts and into neighbourhoods and to involve people in planning their communities, these actions would broaden the material well-being and reduce the alienation of many city residents, but admittedly at the short-term risk of worsening the situation of those still left out.

In part, whether these approaches amount to a progressive direction depends on their forming part of a national programme to improve the situation of working-class people. If urban strategies targeted at middle strata are not part of a general broadening of economic opportunity, they will produce a zero-sum game between the working-class and the poor. A national programme directed at this goal must focus on the creation of full-time, full-benefit employment at a decent wage. Within the United States this means raising of the minimum wage, portability of health and pension benefits, and the regulation of 'temporary' jobs occupied by long-term employees who receive substandard wages and benefits. In Europe it implies a larger government role as direct sponsor of employment combined with a relaxation of private-sector labour regulations so as to introduce more flexibility into labour markets, along the Swedish model.

Such measures would not address issues of cultural conflict, which require actions within the realm of civil society. Racial and ethnic relations and conflicts over lifestyle are less amenable to becoming the object of a political programme than are questions of material distribution, yet they nevertheless must be the constant subject of critical thinking. People often want to live in situations where they do not have to constantly interact with others pursuing radically different life styles. In fact, and contrary to Sennett's prescriptions, the path to greater tolerance may well lie in reduced interaction (although short of the radical separation that has occurred in the US through suburban autonomy). This may mean acceptance of ability grouping (i.e. 'streaming') in schools and neighbourhood homogeneity as long as the larger framework of the school and the city remains heterogeneous.

The task is to figure out and attract people to ways of achieving equality, diversity, and democracy while being sensitive to procedural as well as social norms. Such a programme must not promote the enrichment of the few at the expense of the many and cannot be blind to hostility based on ascriptive attributes. But when one goal precludes another, perhaps there is no alternative to the old Benthamite prescription of 'the greatest good of the greatest number.' Proponents of public policy derived from negotiation express hopes that programmes can be formulated through procedures that accommodate everyone's needs.[72] This viewpoint takes an extraordinarily benign view of social power and social conflict and forgoes the political economist's insight into the substantive content of equality. Nevertheless, it does incorporate Mannheim's thesis concerning the possibility of learning. If it can be combined with a more stringent argument about the material underpinning necessary for participation in the conversation, it can then potentially provide a basis to a governing ethic for the making of an urban agenda.

NOTES

1 Like all such typologies this one does injustice to the refinements of individual thinkers and ignores the degree to which they synthesize arguments across the intellectual boundaries established here.

2 In commenting on an early draft of this chapter, Robert Beauregard remarked that 'it is difficult to link these different perspectives to positions on social justice when they are generally silent about it. . . . In fact, justice is undertheorized, while issues of allocation are used in its stead.' It is, nevertheless, my contention that concepts of social justice are fundamental to the works of the writers discussed here despite their unwillingness to spell out their concepts.

3 Although often not explicitly socialist, urban political economists operate under the vague aegis of 'democratic socialism' in Europe and 'progressive' scholarship in the US. There is also a conservative branch of political economy, which is rooted in the thought of Adam Smith and which, in its modern manifestation, expresses itself in theories of public choice and rational expectations. I am not including this perspective in my discussion.

4 Johns Hopkins University Press, Baltimore, 1973.

5 MIT Press, Cambridge, 1977.

6 David Harvey deals explicitly with the question of 'a just distribution' in chapter 3 of *Social Justice and the City*. He classifies this chapter, however, as a 'liberal formulation', implying that such a normative discussion cannot be accommodated within the Marxist framework of his later work. Despite, however, the seeming inconsistency of seeking to develop a foundational ethic within a Marxist paradigm, Harvey returns to this issue in his recent work (see David Harvey, 'Social justice, postmodernism and the city', *International Journal of Urban and Regional Research* [*IJURR*], 16 [December 1992], pp588–601). The issue of foundationalism will be discussed in the conclusion of this paper.

7 Harvey, *Social Justice and the City*, p233.

8 Neil Smith takes up the theme of uneven employment and elaborates it in his work on gentrification. The most money is to be made when the price differentials between desirably located pieces of land are greatest. Consequently capitalists are impelled to first drive down the value of property occupied by poor people until living conditions become untenable, then to clear them out and raise the value by multiples of the original amount. He goes beyond Harvey's work on ghettos and class monopoly rent by deconstructing the symbolism involved in the takeover of space by gentrifiers and thus employs analytic tools used more often by post-structuralists than political economists. The underlying dynamic that he expounds, however, is fundamentally an economic one and the symbolic level is a legitimation of economic motives rather than an independent causal mechanism. See 'Toward a theory of gentrification: a back to the city movement by capital not people'. *Journal of the American Planning Association*, October 1979, pp538–48, and 'New City, New Frontier: The Lower East Side as Wild, Wild West', in Michael Sorkin (ed), *Variations on a Theme park*, Hill and Wang, New York, 1992.

9 Castells, *The Urban Question*, pp426–7. Mike Davis's *City of Quartz*, Verso, London, 1990, describes contemporary Los Angeles in a way that seems to fulfil Castells's prediction.

10 Henri Lefebvre, *La production de l'espace*, Anthropos, Paris, 1974.

11 Ira Katznelson, *City Trenches*, Pantheon, New York, 1981.

12 *The City and the Grassroots*, University of California Press, Berkeley, 1983, p329.

13 David Harvey, *Consciousness and the Urban Experience*, Johns Hopkins University Press, Baltimore, 1985, p. 274. Italics added.

14 For example, Bennett Harrison and Barry Bluestone, in their critique of the social impacts of conservative economic policy, comment on the particularly disadvantaged position of women and racial minorities within each category of employment. Nevertheless, their concluding chapter, which contains recommendations for 'further redistribution of wealth to root out poverty and promote equality' focuses entirely on economic measures. (*The Great U-Turn*, Basic Books, New York, 1988).

15 *The Sociology of Georg Simmel*, Kurt H. Wolff (ed), Free Press, New York, 1950, p75. Ralf Dahrendorf similarly argues the inevitability of hierarchies of power and social differentiation. See his *Class and Class Conflict in Industrial Society*, Stanford University Press, Stanford, 1959.

16 William Julius Wilson, in his examinations of the causes of ghetto poverty, has made the most significant recent attempt at such a discussion. See *The Declining Significance of Race*, University of Chicago Press, Chicago, 1978, and *The Truly Disadvantaged*, University of Chicago Press, Chicago, 1987.

17 See Nanette Funk and Magda Mueller (eds), *Gender Politics and Post-Communism*, Routledge, New York, 1993.

18 A recent, extensive discussion of the problem of order as dealt with by various social theorists is contained in Dennis Wrong, *The Problem of Order*, Macmillan, New York, 1994.

19 This criticism was most strongly expressed by Max Weber in 'Politics as a Vocation', when he contrasted the ethic of responsibility with the ethic of absolute ends. In Hans Gerth and C. Wright Mills (eds), *From Max Weber*, Oxford University Press, New York, 1958.

21 Harvey, 'Social Justice, Postmodernism, and the City', p600.

21 I have chosen to use the term post-structuralist rather than post-modernist, since the latter approach appears to have faded considerably even while certain elements within it linger and have influenced the work of many thinkers. Post-structuralism incorporates feminist and cultural theories that do not necessarily accept the post-modernist emphasis on discourse and rejection of the Enlightenment tradition. Some of the writers I am including in this category developed their outlooks before the term 'post-structural' came into common usage. Nevertheless, their arguments fit into this general framework.

22 See his discussion in 'What Is Enlightenment?' in Paul Rabinow (ed), *The Foucault Reader*, Pantheon, New York, 1984, pp32–50, where he rejects categorization of his own thought as either structuralist or anti-structuralist, pro- or anti-Enlightenment; see also Hubert L. Dreyfus and Paul Rabinow, 'Introduction', in Dreyfus and Rabinow (eds), Michel Foucault, *Beyond Structuralism and Hermeneutics*, 2nd ed., University of Chicago Press. Chicago, 1983.

23 See especially *Discipline and Punish: The Birth of the Prison*, Vintage, New York, 1979; and 'Space, Knowledge, and Power', in Rabinow, (ed), (1984), pp239–56.

24 Jane Jacobs, *The Death and Life of Great American Cities*, Vintage, New York, 1961; Richard Sennett, *The Uses of Disorder*, Vintage, New York, 1970. These volumes, of course, predate the use of the term post-structuralist and do not involve a theoretical formulation of this philosophic outlook. Nevertheless, their content contains the basic precepts that later became incorporated into post-structuralist urban criticism.

25 *Ibid.*, p162.

26 See Richard Sennett, *The Conscience of the Eye*, Vintage, New York, 1990; Elizabeth Wilson, *The Sphinx in the City*, Virago, London, 1991; Sharon Zukin, *Landscapes of Power*, University of California Press, Berkeley, 1991; Paul Knox (ed), *The Restless Urban Landscape*, Prentice-Hall, Englewood Cliffs, NJ, 1993.

27 M. Sorkin (ed), *Variations on a Theme Park*, pp205–32; Fishman, *Bourgeois Utopias*, Basic Books, New York, 1987.

28 Mike Davis, *op. cit.*

29 See Christine Boyer, 'Cities for Sale: Merchandising History at South Street Seaport', in Sorkin (ed), (1992), pp181–204; Michael Sorkin, 'See You In Disneyland', in *ibid.*

30 Foucault, in contrast to some of his followers, explicitly eschews any golden age mythology, condemning any 'inclination to seek out some cheap form of archaism or some imaginary past forms of happiness that people did not, in fact, have at all'. ('Space, Knowledge, and Power', in Rabinow (ed), *Foucault Reader*, p248).

31 Trevor Boddy, 'Underground and Overhead: Building the Analogous City', in Sorkin (ed), (1992), pp123–53. While an important strand of post-structuralist cultural analysis celebrates artificiality and the spectacle, most specifically urban works denounce the creation of urban space for the purpose of commercial exploitation.

32 In *The City Builders*, (Blackwell, Oxford, 1994), I criticize these critiques for their assumptions concerning a golden age and an authentically urban milieu.

33 Joan Kelly-Gadol, in a well-known essay, argues that the Renaissance did not improve the lives of women and thus did not constitute a liberating turning point of history for at least half of humanity. ('Did Women Have a Renaissance?' in Renate Bridenthal and Claudia Koonz (eds), *Becoming Visible: Women in European History*, Houghton Mifflin, Boston, 1977, pp137–64). Other supposedly liberatory activities likewise have left women unaffected or actually worsened their lot. Many have pointed to the subordination of women by New Left men as an important stimulus for the birth of the second wave of feminism.

34 Iris Marion Young, *Justice and the Politics of Difference*, Princeton University Press, Princeton, 1990. p227.

35 *Ibid.*, p172.

36 *Ibid.*, p166.

37 Harvey, 'Social Justice, Postmodernism and the City', p 599. Italic in original.

38 Young's work seems to allow little leeway for individuals to construct their identities outside their group affiliations. This is not, however, the case for Sennett.

39 Louis Hartz, in his brilliant analysis of the dialectics of political theory, comments: 'The novel proposition [of the monarchical state] that individuals are genuinely autonomous, and equal within the state, meant that human beings could successfully become unraveled from the meaning of their prior associations. Formerly [i.e. in the feudal era] corporate attachment had defined personality; in the breakdown of the old order individuals would be redefined.' (*The Necessity of Choice: Nineteenth-Century Political Thought*, edited, compiled, and prepared by Paul Roazen, Transaction, New Brunswick, NJ, 1990. p29).

40 See Toril Moi, *Sexual/Textual Politics*, Routledge, London, 1985.
41 Hooshang Amirahmadi terms this practice 'reverse Orientalism'.
42 *The Uses of Disorder*, p194.
43 See J. Anthony Lukas, *Common Ground*, Knopf, New York, 1985.
44 Todd Swanstrom, *The Crisis of Growth Politics*, Temple University Press, Philadelphia, 1985, p129.
45 Herbert Gans, *People and Plans*, Basic Books, New York, 1968, p21.
46 *Ibid.*, pp28–9.
47 Harry C. Boyte, *The Backyard Revolution*, Temple University Press, Philadelphia, 1980, Chap. 1.
48 Peter Saunders, 'The Sociological Significance of Private Property Rights in Means of Consumption', *International Journal of Urban and Regional Research*, 8(2)(1984). p219.
49 *Ibid.*, pp218–23.
50 Herbert Gans, *More Equality*, Pantheon, New York, 1973, pp138, 139. Italic in original.
51 Boyte, *op. cit.*, pp38–9.
52 'Beyond Housing Classes', p223.
53 See José Ortega y Gasset, *The Revolt of the Masses*, Norton, New York, 1932.
54 'Social Justice, Postmodernism and the City', p594.
55 Harald Höffding, *A History of Modern Philosophy*, Volume 2, Dover, New York, 1955, p86. Italic in original.
56 John Rawls, *A Theory of Justice*, Harvard University Press, Cambridge, Massachusetts, 1971, pp251–7.
57 Simmel points out that in the nineteenth century the link that Kant made between freedom and equality was severed. 'Crudely . . . these ideals may be identified as the tendencies toward equality without freedom, and toward freedom without equality'. The former was taken up by socialism, the latter by liberal individualism. (*The Sociology of Georg Simmel*, p73).
58 *Ibid.*, p72.
59 'Social Justice, Postmodernism and the City', pp594–600. Harvey does not state this as his purpose. Rather, it is my reading of the article that equates his view of social justice with fairness.
60 Cornel West, *The Ethical Dimensions of Marxist Thought*, Monthly Review Press, New York, 1991, p1.
61 'Social Justice, Postmodernism and the City', p594.
62 West, *op. cit.*
63 *Ibid.*, pp98–9.
64 *Ibid.*, p2.
65 'Social Justice, Postmodernism and the City', p598.
66 Contained in Karl Mannheim, *Ideology and Utopia*, Harcourt, Brace, & World, New York, 1936.
67 *Ibid.*, p281.
68 *Ibid.*, p283.
69 Nancy J. Hirschmann, *Rethinking Obligation: A Feminist Method for Political Theory*, Cornell University Press, Ithaca, 1992, p235.
70 See W.G. Runciman, *Relative Deprivation and Social Justice*, University of California Press, Berkeley, 1966; G. Simmel, *Sociology of Georg Simmel*, 1950, pp73–8.
71 It is widely believed that the British Labour Party lost the 1992 election because it had indicated that it would increase taxes on the middle class.
72 See Patsy Healey, 'Planning through Debate: The Communicative Turn in Planning Theory', *Town Planning Review*, 63 (2), 1992, pp143–62.

The following images are taken from the series *Capital Arcade* (1998–9) by John Goto, from his larger work *Tales of the Twentieth Century*. Each image is a digital manipulation based on a composition derived from a painting in the European tradition, and peopled by contemporary figures. The set of five, including an introductory and concluding work – an entrance from the car park and a closing-down sale – constructs a pictorial equivalent of a shopping mall with its familiar retail chain stores.

The series was exhibited at the Seonam Arts Centre in Seoul in 2000, and at the Andrew Mummery Gallery London, Howard Gardens Gallery, Cardfiff, and Hasselblad Centre at Goteborg in Sweden, during the same year. Goto has exhibited other works at Tate Britain and the National Portrait Gallery, London, Kettles Yard in Cambridge, the Museum of Modern Art, Oxford, and the Holbourne Museum in Bath, as well as in Berlin, Oporto and Edinburgh. A review of *Capital Arcade* can be found in *Creative Camera* (October, 1999) and in *Flux* (April/May, 2000). Goto's work can be seen in full colour on his website: www.johngoto.org.uk; and a book of his work will be published by the Djanogli Gallery, London in 2003.

Plate 11 **'Welcome to Capital Arcade' by John Goto** (Courtesy of the artist).

Plate 12 **'Next' by John Goto** (Courtesy of the artist).

Plate 13 **'Mothercare' by John Goto** (Courtesy of the artist).

Plate 14 **'B & Q' by John Goto** (Courtesy of the artist).

Plate 15 **'McDonald's' by John Goto** (Courtesy of the artist).

PART FOUR

The Culture Industry

INTRODUCTION TO PART FOUR

To take the heading Culture Industry for this section is to adopt a particular position. It is the term used by Theodor W. Adorno and Max Horkheimer in an essay written in 1944 and published in 1947 in *Dialectic of Enlightenment* (London, Verso, 1997, pp. 120–67) to describe the industry which produces mass culture. Adorno and Horkheimer refuse the industry's own term of popular culture as in popular music because such cultural goods are not made *by* but *for* a mass audience, and have an ideological function. Today this industry, which has grown rapidly in the postwar period and engaged with new technologies of sound-recording and image-making, and brought entertainment from the cinema to the domestic interior, might be called the media industry, to include advertising and broadcasting as well as film, video and popular music. This puts emphasis on cultural production rather than reception, but Adorno's point is anyway that the reception of mass culture is passive. Film, for example, uses increasingly sophisticated techniques to create illusions of reality, so that 'the easier it is for the illusion to prevail that the outside world is the straightforward continuation of that presented on the screen' (p. 126). Walter Benjamin, in contrast, sees film as a democratic medium in which the spectator identifies with the actor's alienated labour as well as being able to imagine alternative plot lines and endings, actively working at interpreting the imagery as it unfolds (see Esther Leslie, *Walter Benjamin: Beyond Conformity*, London, Pluto, 2000, pp. 130–67).

But Adorno and Horkheimer are clear: mass culture is a powerful element in an increasingly total manipulation of consciousness, aligned with other kinds of totalitarianism. They write:

> Real life is becoming indistinguishable from the movies. The sound film, far surpassing the theatre of illusion, leaves no room for imagination or reflection on the part of the audience, who is unable to respond within the structure of the film, yet deviate from its precise detail without losing the thread of the story; hence the film forces its victims to equate it directly with reality (*Dialectic of Enlightenment*, p. 126).

Today this might seem a naive or jaded position to adopt, and Benjamin is more frequently cited in the literature of cultural and film studies than Adorno. But Benjamin was thinking of German experimental (Expressionist) film while Adorno retains, in the essay reprinted here, as in the 1944 text, a focus on the Hollywood movies he encountered in north America. This industry offers choice but homogenises entertainment, deceiving those whose cash it takes:

> The culture industry perpetually cheats its consumers of what it perpetually promises. The promissory note which, with its plots and staging, draws on pleasure is endlessly prolonged; the promise, which is actually all the spectacle consists of, is illusory; all it actually confirms is that the real point will never be reached, that the diner must be satisfied with the menu (p. 227).

This could serve as a critique of consumerism, not least of the lifestyle advertising of television commercials and the colour supplements in Sunday newspapers – developments of which Adorno, who died in 1969, saw only the early stages. In another essay, 'The schema of mass culture' (in *The Culture Industry*, pp. 53–84), he notes the industry's ability to absorb even deviant cultural forms: 'Advertising has absorbed surrealism and the champions of this movement have given their blessing to the commercialization of their own murderous attacks on culture' (p. 59). As German exiles, those who reconstituted the Institute for Social Research at Columbia University in New York under Horkheimer's direction in the mid-1930s looked with detachment on north American mass culture's brash commercialism; and as refugees from a country where megalomania ruled, they distrusted the Hollywood cult of the star. But, though Adorno's argument is strong as well as intricately wrought, and consumerism does depend on the consumer never being satisfied, always wanting and paying for more, it is difficult to avoid a sense of nostalgia when Adorno writes that 'The colour film demolishes the genial old tavern to a greater extent than bombs ever could' (p. 89). Were those old taverns really so genial? Perhaps the point most worth excavating from the 1944 text is that the illusion of choice is no choice: the customer is not king or queen, and 'The man [*sic*] with leisure has to accept what the culture manufacturers offer him'; the danger is that the passivity bred by this situation is contagious, so that the 'ruthless unity in the culture industry is evidence of what will happen in politics' (pp. 124, 123).

An alternative term for the culture industry, and one which seeks to let go of its ideological function, is the cultural and creative industries, or cultural industries for short. The cultural industries tend to figure in government reports as a driving force in urban redevelopment, for instance in the designation of cultural quarters. They figure, too, in knowledge economies based on new technologies of communication and imaging. The effect of such policies, on the other hand, may reinforce a sympathy for Adorno's view when they lead to processes of gentrification – the real destroyer of the genial old tavern, if there was one – in an equation of culture and affluence. Then living in a cultural zone is a mark of status attended by conspicuous consumption, property values and rents rise, and residual publics are displaced.

But what exactly are the cultural industries? They include the same components as the culture industry – the recording of popular music, film, broadcasting, and so forth – but also graphic design, advertising, public relations, and now the digital manipulation of images. For the most part these are small- or medium-scale service enterprises using highly portable equipment, with flexible structures, and dependent more on ideas, fashions, and reputations than investment in plant and raw materials. In some cases the visual and performing arts are included as well, so that a cultural quarter in a post-industrial urban zone might include artists' studios, an experimental dance or theatre workshop, design and public relations businesses, and television production studios. There will, today, be a preponderance of new technologies, and very often a range of designer bars, boutiques, florists, and hairdressers to meet the needs of the busy cultural producers.

A wider definition of the cultural industries is also available. Allen Scott, in *The Cultural Economy of Cities* (London, Sage, 2000), written from a perspective of economic geography and urban policy, includes textiles, furnishings, printing and publishing, leather, jewellery and silverware, and toys as well as music recording, film, and advertising media in a comparison of the cultural economies of Los Angeles and Paris. He includes architecture (employing more than 5,000 people) in Los Angeles (pp. 176 and 194; tables 11.3 and 12.1), and writes that 'A noteworthy feature that all these sectors share in common is their disposition to form dense specialized industrial districts within the wider fabric of the city'; and that, in Paris, the locations of older, artisanal industries tend to be in working-class and immigrant neighbourhoods while media industries cluster in more affluent districts (pp. 194–5). Given Scott's broad categorisation of industries producing cultural products, it is not surprising that he finds the sector to be a major employer. He notes, too, from the US cultural economy, that it is closely linked to cities: 'a very significant proportion of employment in the . . . cultural economy is concentrated in large metropolitan areas, and the proportion seems to increase as the cultural content of final products increases' (p. 8). Even when the cultural economy is more narrowly defined as the arts, media, and heritage, it is still a major contributor to employment and, in the United Kingdom, to invisible earnings, as John Myerscough reported in *The Economic Importance of the Arts in Britain* (London, Policy Studies Institute, 1988). But missing from Scott's definition, perhaps in response to the US conditions he mainly studies, is cultural tourism, now a major industry in many European cities, particularly those with an architectural heritage.

If, then, the cultural industries are increasingly central to urban economies, and seen as means to the regeneration of post-industrial zones – using redundant industrial buildings as art galleries and museums – they also have a key influence on the perception of cities. This point was made at length in the previous section on symbolic economies, where cultural consumption and the image of a flourishing cultural sector is used in marketing a city and attracting new investment. In this section, the case of Barcelona examined by Dianne Dodd takes the enquiry further, showing that a city's cultural planning may trade on local culture (in the senses both of cultural production and a surviving, distinctive way of life) as a means to draw in visitors seeking authentic experience. For Adorno, authenticity is the quality lacking in mass culture. But it seems that just as culture became in the nineteenth century (see the introduction to Part One) an antidote to industrialisation, now local culture becomes an antidote to standardisation in a world of globalisation. Whether the local cultures discovered by tourists are authentic, however, remains a question.

John Urry, too, looks at local cultures – in this case in the context of the British tourist industry – in the chapter from *Consuming Places* reprinted in this part. His aim, which contrasts with Adorno's unremittingly bleak outlook, is to question the assumptions that globalisation engenders homogenisation; and that local distinctiveness is in any case so important. At first this might appear confusing, as if one element is eroding another which is not really there or does not really matter, but Urry's concern is with the complex connections *between* local and global cultures: 'It is the *interconnections* between them which account for the particular ways in which an area's local history and culture is made available and transformed into a resource for local economic and social development' (John Urry, *Consuming Places*, London, Routledge, 1995, p. 152). Comparing Urry's work on industrial heritage sites such as Wigan Pier in England with Dodd's on Barcelona is instructive. First, Urry finds that heritage culture tends to grow as a consequence of de-industrialisation, which provides it with sites which it recycles for tourist consumption; Dodd sees Barcelona as expanding an existing economy rather than replacing it, fusing its fairly recent cultural heritage with a presence of contemporary art. Second, Urry sees a lack of interest in contemporary architecture in favour of a reproduction of traditional and vernacular styles, which is specifically British; Dodd notes the commissioning of new buildings as part of a resurgence of Catalan culture, so that the city's modernism is renewed in its postmodernism. Third, Urry laments a lack of funding for the infrastructure which might genuinely rejuvenate a post-industrial area; Dodd describes significant cultural investment, in part through non-profit savings banks, in a cultural infrastructure. But fourth, where the accounts converge, Urry discusses heritage tourism as a service industry, and Dodd regrets Barcelona's failure to support cultural producers such as artists and authors to the same extent as cultural institutions.

Both Dodd and Urry treat culture as the commodity it has been made by the cultural industries. This recognises that cultural experience is the basis of a significant industry. But it may be helpful to remember, too, that just as culture has an anthropological meaning as well as a productive one so there are living cultures, though they tend to be ignored by the cultural industries. Or perhaps, since they are mutable and internally diverse, and cannot be relied on to be unfailingly nice, they are misrepresented in the interests of a homogenised product. An example is the global spread of themed 'Irish' bars with 'Irish' music, which are not re-creations of traditional Irish urban or rural drinking houses, which might be less decorative places – those genial old taverns again. This commodification of Irish drinking experience is contextualised by movies which emphasise the country's lyrical spirit and history of rural hardship and emigration; and by Ireland's own tourism promotion which plays on wild landscapes, medieval monasteries, and 'Celtic' crafts. While Ireland is a modern democracy in the Euro zone, with a thriving film industry, in its packaging by the cultural industries it is a series of twilights.

The linking factor in these representations is community – the *craic* in the bar, the supposed solidarity of an ethnic group. It is this which Mary J. Hickman deconstructs in her chapter 'Differences, Boundaries, Community: The Irish in Britain' (from *Recoveries and Reclamations*, edited by Judith Rugg and Daniel Hinchcliffe, Bristol, Intellect Books, 2002). Although the chapter's aim is to examine the idea of community as constructed in Irish migrant culture, and the chapter is also relevant to Part Seven on contesting identity, it is included here to provide a note of realism as balance to accounts of an industry which deals in myths and dreams. There are parallels between the representation of a culture in tourism and by nation-states. Hickman writes: 'It has been a main function of national cultures to represent what is in fact the ethnic mix of modern nationality as the primordial unity of "one people"' (p. 132). This happened in Britain in the nineteenth century when a need for national traditions to distract attention from social insecurity and a past which included rebellions in Scotland in 1715 and 1745, and a widespread dislike of the monarch, was answered by Scottish tartans and Morris dancing. Yet, as Hickman says later, 'It is important to emphasise another aspect of cultural identity, one which recognises that as well as many points of similarity, there are also critical points of deep and significant difference which constitute "what we really are" or "what we have become"' (p. 135). This point is picked up in the introduction to Hickman's text, but leads to a further question as to what constitutes the space in which 'what we really are' is determined.

Again there is an overlap with the section on Contesting Identities, but the text by Patricia Phillips included below deals with the issue through understandings of public space, and public art's ambiguous relation to democracy. Public art can be seen as a specialism within the cultural industries, commissioned within urban redevelopment schemes to lend them a sense of local distinctiveness – place-making – by referencing local industries, traditions, or aspects of community. It is open to the same commodification of cultural experiences as heritage or cultural tourism, and also has an ideological role when it provides an aesthetic facade to developments which are socially disruptive. Public art, however, operates at an interface between local distinctiveness and an equally formalised notion of democracy as mutual interaction in a public space. It is helpful to interrogate it here because, as Phillips argues, it has tended to make that space a physically bounded entity, not – as she sees in the north American commons of the eighteenth and nineteenth centuries – a psychological space of speech and action, 'the forum where information was shared and public debate occurred: a charged, dynamic coalescence' (p. 94).

So, public art, cultural and heritage tourism, like the mass media, tend to homogenise experience even when selling local diversity. Perhaps Adorno's grim picture is accurate and it is a matter of mass deception, of a past which never was and a future of dreams which have been made for profit. Or not, which is the challenge for cultural planners. It seems appropriate to return to Adorno's analysis. This is not to say his is the only viable position in relation to the culture industry, nor that his hostility to mass culture is supported by all research on consumption. But it does appear necessary to have well-honed critical tools in face of the globalisation of reductive cultural experience as epitomised by the theme park. Writing on Disney World, Michael Sorkin notes that it receives over thirty million visitors a year, is the first urban environment to be copyrighted, and 'America's stand-in for Elysium . . . the utopia of leisure' ('See You in Disneyland', in *Variations on a Theme Park*, edited by Michael Sorkin, New York, Hill & Wang, 1992, p. 205). He sees a continuity as industrial production systems are applied to the leisure industry: 'The industrial army, raised in the nineteenth century and rationalized in the twentieth, is, at Disneyzone, not dispersed but converted to a vast leisure army, sacrificing nothing in regimentation and disciplines as it consumes its Taylorized fun' (p. 223). The reference is to F. Taylor's principles of scientific management published in 1947 in *Scientific Management* (London, Harper and Row). Robert W. Witkin explains in *Adorno on Popular Culture* (London, Routledge, 2003) that, for Adorno, Taylorism breeds a consciousness devoid of moral integrity while art provides insights into the human condition. While Taylorism expropriates the subject, then, as Witkin writes, 'Adorno seeks repeatedly to analyse the ways in which the subject . . . is undermined by the rising tide of popular culture . . . To those key questions, How shall we live? What shall we do next? Adorno's answer was to resist, to refuse identity with oppressive totalitarian forces' (p. 15).

SUGGESTED FURTHER READING

T. W. Adorno, *The Stars Down to Earth and Other Essays on the Irrational in Culture*, London, Routledge, 1994

Franco Bianchini, 'The Relationship between Cultural Resources and Tourism Policies for Cities and Regions', in *Planning Cultural Tourism in Europe*, edited by D. Dodd and A. van Hemel, Amsterdam, Boekmanstichting, 1999

Franco Bianchini and Michael Parkinson, eds, *Cultural Policy and Urban Regeneration*, London, Routledge, 1993

Graeme Evans, *Cultural Planning: An Urban Renaissance?*, London, Routledge, 2001

Robert W. Witkin, *Adorno on Popular Culture*, London, Routledge, 2003

'The Culture Industry Reconsidered'

from *The Culture Industry* (1991) [1975]

Theodor W. Adorno

Editors' Introduction

As Adorno explains in his opening paragraph, he and Max Horkheimer replaced the term 'mass culture' with 'the culture industry' in their 1944 essay, 'The Culture Industry: Enlightenment as Mass Deception' in *Dialectic of Enlightenment* (London, Verso, 1997, pp. 120–67) because this culture is produced for the mass public and not made by them. It does not arise with spontaneity, as a folk culture might be said to, nor does it offer the mass audience their own preferences, but is manufactured in the interest of the producers. As he also reiterates, the culture industry integrates high and low cultural production, for a common and impoverished reception. Hence the seriousness of high art – which includes its capacity for a critical dimension which cannot be reduced to trivia – is destroyed.

Adorno's position may seem outdated, after cultural studies, sub-cultures, and a renewed interest in vernacular cultures. His retention of the separation of high from low culture may seem elitist, and he is not infrequently criticised for his dislike of jazz – though a charge of racism would be difficult to substantiate. His point, however, remains valid: mass culture trades in illusions, mainly now those of globalised consumerism but hitherto those of happy families and other aspects of the liberal state, while the complex criticism of the social order in which Adorno was engaged cannot be reduced to the simplicity of a sound-bite.

Mass culture, in other words, is ideology: like art, it is determined by economic conditions, and in a bourgeois society will tend to affirm bourgeois attitudes, not least in concealing the underlying contradiction between the universal call for Liberty which characterises the bourgeois revolutions of the eighteenth century, and the impossibility of Liberty for all under capitalist mechanisms of exchange. The argument is far more complex, as Adorno's dense prose suggests, than that, but at its core is the idea that mass culture impedes the growth of autonomous subjects (selves). This text could be read, in this respect, with Kimberley Curtis' chapter on Hannah Arendt (like Adorno a migrant to the United States from Nazi Germany in the 1930s) in Part Seven. It could also be read with Adorno's detailed study of a horoscope column in a mass circulation newspaper, in which he sees an ideological dimension, too, in *Stars Down to Earth and Other Essays on the Irrational in Culture* (edited and introduced by S. Crook, London, Routledge, 1994).

If horoscopes feed irrationality, the value of rationality is its defence against the irrationality of fascism. Rationality is flawed by the association of knowledge with power, but, for Adorno and Horkheimer, can be revised from within rather than junked. At the end of *Dialectic of Enlightenment*, in the essay 'Elements of Anti-Semitism', they write that 'Enlightenment which is in possession of itself and coming to power can break the bounds of enlightenment' (p. 208). And when fascism offers lies which are obvious but persistently accepted, then rationality is evident only in seeing the absence of reason this state of affairs makes clear. Although the deception has no place for rationality, only a complete deprivation of the faculty of thought can prevent it being seen for what it is. This is not unlike the case made by Primo Levi: 'There is no rationality in the Nazi hatred: it is a hate that is not in us; it is outside man [*sic*], it is a poison . . . We cannot understand it, but we can and must understand from where it springs, and we must be on our guard.' Levi continues that the way of being on guard is through memory of what happened, but also though reasoning: 'It is, therefore, necessary to be suspicious of those who seek to convince us with means other than reason, and of charismatic leaders' (Primo Levi, from *If This is a Man*, cited in Roger Griffin, ed., *Fascism*, Oxford, Oxford University Press, 1995, pp. 391–2).

Among introductions to Adorno, Robert W. Witkin's *Adorno on Popular Culture* (London, Routledge, 2003) is perhaps the clearest and most relevant here. In a broader way, Simon Jarvis' *Adorno: A Critical Introduction* (Cambridge, Polity, 1998) is also suggested.

The term culture industry was perhaps used for the first time in the book *Dialectic of Enlightenment*, which Horkheimer and I published in Amsterdam in 1947. In our drafts we spoke of 'mass culture'. We replaced that expression with 'culture industry' in order to exclude from the outset the interpretation agreeable to its advocates: that it is a matter of something like a culture that arises spontaneously from the masses themselves, the contemporary form of popular art. From the latter the culture industry must be distinguished in the extreme. The culture industry fuses the old and familiar into a new quality. In all its branches, products which are tailored for consumption by masses, and which to a great extent determine the nature of that consumption, are manufactured more or less according to plan. The individual branches are similar in structure or at least fit into each other, ordering themselves into a system almost without a gap. This is made possible by contemporary technical capabilities as well as by economic and administrative concentration. The culture industry intentionally integrates its consumers from above. To the detriment of both it forces together the spheres of high and low art, separated for thousands of years. The seriousness of high art is destroyed in speculation about its efficacy; the seriousness of the lower perishes with the civilizational constraints imposed on the rebellious resistance inherent within it as long as social control was not yet total. Thus, although the culture industry undeniably speculates on the conscious and unconscious state of the millions towards which it is directed, the masses are not primary, but secondary, they are an object of calculation; an appendage of the machinery. The customer is not king, as the culture industry would have us believe, not its subject but its object. The very word mass-media, specially honed for the culture industry, already shifts the accent onto harmless terrain. Neither is it a question of primary concern for the masses, nor of the techniques of communication as such, but of the spirit which sufflates them, their master's voice. The culture industry misuses its concern for the masses in order to duplicate, reinforce and strengthen their mentality, which it presumes is given and unchangeable. How this mentality might be changed is excluded throughout. The masses are not the measure but the ideology of the culture industry, even though the culture industry itself could scarcely exist without adapting to the masses.

The cultural commodities of the industry are governed, as Brecht and Suhrkamp expressed it thirty years ago, by the principle of their realization as value, and not by their own specific content and harmonious formation. The entire practice of the culture industry transfers the profit motive naked onto cultural forms. Ever since these cultural forms first began to earn a living for their creators as commodities in the marketplace they had already possessed something of this quality. But then they sought after profit only indirectly, over and above their autonomous essence. New on the part of the culture industry is the direct and undisguised primacy of a precisely and thoroughly calculated efficacy in its most typical products. The autonomy of works of art, which of course rarely ever predominated in an entirely pure form, and was always permeated by a constellation of effects, is tendentially eliminated by the culture industry, with or without the conscious will of those in control. The latter include both those who carry out directives as well as those who hold the power. In economic terms they are or were in search of new opportunities for the realization of capital in the most economically developed countries. The old opportunities became increasingly more precarious as a result of the same concentration process which alone makes the culture industry possible as an omnipresent phenomenon. Culture, in the true sense, did not simply accommodate itself to human beings;

but it always simultaneously raised a protest against the petrified relations under which they lived, thereby honouring them. In so far as culture becomes wholly assimilated to and integrated in those petrified relations, human beings are once more debased. Cultural entities typical of the culture industry are no longer *also* commodities, they are commodities through and through. This quantitative shift is so great that it calls forth entirely new phenomena. Ultimately, the culture industry no longer even needs to directly pursue everywhere the profit interests from which it originated. These interests have become objectified in its ideology and have even made themselves independent of the compulsion to sell the cultural commodities which must be swallowed anyway. The culture industry turns into public relations, the manufacturing of 'goodwill' per se, without regard for particular firms or saleable objects. Brought to bear is a general uncritical consensus, advertisements produced for the world, so that each product of the culture industry becomes its own advertisement.

Nevertheless, those characteristics which originally stamped the transformation of literature into a commodity are maintained in this process. More than anything in the world, the culture industry has its ontology, a scaffolding of rigidly conservative basic categories which can be gleaned, for example, from the commercial English novels of the late seventeenth and early eighteenth centuries. What parades as progress in the culture industry, as the incessantly new which it offers up, remains the disguise for an eternal sameness; everywhere the changes mask a skeleton which has changed just as little as the profit motive itself since the time it first gained its predominance over culture.

Thus, the expression 'industry' is not to be taken too literally. It refers to the standardization of the thing itself – such as that of the Western, familiar to every movie-goer – and to the rationalization of distribution techniques, but not strictly to the production process. Although in film, the central sector of the culture industry, the production process resembles technical modes of operation in the extensive division of labour, the employment of machines and the separation of the labourers from the means of production – expressed in the perennial conflict between artists active in the culture industry and those who control it – individual forms of production are nevertheless maintained. Each product affects an individual air; individuality itself serves to reinforce ideology, in so far as the illusion is conjured up that the completely reified and mediated is a sanctuary from immediacy and life. Now, as ever, the culture industry exists in the 'service' of third persons, maintaining its affinity to the declining circulation process of capital, to the commerce from which it came into being. Its ideology above all makes use of the star system, borrowed from individualistic art and its commercial exploitation. The more dehumanized its methods of operation and content, the more diligently and successfully the culture industry propagates supposedly great personalities and operates with heart-throbs. It is industrial more in a sociological sense, in the incorporation of industrial forms of organization even when nothing is manufactured – as in the rationalization of office work – rather than in the sense of anything really and actually produced by technological rationality. Accordingly, the misinvestments of the culture industry are considerable, throwing those branches rendered obsolete by new techniques into crises, which seldom lead to changes for the better.

The concept of technique in the culture industry is only in name identical with technique in works of art. In the latter, technique is concerned with the internal organization of the object itself, with its inner logic. In contrast, the technique of the culture industry is, from the beginning, one of distribution and mechanical reproduction, and therefore always remains external to its object. The culture industry finds ideological support precisely in so far as it carefully shields itself from the full potential of the techniques contained in its products. It lives parasitically from the extra-artistic technique of the material production of goods, without regard for the obligation to the internal artistic whole implied by its functionality (*Sachlichkeit*), but also without concern for the laws of form demanded by aesthetic autonomy. The result for the physiognomy of the culture industry is essentially a mixture of streamlining, photographic hardness and precision on the one hand, and individualistic residues, sentimentality and an already rationally disposed and adapted romanticism on the other. Adopting Benjamin's designation of the traditional work of art by the concept of aura, the presence of that which is not present, the culture industry is defined by the fact that it does not strictly counterpose another principle to that of aura, but rather by the fact that it conserves the decaying aura as a foggy mist. By this means the culture industry betrays its own ideological abuses.

It has recently become customary among cultural officials as well as sociologists to warn against underestimating the culture industry while pointing to its great importance for the development of the consciousness of its consumers. It is to be taken seriously, without cultured snobbism. In actuality the culture industry is important as a moment of the spirit which dominates today. Whoever ignores its influence out of scepticism for what it stuffs into people would be naive. Yet there is a deceptive glitter about the admonition to take it seriously. Because of its social role, disturbing questions about its quality, about truth or untruth, and about the aesthetic niveau of the culture industry's emissions are repressed, or at least excluded from the so-called sociology of communications. The critic is accused of taking refuge in arrogant esoterica. It would be advisable first to indicate the double meaning of importance that slowly worms its way in unnoticed. Even if it touches the lives of innumerable people, the function of something is no guarantee of its particular quality. The blending of aesthetics with its residual communicative aspects leads art, as a social phenomenon, not to its rightful position in opposition to alleged artistic snobbism, but rather in a variety of ways to the defence of its baneful social consequences. The importance of the culture industry in the spiritual constitution of the masses is no dispensation for reflection on its objective legitimation, its essential being, least of all by a science which thinks itself pragmatic. On the contrary: such reflection becomes necessary precisely for this reason. To take the culture industry as seriously as its unquestioned role demands, means to take it seriously critically, and not to cower in the face of its monopolistic character.

Among those intellectuals anxious to reconcile themselves with the phenomenon and eager to find a common formula to express both their reservations against it and their respect for its power, a tone of ironic toleration prevails unless they have already created a new mythos of the twentieth century from the imposed regression. After all, those intellectuals maintain, everyone knows what pocket novels, films off the rack, family television shows rolled out into serials and hit parades, advice to the lovelorn and horoscope columns are all about. All of this, however, is harmless and, according to them, even democratic since it responds to a demand, albeit a stimulated one. It also bestows all kinds of blessings, they point out, for example, through the dissemination of information, advice and stress reducing patterns of behaviour. Of course, as every sociological study measuring something as elementary as how politically informed the public is has proven, the information is meagre or indifferent. Moreover, the advice to be gained from manifestations of the culture industry is vacuous, banal or worse, and the behaviour patterns are shamelessly conformist.

The two-faced irony in the relationship of servile intellectuals to the culture industry is not restricted to them alone. It may also be supposed that the consciousness of the consumers themselves is split between the prescribed fun which is supplied to them by the culture industry and a not particularly well-hidden doubt about its blessings. The phrase, the world wants to be deceived, has become truer than had ever been intended. People are not only, as the saying goes, falling for the swindle; if it guarantees them even the most fleeting gratification they desire a deception which is nonetheless transparent to them. They force their eyes shut and voice approval, in a kind of self-loathing, for what is meted out to them, knowing fully the purpose for which it is manufactured. Without admitting it they sense that their lives would be completely intolerable as soon as they no longer clung to satisfactions which are none at all.

The most ambitious defence of the culture industry today celebrates its spirit, which might be safely called ideology, as an ordering factor. In a supposedly chaotic world it provides human beings with something like standards for orientation, and that alone seems worthy of approval. However, what its defenders imagine is preserved by the culture industry is in fact all the more thoroughly destroyed by it. The colour film demolishes the genial old tavern to a greater extent than bombs ever could: the film exterminates its imago. No homeland can survive being processed by the films which celebrate it, and which thereby turn the unique character on which it thrives into an interchangeable sameness.

That which legitimately could be called culture attempted, as an expression of suffering and contradiction, to maintain a grasp on the idea of the good life. Culture cannot represent either that which merely exists or the conventional and no longer binding categories of order which the culture industry drapes over the idea of the good life as if existing reality were the good life, and as if those categories were its true measure. If the response of the culture industry's representatives is that it does not deliver art at all, this is itself the ideology with which they evade responsibility for that from which the business lives. No misdeed is ever righted by explaining it as such.

The appeal to order alone, without concrete specificity, is futile; the appeal to the dissemination of norms, without these ever proving themselves in reality or before consciousness, is equally futile. The idea of an objectively binding order, huckstered to people because it is so lacking for them, has no claims if it does not prove itself internally and in confrontation with human beings. But this is precisely what no product of the culture industry would engage in. The concepts of order which it hammers into human beings are always those of the status quo. They remain unquestioned, unanalysed and undialectically presupposed, even if they no longer have any substance for those who accept them. In contrast to the Kantian, the categorical imperative of the culture industry no longer has anything in common with freedom. It proclaims: you shall conform, without instruction as to what; conform to that which exists anyway, and to that which everyone thinks anyway as a reflex of its power and omnipresence. The power of the culture industry's ideology is such that conformity has replaced consciousness. The order that springs from it is never confronted with what it claims to be or with the real interests of human beings. Order, however, is not good in itself. It would be so only as a good order. The fact that the culture industry is oblivious to this and extols order *in abstracto*, bears witness to the impotence and untruth of the messages it conveys. While it claims to lead the perplexed, it deludes them with false conflicts which they are to exchange for their own. It solves conflicts for them only in appearance, in a way that they can hardly be solved in their real lives. In the products of the culture industry human beings get into trouble only so that they can be rescued unharmed, usually by representatives of a benevolent collective; and then in empty harmony, they are reconciled with the general, whose demands they had experienced at the outset as irreconcileable with their interests. For this purpose the culture industry has developed formulas which even reach into such non-conceptual areas as light musical entertainment. Here too one gets into a 'jam', into rhythmic problems, which can be instantly disentangled by the triumph of the basic beat.

Even its defenders, however, would hardly contradict Plato openly who maintained that what is objectively and intrinsically untrue cannot also be subjectively good and true for human beings. The concoctions of the culture industry are neither guides for a blissful life, nor a new art of moral responsibility, but rather exhortations to toe the line, behind which stand the most powerful interests. The consensus which it propagates strengthens blind, opaque authority. If the culture industry is measured not by its own substance and logic, but by its efficacy, by its position in reality and its explicit pretensions; if the focus of serious concern is with the efficacy to which it always appeals, the potential of its effect becomes twice as weighty. This potential, however, lies in the promotion and exploitation of the ego-weakness to which the powerless members of contemporary society, with its concentration of power, are condemned. Their consciousness is further developed retrogressively. It is no coincidence that cynical American film producers are heard to say that their pictures must take into consideration the level of eleven-year-olds. In doing so they would very much like to make adults into eleven-year-olds.

It is true that thorough research has not, for the time being, produced an airtight case proving the regressive effects of particular products of the culture industry. No doubt an imaginatively designed experiment could achieve this more successfully than the powerful financial interests concerned would find comfortable. In any case, it can be assumed without hesitation that steady drops hollow the stone, especially since the system of the culture industry that surrounds the masses tolerates hardly any deviation and incessantly drills the same formulas on behaviour. Only their deep unconscious mistrust, the last residue of the difference between art and empirical reality in the spiritual make-up of the masses explains why they have not, to a person, long since perceived and accepted the world as it is constructed for them by the culture industry. Even if its messages were as harmless as they are made out to be – on countless occasions they are obviously not harmless, like the movies which chime in with currently popular hate campaigns against intellectuals by portraying them with the usual stereotypes – the attitudes which the culture industry calls forth are anything but harmless. If an astrologer urges his readers to drive carefully on a particular day, that certainly hurts no one; they will, however, be harmed indeed by the stupefication which lies in the claim that advice which is valid every day and which is therefore idiotic, needs the approval of the stars.

Human dependence and servitude, the vanishing point of the culture industry, could scarcely be more faithfully described than by the American interviewee who was of the opinion that the dilemmas of the contemporary epoch would end if people would simply follow the lead of prominent personalities. In so far as the culture industry arouses a feeling of well-being that the world is precisely in that order suggested by the culture industry, the substitute gratification which it prepares for human beings cheats them out of the same happiness which it deceitfully projects. The total effect of the culture industry is one of anti-enlightenment, in which, as Horkheimer and I have noted, enlightenment, that is the progressive technical domination of nature, becomes mass deception and is turned into a means for fettering consciousness. It impedes the development of autonomous, independent individuals who judge and decide consciously for themselves. These, however, would be the precondition for a democratic society which needs adults who have come of age in order to sustain itself and develop. If the masses have been unjustly reviled from above as masses, the culture industry is not among the least responsible for making them into masses and then despising them, while obstructing the emancipation for which human beings are as ripe as the productive forces of the epoch permit.

'Reinterpreting Local Culture'

from *Consuming Places* (1995)

John Urry

Editors' Introduction

John Urry's work on tourism is possibly best known from his book *The Tourist Gaze* (London, Sage, 1990). In it he first introduced a concept of the tourist consumption of place, using the term gaze. This is derived in its wider use as the masculine gaze, though Urry does not go into this, from film theory, in particular from the work of Laura Mulvey. Urry's approach, however, was that the tourist gaze is constructed in relation to the non-tourist gaze, as a kind of deviation from it. In part, Urry was obliged to argue for the study of tourism as a serious concern. But in part he was also extending a methodology from fields such as anthropology and cultural studies, in which deviance is a route to identification of norms, and shows them to be constructed in specific ways – to be historical not natural. Urry writes: 'Just why various activities are treated as deviant can illuminate how different societies operate' (p. 2).

In *Consuming Places* (London, Routledge, 1995), Urry establishes three main arguments: that places are complex, their investigation requiring multiple tools; that social theory has not been good at dealing with the idea of place because appropriate insights into time, space and nature have not been available; and that places are consumed in ways as yet not fully understood. The growth of cultural and heritage tourism over recent years, and the centrality of the cultural, heritage and associated tourist industries in city marketing and development has not in general been accompanied by a related expansion of a critical literature, and Urry remains one of the few people working in this area. His book could be compared, however, with an essay by Chris Rojek, 'Mass Tourism or the Re-enchantment of the World? Issues and Contradictions in the Study of Travel', in *New Forms of Consumption: Consumers, Culture, and Commodification*, edited by Mark Gottdiener (Lanham, MD, Rowman & Littlefield, 2000, pp. 51–70). Rojek sees tourism as having been industrialised over the past century, with the commodification of experience that involves; but also suggests that tourists, in a postmodern world, are able to play knowingly with tourist promotion. The marketed product, or commodified place, may be seen for what it is – kitsch – while not engendering a compensatory yearning for an authenticity which may not exist anyway. Rojek cites two further texts, both by Dean MacCannell (whose essay on Celebration, Florida, is included in Part Eight): *The Tourist* (New York, Schocken, 1976) and *Empty Meeting Grounds* (London, Routledge, 1992). Urry's chapter below looks at local cultures such as that of post-industrial Lancashire in the north of England. Urry is Professor of Sociology at Lancaster University, so this is local to him. Key issues here are whose past is represented for whom, and the extent to which tourism is a replacement industry for those it packages in museums and heritage sites. Urry's conclusion is that lack of investment means that British cities and towns are unable to compete effectively with those of other European countries. This does not reflect the 'rich reservoir of history' many of them conserve (p. 162).

GLOBAL AND THE LOCAL

In this chapter I will criticise two positions: that which suggests that processes of globalisation are producing economic, political and cultural homogenisation; and that which claims that what is most important about the contemporary world is the surprising emergence of locally distinct cultures. Both of these accounts are one-sided and instead what we need to analyse are the complex interconnections of *both* global and local processes. It is the *interconnections* between them which account for the particular ways in which an area's local history and culture is made available and transformed into a resource for local economic and social development within a globally evolving economy and society.

Deciphering these interconnections is complex and cannot be reduced to an examination of the economy on its own. In particular it involves analysis of the nature of social flows, of people, information, companies, ideas and images. Such flows do not take place in a vacuum but depend partly on cultural processes. These include certain tendencies towards globalised forms of culture but as we will see these do not necessarily produce cultural homogenisation. The following are some of the main forms taken by these global cultural flows: new forms of global communication including satellite technologies and massive media conglomerates which 'collapse space and time' (Brunn and Leinbach 1991); the development of international travel and of 'small worlds' little connected to nation-state relationships; the increasing numbers of international agencies and institutions; the development of global competitions and prizes; the emergence of a small number of languages of communication, most notably English; and the development of more widely shared notions of citizenship and of political democracy (see Lash and Urry 1994, for more detail).

I suggested earlier that what we need to consider are various flows. This thesis has been amplified by Appadurai who attempts to detail five different dimensions of such global cultural flows (1990). These dimensions move in non-isomorphic paths and challenge simple notions of a cultural centre and a subordinate periphery. They constitute building blocks for what Appadurai terms 'imagined worlds', the multiple worlds constituted by the historically situated imaginations of persons and groups spread across the globe. Such worlds are fluid and irregularly shaped.

The five dimensions of such global cultural flows are *ethnoscapes* – the moving landscape of tourists, immigrants, refugees, exiles, guestworkers and so on; *technoscapes* – the movements of technologies, high and low, mechanical and informational, across all kinds of boundaries; *finanscapes* – the movement of vast sums of monies through national turnstiles at bewildering speed, via currency markets, national stock exchanges and commodity speculations; *mediascapes* – the distribution of electronic capabilities to produce and disseminate images and the proliferation of images thereby generated; and *ideoscapes* – concatenations of images often in part linked to the ideologies of states or of movements of opposition (Appadurai 1990: 296–300).

This chapter is concerned with the interconnections between these various flows and the shapes that they take, and with the resulting economic and social organisation of particular towns, cities and regions especially within Britain. I shall presume the following. First, these flows in part derive from very particular places from which in a sense they derive – such as the financial flows which in the British case stem from the history, traditions and spatial form of the 'square mile' of the City of London (Lash and Urry 1994: Ch. 11).

Second, these flows impact upon particular towns and cities in often unexpected and counter-intuitive ways. There are many different kinds of flow as we have seen and the impact of their non-isomorphic shapes can produce distinct non-homogenised outcomes in particular places (see Bagguley *et al.* 1990).

Third, the effect of globalisation is often to increase local distinctiveness for one or more of the following reasons: the increased ability of large companies to subdivide their operations and to locate different activities within different labour markets located in different societies (see discussion in Bagguley *et al.* 1990); the breaking up of previously relatively coherent regional economies; the competition between local states for jobs, the growth of international differences and the localising of regional policy (see discussion in Harloe *et al.* 1990); the decreasing tendency for voting patterns to be nationally determined and the increased importance of 'neighbourhood' effects; the enduring significance of symbols of place and location particularly with the decline in the popularity of the international modern style of architecture and the emergence of local and vernacular styles; and the resurgence of locally oriented culture and politics especially around campaigns for the conservation of the built and physical environment.

Globalization is, in fact, also associated with new dynamics of *re*localization. It is about the achievement of a new global-local nexus, about new and intricate relations between global space and local space. Globalization is like putting together a jigsaw puzzle: it is a matter of inserting a multiplicity of localities into the overall picture of a new global system.

(1991: 34–5)

TRANSFORMING HISTORY AND CULTURE

I am mostly going to refer in this chapter to transformations in the interpretation of the history and culture of industrial Lancashire. Before doing so though I need to make a number of further points about the context in which such transformations took place.

Initially, it is important to note that in terms of employment creation, since most people in Britain work in services, it is the geographical location of service-sector enterprises which is of most significance for the distribution of employment. Further, even in the case of manufactured goods there is an increasing design element built into them. This means that a range of services are particularly pertinent to such manufacturing industry, that is, those services concerned with their design. One particularly important set of industries in contemporary Britain are so-called cultural industries which include music, television, cinema, publishing, leisure and tourism. Cultural industries are concerned in part with the re-presentation of the supposed history and culture of a place. Many authorities have begun to develop a specific strategy with regard to the arts and culture, with designated arts areas or corridors, as in Sheffield, Liverpool and Glasgow.

This in turn is related to the way in which many places in Britain have begun to develop policies designed to attract both tourists and incoming entrepreneurs and their employees to their area. Such policies have mainly involved developing the range of appropriate services available. And this in turn has partly at least involved efforts to re-present the history and culture of their area. The following are some of the reasons why this occurred in the 1980s, so much so that almost every town and city in Britain has now developed a 'tourism strategy'. The re-presentation of an area's history and culture is also seen as relevant to the attraction of new employees and managers.

First, there was the astonishingly rapid de-industrialisation of many towns and cities in Britain in the late 1970s and 1980s. This created a profound sense of loss, both of certain kinds of technology, factories, steam engines, blast furnaces and pit workings, and of the patterns of social life that developed around such technologies. The sense of loss was found in very many places, especially where the work had involved backbreaking and apparently heroic labour by men. The transformation of coal mines into museums in south Wales is perhaps the best example of this development (in 1994 the only pit now in south Wales is a museum).

Second, the costs of job creation in tourist and leisure-related services compared with manufacturing industry are very much lower (up to one-eighth). Moreover, local councils have been willing to engage in leisure and tourism projects because this is one area where there are funds available to initiate projects which may also benefit local residents. Funds have been available both from central government and from the European Community. Such facilities have been important not only in attracting visitors but also prospective employees and employers and in keeping them satisfied once they have relocated. One place in Lancashire where this has clearly happened is Wigan following the establishment of the Wigan Pier Heritage Centre. The Chairman of the North West Tourist Board argues that:

> The growth of the tourism industry has a great deal to do with the growth of every other industry or business: the opening up of the regions as fine places to visit means that they're better places to live in – and thus better places to work . . . a higher quality of life benefits employees.
>
> (cited in Reynolds 1988)

Third, this de-industrialisation mainly took place in the 1980s in northern towns and cities, especially within inner-city areas. It resulted in a paradoxically useful legacy of derelict buildings from the Victorian era. Some of these were intrinsically attractive, such as the Albert Dock in Liverpool, or could be refurbished in a picturesque heritage style, such as the White Cross Mill in Lancaster (on the Albert Dock, see Mellor 1991). A heritage style normally entails sandblasted walls, Victorian street furniture, replaced but 'authentically' appearing windows and brightly painted doors, pipes and balconies. Such derelict buildings have often been suitable for conversion to educational, cultural and leisure uses. Further, the preservation of such vernacular buildings has been particularly marked because of the mostly unappealing character of modern architecture in Britain. Modern buildings have been particularly unpopular, partly because many were built in the 1960s using concrete as the predominant building material (see Coventry city centre for example). The criticisms of such buildings expressed by Prince Charles resonate well with popular sentiment in Britain. Vernacular buildings by contrast appear very attractive to British people and well worth preserving. The novelist Margaret Drabble writes of:

> a sort of stubborn English philistinism about architecture and city life, encouraged by the wilder utterances of the Prince of Wales. Not for us the pride of Paris in its pyramid, in its brave and soaring arch; not for us the multi-coloured panache of Stirling's Staatsgalerie in Stuttgart. . . . One of the reasons for our current architectural timidity lies in the failure of post-war high-rise and deck-access council building.
>
> (1991: 33)

And this in turn raises a more general issue, namely the process of conservation and its relationship to the symbols of history and culture. In Britain there is an extremely well-organised movement for conservation in many rural areas, towns and cities and this can exert a veto over certain kinds of new building development. The strength of such groups varies considerably between different places. In 1980, for example, while there were 5.1 members of such societies per 1000 population in the UK as a whole, the ratio was over 20 per 1000 in Hampshire and over 10 per 1000 in most of the counties around London, in Devon and in Cumbria in the north-west of Britain (Lowe and Goyder 1983: 28–30). Partly this variation varies in response to where there is deemed to be a 'history and culture' worth preserving (see Cowen 1990, on Cheltenham for example). But in some ways it seems that almost anything can be seen as worth conserving, including in Lancashire a slag heap from a coal mine! Preservation also need not be of a building – it can be of a road layout or of the shape of an original building. In Lancaster, for example, there has been a huge protest movement against the re-siting of a Victorian market hall after the original had been destroyed by fire (the distance involved was about half a kilometre).

There are two further points to note about such conservation. First, normally conservation is sought in relationship to some aspect of the built environment which is taken to stand for or represent the locality in question. It is not merely that the object is historical, but that the object signifies the place and that if the object were to be demolished or substantially changed then that would signify a threat to the place itself. This can be seen in debates in Morecambe in Lancashire and the planned refurbishment of the Winter Gardens Theatre. There is an extremely energetic campaign to preserve this theatre which in a sense has come to stand *for the town*. This semi-derelict building has come to symbolise the town itself – if it is demolished then the town itself is thought to have no viable future.

Second, there is nothing inevitable about conservation. It would be perfectly possible to permit many buildings to be destroyed and to build new ones in their place. And this does of course happen, especially in the larger cities in Britain. However, since the planning and building disasters in the 1960s this is often resisted by conservation groups, normally called civic societies, many of which were established in the 1960s. These are normally set up by those in professional and managerial occupations and may then conflict with the interests of large-scale developers. The members of such groups may not necessarily originate from that particular town or city. There appear to be relatively lower rates of geographical mobility among at least the male professionals and managers in Britain. As a result they often develop a strong attachment to 'their place' – what is sometimes called the service class becomes 'localised' and this works its way into the formation of local conservation groups who energetically resist large-scale plans for office and retail development.

Although such civic societies may not literally be 'locals', they often articulate a strong sense of nostalgia for that place. They will suggest that there is a profound sense of loss of one's 'home' resulting from various economic and social changes. This in turn depends upon a particular structuring of the collective memory which is reinforced by various enacted ritual performances. However, much of the 'nostalgia' and 'tradition' of the place may in fact be invented by these conservation groups who articulate a set of particular aesthetic interests often based on the concept of 'community'. This gives rise to attempts to preserve otherwise derelict property such as the Albert Dock in Liverpool, and to favour any new building that is necessary, being built in the local style as opposed to the international architectural style of modernism. This localism or neo-vernacularism in architecture is often developed and encouraged by firms of local architects, the numbers of which have been rapidly growing in recent years.

LANCASHIRE

In this section I will consider some of the ways that history and culture are being used in the efforts to regenerate industrial Lancashire, an area located in the north west of England. This is where the first ever industrial revolution took place. The development of industry worldwide began initially in the cotton textile villages and towns of Lancashire in the late eighteenth century. This area therefore has first, a unique claim as the original site of industrialisation; second, it has the longest history of industrialisation anywhere in the world; and third, it has a culture which is almost literally built around manufacturing industry. For two centuries then the culture of Lancashire has been based upon the work relations of textile mills, coal mines and factories. The best known artist from the area, L. S. Lowry, made his name with his heavily urbanised and industrialised paintings, hardly the typical pastoral scenes of English landscape painting. It should also be noted that these towns and cities of industrial Lancashire were built close to rivers, often at the bottom of quite deeply cut river valleys. Surrounding them is a lot of impressively wild and attractive countryside, especially the Pennine range of hills.

So there is in Lancashire a unique history, an area that for two centuries has been moulded by industry. The main industries have been textiles, coal mining, textile engineering, linoleum and various sorts of engineering including aerospace. More or less all the towns are industrial. The main exceptions consist of those located on the coast. These became in the mid- to late-nineteenth century the very first holiday resorts for the lower middle and skilled working classes. Here developed the first *mass* resorts, including probably the most famous resort in the world, Blackpool. There is then a second kind of history to be found in the area, a history of mass leisure which has in a way become as representative of the area as has the cotton textile industry itself. A symbiotic culture therefore exists in Lancashire, of factory-based industry and of mass leisure, the two being heavily intertwined (see Urry 1990; Ch. 2 for more detail on the work–leisure interconnections in Lancashire).

The history of the twentieth century in Lancashire has been one of long-term decline for many of these textile towns. This has involved: the emergence of various new industries especially between the 1920s and the 1960s; the rapid de-industrialisation of the area from the 1970s onwards; the corresponding weakening of the seaside leisure industry from the 1960s onwards especially as foreign competition emerged from the Mediterranean region; and the unexpected emergence of industrial and urban tourism in the 1980s. What I am going to analyse briefly is the relationship between de-industrialisation on the one hand, and urban and industrial tourism on the other.

It is first worth pointing out that twenty years ago no one in Britain would have contemplated visiting industrial Lancashire *by choice*. It would have been travelled to only for business or for visiting friends. Likewise no one would have considered that it possessed a history that was in any way interesting. It was a 'place on the margin' of British life, a place rendered peripheral by virtue of global economic process (see Shields 1991, on the place-image of 'the north'). The culture of the area was not thought of as worth knowing about. It was 'up there', well away from the supposed centres of British public and artistic life which have for some centuries been based in the south east of the country, in the so-called 'home counties' surrounding London. To the extent that 'southerners' visited Lancashire it was to go to Blackpool and the other resorts – but even this was not that common and mostly undertaken to confirm prejudices about the 'uncivilised' northerners whose tastes were viewed as 'other', as not really English, as irredeemably uncultured.

George Orwell for example in *The Road to Wigan Pier* talked of a 'line north of Birmingham to demarcate the beginning of "the real ugliness of industrialisation"'. As a southerner, Orwell was conscious when travelling north in the late 1930s of 'entering a strange country . . . [which is] partly because of certain real differences which do exist, but still more because of the north–south antithesis which has been rubbed into us for such a long time past' (1959 [1937]: 106–7). Likewise he ridiculed the working class holiday camp. He talked of how muzak would be playing in the background 'to prevent the onset of that dreaded thing – thought' (cited in Hebdige 1988: 52).

The only exception to this generally unfavourable

image of the north and particularly of Lancashire was the belief that most people in fact lived in warm-hearted 'working class communities'. It was believed that these were solidaristic, that they involved a great deal of mutual support and advice in times of trouble, that there were very close-knit contacts between family and neighbours, and that leisure was organised collectively rather than individually.

So these then are some components of the history and culture of Lancashire which have come to be reassessed in the past decade and a half. No longer is Lancashire seen as somewhere merely to pass through, as merely on the margins. The neighbouring city of Liverpool now attracts 20 million visitors a year; Manchester has recently enjoyed a huge revival of fortune, particularly artistically; and almost every town and city in Lancashire seeks both to attract visitors and permanent residents partly through repackaging its history and culture. In other words, what was seen as a set of characteristics which were peripheral to mainstream British life have now been reassessed. As working industry has disappeared so vast numbers of people seem to be fascinated by the memories of that industry and of the forms of life that were associated with it. Britain seems to be engulfed by a vast collective nostalgia in which almost anything from the past, whether an 'old master' or an old cake tin, is viewed as equally interesting and well worth visiting. The Director of the Science Museum in London has said of this growth in heritage that: 'You can't project that sort of rate of growth much further before the whole country becomes one big open air museum, and you just join it as soon as you get off at Heathrow' (quoted in Hewison 1987: 24).

There are now over half a million listed buildings in Britain; a new museum opens every fortnight including many with an industrial theme, more people visit museums and galleries than the cinema and three-quarters of overseas visitors to Britain visit a museum or gallery during their stay (see Urry 1990: 105–6).

One of the most interesting attempts to re-present history has taken place in Wigan in Lancashire, about which Orwell wrote his classic work *The Road to Wigan Pier* in the late 1930s. Incidentally, the pier had been used for loading coal – it was not a seaside pier. I noted above the way in which various local authorities have begun to see in tourist-related developments a way of both generating jobs directly and developing more general publicity about their area. The latter is designed to attract prospective managers and their families. Wigan has attempted to do this via a publicity booklet entitled *I've Never Been to Wigan but I Know What it's Like* (Economic Development, Wigan: undated). The first five pictures in black and white are of back-to-back terraced housing, mines and elderly residents walking along narrow alleyways. But we are then asked if we are sure that this is really what Wigan is now like. The following twelve photos, all in colour, demonstrate contemporary Wigan, which is revealed as possessing countless tourist sites, including the award winning Wigan Pier Heritage Centre, a colourful market and elegant shops, excellent sports facilities, attractive pubs and restaurants, and delightful canalside walkways. Selling Wigan to tourists is then part of the process of selling Wigan to potential investors, who are going to be concerned about the availability of various kinds of services for their employees.

What though is interesting about this is the way in which the industrial past is part of what gets sold. The Wigan Pier Heritage Centre attracts over one million visitors a year. It has unashamedly re-presented the industrial and social history of Wigan albeit in a way which is certainly somewhat sanitised. The set of buildings by the canal have been cleaned up and given a 'heritage' look. Moreover, that history also includes George Orwell and the famous book written about the town. So the history to be re-presented is complex. It is both the industrial and social history of Wigan and it is Orwell's visit to the town. There is for instance a bar in the Heritage Centre called The Orwell.

In Hewison's famous examination of the 'heritage industry' in Britain, Wigan Pier is his first port of call, the first representation of the past to be critiqued (1987). He condemns the way in which the agenda of heritage promotes a mythical English idyll of harmony and community and a romanticised and glamorised industrial past. The effect of this commodification of history systematically distorts attention from the present, from contemporary polarisations and conflicts. He draws a strong distinction between an authentic history, continuing and therefore dangerous, and a packaged heritage, past, dead and safe. The protection of the past conceals the destruction of the present. Indeed Hewison argues that if we are really interested in history then we may need to preserve it from the conservationists. Heritage is for him bogus history.

His arguments do have a a certain plausibility but there are some points to make in opposition, points revealed by the case of the Wigan Pier Heritage Centre. It is educational, even recreating something of the appearance of an old school room; it presents a history of popular struggles against employers and town bosses; it partly blames the employers for mining disasters; and it celebrates a non-elite culture as no longer marginal. Wigan Pier was organised by a Labour local authority and much of the text has been written by professional historians. It also attracts considerable numbers of local people as well as tourists and encourages a degree of active engagement rather than passive acceptance from the visitors. One might also note that most people's understanding of history is rather sketchy and ill-formed. It is not obvious that the Wigan Pier experience is worse than say reading historical novels. Mellor presents a robust defence of sites such as Wigan Pier:

> when you ask other visitors what they are doing there, it turns out many of them . . . are reminiscing. They do so, not simply in passive deference to Wigan Pier's own construction of Wigan life at the turn of the century, but actively using the displays, reconstructions, and discourses of the actors . . . as the point of departure for their own memories of a way of life in which economic hardship and exploited labour were offset by a sense of community, neighbourliness and mutuality.
>
> (1991: 100)

There is however one aspect of the representation of history involved in such sites of industrial heritage which is problematic. This is that heritage history is distorted because of the emphasis upon visualisation, on presenting visitors with an array of artefacts, including buildings, and then encouraging visitors to try to visualise the patterns of life that would have formed around those artefacts. This might then be termed an 'artefactual' history, in which a whole variety of social experiences are trivialised or marginalised (see Urry 1990: 112).

Finally, in this section I will consider the city of Lancaster that has been seeking to turn itself into a heritage city by re-using its history and culture. It seems that there are three preconditions that have to be met for the construction of a heritage city. First, there has to be legacy of a number of attractive and well-preserved buildings from a range of historical periods. In Lancaster's case these are medieval (a castle), eighteenth century (many town houses), nineteenth century (many mills and further town houses) and interwar (art deco hotel). Second, such buildings would have to be used for purposes in some way consistent with their use as tourist sites. Currently much of the prison which possesses a magnificent Norman gate is not open to the public, and in fact is used as a prison, a use that conflicts with its potential as a tourist site. The third condition for Lancaster to become a plausible heritage city is that the buildings should in some sense have been significant historically, that they stand for or signify important historical events, people or processes. In one report Lancaster is thus described as:

> an ancient settlement steeped in history, with Roman origins, an important medieval past. . . . Through the Duchy of Lancaster it has close associations with the Monarchy. . . . The city's many attractions, based on its rich history and fine buildings, together with its royal associations, combine for the promotion and marketing of Lancaster's heritage.
>
> (cited Urry 1990: 118)

CONCLUSION

Thus I have endeavoured to set out some of the ways in which history and culture have been employed within some north-western towns and cities in Britain, employed as part of a strategy of urban regeneration. In conclusion though, it is important to note some of the constraints under which such places have been operating.

First, there have been enormous funding difficulties as the Conservative national government has attempted to minimise the role and scale of local government intervention (for details, see Pickvance 1990). Second, and related so this, have been the efforts by the same government to 'Americanise' urban policy, to find private sector solutions and to minimise the importance of infrastructure, of local government and of public planning (as most clearly shown in the Canary Wharf débâcle in the London Docklands; see Bianchini and Schwengel 1991). Ideas about economic regeneration have 'flowed' to Britain from the US.

The final constraint is that British towns and cities are poorly placed to compete successfully with some European cities which have been able to plan their use of history and culture in a strategic fashion. Public funding has been available to link together different elements particularly via a strategy which has provided extensive support both for the arts and for a public infrastructure. Glasgow in Scotland is probably the British city that has best been able to effect such a 'European' transformation especially through its newsworthy designation as the European City of Culture in 1990 (see Wishart 1991). But in general flows of attractive images of some European cities have weakened the competitive position of many British cities, even those which had a particularly rich reservoir of history and culture to mobilise. Global competition can be a demanding and relentless taskmaster!

NOTE

This was first given as a lecture to The International Conference on Comparative Regional Studies, Tohuku University, Sendai, Japan, 1992. Discussions with Dan Shapiro have helped to develop the argument here.

REFERENCES

Appadurai, A. (1990) 'Disjuncture and difference in the global cultural economy', *Theory, Culture and Society*, 7: 295–310.

Bagguley, P., Mark-Lawson, J., Shapiro, D., Urry, J., Walby, S. and Warde, A. (1990) *Restructuring. Place, Class and Gender*, London: Sage.

Bianchini, F. and Schwengel, H. (1991) 'Re-imagining the city', in J. Corner and S. Harvey (eds) *Enterprise and Heritage*, London: Routledge.

Brunn, S. and Leinbach, T. (eds) (1991) *Collapsing Space and Time*, London: Harper Collins.

Cowen, H. (1990) 'Regency icons; marketing Cheltenham's built environment', in M. Harloe, C. Pickvance and J. Urry (eds) *Place, Policy and Politics: Do Localities Matter?*, London: Unwin Hyman.

Drabble, M. (1991) 'A vision of the real city', in M. Fisher and U. Owen (eds) *Whose Cities?*, Harmondsworth: Penguin.

Economic Development (undated) *I've Never Been to Wigan, But I Know What it's Like*, Wigan: Economic Development.

Harloe, M., Pickvance, C. and Urry, J. (eds) (1990) *Place, Policy, Politics: Do Localities Matter?*, London: Unwin Hyman.

Hebdige, D. (1988) *Hiding in the Light*, London: Routledge.

Hebdige, D. (1990) 'Fax to the future', *Marxism Today*, January: 18–23.

Hefferan, J. (1985) *The Recreation of Landscape*, Hanover: University Press of New England.

Hewison, R. (1987) *The Heritage Industry*, London: Methuen.

Hewison, R. (1993) 'Field of dreams', *Sunday Times*, 3 January.

Lash, S. and Urry, J. (1984) 'The new Marxism of collective action: a critical analysis', *Sociology*, 18: 33–50.

Lash, S. and Urry, J. (1987) *The End of Organized Capitalism*, Cambridge: Polity Press.

Lash, S. and Urry, J. (1994) *Economies of Signs and Space*, London: Sage.

Mellor, A. (1991) 'Enterprise and heritage in the dock', in J. Corner and S. Harvey (eds) *Enterprise and Heritage*, London: Routledge.

Orwell, G. (1938) *Homage to Catalonia*, London: Secker & Warburg.

Orwell, G. (1959) [1937] *The Road to Wigan Pier*, London: Secker & Warburg.

Pickvance, C. (1990) 'Introduction', in M. Harloe, C. Pickvance and J. Urry (eds) *Place, Policy and Politics: Do Localities Matter?*, London: Unwin Hyman.

Reynolds, H. (1988) '"Leisure revolutions" – prime engines of regional recovery'. *Daily Telegraph*, 2 December.

Robins, K. (1991) 'Tradition and translation: National culture in its global context', in J. Corner and S. Harvey (eds) *Enterprise and Heritage*, London: Routledge.

Shields, R. (1991) *Places on the Margin*, London: Routledge.

Urry, J. (1990) *The Tourist Gaze*, London: Sage.

Wishart, R. (1991) 'Fashioning the future – Glasgow', in M. Fisher and U. Owen (eds) *Whose Cities?*, Harmondsworth: Penguin.

'Barcelona: The Making of a Cultural City'

from Dianne Dodd and Annemoon van Hemel (eds), *Planning Cultural Tourism in Europe* (1999)

Dianne Dodd

Editors' Introduction

In contrast to Urry's tale of woe from Lancashire, Dodd presents an up-beat picture of Barcelona. The comments on the city's recent waterfront development which introduce Plates 1–10 could be noted in conjunction with her essay, nonetheless. As Dodd describes, Barcelona's municipal authorities and non-profit savings banks have invested heavily in a local cultural infrastructure. This emphasises its Catalan rather than Spanish (or Castilian) identity, and through that fuses the city's role as a centre of contemporary art, architecture, and fashion with its historical fabric from the nineteenth century and its medieval past (as an independent state until 1714). Although the 1992 Olympics were a key point in the city's move from neglect under the fascist regime (when Catalan was a banned language, and the city's history of resistance to the fascist coup remembered) to a renewed prosperity and cultural life after the liberalisation, investment in culture, and in high-quality public spaces, was a longer process, and continues.

As Dodd explains, the policy which delivered Barcelona's currently thriving tourist industry was initiated in the 1980s as a move away from mass market tourism – as at the beaches of the Costa del Sol – towards a more selective marketing as a cultural destination. This put its promotion in conflict with Spain's national promotion at the time as an inexpensive vacation place with plenty of sun, sand, and cheap alcohol. Barcelona, in contrast, appealed to tourists who saw themselves as travellers, and would be intrepid in seeking out local culture, either as an 'authentic' bar, or as cultural experience in galleries, museums, theatres, and concert halls. The development of a specifically local culture thus served a dual purpose as meeting a local need and, at the same time, meeting a tourist need. The more notices of cultural events were in Catalan the better because the more authentic the events thus became.

Dodd is a freelance researcher working on areas such as new employment patterns in relation to cultural planning. Her essay is in a collection which also includes an examination of the pitfalls of heritage tourism in Bergen by Siri Myrvoll, which is a useful balance to Dodd's enthusiasm; and a piece by Franco Bianchini on cultural resources and tourism in cities and regions. Further useful material, if a little outdated now, is in *Cultural Policy and Urban Regeneration: The West European Experience*, edited by Franco Bianchini and Michael Parkinson (London, Routledge, 1993).

Sun, sand and sangria have ensured that Spain, during the past three decades has become one of the prime package tourism centres in Europe. According to information given by the Autonomous Government of Catalonia, Spain is one of the principle receivers of tourists in the World (second only to France and the United States) attracting 7% of the world market and 12% of the European market.[1] Catalonia, more than any other region in Spain has exploited its natural resources to offer millions of tourists including German, English and Dutch holiday makers a variety of beach and/or skiing holidays. This investment has obviously paid off because Catalonia now attracts 1.7% of the world market and 24.1% of all tourists visiting Spain chose Catalonia.[2]

ESPAÑA – SOL

The principle attraction for tourists visiting these areas was, and still is, sun and beach. The Spanish government has encouraged this type of tourism by developing promotional marketing tools to ensure the word 'sun' is synonymous with Spain 'España – Sol'. Younger tourists were attracted by cheap alcohol, affordable package tours (including flight and accommodation) and the exceptional night life activity expounded by the fact that hundreds of other young tourists shared that 'holiday feeling'. The sun package left little room for cultural activity and so while the economic profit provided by mass package tourism was great, the cultural deficit was high. If culture was evident at all in a holiday itinerary it was usually a very marginal part of the trip i.e. a visit to see Gaudi's cathedral might feature in one afternoon during a two-week holiday. Other than that, tourists might see some flamenco dancing or hear some Spanish music but increasingly in Spanish resorts the novelty of these sights wore thin and tourists opted, on mass, to relax and enjoy their holiday rather than seek educational goals. Package tour operators seeking to keep their profits high would also opt for mainstream attractions, offering cheap thrills to tourists including water parks, beach games and poolside entertainment. While package tourism has served a strong economic purpose in Spain and particularly in Catalonia, it grew so rapidly that vast stretches of the Costa Brava, Costa Dorada and Costa del Sol were compromised by huge tower blocks.

In Catalonia many examples of this phenomenon can be provided from Benidorm to Salou. A saddening example is the case of Tarragona, a large coastal town situated south of Barcelona, which pulled down Roman walls to make room for new hotels (and is presently reconstructing them!). Other traits in these holiday resorts included Spanish restaurants making room for Fish and Chip shops and Andalucian Flamenco music being displayed to gullible tourists who believed that this was the folklore culture of the place in which they were staying.

INDIVIDUAL TOURISTS

It was after witnessing this extreme abuse of local resources and the disrespect evident from tourists visiting the coastal areas of Catalonia, that approximately twelve years ago serious questions about cultural recognition and tourism were being asked in Catalonia. Cultural concerns had grown with increased regional pride that was steadily accumulating in the post-Franco environment. Catalans were fed up with seeing 'Spanish' (Andalucian or Castilliano) cultural life and arts demonstrated to tourists while their own cultural traditions and history were being ignored. Moreover they were sick of poorly behaved tourists taking over their beaches and coast towns.

Not all tourists were to blame, many venturing abroad for the first time put their trust into the hands of package tour operators. In return they got a limited vision of the country in which they were staying, cheap food, basic hotel facilities and conned into believing that the Costa holiday resorts were 'Spain'. More tourists began to venture beyond the package tour; the individual tourist grew in numbers and a new market was found. 'Individual tourists, tired of fighting for increasingly scarce space on the Mediterranean beaches, began to seek less crowded alternatives, often with cultural attractions in place of sun and sand' (Richards 1996, 8).

The individual tourist began to provide an increasingly large market sector, which required a different product. This time the Catalans wanted to control the management of this audience effectively and began to investigate the individual tourist to see who they were. They quickly established that tourists who visit cultural sights are more likely to be tourists with a high level of education and therefore hold a high professional position. Planners in Barcelona began to look towards business people, as well as individual tourists for a future market.

Once the planners in Barcelona had identified their market sector they went about ruthlessly creating the conditions which would attract this type of tourism. A body which has carried out much of the work, is the Barcelona Tourism Consortium (Turisme de Barcelona). This was founded in 1993 by Barcelona City Council (Ajuntament de Barcelona), the Barcelona

Official Chamber of Commerce, Industry and Shipping (Cambra Oficial de Comerç Industria i Navigacio de Barcelona), and the Foundation for the International Promotion of Barcelona (Fundacio por Promocio Internacional de Barcelona).

Turisme de Barcelona is in fact a continuance of the work which was begun by the City Council and the Chamber of Commerce, Industry and Shipping prior to and during the Olympic Games. Turisme de Barcelona clearly state that they want to avoid package sun tourism in their strategy. They instead aim to 'offer specific products and services to the professional tourism sector as well as individual tourists visiting the city'.[3]

THE OLYMPICS: A LAUNCH PAD FOR THE CITY

Their work can be traced back to 1981, but it was the Olympics which was undoubtedly the launching pad for their work. Advantage was taken of the Hellenic tradition to hold a Cultural Olympiad throughout the four years of preparation for the 1992 Olympic Games. A committee called Olympiad Cultural S.A. COOB '92 (COOB) was set up, and managed a whole host of cultural events throughout the four years preceding the Olympic events of 1992 (1988 Gate of the Olympiad, 1999 The Year of Culture and Sports, 1990 Year of the Arts, 1991 Year of the Future and 1992 Year of the Games). COOB's programming not only covered exhibitions, festivals, training and participation programmes, but also included performing arts, music, opera, audio-visual arts (cinema, television, video), scientific and cultural debates, publications (design objects, printed matter), folklore events and the development of artistic heritage. In the words of Pasqual Maragall, the former mayor of Barcelona and president of COOB, it was hoped that the Cultural Olympiad would '(. . .) recover and further the Hellenic ideal, making Olympic Barcelona a focus for the finest expressions of the culture of our time'.[4]

What followed was a re-birth of Catalan cultural life presented in a coordinated effort to regenerate the city and put it at a level to compete with other European cities (particularly Madrid). Tourism of course is an underlining factor of any city intent on marketing its image externally. Rosa Maria Carrasco, the former External Affairs Officer of COOB, claimed that the cultural baseline for the Olympics was a strategy for cultural tourism expansion. However, the development of this strategy was unlike other cities such as Bilboa, which created the Guggenheim specifically for tourist trade, underlining what Greg Richards describes as 'the tendency for culture to be produced for tourist consumption' (Richards 1999). In Barcelona the strategy was different because the local public in Catalonia, due to their general suspicion of package or mass tourism, would not have supported this type of investment. Developing the local audiences interest in cultural life was therefore a priority, because without their support, developing cultural attractions would have been difficult.

CONSTRUCTION WORKS

Cultural facilities have been built in Barcelona and continue to be built (the National Theatre inaugurated in 1997 and the Auditorium inaugurated in spring this year) but these facilities have not been marketed primarily to or for tourists, they are rather presented as a facility for the local citizens. These new cultural infrastructures have been built with the knowledge that they will provide an up beat and modern image of a cosmopolitan city and that tourists visiting these spaces will take away with them the impression that they enjoyed a slice of 'real' Catalan Culture.

This prompts the question in Barcelona: is there a distinction between what is being produced for cultural tourists consumption and what is being produced for the local citizen? The various politicians residing over cultural decisions in Barcelona and Catalonia would prefer the citizens to think that these new cultural infrastructures are being built for the local people. Maybe this is why politicians and public money are giving limited support to the further building developments of the unfinished cathedral and masterpiece of Antoni Gaudi, the Sagrada Familia,[5] or the re-building of a privately owned Opera House, the Gran Teatre del Liceu,[6] which burnt down over six years ago.

COOB was successful in stimulating local public interest in culture. Rosa Maria Carrasco told of estimates suggesting theatre and museum visits had doubled and in some cases tripled,[7] while the profile of the visitor was said to be overwhelmingly local based. This is largely a positive side-effect of the Cultural Olympiad which stimulated Catalan citizens' own pride and interest in Catalan cultural life. The three tiers of local authority (Generalitat, Diputacio and Ajuntament) now compete for prestige and as the cultural image of the city is seen by the citizens to be a priority, cultural life is well supported. Amongst the citizens there is support for public expenditure on cultural facilities as well as the policy which is often used in conjunction with this, to attract business tourists.

It is not just the local authorities that are supporting culture either, the saving banks are non-profit making organisations and are under obligation to invest in culture and social projects by constitution. Therefore public and private purses support many cultural buildings, festivals and arts events and the feeling in the cultural sector is not one of economic difficulties. As money is readily available, there is not the same need for entrepreneurship, which may be experienced in other countries, and certainly no blatant efforts being made to attract visitors and tourists. Investment in cultural infrastructure is high on the agenda, however later in this chapter is explained that there is a deficit of investment in the artists themselves.

Indeed, by not supporting consumable cultural tourist attractions, the politicians safe guard their image, while in actual fact the offer of a variety of cultural activities is a subtler tourist attraction. The feeling is that even if the tourists do not use all the attractions on offer, the knowledge of their presence, by the visitor, will encourage a return visit. In addition it is likely that because these cultural elements are not built for tourist purposes, business or culturally educated tourists will be more interested in them, because they tend to search for authenticity.

Barcelona undoubtedly owes much of its success to the finely tuned and rehearsed Cultural Olympiad and Olympic Games for which the economic debts are still being paid and from which the whole city has profited. Cultural tourism policies prior, during and after the Olympics were concentrated in Barcelona, but not exclusively. Other areas of Catalonia benefited too, but less than Barcelona. In some cases these policies have been elusive, many have been sophisticated and almost all are worth highlighting. The following policies ensured that Barcelona would, during and post-Olympics, attract richer and more culturally minded tourists.

OTHER KEY STRATEGIES

As already outlined, the hypothesis that culturally aware tourists are likely to be decision makers in their professional life means that a strong cultural image would attract business tourists. These tourists, if convinced, would use the city for congresses, conferences, trade fairs and symposiums in the future. This strategy would bring more like-minded people to Barcelona and these like-minded people would appreciate and learn about the Catalan culture as well as spend more money in the city. Therefore, Barcelona was launched as a Cultural City during the Olympics with two principal objectives. First, to promote Catalan regional identity and secondly, to move away from the old 'cheap beach holiday image' and promote the business facilities offered by Barcelona. The Olympics provided world-wide coverage, which, backed by strong marketing policies, offered a strong launch pad. Key strategies adopted during the Olympics included:

- A spectacular, artistic, musical and visual opening ceremony;[8]
- One TV contract awarded which was carefully controlled and supported by pre-filmed highlights of Catalan cultural heritage;
- The order of official languages being English, French, *Catalan* and Spanish;
- The King of Spain giving his speech in Catalan;
- A strong promotion of Catalan landmarks, food and arts.

Other more subtle policies were also employed. A large amount of accommodation was needed to house participants, participants' families, staff, journalists, medical staff etc., not to mention thousands of spectators. Instead of building cheap hotels, which could have been used for the promotion of package tourism after the Olympics, the authorities elected to think long term, by:

- Awarding contracts to thirty architecture prize winners to build new contemporary apartments on old industrial waste land. These apartments housed participants, journalists, staff, etc. The contemporary flats and apartments were later sold on the open market to wealthy house buyers in a bid to create a new affluent neighbourhood;
- Promoting the development of luxury hotels which would attract richer tourists and later provide accommodation for business tourists. This policy included refusing building permission for hotels on the beach front of Barcelona, with the exception of one five star luxury hotel, and refusing the

development of leisure facilities, such as outdoor swimming pools (usually demanded by package tourists);
- Multipurpose and grand conference facilities were built and promoted during and after the Olympics. The most famous being built below the Palace which offered a panoramic view of the city and therefore enjoyed large amounts of television coverage.

A less obvious tactic to safeguard the cosmopolitan city image was to ensure that Public Transport Information was, and currently still is, not available in foreign languages, which makes it difficult for package tourists, staying in the nearby Costa Dorada and Costa Brava resorts, to travel to the city. Package tour operators who profit from offering organised coach tours to the city support this policy.

In striving to attract up-market middle class tourists, the city is intent on promoting the arts and its contemporary image and efforts since the Olympics have not waned. Re-development of the public space to give it a modern and cosmopolitan image has been extensive. In the past twelve years the authorities presiding over Barcelona have built or modernised:

- The Olympic Port and Olympic Village (housing estate);
- The Old Beach Promenade;
- The Old Port (adding a multi-purpose complex with shops, bars, cinemas, night-clubs, an aquarium etc.);
- The Centre de Cultura Contemporania (CCCB);
- The Museu d'Art Contemporani de Barcelona (MACBA);
- The National Theatre;
- The Museu de Catalunya (Catalan History Museum);
- The Auditorium.

Another interesting strategy adopted by the city was in the chosen locations for each of these large cultural venues. These cultural monuments were constructed in the centre of marginal areas, where economic decline was taking its toll and by doing so the city regenerated poorer areas of the city and spread both the wealth and the tourists out, making the latter a more manageable commodity. By encouraging visitors both local and international, to visit previously closed off areas of the city, the offer becomes more appealing to the individual tourist. Also business tourists, wary of being found in the hub of tourism, prefer to feel that they are discovering unturned stones of a city. The local residents were happy because they could provide services for these new visitors and the politicians were happy because they could be credited for taking care of marginal areas.

CREATIVE SIDE OF THE ARTS NEGLECTED

The down side of all this investment in infrastructure is that the creative or production side of the arts has been neglected. There are very few subsidies available for community arts, arts for the disabled or new developing arts, despite great interest in amateur arts associations. The collections housed in the national museums are of poor quality and limited vision. Few plays by new writers get staged and the City Council only gives subsidies to theatre plays using the Catalan language which often encourages the poor quality production of translated classics. As so much attention is given to the overall image of the city, or as Richards describes 'the hardware of structural creativity',[9] it would be easy to believe that artistic production is thriving, vital and innovative but the reality is that little is invested in the derived creativity or the 'software' of the city.

MARKETING POLICY

The marketing of Barcelona has not been problematic despite competing, during the Olympics, with both the Central Spanish government and the Autonomous Government of Catalonia (Generalitat). Fortunately, the Generalitat was in favour of a policy that promoted Catalan culture and because the Spanish Ministry already had a strong publicity campaign that provides ample marketing for the coastal resorts, the Generalitat and Barcelona City Council could afford to target their chosen tourist market: business and individual tourists. Turisme de Barcelona has a marketing policy which targets: 'Public and private bodies from Spain and abroad, businesses and business organisations, public administrations, international organisations and tourism promotion bodies'.[10]

When the tourist arrives in Barcelona, the city also markets the cultural offer in other ways:

- Publicity street flags which line the streets are reserved for cultural event announcements;
- Large sums of money are annually made available for street festivals, processions, national and regional events;
- Theatre and concert electronic box office systems are offered through the most popular Catalan savings banks to ensure that anyone drawing money from a cash machine will see the extensive cultural offer available and be facilitated in buying tickets (this service is available in several languages);
- Even the public transport systems have been modernised to ensure the visitor is met with a contemporary city and can reach the cultural resources housed far from the city centre.

These policies on the surface seem to have been successful. The Autonomous Government of Catalonia (Generalitat de Catalunya) stated in their resume for the Tourist Season 1997, that Barcelona attracted 24.4% of the tourists that entered Catalonia and that every year more foreign tourists were attracted to it. In comparison the whole of the Costa Brava attracted 35.6% and the Costa Dorada (and Garraf) 27.7%. In addition Barcelona was the principal destination for secondary visits (defined as one or two day trips which return to the principal destination) attracting 50% of the tourists that came to Catalonia in 1997.[11]

The number of foreign visitors arriving in Catalonia for business reasons (9.8%) is still far out of reach of the numbers that arrive for sun and beach holidays (81.7%). The principal motivating factor for choosing to visit Catalonia is still beach (7.27) and climate (7.97) with Culture low on the scale in (4.94) in 1997. However if you can compare this with the principal motivating factor for choosing the City of Barcelona as a destination, 43.2% came on business, 11.7% visited fairs and congresses, 36% came for holidays.[12] The global picture offers a more positive view, Barcelona ranks amongst the ten leading congress cities in Europe. In addition Barcelona hosts a number of international fairs with more than sixty-seven celebrated annually, bi-annually or tri-annually.[13]

The statistics prove and public opinion decides that Barcelona has achieved its goal of being an internationally recognised cosmopolitan city. The twist is that Barcelona is hardly associated at all with the package tourist offers lining the shores of Catalonia. While Spain is synonymous with sun, Barcelona has become synonymous with culture. Many inland towns are now copying Barcelona's policy and Turisme de Barcelona is offering its expertise to promote cultural services in the whole of Catalonia. One such example is Tarraco – Barcino – Emporium – la via Romana an important route which moves from Tarragona to Castello de Empuries via Barcelona and maps out the remains of Roman civilisation.[14]

It remains to be seen how successfully Barcelona's strategies can be implemented in the rest of Catalonia, which still lingers with the image of beach holidays.

NOTES

1 www.gencat.es/turisme/tt97/to-pres.htm.
2 www.gencat.es/turisme/tt97/to-pres.htm.
3 www.barcelonaturisme.com/turisme/ing/tb/turio1.htm.
4 Taken from the Olimpiad Cultural S.A. COOB '92 programme brochure printed in 1988 and reprinted in 1990.
5 The Sagrada Familia has been the centre of controversy for years over whether to continue and finish the building or leave it as an unfinished monument. Right now it is one of Barcelona's most visited sights with coach loads coming from nearby holiday resorts to see a glimpse of the temple's facade.
6 The Opera House is also the centre of controversy after plans to rebuild a more spectacular Opera House (which could compete on size and scale with Madrid's own Opera House) revealed that over 200 families would lose their homes.
7 Personal interview with Rosa Maria Carrasco, former External Affairs Director for 'Olimpiada Cultural S.A. COOB '92', 24th March 1999.
8 COPEC is a Promotional Unit of the Regional Government of Catalonia (Generalitat) and they now hire an adapted version of the opening ceremony (complete with balloons and moons and music) to other cities for festival events.
9 See Greg Richards in the introduction of his volume.
10 www.barcelonaturisme.com/turisme/ing/tb/turio4.htm.
11 www.gencat.es/turisme/tt97/pg16.htm.
12 www.bcn.es/catala/barcelon/presenta/anuari96/cturo6.htm.
13 www.barcelonaturisme.com/turisme/ing/ag/agenda05.htm.
14 www.barcelonaturisme.com/turisme/ing/tb/cult04.htm.

'Difference, Boundaries, Community: The Irish in Britain'

from Judith Rugg and Daniel Hinchcliffe (eds), *Recoveries and Reclamations* (2002)

Mary J. Hickman

Editors' Introduction

This essay is included for its critical view of one local culture, that of Irish people in Britain. As indicated in the introduction to this part, cultural realities are seen by the Editors as a necessary antidote to other aspects of the culture industry, to use Adorno's term. It is immediately evident from Mary J. Hickman's text, however, that there is no single such thing as an Irish community in Britain, but rather several overlapping and intersecting, sometimes conflicting, communities each with its own variation of Irish culture. Differences between Catholic and Protestant, Nationalist and Loyalist, or origin from the north or south are some of the operative categories, and the list of pairs should not be taken as repeating the same difference in each case.

Hickman is Professor of Irish Studies and Sociology at London Metropolitan University. She begins by citing Benedict Anderson to the effect that all communities beyond the village level are imagined. Constructed would be another term, and again the question could be asked as to whether the sense of community is constructed by a self-organising group (see Introduction to Iris Marion Young's text in Part Seven) or as object of a marketing campaign, or both, or between the two. One complication Hickman notes is the elevation of rootedness – Irishness – to a birthright in the myths of nationalism, and Hickman refers to the tendency of states in which many migrant elements contribute to the population to construct national myths as representations of a desired national unity. This, too, is an argument which can be compared with Young's on the liberal model of assimilation into a majority society. Hickman looks at culture, in the broad anthropological sense, while Young looks at political organisation, but the cases have much in which they agree. Hickman, however, looks also at a historical dimension in the causes of Irish emigrations, notably the conditions within Ireland under British rule. She observes that although Irish people of many classes came to Britain, the 'problem' of Irish immigration was linked to the mainly Catholic peasant class. The sense of a problem was reiterated with the Prevention of Terrorism Act (1974), as a response to bombings by the IRA. The largest migration, though, has been since the early 1980s, with higher educational and income levels than those of previous generations but also equally strong attachments to religion and tradition. What emerges is that community is only a partially useful term for a complex set of social relations.

The issue of community is an important one for a minority population in Britain because of its relationship to issues of the nation, identity, ethnicity, migration and racism. However, to date, discussions about the Irish community either involve assertions that such an entity exists, or counterarguments suggesting that the degree of differentiation and dispersal of the Irish population negates the idea that we form a community.

It is, however, important to refute the idea that differentiation necessarily negates community. At the same time, I would agree with the detractors of community that it is necessary to do more than just assert that the phenomenon exists. It is necessary to create a framework for understanding the basis of community and within that context establish what is meant by an Irish community.

Benedict Anderson describes nations as 'imagined communities' (Anderson, 1991:56). All communities, he insists, which are larger than primordial villages of face-to-face contact, are imagined. Until recently, the nation represented the largest community that most individuals imagined themselves as belonging to. All communities are distinguished not so much by falsity or genuineness but by the style in which they are imagined. The politics of forming a nation is the process by which the identity of a 'people' or 'community' is forged. The 'people' and its biography are mythical. Most nations, all products of the modern period, are based on imagined histories which posit back a unity, sometimes to antiquity, of its people.

The nationalist myth elevates to a birthright the fantasy of being rooted. For all those who are displaced by migration (frequently forced and structural), or who are refugees, the search for roots becomes inevitable, and often, depending on the context, this can be a poignant and difficult search to accommodate (Feuchtwang, 1992). In this sense then, we can say that the notion of the Irish community in Britain is a myth – it is a myth in just the same way that all nations, or ethnicities, as imagined communities, are based on myth and all migrant groups live the contradictions of maintaining or not maintaining that myth in the diaspora. Thus, Irishness is both a world and a set of representations, carried around in the heads of actual people, and can be displayed across a number of texts and visual representations (Harris, 1991).

All migrant groups from former colonies or, generally, from the 'South' coming to the 'North' (in the Brandt Report's sense of those terms) have to engage with and resolve problems of difference. They migrate bearing the traces of particular cultures, traditions, languages, systems of belief, histories that have shaped them, and are obliged to come to terms with and make something new of the cultures and economic location they come to inhabit, without simply assimilating (Hall, 1991). When the country they migrate to is the former colonising power, how much more acute and sensitive the situation is. Any comparison of the Irish in Britain with the Irish in the USA and Australia in the nineteenth-century will bear this out. The contrast in terms of control of the Catholic Church and open participation in the political system is striking. It is not that the Irish did not face opposition in those two societies, but the response by the Irish was different.

It has been a main function of national cultures to represent what is in fact the ethnic mix of modern nationality as the primordial unity of 'one people'. This has been achieved by centralised nation-states with their incorporating cultures and national identities, implanted and secured by strong cultural institutions, which tend to subsume all differences and diversity into themselves (Hall, 1992). The Irish first came in very large numbers to Britain during the period which was most critical for the successful securing of a national identity and culture in Britain (and by that means a class alliance): i.e., the nineteenth-century. In that period the Irish were both the most sizeable and most visible minority element in the population. In migrant communities in Britain, 'Irishness' as an essentialist notion has shaped itself against other forms of political and cultural identity, especially Englishness. The consequences have been profound for the subsequent history and experience of the Irish in Britain.

The strategy of the British State and the Catholic Church has been incorporation; for example, through the education system. By incorporation I mean the active attempts by the State to regulate the expression and development of separate and distinctive identities by potentially oppositional groups in order to create a single nation-state. The incorporation of the Irish Catholic working class in Britain was based on strategies of denationalisation, and was not the consequence of an inevitable process of assimilation or integration. In Catholic schools, the priority placed on religious instruction, the effort which went into religious instruction, and the manner in which the religious pervaded all the rituals of school life were all part of a strategy for reinforcing the religious identity of the pupils at the expense of their national identity. There was a corresponding silence in the curriculum content of Catholic schools about Ireland (Hickman, 1995).

It is not surprising, therefore, that a contemporary account suggests the pressure experienced by the second generation to marginalise Irish identity. Tom Barclay, in his memoirs of a bottlewasher, recounts his childhood in Leicester in the 1850s and 1860s. After describing his mother's recitation of old bardic legends and laments, he continues:

> But what had I to do with all that? I was becoming English. I did not hate things Irish, but I began to feel that they must be put away; they were inferior to things English. . . . Outside the house everything was English: my catechism, lessons, prayers, songs, tales, games. . . . Presently I began to feel ashamed of the jeers and mockery and criticism (Hollen-Lees, 1979:190).

'Becoming English' was not based on an inevitable process of cultural assimilation but on acquiring a perception of the inferiority of Irishness compared with Englishness. The cultural pressures to become English and reject Irishness that Barclay cites primarily emanated from the Catholic Church. His world outside the home was defined by the Church and the school, and the latter contained textbooks which glorified England and were silent about Ireland. In another example, Hart Kennedy describes his Catholic schooling as being "taught a great deal about the glory of God and the glory of England, and very little about the art of reading and writing. . . . It was a great privilege to be born in England, the teacher said." (Fielding, 1993)

A low public profile for the Irish became characteristic in Britain as a result of these incorporatist strategies. One person I interviewed, when asked what the term Irish community means to him, said, 'hidden people'. This low public profile is the main achievement of the state and institutional response to the Irish presence in nineteenth-century Britain. For example, Catherine Ridgeway, discussing her early years living in England in the late 1920s and early 1930s, commented:

> During that period I didn't mix much with Irish people. Mostly English. I think my uncle and aunt put me off. They said, "Don't get involved in Irish clubs or anything like that", because there was still the political background all the time. As the years went on and I was learning more about the political situation, I still didn't get involved, because you always had at the back of your mind that if anything crops up and you are involved, you might be deported or something like this (Lennon, 1988:50).

This quotation, and there are many others to support it, demonstrates that the low public profile is not just a product of events in Northern Ireland since 1968. The Irish in Britain have been positioned as a potential political and social threat since the Act of Union in 1801 brought Ireland into the United Kingdom.

It was the expulsion of a specific sector of the Irish peasantry, almost exclusively Catholics, that became represented as the problem of Irish migration in Britain. This occurred despite the fact that for over two hundred years a range of social classes have migrated from Ireland to Britain, including both Protestants and Catholics. This process of construction of the Irish 'minority', however, does not solely rest on the fact that historically the structural location of the majority of Irish Catholic migrants has been as part of the casual, unskilled and semi-skilled working class. As important in understanding the 'place' of the problematised 'Irish' is the discursive effects of Anglo-Irish colonial relations and their articulation with the religious signification of British nationalism. In that context it is unsurprising that references to the Irish community in Britain in nineteenth-and twentieth-century discourses usually refer to working-class Irish Catholics, part of whose response has been to construct a community life based on the very features that encapsulated the threat they represented: religion, national politics and class organisation. Obviously, at different times and in different contexts these elements of 'community' are articulated together differently.

The incorporation of the nineteenth-century Irish immigrants was never completely successful because, although the state and its agencies managed to regulate the expression of Irish identity, it was not able to eradicate it from all those of Irish descent. Identity is an arena of contestation, and the result for many was a complex identity with different elements to the fore in different contexts. Both these points are illustrated by Anne Higgins, who was born in Manchester in the 1930s. This is how she described her childhood:

> We were under a kind of siege being Irish Catholics in Manchester in the thirties and forties. We lived initially in a very poor inner-city district where there were many other Irish families. The parish school we went to had mainly Irish teachers and pupils, we knew Irish Catholic families in the street, we met Irish people at the church, and we didn't have to associate with English people if we didn't want to. In point of fact, my mother made friends easily and a next-door neighbour who was a staunch English Protestant became her best friend in no time, but

> we mixed mainly with other Irish people (Lennon, 1988:146).

Reflecting on her own identity at the time of being interviewed in the 1980s, she said:

> My religion, political beliefs and national identity were all inter-related when I was a child. I've had to rethink my position on all of these over the years but I'm glad I have been able to carry with me, much of what was important to me as a child (Lennon, 1988:155).

Anne Higgins speaks for many in this statement. Identity is not fixed, it changes over time and in different circumstances. But the elements she refers to – national identity (Irish), religion (Catholic) and political beliefs (support for the Labour party) – hardly deviate from what clearly emerge as the chief characteristics of the readers of the *Irish Post*, the biggest-selling newspaper for the Irish in Britain, in its recent survey (*Irish Post*, 5th December 1992 to 16th January 1993). The proclaimed Irish identity, Catholicism and to a lesser extent support for the Labour party of the mid-twentieth-century migrants are rooted in the material basis of the 1950s migration and settlement in Britain. The experience of the emigrants of that period can be understood in terms of the co-existence and intersection of their class position (both in Ireland and in Britain) with their ethnicity (as asserted by them, be it in the Countries associations of the 1950s or the welfare or cultural organisations of the 1980s, and as assigned to them by their continuing problematisation as a social and political threat). The imagined community of being Irish in Britain, as so far discussed, is one that has been constituted by the sense of a forced migration and the differences and boundaries which were immanent in the problematisation of Irish immigrants. The making of a sense of community for this generation of migrants at some level has been secured through a common experience of loss (Barber, 1992).This is the concrete reality of a distinct (although not homogeneous) community.

It is important to emphasise another aspect of community or cultural identity, one which recognises that as well as many points of similarity, there are also critical points of deep and significant difference which constitute 'what we really are' or 'what we have become'. Cultural identities, far from being eternally fixed in some essentialised past, are subject to the continuous 'play' of culture and power. Every regime of representation is a regime of power, and the dominant regimes of colonial experience had the power to make us see and experience ourselves as 'Other'. This inner expropriation of cultural identity can cripple and distort if it is not resisted. Cultural identities, therefore, are points of identification which are made within discourses of history and culture, and therefore are not essences but positionings (Hall, 1990).

In Britain, the experience of anti-Irish disadvantage and discrimination has exerted its own influence over the development of Irish identity. Thus the particular articulation of religion, class and national identity that historically has constituted the communal identity of being Irish in Britain can be understood as, in part, an aspect of this resistance of colonial regimes of representation. Irish identity was also formed in resistance to a racist British nationalism, for which Irish migrants were a specific Other. In other words, individuals and collectivities that are prey to racism (its 'objects') find themselves constrained to see themselves as a community (Balibar, 1991). For example, during the past twenty years stereotypes and problematising discourses about the Irish have led to the toleration of the civil liberties abuses, which amount to a form of 'state racism', sustained by Irish people in Britain through the operation of the Prevention of Terrorism Act (PTA). In 1974, after the IRA carried out the Birmingham pub bombings in England, the Prevention of Terrorism Act was rushed through Parliament. It gave the Secretary of State considerable new powers to control the movement of people between Ireland and Great Britain. The Act provided extensive powers to establish a comprehensive system of port controls, and a process of internal exile which gives the Secretary of State the power to remove people who are already living in Great Britain to either Northern Ireland or the Republic of Ireland.

Although the legislation was extended in 1984 to cover international terrorism, the port powers were devised, and have principally been applied, to control Irish people travelling between Britain and Ireland (Hillyard, 1993). The Prevention of Terrorism Act is '. . . a discriminatory piece of law in that it is directed primarily at one section of the travelling public'. In effect it means that Irish people in general have a more restrictive set of rights than other travellers. In this sense, the Irish community as a whole is a 'suspect community' (Hillyard, 1993:13). The evidence suggests that the use of the powers is targeted at two particular groups: principally, young men living in Ireland and Irish people living in Britain. The introduction of the Prevention of Terrorism Act created a dual system of criminal justice in Britain. Of the 7,052 who had been detained under the Act by the end of 1991, 6,097, or 86 per cent, have been released without any action being taken against them (Hillyard, 1993). People are suspects primarily because they are Irish. The usefulness of the PTA has always hinged on the fact that it can suppress political activity, build up information on Irish people and intimidate the whole Irish community.

A nun who was very active in the campaigns to get people like the Guildford Four (four people wrongfully imprisoned for an IRA bombing at Guildford in England) and the Birmingham Six (six people wrongfully imprisoned for the 1974 Birmingham pub bombings) released has recorded the following account of the pressures on Irish people, especially in the 1970s and early 1980s:

> There were widespread arrests. . . . People picked up under the PTA had no rights whatsoever in those early days. They disappeared. Eventually, we found out that they could be held for seven days. Police denied that they were holding people. Detainees were questioned at all hours, day and night, and solicitors were not allowed in. It was a very anxious time for the families of those detained. . . . It was terrible from 1975 to 1981. That was the worst period; I call it the 'bad time'. Police with dogs, guns and vans swooped on houses in the early hours of the morning, frightening young children, damaging property and making innocent law-abiding citizens the targets of suspicion in their streets and neighbourhoods. If they were any way involved, and when I say 'involved', I mean any way Irish at all, they were raided or taken in (Lennon, 1998:196).

The way in which the PTA was implemented fueled anti-Irish racism, with the oft-repeated injunctions of the police after various incidents to 'Keep an eye on Irish neighbours and watch out for Irish accents'. In the campaigns in the late 1980s to obtain the release of the Guildford Four and the Birmingham Six, many Irish people in Britain (and critically some British people) who often had very different views on events in Northern Ireland came together to right these self-evident injustices. In such circumstances, a sense of community is fostered out of particular historical experiences and in response to specific social constructions of the Irish in Britain. Cultural identity, however, also represents hybridity. In this emphasis the diaspora experience necessarily recognises heterogeneity and diversity, because identity lives with and through difference. Compared with the late 1970s and 1980s, in the 1990s there is a greater representation of the Irish 'community' as diverse, if we can take the changes in reportage in the *Irish Post* (the bestselling newspaper for the Irish in Britain) as one gauge of this. In the early 1980s, references to Irish women's groups were at best nervous; nowadays they are routine. The area of sensitivity today, in many Irish arenas, including the *Irish Post*, is much more likely to be acknowledgment of the existence and campaigns of Irish gay and lesbian groups.

These examples, though, still refer to the 1940s–60s rural emigrants from the Republic of Ireland and their children. They are not, however, the only elements in the Irish population in Britain (nor were they ever the only element), although they still remain the largest grouping. The other major constituent elements are those who have migrated from the Northern Ireland, Protestant and Catholic, and the large flow of migrants from the Republic in the past ten years. Compared with the nineteenth-century, the experience of any Protestant from Northern Ireland coming to Britain, but especially to England, is very different. Anyone with a northern accent is viewed as Irish. There is hardly any research published about them as a group, although some studies are now underway. But the numbers from Northern Ireland have increased substantially in the last twenty years, and they form a significant element in what constitutes being Irish in Britain today.

However, the largest augmentation of the Irish population in Britain has come from the Republic of Ireland since the early 1980s. Much has been made of the fact that these migrants are very different from the 1940s–60s generation who left Ireland. The recent migrants have higher levels of educational qualifications and in the main are more likely to come from urban backgrounds. Some of these differences have been exaggerated, but nevertheless, this migration is significantly different from the previous two main phases in the mid-nineteenth and mid-twentieth centuries. For example, it is assumed that attitudes of the new migrants to the Catholic Church are different, and it is expected that this is bound to have an impact on what constitutes 'community' for the Irish in Britain. There have been number of studies of these new migrants in terms of employment, housing, etc., but only a small number which examine attitudes and perspectives, especially about religion and national identity. One study focused on recent migrants of largely working-class origin from the Republic, in their twenties and thirties who left Ireland without a Leaving Certificate. Contradictory sentiments about Catholicism emerge from their responses. Many of the new migrants from the Republic make a direct link between Catholicism and unhappiness, and bemoan the impact they perceive Catholicism to have on their own lives and on Irish society as a whole. But for many it would appear that although they have jettisoned their adherence to Catholic beliefs, they recognise that Catholicism has had a part in shaping their Irish identity. These responses prompt the speculation that the respondents have a strong sense of Irish identity as apart from Catholicism, but that Catholicism touches their lives because of its place in Irish society and politics, and the role it has played historically in the Irish community of which they are now a part (McGlacken, 1992).

Another study indicates that young Irish middle-class migrants comment, whether from the North or the South, that they find Britain 'shockingly secular'. A sense of spirituality, although not necessarily attachment to organised religion, emerges as an important marker that differentiates the Irish from the English. None of these Irish migrants described themselves as an agnostic or an atheist (Kells, 1995). This sample was markedly more middleclass than the other, and although both samples are small, they suggest that further research in this area would be fruitful. Research needs to be carried out on the repercussions for the Irish community in Britain of the changing role of religion as a part of Irish national and cultural identities, against a backdrop of the secularisation of Irish society and the changes in Anglo-Irish relations heralded by the current peace process.

CONCLUSION

I set out at the beginning to indicate a framework for understanding the basis of community and within that context establish what is meant by an Irish community. Broadly, I have situated the discussion within the context of the inevitable problematic that immigrant groups encounter of coming to terms with and making something new of the cultures and economic location they come to inhabit, without simply assimilating. Until the late 1960s, the agenda in Britain was assimilation/incorporation. The strong incorporatist tendencies of British national culture made an indelible mark on the experience of Irish migrants to Britain, and still shape the positioning of the Irish within that national culture.

The agenda, however, is now about plurality; cultural diversity is the hallmark of post-modernity and it is now more apparent that symbols that represent the differences and boundaries that constitute the Irish community in Britain do not necessarily have the same meaning for all Irish people or those of Irish descent. This differentiation is a strength rather than a weakness. The greatest danger surely arises from forms of national and cultural identity that attempt to secure their identity by adopting closed versions of culture and community. The point is that 'community' is highly symbolised, with the consequence that members of the community can invest it with their often very different selves. Its character is sufficiently malleable that it can accommodate all its members' selves. The imagined community which divides the world between 'us' and 'them' is maintained by a whole system of symbolic 'border guards'. These border guards are used as shared cultural resources with shared collective positioning vis-à-vis other collectivities. They can provide the collectivity members with 'imagined communities', but also with 'communicative communities'. Membership in a people consists in the ability to communicate more effectively, and over a wider range of subjects, with members of one large group than with other outsiders (Anthias, 1992). So although people will have different imaginings of the 'community' in their heads, some symbols or practices will unite larger groups of them, effectively forming alliances on an ethnic basis. Question marks remain over Irish identity in Britain in this respect, but there is no doubt it is a more inclusive notion of community than in the past. The essentialised Irish community which was formed in resistance to anti-Irish racism and in opposition to constructions of English/British identity entailed 'silences' which an emphasis on hybridity allows now to be 'voiced'.

REFERENCES

Anderson, Benedict (1991) *Imagined Communities*, London: Verso.

Anthias, F., & Yuval-Davis, N. (1992) *Racialised Boundaries*, London: Routledge.

Balibar, Etienne (1991) 'Is There a Neo-racism?' in Balibar, E. & Wallerstein, I. (1991) (eds) *Race, Nation, Class*, London: Verso.

Barber, Fiona (1991) 'No Great and Recognisable Events: The Representation of Emigration, Gender and Class' in O'Leary, C. (1991) (ed) *Wave/Another Country: Irish Exile and Dispossession*, Huddersfield: Huddersfield Art Gallery.

Feuchtwang, Stephan (1992) 'Where You Belong' in Cambridge, A. & Feuchtwang, S. (1992) (eds) *Where You Belong*, Aldershot: Avebury Press.

Fielding, Steven (1993) *Class and Ethnicity: Irish Catholics in England 1880–1939* Buckingham: Open University Press.

Hall, Stuart (1990) 'Cultural Identity and Diaspora' in Rutherford, J. (1990) (ed) *Identity: Community, Culture, Difference*. London: Lawrence & Wishart.

Hall, Stuart (1991) 'Old and New Identities, Old and New Ethnicities' in King, A. D. (1991) (ed) *Culture, Globalisation and World System*, London: Macmillan.

Hall, Stuart (1992) *Our Mongrel Selves*, New Statesman and Society, 19 June.

Harris, Jonathan (1991) 'Passages: Transportations' in O'Leary, C. (1991) (ed) *Wave/Another Country: Irish Exile and Dispossession*, Huddersfield: Huddersfield Art Gallery.

Hickman, Mary J. (1995) *Religion, Class and Identity*, Aldershot: Avebury Press.

Hillyard, Paddy (1992) *Suspect Community: People's Experience of the Prevention of Terrorism Acts in Britain*, London: Pluto Press.

Hollen-Lees, Lynn (1979) *Exiles of Erin: Irish Immigrants in Victorian London*, Manchester: Manchester University Press.

Irish Post (5th December 1992 – 16th January 1993) *Survey of the Readership*, (published weekly), London.

Kells, Mary (1995) *Ethnic Identity Amongst Young Irish Middle Class Migrants in London*, Irish Studies Centre Occasional Papers Series, No. 7, London: University of North London Press.

Lennon, M., McAdam, M. & O'Brien, J. (1988) *Across the Water: Irish Women's Lives in Britain*, London: Virago.

McGlacken, Sinead (1992) unpublished research dissertation for B.A. Honors Degree, Irish Studies Centre, London: University of North London Press.

'Out of Order: The Public Art Machine',

Artforum (1989)

Patricia Phillips

Editors' Introduction

When this article first appeared in *Artforum* (a leading contemporary art magazine read on both sides of the Atlantic), it was one of very few pieces of writing on public art to adopt a critical position. For the most part, writing on the field, in as much as there was any, consisted of advocacy on the part of publicly funded organisations seeking to promote their role as intermediaries in commissioning public art, or reports from bodies such as the Arts Councils making largely spurious claims for public art as creating place identities or leading economic regeneration. The examples of north American cities such as Seattle were given, but with no understanding of the actualities involved. A critical investigation of projects spawned by this climate is found in Sara Selwood's report for the Policy Studies Institute, *The Benefits of Public Art*, (1995), an ironic title as it turns out. But in the late 1980s, as sites such as New York's Battery Park City were extending commissioning to aesthetically more interesting and experimental areas, Phillips offered an almost unique viewpoint.

Her case revolves around contested definitions of the public sphere. In the literature of public art advocacy, as in most briefs for commissions, this tended to be defined as a physical, geographical space. A work's integration with its site consisted of making visual references to it, or perhaps incorporating some kind of local history. Seldom was there a sense that a public sphere was historically – Phillips has in mind the commons of early north American democracy – a space of multiple meanings and voices. Her argument can be seen in conjunction with the texts included in Part Seven on identity contestation, though it is here for its emphasis on an area of cultural production which is frequently used in place marketing. For Phillips, the public sphere is a mutable realm in which conflict arises but can be constructive.

Further writing by Phillips, who is Dean of Arts at the State University of New York, New Paltz, can be found in *But Is It Art?*, edited by Nina Felshin (Seattle, Bay Press, 1995 – see pp. 165–194 and 283–308). The first page reference is to an essay on Mierle Laderman Ukeles, whose project *Touch Sanitation* is illustrated in Plates 24–31. The second is to an essay on Peggy Diggs, and her project using a commercially made and distributed milk carton to disseminate a message about domestic violence. A more recent text is 'Public Art: A Renewable Resource', in *Urban Futures*, edited by Malcolm Miles and Tim Hall (London, Routledge, 2003). In her last paragraph there, Phillips refers to Ukeles again, and writes 'Negotiation is about navigation – moving to a fruitful, animating idea of common purpose rather than a chilling or dulling compromise. Negotiation occurs on the threshold. Transition is an unerring condition' (p. 133). How remote this seems from the culture industry.

There's good reason to be wary these days when the signs of another specialization start emerging – when one small point is established at the sacrifice of the wide horizon. Contemporary society has become remarkably undisciplined in the ways that it spontaneously endorses new disciplines in almost unimaginable areas of expertise. Those involved in the art world are well accustomed to the coalescences and lightning-like dissipations of style, but a new speciality is not a common notion. In the past 25 years, traditional distinctions between sculpture, painting, drawing, photography, and installation – as well as the idea of art and architecture as independent, exclusive phenomena – have eroded, causing fused and hybrid forms and unusual intersections. And conceptual catholicity, openness, and negotiable categorization have provided the groundwork for the galvanization of a new art: the now very active and hierarchically complex world of public art. Within this arena, there are many players and many productions, some enlightened ideas and little criticism.

Public art – as it is normally understood and encountered today – is a nascent, and perhaps naive, idea. It bears so little resemblance to earlier manifestations – especially the most immediate precedent of civic, elegiac art of the 19th and early 20th centuries – that the idea of a historical progression of uninterrupted continuity seems spurious; there are few instructive models. And so, though public art in the late 20th century has emerged as a full-blown discipline, it is a field without clear definitions, without a constructive theory, and without coherent objectives. When the intentions have been apparent they are usually so modest (amenity) or so obvious (embellishment or camouflage) that they seem to have little to do with art at all. In short, the making of public art has become a profession, whose practitioners are in the business of beautifying, or enlivening, or entertaining the citizens of, modern American and European cities. In effect, public art's mission has been reduced to making people feel good – about themselves and where they live. This may be an acceptable, and it certainly is an agreeable, intention, but it is a profoundly unambitious and often reactionary one. And even these small goals are infrequently satisfied; public art doesn't generally please or placate, or provide any insistent stimulation. Instead, public art today, for the most part, *occupies*. And just at the moment when so much apparatus has been assembled and oiled that might aid in the development of a rigorous critical foundation for public art, there is a growing feeling of – well, why bother? Indeed, an enterprise that emerged with such idealism now feels like a lost opportunity.

Yet many artists, art administrators, and bureaucrats worked hard to promote the current proliferation and professionalization of public art, and did so with the noblest of intentions. Some reflection on the past indicates that those involved had good reason to lobby for "official" policy and protections. For art that appears beyond the configurations and machinations of the gallery and museum encounters different forces and greater risks, and thus should be provided, they believed, with some fundamental assurances and safeguards – for the sake of the artist, as well as the community. And given the very real need for relief from, or challenge to, the loud monotony of the urban landscape, state and federal guidelines for "percent for art" programs were initiated; standards and criteria for selection and review drafted; and bureaucratic procedures codified. But this clarification of operations has ultimately led to a "minimum basic standard" mentality. Not unlike American housing reform in the late 19th century – which was not based on constructive legislation for a sound life, but on the absolute lowest standards of acceptability – the public art "machine" now often encourages mediocrity. To weave one's way through its labyrinthine network of proposal submissions to appropriate agencies, filings and refilings of budget estimates, presentations to juries, and negotiations with government or corporate sponsors, requires a variety of skills that are frequently antithetical to the production of a potent work of art. If the "machine" itself can be put to use as a conduit, rather than as a molder of the art that emerges, then there is still the potential for transforming methodology and materials into positive energy. But more often the result of this process has been what Gordon Matta-Clark, James Wines, and others have referred to as "the turd in the plaza."

Public art operates on a practical as well as a philosophical level, but the contemporary preoccupation has been with the pragmatic. Thus we can find abundant information on the strategies that initiate public art, but we can search far and wide for any compellingly articulated theory of public art. Can provocative art endure the democratic composition of the selection panel and process? Are art and ecumenicism in opposition? Can public art illuminate cultural ideas that other forms frequently cannot? What is it that public art can uniquely do? These are the kinds of questions, I would argue, that must be more vigorously explored. And I would further propose that this discourse will serve to overturn some knee-jerk assumptions about the very nature of the hybrid beast we call public art.

One basic assumption that has underwritten many of the contemporary manifestations of public art is the notion that this art derives its "publicness" from where it is located. But is this really a valid conception? The idea of the public is a difficult, mutable, and perhaps somewhat atrophied one, but the fact remains that the public dimension is a psychological, rather than a physical or environmental, construct. The concept of public spirit is part of every individual's psychic composition: it is that metaphysical site where personal needs and expression meet with collective aspirations and activity. The public is the sphere we share in common; wherever it occurs, it begins in the decidedly "somewhere" of individual consciousness and perception.

Therefore, the public is not only a spatial construct. And thus a truly public are will derive its "publicness" not from its location, but from the nature of its engagement with the congested, cacophonous intersections of personal interests, collective values, social issues, political events, and wider cultural patterns that mark out our civic life. Unfortunately, what we have traditionally seen is a facile definition that links those areas that cities (with private developers) designate as public spaces with the notion of public art. It is presumed that these sites, by virtue of their accessibility or prominence, are the ones where public art *can* and *should* appear. This is a questionable idea of many reasons – not the least of which is that public space, as it is emerging in our time, bears little kinship to the public space of the town square, plaza, or common in which the public art of the past traditionally found its home. Public space, as defined today, is, in fact, the socially acceptable euphemism used to describe the area that developers have "left over," the only "negotiable" space after all of their available commercial and residential space has been rented or sold. The City of New York, for example, has granted many developers the right to upscale the height or bulk of their buildings, contingent upon their agreement to provide a little more "public space" at ground level. But what qualities and characteristics these spaces must offer have been inconsistently interpreted. Thus public space has served as a great new incentive – not to be "public," however, but to satisfy far more profit-motivated market objectives. When public space and public art seem to appear spontaneously, it is usually because some savvy or enlightened developer has discovered that beauty can be profitable, and that offering something to the community (even if no one really understands the nature of the gift) can enhance the corporate image. In the same way that "good fences make good neighbors," the clear delineation of a public space has been packaged as a neighborly gesture, with public art the fence that identifies boundaries.

But a public art that truly explores the rich symbiotic topography of civic, social, and cultural forces can take place anywhere – and for any length of time. It would not have to conform to such formal parameters, for it would not find its meaning through its situation *in* a forum, but would *create* the forum for the poignant and potent dialogue between public ideals and private impulses, between obligation and desire, between being of a community and solitude. Wherever we might find that art, we would be inspired to extend its discourse into the variety of public and private domains we enter. Those two domains are different, of course, but they are interdependent. To define the public as merely that which exists outside the private is to deny the essential and complex relationship between the two.

A major exhibition in lower Manhattan this fall has helped to emblematize these and other disquieting questions about the relationship between so-called public space and public art. In and around a major and unfinished portion of ground-level space in the World Financial Center of the newly emerging Battery Park City complex, the real estate development firm of Olympia & York provided a space for invited artists and architects to install temporary, site-specific works. The Olympia & York assembly of art, entitled "The New Urban Landscape," was an extravaganza – in the best and worst senses of that word. This rich variety of projects announced loudly and emphatically that here lies another public space. And so here, once again, art was defined as public because of its location. Yet there was a particularly shrewd inversion at work. By dangling the bait of abundant and chewy art by some of the "hottest" accomplished and emerging artists from around the world, Olympia & York succeeded in appropriating the notion of public art to entice the public to a new site – that didn't, by any other definition, look or feel very public. And the lure for this consecration was both savory and spicey. The organizers and artists had the courage, and the developer the good sense (and beneficence), to endorse some politically loaded, controversial, and critical work in a corporate-sponsored setting. And yet "The New Urban Landscape" sends out troubling – and by now familiar – messages about public art's application. For "The New Urban Landscape" was a fin-de-siècle enterprise – in some ways, the coda for fifteen years of fervor. And when it all ended, art had served as just one more ingredient in an elaborate coronation that attempted to transform nothing more than a low-ceilinged hallway into a dynamic public space, and a private developer into a public patron.

The involvement of corporations in the sponsorship and support of art is not a new thing. After many years of stimulating the production of private art, it seems quite natural that corporations would eventually find their way to public art, which can now not only boost a corporation's reputation as intelligent and concerned, but can also serve as the vehicle to demonstrate community spirit, a belief in the idea of place in an age of placeless architecture. With this project there was a generous and open sponsor, some very good art, and thoughtful, insightful organizers. So what is the problem? What is it that disturbs?

In fact, some of the answers to these questions will be found in other questions: those that address the implications of the temporary in public art. For in the bureaucratization of public art, there has been a tremendous emphasis on the installation of permanent projects. (Organizations such as the Public Art Fund Inc. and Creative Time, Inc. – dedicated to sponsoring short-lived exhibitions and installations in sites throughout New York – are two of the exceptions.) When evaluating proposals for art that will be commissioned to last "forever," it is not shocking that selection panels have often clammed up and chosen the safe, well-traveled path of caution. When faced with the expanses of eternity, it is not surprising that many artists themselves have tended to propose those cautious, evenhanded solutions. Therefore, the temporary is important because it represents a provocative opportunity to be maverick, or to be focused, or to be urgent about immediate issues in ways that can endure and resonate. But I would argue that the power of the temporary asserts itself productively and genuinely in situations where the pressure of the moment is implicit in the work. Seen in these terms, the temporary is not about an absence of long-term exhibition commitment on the part of any particular sponsor, but about a pledge of a different kind, with more compressed intensity, on the part of the artist.

The nucleated setting and agenda of Olympia & York's endeavour raises serious concerns about the potential for co-opting and institutionalizing even this radical fringe of public art. For what will be the lasting impact of this great event of Olympia & York's? In what significant ways has this exhibition marked this site, or furthered the idea of art as a critical public catalyst, once the gypsy encampment has packed up and moved on? In fact, wasn't this project just another schedule-driven exhibition that had little to do with the present or future of the public life at this site? If a succession of temporary exhibitions might, in fact, animate this public space (something the developers apparently desire) and begin to generate some meaningful dialogue (something the rest of us might like) about space, art, and contemporary urban life, such a possibility is entirely contingent upon some long-range vision as opposed to a shrewd public relations strategy, however magnificently or munificently that strategy is enacted. By "dressing up" (or disingenuously "dressing down" what would be considered even a poorly designed indoor sculpture garden in the garb and lingo of social conscience and inquiry ("The public spaces of The World Finance Center are an ideal context for public art," a four-color brochure tells us. "The works in this exhibition make unusual demands on the viewer," etc.), Olympia & York have, as much as anything, demonstrated to us just how subject to manipulation the concept of public art has become.

Perhaps another one of the great problems of public art today stems from its fundamentally ecumenical intentions. Artists striving to meet the needs of their public audience have too easily subscribed to the notion that these needs can best be met through an art of the widest possible relevance. The ideas of ecumenicism and relevance are not onerous, but they can have – and in the case of public art, often *do* have – insidious and oppressive dimensions. For broad-based appeal and the search for a universal common denominator are not a priori esthetic concepts, but a posteriori results. Reverse that order, and the art's in trouble, for art is an investigation, not an application. So it's disturbing when it looks as if artists are campaigning for public office – going for the majority consensus at all costs.

Not surprisingly, this goal of unanimity has also led to the establishment of what is considered a more democratic composition of public-art selection committees. There has been a generous and well-intentioned effort to include on these committees not only panelists with backgrounds in the arts, but also representatives from the local community in which the public installation will be situated. Yet if followed to its logical conclusion, the concept of "public" that this phenomenon implies reveals itself to be quite ludicrous. For public space is either communal – a part of the *collective* citizenry – or it is not. Somewhere along the line, our democratic process has presumed that the sentiments of one particular community, simply because of its members' propinquity to the prospective installation, should be granted greater significance. What this suggests is that we have arrived at some reliable formula for articulating the precise radius that distinguishes that community's interests from the larger field of public life. Thus the ideas of the local community and of the general public are put into an adversarial relationship, implying a fundamental conflict between those inside a particular neighborhood, area, city, etc., and those outside. This peculiar endorsement of community opinion, sometimes at the expense of larger public concerns, subtly yet effectively affirms a notion of what I would call "psychological ownership," at the same time that it refuses to ground that notion in any terms other than geographic. Thus, because I live, work, or relax near a certain site, I believe that, in a sense, I "have" that site, and am empowered to exert control, regulation, or power over those who are the "have-nots." We have seen the ramifications of this kind of thinking in a variety of controversies. For example, which is the community that should have the most say in "approving" the design of the Vietnam War Memorial in Washington, D.C.? Veterans? The family members of men killed or missing in action? The group of office workers and government bureaucrats who work nearby? The public at large, who might feel a sense of possession of this tragic, poignant space? And which is the community to be consulted when installations are contemplated for City Hall Park in Manhattan? The government employees who work in City Hall and cross the park each day? Or New York City voters for whose civic authority and commitment the site speaks? Or the many homeless who spend their days and their nights in the park? And to which group should the artist throw his or her appeal?

Rather than digging in our heels to examine and analyze the implications of these questions, too many public-art sponsors and makers seem to be trying to sidestep them with a "minimum-risk" art; that is, an art that can be slipped quietly into space and somehow manage to engage everyone but seriously offend or disturb no one. But isn't it ironic that an enterprise aimed, even at the least, at enlivening public life is now running on gears designed to evade controversy? And that so many involved in public art express such dismay – even hurt – if and when controversy occurs? Curiously, Richard Serra's *Tilted Arc*, 1981, in Manhattan's Federal Plaza remains one of the great moments in contemporary public art, not despite but *because* of the conflict its installment generated. Is it offensive? Does it obstruct? Is it public if it does not please? Should the artist's personal vision of site-specificity be permitted to override the desires of the local (specifically professional) community most frequently exposed to the work? In fact, Serra's work achieved its most profound public resonance and significance precisely at the moment when its future seemed most threatened. And that inflexible, somewhat dogmatic object in a deplorable architectural context has been enriched by the color and texture of public debate that continues to surround it. *Tilted Arc* is an important symbol for public art because of the questions it has stimulated – and not because it should not be where it is.

Unfortunately, the avoidance of such controversy has generated an attitude about public art that constrains and segregates thinking. In the 1970s, when troubled cities felt a great vulnerability to the aggressive, often destructive gestures of disenfranchised citizens, the idea of "defensible space" became an important concept. It was Oscar Newman who first proposed that public space could and should be designed in a way that protected it from the onslaughts of graffitists, vandals, and other assaulters. We can see the influence of this proposition in the clunky, immovable concrete benches and barriers in our parks and city streets; the barbed grillwork appearing on ground-level heating ducts to stave off loiterers and the homeless seeking warmth. But we can also see the flip side of this proposition at work in the public-art mentality taking hold today. Public art may not be required to be physically "defensible" but it is, more and more, expected to be *defendable*. So every possible – and ludicrous – objection is raised at the early stages of the artist selection and proposal process, to anticipate and fend off any possible community disfavor. With programs dependent on such tightly woven sieves, it's not surprising that plenty of hefty, powerful projects don't make their way through. And it's not surprising that, over the years, the artists who might propose such projects have ruined their energies elsewhere, while the studios of the artists who have learned the appropriate formula have become mini-factories for the churning out of elegant maquettes for current and future projects.

It is important to consider that the most public and civic space of many early American cities was the common. The common represented the site, the concept, and the enactment of democratic process. This public area, used for everything from the grazing of livestock to the drilling of militia, was the forum where information was shared and public debate occurred: a charged, dynamic coalescence. The common was not a place of absolute conformity, predictability, or acquiescence, but of spirited disagreement, of conflict, of only modest compromises – and of controversy. It was the place where the ongoing dialogue between desire and civility was constantly reenacted, rather than restrained or censored. If it's true that the actual space of the common does not exist as it did two hundred years ago, the idea is still vital. Its problematic shadow image, the idea of an enormous, happy cultural melting pot, was challenged and generally dismissed twenty-five years ago – except by a lot of people involved with public art. So if there is a tragedy here, it is that public art is in the unique position to reconstitute the idea of the common, and yet, by misconstruing the concept – by too often rewarding the timid, the proven, the assuaging – the public art machine has consistently sabotaged its own potential to do so.

Still, it seems, we're returned to the question of where that "common" might be today, or at least where – or how – we might look for the public art to create it. And though I've painted a bleak picture of the contemporary scene, it is not a hopeless one. In fact, if we take that step beyond conceiving of the new urban landscape as a geographic grid of buildings, spaces, and art, to view it instead as an ever-mutating organism sustained by multiple, interrelated vortices and networks and the private trajectories that complicate them, then the horizon line of public art expands to include the "invisible" operations of huge systems and the intimate stories of individual lives. Certainly the artists who choose to work with these polarities where the edges of the public are invented and realized and not the only ones whose projects provide significant stimulation. But their work, by pressing against calcified notions of public art, suggests some fresh visions of the common.

For example, artist Mierle Laderman Ukeles' work with the Department of Sanitation of the City of New York engages city residents in one of the most crucial, life-sustaining but maligned operations of urban life – garbage collection. Her most recent project, *Flow City* (scheduled for completion in 1990), will bring people into a cavernous marine transfer facility at 59th Street and the Hudson River for what is, in effect, a multimedia performance of trucks dumping their loads of household and commercial waste into barges destined for landfills. Ukeles' work proposes that the public in public art is defined by subject rather than object.

At the other end of the spectrum, some of the most fruitful investigations of public life and art are occurring in the most private, sequestered site of all – the home. For just as the public space has become diminished as a civic site, the home has become, in many senses, a more public, open forum. The public world comes into each home as it never has before, through television, radio, and personal computer. So that the rituals that were once shared conspicuously in a group are now still shared – but in isolation. An example of this ambiguous condition is the annual celebration of New Year's Eve in Times Square. Which is the more public event – the throng of people gathering at 42nd Street to watch a lighted apple drop, or the millions of people at home, each watching this congregation on TV? In other words, more and more, the home has become the site for the complex play of social meanings. For this reason, it is a fruitful domain for dialogue about the public/private dialectic. Following in the footsteps of Gent's Museum van Hedendaagse Kunst's 1986 exhibition "*Chambres d'amis*," the Santa Barbara (California) Contemporary Arts Forum organized their 1988 "Home Show," with ten California residents welcoming ten artists into their homes to explore the region of interiority – as it relates to the external, public world.

For this project, the collaborative team of Kate Ericson and Mel Ziegler, with a work entitled *Picture out of Doors*, methodically removed all the doors in Pat and David Farmer's home, including doors from closets, cupboards, and cabinets, even from bedrooms and bathrooms. The tangible evidence of sanctioned voyeurism was stacked in the living room. In a sense, the team's project publicized intimacy by denying privacy. In many ways, the Santa Barbara installation was a tame project for Ericson and Ziegler. For the past ten years they have conducted their own investigations of the private/public dialectic, with much of their work occurring on their own instigation, that is, without the benefit or legitimacy of an arts organization. They have placed advertisements in local newspapers seeking homeowners willing to collaborate on projects. In one project in Hawley, Pennsylvania, for example, called *Half Slave, Half Free*, 1987, the team asked a homeowner to continue to cut only half of his lawn and leave the other portion unmaintained. *Half Slave, Half Free* suggests an expanded and provocative definition of public art, one that has sustained a commitment to independent "guerrilla" activity as an alternative to institutionalized commissioning, and that appeals to and enlists the support of the single vote (the homeowner/collaborator) as opposed to the majority rule in order to explore the half-slave, half-free relationship of personal to public, and vice versa.

Individual vision and independent thinking are possible in the realm of public art; what we've come to expect – or accept with a sigh – is not all we *need* expect. Two years ago, architects Donna Robertson and Robert McAnulty proposed a design of an apartment for an exhibition called "Room in the City." The proposal was spare yet complex. Within the space of a small Manhattan apartment, a procession of five video monitors, hooked up to a satellite communications disk placed outside the window, showed random images of the city, creating an ethereal glow of violated or collaged information. On the other side of a diagonal wall that slashed through the space, a single monitor, unattached to the communications dish and the surrounding city, sat at the foot of the bed. In the traditional site of domesticity and intimacy, this project stands as a metaphor for our new urban landscape; that site where private and public, the intimate and the shared, are fragmented and reconceptualized, where culture both originates and ends, and where the public is permitted to assert itself as an idea of ever-shifting focus and fruitful frustration.

PART FIVE

Cultures and Technologies

INTRODUCTION TO PART FIVE

Technology and the urban are intimately interlinked. First, almost inevitably, most technological advances and new technologies emerge from urban areas, facilitated by the concentration of capital and labour. From the point of view of production, economic geographers have long examined the geography of new technologies and recognised distinctive regional patterns to the emergence of new waves of technology. Recently new technologies have developed very unevenly across space. Hi-tech regions have tended to develop in areas not associated with earlier waves of technology such as manufacturing industry. These new technological clusters have been highly concentrated spatially, for example, in the United Kingdom over 50 per cent of hi-tech employment has long been concentrated in only three regions: the western crescent around London, including the home counties, north west England and 'Silicon Glen' in central Scotland. Similar patterns of concentration occur in other countries. Explanations for this involve locational and structural advantages enjoyed by certain locations. However, a region's position within the cycles of technological development and change has more than just economic impacts, rather it shapes and reverberates throughout its social geography as well. Compare accounts of life in redundant, former industrial regions associated with now long past technologies, for example Nick Danziger's *Danziger's Britain: A Journey to the Edge* (1994), with accounts of life on the frontiers of new technology in the J. G. Ballard novel *Super-Cannes* (2000). This part is concerned with far more than just the economic geography of technology; rather its remit is to introduce a collection of authors who have examined how city cultures are affected by technology, how technology affects city cultures and who imagine possible futures. It introduces these authors and considers a number of themes concerning the relationship between culture and technology that they raise.

Technology is not independent of society. Rather, it is often the case that technologies are the product of a cultural climate and their uses change according to shifts in ideology. In addition, technology has a huge potential to shape the urban and its cultures. This is most powerfully demonstrated by the case of the car, which has curiously been largely overlooked until recently by sociologists compared to other technologies, and is explored in the extract from 'The City and the Car' by Mimi Sheller and John Urry (2000). Sheller and Urry, in a wide-ranging discussion, examine how cars, and provision for cars, shape the physical spaces and environments of cities, patterns of social division and even ways of 'being' in the urban. They argue that cars have created societies of 'automobility' for those who can afford it.

> Mobility is as constitutive of modernity as is urbanity, that civil societies of the West are societies of automobility . . . a civil society of quasi-objects or 'car-drivers' and 'car-passengers' along with disenfranchised 'pedestrians' and others not-in-cars, those that suffer a kind of Lacanian 'lack' (2000: 738–9).

Policies to reduce car use seem doomed to failure through a lack of political will to stop the growth of private transport, lifestyles lived in cities that require a car to access opportunities and the failure of policies to connect with the multiple social and cultural geographies of mobility, Sheller and Urry conclude by outlining some possible solutions.

As the public spaces of the city are shaped and colonised by the car, so the private spaces of the home are becoming increasingly penetrated and shaped by new technologies. Stephen Graham and Simon Marvin write about the changing realities of the home, such as teleworking, in an extract from *Telecommunications and the City* (London, Routledge, 1996). If the city of the nineteenth-century, or twentieth-century Fordism, grew as workforces expanded, now those accumulations of human resource are dispersing, perhaps regaining some control over their working pattern and environment. Graham and Marvin note that in France the Mintel videotex system serves a third of households. Around much of the rest of the world the internet is spreading into homes more rapidly than any other communication technology of the past. Graham and Marvin foresee homes in which all appliances will work on integrated electronic command systems and where shopping is done in virtual reality. Such a course envisages changes in forms of domestic living and will further impact on the denuded public spaces left behind after the car has been provided for. Who can blame us if we buy *The City Cultures Reader* via the internet rather than braving the urban assault course and trying to walk, bus or cycle to our nearest bookshop?

It is worth remembering that access to these new technologies is uneven. There are holes and gaps in the seemingly inexorable spread of the 'net'. What place then for those without cars and without access to new communications technologies in the cities of the future? What relic cities will be left behind for them to inhabit? Technologies can, therefore, be liberating (for the car driver and computer owner) as well as constraining and containing.

The three remaining extracts in the part deal with alternative possibilities offered by technology. The first, 'The People's Radio of Vila Nossa Senhora Aparecida' by Vivian Schelling (1999), explores the impacts of the utilisation of very basic communication technology (tape recorders broadcasting through loudspeakers in public places) in a community of poor rural migrants living on the outskirts of São Paulo in Brazil and the ways in which this basic technology was able to empower the community. Schelling traces the history of radio as a 'means of communication directed towards and used by members of the "popular classes"' (1999: 167) in Latin America, situating it within cultures of resistance, subtle, everyday practices of opposition. The settlement Schelling investigated has a long history and culture of resistance, expressed, for example, in political affiliation. The radio station acted as an appropriate means of communication with a local culture with a long and important oral tradition, a familiarity with the technology being utilised and the high levels of illiteracy within the community. Crucially, the radio station established a space of dialogue for the people in contrast to the mainstream media.

> Ideologically, CEMI (Centre for Popular Education and Communication of São Miguel) adhered to the view that in contrast to the media of the culture industry, 'popular communication' was characterised by the fact that it was dialogic and participatory, enabling the receiver to become the producer and vice-versa (1999: 173).

Schelling does not offer a rose-tinted view of a grassroots movement transforming a community of oppressed rural migrants. In the final third of the paper she deals with the contradictions and problems it threw up as well as its powerful transformative effect on those who experienced it.

Tiziana Terranova considers the relationships between technology, the body and culture in 'Posthuman unbound: artificial evolution and high-tech subcultures' (1996) included here. She considers how technologies and bodies are becoming increasingly interconnected from the mundane (soft contact lens, for example) to the sci-fi imagination of William Gibson, and how the boundaries between the body and technology are becoming blurred. While this subject has generated a huge amount of debate, little of it has considered the ways in which new body-technology formations are implicated in the emergence of new cultural formations as Terranova does here. She talks of the invasion of the body by machines creating new social and cultural divisions and formations that, as yet, have not been considered by the advocates of new technologies. While highlighting possibilities for resistance and subversion of state and corporate power she offers a critique of the 'fragmented anatomy' of the advocates of cyberculture. She notes, for example, that: 'As an expression of a subculture that encourages a "subversive" use of technology, there is little problematization of the ways in which technological change can be and is being shaped by economic and political forces' (1996:173).

Technology has long been central to resistance to state and corporate hegemony. In an age where power appears to have shifted from the streets into cyberspace this is particularly so. However, the pursuit of power and its resistance by counter-hegemonic groups is problematic, as also explored by the Critical Art Ensemble in the final extract here which examines questions of power, resistance and their mediation by new electronic technologies. Here they explore the limitations of what they argue are outdated modes of civil disobedience based on the occupation of physical space. They begin by arguing that 'contemporary activism has had very little effect on military/corporate policy . . . [because] as far as power is concerned, the streets are dead capital' (Critical Art Ensemble, 1996: 10–11).

In contrast to the majority of writing on resistance and protest that has proclaimed the power of the public sphere the Critical Art Ensemble offer a gloomy view of the redundancy of the city as a space with political value, recognised through its abandonment by corporate and political power. Perhaps it is just left behind for the cars? In a key passage they argue:

> Since cities have been abandoned by business and left to rot in a state of bankruptcy, and have become plagued by crime and disease, it seems reasonable to assume that they are no longer useful in the expansion of power. If they were of use, surely they would be continually renewed and defended (Critical Art Ensemble, 1996: 11).

The street has traditionally been seen as the locus where power is contested and expanded. The view of the Critical Art Ensemble then suggests a fundamental shift in our conception of the city as a political space. By contrast they recognise that at the same time as the city has been abandoned by power, cyberspace has become heavily policed and defended. The key to the thesis outlined by the Critical Art Ensemble is that the new frontier through which power might be contested, cyberspace, remains effectively unoccupied by counter-hegemonic forces. They argue that activists remain attached to the idea of the street being the space through which power is unsettled while hackers have failed to wed their technical and criminal skills to a coherent political ideology or project. In summing up the situation they describe activists thus: 'left behind in the dust of history, this political subgroup knows what to do and what to target, but has no effective means to carry out its desires'. They go on to argue: 'Here are two groups motivated to accomplish similar anti-authoritarian ends, but which cannot seem to find a point of intersection' (Critical Art Ensemble, 1996: 20). For meaningful resistance to emerge on this frontier they argue that this hacker/activist schism must be overcome. The extract ends with the sketch of a manifesto for electronic civil disobedience.

'The City and the Car'

International Journal of Urban and Regional Research (2000)

Mimi Sheller and John Urry

Editors' Introduction

'The City and the Car' was a paper published in the *International Journal of Urban and Regional Research*, which emerged out of the work of the Mobilities research group at the University of Lancaster in the late 1990s. The group is multidisciplinary, constituted from across the university, and has been concerned with utilising a variety of humanities and social science approaches to understand the increasing significance of different forms of travel within social and cultural life. The impact of the car on cities has begun to generate a great deal of debate in recent years. Much of this has been concerned with the environmental implications of the increasing growth of car travel across the world. It is only recently that sociologists and others have begun to examine the social and cultural geographies of mobility. It is now apparent that car travel has a number of social impacts and that mobilities are a complex and multiple social and cultural terrain that has been largely ignored by sociological study. Sheller and Urry's paper is significant because it pushes forward the debate on the reciprocity between the car and society in ways few have done before. For both authors an interest in mobility is only one amongst many interests. John Urry's interests are wide ranging and have included the sociology of power and revolution, urban and regional research, the relationship between society and space and the sociology of leisure and tourism. His most influential publications to date have included *The Anatomy of Capitalist Societies* (1981), *Social Relations and Spatial Structures* (1985), *The End of Organized Capitalism* (1987), *Economies of Signs and Space* (1994), and *The Tourist Gaze* (1990). Mimi Sheller's interests also include democracy in post-slavery societies and gender and racial formation. She is widely published in these areas.

INTRODUCTION

According to Heidegger, machinery 'unfolds a specific character of domination . . . a specific kind of discipline and a unique kind of consciousness of conquest' over human beings (quoted in Zimmerman, 1990: 214). In the twentieth century this disciplining and domination through machine technology is most dramatically seen in the system of production, consumption, circulation, location and sociality engendered by the 'motor car', what Barthes describes as 'the exact equivalent of the great Gothic cathedrals' (1972: 88).

Indeed, the car is a particularly good illustration of a putative globalization. One billion cars have been manufactured in the twentieth century. There are currently over 500 million cars roaming the world, a figure that is expected to double by 2015 (Shove, 1998). However, the car is rarely discussed in the 'globalization literature' (see Albrow, 1996), although its specific character of domination is as global as the other great technological cultures of the twentieth century, the cinema, television and the computer, which are seen as constitutive of global cultures. Contemporary 'global cities', and cities in general, remain primarily rooted in and defined by automobility as much as by newer technologies. Thus, to understand the ways in which social life might be reconfigured by new technologies of information and communication will require that social analysts take seriously their relation to the car.

Yet the social sciences have generally ignored the motor car and its awesome consequences for social life. Three 'disciplines' that ought to have examined the social impact of the car are industrial sociology, the analyses of consumption practices and urban studies. Within industrial sociology there has been little examination of how the much-analysed mass production of cars has extraordinarily transformed social life. It did not see how the huge number of cars being produced through 'Fordist' methods, especially within the US, were impacting upon the patterns of social life as car ownership became 'democratized' and generalized. Within the study of consumption there has not been much examination of the use-value of cars in permitting extraordinary modes of mobility, new ways of dwelling in movement and the car culture to develop. The main question for consumption analyses has concerned sign-values, with the ways that car ownership in general or the ownership of particular models does or does not enhance people's status position. The car as the locus of consumption normally remains on the drive of the house (see Shove, 1998).[1]

In this article it is the absence of the car in the analysis of the urban that is our particular concern. It was in the modern city that the founders of sociology first envisioned the contraction of social space, the density of transactions and the compression of 'social distance' that comprised modernity. Yet sociology's view of urban life has failed to consider the overwhelming impact of the automobile in transforming the time-space scapes of the modern urban/suburban dweller. Indeed, urban studies have at best concentrated upon the sociospatial practice of walking and especially upon *flânerie* or 'strolling' the city. It has been presumed that the movement, noise, smell, visual intrusion and environmental hazards of the car are largely irrelevant to deciphering the nature of city life. Many urban analyses have, in fact, been remarkably static and concerned themselves little with the forms of mobility into, across and through the city (although see Lynd and Lynd, 1937; and Hawkins, 1986; Lynch, 1993, on the linearity of traffic). Where such mobilities have been taken into account, it is generally either to lament the effects of the car on the city (Jacobs, 1961; Mumford, 1964) or to argue that a culture of speed replaces older cultures of urbanism (Virilio, 1997).

In general, then, cars have been conceived of either as a neutral technology, permitting social patterns of life that would happen anyway, or as a fiendish interloper that destroyed earlier patterns of urban life. Urban studies has omitted to consider how the car reconfigures urban life, involving, as we shall describe, distinct ways of dwelling, travelling and socializing in, and through, an automobilized time-space. We argue that mobility is as constitutive of modernity as is urbanity, that civil societies of the West are societies of 'automobility' and that automobility should be examined through six interlocking components. It is the *unique combination* of these components that generates the 'specific character of domination' of automobility across most societies across the globe (see Freund, 1993; and Whitelegg, 1997, on many of these). Automobility is:

- The quintessential *manufactured object* produced by the leading industrial sectors and the iconic firms within twentieth-century capitalism (Ford, GM, Rolls-Royce, Mercedes, Toyota, VW and so on); hence, it is the industry from which key concepts such as Fordism and post-Fordism have emerged to analyse the nature of, and changes in, the trajectory of western capitalism.
- The major item of *individual consumption* after housing which (1) provides status to its owner/user through the sign-values with which it is associated (such as speed, home, safety, sexual desire, career success, freedom, family, masculinity, genetic breeding); (2) it is easily anthropomorphized by being given names, having rebellious features, being seen to age and so on; and (3) generates massive amounts of crime (theft, speeding, drunk driving, dangerous driving) and disproportionately pre-occupies each country's criminal justice system.
- An extraordinarily powerful *machinic complex* constituted through the car's technical and social interlinkages with other industries, including car parts and accessories, petrol refining and distribution, road-building and maintenance, hotels, roadside service areas and motels, car sales and repair workshops, suburban house building, new retailing and leisure complexes, advertising and marketing, urban design and planning.
- The predominant global form of 'quasi-private' *mobility* that subordinates other 'public' mobilities of walking, cycling, travelling by rail and so on; and it reorganizes how people negotiate the

opportunities for, and constraints upon, work, family life, leisure and pleasure.

- The dominant *culture* that sustains major discourses of what constitutes the good life, what is necessary for an appropriate citizenship of mobility, and which provides potent literary and artistic images and symbols – such as E.M. Forster's (1931: 191) evocation in *Howard's End* of how cars generate a 'sense of flux', and J.G. Ballard's (1973) *Crash*, which describes the erotics of 'crash culture' (and see Bachmair, 1991; Eyerman and Löfgren, 1995; Graves-Brown, 1997; Creed, 1998).
- The single most important cause of *environmental resource-use* resulting from the exceptional range and scale of material, space and power used in the manufacture of cars, roads and car-only environments, and in coping with the material, air quality, medical, social, ozone, visual, noise and other consequences of pretty well global automobility (see SceneSusTech, 1998).

We use 'automobility' here in order to capture a double-sense. On the one hand, 'auto' refers reflexively to the humanist self, such as the meaning of 'auto' in notions like autobiography or autoerotic. On the other hand, 'auto' often occurs in conjunction with objects or machines that possess a capacity for movement, as expressed by terms such as automatic, automaton and especially automobile, This double resonance of 'auto' is suggestive of the way in which the car-driver is a 'hybrid' assemblage, not simply of autonomous humans, but simultaneously of machines, roads, buildings, signs and entire cultures of mobility (Haraway, 1991; Thrift, 1996: 282–4). In the following we outline a manifesto for the analysis of 'auto' mobility that explores this double resonance, of autonomous humans *and* of autonomous machines only able to roam in certain time-space scapes. Such a manifesto, we argue, will transform current understandings and analyses of contemporary cities (it should be noted that most examples in this article are taken from North America and western Europe).

Automobility is a complex amalgam of interlocking machines, social practices and ways of dwelling, not in a stationary home, but in a mobile, semi-privatized and hugely dangerous capsule. In what follows we argue that automobility has reshaped citizenship and the public sphere via the mobilization of modern civil societies. We begin by tracing urban sociology's paradoxical resistance to cultures of mobility, and the implications this has had for theories of civil society. In particular, we argue that civil society should be reconceptualized as a 'civil society of automobility', a civil society of quasi-objects, or 'car-drivers' and 'car-passengers', along with disenfranchised 'pedestrians' and others not-in-cars, those that suffer a kind of Lacanian 'lack'. There is not a civil society of separate human subjects which can be conceived of as autonomous from these all-conquering machinic complexes. We then explore how automobility makes instantaneous time and the negotiation of extensive space central to how social life is configured. As people dwell in and socially interact through their cars, they become hyphenated car-drivers: at home in movement, transcending distance to complete series of activities within fragmented moments of time. The car is thus not simply an extension of each individual; automobility is not simply an act of consumption since it reconfigures the modes of especially urban sociality. Urban social life has always entailed various mobilities but the car transforms these in a distinct combination of flexibility and coercion. In the conclusion, we will begin to trace a vision of an evolved automobility for the cities of tomorrow in which public space might again be made public.

AUTOMOBILITY, CIVIL SOCIETY AND THE CITY

The material fabric of cities substantiates a static architectonic that is shot through with 'ways' of mobility. Cities are encrusted with ramps and overpasses, bridges and tunnels, expressways and bypasses, roundabouts and 'gyratories' (see Morse, 1998: Chapter 4 on liquidity). Hemmed in by this physical infrastructure of mobility, urban architecture has become a function of movement. City-space is dominated by intersecting quasi-objects in their disorganizing semi-autonomy. Classical urban architecture was designed from the point of view of the pedestrian, using visual perspective to guide the walker through arcades and squares defined by fountains, to open, wider vistas defined by the spatial vanishing points created by judiciously placed obelisks and domes (Sennett, 1990). Post-modern architecture's emphasis on quickly-read signs and surface quotation, in contrast, arose in part out of the experience of driving along the 'strip' with its billboards and neon signs, as epitomized by learning from Las Vegas (Venturi *et al.*, 1972; and see Freund,

1993: 104–6).[2] Yet urban studies clings to the humanist figure of the pedestrian and cars are often seen as the enemy of urbanism, of civility, even of citizenship – an intrusion from the suburban borderlands. Here we will consider two traditions of theorizing the urban that have contributed to the perception of mobility as inimical to civility and citizenship: urban studies in the Chicago School tradition, and theories of civil society.

Urban studies have not ignored mobility altogether; indeed, the diversity, density and stimulus of urban social life have long been associated with forms of mobility. Nevertheless, an excess of mobility has often been blamed for urban degeneration and danger. In the Chicago School's early ecological approach to cities mobility played a central part in diagnosing urban pathologies. Burgess argued that mobility was 'perhaps the best index of the state of metabolism of the city. Mobility may be thought of in more than a fanciful sense, as the "pulse of the community"' (Burgess, 1925: 59). However, too much mobility and the stimuli that it entailed, he suggested, 'tends inevitably to confuse and to demoralize the person', resulting in a host of urban ills that undermined the city as a social organism (*ibid.* 58). Roben Park, even more than Burgess, placed personal mobility at the centre of his understanding of urban social life. He proposed that along with the basic human urge to dwell in a place, 'that other characteristic ambition of mankind [is], namely, to move freely and untrammelled over the surface of mundane things, and to live, like pure spirit, in his mind and in his imagination alone . . . Mind is an incident of locomotion' (Park, 1925: 156).

However, he too saw the need for a balance between locality and mobility. For Park the figure of the Hobo was emblematic of the uprooted, rudderless, footloose human, with no place in the world and no vocation. 'Society is, to be sure, made up of independent, locomoting individuals' he argued:

> But in order that there may be permanence and progress in society the individuals who compose it must be located; they must be located, for one thing, in order to maintain communication, for it is only through communication that the moving equilibrium which we call society can be maintained. All forms of association among human beings rest finally upon locality and local association (Park, 1925: 159).

Modern urban social life – or what might more broadly be called civil society – was therefore seen as *attached* to the city, *located* in neighbourhoods and associational spaces, and *rooted* in places of dwelling. Mobility was a necessary feature of growth and modernization, but had to be stabilized by association and anchored within place.

This understanding of association as an outgrowth of locality, that is, as architecturally located and *opposed* to mobility, is closely related to the conceptualizations of civil society grounded in the history of urbanization, which has also seen automobility as a threat to the public sphere. Civil society is typically defined as a 'realm' of association, mediating either between individuals and the state in the Hegelian tradition, or between economy and state in the Gramscian tradition (Cohen and Arato, 1992). It is understood as located in specific spaces of sociality, rooted in urban architecture rather than in mobilities, at best connected by newspapers and the imagined communities of 'print publics' (Anderson, 1991). This static conception of civil society originates in nineteenth-century notions of urban society, but has been built into late-twentieth-century social theory. Crucially, the civil society concept has been elaborated on a model of refined urban sociality, the realm of *bürgergesellschaft*.

The power of civil society crucially depends on the democratic 'social space' created by the temporal syncopation and movement between two separate 'spheres', the private and the public, through which individuals can develop their deliberative capacities as citizens (Benhabib, 1992; Cohen and Arato, 1992). In the idealized urban public spaces of, for example, Arendt's *polis* (1958), Tocqueville's voluntary associations (1835), or Habermas's coffee-houses (1992), an informed rational debate could supposedly take place – at least among the elite men who could gain entry to such 'public' places (Landes, 1988). Thus, as is often noted, we find the linguistic root of citizen, civil and civic in city (*civitas*), and of politics and polity in *polis*. While these theorists draw a sharp distinction between what is private and what is public, there is an unexamined blurring between the 'public sphere' (of citizenship) and the 'public space' (of the city).

Because of this common slippage, the *spatialization* of 'publicity' has not in itself been systematically examined. Indeed, the word public itself takes on a multitude of overlapping but at times mutually contradictory meanings within different discursive traditions (Weintraub, 1997). Given this confusion of institutional 'spheres' with physical spaces, these models of civil society attend little to how people actually *move* between the private and the public spaces. Indeed, a crucial issue might be how people play with and upon that blurred boundary while moving through urban space (though some theorists of the 'performative' dimensions of the public touch upon such questions; see Keane, 1988). Attention to the mobility of urban publics raises new questions about how such mobility is unequally available, gendered as masculine, or racialized as 'white' (but see Ryan, 1990; and Kelley, 1996). Most importantly, there is an implicit underlying threat that is barely addressed by theorists of civil society: that the very freedom of mobility necessary to publicity somehow also holds the potential to disrupt public space, to interfere with the more stable associational life and to undermine proper politics. Mobility is the enemy of civility.

This fear of mobility (and of the unsettling encounters it may bring) betrays a failure to remember that mobility is in some respects constitutive of democracy – it is a democratic 'right'. From the autonomous public emerged what Habermas describes as 'a sphere of personal freedom, leisure, and freedom of movement' (1992: 129). Yet it was the process of democratization of that freedom of movement – enabled partly by motor cars – which led to the collapse of the very distinction between what is private and what is public through transforming the flows of people in time-space scapes. When many commentators outline the decline of a public sphere of civil society, they often focus on the mass media, but seldom link this 'structural transformation' of the public sphere to this other great transformation of modern civil society, namely, the emergence of various logics of personal mobility/liquidity. In modern western societies, as Turner notes, 'the growth of mass transport systems brought about a democratisation of geographical movement so that the ownership of a motor car became, along with the ownership of a home, a basic objective of modern democracies' (1994: 128), though not one achievable by all. From the railway in the nineteenth century to automobility in the twentieth, the geography of urban public and private space – and with it the networks of civil society – have been fundamentally transformed by mass 'democratic' movement.

Like Burgess and Park, analysts of the public sphere depict a fundamental conflict between urban civility *and* democratized mobility. Both the intimate sphere of the private family and the related public sphere of old urban forms of dwelling are perceived as being drowned-out in the modern urban built environment centred on traffic flow:

> [The] meaningful ordering of the city as a whole . . . has been overtaken, to mention just one factor, by changes in the function of streets and squares due to the technical requirements of traffic flow. The resulting configuration does not afford a spatially protected private sphere, nor does it create free space for public contacts and communications that could bring private people together to form a public (Habermas, 1992: 157–8).

The urban public sphere disintegrates 'into an ill-ordered arena for tyrannical vehicle traffic' (*ibid.*: 159). Sennett, likewise, touches on automobiles in *The fall of public man*, where he suggests that the unfortunate 'idea of public space as derivative from motion parallels exactly the relations of space to motion produced by the private automobile' (1977: 14). People 'take unrestricted motion of the individual to be an absolute right', he argues. 'The private motorcar is the logical instrument for exercising that right, and the effect on public space, especially the space of the urban street, is that the space becomes meaningless or even maddening unless it can be subordinated to free movement. The technology of modern motion replaces being in the street with a desire to erase the constraints of geography' (*ibid.*).

Lefebvre makes a similar point about the negative effects of the car on urban public spaces, and the high social costs of turning over public space to private cars:

> [C]ity life is subtly but profoundly changed, sacrificed to that abstract space where cars circulate like so many atomic particles . . . [T]he driver is concerned only with steering himself to his destination, and in looking about sees only what he needs to see for that purpose; he thus perceives only his route, which has been materialized, mechanized, and technicized, and he sees it from one angle only – that of its functionality: speed, readability, facility (Lefebvre, 1991: 312–13).

This destruction of urban public space and introduction of a superficial built environment reflects a 'strategy of the state bureaucracy', as seen in the Haussmanization of Paris or the building of Brasilia (*ibid.*: 312), married to the interests of oil companies and car manufacturers (*ibid.*: 359, 374). The end result, he argues, is that 'urban space tends to be sliced up, degraded, and eventually destroyed by . . . the proliferation of fast roads and of places to park and garage cars, and their corollary, a reduction of tree-lined streets, green spaces, and parks and gardens' (*ibid.*: 359).

What these widely-accepted formulations of modernity and urbanism thus neglect, and analyses of the decline of publicity only hint at, is what we call the *mobilization* of modern civil societies. The focus on urbanization as a mode of location, of dwelling, of architecture and of association has deflected attention from the ways in which movement also constitutes cities as civil spheres. If *urbanization* leads to the intensification of human habitats, the concentration of places in space, and the unification of condensed temporal flows, then *automobilization*, by contrast, leads to the extension of human habitats, the dispersal of places across space, the opportunities to escape certain locales and to form new socialities, and the fragmentation of temporal flows, especially through suburbanization. On the latter, Morse (1998: 106) notes that the freeway is not so much a place but a vector. Both processes, urbanization and automobilization are *together* characteristic of modernity and of the culture of cities. Meeting places require that people get to them. Mobility cannot simply be conceived as the enemy of *civitas*; however much we may despair of vehicular traffic and busy roads, the auto-freedom of movement is part of what can constitute democratic life.

That this has not been appreciated is because analyses of the modern have normally been in terms of 'structure' rather than 'mobility', and have viewed civil society as essentially 'reflective' and/or 'residual' (Emirbayer and Sheller, 1999). Mobility and the technologies associated with it have generally been understood as arising from capitalism, while non-economic and non-state processes of mobilization have been overlooked. Automobility, the coercive yet flexible freedom of civil actors, is, we suggest, a crucial dimension of processes of democratization and reflexive modernization on the one hand, and of the deeply contested constraints and unintended consequences of modernity on the other. Civil society is thus not to be viewed as simply or principally a 'structure'. Civil societies are sets of mobilities flowing over roads which have only been civilianized as more and more social actors have demanded rights of personal mobility. Civil actors draw on structures such as the road system in ways that may be unexpected and contestatory. This deployment of such 'structured' resources can result in socially innovative practices that are neither residual nor reflective of these larger structures. The logic of communication is also embedded in transportation systems as civic (as well as economic and political) networks. Civil society is thus a site, or perhaps a 'crossroads', where mobilities have been sought by many actors and groups struggling to establish their place of movement within civil society and, in particular, to move within and between cities using the established routeways of capital and the state. In the next section we go on to consider some of the more detailed transformations of time and space that such an automobilized civil society presupposes and produces.

TIME AND SPACE IN THE CITY

Raymond Williams' novels interestingly bring out how twentieth-century social life exists through interconnecting time-space paths linking place with place. Williams elaborates how many socialities of civil society are sustained through technologies of movement which, literally and imaginatively, connect peoples, and especially families, over significant, complexly structured, heterogeneous distances. In *Border country*, Williams is 'fascinated by the networks men and women set up, the trails and territorial structures they make as they move across a region, and the ways these interact or interfere with each other' (Pinkney, 1991: 49; Williams, 1988). Williams mainly considers the connections made possible by the railway. But these are now less significant than those of automobility. 'What was central now was the fact of traffic', as Williams puts it (quoted in Pinkney, 1991: 55).

Automobility permits multiple socialities, of family life, community, leisure, the pleasures of movement and so on, which are interwoven through complex jugglings of time and space that car journeys both allow but also necessitate. These jugglings result from two interdependent features of automobility: that the car is immensely flexible and wholly coercive. We elaborate some of the temporal and spatial implications of this simultaneous flexibility and coercion for the constitutive nature of urban life.

Automobility (in some respects) is a source of freedom, the 'freedom of the road'. Its flexibility enables the car-driver to travel at speed, at any time, in any direction along the complex road systems of western societies that link together most houses, workplaces and leisure sites. Cars, therefore, extend where people can go and hence what as humans they are literally able to do. Much of what many people now think of as 'social life' could not be undertaken without the flexibilities of the car and its availability 24 hours a day. It is possible to leave late by car, to miss connections, to travel in a relatively timeless fashion. People find pleasure in travelling when they want to, along routes that they choose, finding new places unexpectedly, stopping for relatively open-ended periods of time, and moving on when they desire. They are what Shove (1998) terms another of the 'convenience devices' of contemporary society, devices that make complex, harried patterns of social life, especially in the city, just about possible, at least, of course, for those with cars.

But, at the same time, this flexibility and these rights are themselves necessitated by automobility. The 'structure of auto space' (Freund, 1993) forces people to orchestrate in complex and heterogeneous ways their mobilities and socialities across very significant distances. The urban environment built for the convenience of the car has 'unbundled' those territorialities of home, work, business and leisure that have historically been closely integrated (Sassen, 1996).

Automobility has fragmented social practices that occurred in shared public spaces within each city (see SceneSusTech, 1998). In particular, automobility divides workplaces from homes, so producing lengthy commutes into and across the city. It splits homes and business districts, undermining local retail outlets to which one might have walked or cycled, thereby eroding town centres, non-car pathways and public spaces. It also separates homes and various kinds of leisure sites, which are often only available by motorized transport. Automobility turns access zones on urban fringes into wastelands, necessitating ever-further travel to escape the urban prison of cement and pollution. Members of families are split up since they will live in distant places necessarily involving complex travel to meet up even intermittently. People are trapped in congestion, jams, temporal uncertainties and health-threatening city environments, as a consequence of being encapsulated in a privatized, cocooned, moving environment that uses up disproportionate amounts of physical resources. At the same time, automobility disables those who are not car-drivers (children, the sight impaired, those without cars) by making their everyday habitats dangerously non-navigable (Kunstler, 1994).

Automobility thus coerces people into an intense flexibility. It forces people to juggle tiny fragments of time so as to deal with the temporal and spatial constraints that it itself generates. Automobility is a Frankenstein-created monster, extending the individual into realms of freedom and flexibility whereby one's time in the car can be positively viewed, but also structuring and constraining the users of cars to live their lives in particular spatially stretched and time-compressed ways. The car, one might suggest, is Weber's 'iron cage' of modernity, motorized, moving and privatized. It coerces almost everyone to juggle tiny fragments of time in order to put together complex, fragile and contingent patterns of social life, especially within the city. J.G. Ballard in *Crash* describes a car-based infantile world where any desire can be instantly satisfied (1973: 4; Macnaghten and Urry, 1998: Chapter 5).

Automobility thus develops 'instantaneous' or 'timeless' time that is to be juggled and managed in highly complex, heterogeneous and uncertain ways. Automobility involves an individualistic timetabling of many instants or fragments of time. The car-driver thus operates in instantaneous time that contrasts with the official timetabling of mobility that accompanied the railways in the mid-nineteenth century (Lash and Urry, 1994: 228–9). This was modernist clock-time based upon the public timetable, or what Bauman terms 'gardening' rather than 'gamekeeping' (1987). As a car-driver wrote in 1902: 'Travelling means utmost free activity, the train however condemns you to passivity . . . the railway squeezes you into a timetable' (cited in Morse, 1998: 117). The objective clock-time of the modernist railway timetable is replaced by personalized, subjective temporalities, as people live their lives in and through their car(s) (if they have one; Lash and Urry, 1994: 41–2). There is involved here a reflexive monitoring, not of the social, but of the self. People try to sustain 'coherent, yet continuously revised, biographical narratives . . . in the context of multiple choices filtered through abstract systems' (such as that involved in automobility; Giddens, 1991: 6). Thus, automobility coerces urban dwellers to juggle tiny fragments of time in order to put together complex, fragile and contingent patterns of social life, which constitute self-created narratives of the reflexive self.

As personal times are desynchronized, spatial movements are synchronized to the rhythm of the road.

The loose interactions and mobilities of pedestrians are forced to give way to the tightly controlled mobility of machines, especially in the human and machinic density of urban areas. Automobility dominates how both car-users *and* non-car-users organize their lives through time-space. Driving requires 'publics' based on trust, in which mutual strangers are able to follow shared rules, communicate through common sets of visual and aural signals, and interact in a kind of default space or non-place available to all 'citizens of the road' (see Lynch, 1993). Yet car-drivers are excused from the normal etiquette and social coordination of face-to-face interactions. Car travel rudely interrupts the taskscapes of others (pedestrians, children going to school, postmen, garbage collectors, farmers, animals and so on), whose daily routines are merely obstacles to the high-speed traffic that cuts mercilessly through slower-moving pathways and dwellings. Junctions, roundabouts and ramps present moments of carefully scripted intercar action during which non-car users of the road present hazards or obstacles to the drivers intent on returning to their normal cruising speed.

Automobility also generates new scapes that structure the flows of people and goods along particular routes, especially motorways or interstate highways (see Urry, 2000, on scapes and flows). There is a rewarping of time and space by advanced transportation structures, as scapes pass by some towns and cities while connecting other areas along transport-rich 'tunnels'. Such tunnels also shape urban geographies of social exclusion and ghettoization (see Urban Taskforce, 1999). Public housing estates in the UK or so-called 'projects' in US cities are often cut off from bus, rail or subway links to employment-rich business districts within the city and from automobile roadways linked to more desirable (middle-class) residential and leisure areas outside the city. At the same time, tolls and parking fees can deter rural and suburban dwellers from entering the city too easily. Thus, the inequalities among multiple publics are entrenched in urban spaces of unevenly distributed access and exclusion.

DWELLING IN MOBILITY

The shortage of time resulting from the extensive distances that increasingly 'have' to be travelled means that the car remains the only viable means of highly flexibilized mobility. Also, other forms of mobility in the city are, by comparison with the car, relatively inflexible and inconvenient, judged, that is, by criteria that automobility itself generates and generalizes. In particular, the car enables *seamless* journeys from home-away-home. It does away with the stationary pauses necessitated by 'stations', apart from the occasional stop at the gas station. And this is what the contemporary traveller has come to expect. The seamlessness of the car journey makes other modes of travel inflexible and fragmented. So-called public transport rarely provides that kind of seamlessness (except for first-class air travellers with a limousine service to and from the airport). There are many gaps between the various mechanized means of public transport: walking from one's house to the bus stop, waiting at the bus stop, walking through the bus station to the train station, waiting on the station platform, getting off the train and waiting for a taxi, walking though a strange street to the office and so on, until one returns home. These 'structural holes' in semi-public space are sources of inconvenience, danger and uncertainty. And this is especially true for women, older people, those who may be subject to racist attacks, the disabled and so on (see SceneSusTech, 1998). There are gaps for the car-driver involving semi-public spaces, such as entering a multi-storey car park or walking though strange streets to return to one's car or waiting by the side of the road for a breakdown vehicle, but these are much less endemic than for other kinds of travel.

Not only do car-drivers gain the comparative benefits of relative mobility and seamless travel, making older ways of travel seem slow and inflexible, but also the matrix of automobility undermines other forms of mobility. The predominance of the car in government policy and planning afforded seamless car journeys while breaking down those linkages that once made other forms of transport possible. In the US, car manufacturers such as General Motors bought and dismantled electric tramway systems in order to make suburbs car-dependent (see Flink, 1988; Wolf, 1996). Zoning laws and building codes enforced suburban sprawl through the separation of business and residential districts, mandatory large-lot sizes and set-backs from kerbs, which destroyed town centres and the public spaces that they once provided (Kunstler, 1994). Auto-intensive middle-class suburbanization resulted in 'auto sprawl syndrome' in which cars make urban suburbanization/sprawl possible and in so doing they make those living in such areas dependent upon the use of cars (SceneSusTech, 1998: 100). Freund argues that 'Modernist urban landscapes were built to facilitate automobility and to discourage other forms of human movement . . . [Movement between] private worlds is through dead public spaces by car' (1993: 119).

Indeed, large areas of the globe now consist of car-only environments – the quintessential non-places of super-modernity (Augé, 1995). About one-quarter of the land in London and nearly one-half of that in LA is devoted to car-only environments. And they then exert an awesome spatial and temporal dominance over surrounding environments, transforming what can be seen, heard, smelt and even tasted (the spatial and temporal range of which varies for each of the senses). Such car-environments or non-places are neither urban nor rural, local nor cosmopolitan. They are sites of pure mobility within which car-drivers are insulated as they 'dwell-within-the-car'. They represent the victory of liquidity over the urban. One such non-place is the motel, immortalized in the UK by the TV soap *Crossroads*. Clifford notes that the 'motel has no real lobby, and it's tied into a highway network – a relay or node rather than a site of encounter between coherent cultural subjects' (as would, he implies, be found in a hotel; 1997: 32). Motels 'memorialize only movement, speed, and perpetual circulation' since they 'can never be a true place' and one motel is only distinguished from another in 'a high-speed, *empiricist* flash' (Morris. 1988: 3, 5). The motel, like the airport transit lounge or the motorway service station, represents neither arrival nor departure, but the 'pause', consecrated to circulation and movement and demolishing particular senses of place and locale. This 'sense of sameness and placelessness' is accompanied by a 'social organization of space that helps to further auto-dependence and to mask any realistic alternatives to automobility' (Freund, 1993: 11).

As a rolling private-in-public space, automobility affords dwelling inside a mobile capsule that involves punctuated movement 'on the road' from home-away-home. Domesticity is reproduced on the road through social relations such as the 'back-seat driver' or the common dependence upon a partner for navigation and map-reading; the car creates a transpersonal mobile agent. Moreover, a variety of services have become available without leaving the car, as the 'drive-in' becomes a feature of everyday life. Since the 1950s halcyon days of the drive-in movie and the drive-in 'automat' where 'fast food' is served, more recent car-dwellers, especially in the US, have been treated to the conveniences of drive-through banking, drive-through car washes, drive-through safari theme parks, and even drive-through beer distributors in some states (not to mention drive-by shootings and drive-up mail delivery). Thus, fragments of time are increasingly compressed into taskscapes that can keep people inside their cars, while the 'coming together of private citizens in public space' is lost to a privatization of the mechanized self moving through emptied non-places.

Protected by seatbelts, airbags, 'crumple zones', 'roll bars' and 'bull bars', car-dwellers boost their own safety and leave others in the city to fend for themselves in a 'nasty, brutish and short' world. In each car the driver is strapped into a comfortable armchair and surrounded by micro-electronic informational sources, controls and sources of pleasure, what Williams calls the 'mobile privatization' of the car (see Pinkney, 1991: 55). Many aspects involved in directing the machine have been digitized, at the same time that car-drivers are located within a place of dwelling that insulates them from much of the risky and dangerous urban environment that they pass through. The Ford brochure of 1949 declared that 'The 49 Ford is a living room on wheels' (Marsh and Collett, 1986: 11). Features such as automatic gearboxes, cruise control and CD-changers 'free' drivers from direct manipulation of the machinery, while embedding them more deeply in its sociality. Yet it is a room in which the senses are impoverished. The speed at which the car must be driven constrains the driver to always keep moving. Dwelling at speed, people lose the ability to perceive local detail, to talk to strangers, to learn of local ways of life, to stop and sense each different place (see Freund, 1993: 120–1). The sights, sounds, tastes, temperatures and smells of the city are reduced to the two-dimensional view through the car windscreen, something prefigured by railway journeys in the nineteenth century (see Schivelbusch, 1986). The sensing of the world through the screen has, of course, become the dominant way of dwelling in contemporary experience. The environment beyond the windscreen is an alien other, to be kept at bay through the diverse privatizing technologies incorporated within the contemporary car. These technologies ensure a consistent temperature (with the standardization of air-conditioning), large supplies of information, a relatively protected environment, high-quality sounds and sophisticated systems of monitoring. They enable the hybrid of the car-driver to negotiate conditions of intense riskiness on high-speed roads (roads are increasingly risky because of the reduced road-space now available to each car). And as cars have increasingly overwhelmed almost all environments, so everyone is coerced to experience such environments through the protective screen and to abandon urban streets and squares to the metallic cages-on-wheels.

The driver's body is itself fragmented and disciplined to the machine, with eyes, ears, hands and feet all trained to respond instantaneously, while the desire to stretch, to change position, or to look around must be suppressed. The car becomes an extension of the driver's body, creating new urban subjectivities (see Freund, 1993: 99; and Morse, 1998: Chapter 4, on the overlaps between driving, shopping and viewing television). A Californian city planner declared even in 1930 that 'it might be said that Southern Californians have added wheels to their anatomy' (cited in Flink, 1988: 143). The machinic hybridization of the car-driver can be said to extend into the deepest reaches of the psyche. A kind of libidinal economy has developed around the car, in which personalities are deeply invested in the car as object. There is a sexualization of the car itself as an extension of the driver's desirability and fantasy world. The car takes part in the ego-formation of the driver as competent, powerful and able (as advertisers have tapped into). Various 'coming-of-age' rituals revolve around the car, at least since the discovery of bench-seats and 'lover's lanes'. Car-sex has itself become an element of fantasy in everything from music videos to 'crash culture' (see Ballard, 1973). At the same time, the car feeds into our deepest anxieties and frustrations, from fear of accident and death at one extreme, to the less life-threatening but nevertheless intense frustration of being stuck behind a slow vehicle while trying to save precious fragments of time. Within the private cocoon of glass and metal intense emotions can be released in forms that would not otherwise be socially acceptable. Outbursts of 'road rage' represent a breakdown in 'auto-regulation' of the driver, in the double sense of both self-control and of following the rules of the road.

Given these powerful restructurings of time, space and self within the car-driver matrix, urban life and the civil society that arises out of it have to be rethought. Our first step is to reconsider the kinds of inequalities produced by the dominant form of automobility.

INEQUALITIES OF AUTOMOBILITY

Automobility, we have suggested, is a key component of civil society in the contemporary world. Yet, like other aspects of civil society, the automobilized civil society is also productive of new kinds of social inequalities despite efforts at its democratization. Such inequalities are centrally important to the urban experience (as the UK White Paper on *A new deal for transport* brings out: DETR, 1998; and see *Towards an urban renaissance*, Final Report of the Urban Taskforce, 1999). It is both the deficits of mobility in some places, as well as its excess in others, that are symptomatic of contemporary urban inequalities and problems of social exclusion. As Freund argues, although the car 'has been widely hailed as the quintessentially democratic means of transport, the auto actually is not usable by, or available to, large sectors of the population, even in the most auto-saturated societies' (1993: 7; and see his discussion of social inequalities in Chapter 3). Here we discuss two such forms of inequality, those of gender and those of being without a car.

Gender and automobility

Women have a very different relation to cars than do men as a group. In the interwar period automobility was generally organized around a cosiness of family life, both in Europe and the US (Taylor, 1994: Chapter 4). In the latter, this was the period of massive suburbanization that was predicated upon low-density family housing with a sizeable garden, many domestic production goods for the 'wife' to use, and a car to enable the 'husband' to travel quite long distances to get to work. The automobilization of family life not only brought the newest and most expensive car models first to male 'heads of families', while women had to settle for second-hand models or smaller cars, but also led to the uneven gendering of time-space. While working men became enmeshed in the stresses of daily commuter traffic into and out of urban centres, suburban 'housewives' had the greater burden of juggling family time around multiple, often conflicting schedules of mobility epitomized by 'the school run' and mom-as-chauffeur. Once family life is centred within the moving car, social responsibilities tend to push women towards 'safer' cars and 'family' models, while men have greater luxury to indulge in individualistic fantasies of the 'Top Gear' fast sports car or the impractical 'classic car'. Cars were originally designed to suit the average male size and have only recently been designed to be adjustable to drivers of various heights and reaches. The distribution of company cars has also benefited men more than women, due to continuing horizontal and vertical segregation in the job market, which keeps most women out of positions with access to such 'perks'. Actuarial statistics also show that male drivers are more likely to externalize risks onto others through dangerous driving practices (see Meadows and Stradling, 2000).

Given these inequalities, for many women exclusion from automobility has become a crucial political issue, both because it limits their capability to work outside the home and because it makes movement through public spaces difficult. In most countries women became eligible to be licensed drivers later than did men, and in some countries they still face severe restrictions on their ability to drive. Women working in domestic service jobs (often from racialized minority groups or recent immigrants) faced (and still face) a gruelling journey on unreliable public transport between the city and the suburbs. Single mothers without cars are among the groups most dependent on public transport and most likely to find their particular 'taskscapes' fraught with gaps and inconveniences. The male drivers' domination of public space appears in the practice of 'kerb crawling' in the city, one of the most tolerated forms of prostitution, which compounds the difficulties of the female 'street walker' as '*flâneuse*' (Wilson, 1995).

In line with the notion of the 'democratization' of automobility, therefore, greater access to automobility may be seen as empowering women. In some respects women's 'emancipation' has been predicated on the automobile. Cars afford many women a sense of personal freedom and a relatively secure form of travel in which families and objects can be safely transported, and fragmented time-schedules successfully intermeshed. Women's access to cars has had a major impact on the integration of women into the labour market by better enabling the juggling of the conflicting time disciplines of paid and unpaid work (see Wajcman, 1991). Although the sense of freedom of movement that comes with automobility has been largely a masculine prerogative as well as a masculine fantasy, it is nevertheless a 'structure of feeling' that reaches beyond Eurocentrism and androcentrism (Thrift, 1996). Women have therefore struggled to claim automobility, and in doing so have in part re-shaped and paradoxically enhanced car culture.

Carless living

Automobility has also shaped and encoded the physical relation of the city to surrounding suburban and rural environments, of urban-dwellers to non-urban dwellers, and of car-dwellers to others in civil society. Distinctions can be drawn by *not* having a car. Living without a car has become a significant lifestyle choice for both environmentalists and for a small cosmopolitan elite able to live in expensively gentrified city-centres. 'Global cities' (Sassen, 1991), increasingly polarized between ghettos of wealth and of poverty, may no longer have a place for the car-bound middle classes. The carless urban poor and growing population of new immigrants (who may actually want cars) are often cut off from cheaper out-of-town shopping, from many public facilities accessible only by car, and from a host of job opportunities in urban fringes and 'edge-cities'. Freedom of mobility is restricted even further by practices of institutional racism, which lead to racially segregated residential areas and disproportionate stopping of black male drivers by police.

Resisting the coercive features of automobility is nearly impossible, given the extent to which even the lives of non-car-users in the city are transformed by cars. The nostalgic retreat to a romantic lifestyle 'in the country' is an option for some, especially when combined with 'telecommuting' to work, yet country life is especially car-dependent. Automobility has accelerated the collapse of rural and small-town local shops and services, which cannot compete against cheaper 'superstores' on the outskirts of cities. The choice of an alternative lifestyle with voluntary limitations on car use is most feasible in medium-sized regional towns where a mix of cycling, walking and public transport can develop (as in Cambridge, UK). Nevertheless, such towns remain clogged with both moving and stationary cars apart from those small 'pedestrian zones' of civility left to the walker. Attempts to introduce pedal-rickshaws and more bike lanes are still constrained by the imperatives of the car-driver matrix against which they must compete.[3]

New processes of urbanization are also reshaping automobility in the city, and with it new inequalities. Soja, using Los Angeles as his example, has identified several geographies of postmodern urbanization, which have important implications for an automobilized civil society. In the radically restructured postmodern city – described by neologisms such as 'megacities, outer cities, edge cities, metroplex, technoburbs, post-suburbia, technopolis, heteropolis, exopolis' (Soja, 1995: 131) – Soja envisions 'a combination of decentralization and recentralization, the peripheralization of the center and the centralization of the periphery, the city simultaneously being turned inside out and outside in'. Although he does not relate this directly to the system of automobility, it is clear that the car and freeway system is central to the changing geography of Los Angeles. This city, above all others, is what Banham has described as an 'autopia' in which 'the freeway system in its totality is now a single comprehensible place, a coherent state of mind, a complete way of life . . . The freeway is where Angelinos live a large part of their lives' (cited in Jacobs, 1992: 255). As the polymer city reshapes itself, strings of traffic are the most solid structure remaining to fuse the urban sprawl into a bounded entity. This newly emerging urban form, according to Soja, is 'shaped by a very complex redistribution of jobs, affordable housing, and access to mass transit, and modified significantly by income, racial, and ethnic differentiation' (Soja, 1995: 132). Together these processes are leading to two further trends, 'the development of new patterns of social fragmentation, segregation, and polarization', as well as 'an increasingly "carceral" city'.

As the control of mobility into and out of regions of the city becomes a central concern of the state, the 'tactics of mobility' emerge as a potential form of resistance (Thrift, 1996). In accordance with early urban sociology's pessimistic view of the car in the city and political sociology's understanding of the car's effect on the decline of the public sphere, many analysts of contemporary social formations understandably continue to equate automobility with inequalities, exclusions, risk proliferation and, of course, environmental degradation. It is no surprise, then, that automobility involves massive contestation. In the following section we consider some of the ways in which political mobilization and social movements have developed around automobility and have simultaneously been transformed by it.

THE POLITICS OF PROTESTING AUTOMOBILITY

Civil society's mobilization around automobility began in relation to consumer protection. In the US, consumer advocates like Ralph Nader represented 'the public interest' in demanding car safety, road safety, 'lemon laws' to protect consumers against unscrupulous car-salesmen, and industry-wide standards for recalls of defective models and fair pricing (Nader, 1965). The oil-crisis of the 1970s sparked public concern over energy use and the growing demand for 'greener' cars with higher fuel economy and in some instances interest in the recycling of the metals, plastics and rubber that make up the car. As the wider environmental movement developed, the petroleum industry became a target of protest. There were campaigns against the expansion of oil extraction into wilderness areas such as Alaska or various off-shore sites; there were protests against pollution caused by oil extraction, processing and shipping (e.g. the Exxon Valdez oil spill); and there were eventually protests against the transnational corporations that controlled oil production (culminating in the Greenpeace campaign against Shell Oil in the early 1990s). Finally, with the Gulf War, many critics of the car culture recognized the extent to which American and European foreign policies are driven by the petroleum interests driving the global economy. Dependence upon oil and the lengths to which societies would go to protect access to oil were highlighted in public debate.

Furthermore, the unavoidable flexibility necessitated by the fragmented time-space of automobility has become an issue not only of personal management, but also of public policy. In the 1970s, urban quality of life became a crucial political issue as cities were choked with fumes and smog, as well as beset by traffic flow and parking problems. In this period the car began to be viewed as more polluting than the train (Liniado, 1996: 28). Many cities such as Amsterdam, Stockholm and Portland, Oregon, developed explicit policies to upgrade and prioritize cycle lanes and public transport, in order to wean people away from their cars. Later, cities such as Athens attempted to control access of private cars in and out of the city centre, with some success, while others imposed commuter restrictions or incentives such as park-and-ride schemes or enforced car-pool lanes with four occupants required per car. More symbolically, some European cities have instituted an annual 'leave your car at home' day.

The debate over better provision of public transport (and overall urban design) has moved to the fore in a number of countries as various governments wrestle with controlling traffic and many drivers seek to find viable alternative means of transport. Integrating better mass transportation into urban design has been crucial to planning burgeoning new metropolises such as Singapore and Hong Kong (Owen, 1987). Questions of congestion pricing systems, taxation of car use and of petrol have become key aspects of government transport policy from California to Britain. It is only in thinking about what it would take to get people out of their cars that we can see the enormous transformations that automobility has wrought in the social organization of time, space and social life.

Overall, although many people may 'love' their car, the system that it presupposes is often unloved, resisted and raged against. Civil society is significantly remade through contestations over the power, range and impact of the system of automobility. The same people can be both enthusiastic car-drivers, as well as being very active protesters against schemes for new roads (see Macnaghten and Urry, 1998: Chapter 6, on how cars generate intense ambivalence). By 1994, in the UK, the scale of grass-roots protest against the construction of new roads had risen to such a level that it was described as 'the most vigorous new force in British environmentalism' (Lean, 1994). There were by then an estimated 250 anti-road groups, a movement significantly impacting upon civil society. The array of direct actions has also diversified as protesters have become more expert, through the use of mass trespass, squatting in buildings, living in trees threatened by road programmes and digging tunnels (hence the iconic Swampy who tunnelled underground to stop a road from being built in Britain). Stopping traffic has itself become a significant form of symbolic direct action, as practised in 'Reclaim the Streets' events. Protesters also became more sophisticated in the use of new technologies, including mobile phones, video cameras and the internet. These have enabled almost instantaneous dissemination to the media (see Macnaghten and Urry, 1998: Chapter 2). Thus, the politics of automobility is generating new forms of public protest and changing civil society's repertoires of contestation. The recent protests over petrol prices across Europe demonstrate that the traffic-stopping impacts of new styles of direct action are both against and in favour of an automobilized society.

As analysts of the decline of the public sphere have lamented, politics itself has been transformed, 'hollowed out' and 'colonized' due to changes in the media of communication. What they have been less clear about, however, is the way in which changes in communication media are also linked to changes in the spatiality of the urban environment and public space. Urban public spaces and public networks of association and 'legitimate' political communication increasingly occur in virtual forums, rather than in 'real' spaces. Public outdoor gatherings and old-fashioned demonstrations are increasingly criminalized in the 'carceral city', and political protest is framed as riot. Thus, many have connected the new technologies of surveillance and digital information to this carceral aspect of the city (such as Davis, 1990). However, new technologies are not only controlled by the state or by media corporations. We want to suggest that the 'communication revolution' takes on a further dimension when it is inserted into the system of automobility. At the turn of the twenty-first century the simulacra of co-presence created via new information and communication technologies have broken free of hard-wiring and spatial location. Like Park's Hobo, civil society has embraced mobility.

REDESIGNING THE CITY OF AUTOMOBILITY: A VISION OF THE FUTURE

For too long transport has been theorized and planned as if it were a free-standing system disconnected from other technologies and socialities. True, for most of the twentieth century the revolution in communication technology was driven by the separation of information transmission from the physical means of transportation, as implicit in the Chicago School's analysis of urban 'metabolism' at the beginning of the twentieth century. However, the current trend, in contrast, is toward the re-embedding of information and communication technologies (ICT) into moving objects: satellites, mobile phones, palmtop computers and so on. At the same time that information has been digitized and released from location, objects such as cars, roads and buildings have been rewired to send and receive digital information – for example in the building of 'Intelligent Transport Systems' (ITS). We consider very briefly here what effects the convergence of ICT and ITS will have on urban geographies and the culture of cities.

Until now this has been considered a question of interest only to urban planners and car manufacturers – a question of traffic control or car and road safety, for example through computer-assisted operation control systems, dynamic route guidance and traffic information systems (see Sparmann, 1992). However, if civil societies consist of modes of mobility as well as modes of communication, then there may be crucial processes of contention and democratization at work in the reflexive social adaptation of these new technologies, which will contribute to the further (post)modernization of cities. Emerging technologies are not only creating new human-machine hybrids like the car-driver, but are also grafting together existing machines to create new intermodal scapes and hybrid flows: PCs with telephones, televisions with satellites, mobile phones with cars. It is worth noting that highly complex information, communication and simulation systems designed for military vehicles are already making the jump to civilian applications, as have previous military technologies.[4]

Given the environmental pressures on the current system of automobility, including political contention around the control of pollution, traffic reduction, management of risk and addressing social exclusion, we would predict that the politics of urban sustainability will play a major part in the design of new built environments. Thus far, environmentally friendly policy initiatives have focused on reducing car use, stemming the flow of traffic and shifting people onto public transport and non-motorized mobility through penalties for car use and incentives for biking, walking or riding the bus, tram or train. These will still be crucial tools in transforming the existing car-driver matrix and we would not argue against such policies.

However, such policies still fail to address the continuing production of a system of automobility – of cars, roads and drivers – which continues to dominate both the built environment of cities and their hinterlands and the scapes of time-space through which people organize and literally mobilize their lives. Until these fundamental social aspects of mobility are addressed, we will continue in a deadlock between the profit-making imperatives of car manufacturers and the decaying semi-public infrastructure of transportation starved of investment and of the necessary time-horizons and physical space in which to expand. More encompassing approaches to changing cultures of automobility and of cities include the creation of auto-free zones in city centres, fostering 'access by proximity' through denser living patterns and integrated land-use patterns, and promoting greater coordination between both motorized and unmotorized transport systems (see Freund, 1993, Chapter 9).

The question remains, though, how these changes will intersect with changing ICT scapes. Some futurologists have proclaimed the 'electronic cottage' to be the way out of the traffic, as well as out of the city and its urban problems. This would be unfortunate for the public life of cities, and for those marooned in them. Rather than the replacement of physical travel by virtual or 'weightless travel' (see Urry, 2000), what may be starting to occur is a convergence between the two. This is suggestive of another way forward out of the impasse of the limits of ecology, time-space and inequality. Through an interlocking of 'smart' transportation systems and the urban 'info-structure' a new mode of automobilization could be created that would integrate private and public transport, motorized and non-motorized transport, and information transmission and human mobility. Crucial to this detraditionalization of urban transportation will be a redesign of both public mass (and mini) transportation systems and of private or semi-private vehicles. Smaller, smarter, information-rich, communication-enhanced vehicles that are better integrated into the public transport system and public space will be indispensable in the city of the future and to the civil societies that might flourish within them.

Telecommuting will not be the key to transforming urban life because, as Park and others recognized, people do *like* to be physically mobile, to see the world, to meet others and to be bodily proximate, and to engage in 'locomotion' (see Boden and Molotsch, 1994, on the compulsion to proximity). Current developments such as the huge popularity of mobile telephones instead suggest that many people want to engage in communication simultaneously with locomotion – to walk and talk or to drive and jive, as it were. Mobile ICT is also increasingly central to work-practices and information-gathering in contexts of unavoidable time-space distanciation and fragmentation. The introduction of flexitime in order to smooth out and redistribute rush-hour peaks of transportation demand, for example, would be enhanced if communication could occur in transit. It is already possible to check voicemail from a mobile phone, but soon e-mail will be found in the car or train, electronic memos will be sent, and mobile banking and electronic shopping will be commonplace.

The crucial question is how these technological developments can be used to redesign urban public spaces in ways that will address the negative constraints, risks and impacts of automobility. This will require the intermodality of multiple transport and communication flows to allow various mobile publics to switch across a range of spaces and zones easily. Creating a smaller, lighter, fuel-efficient car is a start, but would not be enough. Car manufacturers have already begun production of various micro-cars that are ideal for crowded urban spaces where parking is at a premium and environmental issues are paramount. However, such micro-cars would have to be truly integrated into a mixed transportation system that allowed more room not only for bikes, pedestrians and public transportation, but also for modes of travel that we have only begun to imagine. This would require the redeployment of existing urban zoning laws to exclude or severely delimit 'traditional' cars (as has already begun to occur in many cities; see Owen, 1987) and to place lower speed limits on them (as is currently being suggested in the UK). The aim would be not only to free up space for new kinds of intermodal flows but also *time* for new socialities that would juggle the complex timing of schedules in more flexible ways.

There would also have to be incentives to both car manufacturers and consumers to produce a new culture of automobility (through extensions of already existing legislation, taxation and penalties). Through such means a number of key objectives could be met:

- Reduced energy consumption and polluting emissions through design of smaller vehicles, use of fewer private vehicles, and curtailment of traffic proliferation and road building;
- Redirection of investment to new and better modes of public transportation, bike and 'soft' vehicle lanes, and more diversified multifunctional stations and public spaces;
- Reduced risks to human safety inside and outside the slower and lighter car;
- Minimizing social exclusion through better planning of networks and intermodality.

The key to such a system could be the use of a multi-function 'smart-card' that would transfer information from home to car, to bus, to train, to workplace, to web site, to shop-till, to bank (a system already under development). Cars for cities could then be partially deprivatized by making them available for public hire through using such a smart-card to pay fares on buses, trains, or more flexibly routed collective mini-vans; cards for welfare recipients, students, families with young children and the elderly could be partially subsidized.[5] But all of these vehicles would have to become *more* than technologies of movement – they would also have to be hybridized with the rapidly converging technologies of the mobile telephone, the personal entertainment system and the laptop computer.

The 'carrot' for car manufacturers is that small cars would no longer be at the bottom of the profit scale; the innovation of new ICT applications would provide an endless source of novelty, desirability and profitability. The hook for car-drivers is that the micro-cars and all other forms of transport would be personalized with their own communication links (e-mail addresses, phone numbers, world-wide-web addresses etc.) and entertainment applications (digitally stored music, programmed radio stations etc.), but only when initiated by inserting the smart-card. Thus, any public vehicle could instantly become a home away from home: a link to the reflexive narratives of the private self in motion through public time-space scapes. The streetscapes of global cities could thus be transformed through a more mixed flow of slow-moving, semi-public micro-cars, bike lanes, pedestrians and improved mass transport.

Public-friendly cars would allow people to travel lighter, if not weightlessly (as 'virtual' electronic travel is often depicted), and would restore some of the civility to urban public space that has been destroyed by current traffic flows and by the spatial patterns of segregation and fragmentation generated by automobility. Could such an urban smart-car be the best way to lure twentieth-century speed-obsessed car-drivers to give up their dependence on dinosaur cars and fossil fuels, a system that is unsustainable on every conceivable measure and is really a very old-fashioned Fordist technology? Urban planning that recognizes the need for a radical transformation of transport can use existing legislation and regulation in new ways. to build truly 'integrated' and 'intermodal' public transport systems. Rather than trying to stifle mobility, however, which has been the strategy until now, cities must draw on the power of the democratic urge to be mobile. Civil society can itself be mobilized to demand this radical automobilization. Overcoming the terrific constraints of automobility will require us to recognize and harness its peculiar auto-freedom.

CONCLUSION

We have shown how automobility, as currently constituted, fosters a civil society of hybridized 'car-drivers', accelerates a collapse of movement between the public and the private, generates a new theme and style of political contestation, and points toward a complex interweaving of mobility and communication within the urban infrastructure. Car-drivers dwelling within their cars, and excluding those without cars or without the 'license' to drive such cars, produce the temporal and spatial geographies of cities as a function of motorized mobility. Pedestrians and cyclists, to a significant extent, are confined to small slivers of the urban public, while many public-transport users are relatively disenfranchised and excluded from full citizenship. Only those moving (however slowly) in cars, taxis and trucks are *public* within a system where public spaces have been democratically seized, through notions of individual choice and personal flexibility, and then turned into the 'iron-cages' that deform public roads and the people disciplined within them.

The civil society of automobility that arises within contemporary cities involves the transformation of public space into flows of traffic, coercing, constraining and unfolding an awesome domination which analysts of the urban have barely begun to see even as they sit staring through their own car windscreens. Smith writes of the analogy with modern scientific thought: 'We get into this mode very much as the driver of a car gets into the driving seat. It is true that we do the driving and can choose the direction and destination, but the way in which the car is put together, how it works, and how and where it will travel structure our relation to the world we travel in' (1987: 73). We have suggested some brief ideas at the end of this article that might just save towns and cities from this awesome Frankenstein-created monster of 'auto' mobility.

NOTES

1 A proliferation of recent conferences on automobility and car culture suggest that this is rapidly changing. Just a small selection includes: 'Speed: A Workshop on Space, Time and Mobility' (October 1999), sponsored by the Danish Transport Council; 'Destination Anywhere: The Architecture of Transport in the 20th Century' (8 March 2000), a conference of the Paul Mellon Centre for Studies in British Art, in London; 'Mobilizing Forces: Social and Cultural Aspects of Automobility' (5–6 May 2000), an international workshop organized by researchers at Göteborg University, Sweden, and sponsored by the Swedish Transport and Communications Research Board; 'The New Jersey Turnpike: Exit Into the American Consciousness' (4 November 2000), a conference organized by the New Jersey Historical Society.

2 While some urban architecture tries to reinvent the pedestrian scale (such as the Centre Pompidou in Paris) or avoid the car altogether (such as the Millennium Dome in Greenwich), all buildings exist in relation to a streetscape and a flow of social practices that remain overdetermined by the car.

3 Experiments in restricting car use in cities have been proliferating recently. Car-free days have been promoted throughout Europe, with leadership from Italy in particular. Many town centres have been pedestrianized or have attempted to ban cars from their roads (e.g. Oxford).

4 The Mercedes S-Class, for example, is equipped with voice telemetry, GIS positioning, a television screen, and card-operated ignition. Such machinic hybrids will be increasingly internet-capable given the development of the Wireless Application Protocol (WAP). At the recent Detroit Motor Show both Ford and General Motors introduced new car models that are designed as 'portals to the net'. Ford plan to make 'the internet on wheels' realistic by introducing voice-activated telematics an standard on all new models (see Gow, 2000). Fiat is also developing a dashboard that can access the internet (*The Guardian*, 26 February 2000: 29).

5 On forms of car-sharing already being used in Germany and Switzerland, see Canzler (1999).

REFERENCES

Albrow, M. (1996) *The global age*. Polity, Cambridge.

Anderson, B. (1991) *Imagined communities: reflections on the origin and spread of nationalism*. Verso, London and New York.

Arendt, H. (1958) [1973 edn.] *The human condition*. University of Chicago Press, Chicago.

Augé, M. (1995) *Non-places*. Verso, London.

Bachmair, B. (1991) From the motor-car to television: cultural-historical arguments on the meaning of mobility for communication. *Media, Culture and Society* 13, 521–33.

Ballard, J.G. (1973) [1995 edn.) *Crash*. Vintage, London.

Barthes, R. (1972) *Mythologies*. Cape, London.

Bauman, Z. (1987) *Legislators and interpreters.* Polity Press, Cambridge.

Benhabib. S. (1992) Models of public space: Hannah Arendt, the liberal tradition and Jurgen Habermas. In S. Benhabib, *Situating the self. gender, community and postmodernity in contemporary ethics*, Routledge, New York.

Boden, D. and H. Molotch (1994) The compulsion to proximity. In R. Friedland and D. Boden (eds.), *Now/here. Time, space and modernity*, University of California Press, Berkeley.

Burgess, E. (1925) [1970 edn.] The growth of the city: an introduction to a research project. In R. Park, E. Burgess and R. McKenzie (eds.), *The city*, University of Chicago Press, Chicago and London.

Canzler, W. (1999) Changing speed? From the private car to cash car sharing. In J. Beckmann (ed.), *Speed: a workshop on space, time and mobility*, Danish Transport Council, Copenhagen.

Clifford, J. (1997) *Routes.* Harvard University Press, Cambridge, MA.

Cohen, J. and A. Arato (1992) *Civil society and political theory.* MIT Press, Cambridge, MA and London.

Creed, B. (1998) The *Crash* debate: anal wounds, metallic kisses. *Screen* 39, 175–9.

Davis, M. (1990) *City of quartz: excavating the future of Los Angeles.* Verso, London.

Emirbayer, M. and M. Sheller (1999) Publics in history. *Theory and Society* 28, 145–97.

Eyerman, R. and O. Löfgren (1995) Romancing the road: road movies and images of mobility. *Theory, Culture and Society* 12, 53–79.

Plink, J. (1988) *The automobile age.* MIT Press, Cambridge, MA.

Forster, EM. (1931) *Howard's End.* Penguin, Harmondsworth.

Freund, P. (1993) *The ecology of the automobile.* Black Rose Books, Montreal and New York.

Giddens, A. (1991) *Modernity and self-identity.* Polity, Cambridge.

Gow, D. (2000) Ford unveils the car to surf your way through the traffic. *The Guardian* 11 January.

Graves-Brown, P. (1997) From highway to superhighway: the sustainability, symbolism and situated practices of car culture. *Social Analysis* 41, 64–75.

Habermas, J. (1992) *The structural transformation of the public sphere: an inquiry into a category of bourgeois society.* MIT Press, Cambridge, MA.

Haraway, D. (1991) *Simians, cyborgs, and women. The reinvention of nature.* Free Association Books, London.

Hawkins, R. (1986) A road not taken: sociology and the neglect of the automobile. *California Sociologist* 9, 61–79.

Jacobs, B. (1992) *Fractured cities: capitalism, community and ennpowerment in Britain and America.* Routledge, London.

Jacobs, J. (1961) *The death and life of great American cities.* Random House, New York.

Keane, J. (1988) *Democracy and civil society.* Verso, London and New York.

Kelley, R.D.G. (1996) *Race rebels: culture, politics, and the black working class.* The Free Press, New York.

Kunstler, J. (1994) *The geography of nowhere: the rise and decline of America's man-made landscape.* Touchstone Books, New York.

Landes, J. (1988) *Women and the public sphere in the age of the French revolution.* Cornell University Press, Ithaca and London.

Lash, S. and J. Urry (1994) *Economies of signs and space.* Sage, London.

Lean, G. (1994) New green army rises up against roads. *The Observer* 20 February.

Lefebvre, H. (1991) *The production of space.* Blackwell, Oxford and Cambridge, MA.

Liniado, M. (1996) *Car culture and countryside change.* MSc Dissertation, Department of Geography, University of Bristol.

Lynch, M. (1993) *Scientific practice and ordinary action.* Cambridge University Press, Cambridge.

Lynd, R. and H. Lynd (1937) *Middletown in transition.* Harvest, New York.

Macnaghten, P. and J. Urry (1998) *Contested natures.* Sage, London.

Marsh, P. and P. Collett (1986) *Driving passion.* Jonathan Cape, London.

Meadows, M. and S. Stradling (2000) Are women better drivers than men? Tools for measuring driver behaviour. In J. Hartley and A. Branthwaite (eds.), *The applied psychologist*, Open University Press, Milton Keynes.

Morris, M. (1988) At Henry Parkes Motel. *Cultural Studies* 2, 1–47.

Morse, M. (1998) *Virtualities: television, media art and cyberculture.* Indiana University Press, Indiana.

Munford, L. (1964) *The highway and the city.* Mentor Books, New York.

Owen, W. (1987) *Transportation and world development.* Johns Hopkins University Press, Baltimore and London.

Park, R.E. (1925) [1970 edn.] The mind of the Hobo: reflections upon the relation between mentality and locomotion. In R. Park, E. Burgess and R. McKenzie (eds), *The city*, University of Chicago Press. Chicago and London.

Pinkney, T. (1991) *Raymond Williams.* Seren Books, Bridgend.

Ryan, M. (1990) *Women in public: between banners and ballots, 1825–1880.* Johns Hopkins University Press, Baltimore.

Sassen, S. (1991) *The global city: New York, London, Tokyo.* Princeton University Press, Princeton, NJ.

—— (1996) The spatial organization of information industries: implications for the role of the state. In J.H. Mittelman (ed.), *Globalization: critical reflections*, Lynne Rienner, Boulder, CO and London.

SceneSusTech (1998) *Car-systems in the city. Report 1. Department of Sociology*, Trinity College, Dublin.

Schivelbusch, W. (1986) *The railway journey. Trains and travel in the nineteenth century.* Blackwell, Oxford.

Sennett, R. (1977) *The fall of public man.* Faber and Faber, London and Boston.

—— (1990) *The conscience of the eye: design and social life in cities.* Faber and Faber, London and Boston.

Shove, F. (1998) *Consuming automobility.* SceneSusTech Discussion Paper, Department of Sociology, Trinity College, Dublin.

Smith, D. (1987) *The everyday world as problematic: a feminist sociology.* Northeastern University Press, Boston, MA.

Soja, E. (1995) Postmodern urbanization: the six restructurings of Los Angeles. In S. Watson and K. Gibson (eds.), *Postmodern cities and spaces*, Blackwell, Cambridge, MA and Oxford.

Sparmann, J. (1992) New technologies in public transportation and traffic management. In OECD Report, *Cities and new technologies*, OECD.

Taylor, J. (1994) *A dream of England.* Manchester University Press, Manchester.

Thrift. N. (1996) *Spatial formations.* Sage, London.

Tocqueville, A. de (1835) [1945 edn.] *Democracy in America.* 2 Vols, Vintage Press, New York.

Turner, B. (1994) *Orientalism, postmodernism and globalism.* Routledge, London and New York.

Urban Taskforce (1999) *Towards an urban renaissance.* Final Report of the Urban Taskforce, Department of the Environment and Department of Transport Library, London.

Urry, J. (2000) *Sociology beyond societies.* Routledge, London.

Venturi, R., D. Scott Brown and S. Izenour (1972) *Learning from Las Vegas.* MIT Press, Cambridge, MA.

Virilio, P. (1997) *The open sky.* Verso, London.

Wajcman, J. (1991) *Feminism confronts technology.* Pennsylvania State University Press, University Park.

Weintraub, J. (1997) The theory and politics of the public/private distinction. In J. Weintraub and K. Kumar (eds.), *Public and private in thought and practice: perspectives on a grand dichotomy*, University of Chicago Press, Chicago and London.

Whitelegg, J. (1997) *Critical mass.* Pluto, London.

Williams, R. (1988) *Border country.* Hogarth Press, London.

Wilson, E. (1995) The invisible *flâneur.* In S. Watson and K. Gibson (eds.), *Postmodern cities and spaces*, Blackwell, Oxford and Cambridge.

Wolf, W. (1996) *Car mania.* Pluto, London.

Zimmerman, M. (1990) *Heidegger's confrontation with modernity.* Indiana University Press, Bloomington.

'The Social and Cultural Life of the City'

from *Telecommunications and the City: Electronic Spaces, Urban Places* (1995)

Stephen Graham and Simon Marvin

Editors' Introduction

Stephen Graham and Simon Marvin's *Telecommunications and the City: Electronic Spaces, Urban Places* (1995) was the one of the first substantive attempts to map the ways in which the city was being reshaped by new electronic telecommunications technology. The book is comprehensive in its sweep, looking at implications for the sociology, cultural and economic geographies of the city, along with issues such as infrastructure, transport, urban form and the ways in which planning and governance of the city is affected by these changes. The book's broad sweep coupled with fine grained attention to detail ensured it soon became established as a key text in urban geography in the late twentieth century. Technologies have changed incredibly quickly and to a huge extent since the book was published. It says a great deal about the sound scholarly basis on which the arguments within the book rest that their relevance is little diminished since its publication. While the nature of the subject will ensure that the book will become of historical interest more quickly than others it will remain a key text at a time when urban geography has just begun to get to grips with the implications of advances in electronic communications. Since the publication of *Telecommunications and the City* in 1995 there have been numerous contributions to the emerging field. Some of the more important of these include Manuel Castell's *The Rise of the Network Society* (Oxford, Blackwell, 2000), Martin Dodge and Rob Kitchin's *Mapping Cyberspace* (London, Routledge, 2000) and Saskia Sassen's *Global Networks, Linked Cities* (London, Routledge, 2002).

Telecommunications and the City emerged out of what is now the Centre for Urban Technology in the School of Architecture, Planning and Landscape at the University of Newcastle upon Tyne. Stephen Graham is the Professor of Urban Technology there and Simon Marvin was a member there until moving to the University of Salford. Both have published widely in the area of cities and technology and continue to be research active in the area. Their collaborative work continued to explore the electronic frontiers of urban geography with the publication of *Splintering Urbanism: Networked Infrastructures, Technological Mobilities and the Urban Condition* (London, Routledge, 2001).

THE HOME AS A DOMESTIC 'NETWORK TERMINAL'

In the modern city, the home emerged as 'the last reserve space'. It was, to quote Helga Nowotny, 'a social space of special significance which has come to signify for us the last sanctuary in a bewildering outside world' (Nowotny, 1982; 102). But this sense that the home is isolated from the rest of the social world is changing rapidly. Many homes, as we have seen, are being incorporated into more and more networks on increasingly global scales. These blur the dividing line between what is public and what is private. To Putnam, this means that the best way to consider the home is as a 'terminal'. He writes,

> in speaking about the modern home, we are talking about more than technologized comforts. The modern home is inconceivable except as a terminal, according the benefits of, but also providing legitimate support to a vast infrastructure facilitating flows of energy, goods, people and messages. The most obvious aspect has been a qualitative transformation of the technical specification of houses and their redefinition as terminals of networks.
>
> (Putnam, 1993; 156)

This is not the result of some simple technological 'logic' however. Nor can we attribute this process entirely to broad political and economic forces. Rather, Roger Silverstone has shown that new communications and information technologies are entering homes through complex and diverse processes of social construction and 'domestication' (Silverstone *et al.*, 1992; Silverstone, 1994). The consumption of these new technologies, and the services that are accessible through them, are

> pursued by consumers who seek to manage and control their own electronic spaces, to make mass-produced objects and meanings meaningful, useful and intelligible to them. This is a process of 'domestication' because what is involved is quite literally a taming of the wild and a cultivation of the tame. In this process new technologies and services, unfamiliar, exciting, but also threatening, are brought (or not) under control by domestic users.
>
> (Silverstone, 1994; 221)

Often, though, these technologies create conflicts – particularly gender conflicts – over who has control and access within the household (Mulgan, 1991; 69).

But these processes operate within the context of broader political, social and economic changes. The shift towards information labour, home and telecentre-based teleworking, flexible labour markets and the cultural shifts towards globalisation provide the context for the changes underway. Evidence suggests, however, that beyond the hype, there remains 'deep resistance to electronic interactivity beyond a small number of enthusiasts' (Mulgan, 1991; 69). Most of the development of technology in the home still centres on stand-alone media and entertainment systems such as CD players, personal computers and TVs. Most cable services are not interactive – they merely pump more channels to be passively consumed than terrestrial broadcasting. Telephones remain by far the most important interactive home communications system.

We must also be conscious of the varied prospects for genuine tele-based services to the home In the United States, home tele-services seem to have a greater chance of reaching 'critical mass'. In Europe, resistance seems to be much deeper. A recent study by the Inteco consultancy of 12,000 people across Europe, funded by 30 retailers and telecom companies, found that there is deep consumer resistance to interactive shopping; entertainment and security services were marginally more attractive (Bannister, 1994).

The main problem, then, is that, in mass markets 'the audience does not yet provide the revenue which supports the services' (Curtis and Means, 1991; 24). Thus the focus, as we saw above, is on 'demassification' and the 'cherry picking' of markets by offering higher value-added services for more lucrative consumers and those small markets where genuine prospects exist. The Inteco study also concluded that the only attractive markets at present comprise 'active, relatively high income families where time is at a premium, where someone in the family will be learning to play a musical instrument or be keen on gardening or some form of home study. [These] will be the fertile ground for multimedia seeds to root' (Inteco, 1994; quoted in Bannister, 1994).

Despite uncertainties in demand, commercial efforts to enhance the role of homes as terminals for many technical networks, offering 'pay-per' services (at least for these affluent groups), are intensifying. These are driven by the view that there are huge commercial rewards that maybe in store from commercial services offered over the much-vaunted prospect of multimedia, interactive television or the information superhighway. Because of this, the still rather marginal shifts towards home banking, home- and centre-based teleworking, and tele-based access to services may eventually grow, as new technological capabilities come into the social mainstream – or, perhaps more likely, the financial reach of socially affluent groups.

Already, 'narrowcasted' electronic newspapers have been developed in the United States, accessible by computer and geared to the particular programmed interests of consumers (White, 1994). 'Video on demand' technologies are about to enter the market, offering selected videos played down phone lines as compressed signals. In a few places, cable networks are beginning to offer value added services rather than just TV. A growing range of consumer services such as 'electronic bookstores' are developing, based on the Internet (which is emerging as a domestic service for technology enthusiasts). 'Smart', 'interactive' or 'high definition' TVs have been developed which, when combined with signal compressions technologies, offer the technological potential for user-friendly and interactive services.

A range of full-blown commercial trials are now in the offing for interactive, sophisticated home telematics systems. Time Warner and US West are already operating an interactive TV trial in Orlando, Florida, for 4,000 homes. A system called 'stargazer', developed by Bell Atlantic, will soon offer education, entertainment, information and shopping on a trial basis to 2,000 homes in Northern Virginia. But the commercial and commodified nature of these emerging systems makes the widely used analogy of the electronic highway profoundly misleading.

If the dream is of consumers who 'spend at the touch of a button', those without money are likely to have little to do with these developments. Brody (1993; 32) writes that 'what is emerging has less the character of a highway than a strip mall, focusing on services with proven demand that can rapidly pay for themselves . . . cable operators are seeking to establish a presence in the lush, unregulated territory of enhanced services'. They also seem to be supply-driven – often crude attempts by technology and telecommunications companies to increase the use of existing networks and technologies and so improve profitability (Noll, 1989).

'ELECTRONIC COTTAGE OR NEO-MEDIEVAL MINI-FORTRESS?'

While there are uncertainties over the viability of these systems, there does seem to be an overall trend towards the mediation of more work, travel and consumption via home-based telematics which intensifies the home-centredness of urban society. These trends suggest that homes may be progressively disembedding from their immediate social environment within urban places through their linkage into electronic spaces. Manuel Castells, for example, argues that

> homes . . . are becoming equipped with a self-sufficient world of images, sounds, news, and information exchanges. . . . Homes *could* become disassociated from neighbourhoods and cities and still not be lonely, isolated places. They would be populated by voices, by images, by sounds, by ideas, by genres, by colors, by news.
>
> (Castells, 1985)

In this context, Barry Smart asks the important question: will home-centredness emerge as a benign and positive development as in the Tofflerian 'electronic cottage' scenario or, conversely, will it 'encourage a retreat from public life and public space and serve thereby to increase the sense of insecurity which has become an increasing feature of the "post-modern" urban environment' (Smart, 1992; 53). The trends are complex and varied and belie easy generalisation. But there seems to be a shift, particularly in the middle classes, towards 'cocooning' – their withdrawal from public spaces in cities and the use of home-centred and self-service technologies and network access points in their place (see Gershuny and Miles, 1983). Fear of crime and social alienation with urban life are key supports to this trend. A recent Channel 4 television programme (1994; 5) commented that access to telematics networks for shopping, leisure and work 'looks safe indeed compared to urban decay. Paranoia, violence and pollution are eating away at the soul of America, driving it inward – to the protection of the home, private security, entry codes, and video-surveillance-controlled gated fortresses'. In this vein, Manuel Castells warns of a dystopian (American) urban future where 'secluded individualistic homes across an endless suburban sprawl turn inward to preserve their own logic and values, closing their doors to the immediate surrounding environment and opening their antennas to the sounds and images of the entire galaxy' (Castells, 1985; 19).

There are two key issues here. First, it is becoming increasingly clear that there are dangers that these home-based technological innovations will simply exacerbate existing trends towards individualisation and polarisation within western cities (Robins and Hepworth, 1988). Relying in the telematics field on 'a totally individualised society completely ruled by market mechanisms' (Kubieck, 1988) seems to threaten to undermine the public, civic sense of cities as physical and cultural spaces of social interaction. Certainly, the reliance on technical networks for social interaction and entertainment and the withdrawal and cocooning that goes along with this, encourages fear of crime and a shift toward the 'tribalisation' of socioeconomic groups in cities.

Second, as we have seen, these trends threaten to exclude and marginalise already disadvantaged groups who are unable to afford participation in these technological futures. Women, for example, may become ensnared in domestic space by a pervasive shift towards home-based teleshopping.

REFERENCES

Brody, H. (1993) 'Information highway: the home front', *Technology Review*, August–September, 31–40.

Channel 4 (1994) *Once Upon a Time in Cyberville (programme transcript), London: Channel 4.*

Nowotny, H. (1982) 'The information society: its impact on the home, local community and marginal groups', in H. Bjorn Andersen, M. Earl, O. Holst and E. Mumford (eds), *Information Society, For Richer, For Poorer*, North Holland: Elsevier.

Silverstone, R. (1994) 'Domesticating the revolution – information and communications technologies and everyday life', in R. Mansell (ed.), *Management of Information and Communication Technologies*, London: Aslib, 221–233.

Silverstone, R., Hirsch, E. and Morley, D. (1992) *Consuming Technologies: Media and Information in Domestic Space*, London: Routledge.

'The People's Radio of Vila Nossa Senhora Aparecida: Alternative Communication and Cultures of Resistance in Brazil'

from Tracey Skelton and Tim Allen (eds), *Culture and Global Change* (1999)

Vivian Schelling

Editors' Introduction

'The people's radio of Vila Nossa Senhora Aparecida' was published in the collection *Culture and Global Change*, edited by Tracey Skelton and Tim Allen (1999). The collection examines the importance of recognising and incorporating cultural diversity into development discourse. The book argues that for development to be meaningful it must appreciate the complexities of everyday lives within developing countries. Schelling's work on popular communication in Brazil sits well within this as it recognises the communications traditions within the community she studies that allowed the development of a people's radio and also the social divisions and fractures that predated the intervention and which in some cases were exacerbated by it. Schelling's work demonstrates that development is a contingent and contested process.

Vivian Schelling is a lecturer in third world Development studies in the School of Cultural and Innovation Studies at the University of East London. She has worked mainly on issues surrounding popular culture and globalisation in Latin America, often charting the ways in which the two are closely bound. She has written *Memory and Modernity: Popular Culture in Latin America* (1991) which was co-authored with William Rowe. Her other works include the edited collection *Through the Kaleidoscope: The Experience of Modernity in Latin America* (2000).

INTRODUCTION

The aim of this chapter is to explore, through the study of a poor community of rural migrants from the northeast of Brazil living on the outskirts of São Paulo and the history of its own independent radio station, the much broader theme of the nature and experience of the modern in Brazil. This is a vast topic and the subject of considerable and fascinating debate. Here, however, I would like to illuminate only a particular facet as it relates to specific questions concerning the identity and integration of rural migrants in an urban context. In a society which is marked by uneven development, by the coexistence of ways of life relating to disparate epochs and economic structures, and in which a large percentage of the population has migrated to the cities in search of the fruits of 'development', the ways in which rural migrants negotiate their insertion in the urban context is a rich source of information on the ways in which modernity is experienced and constructed.

RADIO AND 'EMPOWERMENT'

Radio as a means of communication directed towards and used by members of the 'popular classes' of Latin America has a history that goes back to the 1950s with the creation by Bolivian tin miners of several independently controlled radio stations, and the deployment by Cuban guerrillas of *Radio Rebelde* (Rebel Radio) as a means of organising the insurrectionary forces leading to the Cuban Revolution in 1959 (Machado, Magri and Masagao 1987).

However, it is only in the 1970s that a distinctive theory and practice of 'popular' or 'alternative communication' emerged as a result of a variety of experiments taking place simultaneously in different parts of Latin America, using communications media to empower 'the popular classes'. These arose out of a confluence of ideas and social forces prevalent at the time: the election of a socialist government in Chile in 1970; the emergence of new social movements; the critique of modernisation theory and the concomitant belief that Latin America needed to follow an 'alternative' development path which satisfied the basic needs of its people and reflected its own history and experience. An essential precondition for achieving this consisted, according to proponents of 'alternative communication', in an overall democratisation of the media at national level and in the development of a 'popular hegemony' at the level of civil society (Kaplun 1987). This latter aim entailed transforming 'from below', through a variety of forms of popular organisation, the relations of economic, political and cultural domination which characterised the prevailing social order. In this process of transformation 'from below', the media would not only become an instrument of popular organisation, but also establish, through innovative experiments with radio and television, a 'new communicational order' based on participation and on dialogue between receivers and transmitters.

The history of 'The People's Radio', a community radio station set up in Vila Aparecida, a poor neighbourhood in São Miguel Paulista in the outlying areas of São Paulo, was one such experiment.[1] Until the 1930s, when new transport connections with the centre of São Paulo and chemical industries were established, São Miguel Paulista had been a quiet semi-rural district. Following this period, especially from the 1960s onwards, it has become primarily an area where migrants from the countryside, in particular the northeast of Brazil, seek housing or a piece of land (Caldeira 1984). In this process of gradual transformation São Miguel has become known as the *periferia*, an urban area like many others surrounding Brazilian cities, characterised by poor housing and a lack of adequate infrastructural facilities such as electricity, sewage and roads.

For rural labourers or peasants from the northeast of Brazil, migration entails a profound process of re-socialisation. Unskilled and shaped by extremely oppressive labour relations, they form an easily exploitable reserve army of labour employed in construction sites, domestic service, factories and in petty commerce. Integration in the urban context is frequently experienced as a form of cultural invalidation, entailing the loss of skills previously mastered without appropriation of the necessary material and intellectual means to participate in the urban context (Chaui 1987). While in the context of subsistence agriculture the cycle of production and consumption was concrete and transparent, in the city the rural migrant becomes an appendage of the machine and an abstract consumer engaged in the exchange of commodities for which his/her purchasing power is always insufficient. The majority of television programmes represent the industrialised south of Brazil and its 'Westernised' and comparatively affluent lifestyle as the 'true' image of Brazil. Other regions and their ways of life are either hardly represented or, when they are, it is in a stereotypical and derogatory manner. The overall effect of this is the cultural invalidation of the migrant. The force of this process of cultural invalidation, however, also depends on the extent to which 'cultures of resistance' – alternative representations of the migrant condition and positive strategies for constructing a new urban identity – are available.

'CULTURES OF RESISTANCE'

Cultures of resistance are ambiguous and amorphous phenomena, since what distinguishes them from more systematic and confrontational forms of struggle is that, as several writers on the topic have observed, they consist of more subtle and everyday practices of opposition to domination.[2] These can include, for example: 'the prosaic but constant struggle between the peasantry and those who seek to extract labour, food, taxes, rents and interest from them' (Scott 1985: xvi); the sabotage of machinery and work rhythms in off-shore transnational companies; inventive ways of circumventing anti-strike legislation; forms of transgressive popular celebration; as well as millenarian and religious movements in which the imminent overturning of the social order through divine intervention is announced. They may, as Gledhill (1994) notes with regard to neo-Inca Andean rebellions in Peru during the colonial period, leave 'a legacy in historical memory' which generates a 'counter-hegemonic indigenous historical consciousness which changed over centuries and could manifest itself in the form of participation in more conventional political and class-based organisations in recent times' (1994: 89). They may, however, also, by creating a space for 'contained' opposition, facilitate pragmatic adaptation to reality without fundamentally changing the balance of power and possibly even in this way reinforce domination. Movements of resistance are thus frequently shot through with elements of conformism. Consequently, it would seem that decisive in determining the direction in which resistance and the struggle for hegemony which takes place on this terrain move, is the existence of more structured and systematic forms of opposition. For, as Gledhill points out, it is the broader context of power relations which determines 'the precise structural implications of particular counter-hegemonic acts' (Gledhill 1994: 93).

SOCIAL MOVEMENTS IN SÃO MIGUEL PAULISTA

What distinguished São Miguel Paulista from other similar areas was the extent to which a 'culture of resistance' had developed over the years since it had become part of the periphery of São Paulo. Since 1945, its inhabitants have tended to vote for political parties representing the interests of the poor and the working class (Caldeira 1984). It has also been characterised by a variety of local associations involved in campaigning for infra-structural improvements as well as promoting recreational activities, many of which aim to recreate the cultural rituals and forms of sociability of the northeast. Since the early 1980s the inhabitants of São Miguel Paulista have also become organised within local and national social movements, in particular the *Movimento Sem Terra* (Landless Movement) and the *Movimento de Moradia* (Housing Movement), both involved in struggling for a more equal distribution of land. These and other social movements concerned with ethnic and gender issues emerged during the period of the right-wing military régime in Brazil (1964–85) in response to the growth in social inequality and the loss of civil liberties which characterised this period. After the return to civilian government these social movements played a significant role in redefining a political culture based on authoritarianism and clientelism. Mobilising around demands to share in the benefits of a growing modern economy and aiming to achieve this without relying on political parties and professional politicians, but rather on forms of independent popular organisation, the social movements have broadened the concept of democratic participation (Cardoso 1992).

Since the establishment of the military régime, the radical wing of the Catholic Church, tied to Liberation Theology and its vision of religious salvation as a liberation from all forms of economic, political and personal oppression, has been an active promoter of the social movements. Thus the marked presence of these 'Popular Movements' in São Miguel Paulista was partly due to the fact that this area of São Paulo was part of the Catholic Diocese of the Bishop Dom Angelico, a forceful advocate of Liberation Theology. In conjunction with the Labour Movement, the Catholic Church had set up Basic Communities throughout the region, including the neighbourhood of Vila Aparecida. The underlying purpose of the Basic Communities could broadly be defined in terms of two interlinked objectives: first, to carry out group readings of the Bible, developing the poor's own ideas and capacity for reflection through shared reading, discussion and personal interpretation; second, to 'awaken' the poor to the possibility of changing their condition by supporting forms of collective action – participation in cooperatives, local associations and unions which aim to liberate the poor from oppression.

In agreement with the broad aims of Liberation Theology, one of the key features of the grassroots movements in São Miguel Paulista was the adult literacy classes and forms of popular education informed by Paulo Freire's notion of a 'pedagogy of the oppressed' (Freire 1975). According to this pedagogy, any 'cultural action' which aims at liberation from oppression needs to reflect not only on the social and economic causes underlying the immediate and everyday experience of oppression but also on the self and the way it has psychologically internalised these forms of oppression in the form of feelings of, for example, fatalism, inferiority or displaced anger. Only this can ensure that the creation of a more just economic order will also lead to the construction of a democratic culture. As Freire notes, 'Culture as an interiorised product which in turn conditions men's [*sic*] subsequent acts must become the object of men's [*sic*] knowledge so that they can perceive its conditioning power' (Freire 1975: 35).

Informed by these ideas, the introductory section of the Popular Education Manual of São Miguel Paulista states that 'Popular Education aims at strengthening the class consciousness of the popular sectors. Popular Education does not aim as knowing or contemplating reality from outside but as decoding its meaning while intervening in its transformation' (*Catálogo Educação e Comunicação Popular*, undated: 7).

VILA APARECIDA AND 'THE PEOPLE'S RADIO'

The neighbourhood of Vila Aparecida, named after its patron virgin, *Nossa Senhoro Aparecida* (Our Lady the Appeared) stretches across a gently sloping hill and valley at the end of a two-hour bus journey from the centre of São Paulo to the periphery of São Miguel Paulista.

In the 1980s, with the rise of Liberation Theology and the growing participation of São Miguel Paulista in local popular movements, Vila Aparecida was gradually transformed into a Basic Community. At the instigation of Catholic priests and lay Catholic workers, several local groups and institutions emerged: a catechism group and *groupos de rua* (street groups) to organise readings and interpretations of the Bible in the light of the Vila Aparecida inhabitants' experience; youth groups combining vocational and recreational tasks; and an urbanisation and adult literacy group connected to the broader Housing, Education and Landless Movements. Each group was represented in the local 'Council', whose task was to co-ordinate the work of all the groups and advance the process of popular mobilisation. With the help of funds raised locally, material was bought to build a Church and a community centre, and a shared journal recording the events and experiences of the local inhabitants was kept. Its purpose was to act as a memory of the *caminhada* – the way, the path of liberation taken by Vila Aparecida. Following a period of several years after piped water and electricity had been obtained as a result of the pressure exerted by local groups on the state authorities, the idea of setting up a local radio 'station' emerged.

Inspired by the use of loudspeakers as 'popular radios' in Villa El Salvador, on the outskirts of Lima, and the subsequent creation of 'The Day of the People', a local 'Centre for Radiophonic Communication and Production', the adult literacy group of Vila Aparecida, set up a local radio 'station' using a tape recorder, a record player, an amplifier, microphones and loudspeakers wired up outside the local church tower, with a transmission radius of 3 km. It was called 'The People's Radio'. Given the importance of oral culture for rural migrants, their familiarity with the use of loudspeakers in village squares and the rate of illiteracy among the inhabitants of Aparecida, it was hoped that the use of an oral medium would be more effective as a vehicle for popular mobilisation. In the words of the local newspaper *O São Paulo*: 'In the shanty town writing doesn't work, nobody knows how to read, people have an aversion to paper' (*O São Paulo* 1986).

As a result of the success of this first experiment, a further nineteen radio stations of a similar kind, financed by the Catholic Church and European charities were set up in other settlements in the region. They were then transformed into an interconnected network by a co-ordinating regional centre, *Proconel* (Project for Non-Written Communication of the Eastern Zone), made up of the representatives of each of the radio 'stations' in the region. These were in turn divided into the following three subgroups charged with developing the popular communication skills in each locality: a group for the collective production of programmes; a group for the training and formation of new leaders; and a group responsible for the exchange of the 'radio experiences' of different regions. With the growth in the number of radios, *Proconel* became a department of the broader Centre for Popular Education and Communication of São Miguel (CEMI). The task of CEMI was to run courses in popular communication, organise regular meetings with the local representatives of the community radios and promote the exchange of 'radio experiences' by providing the whole area with locally produced taped programmes, audio-visual materials and communications manuals.

Ideologically, CEMI adhered to the view that in contrast to the media of the culture industry, 'popular communication' was characterised by the fact that it was dialogic and participatory, enabling the receiver to become the producer and vice-versa. As such, 'popular communication' was an integral part of the life of the community, present at local activities, creating links between local issues and broader social and political processes and reflecting as well as elaborating in its programmes the needs and interests of the popular classes. In this way 'popular communication' would enable the poor and subaltern classes to 'discover their voice', become aware of their creativity and humanity and thus acquire the experiential precondition for claiming their rights as citizens.

How were the community radios thought to achieve these aims? First, the use of the radio itself by the poor was seen as empowering and as demystifying the medium; hearing their 'own voice', local inhabitants would realise that they were capable of manipulating the required technology and adapting it to their needs. Moreover, as transmitters–receivers speak freely, experiences are shared, giving rise to a language and symbols expressing hope and solidarity pre-figuring a new social order. In that sense, it could be argued, the ultimate purpose of 'popular communication' was to promote the development of a 'popular counter-hegemony' in the sphere of civil society.

In 'The People's Radio' locally distributed booklet the broader theoretical and ideological tenets of 'popular communication' were given concrete expression in terms of the following aims:

1 To recover the voice of the people, in history, religiosity, culture, tradition and legends.
2 To provide basic information, to support the organisation of community struggles and to communicate the hopes and problems of the people.
3 To promote the transformation of society through tasks undertaken in common.
4 To support popular artists as well as organise musical events and festivities.
5 To encourage participation at every level and engage in the formation of new community leaders.

Item 6 states: 'through following this path (*caminhada*) we began to perceive that the radio was no longer a dream and had become part of the history of our people. It is a difficult path but we resist because we believe in the power of the word of our people' (*Radio do Povo*, undated: 21).

The members of 'The People's Radio' were divided into four groups with different tasks designed to advance the process of popular communication: an 'interview group' was responsible for exploring, in conjunction with the inhabitants of Vila Aparecida, the ideas and issues which they would like to see included in the weekly programme as well as evaluating whether the radio was reaching them; a 'visiting group' was responsible for exchanging experiences with other communities; a 'programme dynamics group' was charged with involving the community in the work of the radio and popular struggles using other media such as slides, films and music; finally a 'news group' was responsible for reading daily newspapers and journals and reproducing them together with a critical commentary (*Ante Projeto do Grupo de Comunicação da Vila Nossa Senhora Aparecida* 1984).

The weekly programme prepared beforehand generally ran from two to four in the afternoon on a Saturday; during the rest of the week the radio was at the disposal of the community to transmit messages, for example, to call people together if someone needed help, to come to meetings or to announce events. The weekly programme usually began with a piece of Brazilian urban or rural music. To play Anglo-Saxon rock was frowned upon as 'anti-popular'. This was followed by a reading from the Bible interpreted in such a way as to make sense of an aspect of the local inhabitant's life. An important component of the weekly programme was the 'socio-drama', short pieces invented and recorded by the inhabitants of Aparecida or by other communities and distributed to Aparecida via the Regional Centre for Popular Education and Communication. These pieces dramatised typical aspects of their life which were considered 'conflictive but in favour of the people'. The purpose of the socio-dramas was to act as a mirror of their experience and by objectifying it to understand and invent new solutions to their problems. The manual 'How to Make a Socio-drama' states:

> By acting we are forced to imagine different solutions to the conflicts of our life. We are representing what we live but also preparing ourselves for what we are going to live. In the socio-drama we can *rehearse the future*: better ways of organising, new ways of demanding our rights, new forms of behaviour. We are rehearsing our liberation.
>
> (*Como fazer um socio-drama* 1987: 8)

Included in every programme were announcements relating to the organisation of the grassroots movements at local, regional and national level, announcements brought in by inhabitants of Aparecida as well as information on the price of food, employment opportunities, public services and available courses in literacy, carpentry and sewing. The programme ended with a piece of music. Two representative examples of 'The People's Radio' programmes are provided in Boxes 1 and 2.

Despite the evidently didactic purpose of many of the programmes, such as the ones in the Boxes below, 'The People's Radio' did not, according to one of its members, campaign in support of the preferred Workers Party, defined as *o partido da casa* (the house party). It preferred to set out for listeners the aims of different parties and the current positions regarding single issues. When the new Constitution was being formulated in 1988, 'The People's Radio' walked the streets of Vila Aparecida discussing the items in the Constitution, and socio-dramas on the current anti-inflationary economic plan were enacted. Moreover, the radio changed over time as it was increasingly appropriated and managed by local inhabitants rather than being informed by the guidelines of the Church. As one inhabitant commented:

> In the beginning, the radio played only religious records and there were problems when we played other songs. Then we chose different songs to introduce a subject. If the subject was land then we would play something to do with the land and with news from the northeast. Now it is changing, the radio is no longer so serious – it is better, more fun. To make a programme talking only of problems is difficult, life itself is crushing, so we don't only talk about bad things, but also with more joy, to entertain and distract. Do you have problems? Millions.

'The People's Radio' functioned not only within the premises of Vila Aparecida but also in other locations in the form of a 'mobile radio'. It accompanied representatives from the grassroots movements on demonstrations and expeditions to lobby the state authorities, acting both as witness to and animator of the *caminhada* (the path). According to the testimony of Sonia, one of the main promoters of the radio:

> The radio is carried on a little car and the music and programmes animate the demonstrations while the state authorities make us wait the whole day long. In the (religious) processions and celebrations people carry the radio; the litter with the Saint goes in front and the litter with the radio behind. This year we had a pilgrimage in favour of peace very connected to the question of race and violence. And the radio was there.

Local attachment to the radio also became apparent on the occasion when the radio was stolen. Money was raised from the local inhabitants to buy new equipment, which was then used to exhort the thieves to return the stolen equipment; after some persuasion they obliged.

To celebrate the creation of 'The People's Radio', a festival of locally produced music and poetry was organised annually on a specific theme. 1988 was the centenary of the abolition of slavery and the festival theme was: 'Black Brazilians – A Cry of Justice'.

The effect of the Festival was to consolidate the network of social relations and the sense of belonging to a *pedaço*, a particular neighbourhood. In contrast to the other spaces occupied by the inhabitants of Aparecida – the factory, the bus, the underground, the construction site and the domestic space of others – the *pedaço* embodied a shared identity and a history of struggle which enabled them to resist the experience of marginalisation and invalidation. This was reflected, for example, in one of the poems recited at the festival:

> Who is outside, wants to come in
> Who is in, doesn't want to leave
> Our shanty town is good
> I no longer want to leave
> We have a good church
> And a priest who prays
> We have piped water
> And a small school
> What we still need is a small hospital
> For we already have the school
> To teach us to read
> With the protection of the priest
> and the social worker
> We will struggle for land
> With our rights as voters.

In addition, the Festival itself gave rise to a local 'Black Roots' group and also fostered the development of small bands and duos who played throughout the year at birthdays, weddings and demonstrations.

BOX 1 Typical programme script

The programme begins with a song about the need for agrarian reform.

- Reading from the Gospel of St Luke: a parable about the 'offerings of a poor widow'. She, like the rich man, gives money to someone poorer than herself. The radio commentator highlights that the poor widow gave much more because she gave what she had and not what was left over and not needed, as in the case of the rich man. In that way, he concludes, 'although we ourselves are poor, we all have something to give.'
- A current popular 'hit' by the Brazilian band '*Paralamas do Sucesso*' (The Mudguards of Success).
- Ten Commandments on how to vote consciously (elections being held for the mayor of São Paulo were won unexpectedly by the candidate of the Worker's Party, Luisa Erundina).
- Socio-drama entitled 'which is the party on the side of the people?'
- Poems read by children of Aparecida on 'Children's Day', followed by a reflection on ecological issues and the question 'is nature being loved like the children?' This is accompanied by the jingle of an extremely popular and lucrative children's television programme presented by 'Xuxa', a blonde and blue-eyed young woman.
- Another song entitled: 'It is not necessary to have a degree, a youngster from the country is also of value'.
- The 'afoxé' song 'Oh Faráo' is played.[3]
- The programme ends with a request for help to repair a broken street pipe and the announcement of a radio course which is being run in order to ensure the continuity of 'The People's Radio'.

BOX 2 Typical socio-drama

Frequently used was a taped socio-drama in which the central character was the construction worker 'Pedro'. Blending in with the song *Pedro Pedreiro* by Chico Buarque de Hollanda, a narrator recounts the process of Pedro's awakening to his predicament as worker:

> Pedro built houses where there was nothing but earth; like a bird without wings, he climbed up the houses which sprung forth from his hands . . . but Pedro didn't know that the house he built was both his freedom and his enslavement; he didn't realise that the worker makes things and is in turn made by them.

However, one day Pedro notices that he, the humble worker, makes everything:

> the knife, the bench, the glass, the walls, the window, the house, the city, the nation . . . he looked at his hand and in addition to his profession as worker he acquired a new dimension of poetry. And this event was noted – other workers listened and heard and thus the worker who always said 'yes' began to say 'no' . . . noticing that his pot belonged to his boss, that the hardness of his day was his boss's night's rest, that his great fatigue was his boss's friend. The reasoning of a poor and forgotton man grew, the reasoning which made the construction worker into a worker in the process of construction.

At this point the music blends in with the first words of a very popular 'bossa nova' song of the 1960s *Opinião* (Opinion): 'You can arrest me, you can beat me, you can starve me, but my opinion I will not change'.

'THE PEOPLE'S RADIO' AS RESISTANCE?

To return now to the problem raised earlier concerning the insertion of rural migrants in an urban context, how effective was 'The People's Radio' in developing a counter-hegemonic culture of resistance? And how did it contribute to articulating a new urban identity for the rural migrants? In order to address these questions it is necessary to discuss briefly the concepts of hegemony and counter-hegemony.

Hegemony theory is a refinement of the Marxist concept of ideology; it emphasises that, although acceptance by subordinate classes of belief systems which bind them to the existing power structure is essential to the maintenance of the social order, this does not occur through a process of simple imposition of an articulate system of ideas and values. Rather, it refers to a situation in which, as a result of the unequal relations of cultural power, subordinate groups actively subscribe to values, aspirations, meanings and life styles which secure their adherence to the prevailing social order but which are not in their interest. The contribution of hegemony theory in this sense is to have pointed out that a social order is maintained when particular values and meanings become, to use Raymond Williams' phrase, 'practical consciousness', that is, when they not only give shape to consciously held ideas but 'saturate the whole process of living'. These values and meanings thus come to 'constitute a sense of reality for most people in the society, a sense of absolute because experienced reality beyond which it is very difficult for most members of the society to move, in most areas of their lives' (Williams, undated: 110).

This is achieved significantly through containing the public discussion of issues and the circulation of meanings relating directly or indirectly to the distribution of power and wealth, within a framework which seems to respond to and accommodate the interests of subordinate groups, without, however, altering the prevailing power structure in any significant way. Conflict is covered over or partially resolved and the power of alternative meanings is mitigated through partial incorporation in the hegemonic discourse. In this way hegemonic discourses retain their claim to being the only way of thinking, rendering the prevailing social order 'natural'. Moreover, language is used in such a way that the values implicit in the dominant or preferred discourses are invested with a positive emotional charge, thus promoting subjective identification with them and transforming them into 'practical consciousness'.

By contrast, counter-hegemonic forms either openly or implicitly aim to foster 'structuring of feeling' and ways of thinking outside the parameters of the hegemonic configuration, revealing sources of social conflict so that meanings and values contained in the hegemonic discourse are no longer seen as universally valid but as directly or indirectly benefiting a partial interest, that of the dominant classes, rather than society as a whole. This in turn has the effect of 'de-naturalising' the social order while opening up a space for a counter-hegemonic interpretation and experience of social reality.

The responses of the inhabitants of Vila Aparecida to the work of 'The People's Radio' varied depending on his/her status within the community and how closely involved he/she was with the radio. From the testimonies of those actively involved in the radio's popular communication groups, it was apparent that the public discussion of issues such as racism, the right to land, the Constitution and the right to the product of one's own labour – as in the socio-drama on the construction worker Pedro – had played an important part in building links between their problems, the broader socio-economic structure and the possibility of imagining and bringing about a different future. In addition, working on the weekly programmes in conjunction with participation in the grassroots movements had fostered an awareness of their identity and rights as citizens, while working with a simple form of technology adapted to local needs had demonstrated to them that it was possible to break the communications monopoly of the culture industry, in however limited a form. It thus seemed that they felt they had been given the means to address the contradiction or duality of their predicament in that the 'cultural action' promoted by 'The People's Radio' emphasised the value of rural culture while also providing tools through which to gain access to modernity as creators and knowers. This was reflected in the frequent use of the word 'transformation' and 'path' or 'way' to refer to the changes experienced in living in Vila Aparecida.

In the case of those inhabitants who were within the three kilometre radius of the loudspeakers and who participated intermittently in it, the radio was seen very much as a public utility. In that sense 'The People's Radio' functioned more instrumentally as a vehicle for improving their life chances in the urban environment through the information it provided on the price of food, employment opportunities and the state of negotiations with the authorities over land. Nevertheless, despite its instrumental value in securing some form of social mobility and comments that the radio was noisy and interfered with listening to 'international' music on the official radio stations, the inhabitants' responses also revealed that, since the emergence of the radio and local involvement with the church, Aparecida had been transformed. While it was seen as a place previously characterised predominantly by mutual suspicion, violence, despair and social disintegration, it was now a community with more resources, with an identity which the inhabitants were quite proud of and with a shared memory of struggle and celebration expressed in the community journal, in the photographs of the community's activities which covered the walls of the church, and in a relative degree of political awareness. This view was reflected in the frequent reiteration that the quality of life had improved:

> The atmosphere has changed in the Vila; there is less hostility between the children. There used to be much violence. The community resists and confronts the violence – it has occupied a space and it continues to celebrate [*fazer a festa*], the groups meet and the joy of living encourages people; the community is a small light perhaps.

Nevertheless, in comparison with those consistently involved in the radio and the grassroots movements, this group was far less ideologically coherent, sometimes expressing on the one hand pleasure in the positive changes in Aparecida and on the other dislike of Blacks – this being a fairly marked feature of northeastern rural culture. In fact, the contradiction between a manifestly anti-racist stance and the extent to which Brazil's black inheritance is still devalued and denied is also apparent in 'The People's Radio's' programmes. These included discussions of questions of race and the fate of Zumbi, the leader of a major slave rebellion in the seventeenth century, as well as an uncritical acceptance of the popularity of the children's television presenter 'Xuxa'. Young, white, blonde and blue-eyed, she is the embodiment of the stereotypical North American ideal of female attractiveness and a glaring example of the hegemonic aspiration to 'whiten' Brazil, despite the widespread claim that Brazil constitutes a 'racial democracy'.

In the testimonies of the women, who, in contrast to the men, spent a greater amount of time in Aparecida, the radio had to a greater extent, through naming and symbolising their experience, fostered what could be defined as a 'counter-hegemonic sensibility':

> The Radio plays a different kind of music; it tells the story of unemployment; it talks of bread, flour and the rise in prices and it is all so true – I think it is beautiful – people like it. When the light or water bill arrives they call you; they invite people to come and help when someone is ill, when there are disasters; when someone needs a document they help; they call people for meetings. All that helps a lot. And it has gotten better – every year there is a festivity – I think its beautiful, the children dancing in a circle. I wash and listen to the music they play, which is about what we go through: one day you go to the supermarket and the milk has gone up; it talks about bread, flour, salaries; I think that is important.

There was also tension and conflict surrounding the work of the radio: some of the core members of the radio were seen as having acquired élite status in the community and within this group itself criticisms were voiced that 'the people have to come to the radio not the radio to the people'. In that sense 'The People's Radio', it could be argued, had predominantly transformed a small number of the inhabitants of Aparecida into transmitters–receivers. Although there was no censorship, there were differences in the selection criteria of material to be used in the weekly programmes between church workers and local inhabitants, as exemplified in the above quotation on how the radio had become less serious. This difference also became apparent when comparing the radio programmes with the repertoire of the festivals celebrating the anniversary of the radio, which were more profane, melodramatic and influenced by the language of the culture industry. However, these were differences and tensions which the church workers, in accordance with their radical democratic ideological position, appeared to want to deal with through dialogue rather than the foreclosing of difference.

Notwithstanding these limitations, in the context of Brazilian society, where the view of the popular classes as unfit for self-government and incapable of thinking is a central element in the hegemonic discourse, cultural practices such as 'The People's Radio' constitute, despite their shortcomings, a counter-hegemonic force. For a subordinate social group to validate its needs and interests, to experience relationships and create symbols and meanings not characterised by domination, and to relate this with whatever degree of clarity to how the social order is constructed is the precondition for the development of a counter-hegemonic cultural force. In that sense it could be argued that 'The People's Radio' contributed towards the development of an alternative modernity privileging a form of citizenship in which the moral, political and cultural resources of the poor form a central part of the way modernity is defined.

NOTES

1 The research on 'The People's Radio' in Vila Nossa Senhora Aparecida was carried out between 1988–9 and funded by the Brazilian National Research Council (Conselho Nacional de Pequisa). It was part of a broader research project which included, in addition to the study of 'The People's Radio', investigating other aspects of northeastern culture in São Paulo and the way it is transformed in the urban context, for example, oral poetry and northeastern dance halls.

2 See in this respect Scott J C (1985) *Weapons of the Weak: Everyday Forms of Peasant Resistance*, New Haven, Yale; Gledhill J (1994) *Power and its Disguises*, London, Pluto; Chaui M (1987) *Conformismo e Resistencia: aspectos da cultura popular no Brasil*, São Paulo, Brasiliense.

3 'Afoxé' groups are groups of musicians, dancers and percussionists connected to the Afro-Brazilian religion of *Candomblé* in Salvador, Bahia who perform and enact aspects of their religious ritual during the Carnival period. The song 'Oh Faraó' celebrates black culture.

‘Posthuman Unbounded: Artificial Evolution and High-tech Subcultures’

from George Robertson *et al.* (eds), *FutureNatural: Nature, Science, Culture* (1996)

Tiziana Terranova

Editors’ Introduction

This essay on high-tech subcultures is included in *FutureNatural*, a collection edited by a group of academics in cultural studies and art history at Middlesex University. It also includes contributions by Kate Soper, Trinh T. Minh-ha, Sadie Plant, and Slavoj Žižek. The book’s theme is the impact of new scientific work on how possible futures are imagined, alongside the more familiar areas of economic, social, and cultural change. In the Introduction, the editors state: ‘The question of what remains to the “natural” when nature is cultural – a product of discourse – when what has been the territory of the “natural” is taken over by the intervention of human engineering, is a pressing one’ (p. 2). This text and the other essays included in *FutureNatural* could be compared with Critical Art Ensemble’s text which follows. It should also be seen, through its concern with the boundaries between discourses, in conjunction with the texts in Part Eight.

Terranova lectures in the sociology of media and film at the University of Essex, and was previously in the Department of Cultural Studies at the University of East London. Her doctoral research concerned ‘The Construction of Science and Technology within Internet Culture and Cyberculture’. In the text reprinted here she addresses ‘posthumanism’, as artificially enhanced evolution. When the essay was first published this may have been more future-based, almost like science fiction, than it is today, after the cloning of Dolly the sheep, and reports that human cloning is either about to happen or already has. But the boundaries she cites are not those between human bodies and artificial elements, but between progressive political argument and a conservative-futurist position in which change is seen as conservation by other means. As Terranova says: ‘The boundaries between electronic liberalism, anarchism, socialism and conservatism are . . . the places where I have chosen to tell my stories about the monstrous mutations of the cybercultural discourse’ (p. 165). In that telling she cites several cyberculture journals from Italy and the anglophone world. Her new book is titled *Network Culture: Cultural Politics and Cybernetic Communication* (forthcoming).

My story about ‘posthumanism’ (the belief in artificially enhanced evolution) in the high-tech subcultures is, in a certain sense, a tale whose main plot-line has been repeated and rehearsed many times before, another story about a seemingly unstoppable confusion of boundaries.

The kind of boundaries I have decided to focus on, however, are not simply the deliciously fluid or earnestly hard boundaries between human flesh, electronic chip, gene-splicing chemical sequences, and stainless steel. If the postmodern imagination seems somehow to have come to terms with, rejoiced in, or simply taken for granted such unions, there are other boundaries, and other interchanges that still make us uncomfortable, boundaries we would like to see not disrupted, but sealed, separated, tidily split in a dialectical antagonism where action is still possible.

The boundaries that these self-reflexive stories seem to disrupt are those between 'progressive' political discourses and the frightfully lively imagination of a 'conservative futurist' view of the world where, to quote the Sicilian novelist Tomasi di Lampedusa, 'everything has to change so that everything can stay the same'.

This is a story at whose near horizon looms the nightmare of the truly uncanny cyborgism of the Republican House Speaker, Newt Gingrich, whose monstrous futurist conservatism makes the amused headlines of American newspapers in these early days of 1995:[1] an utopic cyberfuture of 'third-wave' technologies and wired orphanages; a virtual townhall for direct decision-making by local communities deliberating about access to health care by 'alien' citizens. The boundaries between electronic liberalism, anarchism, socialism and conservatism are therefore the places where I have chosen to tell my story about the monstrous mutations of the cybercultural discourse.

ORIGINARY STORIES

In a certain sense, rudimentary versions of 'posthumanity' have circulated in western culture at least since Friedrich Nietszche and the high-tech modernist avant-garde represented by the Italian Futurists and the German Dada. A rhetoric of simulation, media-originated epochal shifts, and enhanced evolutionism by artificial means are also available in the writings of people like Jean Baudrillard, Marshall McLuhan and Timothy Leary. In the sphere of popular culture, mutated bodies and superhuman intelligences have populated the SF imagination since its very beginnings, but it is only through the activity of the cyberpunk SF group in the mid1980s that a synergetic movement between developing subcultures interested in the new developments in biotechnology and computer-mediated communication (CMC) really took off.

The American West Coast mutation of the cyberpunk imagination as expressed by *Mondo 2000*'s New Edge, the socially varied experience of the new on-line population, the complex national varieties of subcultures and political movements that grafted themselves on images and tropes vividly illustrated in the cyberpunks' fiction, are all part of a cultural landscape in many ways still in the making.

In *Mirrorshades*, the mythical anthology of 'originary' cyberpunk fiction, Bruce Sterling made a significant contribution to the propagation of a new, wired variety of 'posthumanism'. Beyond turning what had mostly been 'computer nerds' into 'cyberpunks', in his manifesto Sterling literally produced a new 'body', one thoroughly invaded and colonized by invisible technologies: 'Eighties tech' Sterling opened our eyes to [*sic*] almost ten years ago, 'sticks to the skin, responds to the touch: the personal computer, the Sony Walkman, the portable telephone, the soft contact lenses' (Sterling 1986: xiii).

In Sterling's preface, such harmless devices are reconstructed so as to become the evident truth of a 'redefinition of the nature of humanity' and 'the nature of the self'. The year before, Sterling had published the novel *Schismatrix* (1985), representing a far future where humanity will have expanded in space and learned to alter its biological frame for efficiency and longevity; the term 'posthuman' is used here to describe a particular philosophy evolving into the *Schismatrix*, the solar system, but also an existing race of mutated human beings; the 'posthuman' or 'shaper/mechanist' theme is also the object of a series of short stories collected in the anthology *Crystal Express* (1989).

The extraordinary popularity of some of William Gibson's most imaginative icons (the famous 'sockets' connecting human brains to computers, the grafted microtechnology remoulding the highly efficient body of his favourite heroine, Molly Millions, the virtual existence of his millionaires) has turned his fiction into an almost 'biblical' repertoire of images and cultural references whose contribution to the creation of the electronic culture as a whole cannot be underestimated.

If the meaning and the implications of these cybernetic bodies, both in their literary manifestations, and in the huge popularity of Hollywood cybericons, have been object of intense critical attention, other places where the posthuman body has implanted and mutated itself are perhaps of greater interest at this stage.

I am thinking in particular of the variations of posthumanism represented by *Mondo 2000*'s New Edge, the esoteric technocults practised by the Extropians, or the discussions on-line on the presence/absence/mutation of the body following intensive exposure to electronic communication. Beyond offering an interesting point of view on the high-tech imagination of some of the contemporary techno-tribes, these stories might help us also to understand, almost by the back door, important ideas about access and social change as developed by these strategically crucial subcultures.

POSTHUMANISM

The first echoes of the 'posthuman' theme in the electronic culture can be traced to the postulation of a postbiological age by *Omni* magazine in 1989 ('Interview to Hans Moravec'); and the publication on the pages of the *Whole Earth Review* in the 1988/9 issue of a forum entitled 'Is the Body Obsolete?'. In 1988 'Max More' and 'Tom Morrow' founded the *Extropy* journal and in 1990 the transformation of hip magazine *Reality Hackers* into *Mondo 2000* gave posthumanism its ultimate edge.

Both in recent academic critical theory and in various magazines interested in high tech, descriptions of posthuman evolutionism have circulated as a kind of acquired common sense for the insiders, and a potentially revolutionary revelation for the uninitiated.[2]

The story-line underlying most of these statements can be summarized in this way: there has been a huge ontological shift not only in the nature of human society, but in that of our very bodies. This mutation has been brought about, on the one hand, by the exposure to simulated images in the most traditional media, and, on the other, by the slow penetration into our daily life of almost invisible technological gadgets, from contact lenses to personal computers. This process of 'invasion' of the human body and psyche by the machine is destined to increase over the years (it is already doing it spectacularly) and give rise to a potentially new race of human beings whose symbiosis with the machine will be total. Most of the commentary by the high-tech subcultures about this phenomenon has been positive, in spite of the fact that most of cyberpunk fiction is far from optimistic on this turn of 'technology as history as destiny' (which seems to have replaced 'natural biology' as destiny).

The nature and the tone used by the groups who adopted the posthuman motif is far from homogeneous: it ranges from total commitment to the cyberpunk ethos mixed up with leftovers from the psychedelic movement in *Mondo 2000*, to an anti-consumerist, situationist magazine such as *Adbusters*.[3] The posthuman pops up in the autonomist Italian *Decoder*, and it is at the centre of the philosophy advocated by the Extropy Institute.

For reasons of space I have chosen to focus on two examples: one is an extract from one of the first issues of the California-run magazine *Mondo 2000*. The other comes from the various statements expressed in different publications, on-line and printed, by the Extropian group, a loose organization devoted to the discussion of artificially enhanced forms of evolution.

Mondo 2000, born from the ashes of *Reality Hackers* under the guide of R.U. Sirius and Queen Mu, is the glossiest and the hippest of the cybermagazines and possibly the most famous – certainly, as the editor of *bOING bOING* has noticed, the one with the heaviest financial back-up. It is the one which has attracted the widest range of writers from the cyberpunk ranks (Sterling, of course, and the mathematician Rudy Rucker, a regular contributor and almost one of its editors); *Mondo 2000*'s rhetoric of the New Edge is an ambitious attempt to fuse psychedelic sixties counterculture (Timothy Leary is one of its godfathers), New Age rhetoric and cyberpunk.[4]

The following feature is collected in *Mondo 2000: A User's Guide to the New Edge*, and is part of a section on 'Evolutionary mutation' written by the chief editor R.U. Sirius:

> Ultimately, the New Edge is an attempt to evolve a new species of human being through a marriage of humans and technology. We are ALREADY cyborgs. My mother, for instance, leads a relatively normal life thanks to a pace-maker. As a species, we are moving toward replaceable parts. Beyond that, genetic engineering and nanotechnology . . . offer us the possibility of literally being able to change our bodies into new and different forms. . . . Hans Moravec, director of the Mobile Robot Lab at Carnegie-Mellon University in Pittsburgh, has investigated three possibilities, and he believes that a form of post-biological humanity can be achieved within the next fifty years.
>
> Think about it. The entire thrust of modern technology has been to move us away from solid

> objects and into information space (or cyberspace). Man the farmer and man the industrial worker are quickly being replaced by man the knowledge worker. . . . We are less and less creatures of flesh, bone, and blood pushing boulders uphill; we are more and more creatures of mind-zapping bits and bytes moving around at the speed of light.[5]

The flamboyant rhetoric of this piece is remarkably reminiscent of Sterling's manifesto: the posthuman, on the one hand is already here – the homely image of mum with the pace-maker – and at the same time still in the future – creatures of mind-zapping bits and bytes moving around at the speed of light, Supermen for the electronic age. This collapse of present and future, domesticity and science fiction, is characteristic of the style of *Mondo 2000* as a whole. It is also interesting to note another of *Mondo*'s characteristic touches, the quote of a 'legitimate' scientist working either for a university or some established lab, in this case cyber-cultural superhero Hans Moravec.

As the Mechanoids and the Shapers in *Schismatrix*, or as Molly and Case in *Neuromancer*, human species will move either in the direction of an intensification of bodily performativity or towards the ultimate flight from the body cage.[6] In either way the human body will undergo a total ontological transformation that *Mondo 2000* feels justified in calling the New Edge. Noticeably the only differentiating factor accompanying the un-adulterated, obsolete, universalized human body in this transformation, is his/her ability to 'surf' the New Edge. This ability, as we will explore in some detail later on, is of course directly proportional to one's skills at manipulating these new technologies – something that should tip us off, if we do not consider ourselves as already safely part of the technosavant crowd, that not everybody will go through the posthuman magic gate at the same pace. The 'edge' that the aspiring posthumans have to learn to surf, can be (and it is) uncannily doubled into an 'evolutionary' edge whose possible origins might be themselves 'genetic'.

Mondo 2000 portrays its readers, the 'surfers' of the New Edge, not only as people characterized by an interest for new technologies, but also as possessing qualities such as 'an independent spirit, a wildly speculating mind, limitless imagination and daring' – a description that could have been given at the beginning of the century for the first, faboulously rich, self-made tycoons, and today quite a good self-portrait of the new 'creative', 'enterpreneurial' crowd. This aggressively future-oriented group is often contemptuously opposed to the 'talk show' crowd, addicted to an old-fashioned technology, television. Quite familiarly and providing a weird resonance to the other image of 'my mother with a pace maker', in an interview to *Ben Is Dead*, R.U. Sirius refers to 'this whole talk-show-as-new-electronic-forum thing' as a scary phenomenon, 'given the preponderance of bored housewives with nothing better to do than get all twitchy over different people's sexual difference'.[7] The association television/femininity/dumbness/addiction is one that is found early on in the portrait of the addiction of Bobby Newmark's mum to soap operas in Gibson's *Count Zero*, and is part of a significant strategy of oppositions and analogies defining the identity both of the new 'interactive' technologies (as opposed to television) and their 'active' users (as opposed to feminine passivity).

It is part of *Mondo 2000*'s ideological foundations, in fact, that access to technology and technological enhancement is provided by certain personality types. R.U. Sirius, again, describes one of the goals of the magazine as the amplification of the 'productive' myth of the 'sophisticated, high-complexity, fast-lane/real-time, intelligent, active, and creative reality hacker', the social subject able to catch up with a technological development vastly beyond the reach of the 'dulled, prosaic, practically-minded, middle-of-the-road public'.[8]

There is no appreciation of, or consideration for, social or economic limitations which impede access to different technologies. In *Mondo 2000*'s rhetoric of avant-gardism there is no space for social, economic or cultural handicaps in the race toward technological appropriation and post-humanism, but only the individual 'fitness' for technological survival.

One of the conditions that makes possible and explains this neglect of technoliteracy questions on the side of *Mondo 2000*, is, undoubtedly, the constitution of its audience, new professionals working in the electronic businesses (graphic, multimedia, music), people whose relationship to this new technology is both lucrative and creative, as *Mondo*'s editor very well knows, 'A large portion of our audience is successful business people in the computer industry, and in industry in general, because industry in the United States is high-tech'.[9]

Of course, as Andrew Ross suggests, we should be careful in giving a 'literal interpretation of what is essentially a utopian injuction on the part of an experimental counterculture'.[10] On the other hand, the tech-counterculture is not the only group eager to embrace the posthuman faith, and the permutations on the theme are too widespread to be taken as isolated, provocative statements.

The activity of the Extropy Institute, in California, is another example of posthumanism which takes the technological liberating promises made by the dominant culture ways more seriously. The Extropians include a different segment of that same professionalized group catered to by *Mondo 2000*'s countercultural rhetorics. As their FAQ (frequently asked questions) file tells us,

> Extropians have made career choices based on their extropian ideas; many are software engineers, neuroscientists, aerospace engineers, cryptologists, privacy consultants, designers of institutions, mathematicians, philosophers, and medical doctors researching life-extension techniques. Some extropians are very active in libertarian politics, and in legal challenges to abuse of government power.

Extropians, in a certain sense, belong in a different kind of subculture, even more heavily professionalized than *Mondo 2000*'s and with less sympathy for music and art or the Burroughsian rhetoric taken for granted by *Mondo 2000*'s audience.

Extropians publish their own magazine (*Extropy*), have their own newsgroup on Usenet, and their electronic mailing list.[11] The *Extropian Manifesto* is available on the Internet, while the official Extropy Institute commercially runs the electronic mailing list. The commercialization of the mailing list is, according to them, a way to keep it a place of 'discussion *among* extropians, and not for constant debating with outsiders' (Extropians FAQ).

The main tenets of Extropianism are 'boundless expansion, self-transformation, dynamic optimism, intelligent technology, and spontaneous order'. The goal of Extropianism is to enable those who want to be at the vanguard of the incumbent transformation into a 'posthuman' age. In the file 'Extropians FAQ', 'transhumanism' and 'posthumanism' are so identified:

> Q3. What do 'transhuman' and 'posthuman' mean?
>
> A3. TRANSHUMAN: We are transhuman to the extent that we seek to become posthuman and take action to prepare for a posthuman future.
>
> This involves learning about and making use of new technologies that can increase our capacities and life expectancy, questioning common assumptions, and transforming ourselves ready for the future, rising above outmoded human beliefs and behaviours.
>
> TRANSHUMANISM: Philosophies of life (such as the Extropian philosophy) that seek the continuation and acceleration of the evolution of intelligent life beyond its currently human form and limits by means of science and technology, guided by life-promoting principles and values, while avoiding religion and dogma.
>
> POSTHUMAN: Posthumans will be persons of unprecedented physical, intellectual, and psychological ability, self-programming and self-defining, potentially immortal, unlimited individuals. Posthumans have overcome the biological, neurological, and psychological constraints evolved into humans. Extropians believe that the best strategy for attaining posthumanity to be a combination of technology and determination, rather than looking for it through psychic contacts, or extraterrestrial or divine gift. [sic]
>
> Posthumans may be partly or mostly biological in form, but will likely be partly or wholly postbiological – our personalities having been transferred 'into' more durable, modifiable, and faster, and more powerful bodies and thinking hardware. Some of the technologies that we currently expect to play a role in allowing us to become posthuman include genetic engineering, neural-computer integration, molecular nanotechnology, and cognitive science.

This version of 'posthumanism' is quite distant from the popular iconography of the cyberpunk, and the affinity between the two groups is quite limited. The Extropians are, therefore, a social group quite different both from the hard cyberpunk of the hacker magazine *2600*, or the techno-artistic crowd of *Mondo 2000*. It would be possible to recognize in the Extropians the result of the convergence of other traditions, like the technological utopianism and the technocratic movement of the early twentieth century whose belief that there are technological solutions to social problems is shared by the Extropians.[12] On the other hand their emphasis both on the 'voluntary' and 'individually-planned' nature of mutation, and on physical evolution beyond the boundaries of humanity as we know it, are new developments that link them more closely to the variation of posthuman evolutionism represented by *Mondo 2000*.

Most of the evolutionary mutations envisaged by the Extropians are part of the cyberpunk folklore (neural interfaces, nanotechnology, genetic engineering), and the same label 'posthuman' is a legacy of cyberpunk fiction. Nevertheless the future visions of Extropians are surprisingly empty of the dingy dystopianism of cyberpunk fiction, which still lingers somehow even around the propagandistic hype of a publication like *Mondo 2000* (what Vivian Sobchack has defined its 'utopian cynicism').[13]

Another manifestation of this uneasy confusion of boundaries inside cyberculture between differently inspired interpretations of technological progress is the publication of the Extropian Manifesto on the pages of the Italian independent magazine *Decoder*.[14] *Decoder* is fairly representative of the enthusiasm and the receptivity to cyberpunk among European autonomist and extreme leftist groups. Although often critical of *Mondo 2000*'s politics, *Decoder* has been fairly active in keeping up international cooperation between different groups involved in computer counterpolitics. They have translated and published from French, British, German, Dutch and American sources.

Decoder sees in the electronic revolution taking place in the western world a possibility of positive, radical mutation of the social system in ways that overlap but differ also considerably from those of their American colleagues. The political affiliations of *Decoder* are clearly to the left of the Italian ex-Communist party, rooted in the tradition of the anarcho-marxism of the extra-parliamentary groups. The specific political and cultural problems of this milieu have shaped the Italian interpretation of cyberpunk as heralded by *Decoder*: in this context, cyberpunk has been turned into a political commentary on technological development and social change alternative to both the 'industrialist', orthodox marxist paradigm and the 'postmodern' socialism of the Italian ex-Communist party. The publication of the *Extropian Manifesto* is part of their politics of diffusion of international, provocative publications on the subject of technology.

In the editor Raffaele Scelsi's words, Italian cyberpunk's 'political tension is oriented towards the reappropriation of communication by social movements, through the constitution of alternative electronic networks which could finally affect the excessive power of multinationals in the field'.[15] The Extropians, on the other hand, seem to be rooted in a different kind of libertarian anarchism, which expresses itself in their faith in the self-regulatory nature of the 'free market':

> The principle of spontaneous order is embodied in the free market system – a system that does not yet exist in a pure form . . . The free market allows complex institutions to develop, encourages innovation, rewards individual initiative, cultivates personal responsibility, fosters diversity, and decentralizes power. Market economies spur the technological and social progress essential to the Extropian philosophy . . . Expert knowledge is best harnessed and transmitted through the superbly efficient mediation of the free market's price signals – signals that embody more information than any person or organization could ever gather.

The publication of the *Extropian Manifesto* on the pages of *Decoder* poses therefore some puzzling questions that could be explained by the partial ignorance by European cyberpunk publications of the American scene or their desire to construct an international cyberpunk network that would validate and reinforce the cultural and political position they are trying to build for themselves. On the other hand, it is more interesting to consider it as a signal of the unresolved contradictions embedded at the heart of the high-tech subculture. Beyond the common debt to cyberpunk fiction and their interest in the same cutting-edge technologies there seems to be no clear and elaborate articulation of differences at this stage. This absence of a constructive and articulated internal debate might be the result of the cohesive effect produced by the widespread hostility still perceived coming from the 'outside'. Too many energies, perhaps, are still spent rejecting outdated ethical and religious objections to technology. Cyberculture is still too busy building its defensive walls to care to look too closely at its own increasingly complex and fragmented anatomy.

THE PROBLEM WITH EVOLUTION

The most distinguishing common trait between the cyberpunk, Extropian, and *Mondo 2000*'s treatment of the 'posthuman' theme is the inability to introduce patterns of differentiations and complexity in their portrait of the future as teleological evolution. As an expression of a subculture that encourages a 'subversive' use of technology, there is little problematization of the ways in which technological change can be and is being shaped by economic and political forces. In particular, this evolutionary trend is represented as existing 'outside' the contentious history of the social uses of 'evolutionary' principles in the United States.[16] It is worth remembering here, while boundaries keep blurring fairly disturbingly, the popularity of mass IQ testing at the beginning of the century as a way to reinforce racial differences within the ideological frame given by social Darwinism. A disturbing pattern of historical continuity has recently brought to the centre of widespread discussion these same patterns of interpretation in the infamous bestseller *The Bell Curve*.[17] Murray and Herrnstein's focus on 'smartness' and 'cognitive stratification' creates the fiction of a 'cognitive disadvantage' distributed across racial coordinates as an elegant way out from embarrassing discussions about the effect of historically specific economic and cultural policies.[18] IQ testing, race discrimination and cognitive fatalism live together uneasily, dormant in the unconscious of some of these high-tech subcultures, which certainly would not recognize them as part of their 'radical' and hopefully 'progressive' imagination.

On the other hand, biological determinism, which is more or less the message given by works such as *The Bell Curve*, is a label that is too narrow and unfair to describe the belief in individual fitness in techno-evolution: it is, nevertheless, important to examine carefully these voices 'uneasy and hidden connections. Believers in posthumanism are not so much saying, 'we are what our genes say we are' but 'we are what we want to be', and, 'thanks to technology, there are no limits to what we can be'. In this triumph of the will society is erased, and the social universe emerges as a fragmented aggregate of individuals in a void without historical and material constraints. Not primarily biological determinism, then, but *rampant super-voluntarism* is the problem with cybernetic posthumanism.

It is necessary at this point to perform the dangerous jump from these limited uses of a provocative trope such as 'posthumanism' to the textual oceans of electronic networks at large. If this jump does not produce an exhaustive and conclusive account of what the Internet and associated networks think exactly, it will try, on the other hand, to sketch the contours of a general consensus (if any such thing is possible) in the most self-reflexive and argumentative Usenet newsgroups.

IN THE NET

The opinions and the arguments developed in the Usenet newsgroups considered here (alt.cyberpunk, alt.cyberpunk.movement, alt.internet. media-coverage, alt.cyberspace, alt.politics.datahighway, alt.current-events.net-abuse) are not homogenous and represent only a portion of the discussions going on among Internet aficionados all over the world. Regular contributors to these newsgroups, on the other hand, are careful readers and heated debaters, and the ideas thrown around in these weeks-long arguments feed on, and very often feed back to, the writings of more public voices, all of which contribute to the creation of an unstable consensus among 'netsurfers' about what electronic communication is about.

In this context, 'posthumanism' is not only one of the excessive manifestations of the cybercultural spirit, it is also a useful gateway into cybercultural discourse at large, intersecting with one of its most ferociously preserved opinions: computer-mediated communication will change significantly the world as we know it in as much as it will involve a collective evolution from the passive consumption of corporation-dominated media to interactive, symbiotic relationships with intimate machines. This belief is not coincident with posthumanism (of which it could be said to represent the utopic side) but is related to it and is itself fraught with disturbing contradictions.

The more generally shared faith in the power of individual self-transformation into 'cyborgian hybrids of technology and biology' overlaps, often disturbingly, with the evolutionary posthumanism described above. Here is a fairly typical statement about on-line cyborgism as expressed by one of the participants in alt.cyberpunk;

> I think that the boundary of 'self' is growing larger and approaching undefinable. Right now with text we are on a lower level, not quite in total tele-presence and virtual reality, but still you can see our little buds of flesh growing behind our ears where we will be ready for such things. People who participate now in this, will have much less trouble when we reach total immersion VR and tele-presence, as they are used to their self being [sic]

mutated and expanded. Not virus, but a fuzzying of physical boundaries, not only of nations, but also the individual body adn [sic] self.

In this 'weak' version, 'posthumanism' is not even named as such, but it resurfaces in the vision of a collective 'cyborgization' of society resulting from the individual act of tuning in: a transformation 'into cyborged hybrids of technology and biology through our ever-more-frequent interactions with machines, or with one another *through* technological interface'.[19]

These statements, although expressed in restricted circles of like-minded individuals, should not prevent us from remembering the fact that they are an expression of a widely felt belief. The idea that current regulars of the virtual communities are the avant-garde of a historical process that will soon be universal does certainly possess wider political currency at this stage. This vision of a wave of cognitive change spreading steadily, at viruslike speed, is expressed also through the evocative use of statistics describing the rate of growth of the Net population, whose magnitude cannot but strike a chord in these 'scarcity', 'budget-obsessed' 1990s.[20]

The on-line population is extremely aware of being still, but not for long; the first wave of the 'wiring' of society, where their own lived experience of 'cyborgization' will be extended to the mass. In Mark Dery's words, 'these subcultural practices offer a precognitive glimpse of mainstream culture a few years from now, when ever-greater numbers of Americans will be part-time residents in virtual communities.'[21] Dery is certainly in touch with the general mood of the electronic tribes (and the communication companies commercial hype) in envisaging the current experience of these electronic groups as the general experience of the masses in the near future.

This is where the 'posthumanist' theme becomes a variation of a more diffused individualist, voluntarist, avant-garde attitude widely shared by the electronically wired communities. As in *Mondo 2000*'s rhetoric, this is a language of action where mainstream, 'old' media such as television and newspapers play, more often than not, a villainous role in propagating one-way indoctrination and passivity in the mass public, while the Internet two-ways, active experience of logging 'in' provides the way 'out'.

Well resistance implies opposition, but i [sic] see it more like moving sideways. For me it is using tech to help myself, and those around me. Wether [sic] it is making art form [sic] it, or helping people learn to use it. *It takes away the dependence peole [sic] have upon normal media, and the monopolies that control it. Which in turn eventually corrodes the hold the media has on people, and after that, it's only small step until they become really free. [sic]* (emphasis added)

the only dirrection [*sic*] I could think of would be towards individual freedom thru [*sic*] the use of technology to breakt [*sic*] he [*sic*] mono[oly] [*sic*] of information held by the media and the control structure. [*sic*]

Of course there's nothing wrong in believing in the superiority of one medium of communication over another, especially when one (like TV) is heavily dominated by a commercial, monopolistic culture and the other is perceived as a free, anarchic universe almost uncontaminated by the powers-that-be. The most told and retold horror stories in the wired universe are stories about the corporate takeover of the Internet, the fatal 'clipping' of its wings in a nightmare of total surveillance;[22] the most repeated exorcism is that performed by the repeated re-enactment of the rhizomatic nature of the Net and its impermeability to any complete form of censorship – 'the net can't be bombed or censored out of existence', 'is not a single entity, solid, it is not an object or a controllable resource'.

In a certain way, there are also some beautiful stories being told here, stories we like to hear and believe in, stories that might come true if virtual communities hold on tightly enough against the various attempts by arch-villains 'big government' and 'big money' to colonize an electronic frontier where 'we' might be luckier natives this time. In this utopic sense, posthumanism could be really a 'cyborg dream'. The mobilization of the strategic fiction of an 'open-to-all', 'universally-accessed' or 'in-theprocess-to-be-universally-accessed' Internet is itself useful in as much as it is at this point an enabling strategy. These fictions allow collective mobilization in the form of electronic activism against imperialist projects of colonization such as those advocated by the Clinton administration with the infamous 'Clipper-chip' plan.

In spite of the strategic needs that these fictions serve, it is worth remembering in a longer-term perspective, the necessity to constantly critique 'the things that are extremely useful, things without which we cannot live on'.[23] One of these myths is, exactly, the strategically useful myth that if technologies (and especially CMC technologies) would be left on their own, they would naturally and spontaneously extend harmoniously to the entire population, making the experience of 'technologically enabled, postmulticultural' identities 'disengaged from gender, ethnicity and other problematic construction', universally achievable.[24]

These are the points where cyberculture offers the possibility of accommodation to various variations on the theme of 'conservative futurism' of which Newt Gingrich at the moment is just the most public version. His statement about favouring tax deductions for poor kids to buy laptops (subsequently named a 'dumb' idea) fits in with the New Right's call to individual empowerment. It is perhaps superfluous to remember here its rhetorical value in conservative rhetorics as the other face of the savage demolition of public mechanisms of social solidarity. Both cybercultural discourse and the new 'futurist conservatism' fall prey to the erasure of the cultural, economic, and political conditions enabling or restricting the social subject. Although fully supporting a policy of popularization of computer-access through the reduction of prizes, we should be aware of statements that link simplistically the accessibility of CMC, the jump into collective cyborgism, and the democratization of society as a whole only to market-availability of cheap hardware, software and E-mail accounts.

The establishment of a possible mass-use of electronic communication devices in the context of a political project aimed at overcoming current social inequalities, should be more than simply encouraging individuals to personal empowerment through technology. Any 'radical' use of CMC has to pass through the recuperation of the collective act of disavowal performed by most virtual communities regarding their current position in the jobless, two-tier, 'flexible' homework economy rampant in the mid-1990s.[25] It is this economy which has created the variety of occupations which is the much vaunted constituency of the Internet, a variety which is 'structurally dependent today on the primitive labour-regimes of minimally educated immigrant minorities in the dense metropolitan centers'.[26]

This article is not intended to be a condemnation of the virtual communities as a whole on the basis of the economic status of their dwellers: the point that I have tried to make is different. My critique is directed at those portions of the electronic community which earnestly believe in a possible contribution given by CMC to the establishment of a world of more equally distributed material and cultural resources. The political activism of these communities is currently fundamental in limiting the consequences of the focusing of corporate and governmental interests on electronic communication.

If, as Howard Rheingold would like it to be, electronic communication can develop into a new, more competent form of citizenship, then the conditions producing the Internet citizen cannot be underestimated. The process of normalization of cyberspace, today taking place in the New Right's recent endorsements and in the boosterism of major communication companies, should not just be rejected as the external appropriation by interested parties of spontaneous and 'genuine' principles: at some level their rhetoric fits in smugly with the unacknowledged or willingly ignored contradictions affecting the most radical of the virtual communities. Haraway's cherished dream of cyborgism was placed against an economic, political and social order whose monstrous character needed desperately the exorcism of a possible utopian imagination. If the cyborg dream has any chance of being delivered, maybe we should learn to sleep less peacefully and, sometimes, even with both of our screen-reddened eyes wide open.

NOTES

1 Gingrich's version of 'third-wave' conservatism has been the object of the attention of several leading national newspapers and magazines since his nomination to House Speaker at the beginning of December 1994. Journalists have underlined the discontinuity and the potential schism between Gingrich's futurism and more conservative Republicans (see David E. Rosenbaum 'Time Warp Republicans like both previews and reruns', *New York Times*, 11 December 1994). The presence of Heidi and Alvin Toffler as close friends and counsellors of the House Speaker has also led somebody to suggest the existence of a 'conservative counterculture' in the making (cf. Steven Waldman 'Creating a Congressional Counterculture', *Newsweek*, 16 January 1995). Gingrich has also been recently patron to a conference called 'Democracy in Virtual America' where the most distinguished speakers have been the Tofflers and Arianna Huffington (author of the new-ageish work *The*

Fourth Instinct)(1994), New York: Simon & Schuster. The subject of course was the transition to the information age. Gingrich (stealing a classic SF motif amplified by cyberpunk) argued that 'we are not at a new place. It is just becoming harder and harder to avoid the place we are'. (cf. Maureen Dowd 'Capital's virtual reality: Gingrich rides a 3rd Wave', *New York Times*, 11 January 1995.)

2 See for variously nuanced versions of this story: Larry McCaffery (1991) *Storming the Reality Studio*, Durham and London; Duke University Press; Arthur and Marilouis Kroker (eds) (1988) *Body Invaders: Sexuality and the Postmodern Condition*, London: Macmillan; Arthur Kroker (1994) *Data Trash*, New York; St Martin's Press; Scott Bukatman (1993) *Terminal Identity*, Durham and London: Duke University Press.

3 See Jeffery Deitch (1994) 'Posthuman' in *Adbusters*, 3 (1), winter.

4 See Douglas Rushkoff, *Cyberia: Life in the Trenches of Hyperspace*, San Francisco: Harper, 1994.

5 See R.U. Sirius, 'Evolutionary mutation', in *Mondo 2000: A User's Guide to the New Edge*, London: Thames & Hudson, 1992, 100.

6 A useful analysis of the variety of cybernetic bodies in William Gibson is given by David Tomas, 'The technophilic body: On technicity in William Gibson's cyborg culture', *New Formations* 8, summer 1989, 113–129)

7 Ethan Port 'Perpetual cyber jack-off', in *Ben Is Dead*, summer 1993, 107–113, especially 113.

8 R.U. Sirius, 'The new edge', in *Mondo 2000: A User's Guide to the New Edge*: 195.

9 R.U. Sirius, quoted in 'Sex, drugs, & cyberspace', *Express: The East Bay's Free Weekly*, 28 September 1990: 12.

10 Andrew Ross, 'The new smartness', in Gretchen Bender and Timothy Druckrey (eds), *Culture on the Brink: Ideologies of Technology*, Seattle: Bay Press, 1994, 329–41, especially 335.

11 See also Ed Regis, 'Meet the Extropians', in *Wired*, October 1994: 102

12 On the technocratic movement of the 1930s, see Andrew Ross, *Strange Weather: Culture, Science and Technology in the Age of Limits*, London and New York: Verso, 1991.

13 See Vivian Sobchack 'New Age Mutant Ninja hackers: reading *Mondo 2000*', *South Atlantic Quarterly* 92 (4) (Fall 1993), 569–84.

14 See *Decoder* 8 September 1993. *Decoder* is published three times a year by the independent publishing house Shake and much of its staff comes from the ranks of La Conchetta, a squatted alternative centre in Milan.

15 See Raffaele Scelsi (ed.) Scelsi, Raffaele, *Cyberpunk: Antologia*, Milan, Shake Edizioni, 1990, 9.

16 On this subject, see Carl Degler, *In Search of Human Nature: The Decline and Revival of Darwinism in American Social Thought*, New York and Oxford: Oxford University Press, 1991.

17 Richard J. Herrnstein and Charles Murray, *The Bell Curve: Intelligence and Class Structure in American Life*, New York, Free Press, 1994.

18 See Andrew Ross 'Demography Is Destiny', *Village Voice*, 29 November 1994: p. 95–6.

19 See Mark Dery, 'Flame Wars' *South Atlantic Quarterly 92 (4)* (Fall 1993), 564.

20 Here is Howard Rheingold's typical rhetoric of utopic, electronic communication boosterism: 'If a citizen today can have the telecomputing power only the Pentagon could afford twenty years ago, what will citizens be able to afford in telecommunication power five or ten years into the future? . . . In terms of population growth, the original ARPANET community numbered around a thousand in 1969. A little over twenty years later, the Internet population is estimated at five to ten million people. The rate of growth is too rapid for accurate measurement at this point.' (Howard Rheingold, *Virtual Communities*, New York: Harper, 1993, 80).

21 Dery, op. cit., p.564.

22 For a nightmarish picture of the total electronic panopticon see Charles Ostman, 'Total surveillance', *Mondo 2000* 13, winter 1995, 16–20; in the same paranoid tone see also most of the articles published by members of the EFF (electronic frontier foundation) at the time of the 'Clipper chip' debate: in this case, conspiracy theory and all-out paranoia proved to be effective strategies in mobilizing the electronic communities towards legal action (see John Perry Barlow 'Jackboots on the Infobahn'. *Wired*, 2.04, April 1994, 40–9; Brock N. Meeks, 'The end of privacy', *ibid.*: 40).

23 See Gayatri Spivak, *Outside in the Teaching Machine*, New York and London: Routledge, 1993, 4.

24 see Dery: 561.

25 See Richard Gordon, 'The computerization of daily life, the sexual division of labor, and the homework economy', Silicon Valley Workshop Conference, University of California at Santa Cruz, 1983; Donna Haraway 'A cyborg manifesto: science, technology, and socialist-feminism in the late twentieth century', in *Simians, Cyborgs, and Women: The Reinvention of Nature*, London: Free Association Books, 1991, 149–81; Stanley Aronowitz, 'Technology and the future of work' in G. Bender and T. Druckrey (eds), *Culture on the Brink: Ideologies of*

Technology, Seattle: Bay Press, 1994, 15–29. See, also the rich material collected in Chris Carlsson and Mark Leger (eds), *Bad Attitude: The Processed World Anthology*, London: Verso, 1990, about conditions of work in the new, computerized, flexible marketplace.

26 See Andrew Ross, *The Chicago Gangster Theory of Life: Nature's Debt to Society*, London and New York: Verso, 1994, 149.

'Electronic Civil Disobedience'

from *Electronic Civil Disobedience and Other Unpopular Ideas* (1996)

Critical Art Ensemble

Editors' Introduction

Critical Art Ensemble is a collective of five artists based in various American cities who have been working together since 1987. Their projects include film and video, photography, books, performances and webpages critically exploring the relationships between art, technology, critical theory and activism. Their inspiration comes primarily from historical movements of resistance such as the Situationists and Radical American Theatre. The extract from 'Electronic civil disobedience and other unpopular ideas' included here was central to much of their work during the 1990s which was concerned with developing a critique of the electronic economy. It was published in 1996 and was followed by a companion volume *The Electronic Disturbance* (2000). The project from which this extract is taken is concerned with using the tools of the electronic economic to foster forms of nomadic resistance. They argue that as the nature of the enemy changes so too must the nature of resistance. The politics of resistance and single-issue politics has attracted a great deal of media and political attention recently. George Mackey's *Senseless Acts of Beauty* (London, Verso, 1996), for example, charts road protests in the United Kingdom detailing how these activities involved a highly spatial form of cultural protest. By contrast the Critical Art Ensemble claim 'as far as power is concerned the streets are dead capital!' (1996: 11). A claim that has serious repercussions for political geography which has seen the streets as the arena within which power is exercised, extended and contested.

Recently the work of Critical Art Ensemble has moved on to include critiques of the manifestation of the body within biotechnology and genetics research. These works have included the performances *Flesh Machine* (1997–98), *Society for Reproductive Anachronisms* (1999) and *Cult of the New Eve* (2000). Each of these performances has resulted in books of essays and CD-roms.

One essential characteristic that sets late capitalism apart from other political and economic forms is its mode of representing power: What was once a sedentary concrete mass has now become a nomadic electronic flow. Before computerized information management, the heart of institutional command and control was easy to locate. In fact, the conspicuous appearance of the halls of power was used by regimes to maintain their hegemony. Castles, palaces, government bureaucracies, corporate home offices, and other architectural structures stood looming in city centers, daring malcontents and underground forces to challenge their fortifications. These structures, bespeaking an impregnable and everlasting solidity, could stop or demoralize contestational movements before they started. Indeed, the prominence of this spectacle was a double-edged sword; once the opposition became desperate enough (due to material privation or to symbolic collapse of a given regime's legitimacy), its revolutionary force had no problem finding and confronting the powerholders. If the fortifications were breached, the regime would most likely collapse. Within this broad historical context emerged the general strategy for civil disobedience.

This strategy was unusual because the contestational groups decided they did not need to act violently toward those who occupied the bunkers of power, and chose instead to use various tactics to disrupt the institutions to such an extent that the occupants became disempowered. Although the smiley face of moral force was the pretext for using this approach, it was economic disruption and symbolic disturbance that made the overall strategy effective. Today acts of civil disobedience (CD) are generally intended to hasten institutional reform rather than bring about national collapse, since this style of resistance allows the possibility for negotiation. For this reason, modern first-world governments tend to be more tolerant of these acts, since they do not necessarily threaten the continued existence of a nation or its ruling class. While civil disobedience does not go unpunished, it is generally not met with extreme violence from the state, nor are participants in CD ordinarily labeled as revolutionaries and treated as political prisoners when arrested. (There have of course been some notable exceptions to this policy in the first world, such as the persecution of American civil rights activists in the deep South).

Although CD is still effective as originally conceived (particularly at local levels), its efficacy fades with each passing decade. This decline is due primarily to the increasing ability of power to evade the provocations of CD participants. Even though the monuments of power still stand, visibly present in stable locations, the agency that maintains power is neither visible nor stable. Power no longer permanently resides in these monuments, and command and control now move about as desired. If mechanisms of control are challenged in one spatial location, they simply move to another location. As a result, CD groups are prevented from establishing a theater of operations by which they can actually disrupt a given institution. Blocking the entrances to a building, or some other resistant action in physical space, can prevent reoccupation (the flow of personnel), but this is of little consequence so long as information-capital continues to flow.

These outdated methods of resistance must be refined, and new methods of disruption invented that attack power (non)centers on the electronic level. The strategy and tactics of CD can still be useful beyond local actions, but only if they are used to block the flow of information rather than the flow of personnel. Unfortunately, the left is its own worst enemy in developing ways to revise CD models. This situation is particularly ironic, since the left has always prided itself on using history in critical analysis. Now, rather than acknowledge the present shift in historical forces when constructing strategies for political activism, members of the left continue to act as if they still live in the age of early capital. This is particularly strange because contestational theory always stresses the importance of dramatic shifts in political economy (early capital to late capital, industrial economy to service economy, production culture to consumption culture, etc). Indeed, the left's lapse of insight on this matter indicates that the schism between theory and practice is as bad as (or worse than) it has ever been.

This particular form of cultural lag prevents activists from devising new strategies for reasons that are difficult to pinpoint. At least one factor responsible is the continued presence of the remnants of the 60s New Left within the ranks of activist groups. Preoccupied as they are with the means used to achieve past victories (primarily the contribution that the New Left made to the withdrawal of American troops from VietNam), members of these groups see no need to invent new approaches. Nostalgia for 60s activism endlessly replays the past as the present, and unfortunately this nostalgia has also infected a new generation of activists who have no living memory of the 60s. Out of this sentimentality has arisen the belief that the "take to the streets" strategy worked then, and will work now on current issues. Meanwhile, as wealth and education continue to be increasingly distributed in favor of the wealthy, as the security state continues to invade private life, as the AIDS crisis still meets with government inaction, and as the homeless population continues to expand, CAE is willing to go out on a limb and say that perhaps an error in judgment has occurred. This claim is not intended to undermine what has been accomplished on local levels; it is intended only to point out that contemporary activism has had very little effect on military/corporate policy.

CAE has said it before, and we will say it again: as far as power is concerned, the streets are dead capital! Nothing of value to the power elite can be found on the streets, nor does this class need control of the streets to efficiently run and maintain state institutions. For CD to have any meaningful effect, the resisters must appropriate something of value to the state. Once they have an object of value, the resisters have a platform from which they may bargain for (or perhaps demand) change.

At one time the control of the street was a valued item. In 19th century Paris the streets were the conduits for the mobility of power, whether it was economic or military in nature. If the streets were blocked, and key political fortresses were occupied, the state became inert, and in some cases collapsed under its own weight. This method of resistance was still useful up through the 60s, but since the end of the 19th century it has yielded diminishing returns, and has drifted from being a radical practice to a liberal one. This strategy is grounded in the necessity of centralizing capital within cities; as capital has become increasingly decentralized, breaking through national boundaries and abandoning the cities, street action has become increasingly useless. Since cities have been abandoned by business and left to rot in a state of bankruptcy, and have become plagued by crime and disease, it seems reasonable to assume that they are no longer useful in the expansion of power. If they were of use, surely they would be continually renewed and defended.

Dangers do lie in this often tautological line of argument. Is the city of no value because it is not maintained, or is it not maintained because it is of no value? This error in logic is inescapable, since the question of who or what is in control cannot be answered. Power itself cannot be seen; only its representation appears. What lies behind the representation is lost. The location and nature of cynical power is purely a matter of speculation. Macro power is known only as a series of abstractions such as "straight white males," "the ruling class," or best of all, "the powers that be." Macro power is experienced only by its effects, and never as a cause. Consequently, certain indicators must be used to determine what is of value to power, or to find the (non)location of power. The assumption here is that key indicators of power-value are the extent to which a location or a commodity is defended, and the extent to which trespassers are punished. The greater the intensity of defense and punishment, the greater the power-value. These indicators have been derived from experience, but they cannot be given theoretical justification, since a second principle will eventually have to be used to explain a first principle.

If the traditional location for deploying power has been abandoned, where has power moved? If we assume that the flow of capital is still crucial to the present system, then there is a trail to follow. (Un)common sense tells us that we can follow the money to find power; however, since money has no point of origin but is part of a circular or spiraling flow, the best we can expect to find is the flow itself. Capital rarely takes a hard form; like power, it exists as an abstraction. An abstract form will probably be found in an abstract place, or to be more specific, in cyberspace. Cyberspace may be defined as a virtual informational landscape that is accessed through the phone system. (For the purposes of this essay, the association between cyberspace and VR proper should be ignored). The degree of access to the information located in cyberspace suggests how institutions are configured in real space. In complex society, the division of labor has become so differentiated that the organizational speed necessary to keep the many segments synchronized can only be achieved by using electronic communication networks. In turn, the controlled deployment of information and access to it becomes a central clue in solving the puzzle of social organization. When access to information is denied, the organizational properties of the institution from which it is withheld become unstable, and – should this condition be maintained for too long – the institution will eventually collapse because of a communication gap. The various segments will have no idea if they are working at cross purposes against each other or if they are working in unison against competing institutions. Blocking information access is the best means to disrupt any institution, whether it is military, corporate, or governmental. When such action is successfully carried out, all segments of the institution are damaged.

The problem with CD as it is now understood is that it has no effect on the core of organization; instead, it tends to concentrate on one localized sedentary structure. In the case of national or multinational institutions, such actions are no more disruptive than a fly biting an elephant. Back when power was centralized in sedentary locations, this strategy made sense, but it is vain now that power is decentralized. To dominate strategic sites in physical space was once the key source of power, but now domination rests on the ability of an institution to move where resistance is absent, in conjunction with the ability to temporarily appropriate a given physical space as needed. For an oppositional force to conquer key points in physical space in no way threatens an institution. Let us assume that a group of dissidents managed to occupy the

White House. It might prove embarrassing for the administration in power and for the Secret Service, but in no way would this occupation actually disrupt the efficient functioning of executive power. The presidential office would simply move to another location. The physical space of the White House is only a hollow representation of presidential authority; it is not essential to it.

In measuring power-value by the extent to which actions are punished and sites are defended, it is readily apparent that cyberspace ranks high on the scale. Defense systems in cyberspace are as well-developed as they can be. The Secret Service (previously an agency whose job was to protect individuals connected with the office of the President and to investigate counterfeiting rackets) has become increasingly swept up in its role as cyberpolice. At the same time, private corporations have developed their own electronic police forces, which function in two ways: First, they act as security forces, installing information surveillance and defense systems, and second, they act as a posse of bounty hunters to physically capture any person who breaks through the security systems. These forces, like the legal system, do not distinguish between actions in cyberspace on the basis of intent. Whether private information sources are accessed simply to examine the system, or whether the purpose is to steal or damage the source, these forces always assume that unauthorized access is an act of extreme hostility, and should receive maximum punishment. In spite of all this security, cyberspace is far from secure. It has expanded and mutated at such a rapid rate that security systems are unable to reconfigure and deploy themselves with equal speed. At present, the gate is still open for information resistance, but it is closing.

Who is attempting to hold the gate open? This is perhaps one of the saddest chapters in the history of resistance in the US. Right now the finest political activists are children. Teen hackers work out of their parents' homes and college dormitories to breach corporate and governmental security systems. Their intentions are vague. Some seem to know that their actions are political in nature. As Dr. Crash has said: "Whether you know it or not, if you are a hacker you are a revolutionary." The question is, a revolutionary for what cause? After poring through issues of *Phrack* and surfing the internet, one can find no cause mentioned other than the first step: free access to information. How this information would be applied is never discussed. The problem of letting children act as the avant-garde of activism is that they have not yet developed a critical sensibility that would guide them beyond their first political encounter. Ironically enough, they do have the intelligence to realize where political action must begin if it is to be effective – a realization that seems to have eluded leftist sophisticates. Another problem is the youthful sense of immortality. According to Bruce Sterling, their youthful fearlessness tends to get them arrested. A number of these young activists – the Atlanta Three, for example – have served time in what has to be recognized as political imprisonment. With only the charge of trespass against them, jailing these individuals seems a little extreme; however, when considering the value of order and private property in cyberspace, extreme punishment for the smallest of crimes should be expected.

Applying the maximum punishment for a minimal offense must be justified in some way. Either the system of punishment must be kept hidden from the public, or the offense must be perceived by the public as a horrific disruption of the social order. Currently, the situation in regard to crime and cyberspace seems neutral, as there is no solid commitment by the state to either path. The arrest and punishment of hackers does not make headlines, and yet the law and order alarm has started to ring. Operation Sundevil, a thorough sweep of hacker operations in 1990 by the Secret Service and corporate posses, received minimal attention from the media. It was well publicized amongst the groups affected by such activities, but it was hardly the material needed for a "60 Minutes" investigation or even a Phil Donahue show. Whether this lack of publicity was intentional or not on the part of the Secret Service is difficult to say. Certainly corporations do not like to call attention to their posses, nor does the Secret Service want to advertise its Gestapo tactics of confiscating the property of citizens not charged with any crime, and neither of the two want to encourage hacker behavior by openly revealing the power that can be gained through "criminal" access to cyberspace. From the point of view of the state, it makes strategic sense to limit the various threats of punishment to the technocracy, until electronic dissidents can be presented to the public as the incarnation of evil bent on the destruction of civilization. However, it is difficult for the state to designate a techno-child as the villain of the week along the lines of Noriega, Saddam Hussein, Khadafy, Khomeny, or anyone involved with drugs from users to cartel leaders. To go public will require something more than just a charge of trespass; it will have to be something that the public can really panic about.

Hollywood has begun to make some suggestions in films such as *Die Hard II* and *Sneakers*. In *Die Hard II*, for example, terrorist hackers appropriate airport computers and use them to hold planes hostage, and even crash one. Fortunately these scenarios are still perceived by the public as science fiction, but it is precisely this kind of imaging which will eventually be used to suspend individual rights, not just to catch computer criminals, but to capture political dissidents as well. Legal agencies are just as able to persecute and prosecute political factions when what they could do arouses fear in others.

Herein lies the distinction between computer criminality and electronic civil disobedience. While the computer criminal seeks profit from actions that damage an individual, the person involved in electronic resistance only attacks institutions. Under the rubric of electronic resistance, the value system of the state (to which information is of higher value than the individual) is inverted, placing information back in the service of people rather than using it to benefit institutions. The authoritarian goal is to prevent this distinction from being perceived; all electronic resistance must fall under the totalizing sign of criminality. Conflating electronic civil disobedience (ECD) with criminal acts makes it possible to seal off cyberspace from resistant political activity. Attacks in cyberspace will carry penalties equivalent to those merited by violent attacks in physical space. Some leftist legal agencies, such as the Electronic Frontier Foundation, have already realized that basic freedoms (of speech, assembly, and press) are denied in cyberspace and are acting accordingly, but they have yet to start work on legitimizing the distinction between political and criminal action. The same legal penalties that apply to CD should also apply to ECD. However, state and corporate agencies should be expected to offer maximum resistance to legal activities aimed at legitimizing ECD. If these authoritarian structures are unwilling to grant basic rights in cyberspace to individuals, it seems safe to assume that a pseudo-legitimized resistance will not be tolerated either.

The strategy and tactics of ECD should not be a mystery to any activists. They are the same as traditional CD. ECD is a nonviolent activity by its very nature, since the oppositional forces never physically confront one another. As in CD, the primary tactics in ECD are trespass and blockage. Exits, entrances, conduits, and other key spaces must be occupied by the contestational force in order to bring pressure on legitimized institutions engaged in unethical or criminal actions. Blocking information conduits is analogous to blocking physical locations; however, electronic blockage can cause financial stress that physical blockage cannot, and it can be used beyond the local level. ECD is CD reinvigorated. What CD once was, ECD is now.

Activists must remember that ECD can easily be abused. The sites for disturbance must be carefully selected. Just as an activist group would not block access to a hospital emergency room, electronic activists must avoid blocking access to an electronic site that may have similar humanitarian functions. For example, let us assume that a profiteering pharmaceutical company is targeted. Care will have to be taken not to block the data controlling the manufacture and distribution of life-saving medications (no matter how bad the extortion profits might be from the drugs). Rather, once the company is targeted, activists would be wiser to select research or consumption-pattern data bases as sites for occupation. Having the R&D or marketing division shut down is one of the most expensive setbacks that a company can suffer. The blockage of this data will give the resistant group a foundation from which to bargain without hurting those who are in need of the medications. Further, if terms are not met, or if there is an attempt to recapture the data, ethical behavior requires that data must not be destroyed or damaged. Finally, no matter how tempting it might be, do not electronically attack individuals (electronic assassination) in the company – not CEOs, not managers, not workers. Don't erase or occupy their bank accounts or destroy their credit. Stick to attacks on the institutions. Attacking individuals only satisfies an urge for revenge without having any effect on corporate or government policy.

This model, although it seems so easy to grasp, is still science fiction. No alliance exists between hackers and specific political organizations. In spite of the fact that each would benefit through interaction and cooperation, the alienating structure of a complex division of labor keeps these two social segments separated more successfully than could the best police force. Hacking requires a continuous technical education in order to keep skills up to date and razor sharp. This educational need has two consequences: First, it is time-consuming, leaving little or no leisure time for collecting information about specific political causes, building critical perspective, or designating contestational sites. Without such information, hacker politics will continue to be extraordinarily vague. Second, continuous reeducation keeps hackers tied into their own hermetically-sealed classroom. Little interaction occurs with others outside this technocratic subclass. Traditional political activists do not fare any better. Left behind in the dust of history, this political subgroup knows what to do and what to target, but has no effective means to carry out its desires. Political activists, as knowledgeable as they might be about their causes, are too often stuck in assembly meetings debating which monument to dead capital they should strike next. Here are two groups motivated to accomplish similar anti-authoritarian ends, but which cannot seem to find a point of intersection. While the former group lives on-line, the latter group lives in the street, and both are unknowingly being defeated by a communication gap for which neither is responsible. The schism between knowledge and technical skill has to be closed, to eliminate the prejudices held by each side (hacker intolerance for the technologically impaired, and activist intolerance for those who are not politically correct).

The hacker/activist schism is not the only difficulty that keeps the idea of ECD in the realm of science fiction. The problem of how to organize potential alliances is also significant. Leftist activism has traditionally been based on principles of democracy – that is, on a belief in the necessity of inclusion. They believe that with no other bargaining power besides sheer number, the populist mass must be organized so that its collective will can be asserted. The weaknesses of this strategy are rather obvious. The first weakness is the belief in a collective will itself. Since the populist mass is divided by so many sociological variables – race/ethnicity, gender, sexual preference, class, education, occupation, language, etc. – it is readily apparent that viewing "the people" as a monolith of consensus is absurd. What fulfills the needs of one group can be repressive or oppressive to another. Centralized organizations attempting to flex their political muscles through the power of numbers find themselves in a peculiar position: Either the group size is relatively large, but it cannot move en masse, or the group advocates an ideological position useful only to a limited sociological set, thereby shrinking their number. In addition, in order for the most simple organization to exist, there must also be bureaucracy. Bureaucracy requires leadership, hence hierarchy. Leadership structures are generally benevolent in these situations, since the leadership is often based on talent and motivation rather than on ascriptive characteristics, and it fluctuates among the membership; however, bureaucratic structure, regardless of how relentlessly it strains toward justice, still erodes the possibility of community (in its proper sense). Within such an organizational pattern, individuals are forced to trust an impersonal process over which they have no real control.

The use of democratic principles of centralization, when analyzed on a global scale, becomes even more depressing. As yet, no democratic organization exists that comes even remotely close to constructing a multinational resistance. Since power has gone global, avoiding attack is merely a matter of moving operations to a location where resistance is absent. Further, in regard to the condition of pluralism, national interest becomes a variable – a policy that is useful within one national situation becomes repressive or oppressive in another. Collective democratic action may be weakly effective on the local (micro) level, but it becomes next to useless on a macro scale; the complexity of the division of labor prevents consensus, and there is no apparatus through which *to organize*.

The option of realizing hacker fantasies of a new avant-garde, in which a technocratic class of resistors acts on behalf of "the people," seems every bit as suspect, although it is not as fantastic as thinking that the people of the world will unite. A technocratic avant-garde is theoretically possible, since an apparatus is in place for such a development. However, since the technocracy consists overwhelmingly of young white first-world males, one has to wonder just what issues would be addressed. That dreaded question of "who speaks for whom?" looms large whenever the idea of avant-gardism is shuffled about.

The question of resistance then becomes threefold: First, how can the notion of an avant-garde be recombined with notions of pluralism? Second, what are the strategies and tactics needed to fight a decentralized power that is constantly in a state of flux? Finally, how are the units of resistance to be organized? Without question, no certain answers are available, but CAE would like to offer the following proposals. The use of power through number – from labor unions to activist organizations – is bankrupt, because such a strategy requires consensus within the resisting party and the existence of a centralized present enemy. However, in spite of the lack of consensus on what to do, most organizations do share a common goal – that is, resistance to authoritarian power. Yet even in terms of goals there is no consensus about the practical basis of authoritarian power. The perception of authoritarianism shifts depending on the coordinates from which a given sociological group chooses to resist authoritarian discourse and practice. How then can this situation be redefined in constructive terms? An anti-authoritarian predisposition becomes useful only when the idea of the democratic monolith is surrendered. To fight a decentralized power requires the use of a decentralized means. Let each group resist from the coordinates that it perceives to be the most fruitful. This means that leftist political action must reorganize itself in terms of anarchist cells, an arrangement that allows resistance to originate from many different points, instead of focusing on one (perhaps biased) point of attack. Within such a micro structure, individuals can reach a meaningful consensus based on trust in the other individuals (real community) in the cell, rather than one based on trust in a bureaucratic process. Each cell can construct its own identity, and can do so without the loss of individual identity; each individual within the cell maintains at all times a multidimensional persona that cannot be reduced to the sign of a particular practice.

How can a small group (four to ten people) have any type of political effect? This is the most difficult question, but the answer lies in the construction of the cell. The cell must be organic; that is, it must consist of interrelated parts working together to form a whole that is greater than the sum of the parts. To be effective, the schism between knowledge and technical ability in the cell must be closed. A shared political perspective should be the glue that binds the parts, rather than interdependence through need. Avoid consensus through similarity of skills, since in order for the cell to be useful, different skills must be represented. Activist, theorist, artist, hacker, and even a lawyer would be a good combination of talents – knowledge and practice should mix. With the cell in place, ECD is now a viable option, and as explained earlier in the essay, with ECD, demands will at least be recognized. Another advantage is that the cell has the option of pooling financial resources, so the minimal equipment needed for ECD can be purchased. The problem of potential legal fees is an argument for centralization – cells may not have a long lifespan. Admittedly, the toxic illegality of electronic political action is one of the key variables that relegates this narrative to the realm of science fiction.

For more radical cells ECD is only the first step. Electronic violence, such as data hostages and system crashes, are also an option. Are such strategies and tactics a misguided nihilism? CAE thinks not. Since revolution is not a viable option, the negation of negation is the only realistic course of action. After two centuries of revolution and near-revolution, one historical lesson continually appears – authoritarian structure cannot be smashed; it can only be resisted. Every time we have opened our eyes after wandering the shining path of a glorious revolution, we find that the bureaucracy is still standing. We find Coca-Cola gone and Pepsi-Cola in its place – looks different, tastes the same. This is why there is no need to fear that we will one day wake up and find civilization destroyed by mad anarchists. This mythic fiction is one that originates in the security state to instill in the public a fear of effective action.

Do centralized programs still have a role in this resistance? Centralized organizations have three functions. The first is to distribute information. Consciousness raising and spectacle production should be carried out by centralized counter-bureaucracies. Cash and labor pools are needed in order to research, construct, design, and distribute information contrary to the aims of the state. The second function is for recruitment and training. It cannot be emphasized enough that there must be more bases for training technologically literate people. To rely only on the chance that enough people will have the right inclination and aptitude to become technically-literate resisters means that there will be a shortage of resistant technocrats to fill the cellular ranks, and that the sociological base for the technocratic resistance not be broad enough. (If technical education continues to be distributed as it is today, the attack on authority will be horribly skewed in favor of a select group of issues). Finally, centralized organizations can act as consultants on the off chance that an authoritarian institution has decided to reform itself in some way. This can happen in a realistic sense, not because of a corporate-military ideological shift, but because it would be cheaper to reform than to continue the battle. The authoritarian fetish for efficiency is an ally that cannot be underestimated.

All that centralized organizations must do – in a negative sense – is to stay out of direct action. Leave confrontation to the cells. Infiltrating cellular activity is very difficult, unlike infiltrating centralized structures. (This is not to say that cellular activities are difficult to monitor, although the degree of difficulty does rise as more cells proliferate). If the cells are working in double blind activities in a large enough number, and are effective in and of themselves, authority can be challenged. The fundamental strategy for resistance remains the same – appropriate authoritarian means and turn them against themselves. However, for this strategy to take on meaning, resistance – like power – must withdraw from the street. Cyberspace as a location and apparatus for resistance has yet to be realized. Now is the time to bring a new model of resistant practice into action.

ADDENDUM: THE NEW AVANT-GARDE

CAE fears that some of our readers might be getting a bit squeamish about the use of the term "avant-garde" in the above essay. After all, an avalanche of literature from very fine postmodern critics has for the past two decades consistently told us that the avant-garde is dead and has been placed in a suitable resting plot in the Modernist cemetery alongside its siblings, originality and the author. In the case of the avant-garde, however, perhaps a magic elixir exists that can reanimate this corpse. The notion has decayed quite a bit, so one would not expect this zombie to look as it once did, but it may still have a place in the world of the living.

The avant-garde today cannot be the mythic entity it once was. No longer can we believe that artists, revolutionaries, and visionaries are able to step outside of culture to catch a glimpse of the necessities of history as well as the future. Nor would it be realistic to think that a party of individuals of enlightened social consciousness (beyond ideology) has arrived to lead the people into a glorious tomorrow. However, a less appealing (in the utopian sense) form of the avant-garde does exist. To simplify the matter, let us assume that within the present social context, there are individuals who object to various authoritarian institutions, and each has allied h/erself with other individuals based on identification solidarity (race/ethnicity, sexual orientation, class, gender, religion, political beliefs, etc.) to form groups/organizations to combat the mechanisms and institutions that are deemed oppressive, repressive, exploitive, and so on. From a theoretical perspective, each of these alliances has a contestational role to play that should be respected and appreciated; however, in terms of practice, there is no basis to view them all as equals. Unquestionably, some groups will have greater resource power than others; that is, some will have greater access to wealth, prestige, hardware, education, and technical skills. Typically, the greater the resources, the greater the effect the group can have. However, the configuration of access in conjunction with the groups' placement along political, numerical, and spatial/geographic continuums will also greatly alter the effectiveness of the group. (A full catalogue of possibilities cannot be listed within the parameters of this discussion). For example, a large, very visible group that is on the radical fringe, which works to change national policy, and which has reasonably good access to resources will also receive stiff counter resistance from the state, thereby neutralizing its potential power. The rapid destruction of the Black Panther Party by the FBI is an example of this vulnerability. A relatively large liberal group with strong resources that acts locally will receive less counter resistance. (Hence the misguided belief that if everyone acts locally for reform, policy will change globally and peacefully. Unfortunately local action does not affect global or national policies since the sum of local issues does not equal national issues). For example, an alliance of various green groups in North Florida has been very successful at keeping oil companies off the Gulf coast line and protecting the local national forests and preserves from logging companies and land speculators; however, such success is by no means representative of the national or international situation in regard to the Green movement.

Then what kind of group configuration *will* gain the most far-ranging results, in terms of disturbing the political/cultural landscape? This is the question that CAE tried to answer in this essay. To repeat: cellular constructions aimed at information disruption in cyberspace. The problem is access. The education and technical skills needed are not widely distributed, and moreover are monopolized (though not through individual intentionality) by a very specific group (young white men). Education activists should be and in many cases are working as hard as possible to correct this problem of access, even though it does seem almost insurmountable. At the same time, contestational forces cannot wait to act until this access problem is corrected. Only in theory can we live by what ought to be; in practice we must work in terms of what is. Those who are trained and ready now need to start building the model of electronic resistance. Those who are ready and willing to begin to form the models of electronic resistance in the new frontier of cyberspace are the ones CAE views as a new avant-garde.

The technocratic avant-garde offers one slim hope of effective resistance on a national and international scale; and, in its favor, in terms of efficiency, and unlike its Modernist predecessors, the intelligentsia, this group does not have to organize "the people." Much like the problems of resource access, this necessity or desire has always bothered the forces of democracy. Avant-gardism is grounded in the dangerous notion that there exists an elite class possessing enlightened consciousness. The fear that one tyrant will simply be replaced by another is what makes avant-gardism so suspect among egalitarians, who in turn always return to more inclusive local strategies. While CAE does not want to discourage or disparage the many possible configurations of (democratic) resistance, the only groups that will successfully confront power are those that locate the arena of contestation in cyberspace, and hence an elite force seems to be the best possibility. The increased success of local and regional resistant configurations, in part, depends upon the success of the avant-garde in the causal domain of the virtual. As for "enlightened consciousness," CAE believes blind groping is a more accurate description. Avant-gardism is a gamble, and the odds are not good, but at present, it's the only game in town.

ADDENDUM II: A NOTE ON ABSENCE, TERROR, AND NOMADIC RESISTANCE

In *The Electronic Disturbance*, CAE argued that a major change in the representation of power had occurred over the past twenty years. Power once represented itself as a visible sedentary force through various types of spectacle (media, architecture, etc.), but it has instead retreated into cyberspace where it can nomadically wander the globe, always absent to counterforces, always present whenever and wherever opportunity knocks. In "Electronic Civil Disobedience," CAE notes that for every strategy there is a counter-strategy. Since cyberspace is accessible to all of the technocratic class, the resistant within this class can also use nomadic strategies and tactics. Indeed, the primary concern among the military/corporate cyber police (Computer Emergency Response Team, the Secret Service, and the FBI's National Computer Crime Squad) is that nomadic strategy and tactics are being employed at this very moment by contestational groups and individuals (in the words of authority, "criminal" groups). The cyberpolice and their elite masters are living under the sign of virtual catastrophe (that is, anticipating the electronic disaster that *could* happen) in much the same way that the oppressed have lived under the signs of virtual war (the war that we are forever preparing for but never comes) and virtual surveillance (the knowledge that we may be watched by the eye of authority).

The current wave of paranoia began in early 1994 with the discovery of "sniffer" programs. Apparently some adept crackers are collecting passwords for unknown purposes. The reaction of the cyberpolice was predictable: They are convinced that this could only be done for criminal intent. Of prime concern is the development of the tactic of data hostaging, where criminals hold precious research data for ransom. Motivations for such an activity are construed solely as criminal. (This is typical of US policy – criminalize alternative political action, arrest the guilty, and then claim with a clear conscience that the US has no political prisoners). CERT, the FBI, and the SS seem convinced that teen crackers have matured and are evolving past information curiosity into information criminality. But something else of greater interest is beginning to occur. The terror of nomadic power is being exposed. The global elite are having to look into the mirror and see their strategies turned against them – terror reflecting back on itself. The threat is a virtual one. There could be cells of crackers hovering unseen, yet poised for a coordinated attack on the net – not to attack a particular institution, but to attack the net itself (which is to say, the world). A coordinated attack on the routers could bring down the whole electronic power apparatus. The vulnerability of the cyber apparatus is known, and now the sign of virtual catastrophe tortures those who created it. As James C. Settle, founder and head of the FBI's National Computer Crime Squad, has said: "I don't think the stuff we are *seeing* is the stuff we need to be worried about. What that activity we do see is indicative of, however, is that we have a really big problem. . . . Something is cooking but no one really knows what." The motto of the sight machine reverberates out of Settle's rhetoric: "If I can see it, it's already dead." At the same time, the opposite – what Settle calls "the dark side" – is out there, planning and scheming. Nomadic power has created its own nemesis – its own image. This brings up the possibility that as a tactic for exposing the nature of nomadic power, ECD is already outdated without having ever been tried. No real "illegal" action needs to be taken. From the point of view of traditional terrorism, action that can reveal the cruelty of nomadic power need only exist in hyperreality, that is, as activities that merely indicate a possibility of electronic disaster. From this moment forward, strategies of the hyperreal will have to be downgraded into the real, meaning the technocratic class (those with the skill to mount a powerful resistance) will have to *act* on behalf of liberation from electronic control under the nomadic elite. The reason: They are not going to have a choice. Since the individuals in this class are the agents of vulnerability within the realm of cyberspace, repression in this class will be formidable. Since "the dark side" has no image, the police state will have no problem inscribing it with its own paranoid projections, thus doubling the amounts of repression, and pushing the situation into a McCarthyist frenzy. To be sure, each technocrat will be paid well to sell h/er sovereignty, but CAE finds it hard to believe that all will live happily under the microscope of repression and accusation. There will always be a healthy contingent who will want to die free rather than live constrained and controlled in a golden prison.

A second problem for nomadic power, as it finds itself suddenly caught in the predicament of sedentary visibility and geographic space, is that not only could an attack on cyberspace bring about the collapse of the apparatus of power, but the possibility also exists for attacking particular domains. This means that ECD could be used effectively. Even though nomadic power has avoided the possibility of a theater of operations emerging contrary to its needs and goals in physical space, once a resistant group enters cyberspace, elite domains can be found and placed under siege.

Whether or not the barbarian hordes – the true nomads of cyberspace – are ready to sweep through the orderly domains of electronic civilization remains to be seen. (If the hordes do their jobs well, they never *will* be seen. The domains will not report them, as they cannot expose their own insecurity, in much the same way a failing bank will not make its debts public). The hordes do have one advantage: They are without a domain, completely deterritorialized, and invisible. In the realm of the invisible what's real and what's hyperreal? Not even the police state knows for sure.

NOTE

"Electronic Civil Disobedience" was originally written as part of a window installation for the *Anti-work Show* at Printed Matter at Dia in the Spring of 1994. It was then reprinted by Threadwaxing Space in *Crash: Nostalgia for the Absence of Cyberspace*. The version presented here is the original form with only a few modifications. The addendums were written the following summer before the article was presented at the *Terminal Futures* conference at the Institute of Contemporary Art in London.

Two sets of images are included here of work by Slovenian artist Marjetica Potrč, Associate Professor at the Academy of Fine Arts, Ljubljiana and working also in Berlin and the United States, represented by the Max Protetch Gallery in New York. The first five are from a series of large ink-jet prints made in 2001. The next three are installation shots from an exhibition at the Kunstlerhaus Bethanien, Berlin, in 2001 (Potrč also showed some of this work at the Guggenheim Museum, New York in the same year, where she was awarded the Hugo Boss Prize, and at the Max Protetch Gallery). These show sightings of animals such as bears and a coyote – another in the series shows an alligator – in urban and suburban sites from streets to elevators and pools. Such sightings are now common in north America and draw 'wild' creatures from their habitats in search of food. Where the wilderness ends no longer seems an answerable question. The second group of images show material culture's response to the uncanny relation of human settlement and nature encapsulated in the sightings. They take the forms of a pepper spray for use on bears (which comes in small and large sizes, there is always a choice), a container of emergency water, and a survival kit for unauthorised immigrants. These images are stark, suggesting that the survivor mentality previously aligned with stories of post-apocalyptic (nuclear or other) destruction has permeated everyday lives, and that survival leads to border crossings of various kinds – from wilderness to city, from one state to another.

Potrč writes:

> For me, animal sightings prove that city space is constantly negotiated . . . they draw attention to the creativity of border spaces in general. The US/Mexican border, for instance, is a good example. Border cities like Tijuana and San Diego flourish because of the border. As for actual border barriers, no matter how high you build a wall, immigrants will always find a way to cross . . . (Marjetica Potrč, 'The Pursuit of Happiness', in exhibition catalogue, Philip Morris Kunstförderung, Berlin, 2001, p. 23).

She adds that 'I read someplace that by 2100 national states will have yielded to city states. If you can imagine that, . . . the idea of border space will change too. It will become a close proximity experience, all due to survival instincts, and of course, the selfish pursuit of happiness' (*ibid.*).

Plate 16 **'Bear Falling' by Marjetica Potrč** (ink-jet print). (Courtesy of the artist and Max Protetch Gallery.)

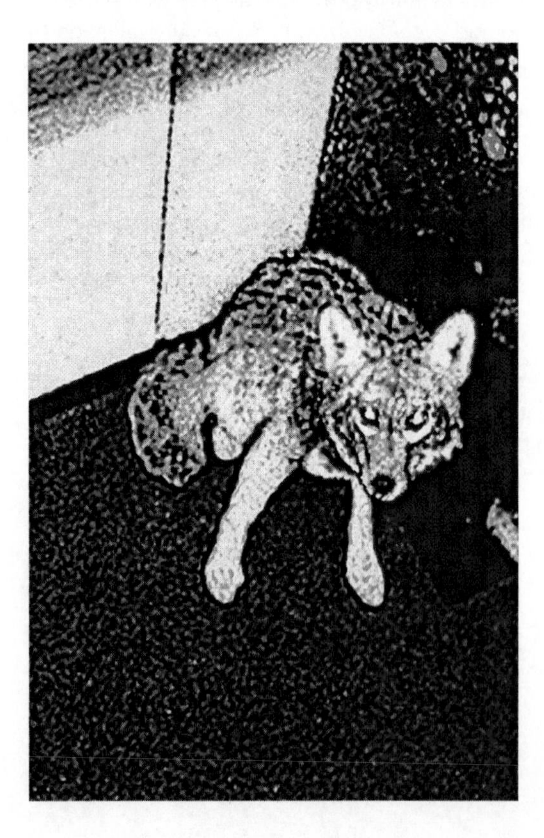

Plate 17 **'Coyote' by Marjetica Potrč** (ink-jet print). (Courtesy of the artist and Max Protetch Gallery.)

Plate 18 **'Pool Bear' by Marjetica Potrč** (ink-jet print). (Courtesy of the artist and Max Protetch Gallery.)

Plate 19 **'Racoon' by Marjetica Potrč** (ink-jet print). (Courtesy of the artist and Max Protetch Gallery.)

Plate 20 'Urban Bear' by Marjetica Potrč (ink-jet print). (Courtesy of the artist and Max Protetch Gallery.)

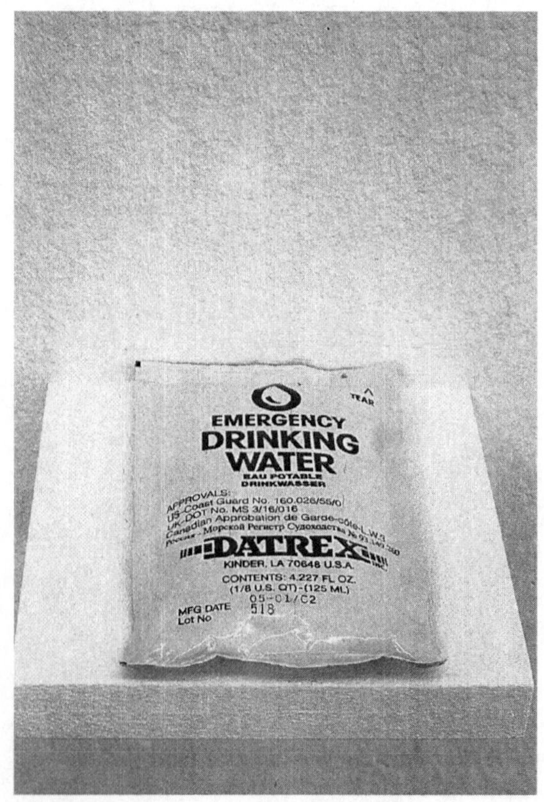

Plate 21 'Emergency Drinking Water' by Marjetica Potrč (installation). (Courtesy of the artist and Max Protetch Gallery.)

Plate 22 **'Survival Kit' by Marjetica Potrč** (installation). (Courtesy of the artist and Max Protetch Gallery.)

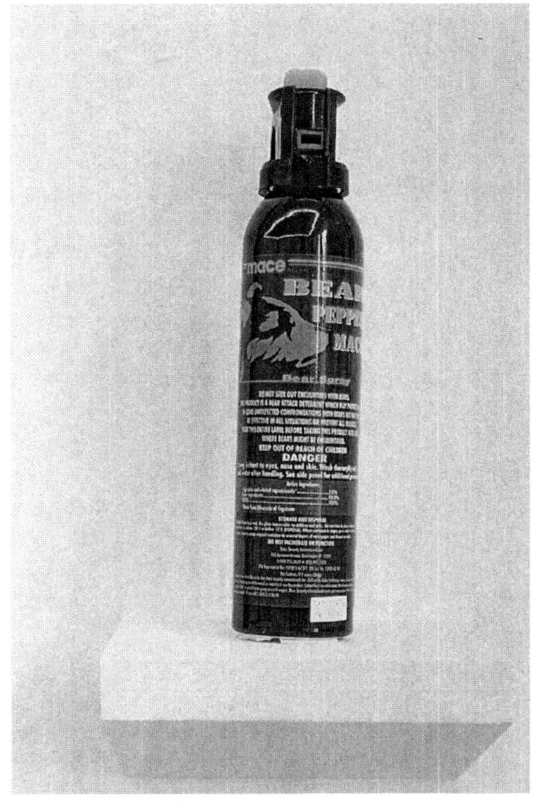

Plate 23 **'Pepper Spray for Bears' by Marjetica Potrč** (installation). (Courtesy of the artist and Max Protetch Gallery.)

PART SIX

Everyday Lives

INTRODUCTION TO PART SIX

The city exists on paper, in plans and drawings, in the minds of architects, planners and the state. The city is spectacular – cover shot architecture, political leaders, famous people, the super-rich, big events, Olympic Games. But it is also mundane, everyday. The city of people, of everyday life, common occurrences, small shops, bus stops, allotments and waste ground is every bit as much 'the city' as the more visible and high profile. More so, as it provides the backdrop and the context within which most of us shape our everyday lives.

The everyday life of cities is often hidden away. It does not feature in the glossy brochures advertising the city to investors and tourists – except occasionally in some sanitised, saleable form ('vibrant, multicultural street life') – and only facsimiles of it find their way into the plans and imaginations of those who shape and manage the city. If you want to find out about the everyday lives of cities, read detective novels, the local newspaper, the notices pasted to lampposts and corrugated iron fences telling that someone's pet cat has gone missing, and the graffiti that obscures these traces of everyday life.

Despite the fact that an ambivalence towards the everyday lives of the city has been shared by cultural commentators, some of the most revealing and most influential writing on the city has concerned aspects of its hidden, everyday lives. From those of the Chicago School sociologies of Park, Burgess and Thrasher, the theories of everyday life of Simmel, Lefebvre, and de Certeau, to the ethnographies of the 1960s and 1970s by Brian Jackson (*Working Class Community*, 1968) or Lee Rainwater (*Behind Ghetto Walls*, 1970), to the writings of Jane Jacobs and William Whyte and beyond. A number of these writing are collected here.

Henri Lefebvre is arguably the pre-eminent and most influential theorist of space of the twentieth century. In the extract included here 'Plan of the present work' from *The Production of Space* (1974) he outlines his central idea of space as socially and historically produced rather than natural. Lefebvre provides a total theory of space, a metaphorical schema that accounts for all kinds of space-production in all cities at all times. Lefebvre argues that space is made up of spatial practices (buildings and actions), representations of space (conscious theories and figures) and representational spaces (imaginations, experiences). The relevance of Lefebvre's ideas to everyday life are that his schema encompass both the production of space through professions such as architecture and planning and its experience and negotiation through the practices of everyday life. Lefebvre's work has had a huge influence on geographically inflected readings of the city since and has been especially important in the theorisation of space by geographers and sociologists in the 1990s. However, despite its undoubted influence and the frequency with which it is cited, doubts remain about the extent to which academics have been able to empirically action Lefebvre's project. Savage and Wards argue, for example, that:

> Lefebvre's laudable project to find a bridge between experienced space, representations of space, and spaces of representation has proved too hard to put into operation empirically. The crucial link between the construction of place in representation and at the level of everyday experience has not been demonstrated (1993:132).

However, this failure to extend and apply Lefebvre's ideas does not diminish their importance to the study of everyday life.

Everyday life is lived within the spaces shaped by the powerful, or left behind by their shaping, but it is not determined by them. A major theoretical contribution to this relationship between the production and experience of space is provided by Michel de Certeau in the extract included here from *The Practice of Everyday Life* (1984). De Certeau recognises 'strategies' by which the powerful produce instrumental or functional space but argues that through the 'tactics' employed in the experience and negotiation of these spaces everyday life is able to subvert or elude the imposition of dominant meanings of space. De Certeau argues that these tactics 'trace out the ruses of other interests and desires that are neither determined nor captured by the systems in which they develop' (1984: xviii). Tim Edensor, in his application of de Certeau's ideas to the culture of the Indian street outlines his contribution:

> His evocation of the plethora of desires that are stimulated through the relationship between sensual bodily movement, fantasy and reverie convincingly refute deterministic notions of pedestrians being shaped mentally and physically by urban space and its control (1998: 219).

De Certeau's ideas have a wider relevance to recent debates within the social sciences concerning the meanings of urban space. Much writing on the meanings of urban space prior to the 1990s was informed by some version of the semiotic perspective in which meanings were investigated through the unpacking of the processes whereby spaces, typically instrumental spaces such as the shopping mall, were produced. However, critics of this approach have highlighted the determinisms inherent and implicit in this approach whereby uses of space was largely assumed to mirror the intentions of the producers of space. Drawing both implicitly and explicitly on the work of de Certeau and Lefebvre, critical social scientists have begun to unearth the performity of meaning and the complex hidden geographies of the everyday use of space.

The theme of the performity is picked up in Iain Borden's discussion of skateboarding in the city in 'A Performative Critique of the City: The Urban Practice of Skateboarding, 1958–1998'. Drawing on Lefebrve, Borden considers skateboarding as one spatial practice through which dominant meanings of the city are subverted, or ignored. As Borden argues:

> Skateboarding here resists the standardisation and repetition of the city as a serial production of building types, functions and discrete objects; it decentres building-objects in time and space in order to recompose them as a strung-out yet newly synchronous arrangement. Thus while many conceive of cities as comprehensive urban plans, monuments or *grands projets*, skateboarding suggests that cities can be thought of as series of micro-spaces. Consequently, architecture is seen to lie beyond the province of the architect and is thrown instead into the turbulent nexus of reproduction (1999: 40).

Or, as the skate magazine *Thrasher* more succinctly puts it: 'Find it. Grind it. Leave it behind'.

Borden's work fits into an emergent body of work that examines this performative aspect of the city and the ways in which spaces are used, experienced and negotiated by their users. Recent examples of the rich geographies of everyday urban life include, in addition to Tim Edensor's discussion of the Indian street already cited, work which has uncovered the multiple meanings of that apparently most sterile urban space the shopping mall (Miller *et al.* 1998) and Loretta Lees' discussion of the new public library building in Vancouver (2001).

Lucy Lippard in 'Home in the Weeds', an extract from her book *The Lure of the Local* (1997), talks of other urban everydays, those of homeless people and residents of neglected neighbourhoods. She talks, not as they are encountered, by wealthy residents or by tourists, but of their everyday worlds and attempts by artists, activists and local residents to empower, articulate and represent their lives. She talks, for example, of the 'photoscalps' taken by homeless residents of the Hill, a camp in New York City. The artists Gabriele Schafer and Nick Fracaro, as part of an installation with the residents of the Hill, provided them with disposable cameras for them to photograph the tourists who normally photographed them. Lippard's project, in this extract, is to make visible the oppressions and histories that lead to these 'invisible' and largely ignored populations, but not in ways that are patronising or racist. She seeks to expose the ways in which the city is talked about and represented, as a battleground and frontier, for example, by property developers and in films such as *Fort Apache the Bronx* (1981). This naturalises the actions of those who 'cleanse' the city, she argues, be they developers selling loft space or the police and civic authorities. Lippard shows how everyday life in the revanchist city is a struggle for space, belonging and identity, but like de Certeau suggests, and those discussed in the extract demonstrate, incarceration and exclusion are not inevitable or natural.

In the final extract, from 'Tango: A Choreography of Urban Displacement', Ana Betancour and Peter Hasdell write of the mythologisation of the everyday in the tango. They trace the origins of the dance to the barrios of Buenos Aires as they were populated by immigrants entering Argentina, to fuel the country's industrial expansion, during the early twentieth century. In their discussion Betancour and Hasdell argue that the unfolding of tango as a cultural form is intertwined with the unfolding of Buenos Aires. They see tango as evidence of the subtle processes of becoming where cities expand and non-places become places. In this narrative, tango becomes a trace of the unfolding of a place, for example, telling through lyrics and dance, stories of place, home and displacement, and a tactic of resistance that allows its practitioners, as Tim Edensor says of the tactics recognised by de Certeau, 'to escape [these] carceral networks' (1998: 219).

> To understand tango, therefore, only in terms of its legitimised form as a national cultural expression and a specific marketable cultural commodity is to profoundly misunderstand its significance as a resistance to the rapidly expanding urban periphery and the marginal existence this had in relation to the legitimate culture capital of the city centre . . . it can be seen as a model for understanding the massive influx of immigrant cultures into peripheral areas of cities and the emergence of cultures within ghettos. It may also point to the subtle transformations of such areas and non-places into places with distinct identities, where dreams and memories – and everyday lives – touch the physical, official world of the city (2000: 149).

And we are back where we started . . .

SUGGESTED FURTHER READING

I. Borden, *Skateboarding, Space and the City*, Oxford, Berg, 2001

Tim Edensor, 'The Culture of the Indian Street', in *Images of the Street: Planning, Identity and Control in Public Space*, edited by N. R. Fyfe, London, Routledge, 1998

Lorretta Lees (2001) 'Towards a Critical Geography of Architecture: The Case of an Ersatz Colosseum', *Ecumene*, 8,1: 51–86

D. Miller, P. Jackson, and N. Thrift, *Shopping, Place and Identity*, London, Routledge, 1998

M. Savage and A. Warde, *Urban Sociology, Capitalism and Modernity*, Basingstoke, Macmillan, 1993

'Plan of the Present Work'

from *The Production of Space* (1991)

Henri Lefebvre

Editors' Introduction

Henri Lefebvre was one of the most important French thinkers of the twentieth century. Lefebvre was a philosopher and sociologist born in the south west of France in 1901 who was concerned with providing an account of alienation in everyday life. In doing this he drew heavily upon and extended Marx's explanation of economic exploitation. His work has had a huge influence on the upsurge in interest in the ways that space is theorised by academics, and was, for example, a central influence on significant publications by such notable academics as Manuel Castells (*La Question Urbane*, 1972), David Harvey (*Social Justice and the City*, 1973 and *The Condition of Postmodernity*, 1989) and Ed Soja (*Postmodern Geographies: The Reassertion of Space in Critical Social Theory*, 1989). Debates about the contribution of Lefebvre and the application of his ideas have hardly gone away since these publications, indeed since his book *The Production of Space* was translated into English in 1991 they have intensified, in the English language academy at least. Lefebvre's attempt to develop a total theory of space, mental, physical and social, as socially constructed and historically conditioned has provided a blueprint within which many geographers and sociologists have operated since. Lefebvre in *The Production of Space* provides perhaps the most well worked through, and certainly most influential, geographical contribution to social theory and analysis.

In addition to *The Production of Space*, Lefebvre's many other influential publications include *Critique of Everyday Life* (1947), which is arguably the founding text of what became cultural studies in the importance it attaches to the meanings and significance of popular culture. Lefebvre's work has also provoked a huge amount of interpretation, appreciation and criticism, one of the best of these is *Lefebvre: Love and Struggle* by Rob Shields (London, Routledge, 1998). Many collections of Lefebvre's writings are available. Two of the most interesting are *Writings on Cities* (Oxford, Blackwell, 1995) and *Henri Lefebvre: Key Writings* (London, Continuum, 2003), edited by Eleonore Kofman, Elizabeth Lebas and Stuart Elden.

XVII

The third implication of our initial hypothesis will take an even greater effort to elaborate on. If space is a product, our knowledge of it must be expected to reproduce and expound the process of production. The 'object' of interest must be expected to shift from *things in space* to the actual *production of space*, but this formulation itself calls for much additional explanation. Both partial products located *in space* – that is, things – and discourse *on space* can henceforth do no more than supply clues to, and testimony about, this productive process – a process which subsumes signifying processes without being reducible to them. It is no longer a matter of the space of this or the space of that: rather, it is space in its totality or global aspect that needs not only to be subjected to analytic scrutiny (a procedure which is liable to furnish merely an infinite series of fragments and cross-sections subordinate to the analytic project), but also to be *engendered* by and within theoretical understanding. Theory *reproduces* the generative process – by means of a concatenation of concepts, to be sure, but in a very strong sense of the word: from within, not just from without (descriptively), and globally – that is, moving continually back and forth between past and present. The historical and its consequences, the 'diachronic', the 'etymology' of locations in the sense of what happened at a particular spot or place and thereby changed it – all of this becomes inscribed in space. The past leaves its traces; time has its own script. Yet this space is always, now and formerly, a *present* space, given as an immediate whole, complete with its associations and connections in their actuality. Thus production process and product present themselves as two inseparable aspects, not as two separable ideas.

It might be objected that at such and such a period, in such and such a society (ancient/slave, medieval/feudal, etc.), the active groups did not 'produce' space in the sense in which a vase, a piece of furniture, a house, or a fruit tree is 'produced'. So how exactly did those groups contrive to produce their space? The question is a highly pertinent one and covers all 'fields' under consideration. Even neocapitalism or 'organized' capitalism, even technocratic planners and programmers, cannot produce a space with a perfectly clear understanding of cause and effect, motive and implication.

Specialists in a number of 'disciplines' might answer or try to answer the question. Ecologists, for example, would very likely take natural ecosystems as a point of departure. They would show how the actions of human groups upset the balance of these systems, and how in most cases, where 'pre-technological' or 'archaeo-technological' societies are concerned, the balance is subsequently restored. They would then examine the development of the relationship between town and country, the perturbing effects of the town, and the possibility or impossibility of a new balance being established. Then, from their point of view, they would adequately have clarified and even explained the genesis of modern social space. Historians, for their part, would doubtless take a different approach, or rather a number of different approaches according to the individual's method or orientation. Those who concern themselves chiefly with events might be inclined to establish a chronology of decisions affecting the relations between cities and their territorial dependencies, or to study the construction of monumental buildings. Others might seek to reconstitute the rise and fall of the institutions which underwrote those monuments. Still others would lean toward an economic study of exchange between city and territory, town and town, state and town, and so on.

To follow this up further, let us return to the three concepts introduced earlier.

1 *Spatial practice*: The spatial practice of a society secretes that society's space; it propounds and presupposes it, in a dialectical interaction; it produces it slowly and surely as it masters and appropriates it. From the analytic standpoint, the spatial practice of a society is revealed through the deciphering of its space.

   What is spatial practice under neocapitalism? It embodies a close association, within perceived space, between daily reality (daily routine) and urban reality (the routes and networks which link up the places set aside for work, 'private' life and leisure). This association is a paradoxical one, because it includes the most extreme separation between the places it links together. The specific spatial competence and performance of every society member can only be evaluated empirically. 'Modern' spatial practice might thus be defined – to take an extreme but significant case – by the daily life of a tenant in a government-subsidized high-rise housing project. Which should not be taken to mean that motorways or the politics of air transport can be left out of the picture. A spatial practice must have a certain cohesiveness, but this does not imply that it is coherent (in the sense of intellectually worked out or logically conceived).
2 *Representations of space*: conceptualized space, the space of scientists, planners, urbanists, technocratic subdividers and social engineers, as of a certain type of artist with a scientific bent – all of whom identify what is lived and what is perceived with what is conceived. (Arcane speculation about Numbers, with its talk of the golden number, moduli and 'canons', tends to perpetuate this view of matters.) This is the dominant space in any society (or mode of production). Conceptions of space tend, with certain exceptions to which I shall return, towards a system of verbal (and therefore intellectually worked out) signs.
3 *Representational spaces*: space as directly *lived* through its associated images and symbols, and

hence the space of 'inhabitants' and 'users', but also of some artists and perhaps of those, such as a few writers and philosophers, who *describe* and aspire to do no more than describe. This is the dominated – and hence passively experienced – space which the imagination seeks to change and appropriate. It overlays physical space, making symbolic use of its objects. Thus representational spaces may be said, though again with certain exceptions, to tend towards more or less coherent systems of non-verbal symbols and signs.

The (relative) autonomy achieved by space *qua* 'reality' during a long process which has occurred especially under capitalism or neocapitalism has brought new contradictions into play. The contradictions within space itself will be explored later. For the moment I merely wish to point up the dialectical relationship which exists within the triad of the perceived, the conceived, and the lived.

A triad: that is, three elements and not two. Relations with two elements boil down to oppositions, contrasts or antagonisms. They are defined by significant effects: echoes, repercussions, mirror effects. Philosophy has found it very difficult to get beyond such dualisms as subject and object, Descartes's *res cogitans* and *res extensa*, and the Ego and non-Ego of the Kantians, post-Kantians and neo-Kantians. 'Binary' theories of this sort no longer have anything whatsoever in common with the Manichaean conception of a bitter struggle between two cosmic principles; their dualism is entirely mental, and strips everything which makes for living activity from life, thought and society (i.e. from the physical, mental and social, as from the lived, perceived and conceived). After the titanic effects of Hegel and Marx to free it from this straitjacket, philosophy reverted to supposedly 'relevant' dualities, drawing with it – or perhaps being drawn by – several specialized sciences, and proceeding, in the name of transparency, to define intelligibility in terms of opposites and systems of opposites. Such a system can have neither materiality nor loose ends: it is a 'perfect' system whose rationality is supposed, when subjected to mental scrutiny, to be self-evident. This paradigm apparently has the magic power to turn obscurity into transparency and to move the 'object' out of the shadows into the light merely by articulating it. In short, it has the power to *decrypt*. Thus knowledge (*savoir*), with a remarkable absence of consciousness, put itself in thrall to power, suppressing all resistance, all obscurity, in its very being.

In seeking to understand the three moments of social space, it may help to consider the *body*. All the more so inasmuch as the relationship to space of a 'subject' who is a member of a group or society implies his relationship to his own body and vice versa. Considered overall, social practice presupposes the use of the body: the use of the hands, members and sensory organs, and the gestures of work as of activity unrelated to work. This is the realm of the *perceived* (the practical basis of the perception of the outside world, to put it in psychology's terms). As for *representations of the body*, they derive from accumulated scientific knowledge, disseminated with an admixture of ideology: from knowledge of anatomy, of physiology, of sickness and its cure, and of the body's relations with nature and with its surroundings or 'milieu'. Bodily *lived* experience, for its part, maybe both highly complex and quite peculiar, because 'culture' intervenes here, with its illusory immediacy, via symbolisms and via the long Judaeo-Christian tradition, certain aspects of which are uncovered by psychoanalysis. The 'heart' as *lived* is strangely different from the heart as *thought* and *perceived*. The same holds *a fortiori* for the sexual organs. Localizations can absolutely not be taken for granted where the lived experience of the body is concerned: under the pressure of morality, it is even possible to achieve the strange result of a body without organs – a body chastised, as it were, to the point of being castrated.

The perceived – conceived – lived triad (in spatial terms: spatial practice, representations of space, representational spaces) loses all force if it is treated as an abstract 'model'. If it cannot grasp the concrete (as distinct from the 'immediate'), then its import is severely limited, amounting to no more than that of one ideological mediation among others.

That the lived, conceived and perceived realms should be interconnected, so that the 'subject', the individual member of a given social group, may move from one to another without confusion – so much is a logical necessity. Whether they constitute a coherent whole is another matter. They probably do so only in favourable circumstances, when a common language, a consensus and a code can be established. It is reasonable to assume that the Western town, from the Italian Renaissance to the nineteenth century, was fortunate enough to enjoy such auspicious conditions. During this period the representation of space tended to dominate and subordinate a representational space, of religious origin, which was now reduced to symbolic figures, to images of Heaven and Hell, of the Devil and the angels, and so on. Tuscan painters, architects and theorists developed a representation of space – perspective – on the basis of a social practice which was itself, as we shall see, the result of a historic change in the relationship between town and country. Common sense meanwhile, though more or less reduced to silence, was still preserving virtually intact a representational space, inherited from the Etruscans, which had survived all the centuries of Roman and Christian dominance. The vanishing line, the vanishing point and the meeting of parallel lines 'at infinity' were the determinants of a representation, at once intellectual and visual, which promoted the primacy of the gaze in a kind of 'logic of visualization'. This representation, which had been in the making for centuries, now became enshrined in architectural and urbanistic practice as the *code* of linear perspective.

For the present investigation to be brought to a satisfactory conclusion, for the theory I am proposing to be confirmed as far as is possible, the distinctions drawn above would have to be generalized in their application to cover all societies, all periods, all 'modes of production'. That is too tall an order for now, however, and I shall at this point merely advance a number of preliminary arguments. I would argue, for example, that representations of space are shot through with a knowledge (*savoir*) – i.e. a mixture of understanding (*connaissance*) and ideology – which is always relative and in the process of change. Such representations are thus objective, though subject to revision. Are they then true or false? The question does not always have a clear meaning: what does it mean, for example, to ask whether perspective is true or false? Representations of space are certainly abstract, but they also play a part in social and political practice: established relations between objects and people in represented space are subordinate to a logic which will sooner or later break them up because of their lack of consistency. Representational spaces, on the other hand, need obey no rules of consistency or cohesiveness. Redolent with imaginary and symbolic elements, they have their source in history – in the history of a people as well as in the history of each individual belonging to that people. Ethnologists, anthropologists and psychoanalysts are students of such representational spaces, whether they are aware of it or not, but they nearly always forget to set them alongside those representations of space which coexist, concord or interfere with them; they even more frequently ignore social practice. By contrast, these experts have no difficulty discerning those aspects of representational spaces which interest them: childhood memories, dreams, or uterine images and symbols (holes, passages, labyrinths). Representational space is alive: it speaks. It has an affective kernel or centre: Ego, bed, bedroom, dwelling, house; or: square, church, graveyard. It embraces the loci of passion, of action and of lived situations, and thus immediately implies time. Consequently it may be qualified in various ways: it may be directional, situational or relational, because it is essentially qualitative, fluid and dynamic.

If this distinction were generally applied, we should have to look at history itself in a new light. We should have to study not only the history of space, but also the history of representations, along with that of their relationships – with each other, with practice, and with ideology. History would have to take in not only the genesis of these spaces but also, and especially, their interconnections, distortions, displacements, mutual interactions, and their links with the spatial practice of the particular society or mode of production under consideration.

We may be sure that representations of space have a practical impact, that they intervene in and modify spatial *textures* which are informed by effective knowledge and ideology. Representations of space must therefore have a substantial role and a specific influence in the production of space. Their intervention occurs by way of construction – in other words, by way of architecture, conceived of not as the building of a particular structure, palace or monument, but rather as a project embedded in a spatial context and a texture which call for 'representations' that will not vanish into the symbolic or imaginary realms.

By contrast, the only products of representational spaces are symbolic works. These are often unique; sometimes they set in train 'aesthetic' trends and, after a time, having provoked a series of manifestations and incursions into the imaginary, run out of steam.

This distinction must, however, be handled with considerable caution. For one thing, there is a danger of its introducing divisions and so defeating the object of the exercise, which is to rediscover the unity of the productive process. Furthermore, it is not at all clear *a priori* that it can legitimately be generalized. Whether the East, specifically China, has experienced a contrast between representations of space and representational spaces is doubtful in the extreme. It is indeed quite possible that the Chinese characters combine two functions in an inextricable way, that on the one hand they convey the order of the world (space–time), while on the other hand they lay hold of that concrete (practical and social) space–time wherein symbolisms hold sway, where works of art are created, and where buildings, palaces and temples are built. I shall return to this question later – although, lacking adequate knowledge of the Orient, I shall offer no definite answer to it. On the other hand, apropos of the West, and of Western practice from ancient Greece and Rome onwards, I shall be seeking to show the development of this distinction, its import and meaning. Not, be it said right away, that the distinction has necessarily remained unchanged in the West right up until the modern period, or that there have never been role reversals (representational spaces becoming responsible for productive activity, for example).

There have been societies – the Chavin of the Peruvian Andes are a case in point[1] – whose representation of space is attested to by the plans of their temples and palaces, while their representational space appears in their art works, writing-systems, fabrics, and so on. What would be the relationship between two such aspects of a particular period? A problem confronting us here is that we are endeavouring with conceptual means to reconstruct a connection which originally in no way resembled the application of a pre-existing knowledge to 'reality'. Things become very difficult for us in that symbols which we can readily conceive and intuit are inaccessible as such to our abstract knowledge – a knowledge that is bodiless and timeless, sophisticated and efficacious, yet 'unrealistic' with respect to certain 'realities'. The question is what intervenes, what occupies the interstices between representations of space and representational spaces. A culture, perhaps? Certainly – but the word has less content than it seems to have. The work of artistic creation? No doubt – but that leaves unanswered the queries 'By whom?' and 'How?' Imagination? Perhaps – but why? and for whom?

The distinction would be even more useful if it could be shown that today's theoreticians and practitioners worked either for one side of it or the other, some developing representational spaces and the remainder working out representations of space. It is arguable, for instance, that Frank Lloyd Wright endorsed a communitarian representational space deriving from a biblical and Protestant tradition, whereas Le Corbusier was working towards a technicist, scientific and intellectualized representation of space.

Perhaps we shall have to go further, and conclude that the producers of space have always acted in accordance with a representation, while the 'users' passively experienced whatever was imposed upon them inasmuch as it was more or less thoroughly inserted into, or justified by, their representational space. How such manipulation might occur is a matter for our analysis to determine. If architects (and urban planners) do indeed have a representation of space, whence does it derive? Whose interests are served when it becomes 'operational'? As to whether or not 'inhabitants' possess a representational space, if we arrive at an affirmative answer, we shall be well on the way to dispelling a curious misunderstanding (which is not to say that this misunderstanding will disappear in social and political practice).

The fact is that the long-obsolescent notion of ideology is now truly on its last legs, even if critical theory still holds it to be necessary. At no time has this concept been clear. It has been much abused by evocations of Marxist, bourgeois, proletarian, revolutionary or socialist ideology; and by incongruous distinctions between ideology in general and specific ideologies, between 'ideological apparatuses' and institutions of knowledge, and so forth.

What is an ideology without a space to which it refers, a space which it describes, whose vocabulary and links it makes use of, and whose code it embodies? What would remain of a religious ideology – the Judaeo-Christian one, say – if it were not based on places and their names: church, confessional, altar, sanctuary, tabernacle? What would remain of the Church if there were no churches? The Christian ideology, carrier of a recognizable if disregarded Judaism (God the Father, etc.), has created the spaces which guarantee that it endures. More generally speaking, what we call ideology only achieves consistency by intervening in social space and in its production, and by thus taking on body therein. Ideology *per se* might well be said to consist primarily in a discourse upon social space.

According to a well-known formulation of Marx's, knowledge (*connaissance*) becomes a productive force immediately, and no longer through any mediation, as soon as the capitalist mode of production takes over.[2] If so, a definite change in the relationship between ideology and knowledge must occur: knowledge must replace ideology. Ideology, to the extent that it remains distinct from knowledge, is characterized by rhetoric, by metalanguage, hence by verbiage and lucubration (and no longer by philosophico-metaphysical systematizing, by 'culture' and 'values'). Ideology and logic may even become indistinguishable – at least to the extent that a stubborn demand for coherence and cohesion manages to erase countervailing factors proceeding either from above (information and knowledge [*savoir*]) or from below (the space of daily life).

Representations of space have at times combined ideology and knowledge within a (social-spatial) practice. Classical perspective is the perfect illustration of this. The space of today's planners, whose system of localization assigns an exact spot to each activity, is another case in point.

The area where ideology and knowledge are barely distinguishable is subsumed under the broader notion of *representation*, which thus supplants the concept of ideology and becomes a serviceable (operational) tool for the analysis of spaces, as of those societies which have given rise to them and recognized themselves in them.

In the Middle Ages, spatial practice embraced not only the network of local roads close to peasant communities, monasteries and castles, but also the main roads between towns and the great pilgrims' and crusaders' ways. As for representations of space, these were borrowed from Aristotelian and Ptolemaic conceptions, as modified by Christianity: the Earth, the underground 'world', and the luminous Cosmos, Heaven of the just and of the angels, inhabited by God the Father, God the Son, and God the Holy Ghost. A fixed sphere within a finite space, diametrically bisected by the surface of the Earth; below this surface, the fires of Hell; above it, in the upper half of the sphere, the Firmament – a cupola bearing the fixed stars and the circling planets – and a space criss-crossed by divine messages and messengers and filled by the radiant Glory of the Trinity. Such is the conception of space found in Thomas Aquinas and in the *Divine Comedy*. Representational spaces, for their part, determined the foci of a vicinity: the village church, graveyard, hall and fields, or the square and the belfry. Such spaces were interpretations, sometimes marvellously successful ones, of cosmological representations. Thus the road to Santiago de Compostela was the equivalent, on the earth's surface, of the Way that led from Cancer to Capricorn on the vault of the heavens, a route otherwise known as the Milky Way – a trail of divine sperm where souls are born before following its downward trajectory and falling to earth, there to seek as best they may the path of redemption – namely, the pilgrimage that will bring them to Compostela ('the field of stars'). The body too, unsurprisingly, had a role in the interplay between representations relating to space. 'Taurus rules over the neck', wrote Albertus Magnus, 'Gemini over the shoulders; Cancer over the hands and arms; Leo over the breast, the heart and the diaphragm; Virgo over the stomach; Libra takes care of the second part of the back; Scorpio is responsible for those parts that belong to lust. . . .'

It is reasonable to assume that spatial practice, representations of space and representational spaces contribute in different ways to the production of space according to their qualities and attributes, according to the society or mode of production in question, and according to the historical period. Relations between the three moments of the perceived, the conceived and the lived are never either simple or stable, nor are they 'positive' in the sense in which this term might be opposed to 'negative', to the indecipherable, the unsaid, the prohibited, or the unconscious. Are these moments and their interconnections in fact conscious? Yes – but at the same time they are disregarded or misconstrued. Can they be described as 'unconscious'? Yes again, because they are generally unknown, and because analysis is able – though not always without error – to rescue them from obscurity. The fact is, however, that these relationships have always had to be given utterance, which is not the same thing as being known – even 'unconsciously'.

NOTES

1 See François Hébert-Stevens, *L'art de l'Amérique du Sud* (Paris: Arthaud, 1973), pp. 55ff. For a sense of medieval space – both the representation of space and representational space – see *Le Grand et le Petit Albert* (Paris: Albin Michel, 1972), particularly 'Le traité des influences astrales'. Another edn: *Le Grand et le Petit Albert: les secrets de la magie* (Paris: Belfond, 1972).

2 Karl Marx, *Grundrisse*, tr. Martin Nicolaus (Harmondsworth, Middx: Penguin, 1973).

'Railway Navigation and Incarceration'

from *The Practice of Everyday Life* (1984)

Michel de Certeau

Editors' Introduction

Michel de Certeau is regarded as one of the most influential and significant theorists of everyday life. His work has become increasingly widely read and influential since the translation into English of his most famous text *The Practice of Everyday Life* in 1984. Although often described as a historian and cultural theorist, much of de Certeau's writing defies easy classification given the enormous range of disciplines, from history, sociology, philosophy, linguistics, semiotics, to anthropology and psychoanalysis that he draws upon in his writing.

Much of de Certeau's most influential work was concerned with the distinction between the strategies employed by the powerful in the reproduction of carceral spaces, instrumental and functional space, for example, and the tactics of pedestrians through which, albeit temporally in some cases, these constructions of space are subverted or escaped from. This work has both explicitly and implicitly fed into an increasingly powerful critique of the semiotic approach to the unpacking of meanings of urban spaces. Nigel Thrift, for example, has called for social scientists to adopt non-representational theory and to consider the ways in which meanings of space are performative and contingent, an approach that is derived in part from the ground-breaking work of de Certeau in *The Practice of Everyday Life*.

Two excellent guides to the life and work of Michel de Certeau are Jeremy Aherne's *Michel de Certeau: Interpretation and Its Other* (Cambridge, Polity Press, 1995) and Ian Buchanan's *Michel de Certeau: Cultural Theorist* (London, Sage, 2000).

A travelling incarceration. Immobile inside the train, seeing immobile things slip by. What is happening? Nothing is moving inside or outside the train.

The unchanging traveller is pigeonholed, numbered, and regulated in the grid of the railway car, which is a perfect actualization of the rational utopia. Control and food move from pigeonhole to pigeonhole: "Tickets, please . . ." "Sandwiches? Beer? Coffee? . . ." Only the restrooms offer an escape from the closed system. They are a lovers' phantasm, a way out for the ill, an escapade for children ("Wee-wee!") – a little space of irrationality, like love affairs and sewers in the *Utopias* of earlier times. Except for this lapse given over to excesses, everything has its place in a gridwork. Only a rationalized cell travels. A bubble of panoptic and classifying power, a module of imprisonment that makes possible the production of an order, a closed and autonomous insularity – that is what can traverse space and make itself independent of local roots.

Inside, there is the immobility of an order. Here rest and dreams reign supreme. There is nothing to do, one is in the *state* of reason. Everything is in its place, as in Hegel's *Philosophy of Right*. Every being is placed there like a piece of printer's type on a page arranged in military order. This order, an organizational system, the quietude of a certain reason, is the condition of both a railway car's and a text's movement from one place to another.

Outside, there is another immobility, that of things, towering mountains, stretches of green field and forest, arrested villages, colonnades of buildings, black urban silhouettes against the pink evening sky, the twinkling of nocturnal lights on a sea that precedes or succeeds our histories. The train generalizes Dürer's *Melancholia*, a speculative experience of the world: being outside of these things that stay there, detached and absolute, that leave us without having anything to do with this departure themselves; being deprived of them, surprised by their ephemeral and quiet strangeness. Astonishment in abandonment. However, these things do not move. They have only the movement that is brought about from moment to moment by changes in perspective among their bulky figures. They have only *trompe-l'oeil* movements. They do not change their place any more than I do; vision alone continually undoes and remakes the relationships between these fixed elements.

Between the immobility of the inside and that of the outside a certain *quid pro quo* is introduced, a slender blade that inverts their stability. The chiasm is produced by the windowpane and the rail. These are two themes found in Jules Verne, the Victor Hugo of travel literature: the porthole of the *Nautilus*, a transparent caesura between the fluctuating feelings of the observer and the moving about of an oceanic reality; the iron rail whose straight line cuts through space and transforms the serene identities of the soil into the speed with which they slip away into the distance. The windowpane is what allows us to *see*, and the rail, what allows us to *move through*. These are two complementary modes of separation. The first creates the spectator's distance: You shall not touch; the more you see, the less you hold – a dispossession of the hand in favor of a greater trajectory for the eye. The second inscribes, indefinitely, the injunction to pass on; it is its order written in a single but endless line: go, leave, this is not your country, and neither is that – an imperative of separation which obliges one to pay for an abstract ocular domination of space by leaving behind any proper place, by losing one's footing.

The windowglass and the iron (rail) line divide, on the one hand, the traveller's (the putative narrator's) interiority and, on the other, the power of being, constituted as an object without discourse, the strength of an exterior silence. But paradoxically it is the silence of these things put at a distance, behind the windowpane, which, from a great distance, makes our memories speak or draws out of the shadows the dreams of our secrets. The isolation of the voting booth produces thoughts as well as separations. Glass and iron produce speculative thinkers or gnostics. This cutting-off is necessary for the birth, outside of these things but not without them, of unknown landscapes and the strange fables of our private stories.

Only the partition makes noise. As it moves forward and creates two inverted silences, it taps out a rhythm, it whistles or moans. There is a beating of the rails, a vibrato of the windowpanes – a sort of rubbing together of spaces at the vanishing points of their frontier. These junctions have no place. They indicate themselves by passing cries and momentary noises. These frontiers are illegible; they can only be heard as a single stream of sounds, so continuous is the tearing off that annihilates the points through which it passes.

These sounds also indicate, however, as do their results, the Principle responsible for all the action taken away from both travellers and nature: the machine. As invisible as all theatrical machinery, the locomotive organizes from afar all the echoes of its work. Even if it is discreet and indirect, its orchestra indicates what makes history, and, like a rumor, guarantees that there is still some history. There is also an accidental element in it. Jolts, brakings, surprises arise from this motor of the system. This residue of events depend on an invisible and single actor, recognizable only by the regularity of the rumbling or by the sudden miracles that disturb the order. The machine is the *primum mobile*, the solitary god from which all the action proceeds. It not only divides spectators and beings, but also connects them; it is a mobile symbol between them, a tireless shifter, producing changes in the relationships between immobile elements.

There is something at once incarcerational and navigational about railroad travel; like Jules Verne's ships and submarines, it combines dreams with technology. The "speculative" returns, located in the very heart of the mechanical order. Contraries coincide for the duration of a journey. A strange moment in which a society fabricates spectators and transgressors of spaces, with saints and blessed souls placed in the halos-holes (*auréoles-alvéoles*) of its railway cars. In these places of laziness and thoughtfulness, paradisiacal ships sailing between two social meeting-points (business deals and families, drab, almost imperceptible violences), atopical liturgies are pronounced, parentheses of prayers to no one (to whom are all these travelling dreams addressed?). Assemblies no longer obey hierarchies of dogmatic orders; they are organized by the gridwork of technocratic discipline, a mute rationalization of laissez-faire individualism.

To get in, as always, there was a price to be paid. The historical threshold of beatitude: history exists where there is a price to be paid. Repose can be obtained only through payment of this tax. In any case the blessed in trains are humble, compared to those in airplanes, to whom it is granted, for a few dollars more, a position that is more abstract (a cleaning-up of the countryside and filmed simulacra of the world) and more perfect (statues sitting in an aerial museum), but enjoying an excess that is penalized by a diminution of the ("melancholy") pleasure of seeing what one is separated from.

And, also as always, one has to get out: there are only lost paradises. Is the terminal the end of an illusion? There is another threshold, composed of momentary bewilderments in the airlock constituted by the train station. History begins again, feverishly, enveloping the motionless framework of the wagon: the blows of his hammer make the inspector aware of cracks in the wheels, the porter lifts the bags, the conductors move back and forth. Visored caps and uniforms restore the network of an order of work within the mass of people, while the wave of travellers/dreamers flows into the net composed of marvellously expectant or preventively justiciary faces. Angry cries. Calls. Joys. In the mobile world of the train station, the immobile machine suddenly seems monumental and almost incongruous in its mute, idol-like inertia, a sort of god undone.

Everyone goes back to work at the place he has been given, in the office or the workshop. The incarceration-vacation is over. For the beautiful abstraction of the prison are substituted the compromises, opacities and dependencies of a workplace. Hand-to-hand combat begins again with a reality that dislodges the spectator without rails or windowpanes. There comes to an end the Robinson Crusoe adventure of the travelling noble soul that could believe itself *intact* because it was surrounded by glass and iron.

'Home in the Weeds'

from *Lure of the Local* (1997)

Lucy Lippard

Editors' Introduction

Lucy Lippard is an activist and art critic who has written for *Art in America*, *Z Magazine* and *The Village Voice* amongst many other publications such as newspapers, periodicals and exhibition catalogues. In addition she has founded the journal *Upfront* and produced more than fifteen books on a wide range of art subjects. She has been particularly interested in uncovering the social and political contexts within which art is produced. Lippard was a key commentator on conceptual art in the late 1960s, and an influential early feminist art critic who, along with writers such as Suzi Gablik and Suzanne Lacy, introduced the concerns of women artists into art history and who wrote extensively about, and advocated, art for social change. She was also a critic of the failures of modernist art, arguing that it was a masculinist genre that had failed to provide any means for social or political emancipation or improvement.

Lippard's more recent work has focused on various aspects of the relationship between place, power and identity. These themes are explored in the extract included here, 'Home in the Weeds' from the *Lure of the Local* (1997) and her 1999 book *On the Beaten Track: Tourism Art and Place*. In the *Lure of the Local* she considers the notions of place and the local and their relationship to identity. In *On the Beaten Track* she traces the power relations and erasures implicit in the tourist gaze and its associated practices.

> I am not the garbage, the booze, the guns, the dirt,
> I am the song, the baptism, the wedding.
> I am the newborn child that grows among rank weeds.
> (Miguel Algarin)

Deprived of "knowing their place," the homeless are adamantly "out of place," confusing the boundaries maintained by those who think they know their places. Their poverty forced into public view, the homeless remind everyone of the hypocrisy and greed that underlies a city's structures. Many are working people with families, brought down by a single stroke of bad luck – eviction, fire, job loss, illness, the kinds of things that can happen to anyone. Some have AIDS; some are addicts; some are mentally ill some suffer a whole litany of afflictions for which society takes less and less responsibility. Even in pathos, even when their anger is dulled and they seem almost saintly in their endurance, the homeless threaten a society that brags it's the best in the world.

Homelessness runs counter to the wider population's illusions of what their public spaces are or should be. Yet the public domain is a place where no one has or takes responsibility, so the homeless are people for whom no one is responsible. One of the best overviews of the city from viewpoints sympathetic to the homeless is a book based on an art exhibition organized by artist Martha Rosler: *If You Lived Here: The City in Art, Theory, and Social Activism*. Rosler contends that art functions socially "to crystallize an image or a response to a blurred social picture, bringing its outlines into focus." Yet for the most part, art about the

"I've always told people who come here, 'It's a wonderful place to live, it's a horrible place to make a living.' You've got to know that your lifestyle's going to include getting along on what you get." – Ed Walden, great great great grandson of the founder of Greenville, Maine.

Maine enjoyed, or suffered, a population surge in the 1970s and 1980s, with southern towns doubling and tripling in size. Younger and better educated newcomers caused property values to escalate and a 'circuit breaker bill' was enacted to offset higher taxes based on ability to pay. Affluence and growth in mid-coast Maine had little affect on more isolated rural areas. The gap between rich and poor is widening. For most residents the changes came too fast, and voters mandated a comprehensive planning bill that encouraged communities to think about where they were going.

In Maine, as across rural America, "low road" economics, based on cheap labor and cheap natural resources, is nearing a dead end. Global competition, branch plant mobility, and technological change are making old economics obsolete. Despite state subsidies, industry is being clobbered by imports. Maine lost 45,0000 jobs between 1989 and 1994. Around 1990 the number of service jobs began to surpass manufacturing jobs for the first time. Some predict that the recession will last another twenty years.

Over New Years of 1990, Natasha Mayers, with some 30 members of the Union of Maine Visual Artists, organized "Artists for the Homeless" to make art in 20 store windows on Portland's Congress Street. She thought support for the homeless (some made window works of their own) would be relatively uncontroversial, but five works (including her own) were censored by landlords. One "expected a nativity scene"; she had in fact made one, based on the line "You never know who you might be turning away . . ." Participating artist Abby Shahn observed, "It's impossible to raise these issues without being controversial. The distribution of wealth along Congress Street is glaring. The contradictions are very powerful." Mayers added, "The act of censorship doesn't get rid of an image, it draws attention to it." Homelessness continues to rise in Maine, thanks to the lack of affordable housing, lack of support for the mentally ill, domestic abuse and violence, government cuts, and tough economic conditions. In the first half of 1995, it increased by more than 10 percent. The Tedford Shelter in Brunswick often has no beds. Gary Lawless, a respected radical poet and bookstore owner who teaches writing to the homeless, says, "The greatest gift you can give someone is to listen to their stories . . . Why should [their talent] be a surprise? Because you have one leg or no paycheck, does that mean you can't write something beautiful?"

From around 1870 to 1920 a wave of Acadians migrated from Quebec to Maine, often in covered wagons, becoming "Franco Americans" in the process. "They spoke only French, worshipped exclusively in the Roman Catholic faith, held politically conservative values, and did not assimilate in Yankee society until after World War II." When dance critic June Vail came to live in Brunswick in 1970, she recalls hearing "French every day in supermarket aisles, shops and restaurants. Today I only rarely pick up those distinctively nasal tones." Nevertheless, Franco Americans are roughly 25% of Maine's population; their cultural centers are Fort Kent and the milltowns of Biddeford and Lewiston, which still hold annual French festivals and once had French newspapers.

In 1971 the University of Maine opened a Franco American Center and later a Franco American student newspaper. In 1993 both houses of the state legislature were headed by Franco Americans.

Rhea J. Côté Robbins, who edited *Le Forum* at the University of Maine for ten years, is researching Franco American women and has begun The Franco American Women's Initiative to publish material on Quebecois and Acadian women. She decided to initiate dialogues around Franco American women's issues because she herself was looking for "a net that doesn't let the Franco American woman's soul fall through." "I am from Waterville," she writes, "and my *maman* was from Wallagrass. In Franco American lore, that's North and South." She writes about eating *tourtières* (pork pies) and *creton*, and dancing jigs, about birthing practices and knitting patterns, the arts and the church, about losing the language, and ties to France. Artist Celeste Roberge has subtly addressed her Franco American identity in her art, recalling her education by nuns from Quebec: "It made me comfortable depicting surreal things. . . . If you remove Catholicism from Franco Americanism, you don't have Franco Americanism."

Georgetown has a few French names on its mailboxes; most of these families have been in the area for generations and have assimilated, though most remain Catholics. The 300-year alliance between French and Indians was recalled in the 1995 ordination procession of Reverend Michael R. Cote, Maine's newest Catholic bishop, of French descent. It was led by a Passamaquoddy and two Penobscot men, representing "the first Catholic Mainers."

Despite its demographics, Maine is not free of racism, which is always encouraged by economic recession. The summer of 1996 was marked by several incidents of vandalism, insults, spray-painted racial slurs combatted by anti-racism rallies and much handwringing. Columnist Nancy Grape pointed out that not one speech by the 31 candidates for Congress in 1995 mentioned racial inequality: "Apparently it's not our problem, despite demographic changes and tourists, college students and residents who expect to be treated with respect. . . ."

The Ku Klux Klan was briefly powerful in Maine in the 1920s, as a backlash against mostly Catholic immigrant labor, although it was also aimed at African Americans and Jews. Mississippi editor Hodding Carter came to study at Bowdoin College in 1923, just in time for the first KKK parade in New England; he saw his first burning cross in Camden, Maine.

An increasing immigrant population challenges Maine's monoculturalism. The local paper recently showed the top winners of the New England Bluefish Open on the Kennebec receiving their prizes in Bath's Waterfront Park. They were all from Freeport, and they were named Cuong Ly, Cuong Tang, and Hong San.

homeless suffers from contextual oversimplifications and lack of understanding of the specific audiences and geographic subjectivities that mediate our relationships to places.

James Lardner elucidated one specific context in a *New Yorker* profile of "the Hill" – a homeless encampment under the Canal Street end of the Manhattan Bridge. In 1909 the site boasted a grandiose neoclassical arch and a pretty little park with a fountain. The seventy-five or so intervening years have been lost in obscurity, since "historians have yet to give to the neglect and deterioration of the city's great monuments of architecture and engineering anything like the attention lavished on their design and construction." By the late eighties, the place had become a weed-strewn vacant lot – a refuge for those who eschewed cardboard boxes for more permanent jerry-built shacks, some of them wildly innovative in their construction. One, built by Juan Samuel Ramirez, adopted the frontier theme with a wagon wheel labeled LA PONDEROSA, referring to a ranch on the TV series *Bonanza*.

"People fight a lot on the Hill," wrote Lardner, "but they exhibit a degree of neighborliness not to be found in every high-rent apartment building." The Hill's inhabitants "were living on a site that was all but abandoned before they got there. They have as much right as most non-homeless people, perhaps, to feel that they have improved on what came before, and more right than most to think of themselves as having built a community from the ground up." For a while, even the police acknowledged this by keeping a protective eye on the people living there.

In 1991, the Hill became art. Gabriele Schafer (from West Germany) and Nick Fracaro (from Illinois), came to the Hill with a vague idea of doing a theater piece with the homeless – "something that would link them with the nomadic tradition stretching back through the centuries" and assert homelessness as a choice. They built a tipi and moved in "without a script and without any performance dates." They drew portraits of the residents, did some gardening, and, most acutely, handed out disposable cameras with which residents could shoot back at the tourists photographing them. ("Do unto others what they shouldn't be doing unto you," is how Ace, a resident, described this process.) The results were exhibited on the site (in the tipi, by then called "The Living Museum of the Nomad Monad") as "photoscalps" captured in a "last stand" – yet another phantom of the frontier. A year or so after Lardner published his article, the tipi was burned and a rash of suspicious fires caused police to close the community down.

> The photographer is the philosopher of the shards of gloss that sprout between the crocks of the concrete sidewalk.
> (Miguel Algarin)

Inumerable photographers have documented the plight of the homeless over the last decade, with results ranging from condescending, sentimental, even racist images, to moving, empowering, and genuinely informative ones. From Los Angeles, the national capital of homelessness, Anthony Hernandez's "Landscapes for the Homeless" are large Cibachromes of the detritus of nomadic life, omitting the human presence. A more overtly engaged approach was that of Oakland photographer Scott Braley, who first spent eight months photographing Rufus Hockenhull's life as a homeless black Vietnam vet and then taught Hockenhull to photograph his own landscape of homelessness, alcoholism, and mental illness.

Around the same time, in Washington D.C., Jim Hubbard embarked on a similar but much larger scale project called "Shooting Back" – after a homeless child explained why he was photographing his world. The participants, who lived in shelters, ranged in age from seven to eighteen; they photographed their immediate surroundings, children's views of the society that has condemned them to that place. (According to Jonathan Kozol, 40 percent of the poor in America are children.) Poet Carolyn Forché described the images as "the landscape of those who suffer from exposure to anonymous space and the practice of human warehousing"; its recurring motifs, she says, are "incarceration and ruin." One child photographed a handlettered sign: I WILL KILL ANYONE DESTROYING MY PROERTY (*sic*).

The most impressive, and ongoing, artwork done with (rather than about) the homeless is the Los Angeles Poverty Department (LAPD), founded by performance artist John Malpede, who tired of the New York artworld, moved to L.A. to work as a paralegal on Skid Row, and found himself using performance techniques to find out what was going down on the streets. His workshops developed into a unique improvisational theater troupe made up primarily of homeless people. In its dozen years of existence, LAPD has traveled around the country and inspired similar work among artists and street people. I saw LAPD perform in Boulder, Colorado – a university town not known for its squalid side. One local participant described himself as "an individual far away reaching for a space in life." In a related discussion group, a Boulder resident made it clear that people didn't realize how difficult it is to live in a town with little lower-income housing, where college students snap up the minimum-wage jobs, where "quality of life" and open space are favored over poor people. As a result of LAPD's visit, a homeless theater program was set up in Boulder. (Some money was stolen, though, and the program was subsequently closed.) For all its pitfalls, art like this helps the public to perceive the places where the homeless live as real places inhabited by real people with lives, however chaotic, rather than as "objects" left on the corners of other people's places.

In the eighties I read a newspaper article about a homeless woman, a carpenter by trade, named Kea Tawana (she looked to be white and in her fifties or sixties) who spent five years building from salvaged lumber a remarkable three-story, ninety-foot-tall ark in the center of Newark, New Jersey, because she "needed a decent place to live"; on the eve of the ark's completion, authorities condemned it to make room for a high-rise. In response to this kind of brutality, which is common nationwide, a group in Atlanta called the Mad Housers began helping to build rudimentary shelters for the homeless, "in recognition of the inability of the existing social order to meet the requirements of those citizens who are unable to compete effectively." Like the homeless themselves, the Mad Housers operate outside of regulatory frameworks. Bailey Pope described the project in terms of the "physical division of here from there" (walls) and extension of self to other (doors, which open to dialogue and the development of community).

The Mad Housers spread to Chicago, where the story of their wildly successful "productive protest" is instructive. Having constructed a group of huts along a commuter railroad track, the homeless residents were praised for replacing an ugly abandoned space with a clean, green community. They attracted much attention, first from train passengers and employees, who donated food and clothing, and then from the media and a much broader public. As its fame grew, Chicago Mayor Daley decided the shantytown was too visible and had to go. Residents, activists, and a large support base girded for battle, vowing to block demolition crews. The city finally won by moving the hut dwellers to the top of the city housing waiting lists (over thousands of people, some of whom had been waiting for years), and the huts came down. Thereafter, the Mad Housers took care to build discretely, keep their sites secret, avoid media attention, and keep up the good work.

The antiregulatory approach has proved appealing to artists, among them: Jon Peterson, who made fiberglass street sculptures in L.A. and New York City in the late seventies, which he called "Bum Shelters"; Athena Tacha, who proposed a very simple but structurally handsome honeycombed movable homeless shelter in the eighties; Donald McDonald, architects from San Francisco who have constructed plastic sleeping boxes with tiny windows; and Brad McCallum, who designed a tent that fit over a park bench. Krzysztof Wodiczko, best known for his giant projections onto the sides of building, expanded on homeless people's own adaptions of the shopping cart as mobile home by inventing an ingenious, if ingenuous, "homeless vehicle" – a body-sized, missile-shaped expandable bed/cart (complete with metal basin and returnables receptacle). Conceived in 1988 literally as a weapon against social neglect and a survival strategy, the vehicle provided "both emergency equipment and an emergency form of address" for the evicted. Another version of the mini-mobile home has been seen recently on the streets of Manhattan's Lower East Side: a little house on wheels dragged around town to various sites by the artist couple who live in it and exhibit their raunchy drawings in the windows. Such fragile "housing" proposed or executed by artists raises questions. Some argue that such stopgap measures accept the necessity for homelessness and fail to deal with its social causes; others insist that since homelessness is a fact, anything that can be done to alleviate the suffering or call attention to it is worth doing.

Like indigenous people in the nineteenth-century West, the homeless are either perceived as a negligible part of the "landscape" or slated for internment. The connections with the myths and military ideologies of the old West lie in the issues that surround any contested territory: land and housing, displacement of the "savages" (poor residents), homesteaders and squatters, and "civilization" (wealthy newcomers). Violence, both economic and physical, is taken for granted. As Neil Smith explains, "the frontier motif makes the new city explicable in terms of old ideologies. . . . [It] rationalizes social differentiation and exclusion as natural and inevitable." Smith calls developers "real estate cowboys," but "ranchers" would be a more accurate description, given the profits involved. He quotes one developer who illogically and ahistorically denied responsibility for the homelessness he had promulgated: "To hold us accountable for it is like blaming the development of a high-rise building in Houston for the displacement of the Indians a hundred years before."

A classic New York version of the Wild West show came in 1981 with the much-publicized making of the movie *Fort Apache: The Bronx*, an all-out media attack on the South Bronx, starring Paul Newman. It became a rallying point for the Committee Against Fort Apache (CAFA), a broad coalition of roughly one-hundred local organizations, ranging from former Young Lords to the Catholic Archdiocese. The film took a struggling neighborhood – beleaguered by civic neglect, bad press, landlord arson and redlining – and presented it as an inferno of crime and abnormality: "a 40 block area with the highest crime rate in New York. Youth gangs, winos, junkies, pimps, hookers, maniacs, cop killers," according to the movie's advertising campaign.

The cavalry was played by the cops, who saw themselves holed up in their "fort," a "thin blue line" maintaining "civilization" against the local "savages." In the movie, one of them describes the community as "seventy thousand people packed in like sardines, smelling each others' farts, living like cockroaches," including "fifty thousand potential cop killers." The film's final solution was "bulldozers . . . that's the only way. Just tear it down and push it into the river." Life then imitated Hollywood: most of the housing in the forty blocks was in fact demolished, and the remodeled precinct rising from the ruins got a new nickname: "little house on the prairie." Richard, the voice of East 13th Street in Manhattan, predicts a similar fate for his neighborhood: "We've all seen neighborhoods rise and fall like little countries. In five years I bet you won't even see our asses on this block. Only this time, I don't know where people will go. There's always the Bronx. Lots of family up there. You know, you'll be walking on the East Side sometime and you'll see Richard got pushed right into the East River! Chased by fucking bulldozers, man!"

In the late seventies and early eighties, the Lower East Side and East Village became a hotbed of punk, new wave, avant garde art as groups of dynamically disenchanted young visual artists rebelled against the expectations of an artworld career and identified to an unprecedented extent with the disenfranchised in whose midst they lived. This was a political and an aesthetic choice, and the art that resulted was more profoundly influenced by the city as place than any other modern movement I can think of. Groups like Group Material, Fashion Moda (the "cultural concept," or alternative space, in the South Bronx), Collaborative Projects (CoLab), Political Documentation/Distribution (PADD), Contemporary Urbicultural Documentation (CUD), and the loose-knit ABC No Rio artspace on the Lower East Side, *World War 3* comics, community muralists such as Cityarts and Artmakers, and individuals like Robbie McCauley and Ed Montgomery, Christy Rupp, Rebecca Howland, Tim Rollins, David Wojnarowicz, Kiki Smith, Justen Ladda, John Ahearn and Rigoberto Torres, Seth Tobocman, Jenny Holzer, Lisa Cahane, Mike Glier, Julie Ault, Bobby Gee, Alan Moore, John Fekner and Don Leicht, and David Wells – all lived, worked, and participated in seamy urban places. Young, cool, into "new contexts," they came to know these "ghettos" and the people economically confined there more intimately than most artists know the places they live in for economic convenience. Many of their projects were collaborations within the neighborhoods. In those early years, Loisaida and the South Bronx were the hottest art spots in town. Downtown, gentrification was already well on its way. The hype around the so-called Lower East Side Art Scene soon polarized the area, breaking all but the strongest alliances.

Photographer Geoffrey Biddles book *Alphabet City*, on the Puerto Rican Lower East Side, was begun in 1978, then resurrected "after the scene" in 1988 with the addition of brief, compelling interviews recorded as his subjects looked at their earlier photos. Thus, the book is a "retrospective": it is about memory, time passing, change for better and worse. With a startling intimacy, and without condescension, Biddle chronicles the dying spaces of the bleak avenues from A to D and the resistent vitality of those who have been dumped there. Many people are photographed in their homes, squeezed by gentrification to double and triple up in the housing projects "in numbers the census won't show but that the city guesses at by monitoring gas and electric use." The neighborhood is epitomized by its harsh, often fearful streets, leading to the East River through tenements and small stores to public housing projects built in the forties and fifties to a strip of park reached by bridges over FDR Drive.

Here are four residents' views of a changing "Alphabet City":

> They're charging. . . . a thousand dollars, That's not for the Black or the Puerto Rican. The only reason you makin that kinda money is because you're involved in drugs. So automatically they're pushing the good people out.
>
> (Evalene Claudio)

> I see a change for the better, because I see a lot of preppies that are coming down, fixing old buildings. There's a new vibration, putting the neighborhood back. There's uniformed cops on every corner. It was a shooting gallery, now it's an art gallery.
>
> (David Garcia)

> There ain't no good future in this neighborhood. If you lived here for like twenty years, best thing is just to get out now. . . . there's nuttin here to advance to. Move to another building that got more mice or something? Nah.
>
> (Howie Wheeler)

> Lately, I see a lot more white people in the neighborhood. I'm not gonna let them take over. Not my home. . . . If they try to move in, we're gonna beat them up, make them think about it twice. That's how it is. They already look at us like we nobodies, and now they gonna try to take away what we got. . . . One day, we're gonna get our sunlight.
>
> (Richard Morales)

In 1988, within this context, the frontier melodrama – real estate versus community, racism versus respect – enjoyed a downtown rerun. The Lower East Side "community," centered on Tompkins Square Park between Avenues A and B, was smaller than the South Bronx and already a veteran of gentrification battles. It was more racially mixed, and much more visible. The battles in Loisaida took place in view of all downtown; the media did not have to be cajoled into "foreign" territory.

Loisaida also had a very different history from the South Bronx. In the fifties, poets and artists of the Beat Generation (as well as working-class gays and lesbians) trickled into what was then dubbed "the East Village," Greenwich Village having become too expensive and commercial for the avant-garde, The 10th Street galleries – New York's first "alternative spaces" – bloomed on the western border. From the mid sixties on, the area provided cheap housing for the creative and countercultural young who mixed cheerfully with diminishing Ukrainian and Jewish communities, an increasing African American and Latino population, and some overflow from nearby Chinatown. Despite a heavy drug scene and police presence, a certain community renaissance also took place, in which some artists were deeply involved. In the seventies, El Bohio, an old public school on 9th Street between Avenues B and C, became the cultural heart of Loisaida. Charas, the organization running the building, coexisted and often collaborated with the burgeoning alternative "punk" art and club scene, which almost immediately attracted hyperbolic media attention and gave way to a rising tide of proto-Yuppies. Around 1983 some artists began to organize against "yupper-income housing," including a PADD project called "The Lower East Side Is Not for Sale," with site-specific works around the neighborhood and exhibitions at Charas and at ABC No Rio.

> It's like a lot of bored people from good backgrounds getting into the bad of the neighborhood. And here we are struggling like hell to get rid of the bad, you know? We find no romance in junk and shit. . . . I mean we got to look at all the shit on the streets everyday. Who wants to see it in a gallery too? Know what I mean? You might think the art is for someone, Who? I can't honestly say it looks like it's for us.
>
> (Richard from East 13th Street)

In 1984, a city-sponsored reconstruction project was rejected by local residents as a gentrification ploy. By then, gentrification had taken hold in earnest. The SoHo model was applied, though with less saturation: small shopkeepers were ousted in favor of boutiques, galleries, restaurants, and suddenly expensive housing; classy spacious white galleries replaced local small businesses; European art dealers arrived; crack dealers hung in, despite a drug crackdown intended to clean the place up for the middle class. Loisaida was also a center for the city's small but lively squatters movement. (Neil Smith has pointed out that "before 1862, when the Homesteading Act was passed, the majority of rugged frontier heroes [moving west] were illegal squatters.") Miguel Algarin described the process: "You can either comply with the law or grab the moment. Take over a building. Go downtown and argue for the deed of ownership. The squatters of Loisaida and Harlem are doing it. They risk having to learn how to pipe a building, how to gut it, how to build a roof. They risk in order to construct the life that is happening to them."

By 1988, the lines were drawn between the gentrifying newcomers and developers (backed by the police department) and the earlier residents, artists, squatters, and homeless. The battleground was Tompkins Square – a one-by-three-block city park created from swampland in the 1830s. It has tall old elm trees, not much grass, benches, winding walks, a playground, and some basketball courts; it *had* a bandshell where the Grateful Dead played in the sixties, and a large homeless population. The square has a protest history that goes back to the 1850s. In 1874 a workers demonstration, seven thousand strong, was brutally broken up by police; an Irish worker asked "Is the Square private, police, or public property?" – a question that might have been asked in 1988.

In the sixties, police brutality was aimed at hippies who refused to keep off the grass. By 1988, people were living in the weeds. Teaming up with progressive elements of the local working communities, young lefties, anarchists, and artists, many of whom were also locals, specifically opposed a pricey condominium on Avenue B, formerly a settlement house and community center (the value of which went from $62,500 in the late seventies to $1.3 million in 1983 to several million in 1987, when a single penthouse apartment was listed at over $1 million.) The protesters introduced a countercultural ingredient that made the struggle reminiscent of land rights battles in rural areas or the Berkeley People's Park confrontations, in which the struggle over a patch of "liberated" land between students, lefties, and street people versus the University of California, the state, and "the Establishment" made sixties history.

"Housing is a human right," "Gentrification = Class War," "Out Yuppie Scum" were among the slogans. The tinder was sparked by exaggerated police hostility during raids to destroy "Tent City," the shanties erected by the homeless in this last of city parks to resist a midnight curfew. It ignited on August 6, 1988, with a police riot: mounted police backed up by hundreds of footsoldiers were later called "out of control" by an investigative commission. The overkill was ineffective: the cavalry beat a retreat and eventually lost the media battle as well. Tompkins Square became a "liberated space" – briefly. In December, more than three hundred homeless people were evicted, their belongings hauled off in garbage trucks. Two homeless people froze to death. The squatters movement was targeted for legal harassment; several squats were demolished, to be replaced by nothing; others fell to suspicious arson, blamed on developers. Some of the homeless were taken in by those squats remaining. Others eventually set up a new tent city in a vacant lot nearby, dubbed "Dinkinsville" after New York's new African American and supposedly progressive mayor. They were in turn evicted in January 1990. In September 1990, another "riot" ensued when police interrupted a benefit concert for Tompkins Square squatters.

The homeless community was evicted again in 1991, but in the process the homeless burned their own tents in protest and the City backed down for a while; the Police Department declined to keep evicting squatters who just moved back in when the City did nothing with the empty buildings and the City scrambled for funding to plan ahead. (One of several ironic turns is the role of well-intentioned non-profit groups who, by supporting the City, help to evict the squatters.) During another riot on Memorial Day 1991, sparked by a still-unsolved murder, a local business was destroyed (some say it was police provocation) and Tompkins Square Park was shut down, only to be reopened with a curfew and strong police presence. Now the homeless stay up all night and come to the park to sleep during the day. In August 1996, the courts ruled against a group of 13th-Street squatters and rebellion is simmering again, "Whatever else happens," says a longtime Tompkins Square activist, "A lot of people have gotten apartments for around $75 a month in New York City. Kids have grown up in these houses – though not as many as we'd hoped."

'Tango: A Choreography of Urban Displacement'

from Lesley N. N. Lokko (ed.), *White Papers, Black Marks: Architecture, Race, Culture* (2000)

Ana Betancour and Peter Hasdell

Editors' Introduction

Ana Betancour and Peter Hasdell are architects, writers, and co-founders of P.H.A.B. Associates, an architectural design, cultural, and urban planning practice in London, which has worked in Sweden, Japan, Australia, and the United Kingdom. Betancour comes originally from Montevideo, and teaches architecture at the Royal Institute of Technology in Stockholm. Her research interests include borders, marginal settlements, and social exclusion. Hasdell, from Australia, shares some of these concerns and teaches at the Architectural Association in London and the Bartlett School of Architecure, University College London. Their essay on tango is included in *White Papers, Black Marks: Architecture, Race, Culture*, a collection described by its editor in terms of a set of questions on the importance of race in architectural and spatial thinking, and in the shaping of the built environment. Other contributors to the volume look at historical and contemporary realities in a range of cities mainly outside Europe and north America.

The text reprinted here is relevant to Parts Seven and Eight, on identity and borders respectively, but is included here because its main interest is in the everyday construction of a cultural tradition – tango – originating in migration but eventually becoming part of a national identity. Tango's development as a genuinely popular art form contrasts with the Haussmannesque plans for the city, as the authors say, produced by the dictatorship of the 1970s. This gives a political slant to everyday culture – not so much as resistance, more as a positively practised other to official culture's posturing.

Please note that there is a discrepancy in the original numbering of the notes (from number 27 onwards). The Editors of this volume have decided to leave it unchanged.

THE TOUCHING LINE

Tango, it is said, originates from the Afro-Argentinean word *zango* or *tambor*, the word for the drums of the arriving slaves from Africa, calling up memories and rituals of their homelands. Perhaps it means the place for the forbidden celebrations of the black population on the edges of the cities of the South Cone? A song, a dance, a choreography, a noun without etymology, no origin other than naming that without name, describing a place, a sentiment, an atmosphere or a certain time or place. Or perhaps it is linked to the word *tangere*, meaning touch, tangent – that which barely touches on a marginal point of the circumference of a circle . . . its edge, the periphery.

A tangent, the touching of two lines that do not intersect. An infinitesimally delicate act, like kissing – touching at a point that tends to nothing mathematically. A definition that speaks of the tension between two lines, two bodies . . . and their displacements. The touching line, linea tangens. Into the abstract, rarefied language of geometry[1] and line enters a beautifully descriptive definition that speaks as though the line had emotive power capable of touching, of being touched or of having a physical existence. A movement of the interior soul of the line that reaches out, without deflecting and touches – only just – another.

To go off on a tangent we might speak of completely different or divergent courses. That play in the mind, digressing, so that one might think oneself moving on in a particular direction only to be displaced, elsewhere. A tangential existence. In tango's obscure origins, the nature of these tangents, digressions and displacements are fragments, journeys, displacements and digressions. They are bits of Paris or Napoli, of detours made, trivial moments in everyday life: the *barrio*;[2] a photograph carried in a suitcase; broken lineages; oceans that separate; colonies claimed[3] – the hope that springs from a particular sound; one's presence in a city that has no collective past, like the sudden changes of direction in a dance. They circumscribe, in a way that can only ever be incomplete, their subject.

One way of describing culture is to see it as the 'unfolding' of a place. In these terms, the relationship between tango as an emerging culture and the city in which it 'unfolded' is a complex one. Factors of geography, heterogeneity, displacement and migration all play a role in this, as do ethnic origins, 'race', colonialism and slavery. Tango's origins, analogous to oral cultures whose origins and expressions are quasi-mythological, become hazy and indistinct, ascribed at a later time and written into the history of the city. Tango in its written and popular history becomes an 'after-the-fact' cultural expression that sublimates most of its hazy and complex beginnings, and by implication, those of the city. Parallel to this, key aspects of its milieu – the newly urbanized cities of Buenos Aires and Montevideo – become similarly indistinct, their origins reconstructed post-factum.

To understand tango, therefore, only in terms of its legitimized form as a national cultural expression and a specific, marketable cultural commodity is to profoundly misunderstand its significance as a resistance to the rapidly expanding urban periphery and the marginal existence this had in relation to the legitimate culture capital of the city centre. In the general context of the history of urban change it can be seen as a possible model for understanding the massive influx of immigrant cultures into peripheral areas of cities and the emergence of cultures within ghettos. It may also point to the subtle transformations of such areas and non-places into places with distinct identities, where dreams and memories – and everyday lives – touch the physical, official world of the city. Describing, in other words, places that are open to interpretation.

Three such descriptions are offered here. They offer neither closure nor finite interpretation, but simultaneous existence in the space of the city. Through the use of ellipses, tropes, analogies and digressions, they rely on tango's obscure origins, allowing slippages and tangents to occur such that anecdotes can enter grand narratives and private, inner emotional spaces may be read in terms of local or national geographies. The tangential and the circumstantial both exist in the emerging culture of Buenos Aires, evident as a state-run enterprise that expanded its rational grid according to a larger ambition of the Modern Project.

CITY OF FRAGMENTS

> When accelerated changes in society arouse feelings of uncertainty . . . an old remembered or imagined order is reconstructed by the memory as past. Against this horizon the present is placed and evaluated.[4]

The massive migrations of disparate peoples from Europe to the New World brought an unfamiliar existence to the milieu of Buenos Aires, perceived as a loss of cultural referents and a poverty of existence. By the turn of this century, up to 70 per cent of its working inhabitants were immigrants, flooding into the country to provide labour for Argentina's rapidly expanding export market to Europe.

> More than fifty thousand workers come each year to the River Plata – Europeans washed up by desperation on these coasts. Eight of every ten workers are foreigners, and among them are Italian socialists and anarchists, Frenchmen of the Commune, Spaniards of the first republic and revolutionaries from Germany and Central Europe.[5]

The enormous flood of immigrants into Buenos Aires, commencing in the 1870s, meant, for example, that the city expanded by over 300 per cent from 187,000 inhabitants to 650,000 in the 25 years[6] to 1895, later swelling to 1.5 million by 1914 and to 2.5 million by the 1930s. By then, the city had grown 10-fold in a matter of 50 years.

> The numbers given in the official census tell that 184,427 immigrants arrived to the country during 1924. An amount really full of promise because it represents an average of more than 15,000 immigrants monthly, an important current of human energy flowing into our country.[7]

These numbers, naturally enough, had drastic effects on the city's fabric, on its public spaces and also on its culture and ethnicity. That the immigrant population formed the majority of the city's citizens meant that a very particular common condition could be said to prevail: the city effectively became the place in which a culture of mixture emerged, one in which the heterogeneity and diversity of people made for an incredible ferment. Given the scale and rapidity of change in this context, the emergence in this context of tango as a specific culture of 'otherness' seems a natural consequence.

One could say that the implicit understanding shared by immigrant cultures, irrespective of origin, drew together their differences, inasmuch as the lyrics of the songs speak of the laments of individuals, whose common plight is migration – to Argentina and Uruguay in the nineteenth century – and whose origins are left behind or only carried as memories, fragments, rhythms and songs.[8] Many tangos specifically tell of the *barrio* and *arrabal*, places of hope and longing, laments for a place where dreams have a patina of expectation and a mythos of home that replaces, substitutes perhaps, for an originary displacement to the far end of the world. Laments on the absence of a loved one, machismo lyrics about the grim or tough conditions of street life in Buenos Aires are also found in tango, mythologizing the present moment, the everyday, in a city that apparently had no past for the newly arrived immigrant other than their carried memories.

Yet what is not commonly understood is that tango was, in essence, a form of cultural resistance; akin – whether explicitly or implicitly, to a masked critique of the larger forces that contributed to this condition, and therefore also an implicit critique of the state. In the specific context of Buenos Aires or Montevideo, this allows us to understand the urban peripheries and their growth not only as physical or manifest architectural form but also as factors in the emergence of cultural identities, offering a critique of the structures of power of such cities, even if this critique is masked (and in a sense impotent) as a lament on the nature of the powerlessness of the individual, akin to many tango lyrics. In these terms, tango is not a pure cultural commodity but also a marker of spatial territory and of very specific conditions. It is no wonder that the authorities of the time regarded tango lyrics and dance as lascivious and threatening, subjecting them to censorship (as did the military dictatorship of the 1970s). Only later, during the 1930s, was tango incorporated into the Argentinean national identity via its acceptance in Europe. The 1930s and later were also periods of heightened nationalism, epitomized in the 1950s by Peronism – dubious arbiter of national identity – and, in some senses, a reaction to the earlier fervour of immigration. During this period the construction of a national identity via an increased nationalism appropriated tango as a distinctly Argentinean characteristic. Beneath the surface, the discontent and underworld image of tango threatened the hegemony of the official and authorized identity. Public space in Buenos Aires during this time was a civic and urbane gesture, modelled on the European, French and Anglo-Saxon ideals. Tango, in contrast, was regarded as part of the hidden or shameful world of the night. Tango took place in the seedy darkened corners of the city peripheries, in the barrios and harbours, beyond those places where the acceptable and recognizable European high culture resided, thus territorializing the extent to which the body politic and an emotional inner landscape of the immigrant masses could be controlled. Where the cultural 'landscape' and the political met was, in essence, the street, the newly constructed urban fabric, the physical city.

Tango, it may be argued, results from the conjunction of two key factors: geographic displacement and an inner emotional condition. The restrained passion of tango, in dance and song alike, is only revealed at certain controlled moments, giving a glimpse of a deeper passion that surfaces where it touches others and is understood by many. Tango's appearance is two-fold: the tension, sexuality and hidden passion of the dance and the withholding and control – these characterize tango. In these terms, it can be said to reveal, by suggestion and expression, an inner landscape, resulting from the wider geographical displacements, a consequence of migration. The inner landscape is unspoken, has – *is* no place recognizable – yet reveals itself in the words, lyrics, movements of tango trying to make sense of a fragmentary existence. It expresses the topographies of hope, despair, depression, relief and elation; the oceans, rivers, mountains, chasms and fields that give shape to a city without relief, existing frenetically in the present and future tenses only and in the numbing horizontality of Buenos Aires.[9] The inner landscape is illustrated by Galeano, in exile himself during the 'difficult' years, who constructed his life in fictions, anecdotes and commentaries from afar, analogous in principle to the songs of tango. Of the immigrants' arrival in Buenos Aires, Galeano writes:

> One fire less, they say in the villages of Galicia when someone emigrates. Over here, he was excess population . . . in a foreign city he takes up less room than a dog. Here they make fun of him and treat him with contempt, because he can't even sign his name, and manual labour is for inferior species. He gets little sleep, the lonely immigrant, but no sooner does he close his eyes than some fairy or witch comes to love him on green mountains and snowy precipices. Sometimes he has nightmares. Then, he drowns in the river. Not just any river but a particular river over there. Whoever crosses it, they say, loses his memory.[10]

Galeano's text is concerned with the differences between here and there, the 'otherness' and alterity that results from a displacement. Analogous to the hero of Cortázar's *Rayuela*, Hopscotch (1963), who 'wavers' irresolutely between Buenos Aires and Paris in a permanent alternation of 'now here' and 'now there'. Between the manifest brutal reality of Buenos Aires and the familiar, left behind, can be located a glimmer of a politic, concerned with factors of alienation, and an inner imagined or remembered landscape. The imaginary and the manifest begin to construct a space around which tango can be situated.

Buenos Aires during this period is likewise afflicted by similar contradictory factors. The city is in essence an elaborate construct. It does not exist in the sense that it is constituted from citizens. What binds the immigrants together, therefore, is only a common displacement from Europe. Society exists as an intention, an imagined utopia governed under the guise of the modern project, a state-run and controlled enterprise that imagines Buenos Aires to be a city such as Paris: European, civilized, dignified. The city as an imaginary construct or even fictional construct by no means stems wholly from this period. Its foundation is based on a similar gap or rupture.

Searching for a veritable fortune, Juan de Solis sailed into the Rio de la Plata,[11] the River of Silver, in 1516. A vast flat sea, which, upon tasting, he named the Mar Dulce, 'freshwater sea'. Solis noted in his journal that the endless adjoining plains were an 'ocean of grass', the pampas that were later to become the fount of wealth for Argentina. Upon landing, native Indians attacked de Solis and his crew, and, apparently, devoured them. Some time later, Sebastian Cabot,

> sought the treasure of King Solomon sailing up this Plate River – so innocent of its silvery name – which has only mud on one bank and sand on the other. While his soldiers, maddened by hunger, ate each other, the captain read Virgil and Erasmus and made pronouncements for immortality.[12]

Claiming to have discovered El Dorado on the basis of a few trinkets of silver, Cabot could easily appeal to the greed of the Spanish kingdom.

In 1536, on royal command and with an army of 1,600 men, Don Pedro Mendoza, whilst searching for a cure for syphilis in *guaiacum* (a plant he mistakenly thought grew in the River Plata area) 'founded a city, a fortress surrounded by huts, and upriver from here he went hunting for the silver mountain and the mysterious lake where the sun sleeps'. The city was called Buenos Aires, 'good air' and was located near what is now Barrio La Boca. The map of Buenos Aires drawn by its founder, Mendoza shows a colonial grid surrounded by written script on all four sides – closer perhaps to a rhetorical or literary device. His map is emblematic of the gap between the inner landscape, an imagined utopia of riches and civilization, and a city that did not yet exist. Mendoza never found the cure for syphilis, silver or the lake and subsequently died from his illness. The city, similarly; was not as enduring as Mendoza wished and lasted for just five years, although the name endures until this day Phoenix-like, in 1580, Buenos Aires was reborn, founded again, a foothold on a continent perched between the flat plane of the sea and the endless plain of grass, and its second founder Juan de Garay drew the plan of the city, a Spanish colonial grid, on cow hide. Angel Rama in *The Lettered City* describes the relationship of literature to the origin of South American cities, in which the pre-existence of the city was in lettered form as edicts, laws, desires and codes, often issued from afar in colonial Spain. Looking at a map depicting Buenos Aires and surroundings in the beginning of the nineteenth century shows the grid laid upon the land and the existing buildings. The legal boundaries for the city settlement evolved in accordance with the planning ordinances of the Law of the Indies, and the Iberian conquerors created, in Rama's clear words: 'a supposedly "blank slate", though the outright denial of impressive indigenous cultures.'[13] The implementation in the colonies of the Spanish cuadricula, and the rational urban grid from the late 19th century are very much alike. The lines drawn upon the supposed 'blank sheet of paper' are an imagined landscape not yet built, and not yet inhabited.

The city in its inception could be described more closely as an apparition, given apparent substance as if through a state of delirium. The equivocation of origin of the city, a precursor to the immense gap between the apparition and the manifest reality of the city – as Corbusier would discover much later – is evident. The dreams of immortality for a presence that is manifestly solid embody themselves in the words and actions of its founder Mendoza, and in the literature of Borges, who wrote: 'it seems impossible to me that Buenos Aires ever had a beginning. It is eternal, like the water and air.'[14] The city effectively was represented before existing in reality. It therefore had an *a priori* existence in the imaginary.

Yet as the names of the nation and its key city never managed to match the reality that promised 'silver' and 'good air', *Argentina* and *Buenos Aires*, so too is there an apparent gap, an absence between the image and the reality of the city. From this lack of substantial reality, of authority, one can understand why Buenos Aires, for example, desired to be the 'Paris of the South': 'in Buenos Aires . . . Sarmiento . . . imported sparrows so the city would seem more like Paris'[15] – Paris being deemed to be an authentic origin and repository of culture, as if to make up for the absence of reality that the fictional identity could never attain. In terms of its cultural identity, the city appealed to elsewhere, to England for its model of society; to Europe for its authenticity; to Paris for an image of its desire and later to the *gauchos* of the *pampas* as a source of its mythology. The cutting of Haussmannesque *grands boulevards* through the colonial Spanish grid during the mid-nineteenth century, by Torcuato de Alvear, shifted the perception of the city from a former outpost of the Spanish colony to a centre of culture, belonging, therefore, to the same sphere as historic and 'civilized' Europe. Coupled with the construction of grand civic institutions such as the Congress Hall and the central plazas, the city was envisioned as exemplary and necessary in the on-going construction of the identity of the nation. Corbusier's invitation to Buenos Aires in 1929 can be understood as part of the same, using European culture as the model, and architecture as the emblematic representation of the new, modern metropolis.

Corbusier approached the city from the sea at night. He saw a southern sky full of stars, glittering mirage-like on the surface of the River Plata. The city 'hovered between these, suturing both, but not quite rooted – not quite of the earth.'[16] His lyrical description suggests a complete seduction by the appearance of things. However, Buenos Aires was duplicitous for Corbusier: a city that elicited both mirage-like apparitions and abject horror. He wrote: 'Buenos Aires is one of the most inhumane cities I have known; really, one's heart is martyred. For weeks I walked its streets without hope like a madman, oppressed, depressed, furious, desperate.'[17] In these terms his reactions mirror those of his surroundings. We can retrospectively view Corbusier's reaction as being an intrinsically European one, looking outwards, away from the pampas and metaphorically towards the origin of culture and of urbanism. As Collins writes:

> in his ninth lecture (18 October, 1929) he advanced the possibility of adapting the Plan Voisin of Paris to Buenos Aires – or one might say, Buenos Aires to the Plan Voisin – affirming that Buenos Aires can transform itself into one of the great (dignas) cities of the world.

Corbusier's plans for the city also embody a desire to shift Buenos Aires towards that which was emblematic of a European modernism, one which implicitly homogenized differences. In other words, the compass of the Plan Voisin pointed towards an implicit European centre, away from the intrinsic differences and cultural 'otherness' that were manifest in the social matrix of the city.[18]

It was to this city that immigrants arrived in the 1900s from Europe. They passed through the mouth of the harbour, across the threshold to the city, then flooded into the city itself. En route, most ships passed by La Boca, a harbour-side *barrio* that formed one edge of the port area. La Boca is the 'mouth' of the city.[19] La Boca, it is held, is the 'birthplace' of tango, according to its reconstructed and legitimized history. La Boca was built as a shanty town by immigrants from Liguria in the 1840s–50s,[20] but inhabited previously by the French Basques. It is situated on low-lying flood-lands where the Riachuelo river meets the Rio de la Plata.

An area distinct from the unrelenting grid of the city, it evolved a particularly colourful corrugated-iron architecture and raised sidewalks to counter the frequent floods. La Boca is, in effect, a transitional area, half river and half land, inextricable from the port and its history of immigration, import and export, a threshold or liminal condition that characterizes Buenos Aires.

Thus during this period, it is possible to see the whole city spatially, culturally, historically and in terms of identity as a threshold between Europe and Argentina's interior,[21] between the Rio de la Plata and the pampas. The port and the city therefore link between two vast flat planes, one a body of water and the other a body of grass. The flatness of the city and its surroundings, which extend into the sea, meant that 'the bay's bottom is essentially a continuation of the *pampas*.'[22] Because of this, the sea is able to flow into the city in innumerable floods. Analogously, the city extends across the *pampas* like water. This duplicity marks the city equivocation of import and export, sea and *pampas*, Europe and the interior, and, in essence, registers the spatial duality of the city in the form of a horizon. The flat plane on which the city is built, horizontal and without relief, is a specific characteristic that historically allowed for speculative development and expansion of the city without physical constraints.

The modernist photographer Horacio Coppola's images of Buenos Aires of the 1930s show the ubiquitous grid receding into the *pampas*. The road is just one tentacle of the city grid, and is depicted as being wide as the land. No mere farmers' lane, but a road laid down as progress. The periphery here is not the left over nether-spaces of the city, but a frontier whose endurance in the beginning of the century is temporal, progressive, and as necessary as the influx of immigrants who flooded into the city from the sea. The flat *pampas* is the hinterland on which the economic future of Argentina was assured. The *pampas* was largely unwritten about until the nineteenth century. Upon this blank sheet, the speculations of the Modern project occurred. In these terms the Modern project that aimed to construct an Argentinean national identity is problematic against the background of massive change, urban growth and mass immigration that altered the demographic constitution.

The reality of Buenos Aires during this time – a city of fragments – is in essence counterpoised against a desired image of the city and a desire to appropriate the 'interior' of the *pampas*. National identity, an authenticity as origin and foundation, therefore, became a fictionalized construct, and at times an imaginary entity, used to make up for the lack of stable referents, a projection put on a less than coherent body of immigrants. And parallel with this, the imaginary landscape of the immigrant is a solace from the brutality of everyday life in the raw city, through which their condition of displacement becomes evident, manifest in tango.

WHITENED CITY

> America, instead of remaining abandoned to the savages, incapable of progress, is today occupied by the Caucasian race – the most perfect, the most intelligent, the most beautiful and the most progressive of those that people the earth.[23]

> Sarmiento, founder of the Animal Protection Society, preaches pure unabashed racism and practices it with untrembling hand. He admires the North Americans, free from any mixture of inferior races, but from Mexico southwards he sees only barbarism, dirt, superstition, chaos and madness. Those dark shadows terrify and fascinate him. He goes for them with sword in one hand, lamp in the other. He publishes prose works of great talent in favour of the extermination of gauchos, Indians, and blacks and their replacement by white laborers from northern Europe.[24]

The history of slavery to the region pre-dates the founding of Buenos Aires by two years. Slavery was set up by royal decree – an *asiento* – granted in 1534, by the expansionist policy of the Spanish Empire. The decree assumed the necessity of African slaves who would work in the gold or silver mines they expected to find in the River Plate area. The implications of this decree in relation to the origins of Buenos Aires are astonishing: the 'black' city, at least on paper, intention and in the imagination, is the precursor to the city. It is significant that during the early years of Buenos Aires, in 1778 or thereabouts, African slaves and indigenous Indians[25] constituted about 30 per cent of the population, a figure that declined to less than 2 per cent of the total population by 1887.[26] This was due, in part, to *mestizaje* (miscegenation), but also due to the burgeoning white immigration policies of the Argentinean government. This miscegenation, which

Andrews refers to as the 'whitening of Buenos Aires'[27] occurred primarily after the ideologue, forefather and future president of modern Argentina (1868–74), Domingo Sarmiento, came to power. In essence, Sarmiento was poised to change Argentina, to usher out colonialism as dominated by the Spanish and bring about a grand project of modernization in an era on the verge of independence. Sarmiento and his 'social Darwinist' predecessors, Bunge, Alberdi and Ingenieros, adopted a 'progressive' set of beliefs, based on 'scientific' racism and on the assumption that European models of society and progress were the desirable evolution of Argentinean society. Sarmiento's vision of an urban and 'civilized' cultivated utopia proclaimed that 'the *pampas* is an immense sheet of paper upon which a poem of prosperity and culture will be imposed.'[28] In the banishing or whitewashing of the 'barbaric' elements, Sarmiento played upon the dramaturgy of this tableau in describing the pre-modern *pampas* as 'shadow' and the River Plata as 'light.'[29]

The project for the modern independent Argentina required a civilized European population that would evolve into a policy-aligned, market-driven economy with a white national identity and white immigration policies to match. The immigration policy aimed at erasing the tensions between the country and the colonial city that clung to the edge of the continent,[30] 'progressively annihilating ethnic differences between the people of the interior and those of the littoral.'[31] This policy was effected deliberately and carefully, attempting to fulfil the desire that, 'there will be only one unique Argentinean type, as imaginative as the aborigine of the tropics and as practical as the dweller of the cold climates, one complex and complete type, which could appear to be the total man, the model of the modern man: *ecce homo*!'[32] as Bunge wrote. The ideologues wanted a (pure) national identity, set apart from their Latin American neighbours who were 'mongrelised' – the flooding of Argentina with immigrants from the North would alter the balance of skin colour and assimilation and intermarriage would take care of the rest, to the extent that Argentina could fulfil for Ingenieros 'a tutelary function over the other republics of the continent.'[33]

The modern project here was the construction of identity, scaled appropriately to 'blank' out the past. Considered in the context of other colonial countries of the time (or, indeed, the future), the fact of the drive towards independence and modernization was not surprising – the scale and deliberation of the construction of a post-colonial identity predicated on whiteness, was: 'by the 1880s, Domingo Sarmiento could write that the banners of the African nations that one used to see at the old Carnival celebrations had been replaced by the flags of the various French, Italian and Spanish clubs and societies.'[34] Curiously, though, as the whitewashing of the city and of the black population was occurring, the disappearance of the Indian and African people was puzzling to the ideologues and ascribed as 'natural.' As Helg tells us:

> Sarmiento, in his desire to see a white America, attributed the process to the secret action of affinity and repulsion. Bunge imputed it to Buenos Aires' climate, to the inability of the blacks' lungs to resist the pampas winds, to intermarrying with whites, and to the waves of European immigration; elsewhere he blessed alcoholism, smallpox, and tuberculosis for having decimated the capital's non-white population. Ingenieros added the devastating effects of the wars of independence and the civil wars of the nineteenth century.[35]

Henceforth, tango's origins are, as Savigliano points out, a 'popular and controversial topic',[36] not least because of tango's early illegitimacy – bastard offspring of a polygamous union, regarded initially as threatening and provocative, resulting in an unwritten history. The origin of tango as a construct of Argentinean national identity is said to have occurred in the 1880s, that this date coincides with the year of independence is hardly surprising. Tango is presented as a European phenomena, portrayed as the struggles of working class immigrants in the burgeoning 'wild west' atmosphere of the city in the 1900s, coupled with the cross-cultural fusion of the differences of those arriving from Italy, France or Spain. In its legitimized form, tango is presented as miscegenation occurring primarily between *white* and *white*, European and Rio Platense – a miscegenation that grew out of the *conventillos*[37] and *barrios*, peripheral and immigrant areas of the city, moving into the centre where it could be adopted as a national identity and presented as such. However, the construction of national identity is nothing other than a mask, as Andrews in *The Afro-Argentineans of Buenos Aires* elaborates, one that reveals as much as it conceals. Miscegenation, in the terms of Argentina and Buenos Aires, is not a simple matter.

The immigration waves of the nineteenth century did not consist solely of a bountiful supply of able and skilled labour, part of the greater economic imperatives of the Argentine government. In essence, the relationships between the black (African and native Indian) population and the later white Europeans is inextricable, their histories and lineages touch each other, similar to the relationship between *candombé* and tango. Tango, as its untraceable etymology reveals in the various Spanish and African claims, cannot easily be pinned down to a specific time, place, cultural group or race. That tango originated in the 1880s seems fallacious, borne out by the following: 'an 1818 will [for a house and tango site] in which the property is mentioned refers to the "said lot situated in the neighborhood of the Parish of Concepcion Tango of the Blacks, by which name it is known"', seemingly locating 'tango' as a place name of the black population. Similarly, as Andrews points out, an ordnance considered by the Montevideo town council in 1807 proposed to 'prohibit the tango of the blacks, their weekly dances.'[38] This, a full sixty years before the historical claim for the 'white' origin of tango. This historical discrepancy is further substantiated by Rossi and Savigliano elsewhere. However, the point here is not to claim authenticity of origin, but rather to raise issues that problematize what we might term the 'cleansing' of this particular aspect of the history of tango and, by implication, of the city.

Andrews, writing of the participation of Afro-Argentines in cultural life in Buenos Aires, traces the relationship between *candombé*, the public dances of the African slave population, from the 1760s onwards and the later emergence of tango. He writes: 'whites were aware that the *candombé* were occasions at which the Africans performed their national dances, calling up memories of their homeland and recreating, even if only for an afternoon, a simulacrum of African society in the New World.'[39] During these times (1780s) the city council viewed these acts as being against Catholicism and against the republic as a whole. However, through the *candombé*, the Afro-Argentines kept a part of their lives free of the absolute control that a slave-owning society sought to exercise over them – an act of remembrance and resistance that was geographic, the dances were a way to maintain a sense of their identity despite their displacements.

As emancipation allowed the black population to have a second class existence, gradually the white and black populations intermingled during the mid-1800s. They met in dance halls called *academias de baile* which were located in lower-class, peripheral areas of the city. In these place 'poor whites and poor blacks met to drink and gamble, to fight and dance'. Out of this cross-racial contact was born the *milonga*, 'a dance created by the young white toughs in mocking imitation of the *candombé*.'[40] This was a precursor to tango as we know it now, the *milonga* being described as a 'slow tango'. This cross-over, occurring on the eve of the death of the *candombé* and the whitewashing of the culture of the Afro-Argentines is ironically the birth of the *milonga* (and therefore tango) as a syncretic culture in its own right. Through 'simulation' or imitation across a racial divide this territorial process occurred.

The predominantly black areas of Buenos Aires were located in the parishes or *barrios* of Monserrat (*'barrio del tambor'* – the *barrio* of the drums) and San Telmo, low-lying lands along creeks adjacent to the river edge of Rio de la Plata. During the 1800s these areas were south of and peripheral to the gridded city centre. Later, as the harbour was constructed and grew, the river edge became incorporated into the more controlled city grid. The black populations of the time living in these areas predominantly served the arriving ships as porters or alternatively as washerwomen, until a yellow fever outbreak in 1871 decimated the population. In this area, excluded from any provision by the city, the blacks established their own community structures such as *La Nación Conga*, a significant organization that was responsible for the annual *candombé* carnival for the local population, a celebration of the African 'nations' that recalled their tribal and geographic roots. Spatially, San Telmo was sandwiched between the official city to the north and the slave market.

Immediately to the south of this area lies Parque Lezama, the site of Mendoza's first disastrous settlement, the short-lived 'first' or originary Buenos Aires. It was also the site of the first slave-market built by the French Guinea Company. Topographically, the park sits on an escarpment, perched just above La Boca where the bank of the Rio de la Plata curves inland as the Riachuelo river joins it, serving as a gateway between the formal grid city and low-lying river flats of La Boca with their different urban pattern. It also separates distinct layers of immigration into the city across this topographic line, delimiting the African *barrios* of the 1830s from the Ligurian and French–Basquean *barrios*. The complex relationship between tango and the city throws up a spatial coincidence of a tangled set of factors: race, slavery, the threshold nature of the city, the official city versus the periphery, and the 'natural' topography.[41] Conjecturally, therefore, one may speculate that from this milieu, the touching line between the black and white population, between San Telmo or Monserrat and La Boca across the space of Parque Lezama, has led to the imitation and cross-over alluded to by Andrews that gave rise to tango, as we now know it occurred. Such touching lines are resonant with significance[42] yet, given the influx of the European immigrants, the social Darwinist policies of Sarmiento and his successors and the miscegenation of black and white, these have become sublimated in the whitening of Buenos Aires. The decline of the black population during these times and the supplantation by the incoming European white immigrants constitutes a tangential line, a cultural, racial and spatial tangent of differences. Tango, therefore, becomes a touching point of two races at which the two lineages of tango – black and white – meet.

DISPLACED CITY

Argentina's modern project depended, as did the city's growth, on expansionist trade. The rapid modernization programme of Buenos Aires as a modern state-sanctioned development generated the periphery as a progressive frontier, a grid of urbanity spreading over the *pampas*. The grid during this period served as a promissory note for an expansionist future envisaged by the planners of the modern project, a tool that, with civic and state impetus and drive, is enmeshed in a 'dense public, metropolitan space.'[43] The grid, like the periphery, was 'the promise of equality and integration',[44] projected, planned and unrolled, only to be quickly – if not immediately – filled with a diverse population. It served effectively to integrate cultural and social differences within the construct of the *arrabals* and *barrios*.[45] It was from the *barrio*,[46] the physical manifestation of the modern project – both culturally, socially, politically and in terms of urban form – that tango, in its modern form, emerged. The street thus became, in the absence of open space in a rapidly expanding city during this time, the provisional public space. This notwithstanding, since the Second World War commencing with the Peronist era, and in many ways exacerbated by the dictatorship of the 1970s, a shift occurred. The periphery became 'the gate to the city' with people flooding into the city from the interior. This shift coincided with the end of the expansionist relationship between the *pampas*, the city and the northern hemisphere. The external markets no longer required Argentina to produce to the same extent it once did. And the city as an import–export threshold that once extended into the *pampas* turns, in effect, inwards on itself, corresponding too with growing political instability that culminated in the dictatorship years during the 1970s. The failed 'promise' of the periphery has also led, in contemporary terms, to a crisis due to the disappearance or collapse of the state and all its manifestations. The decline of the dream of the *pampas*, a victim to global and historical shifts in the economy, has therefore had major repercussions on a city that depended solely on this ambition. Gorelik refers to this as the development of the 'second periphery', a periphery that is 'no longer inhabited by immigrants coming from Europe but by migrants coming from the provinces and neighboring countries.'[47] This, he indicates, has had a 'LatinAmerican-ising' effect on Buenos Aires since the 1970s, on what was once the most European of Latin American cities, whose external immigration from Europe 'staved off' the effects that have characterized other South American cities such as Lima or Mexico City, but is now witnessing the emergence of a migration from the interior.

As this dream became increasingly unfulfilled, the periphery of the city became a receptacle for the increasingly poverty stricken, those fleeing the land. Similarly, the harbour edge became derelict. The modern project, built on the 'utopian' vision of Sarmiento and others half a century before, thus remains unfinished. As a consequence, the resulting urbanity now appears fragmentary 'like postcards containing unkept promises, in which *objets trouvés* and insignificant gaps alternate on the unvarying expanse of the *pampas*.'[48]

As Gorelik further elaborates, the idea of the periphery since the 1970s no longer corresponds to a notion of equality, progress and modernity as it once did, but develops in a fragmentary or island-like way. The effect of this is evident in the wide disparities that have emerged between neighbouring areas and accordingly, the emblems of this shift have also been subject to change. If the street was once the domain within the grid system of the *barrio*, immortalized in tango or in literature by Borges (for example, 'The Street Corner-Man') as that place within a city where the charades of *machismo*[49] and ritual dance are enacted, then the emergence of shopping malls, enclosed and isolated, separates people from the clamorous and chaotic street where differences are constantly met. 'By cultural tradition, Argentineans have always been instinctively against supermarkets and closed spaces generally which can't be traversed [or] are cut off from the streets. All the same, the grave economic circumstances of the early 80s forced certain social classes to make use of mass-distribution stores.'[50] This retreat of public space under the dictatorship, out of fear and imposed curfews, has meant that the transition to a post-dictatorship free-market economy,[51] albeit under economic sanctions imposed from North America, naturally privatizes public space in a move that is almost a seamless transition between dictatorship and democratic process.

The Hausmannesque plans of the dictatorship cut through the fabric of the city grid and are part of a massive but aborted highways project during the 1970s. These highways, in particular, run through the outer-lying *barrio* areas, in what must surely have been part of a system of political control. One such highway cut is *Autopista 25 de Mayo* between San Telmo and La Boca, just north of Parque Lezama, running parallel to the route of an old creek that marks the southern edge of the city centre grid. This highway effectively separates the poorer barrios of La Boca from the more official city centre which had long since incorporated San Telmo in its official grid structure. New tensions across this line are now established that repeat, to some extent, the lines written in the urban fabric of earlier times.

The rubble from these cuts was dumped into the river on the Banco de la Boca, a sandbank just outside of the Puerto Madero harbour, creating new islands in the Rio de la Plato and accidentally manifesting Corbusier's plan of fifty years earlier for an island in the river. Over time, the new islands have become a natural wasteland, attracting all kinds of wildlife who dwell amongst the bricks and rubble. This new land separates the city and its port from the Rio de la Plata. At the same time it grafts a strange new mask on to the city in the form of beaches and ecological habitats. This is now an area witnessing transformation in the form of the construction of enclaves for the economic élite, and in effect privatizing public space, as they have done globally in various cities in the past decade or so. Speculative developments, removed from the city, are proposed for this displaced land that have more to do with globalization than local factors. At the same time, in conjunction with these developments, government policy is encouraging economic immigration as a possible panacea to Argentina's post-dictatorship economic situation, whilst controlling internal migration. This migration policy is exemplified in an advertisement published in the following London newspaper:

> The National State Ministry of the Interior announcement for public, national and international bid (Number 01/96) for the full, indivisible contract of a service for the design, start-up and support of a System of Migration Control and Identification of Individuals and of Electoral Information.
>
> (Ministry of the Interior, Advertisement in the *Guardian* newspaper, 26 August 1996, London)

A form perhaps of economic whitening coupled with the implementation of the generic city, a city that is removed from place and ground.[52]

These actions are emblematic of the incompleteness of the Buenos Aires modern project, and part of the '*tradición de ruptura*',[53] that Octavio Paz has written about. The rupture between modernity and centralized economic and cultural policy that has occurred implies a shift of the order of things, leading to new displacements and disruptions. The ideological neo-conservative privatization of the city and of public spaces implemented under the present government[54] borrows heavily on the one hand from Thatcher's Britain of the 1980s, as indicative in the calling for tenders for control mechanisms by the Ministry of the Interior, whilst on the other hand, as in other Latin American countries, national policy becomes increasingly dictated by external North American agencies such as the Brady Plan and the IMF,[55] with its now familiar austerity or restructuring impositions of free-market economy, biased towards a North–South free trade zone. These are factors which, although derived from global extraneous influences, have transformative effects on cities and similarly on urban culture. They are factors that are by no means new to the Rio de la Plata region, whose culture has always been geographic in some way and influenced by sources outside its physical domain, but the repercussions become manifested locally as physical ruptures in the city fabric, the contraction and privatization of public space. We can view the current large structurally changing North–South emphasis of USA–Latin America and the effects of this (structural economic adjustment, internal migration, etc.) in terms of a politic of memory that constructs and overrides the old order: the execution of a new set of global forces impacting upon the urban, physical and cultural, social structures, and, moreover, beginning to touch on the inner conditions of those whom it effects.

Thus, if Buenos Aires is becoming Latin-Americanised, then the proposed global city, on the new island of displaced land and rubble, is somehow symbolic of the same factors that marginalized the areas of La Boca and San Telmo previously, given by immigration in relation to the state. It becomes external to the city yet connected spatially across a touching line, a part of the city that desires to be elsewhere.

The touching lines between each of these areas, these cultures, and the city itself manifest similar patterns. The same figures reoccur in the unfolding of the city; the erasure or displacement of one body (one cultural layer) within, coupled with an appeal to an external body to provide impetus or identity; the spatial connections between the geographic and the localized across the specific thresholds of Buenos Aires. Bound in an intertwined relationship, this dynamic is a tangoing movement of two bodies.

LIBERTANGO[56]

The touching line, we postulate, enables us to understand a horizon of identity located in a specific place at a particular time, where cultural factors meet with spatial; where two radically incommensurable, different figures, entities or bodies touch and an emotional landscape, buried under the surface, emerges. At these points, assumptions of raw emotions, ill-defined identities, cultures and raw urbanity give rise to new form as ephemeral phenomena that may well be erased, forgotten, disappeared, blanked-out. It results in spatial stories, rather than 'history' or hard-line urbanism, that deal with the affect and effect of the geographic and large-scale factors (the modern project and post-colonial issues, for example) meeting with everyday conditions, the interior and the imaginary.

Tango's legitimization, we argue, is therefore not based on part of a reinterpretation of history or a revisionist understanding, nor as authenticity or lament. Instead it is more potent to see it as tango-ing, with the mechanic of the touching line as the understanding of submerged emotional conditions and allied with specific urban place and conditions in the development of a cultural identity, attempting to understand the touching of the local, the barrios, daily life (everyday life) with immense geographic or global phenomena. It is an approach open to interpretation that considers the potential of place whose emotional intensity is a dynamic and potent urban condition that manifests a collective understanding and is therefore a constituent of a possible new culture. Intervening in such a locale then may articulate strange links, allow heterogeneous bodies to touch, choreograph their tensions and make their qualities into a collective expression; an urbanity.

What is touched upon, *taconeando*, a footnote[57] of the dance, is a choreography of bodies held in tension.

> Gone is the arrabal with all its lustre,
> perhaps its history is the crucifix of the dagger
> The arrabal is gone that spoke of love,
> and the stamp of the heels is gone.
>
> Toconeando (Tango song from 1930)[58]

NOTES

1 Edmund Husserl, who wrote *The Origin of Geometry*, (as Michel Serres reminds us) at a time when geometry is 'disappearing' or 'dying', calls the presence of an unknown a 'horizon', for that which is outside of our perception and knowledge. The tangent reminds us of the limitations of Euclidean geometry's parallel lines, an axiom that is the Achilles heel of geometry for it cannot be proved that two parallel lines do not touch somewhere, over a horizon.

2 'In 1930, the term "periphery" was not in common use and one referred to such areas as *arrabal*, or *barrio*, i.e. as "suburbs" or "outskirts"'. Gorelik, A., 'A Place of Time: the Periphery of Buenos Aires', *Daidalos* 50, December 1993, p112. *Arrabal* means 'slum-suburb', or 'outskirts', the edge where the *pampas* meets the city.

3 The conferring of authority is passed on through lineage, embodied, claimed, discovered and appropriated as Michel de Certeau, in *The Writing of History*, Columbia University Press, New York, 1988, pxxv, tells us: 'Amerigo Vespucci the voyager arrives from the sea. A crusader standing erect in his body armor, he bears the European weapons of meaning. Behind him are the vessels that will bring back to the European West the spoils of a paradise. Before him is the Indian "America"', a nude woman reclining in her hammock, an un-named presence of difference, a body which awakens within a space of exotic fauna and flora. An inaugural scene: after a moment of stupor, on this threshold dotted with colonnades of trees, the conqueror will write the body of the other and trace

there his own history. From her he will make a historicised body – a blazon – of his labors and phantasms. She will be "Latin" America. . . . she is *nuova terra* not yet existing on maps – an unknown body destined to hear the name, *Amerigo*, of her inventor. But what is really initiated here is a colonialization of the body by the discourse of power. This is writing that conquers. It will use the New World as if it were a blank, "savage" page on which Western desire will be written.'

4 Sarlo, Beatriz, '*Una Modernidad Periferica: Buenos Aires 1920 y 30', in Buenos Aires: Ediciones Nueva Vision*, 1988, p31. 'Cuandos cambios acelerados en la sociedad suscitan sentimientos di incertidumbre. Un viejo orden recordado fantaseado es reconstruido por la memoria coma pasado. Contra este horizonte se coloca y se evalua el presente.' Translation from Spanish by authors.

5 Galeano, Eduardo, '1890 River Plata: Comrades', *Memory of Fire*, Quartet Books, London, 1995, p555.

6 Refer to James Scobie's classic two-part study, *Buenos Aires: Plaza to Suburb, 1870–1910*, Oxford University Press, New York, 1974, for further elaboration.

7 '*La Argentina recibia a 15 mil immigrantes por mes*,' *El Popular* newspaper, Buenos Aires, 9 January, 1925.

8 'The tango comes from *gaucho* tunes of the interior and comes from the sea, the shanties of sailors. It comes from the slaves of Africa and the gypsies of Andalusia. Spain contributes with guitar, Germany its concertina, Italy its mandolin. The driver of the horse-drawn streetcar contributed his trumpet and the immigrant worker his harmonica, comrade of lonely moments.' Galeano, op cit., 'Faces and Masks', *Tangoing*, p557.

9 'The city is roofed over by a vast horizontal sky: yes, the sole great consolation. For I have seen it, this sky, on the endless plain of grasslands, punctuated by a few weeping willows. It is unlimited, as sparkling by day as by night with a transparent blue light or with myriads of stars; it spreads to all four horizons . . . to tell the truth, all this landscape is one single and same straight line – the horizon,' wrote Le Corbusier in *Precisions*, MIT Press, Cambridge, Massachusetts, 1991, p4.

10 Galeano, op. cit. 'Man Alone', p556.

11 '*Plata*' as in '*Rio de la Plata*' means 'silver', but is commonly used as the word for 'money' in the region.

12 Galeano, op. cit., 'The Founders', p170.

13 Rama, Angel, *The Lettered City*, translation by John Charles Chasteen, Duke University Press, Durham, 1996, p2.

14 Borges, Jorge Luis, 'The Mythical Founding of Buenos Aires', *Selected Poems 1923–1967*, Penguin, London, 1972, p63.

15 Tuer, Dot, 'Cartographies of Memory', *Parachute*, no. 83, Jul/Aug/Sept 1996, p25.

16 'All at once, above the first illuminated beacons, I saw Buenos Aires. The uniform river, flat, without limits to the left and to the right; above your Argentine sky so filled with stars; a Buenos Aires, this phenomenal line of light beginning on the right at infinity and fleeting to the left towards infinity. Nothing else, except, at the centre of the line of light, the electric glitter which announces the heart of the city. The simple meeting of the *pampas* and the river in one line, illuminated the night from one end to the other. Mirage, miracle of the night, the simple punctuation regular and infinite of the lights of the city describes what Buenos Aires is in the eyes of the voyageur. This vision remained for one instance and imperious I thought: nothing exists in Buenos Aires – but what a strong and majestic line.' Le Corbusier, op. cit., p201.

17 *Ibid.*, p.202.

18 Refer to Collins, Christine Crasemann, 'Urban Interchange in the Southern Cone: Le Corbusier (1929) and Werner Hegemann (1931) in Argentina,' *Journal of the Society of Architectural Historians*, No. 542, June 1995. At around the same time, Werner Hegemann, a German-American architect and planner visited Buenos Aires and made a number of urban proposals for Buenos Aires. Hegemann's more in-depth proposal for Buenos Aires focused instead on the connection of the city to the *pampas*. In a sense, this further articulated the modern project where the periphery effectively became the centre. Hegemann, unlike Corbusier, based the premise of his plan on an in-depth study and protracted fieldwork, rather than on immediate impressions. He immersed himself in the culture of the city and attempted to comprehend its complex dynamic, as Collins writes: 'anchoring his recommendations in the local context, Hegemann reinforced the uniqueness of the South American city as it was striving for its own identity.' In this focus on the periphery, Hegemann specifically considered the role of the *barrio* and presented his findings in a film entitled *La Cuidad del Mañana*, (City of Tomorrow). In part, Hegemann recognised the heterogeneity and ad-hoc nature of the outer edges of the city as 'democratic in spirit . . . [he] saw in these unpretentious buildings, as well as in the self-built dwellings on the periphery of the capital, an incipient indigenous style.' Collins, drawing on della Paolera, points out the crucial differences between these two imported authorities, Corbusier and Hegemann, as 'emblematic' of the difference between a *plano* regulador and a *plan* regulador. The former is a 'graphic prescription for an urban

intervention, which, at the moment of its creation, may be the correct prescription: but already while it is being implemented, the complex city organism has transformed itself.' The latter is comprised of 'many planos in a continuous, energetic planning politic that commands the momentum and flexibility required to organise a major city.'

19 *La boca* means 'mouth' in Spanish.

20 Apart from the Ligurians, all other immigration was stopped at the time by the then president Rosas. 'The man of the interior has stripped Buenos Aires of any materiality and transformed her into a formidable emporium of the best that exists in our reality and in our imagination. Thus Buenos Aires is the centre of a circumference formed by the most populated points and cultivated by the interior. They are all at the same distance. They are periphery as she is centre.' Martinez Estrada, *Civitas*, 1963.

21 Andrews, George Reid, *The Afro-Argentines of Buenos Aires 1800–1900*, The University of Wisconsin Press, Madison, Wisconsin, 1980, p11. During the 1880s, when ships were forced to anchor five or six miles off the coast, horses and carts were used to traverse the land and the water, due to the shallowness of the bay and lack of an adequate harbor: 'a base of planks two or three inches apart, through which water splashes at every wave, mounted on a big heavy wooden axle between a pair of gigantic wheels. To this ungovernable machine is tied a horse. The wild brutish appearance of the sun-tanned cart drivers who, half naked, swear and scream and shove one another and whip their poor exhausted horses into the water.' Suarez, Odilia E., 'Buenos Aires: La Story Urbanistica', *Abitare*, No. 342, July/August 1995, p87.

22 Savigliano, Marta E., *Tango and the Political Economy of Passion*, Westview Press, San Francisco, 1995, p24.

23 Galeano, op. cit., '1870: Buenos Aires: Sarmiento', p530. As Aline Helg writes concerning the writings of Sarmiento and his ideologues in 'Race in Argentina and Cuba, 1880–1930', *Theory, Policies and Popular Reaction in The Idea of Race in Latin America, 1870–1940*, (ed.) Richard Graham, University of Texas Press, Texas, 1990, 'Ultimately, then, the difficult relationship between scientific racism and social environment brings up the question of the audience. For whom were Hispanic American intellectuals writing? For their fellow countrymen in order to redeem them, as they claimed? Or for European colleagues, in order to be published in their Journals and welcomed in their circles? Or for themselves, to relieve anxiety and guilt? The last hypothesis seems most likely. They tried to build up an imaginary and stereotyped world that would function according to permanent and logical biopsychological laws. It would be a world with acknowledged enemies and myths. A 'scientific,' 'rational' world, when a confusing and rapidly changing reality made it too, difficult to find one's identity.'

24 Helg, op. cit., p41: 'The Indians, however, were considered the most challenging enemy of Argentinean civilisation until the early 1880s. In 1879, Gen. Julio A. Roca initiated his 'Conquest of the Desert', a misnamed war that expanded the Argentinean frontier southward by subduing or exterminating an entire aboriginal group . . . in brief, the policies implemented converged on a single goal: to eliminate the aborigines in order to direct new European immigrants to the exploitation of interior lands.'

25 *Ibid.*, p43: 'In reality, the whitening of Argentina through immigration had been a fast process. In 1869 Indians represented 5 percent of the population, but by 1895, only 0.7 percent of a total of 3,955,000. The blacks, centered in Buenos Aires, had composed 25 percent of the capital's population in 1838, but had dropped to 2 percent by 1887. Between 1880 and 1930, total net immigration added nearly 3,225,000 inhabitants to Argentina. Among the immigrants, 43 percent were Italian and 34 percent Spanish; far behind came the highly valued Anglo-Saxons. As for the Jews, during the peak years of 1907 to 1914, they represented between 2 and 6 percent of total net immigration.'

26 Barrenechea, Ana María, 'Sarmiento and the 'Buenos Aires/Córdoba' Duality', in Donghi, T H., Jaksic, I. & Kirkpatrick, G., (eds.), Masiello, F., *Sarmiento: Author of a Nation*, University of California Press, Berkeley, 1994, p106.

27 *Ibid.*, p63.

28 *Ibid.*, pp62–63. Donghi, Jaksic, Kirkpatrick & Masiello elaborate these issues in their biography of Sarmiento.

29 *Ibid.*, p62.

30 Helg, op. cit., p41.

31 *Ibid.*, p41.

32 *Ibid.*, pp42–43.

33 Andrew, op. cit., p106. The slaves and Indians effectively become earlier versions of *los desaparecidos* of the later dictatorship (1976–1983), during which 15,000 people 'disappeared,' of whom no trace other than *memorias de sangre* (memories of blood) remain. The city turns red.

34 Helg, op. cit., p43.

35 Savigliano, op. cit., p199.

36 *Conventillos* were tenements that housed immigrants in areas such as San Telmo, the former black area and other harbour-side areas. Often larger houses originally, they

were subdivided into which newly arriving immigrants were crowded.

37 Andrews, op. cit., p165.

38 *Ibid.*, p162.

39 *Ibid.*, p166.

40 By no means a pattern unique to Buenos Aires, this phenomenon can be found in other cities such as London's East End where the relationships between the harbor area and immigrant cultures are inextricable: the immigrant in effect is almost commodified as import or export, being somehow just outside of the city, often in the floodplains or low lying lands and yet giving new input to the culture of the city. Spitalfields and Brick Lane in London are clear examples of this, the areas having a richness of layers from the various cultural groups. Similarly, a touching line could be said to exist between Spitalfields and the City of London across Bishopsgate.

41 Akin to what Michel de Certeau calls 'Local Authorities', a 'crack' in a place pregnant with signification, from which many spatial stories may be unfolded as the emergent potential of place to speak of emotion (the soul of a place) as opposed to the imposition of this from above as historic narrative. See Michel de Certeau, *The Practice of Everyday Life*, translated by Steven Rendall, University of California Press, California, 1984.

42 Gorelik, op. cit., p110.

43 See Paul Carter's writings on the colonial city in *Australia: The Road to Botony Bay*, Faber & Faber, London, 1987, p204: '. . . land there could be regarded very much like land here. Empty spaces on the map, accountable and equally subdivided, should yield returns that could more or less be computed in advance. Located against the imaginary grid, the blankness of unexplored country was translated into a blueprint for colonisation: it could be divided up into blocks, the blocks numbered and the land auctioned without ever having to leave their London offices.'

44 The *barrio* is not a measurable entity: it is a vicinity – a set of invisible and visible relations constituted by approximation in social contacts; distance and access to the centre of the city; density of inhabitants and buildings; ethnicity; class and history. A result of this is that crossing over a street might mean passing by doorways in silence – in the previous *cuadra* (block), one would always be greeted. For further definitions of the *barrio* see James Scobie, *Buenos Aires – From Plaza to Suburb*, 1870–1910, Oxford University Press, New York, 1974.

45 Tracing the development of the city of Buenos Aires, the historian James Scobie writes: 'terrain, transportation, and land use served to differentiate suburbs, but running throughout the outward expansion of the city towards the suburbs was the common experience and unifying theme of the *barrio*, or local neighborhood. Although nowhere precisely defined or recorded as a unit of measurement, the *barrio*, along with its smallest component, the *cuadra*, was integral to the city's formation. The *barrio* and the *cuadra* were developed principally through a sense of attachment and social contact between inhabitants . . . in heavily built-up zones, the *barrio* might consist of a single *cuadra*, while in outlying, sparsely inhabited areas it might include as many as a dozen *cuadras*.' Scobie, op. cit., pp201–202.

46 *Ibid.*, p112.

47 *Ibid.*, p109.

48 For a discussion of gender-related issues in tango and Argentinean culture, refer to Savigliano. It is worth bearing in mind that during the early years of expansion, both as colony and during the nineteenth century, the predominance of males meant, for example, that tango was often danced between males, with one male pretending to be the female partner.

49 Fiorenza, Dante, *Arbitare*, op. cit., 'I Mall d'Argentina', p93–94.

50 As Jean Franco has elaborated in her detailed article 'From Public Space to the Fortified Enclave: Neoliberalism's effect on the Latin American City' in *Anybody*.

51 This is similar to cities currently being built in the Far East, not to mention London's Canary Wharf.

52 Paz, Octavio, 'The Labyrinth of Solitude', quoted in George Yúdice, 'Postmodernity and Transnational Capitalism', Yúdice, G., Franco, J. & Flores, J. (eds.) *On Edge: The Crisis of Contemporary Latin American Culture*, University of Minnesota Press, Minneapolis, 1992, p5. '. . . what does it really mean to be Argentine? Who has acquired the rights to define the still un-delimited field of Argentine culture?' Beatriz Sarlo, 'Xul Solar', in *Argentina 1924–1994*, Exhibition catalogue: MOMA, Oxford, 1994, p34.

53 Here referring to the Menem administration.

54 See Rex Butler, 'Buying Time' in *FutureFall: Excursions into Post-Modernity*, (ed.) Grosz, E. A., Threadgold, T., Kelly, D., Cholodenko, A., and Colless, E., Power Institute Publications, Sydney, 1986.

55 *Libertango* is a song composed by Astor Piazzola, during the 1970s. The name comes from *libertad* and *tango*.

56 A footnote is also the sound the foot makes on the floor. *The Authors' and Printers' Dictionary* by F. Howard Collins (1948) indexes good practice in the art of books. Collins writes of footnotes that 'a white line, or rule, should separate footnotes from the text (see also authorities, reference marks).' p135.

58 Translation from Spanish by authors.

'A Performative Critique of the City: The Urban Practice of Skateboarding, 1958–98'

Everything (2003)

Iain Borden

Editors' Introduction

Iain Borden is Director of the Bartlett School of Architecture, University College London, where he is Professor of Architecture and Urban Culture. An architectural historian and urban commentator, his historical and theoretical interests have led to publications on critical theory and architectural historical methodology, the history of skateboarding as an urban practice, boundaries and surveillance, Henri Lefebvre and Georg Simmel, Renaissance urban space, architectural modernism and modernity, film and architecture, gender and architecture, body spaces and the experience of space. His photographs have been published both in his own publications and those by other historians and architects. He is currently working on a history of different kinds of movement and architectural space, and on a television documentary and international touring exhibition about skateboarding and urban space.

His authored and co-edited books include *Manual: the Architecture and Office of Allford Hall Monaghan Morris* (2003), *Skateboarding, Space and the City* (2001), *The Unknown City* (2001), *New Babylonians* (2001), *The Dissertation* (2000), *InterSections* (2000), *Gender Space Architecture* (1999), *Strangely Familiar* (1996) and *Architecture and the Sites of History* (1995).

> Surely it is the supreme illusion to defer to architects, urbanists or planners as being experts or ultimate authorities in matters relating to space.[1]

Lefebvre's attitude toward space has come to be widely held across different disciplines and discourses. Architects and planners may be the functionaries and ideologists of urban space, but their schema and drawings, their buildings and planned spaces, do not themselves constitute urban space. Rather, urban space is a continual reproduction, involving not just material objects and practices, not just codified texts and representations, but also imaginations and experiences of space.

What I want to do here is to explore a particular kind of urban space production, one which utilises the objects and spaces of the city, but which does not itself produce any objectival thing. In particular, I want to focus on the compositional and representational mode of skateboarding, to consider *how it represents* the city without maps, and *how it speaks* something of the city, without recourse to theory or texts.

FROM SURF TO STREETS

Skateboarding began in the beach cities of California, first in the late 1950s through to the early 1970s as a surfer's activity, emulating the surf moves on the hard surfaces of urban subdivisions and rolling tarmac.

By the mid-1970s skaters had located a variety of what I call found terrains, on which they further extended their surf-related moves. These ranged from schoolyard banks, such as those at Kenter School in the Brentwood area of LA, to drainage ditches, such as Stoker Hill, to concrete pipes found out in the desert. Most importantly of all, skaters discovered that once drained of water, the round, keyhole or kidney shaped swimming pools favoured in many of the more moneyed LA residences offered a curved transition from floor to wall. Skaters carved up the walls, explored the limits of the tile and coping, and even the space beyond the wall with "aerial" moves in which the skater turns around in mid-air before returning to the pool surface below.

In the late 1970s, such moves became even more dramatic in the new skateparks – purpose-built, commercial facilities built all over the US, UK and other countries worldwide. Such skateparks typically offered a range of elements, including dramatically exaggerated pools, replete with tiles and coping – some of the most famous include the skatepark at Marina del Rey (Los Angeles) and "Pipeline" (Uplands, Los Angeles) in the USA and "The Rom" (Romford) and "Solid Surf" (Harrow) in the UK.

But from the early 1980s onward, skateboarding has increasingly gone back to the streets, not so much to the suburban drives of California but to the inner city cores of other cities worldwide. The urban practice of skateboarding has become a global phenomenon, with I estimate around 20–40 million dedicated practitioners dispersed through just about every modern city worldwide. The focus of it all remains, however, the USA, with new centres like Philadelphia, Chicago, New York and San Francisco joining Los Angeles as major concentrations of skate activity. Here, in the modernist city, skaters ride on to the walls, benches, ledges, railings, fire hydrants and other paraphernalia of the urban street.

There is also a social and cultural dimension to this, for skateboarding in fact represents a totalising urban subculture, complete with its own graphic design, language, music, magazines, junk food and codes of behaviour. It postulates certain attitudes towards matters of gender relations, race, sexuality and masculinity. Above all, this is a subculture which rejects work, the family and normative American values. As one American skater put it,

> Baseball, hotdogs, apple pie, weed, beer, pills, needles, alcohol etc., etc., are all typical hobbies of all the typical people in all the typical states in the typical country of the United States of Amerika [. . .] Why be a clone? Why be typical?[2]

This is a totalising subculture, in which partial allegiance is to miss the point and which ultimately presents the skater with a single binary choice: skate or be stupid.

Skateboarding thus brings together a concern to live out an idealised present, trying to live outside of society while being simultaneously within its very heart.[3] But for skateboarders to produce themselves in this way, their activity must take place in the streets of the city. Its representational mode is not that of writing, drawing or theorising, but of performing – of speaking their meanings and critiques of the city through their urban actions. Here in the movement of the body across urban space, and in its direct interaction with the modern architecture of the city, lies the central critique of skateboarding – a rejection both of the values and of the spatio-temporal modes of living in the contemporary capitalist city.

The skater's engagement with the city is, in particular, a run across its terrains, with momentary settlings and encounters with all manner of diverse objects and spaces: ledges, walls, hydrants, rails, steps, benches, planters, bins, kerbs, banks and so on. In the words of Stacy Peralta, '[S]katers can exist on the essentials of what is out there. Any terrain. For urban skaters the city is the hardware on their trip.'[4]

In this sense, skaters see the city as a set of objects. Yet cities are not things, but the apparent form of the urbanisation process,[5] and are in fact filled with ideas, culture and memories, with flows of money, information and ideologies, and are dynamically constitutive of the continual reproduction of the urban. To see the city as a collection of objects is then to fail to see its real character. And this is exactly the failure one could say of skateboarding, which does little or nothing to *analyse* the processes which form the urban; instead, the phenomenal procedures of skateboarding rely entirely on the objectival nature of the city, treating its surfaces – horizontal, vertical, diagonal, curved – as the physical ground on which to operate.

Yet within this failure lies a profound critique of the city *qua* object-thing. Capitalism has replaced the city as *oeuvre* – the unintentional and collective work of art, richly significant yet embedded in everyday life[6]

– with "repetitive spaces", "repetitive gestures" and standardised *things* of all kinds to be exchanged and reproduced, differentiated only by money.[7] Skateboarding, however, at once accepts and denies this presentation of cities as collections of repetitive things. On the one hand, skateboarders accept it, by focusing purely on the phenomenal characteristics of architecture, on its compositions of planes, surfaces and textures as accessible to the skateboarder. 'Look around. Look at a world full of skate shapes [. . .] shapes left there by architects for you to skate.'[8]

Here the city and its architecture is undoubtedly a thing. On the other hand, it is also through this very focus on the phenomenal that a change is made. When skateboarders ride along a wall, over a fire hydrant or up a building, they are entirely indifferent to its function or ideological content. They are therefore no longer even concerned with its presence as a *building*, as a composition of spaces and materials logically disposed to create a coherent urban entity. By focusing only on certain elements (ledges, walls, banks, rails) of the building, skateboarders deny architecture's existence as a discrete three-dimensional indivisible thing, knowable only as a totality, and treat it instead as a set of floating, detached, physical elements isolated from each other; where architects' considerations of building "users"[9] imply a quantification of the body subordinate to space and design, the skater's performative body has, "the ability to deal with a given set of pre-determined circumstances and to extract what you want and to discard the rest".[10] Skateboarding reproduces architecture in its own measure, re-editing it as series of surfaces, textures and micro-objects.

> Buildings are building blocks for the open minded.[11]

Architecture (following here Lefebvre's body-centric formulations) "reproduces itself within those who *use* the space in question, within their lived experience".[12] This occurs in skateboarding through architecture being encountered in relation to height, tactility, transition, slipperiness, roughness, damage to skin on touching, damage to body from a fall, angle and verticality, sequencing, drops (stairs and ramps), kinks and shape (hand-rails), profiles (edges), materials, lengths and so on. And only a very small part of the architecture is used – the "building" for a skater only an extracted edit of its total existence.

For example, a particular English school in Ipswich is known by skaters not as a building or function, but for its handrails. '[T]ravel to Ipswich and ask to check out the school with the handrails, they'll know which one and it's sick.'[13] Also in Ipswich, Suffolk College was known primarily for its roof, stairs and ledges, a specific church was known for the wooden benches outside, another school for some steps, and an entire US air base for a single, yellow fire hydrant.[14]

Similarly, on the other side of the Atlantic, the Marriott Marquis Hotel in New York, (1985, architect John Portman), and offering the usual Portman features of vast glass elevations, spectacular atrium, rocket ship elevators and internal glitz,[15] was reconceived by skaters as "modern day skate architecture" and identified for its "tight transitions", "black walls", street-level walkway and for its planters.[16] Similarly, New York's Museum of Natural History became "100 yards of Italian marble, marble benches curbed for frontside and backside rails, six steps, and statues of famous dudes with marble bases [. . .] basically an awesome skate arena".[17]

What ties these elements together is neither compositional, structural, servicing or functional logic, but the entirely separate logic composed from the skateboarder's moving rapidly from one building or urban element to another. Such "strategies embracing architecture"[18] select what in design-architectural terms are a discontinuous series of walls, surfaces, steps and boundaries, but which in skateboarding's space–time become a flow of encounters and engagements between board, body and terrain.

> Find it. Grind it. Leave it behind.[19]

Skateboarding here resists the standardisation and repetition of the city as a serial production of building types, functions and discrete objects; it decentres building-objects in time and space in order to recompose them as a strung-out yet newly synchronous arrangement. Thus while many conceive of cities as comprehensive urban plans, monuments or *grands projets*, skateboarding suggests that cities can be thought of as series of micro-spaces. Consequently, architecture is seen to lie beyond the province of the architect and is thrown instead into the turbulent nexus of reproduction.[20] 'On the street the urban blight is being reworked to new specifications. The man on the avenue is the architect of the future [. . .] There are now no formalized plans. Invent your own life.'[21] Through such compositions, skateboarding brings back that which strictly economistic Marxism evacuates – it brings back the dream, imaginary and "poetic being",[22] what one skater called the "skate of the art".[23] Skateboarding points to the resurrection of the urban not as a product, but as a way of living.

PERFORMING CITIES

Skateboarding is, then, at one level an aesthetic rather than ethical practice, using the "formants" at its disposal to create an alternative reality.[24] Skateboarders analyse architecture not for historical, symbolic or authorial content but for how surfaces present themselves as skateable surfaces. This is what *Thrasher* skateboard magazine calls the "skater's eye": 'People who ride skateboards look at the world in a very different way. Angles, spots, lurkers and cops all dot the landscape that we all travel.'[25] How then does this aesthetic activity take place? What techniques or modes of representation are involved?

As already noted, skateboarders undertake a discontinuous edit of architecture and urban space, recomposing their own city from different places, locations, urban elements, routes and times. The city for the skateboarder becomes a kind of *capriccio*, the tourist's postcard where various architectural sites are compressed into an irrational (in time and space) view,[26] except the editing tool is here not eye, camera or tourist coach but motile body.

One effect of this is that a different kind of canon of city architecture is drawn up – substituting everyday architecture for great monuments and buildings by famous architects. The city for skateboarders is not buildings but a set of ledges, window sills, walls, roofs, railings, porches, steps, salt bins, fire hydrants, bus benches, water tanks, newspaper stands, pavements, planters, kerbs, handrails, barriers, fences, banks, skips, posts, tables and so on: "To us these things are more."[27] New York, for example, is for skaters not the New York of the Statue of Liberty, Times Square, 42nd Street, Central Park and the Chrysler Building, but of the Bear Stearns Building (46th and 47th, Park and Lexington), "Bubble banks" (south side of 747 3rd Avenue), "Harlem banks" (Malcolm X Avenue and 139th), "Brooklyn banks" (Manhattan end of Brooklyn Bridge), Washington Square Park, Mullaly Park in Brooklyn, Marriott Marquis Hotel (45th and Broadway), Bell Plaza banks etc.[28] Washington, by the same process, became known architecturally to skaters as Pulaski Park, National Geographic Building, Federal Welfare Archives, Georgetown School banks, "Gold Rail" and "White Steps".[29] Other cities receive the same treatment; Tokyo, for example, becomes Akihabara Park, 'jabu jabu" banks in Shinjuku, ledges at Tokyo Station, curbs at Yotsuya Station, banks at Tokyo Taikan and so on.

What is the mode involved in such an recomposition? Occasionally, this takes the form of a nap or geographic list, such as alternative routes through Bristol[30] or the *Knowhere* internet site, where nearly every skate location in the UK is identified.[31] But more usually a more localised kind of mapping takes place. Skate magazines in the 1990s have tended to focus less on professional skaters, major cities and well-known skate places and more on local skate scenes – the "streets and back yards of Anytown"[32] – like those in Oxted, Ipswich, Oxford, Milton Keynes, High Wycombe, Stroud, Cirencester and Cardiff; in the US, a single issue of *Slap*, for example, covered not just LA (the oldest centre of skateboarding) but also places like Sacramento (California), Fort Lauderdale (Florida) and the urban backwaters of Nevada, Utah, Iowa, Kentucky, Connecticut, and New Jersey.[33] In such articles, the reader-skater finds descriptions of local banks, rails, curbs etc., not just to encourage a visit, but to generally demonstrate that such locations are to be found in all urban centres, and so available to all urban skaters.

> Here are more pictures of Everyman skating in Everytown. It could be your town. It could be you.[34]

This is a communication which engenders empathy and similarity between towns and skaters, not a spectacularised Other of terrain and personalities.

In their own locality, therefore, the skateboarder's cognitive representation is neither map nor directory, for skateboarding is "hard to put onto paper",[35] nor of a spectacularised centre-point, but a mental knowledge composed of highly detailed local knowledge about dispersed places, micro-architectures and accessible times.

> ALWAYS be on the alert for a possible spot [. . .] Be alert [. . .] keep your eyes open and your head oscillating.[36]

Skaters' representations thus have more in common with the Situationist tactics of the *dérive, détournement* and psychogeography – "maps" composed from the opportunities offered by the physical and emotional contours of the city, and, above all, enacted through a run across different spaces and moments.[37]

> I'm directed most to movements, the way I travel, the directions I move in. I follow my feelings.[38]

> Skating is a continual search for the unknown.[39]

Skateboarders' representational maps are thus always situated through a continual re-living of the city – "an open mind always seeking out new lines and possibilities".[40] Skaters attempt neither to "see" the city or comprehend it as a totality, but to live it as simultaneously representation and physicality.

> Walls aren't just walls, banks aren't just banks, curbs aren't just curbs and so on [. . .] mapping cities out in your head according to the distribution of blocks and stairs, twisting the meaning of your environment around to fit your own needs and imagination. It's brilliant being a skateboarder isn't it?[41]

Another distinction from conventional maps concerns temporality. In the aerial form of map, the entire city is understood simultaneously within a single glance – but in skateboarders' cognitive mapping the time is that of the run, composed of disparate objects in a sequence (linear time), with some objects "read" once (isolated time), others encountered several times (repeated time) and still others returned to again and again on different occasions (cyclical time). The whole run can also be repeated the same or differently (differential time). As one skater described the experience of skateboarding among traffic:

> Ridin' from spot to spot, at high speed, during rush hour is my version of the ultimate test for any urban "street skater". On a good day, when all the stop lights are working in my favour, I feel like I've figured out where my place is in this fucked-up world. That lasts for maybe a minute, then the feeling disappears and I'm lost again. So it goes.[42]

Skateboarders are thus more concerned with temporal distance as proximity (temporal closeness of things, temporal locality), and its repetition, than with time as a valuable resource or measure of efficiency; time for skaters is what is lived, experienced and produced, not what is required.

> It's about time, it's about space, it's about time to skate someplace.[43]

Another aspect of this sense of adaptive temporality concerns memory and documentation, for the skateboarder's is not a historical but everyday memory, often surviving only for the period in which a set of places are skated. Skateboarders thus negate the "historical" time of the city, being wholly unconcerned with the many decades and processes of its construction, so that the city appears out-of-the-blue with no temporal past. "I've always lived for the present. I live for the present."[44]

Nor is the city *recorded* by skateboarders, but is that of the here-and-now, the immediate object, reborn each day of the skater's run. "This isn't art, it isn't business, it's life."[45] Just, then, as skateboarders do not attempt to understand the city, nor do they try to document it. Skateboarding leaves almost no text to be read; its marks and assaults leave virtually no discernible script for others to translate and comprehend. These kinds of marks are about the only "text" left by the activity of skateboarding itself.

Skateboarding is, then, less a mode of writing or drawing, and more a mode of *speaking* of the city that "speech doubling"[46] which at once interrogates and increases the meaning of the city, while leaving its original text intact. Above all, speech requires the actual presence of the subject, the active speaker of the city. Speaking-skateboarding is not a mimicking of the city, an oration of a pre-given text, but a performative utterance wherein the speaker forms anew themselves and the city.

> The new urban strategist realizes that while it may not pay to be different, no one can really afford the price of being the same. In the new master plan, conformation has been replaced by confrontation. Act, don't react, turn off the air conditioner go outside and move.[47]

It is, therefore, in the continual performance of skateboarding that its meaning and actions are manifested; as one skateboard maxim puts it, "shut up and skate." These are not things which can be simply seen or understood through pure abstraction; like any socio-spatial rhythm, skateboarding requires a multiplicity of senses, thoughts and activities to be represented and comprehended. Above all, because the experiencer relates the fundamental conditions of their own temporality to that of the world outside, they create a subject–object engagement that is ultimately a lived form of dialectical thought. They produce themselves bodily and socially, and they produce the city in terms of their own specific bodily encounter with it.

NOTES

1 Henri Lefebvre, *The Production of Space*, (Oxford: Blackwell, 1991), p. 94.
2 Gary Davis, 'Steep Slopes," *Thrasher*, v. 3 n. 5 (May 1983), p. 8.
3 Henri Lefebvre, *Introduction to Modernity: Twelve Preludes September 1959 – May 1961*, (London: Verso, 1995), p. 301–2.
4 Stacy Peralta, interview, *Interview*, n.17, (July 1987), pp. 102–3.
5 David Harvey, *Justice, Nature and the Geography of Difference*, (Oxford: Blackwell, 1996), p. 418.
6 Henri Lefebvre, *Writings on Cities*, (Oxford: Blackwell, 1996), Eleonore Kofman and Elizabeth Lebas (eds.), p. 101.
7 Lefebvre, *Production of Space*, p. 75.
8 "Where?," *R.A.D.*, n.79 (September 1989), p. 18.
9 Lefebvre, *Production of Space*, pp. 338–9.
10 John Smythe, "The History of the World and Other Short Subjects, or, From Jan and Dean to Joe Jackson Unabridged," *SkateBoarder*, v. 6 n. 10 (May 1980), p. 29.
11 "Searching, Finding, Living, Sharing", *R.A.D.*, n.79 (September 1989), p. 15.
12 Lefebvre, *Production of Space*, p. 137.
13 "Fire and Friends," *Sidewalk Surfer*, n. 3 (January–February 1996), unpaginated.
14 "Fire and Friends," unpaginated.
15 Elliott Willensky and Norval White, *AIA Guide to New York*, (Hew York: Harcourt Brace Jovanovich, third edition, 1988), p. 230.
16 Kevin Wilkins, "New England Hot Spots," *TransWorld Skateboarding*, v. 9 n. 11 (November 1991), p. 43.
17 Pete and the Posse, letter, *Thrasher*, v.11 n.9 (September 1991), p. 6.
18 Santa Cruz, advertisement, *Action Now*, v.7 n.12 (July 1981), p. 53.
19 "Blast From the Past," *Thrasher*, v.17 n.9 (September 1997), p. 56.
20 Iain Borden, Joe Kerr, Alicia Pivaro and Jane Rendall, "Narratives of Architecture in the City," Iain Borden, Joe Kerr, Alicia Pivaro and Jane Rendall (eds.), *Strangely Familiar: Narratives of Architecture in the City*, (London: Routledge 1996), p. 9.
21 John Smythe, "No Parking," *Action Now*, v.8 n.2 (September 1981), p. 55.
22 Henri Lefebvre, *Espaces et Sociétés*, v.4 (1976–8), p. 270, quoted in Eleonore Kofman and Elizabeth Lebas, "Lost in Transposition," Lefebvre, *Writings on Cities*, p. 23.
23 Stacy Peralta, "Skate of the Art, '85" *Thrasher*, v.5 n.8 (August 1985), pp. 38–40.
24 Lefebvre, *Introduction to Modernity*, p. 321.
25 "Skater's Eye," *Thrasher*, v.17 n.1 (January 1997), p. 71.
26 Barry Curtis, "Venice Metro," Borden, Kerr, Pivaro and Rendall (eds.), *Strangely Familiar*, p. 45.
27 "Searching, Finding, Living, Sharing", p. 15.
28 Marco Contati, "New York, New York," *Skateboard!*, (second series), n.43 (June 1990), pp. 32–41; "Skatetown: New York City," *Thrasher*, v.9 n.10 (October 1989), pp. 58–65 and 106; and Wilkins, "New England," p. 43.
29 Pete Thompson, "Washington DC," *TransWorld Skateboarding*, v.13 n.5 (May 1995), pp. 86–9; and Andy Stone, interview, *TransWorld Skateboarding*, v.13 n.5 (May 1995), pp. 90–3.
30 Steve Kane, "Street Life: Bristol," *Skateboard!*, n.15 (November 1978), pp. 36–9.
31 *Knowhere* internet site, URL http://www.state51.co.uk/state51/knowhere/skindex.html, (accessed 7 February 1997).
32 "From Surf to Hellbows: the Styling of Street," *R.A.D.*, n.75, (May 1989), p. 60.
33 *Slap*, v.6 n.1 (January 1997). See also Jerry Mander, "Sacto Locals," *TransWorld Skateboarding*, v.9 n.10 (October 1991), pp. 80–5.
34 "Scary Places," *R.A.D.*, n.82, (December 1989), p. 20.
35 Ewan Bowman, "Comment," *Sidewalk Surfer*, n.13 (January-February 1997), unpaginated.
36 Gary Davis, "Radical Manifesto," *Thrasher*, v.2 n.2 (February 1982), p. 18, reprinted from *Skate Fate*, (Cincinnati, Ohio).
37 Guy Debord, "Introduction to a Critique of Urban Geography," and "Theory of the Dérive," Ken Knabb (ed.), *Situationist International Anthology*, (Berkeley: Bureau of Public Secrets, 1981), pp. 5–8 and 50–4.
38 Rodney Mullen, interview, *R.A.D.*, n.74, (April 1989), p. 28.
39 Caine Gayle, "Multiple choice Through Words and Pictures," *Slap*, v.4 n.9 (September 1995), p. 33.

40 Christopher James Pulman, "An Environmental Issue," *Sidewalk Surfer*, n.1 (September–October 1995), unpaginated.
41 "Twisted," *Sidewalk Surfer*, n.14 (March 1997), unpaginated.
42 Jesse Driggs, "Swamp Trogs from Outer Space," *Thrasher*, v.15 n.9 (September 1995), p. 43.
43 Rick Blackhart, "Ask the Doctor," *Thrasher*, v.11 n.11 (November 1991), p. 24.
44 Rune Glifberg, interview, *Sidewalk Surfer*, n.14 (March 1997), unpaginated.
45 Mark Gonzales, in "Trash," *Thrasher*, v.16 n.2 (February 1996), p. 139.
46 Henri Lefebvre, *Everyday Life in the Modern World*, (London: Transaction Publishers, 1984), p. 176.
47 Smythe, "No Parking," p. 57.

PART SEVEN

Contesting Identity

INTRODUCTION TO PART SEVEN

During the cold war it was viable for an American President to stand at the Berlin Wall and announce in borrowed German that he was a Berliner. There is an ambiguity as to whether the word used means a citizen of Berlin or a type of bun – in Frankfurt he might be a sausage – but this was not the point. It was that the world consisted of two opposed power blocs, one of which identified itself as the free world and identified the other as tyranny. This, after all, was the President who won the Cuban missile crisis. The fault line between the two worlds was the Wall. The Wall marked a division between a free-market economy (though one prepared to resort to controls, as in the prohibition of imports from Cuba) and a command economy. The difference is illusory, since all economies are commanded by the interests of either capital or the state, but the rhetoric persists despite its irrationality. Standing at the Wall, the President sought to identify himself as belonging to the fault-line city, or more exactly to that part of it situated in the free sector, surrounded by the territory of the other side. The beleaguered city offered the President the role of protector, though whom he protected was not stated in the sound-bite. If this act of identification extends a development which begins with the unveiling of the Statue of Liberty in 1886, there were ambiguities then, too. Shortly before the unveiling a law was passed to limit immigration into the United States by Chinese labourers.

Some of the ambiguities are present in Frédéric-Auguste Bartholdi's design for the statue – fabricated in France, shipped in parts to New York as a gift in 1876, stored while the cost of the plinth was raised by subscription, finally bolted on to a cast-iron armature designed for it by Gustave Eiffel and opened to public visits – and in its inscription (added in 1903). Liberty's form echoes the iconography of the French Revolution of 1789, transmuted after the fall of the Paris Commune to a more sober image of the Republic, fused with that of the American War of Independence of 1776–81. In place of a liberty bonnet – used in both revolts and derived from two sources in antiquity: the cap of a Phrygian and the red hat of a freed slave in Rome – are the rays of the sun-god Helios, and in Liberty's hand is a book bearing the date July 4th, 1776. These are conventionally masculine traits, and Marina Warner (*Monuments and Maidens: The Allegory of the Female Form*, London, Picador, 1987) sees the inversion as an adaptation of a phallocratic monument to one of universal motherhood, suggesting that Bartholdi's mother was the model for its features. Warner writes that 'Containing a sequence of inversions, the iconography of Liberty thus operates on the central premise that signs of Otherness can be recuperated to express an ideal' (p. 277). Warner's immediate reference is Delacroix's painting *Liberty Guiding the People* (1831) but the comment could apply to Bartholdi's Liberty, especially given the verses beneath it: 'Give me your tired, your poor/Your huddled masses yearning to breathe free/The wretched refuse of your teeming shore' (from *The Mother of Exiles* by Emma Lazarus, cited in Warner, 1987, p. 10).

Writing in 1883 after the Tsarist pogroms, Lazarus expressed a sentiment easily appropriated by a nation reconstructed out of immigration, many of whose citizens had fled poverty or persecution in Europe. There is a strange ambiguity: is their identification with a strong national myth – perpetuated in the cold war as the defence of freedom, and again in the twenty-first century in a war on terror – an expression of a need produced in a history of otherness (which the mythicised identity represses)? John Adams writes that 'The poor man's conscience is clear; yet he is ashamed' (cited in Kimberly Curtis, *Our Sense of the Real: Aesthetic Experience and Arendtian Politics*, Ithaca, Cornell University Press, 1999, p. 67), and it seems that an inability to cope with otherness does run through the free world's myth of itself, which is an issue separate from the guilt of genocide which has only recently dawned on the white north American majority faction. It is seen, for instance, as Richard Sennett points out in what remains arguably his most incisive contribution to discussion of urban cultures, *The Uses of Disorder* (New York, Norton, 1970), in the suspicion of outsiders such as people of colour felt by white, middle-income suburbanites. Unable to cope with difference, the suburban guardians of a social ordering project the content of their fears onto others, demonising them. In contemporary city governance this continues to produce a notion of an underclass (see the text by Rosalyn Deutsche reprinted in Part Nine, citing Mayor Koch on homeless people in New York), and defensive design strategies to prevent them gathering, or being visible in public, downtown spaces (see Mike Davis, *City of Quartz*, London, Verso, 1990).

Returning to the Wall and the President's identification with Berlin: just as the poor and oppressed identify with Liberty, so he, in an opposite movement, becomes one of the oppressed for a day.

The President, until his death, inhabited a world of certainty which no longer exists, replaced by one of contingency. The world is fragmented rather than divided. One of its power blocs has dissolved but other divisions than those of west and east, capitalism and state communism, are now more recognised. These include the divisions of the affluent north from the non-affluent south (though enclaves of each exist inside the terrains of the other), of majorities from minorities, of sacred from secular associations within and between states. In particular, the question 'In whose voice . . .?' and its content of a problematics of representation is now unavoidable. Yet if the statement 'I am a Berliner' cannot now be made without raising complex difficulties, perhaps it was already a sham; not because the President was palpably not a Berliner – the audience knew that – but because the state of which he was the figurehead was internally divided, not least by a continuing concern for civil rights and what was to become mass protest over a colonial war in Indochina initiated by the same President. But if the civil rights movement added race to class in a taxonomy of the social, feminism added gender, and further categories including age, mobility, and sexual orientation have followed.

After feminism and post-colonialism, then, representation is seen as contested, and identity seen as something formed not given. There are two aspects of this which need to be noted at this stage: that the *conditions* in which identity formation takes place are complex; and that the awareness of *formations* itself raises questions about what constitutes the human subject. The emphasis in the selection of texts for this section is on the process of identity formation, or contestation, so it may be helpful to say something here about the conditions.

A frequently rehearsed trope of the city is its moulding by globalisation. Saskia Sassen's *The Global City* (Princeton, Princeton University Press, 1991) introduced the idea of a single global city of financial services enclaves around the world, bound by information super-highways. Manuel Castells has written extensively on the scape of flows which this incorporates. But it is not only capital which is globalised but also communications and resistance to the deregulated economy of the free-market world. The flows are in many directions: capital to sites of low-cost production; labour to sites of employment; refugees to asylum. This changes patterns of urban sociation as much as the inception of metropolitan cities in the nineteenth century. John Eade writes in his Introduction to *Living the Global City: Globalization as Local Process* (London, Routledge, 1997) that people's lives are not shaped by global processes, 'rather they are participating in diverse global processes which are occurring throughout the world but can be analyzed in specific contexts' such as a city (p. 11). Through mobility and migration, patterns of urban sociation are now complex, global and mutable in the short term.

In these conditions, and figured particularly in postmodern cultural theory, the process of identity formation is as complex as the patterns of sociation, and as much liable to draw on diverse elements. Juliet Steyn writes:

> Rethinking identity entails a demand: to split the traditional link between self and identity. Identity is presented as akin to culture, which allows identity but is not a guarantor of the subject. Likewise the assumptions of the subject are no guarantors of subjectivity. Nowadays it has

become commonplace in cultural theory to describe ourselves as 'split', in between, fragmented, ceaselessly losing our identity, destabilised by changes in our relationships to the other (Juliet Steyn, *Other Than Identity: The Subject, Politics and Art*, Manchester, Manchester University Press, 1997, p. 3).

In these shifting sands the term 'subject' replaces 'self', as the notion of a unified and coherent self-identity gives way to something negotiated and further negotiable. Dani Cavallaro explains, in *Critical and Cultural Theory* (London, Athlone, 2001, p. 86): 'the word "self" traditionally evokes the idea of identity as a private possession and a notion of the individual as autonomous' while 'A subject is both active and passive.' He gives the example of the subject of a sentence as denoting a person doing and/or done-to, and notes the use as subject of a state. He continues: 'Poststructuralism has emphasized that the subject is not a free consciousness or a stable human essence but rather a construction of language, politics and culture' (*ibid.*). The subject, then, can be contested. The process and performance of that contestation became a pervasive element of socio-cultural debate in the 1980s, through what has been called a second-wave feminism. The Guerrilla Girls, for instance, a group of women artists preserving anonymity by wearing gorilla masks while undertaking cultural guerrilla warfare in cities such as New York, used billboard posters and fliers to draw attention to discrimination against people of colour as well as women in the art world. One of their posters reads 'Do women have to be naked to get into the Met. Museum' with a subtext that less than 5 per cent of the artists represented in the collection are women while 85 per cent of the nudes are female (see *Confessions of the Guerrilla Girls*, New York, Pandora, 1995). Notices were affixed to the facades of contemporary art galleries exhibiting a prejudice against women and people of colour in their selection of artists. The aim was not to be assimilated to the white men's artworld with its values and practices rooted in patriarchy, but to change those structures. This does not suppose an essentialism in which women's art is by nature different from men's, but questions the criteria of admission to the artworld.

Among the memorable phrases of this period was 'The personal is political', interpreted by many women academics as a rejection of the conventional third-person voice (used here), and introduction into academic writing of personal experiences. Doreen Massey, whose chapter on 'Space, Place and Gender' is reprinted below, writes, for instance, of childhood and adolescent experiences of gendered spaces. This is not a gesture of informality but a claim to the validity of such experiences as a basis of knowledge.

Arguments on difference were clarified within feminism when categories of class, race, and sexual orientation were seen to intersect gender. In an address to a panel on 'The Personal and the Political' at a conference at the New York University Institute for the Humanities in 1979, Audre Lorde argued that:

> Those of us who stand outside the circle of this society's definition of acceptable women; those of us who have been forged in the crucibles of difference – those of us who are poor, who are lesbians, who are Black, who are older – know that *survival is not an academic skill*. It is learning how to stand alone, unpopular and sometimes reviled, and how to make common cause with those others identified as outside the structures in order to define and seek a world in which we can all flourish. It is learning how to take our differences and make them strengths. *For the master's tools will never dismantle the master's house*. They may allow us temporarily to beat him at his own game, but they will never enable us to bring about genuine change (Audre Lorde 'The Master's Tools will Never Dismantle the Master's House' (1979) in Jane Rendell, Barbara Penner and Iain Borden (eds), *Gender Space Architecture*, London, Routledge, 2000, p. 54).

In second-wave feminist thought emphasis moves from a reform of liberalism to a radical avowal of differences between feminists. Lorde asks how her audience of women academics deals with the fact that the women who clean their houses while they attend feminist conferences are poor and black, seeing the ignorance of difference this implies as a repetition by other means of racism and patriarchy.

Lorde's argument is, in effect, restated by Griselda Pollock in the specialist terms of art history in *Vision and Difference* (1988):

> Is adding women to art history the same as producing feminist art history? Demanding that women be considered not only changes what is studied and what becomes relevant to investigate but it challenges the existing disciplines politically. Women have not been omitted through forgetfulness (Griselda Pollock, *Vision and Difference: Femininity, Feminism and the Histories of Art*, London, Routledge, 1988, p. 1).

Women have been left out through a structural sexism which diminishes what can be learned about the world, or about art history. The same structural sexism denies women access to public space in conventional arrangements of the city, in which the street is a space of masculine gazing and women, particularly if not accompanied by men, are objects of that gaze and its possessive tendency. This patriarchy is reinforced by limits (not only historically) to women's access to professions from medicine and law to architecture and planning, but operates, too, as Doreen Massey argues, in terms of what is included within academic disciplines. Massey says that when she began to study geography, personal experiences of spatial gendering were 'just not talked about' (Doreen Massey, *Space, Place and Gender*, Cambridge, Polity, 1994, p. 186). From a study of labour trends, however, Massey shows that when gender is a factor in geographical analysis it produces a more nuanced understanding of use to all. In the previous chapter, Massey applies a similar framework to the question of place identity, arguing that concepts such as home are mutable and socially produced.

In a similar vein, but looking to urban planning, Leonie Sandercock writes in *Towards Cosmopolis* (1998) that:

> The enterprise of planning theory . . . had been an almost exclusively male and white domain. The handful of women who had engaged in these debates had not necessarily done so from a feminist perspective, and those few scholars of color, likewise, for the most part had not insisted on 'race' and racism as necessary categories of analysis (Leonie Sandercock, *Towards Cosmopolis*, Chichester, Wiley, 1998, p. 109).

She contends that in a feminist analysis planning cannot be theorised as if neutral in its impact on women, requiring a shift from the difference that theory makes to, as she puts it, the theory that difference makes. In 'The Death of Modernist Planning' (included in Part Nine for its commentary on the Modernist utopia), Sandercock notes that questions were asked in the 1980s as to who constituted the 'we' of feminist urban analysis. This echoes Lorde's view cited above. Sandercock writes that women of color, lesbians and the physically challenged began to voice feelings of exclusion from the society of white, middle-class, able-bodied, heterosexual women who lived in conventional family structures in metropolitan cities. One of her aims in re-visioning planning theory is to introduce into it awareness of voices from the margins. What this means is not an enlarged centre which gradually draws in marginalised elements, but a new map with neither centres nor margins.

The purpose of the above outline of a second-wave feminist position is to draw attention to some prerequisites for a discussion of identity formation. These can be summarised as: that identities are complex and do not correspond to unifying typologies; that they are contested both between and within social categories; and that the categories which have generally been used in academic and professional debate require renegotiation. The contestation of identities follows from a multi-dimensional *intersection* of categories, and more or less abolishes the dream of a future social stability to be gained by movement towards homogeneity. This does not mean societies are inevitably unstable, of course, but that stability, or equilibrium, is more difficult.

Iris Marion Young's *Justice and the Politics of Difference* (Princeton, Princeton University Press, 1990) is a seminal book in this field. Its key argument is that assimilation is not the route to social justice, and that for members of marginalised areas of a society, group identities are the means to recognition of rights of difference, as well as providing a necessary solidarity. As she writes, the parts of society not included in the dominant model are diverse: 'If "cultural minority" is interpreted to mean any group subject to cultural imperialism, then this statement applies to women, old people, disabled people, gay men and lesbians, and working-class people as much as it applies to ethnic or national groups' (p. 175). These, taken together, would comprise a majority in any industrialised (or post-industrial) society. But that is the point, the so-called majority is a majority only when measured in terms of its ability to determine the terms of debate.

Young's argument deserves a lengthier commentary, and is discussed again in the introduction to her text below. One further point can be made here, which is Young's support for positive action for social justice, and the possibility for self-organisation she sees following from this. A consequence of that self-organisation, though she does not discard the politics of coalition, is that new exclusions necessarily occur: 'This politics of group assertion . . . takes as a basic principle that members of oppressed groups need separate organizations that exclude others, especially those from more privileged groups' (p. 167). This flies in the face of liberal humanism, and challenges more radical or revolutionary theories such as those of the mass protest or movement. It is not, on the other hand, incompatible with an older, more classically oriented, concept of self-realisation through immersion in the society of others.

Hannah Arendt – writing before the postmodern turn to subjects – terms this a process of natality in which a mature self is formed through publicity, that is, through exposure to difference in public. She states this in *The Human Condition* (Chicago, Chicago University Press, 1955), but the text included here is a recent commentary on Arendt's work and politics by Kimberly Curtis. The chapter begins with John Adams' remark cited above, and examines identity formation through its inverse, a denial of that publicity which is its precondition – a particular form of which afflicted Europe in the 1930s.

Arendt's delineation of a light outside and a dark inside is based in classicism – see Richard Sennett's exposition of Greek thought on the body, men's exterior spaces, and the domestic interior, in *Flesh and Stone* (London, Faber and Faber, 1995). But, as Curtis argues, she saw it, too, as produced by modern poverty. The deprivations of a consumerist society are obscured, it could be added, as an unwelcome contradiction, but are real and experiential for those afflicted. Whether economic or social in origin, the phenomena of deprivation cast real shadows: 'Those relegated to oblivion suffer a loss of feeling for their own existence, their own reality, as well as for the larger world and their relationship to it' (Curtis, 1999, p. 69). The contemporary world is not short of cases of violence which could exemplify this.

The remaining two texts included in this part both concern cases of such violence. Susana Torre considers the women who, in Argentina in the mid-1970s, began to publicly and collectively refuse to accept the 'disappearance' of their sons and daughters in the military government's suppression of political life. As Torre writes: 'Because of the clandestine, unrecorded activities of the paramilitary groups charged with these deeds, and because many burial sites still remain undisclosed, agreement as to the exact number of "disappeared" may never be achieved' ('Claiming the Public Space: The Mothers of the Plaza de Mayo', in Diane Agrest *et al.*, *The Sex of Architecture*, New York, Abrams, 1996, pp. 241–50). Her aim is not to write a history of Argentina, but to draw attention to the role of women, in particular through self-organisation, in changing political histories. She sees this as necessary because much feminist scholarship to date has directed its focus at past rather than present situations, and on academic arguments around the masculine gaze. Feminist critics, she claims, have seen women who had a transformative impact on the society in which they lived as exceptional, or bohemian, and hence outside the scope of contemporary struggles for visibility. Torre's concern is how women today inscribe their stories within an urban palimpsest.

Farha Ghannam, in a chapter in *Space, Culture and Power* (edited by Ayşe Öncü and Petra Weyland, London, Zed Books, 1997, pp. 119–39), writes of women's displacement from one part of Cairo to another through development. This text emphasises that global change has diverse local impacts, and complements Eade's work cited above. Ghannam writes that 'People in al-Zawiya al-Hamra [a social housing zone in the northern suburbs of Cairo] not only experience the American culture that is transmitted to them in movies . . . but they also experience the global through oil-producing countries where their children and male relatives work as well as through the mixture of people who visit and work in Cairo from different Arab countries' (p. 125). Beside these global currents are others specific to Islam, in particular that which permeates the space of the mosque in which women find a collectivity, and specific to geographical locations such as the older neighbourhood from which the population in question were moved. Ghannam concludes – topically at the time of writing in January, 2003: 'When people experience the global as a violent attack on their cultural identities and self-images, it is not strange that they do not embrace global discourses . . . In short, more attention should be devoted to those who live on the margin of the marginal: those who are displaced in their own "culture"' (p. 137). This reinforces the concept of a margin, rather than deconstruct the concept of a centre, but the accounts by Torre and Ghannam of realities at two of the edges of oblivion are a vital balance to more theoretical texts such as those reprinted alongside Torre's and Ghannam's below.

SUGGESTED FURTHER READING

Jo Beall, ed., *A City for All: Valuing Difference and Working with Diversity*, London, Zed Books, 1997, pp. 2–37

Mike Douglass and John Friedman, eds, *Cities for Citizens*, Chichester, Wiley, 1998

John Eade, ed., *Living the Global City: Globalization as Local Process*, London, Routledge, 1997

Hal Foster, *Recordings: Art, Spectacle, Cultural Politics*, Seattle, Bay Press, 1985

Sally R. Munt, 'The Lesbian Flâneur', in *The Unknown City: Contesting Architecture and Social Space*, edited by Iain Borden *et al.*, Cambridge, MA, MIT, 2001, pp. 246–61

Leonie Sandercock, *Towards Cosmopolis*, Chichester, Wiley, 1998

Juliet Steyn, *Other Than Identity: The Subject, Politics and Art*, Manchester, Manchester University Press, 1997

‘Space, Place and Gender’

from *Space, Place and Gender* (1994)

Doreen Massey

Editors’ Introduction

In an earlier chapter of *Space, Place and Gender*, titled ‘Uneven Development: Social Change and Spatial Divisions of Labour’ (pp. 86–114), Massey’s emphasis is on class relations within the UK economy during the 1980s. Citing a newspaper cartoon by Steve Bell showing yuppies floating into space saying they don’t need dustbins, she says ‘They are wrong. They do. And they need people to empty them’ (p. 87). Similarly, she argues there cannot be a knowledge economy, as it is called, without manual labour to support it. The chapter charts the uneven development of opportunities within the economy and their spatial distribution, but towards the end Massey moves to another point, the lack of research on different levels of economic opportunity which affect (in this case professional) men and women in the same region.

Regional employment patterns are the subject, too, of the chapter reprinted here. But Massey, Professor of Geography at the Open University, prefaces the results of an investigation into regional employment patterns with a recollection of her experiences as a nine- or ten-year old in outer Manchester, seeing acres of playing fields from the top of a bus and knowing they were not for her. She remembers also seeing paintings of female nudes in an art gallery and feeling objectified by the gaze implicit in them. Massey has used personal memories, usually associated with specific places, in other writing – such as ‘Living in Wythenshaw’, in *The Unknown City: Contesting Architecture and Social Space*, edited by Iain Borden, Joe Kerr, and Jane Rendell, with Alicia Pivaro (Cambridge, MA, MIT, 2001, pp. 458–75). It is, as the introduction to this part stated, more than a detail of style, affirming the importance of personal knowledge in geography, and of the gendered experiences which are one way in which meanings are constructed in spaces.

Part of Massey’s aim is to widen geographical insights through a more diverse range of ways of finding things out, and part to deepen them through a closer attention to the nuances of meaning which different knowledge give. Among her findings in the labour market study was that while disappearing jobs tended to be men’s, new jobs with lower pay and less security tended to be allocated to women, although women’s economic activity in the region had traditionally been low. Massey is careful to avoid essentialism, seeing gender roles as socially and economically constructed, hence contestable. A difficulty, however, which Massey notes at the end, is that even in new economic structures new masculinities tend to be constructed.

I can remember very clearly a sight which often used to strike me when I was nine or ten years old. I lived then on the outskirts of Manchester, and ‘Going into Town’ was a relatively big occasion; it took over half an hour and we went on the top deck of a bus. On the way into town we would cross the wide shallow valley of the River Mersey, and my memory is of dank, muddy fields spreading away into a cold, misty distance. And all of it – all of these acres of Manchester – was divided up into football pitches and rugby pitches. And on

Saturdays, which was when we went into Town, the whole vast area would be covered with hundreds of little people, all running around after balls, as far as the eye could see. (It seemed from the top of the bus like a vast, animated Lowry painting, with all the little people in rather brighter colours than Lowry used to paint them, and with cold red legs.)

I remember all this very sharply. And I remember, too, it striking me very clearly – even then as a puzzled, slightly thoughtful little girl – that all this huge stretch of the Mersey flood plain had been entirely given over to boys.

I did not go to those playing fields – they seemed barred, another world (though today, with more nerve and some consciousness of being a space-invader, I do stand on football terraces – and love it). But there were other places to which I did go, and yet where I still felt that they were not mine or at least that they were designed to, or had the effect of, firmly letting me know my conventional subordination. I remember, for instance, in my late teens being in an Art Gallery (capital A capital G) in some town across the Channel. I was with two young men, and we were hitching around 'the Continent'. And this Temple of High Culture, which was one of The Places To Be Visited, was full of paintings, a high proportion of which were of naked women. They were pictures of naked women painted by men, and thus of women seen through the eyes of men. So I stood there with these two young friends, and they looked at these pictures which were of women seen through the eyes of men, and I looked at them, my two young friends, looking at pictures of naked women as seen through the eyes of men. And I felt objectified. This was a 'space' that clearly let me know something, and something ignominious, about what High Culture thought was my place in Society. The effect on me of being in that space/place was quite different from the effect it had on my male friends. (I remember that we went off to a café afterwards and had an argument about it. And I lost that argument, largely on the grounds that I was 'being silly'. I had not then had the benefit of reading Griselda Pollock, or Janet Wolff, or Whitney Chadwick . . . maybe I really *was* the only person who felt like that . . .)

I could multiply such examples, and so I am sure could anyone here today, whether woman or man. The only point I want to make is that space and place, spaces and places, and our senses of them (and such related things as our degrees of mobility) are gendered through and through. Moreover they are gendered in a myriad different ways, which vary between cultures and over time. And this gendering of space and place both reflects *and has effects back on* the ways in which gender is constructed and understood in the societies in which we live.

When I first started 'doing geography' these things were just not talked about. What I want to do here is simply to give one example of how issues of gender began to creep into our subject matter. The example is perhaps quite mundane; it concerns empirical issues of regional development which are now well established in debate; but in spite of that some interesting lessons can be drawn.

The example, then, is from studies of regional employment in the United Kingdom. It concerns the story of the regional decentralization of jobs which took place in this country between the mid-1960s and the early 1970s. There are some facts which ought to be known before the story begins. This was a period largely of Labour government, with Harold Wilson as Prime Minister. There were major losses of jobs in coal mining, in the north-east of England, in south Wales and in central Scotland. It was the great era of regional policy, when there were numerous incentives and inducements to firms to invest in the regions where job loss was taking place. And it was also an era of the decentralization of jobs from the high employment areas of the south-east and the west midlands to these 'northern' regions of high *un*employment. And the question which preoccupied many of us at that time was: how were we to put these facts together? Or, specifically, how were we to explain the decentralization of jobs to the regions of the north and the west?

The argument went through a series of stages. Or, at least, I shall present it as a series of stages – there are many occupants in what I label as the early stages who will doubtless disagree with what I say. Intellectual change is just not as linear as that.

The analysis, then, in 'stage one' was led primarily by people with computers and statistical packages, who correlated the timing and size of the decentralization of employment with the timing and distribution of regional policy. They found a high correlation between the two, and deduced that they were causally related: namely (although this was of course not directly shown by the statistics themselves) that regional policy was the cause of the decentralization of jobs. Thus regional policy, on this reading, was seen as having been quite successful.

But then came stage two. It was provoked by political rumblings of discontent, from male-dominated trade unions and local councils, and from evidence given to a parliamentary sub-committee. For jobs were not just jobs, it seemed: they were gendered. While the jobs which had been lost had been men's, the new jobs, arriving on the wave of decentralization, were largely being taken by women. And within academe, a whole new line of inquiry started as to *why* these jobs were for women. The answers which were found are now well known. Women workers were cheap; they were prepared to accept low wages, the result of years of negotiating in terms of 'the family wage'. Women were also more available than men for part-time work, an effect of the long established domestic division of labour within the household. Both of these reasons were characteristic of male/female relations, within the home and within the employment market, across the country. But some reasons were more specific, or at least more important, to these particular regions to which the jobs had been decentralized. Thus, the women in these regions had very low rates of organization into trade unions, a result of the very low levels of their previous incorporation into paid employment. The female economic activity rates there were indeed amongst the lowest in the country. These women, in other words, were classic 'green labour'.

With this development of the argument a slightly more complex story evolved which recognized some differences within the labour market, which recognized certain constraints and specificities of women as potential employees, which, in brief, recognized that women and women's jobs were different. Such a revised understanding led also to a revised evaluation of the effectivity of regional policy. It was now clearly necessary to be more muted in any claims for its success. There were two versions of this re-evaluation. One, clearly sexist, persisted in its claim that the new jobs being made available in the regions should be criticized for being 'not real jobs', or for being 'only for women'. There was, however, also another form of re-evaluation, more academically respectable although still worrying in its implications: that the fact that the new jobs were for women was unfortunate in the sense that, because women's jobs were less well paid than were men's, aggregate regional income was still lower.

And yet there was a further stage in the development of this argument: stage three. For the more that one thought about it, the more the story seemed more complicated than that. Why, for example, had the economic activity rate for women in these regions been historically so low? This raised the whole question of local gender cultures. Many people, writing in both geography and sociology, commented upon the domestic labour burden of being a wife or mother to miners. They commented also on how the length and irregularity of shift-work made it problematical for the other partner in a couple also to seek paid employment outside the home. There was much detailed investigation of the construction of particular forms of masculinity around jobs such as mining. And all these investigations, and others besides, pointed to a deeper explanation of why, more than in most other regions of the country, there was in these areas a culture of the man being the breadwinner and of the women being the homemaker.

We had, in other words, moved through a series of approaches; from not taking gender into account at all, we had moved first to looking at women, and from there to looking at gender roles, men, and locally constructed gender relations. Moreover this gave us, once again, both a different story of what had happened and a different evaluation of regional policy. The new story was again more complicated and more nuanced. Harold Wilson had come to power in 1964 on a programme of modernizing social democracy, part of which centred on the rationalization of old industries such as coal mining. Contradictorily for him, however, the loss of jobs which would be consequent upon that rationalization would occur precisely in the regions which were his main geographical power base – regions such as the north-east of England, south Wales, and the central area of Scotland. In order, therefore, to proceed with this reconstruction of the old basic sectors of these regions, it was necessary to have as the other side of the deal a strong regional policy. Given this, acquiescence might be won from the trade unions and their members. However, it was the very fact that the men in the region were being made redundant which was important in creating the availability of female labour. For women were now for the first time in decades 'freed' on to the labour market. They needed paid employment, most particularly now in the absence of work for men, and there was less of a domestic labour burden upon them restraining them from taking it. Moreover these women had been constructed over the years, precisely by the specificity of the local gender culture, into just the kind of workforce the decentralizing industries were looking for.

Moreover, there was yet again a different evaluation of regional policy. For regional policy could no longer be accepted as the single dominant factor in the explanation of decentralization of employment because the labour-force which had been part of the attraction to the incoming industries had been created not by regional policy but by the simultaneous decline of men's jobs and as a result of the previous gender culture. It certainly remained true that regional policy had brought with it only low-paid jobs, but on the other hand there were some positive aspects to the jobs it did bring, which previously had been unrecognized. Most importantly, it did bring some independent income for women, and for the first time in decades. Moreover, as the very fact of the initial complaints indicated, precisely by bringing in those jobs it began to disrupt some of the old gender relations. In other words, on this score (though not on many others) regional policy can be seen to have had some quite positive effects – though in a wholly different way from that initially claimed in stage one of the development of the argument.

There are a number of reflections which can be drawn from this story of a developing analysis. First, and most obviously, taking gender seriously produced a more nuanced evaluation of regional policy, a far better understanding of the organization and reorganization of our national economic space, and indeed – since these decentralizing industries were moving north to cut costs in the face of increasing international competition – it has shown us how British industry was actively *using* regional differences in systems of gender relations in an early attempt to get out of what has become the crisis of the British economy. Second, this understanding was arrived at not just by looking at women – although that was a start – but by investigating geographical variations in the construction of masculinity and femininity and the relations between the two. Feminist geography is (or should be) as much about men as it is about women. Third, moreover, the very focus on geographical variation means that we are not here dealing with some essentialism of men and women, but with how they are constructed as such.

The fourth reflection is a rather different one. It is easy now to look back and criticize this old-time patriarchy in the coalfields. Indeed it has become a stick with which to beat 'the old labour movement'. But that should not let us slide into an assumption that because the old was bad the new is somehow unproblematical. So, partly in response to the last three reflections (the need to look at men and masculinity, the importance of recognizing geographical variations and of constructing a nonessentialist analysis, and the feeling that it is important to look at new jobs as well as at old) I am now involved in research on a 'new' region of economic growth – Cambridge. Cambridge: the very name of the place gives rise to thoughts of 'the Cambridge phenomenon' of high-technology growth, of science and innovation, and of white-collar work. It is all a million miles from coal mines, geographically, technologically, and – you would think – socially. In fact the picture is not as clear as that.

It is the highly qualified workers in high technology sectors on which this new research is concentrating. Well over 90 per cent of these scientists and technologists are men. They frequently love their work. This is no bad thing, until one comes across statements like 'the boundary between work and play disappears', which immediately gives pause for thought. Is the only thing outside paid employment 'play'? Who does the domestic labour? These employees work long hours on knotty problems, and construct their image of themselves as people around the paid work that they do. But those long hours, and the flexibility of their organization, is someone else's constraint. Who goes to the launderette? Who picks up the children from school? In a previous project, from which this one derived, and from which we have some initial information, only one of these employees, and that one of the few women whom we found, mentioned using the flexibility of work hours in any relation to domestic labour – in this case she said that on occasions she left work at six o'clock to nip home to feed the cat![1] The point is that the whole design of these jobs requires that such employees do not do the work of reproduction and of caring for other people; indeed it implies that, best of all, they have someone to look after *them*. It is not therefore just the old labour movement, it is also the regions of the 'new man' which have their problems in terms of the construction of gender relations. What is being constructed in this region of new economic growth is a new version of masculinity, and a new – and still highly problematical – set of gender roles and gender relations.

NOTE

1 See Doreen Massey, Paul Quintas and David Wield, *High-Tech Fantasies: Science Parks in Society, Science and Space*, London, Routledge, 1992.

'Social Movements and the Politics of Difference'

from *Justice and the Politics of Difference* (1990)

Iris Marion Young

Editors' Introduction

The book from which this chapter is reprinted takes a position against the conventions of liberal humanism. Young, Professor of Political Science at the University of Chicago, argues that assimilation – the aim of the liberal social model which categorises groups as having majority or minority status – means only assimilation to a majority, and that this is not a form of social justice but a denial of identity on the terms of the majority. At the same time, Young does not take an extreme individualist, or atomist, position, believing that group identity, and the contestation of claims by groups, is the most viable means by which citizens included in minority categories (of race, class, gender, or sexual orientation) will gain justice and recognition. The chapter after that reprinted here deals with affirmative action, and what Young calls the myth of merit. That myth is one legacy from the modern faith in disinterested professional judgements, in a way a Kantian idea, which is epitomised by the rational planning model criticised by Leonie Sandercock in her text in Part Nine. The two chapters could be read together, giving insights from political science and urban planning which together redefine the concept of civil society.

In a more recent book, *Inclusion and Democracy* (Oxford, Oxford University Press, 2000), Young argues that the construction of civil society has three levels of associational activity: private; civic; and political. Of these the third focuses on what the social collective should do, that is, how it shapes society (p. 162). This leads to a discussion of self-organisation within civil society as the means to end the silencing of minority groups within a majority society. This follows Young's conclusions in the chapter reprinted below from *Justice and the Politics of Difference*, for instance that group interests are better represented in a coalition of difference than in a conventional coherence of interests or notion of a unified public. This, in Young's view, will not worsen conflict between groups, but may be conducive to recognition (of groups and their sense of difference). She states 'A principle of representation for oppressed or disadvantaged groups has been implemented most frequently in organizations and movements that challenge politics as usual in the welfare capitalist state' (p. 189). Today that welfare capitalist state is in danger of becoming simply a capitalist, or free-market, state. The needs for representation of those disadvantaged grows, and with it a strange realisation, perhaps, that if the term minority applies to any disadvantaged group, it is possible a heterogeneous majority could become such a minority.

> The idea that I think we need today in order to make decisions in political matters cannot be the idea of a totality, or of the unity, of a body. It can only be the idea of a multiplicity or a diversity. . . . To state that one must draw a critique of political judgment means today to do a politics of opinions that at the same time is a politics of Ideas . . . in which justice is not placed under a rule of convergence but rather a rule of divergence. I believe that this is the theme that one finds constantly in present day writing under the name "minority."
>
> (Jean-François Lyotard)

There was once a time of caste and class, when tradition decreed that each group had its place, and that some are born to rule and others to serve. In this time of darkness, law and social norms defined rights, privileges, and obligations differently for different groups, distinguished by characteristics of sex, race, religion, class, or occupation. Social inequality was justified by church and state on the grounds that people have different natures, and some natures are better than others.

Then one day Enlightenment dawned, heralding a revolutionary conception of humanity and society. All people are equal, the revolutionaries declared, inasmuch as all have a capacity for reason and moral sense. Law and politics should therefore grant to everyone equal political and civil rights. With these bold ideas the battle lines of modern political struggle were drawn.

For over two hundred years since those voices of Reason first rang out, the forces of light have struggled for liberty and political equality against the dark forces of irrational prejudice, arbitrary metaphysics, and the crumbling towers of patriarchal church, state, and family. In the New World we had a head start in this fight, since the American War of Independence was fought on these Enlightenment principles, and our Constitution stood for liberty and equality. So we did not have to throw off the yokes of class and religious privilege, as did our Old World comrades. Yet the United States had its own oligarchic horrors in the form of slavery and the exclusion of women from public life. In protracted and bitter struggles these bastions of privilege based on group difference began to give way, finally to topple in the 1960s.

Today in our society a few vestiges of prejudice and discrimination remain, but we are working on them, and have nearly realized the dream those Enlightenment fathers dared to propound. The state and law should express rights only in universal terms applied equally to all, and differences among persons and groups should be a purely accidental and private matter. We seek a society in which differences of race, sex, religion, and ethnicity no longer make a difference to people's rights and opportunities. People should be treated as individuals, not as members of groups; their life options and rewards should be based solely on their individual achievement. All persons should have the liberty to be and do anything they want, to choose their own lives and not be hampered by traditional expectations and stereotypes.

We tell each other this story and make our children perform it for our sacred holidays – Thanksgiving Day, the Fourth of July, Memorial Day, Lincoln's Birthday. We have constructed Martin Luther King Day to fit the narrative so well that we have already forgotten that it took a fight to get it included in the canon year. There is much truth to this story. Enlightenment ideals of liberty and political equality did and do inspire movements against oppression and domination, whose success has created social values and institutions we would not want to lose. A people could do worse than tell this story after big meals and occasionally call upon one another to live up to it.

The very worthiness of the narrative, however, and the achievement of political equality that it recounts, now inspires new heretics. In recent years the ideal of liberation as the elimination of group difference has been challenged by movements of the oppressed. The very success of political movements against differential privilege and for political equality has generated movements of group specificity and cultural pride.

In this chapter I criticize an ideal of justice that defines liberation as the transcendence of group difference, which I refer to as an ideal of assimilation. This ideal usually promotes equal treatment as a primary principle of justice. Recent social movements of oppressed groups challenge this ideal. Many in these movements argue that a positive self-definition of group difference is in fact more liberatory.

I endorse this politics of difference, and argue that at stake is the meaning of social difference itself. Traditional politics that excludes or devalues some persons on account of their group attributes assumes an essentialist meaning of difference; it defines groups as having different natures. An egalitarian politics of difference, on the other hand, defines difference more fluidly and relationally as the product of social processes.

An emancipatory politics that affirms group difference involves a reconception of the meaning of equality. The assimilationist ideal assumes that equal social status for all persons requires treating everyone according to the same principles, rules, and standards. A politics of difference argues, on the other hand, that equality as the participation and inclusion of all groups sometimes requires different treatment for oppressed or disadvantaged groups. To promote social justice, I argue, social policy should sometimes accord special treatment to groups. I explore pregnancy and birthing rights for workers, bilingual-bicultural rights, and American Indian rights as three cases of such special treatment. Finally, I expand the idea of a heterogeneous public here by arguing for a principle of representation for oppressed groups in democratic decisionmaking bodies.

COMPETING PARADIGMS OF LIBERATION

In "On Racism and Sexism," Richard Wasserstrom (1980a) develops a classic statement of the ideal of liberation from group-based oppression as involving the elimination of group-based difference itself. A truly nonracist, nonsexist society, he suggests, would be one in which the race or sex of an individual would be the functional equivalent of eye color in our society today. While physiological differences in skin color or genitals would remain, they would have no significance for a person's sense of identity or how others regard him or her. No political rights or obligations would be connected to race or sex, and no important institutional benefits would be associated with either. People would see no reason to consider race or gender in policy or everyday interactions. In such a society, social group differences would have ceased to exist.

Wasserstrom contrasts this ideal of assimilation with an ideal of diversity much like the one I will argue for, which he agrees is compelling. He offers three primary reasons, however, for choosing the assimilationist ideal of liberation over the ideal of diversity. First, the assimilationist ideal exposes the arbitrariness of group-based social distinctions which are thought natural and necessary. By imagining a society in which race and sex have no social significance, one sees more clearly how pervasively these group categories unnecessarily limit possibilities for some in existing society. Second, the assimilationist ideal presents a clear and unambiguous standard of equality and justice. According to such a standard, any group-related differentiation or discrimination is suspect. Whenever laws or rules, the division of labor, or other social practices allocate benefits differently according to group membership, this is a sign of injustice. The principle of justice is simple: treat everyone according to the same principles, rules, and standards. Third, the assimilationist ideal maximizes choice. In a society where differences make no social difference people can develop themselves as individuals, unconstrained by group norms and expectations.

There is no question that the ideal of liberation as the elimination of group difference has been enormously important in the history of emancipatory politics. The ideal of universal humanity that denies natural differences has been a crucial historical development in the struggle against exclusion and status differentiation. It has made possible the assertion of the equal moral worth of all persons, and thus the right of all to participate and be included in all institutions and positions of power and privilege. The assimilationist ideal retains significant rhetorical power in the face of continued beliefs in the essentially different and inferior natures of women, Blacks, and other groups.

The power of this assimilationist ideal has inspired the struggle of oppressed groups and the supporters against the exclusion and denigration of these groups, and continues to inspire many. Periodically in American history, however, movements of the oppressed have questioned and rejected this "path to belonging" (Karst, 1986). Instead they have seen self-organization and the assertion of a positive group cultural identity as a better strategy for achieving power and participation in dominant institutions. Recent decades have witnessed a resurgence of this "politics of difference" not only among racial and ethnic groups, but also among women, gay men and lesbians, old people, and the disabled.

Not long after the passage of the Civil Rights Act and the Voting Rights Act, many white and Black supporters of the Black civil rights movement were surprised, confused, and angered by the emergence of the Black Power movement. Black Power advocates criticized the integrationist goal and reliance on the support of white liberals that characterized the civil rights movement. They encouraged Blacks to break their alliance with whites and assert the specificity of their own culture, political organization, and goals. Instead of integration, they encouraged Blacks to seek economic and political empowerment in their separate neighborhoods (Carmichael and Hamilton, 1967; Bayes, 1982, chap. 3; Lader, 1979, chap. 5; Omi and Winant, 1986, chap. 6). Since the late 1960s many Blacks have claimed that the integration successes of the civil rights movement have had the effect of dismantling the bases of Black-organized social and economic institutions at least as much as they have lessened Black-white animosity and opened doors of opportunity (Cruse, 1987). While some individual Blacks may be better off than they would have been if these changes had not occurred, as a group, Blacks are no better off and may be worse off, because the Blacks who have succeeded in assimilating into the American middle class no longer associate as closely with lower-class Blacks (cf. Wilson, 1978).

While much Black politics has questioned the ideal of assimilation in economic and political terms, the past twenty years have also seen the assertion and celebration by Blacks of a distinct Afro-American culture, both as a recovery and revaluation of an Afro-American history and in the creation of new cultural forms. The slogan "Black is beautiful" pierced American consciousness, deeply unsettling the received body aesthetic which I argued in Chapter 5 continues to be a powerful reproducer of racism. Afro-American hairstyles pronounced themselves differently stylish, not less stylish. Linguistic theorists asserted that Black English is English differently constructed, not bad English, and Black poets and novelists exploited and explored its particular nuances.

In the late 1960s Red Power came fast on the heels of Black Power. The American Indian Movement and other radical organizations of American Indians rejected perhaps even more vehemently than Blacks the goal of assimilation which has dominated white-Indian relations for most of the twentieth century. They asserted a right to self-government on Indian lands and fought to gain and maintain a dominant Indian voice in the Bureau of Indian Affairs. American Indians have sought to recover and preserve their language, rituals, and crafts, and this renewal of pride in traditional culture has also fostered a separatist political movement. The desire to pursue land rights claims and to fight for control over resources on reservations arises from what has become a fierce commitment to tribal self-determination, the desire to develop and maintain Indian political and economic bases in but not of white society (Deloria and Lytle, 1983; Ortiz, 1984, pt. 3; Cornell, 1988, pt. 2).

These are but two examples of a widespread tendency in the politics of the 1970s and 1980s for oppressed, disadvantaged, or specially marked groups to organize autonomously and assert a positive sense of their cultural and experiential specificity. Many Spanish-speaking Americans have rejected the traditional assumption that full participation in American society requires linguistic and cultural assimilation. In the last twenty years many have developed a renewed interest and pride in their Puerto Rican, Chicano, Mexican, or other Latin American heritage. They have asserted the right to maintain their specific culture and speak their language and still receive the benefits of citizenship, such as voting rights, decent education, and job opportunities. Many Jewish Americans have similarly rejected the ideal of assimilation, instead asserting the specificity and positive meaning of Jewish identity, often insisting publicly that Christian culture cease to be taken as the norm.

Since the late 1960s the blossoming of gay cultural expression, gay organization, and the public presence of gays in marches and other forums have radically altered the environment in which young people come to sexual identity, and changed many people's perceptions of homosexuality. Early gay rights advocacy had a distinctly assimilationist and universalist orientation. The goal was to remove the stigma of being homosexual, to prevent institutional discrimination, and to achieve societal recognition that gay people are "no different" from anyone else. The very process of political organization against discrimination and police harassment and for the achievement of civil rights, however, fostered the development of gay and lesbian communities and cultural expression, which by the mid 1970s flowered in meeting places, organizations, literature, music, and massive street celebrations (Altman, 1982; D'Emilio, 1983; Epstein, 1987).

Today most gay and lesbian liberation advocates seek not merely civil rights, but the affirmation of gay men and lesbians as social groups with specific experiences and perspectives. Refusing to accept the dominant culture's definition of healthy sexuality and respectable family life and social practices, gay and lesbian liberation movements have proudly created and displayed a distinctive self-definition and culture. For gay men and lesbians the analogue to racial integration is the typical liberal approach to sexuality, which tolerates any behavior as long as it is kept private. Gay pride asserts that sexual identity is a matter of culture and politics, and not merely "behavior" to be tolerated or forbidden.

The women's movement has also generated its own versions of a politics of difference. Humanist feminism, which predominated in the nineteenth century and in the contemporary women's movement until the late 1970s, finds in any assertion of difference between women and men only a legacy of female oppression and an ideology to legitimate continued exclusion of women from socially valued human activity. Humanist feminism is thus analogous to an ideal of assimilation in identifying sexual equality with gender blindness, with measuring women and men according to the same standards and treating them in the same way. Indeed, for many feminists, androgyny names the ideal of sexual liberation – a society in which gender difference itself would be eliminated. Given the strength and plausibility of this vision of sexual equality, it was confusing when feminists too began taking the turn to difference, asserting the positivity and specificity of female experience and values (see Young, 1985; Miles, 1985).

Feminist separatism was the earliest expression of such gynocentric feminism. Feminist separatism rejected wholly or partly the goal of entering the male-dominated world, because it requires playing according to rules that men have made and that have been used against women, and because trying to measure up to male-defined standards inevitably involves accommodating or pleasing the men who continue to dominate socially valued institutions and activities. Separatism promoted the empowerment of women through self-organization, the creation of separate and safe spaces where women could share and analyze their experiences, voice their anger, play with and create bonds with one another, and develop new and better institutions and practices.

Most elements of the contemporary women's movement have been separatist to some degree. Separatists seeking to live as much of their lives as possible in women-only institutions were largely responsible for the creation of the women's culture that burst forth all over the United States by the mid 1970s, and continues to claim the loyalty of millions of women – in the form of music, poetry, spirituality, literature, celebrations, festivals, and dances (see Jaggar, 1983, pp. 275–86). Whether drawing on images of Amazonian grandeur, recovering and revaluing traditional women's arts, like quilting and weaving, or inventing new rituals based on medieval witchcraft, the development of such expressions of women's culture gave many feminists images of a female-centered beauty and strength entirely outside capitalist patriarchal definitions of feminine pulchritude. The separatist impulse also fostered the development of the many autonomous women's institutions and services that have concretely improved the lives of many women, whether feminists or not – such as health clinics, battered women's shelters, rape crisis centers, and women's coffeehouses and bookstores.

Beginning in the late 1970s much feminist theory and political analysis also took a turn away from humanist feminism, to question the assumption that traditional female activity expresses primarily the victimization of women and the distortion of their human potential and that the goal of women's liberation is the participation of women as equals in public institutions now dominated by men. Instead of understanding the activities and values associated with traditional femininity as largely distortions and inhibitions of women's truly human potentialities, this gynocentric analysis sought to revalue the caring, nurturing, and cooperative approach to social relations they found associated with feminine socialization, and sought in women's specific experiences the bases for an attitude toward the body and nature healthier than that predominant in male-dominated Western capitalist culture.

None of the social movements asserting positive group specificity is in fact a unity. All have group differences within them. The Black movement, for example, includes middle-class Blacks and working-class Blacks, gays and straight people, men and women, and so it is with any other group. The implications of group differences within a social group have been most systematically discussed in the women's movement. Feminist conferences and publications have generated particularly fruitful, though often emotionally wrenching, discussions of the oppression of racial and ethnic blindness and the importance of attending to group differences among women (Bulkin, Pratt, and Smith, 1984). From such discussions emerged principled efforts to provide autonomously organized forums for Black women, Latinas, Jewish women, lesbians, differently abled women, old women, and any other women who see reason for claiming that they have as a group a distinctive voice that might be silenced in a general feminist discourse. Those discussions, along with the practices feminists instituted to structure discussion and interaction among differently identifying groups of women, offer some beginning models for the development of a heterogeneous public. Each of the other social movements has also generated discussion of group differences that cut across their identities, leading to other possibilities of coalition and alliance.

EMANCIPATION THROUGH THE POLITICS OF DIFFERENCE

Implicit in emancipatory movements asserting a positive sense of group difference is a different ideal of liberation, which might be called democratic cultural pluralism (cf. Laclau and Mouffe, 1985, pp. 166–71; Cunningham, 1987, pp. 186–99; Nickel, 1987). In this vision the good society does not eliminate or transcend group difference. Rather, there is equality among socially and culturally differentiated groups, who mutually respect one another and affirm one another in their differences. What are the reasons for rejecting the assimilationist ideal and promoting a politics of difference?

As I discussed in Chapter 2, some deny the reality of social groups. For them, group difference is an invidious fiction produced and perpetuated in order to preserve the privilege of the few. Others, such as Wasserstrom, may agree that social groups do now exist and have real social consequences for the way people identify themselves and one another, but assert that such social group differences are undesirable. The assimilationist ideal involves denying either the reality or the desirability of social groups.

Those promoting a politics of difference doubt that a society without group differences is either possible or desirable. Contrary to the assumption of modernization theory, increased urbanization and the extension of equal formal rights to all groups has not led to a decline in particularist affiliations. If anything, the urban concentration and interactions among groups that modernizing social processes introduce tend to reinforce group solidarity and differentiation (Rothschild, 1981; Ross, 1980; Fischer, 1982). Attachment to specific traditions, practices, language, and other culturally specific forms is a crucial aspect of social existence. People do not usually give up their social group identifications, even when they are oppressed.

Whether eliminating social group difference is possible or desirable in the long run, however, is an academic issue. Today and for the foreseeable future societies are certainly structured by groups, and some are privileged while others are oppressed. New social movements of group specificity do not deny the official story's claim that the ideal of liberation as eliminating difference and treating everyone the same has brought significant improvement in the status of excluded groups. Its main quarrel is with the story's conclusion, namely, that since we have achieved formal equality, only vestiges and holdovers of differential privilege remain, which will die out with the continued persistent assertion of an ideal of social relations that make differences irrelevant to a person's life prospects. The achievement of formal equality does not eliminate social differences, and rhetorical commitment to the sameness of persons makes it impossible even to name how those differences presently structure privilege and oppression.

Though in many respects the law is now blind to group differences, some groups continue to be marked as deviant, as the Other. In everyday interactions, images, and decisions, assumptions about women, Blacks, Hispanics, gay men and lesbians, old people, and other marked groups continue to justify exclusion, avoidance, paternalism, and authoritarian treatment. Continued racist, sexist, homophobic, ageist, and ableist institutions and behavior create particular circumstances for these groups, usually disadvantaging them in their opportunity to develop their capacities. Finally, in part because they have been segregated from one another, and in part because they have particular histories and traditions, there are cultural differences among social groups – differences in language, style of living, body comportment and gestures, values, and perspectives on society.

Today in American society, as in many other societies, there is widespread agreement that no person should be excluded from political and economic activities because of ascribed characteristics. Group differences nevertheless continue to exist, and certain groups continue to be privileged. Under these circumstances, insisting that equality and liberation entail ignoring difference has oppressive consequences in three respects.

First, blindness to difference disadvantages groups whose experience, culture, and socialized capacities differ from those of privileged groups. The strategy of assimilation aims to bring formerly excluded groups into the mainstream. So assimilation always implies coming into the game after it is already begun, after the rules and standards have already been set, and having to prove oneself according to those rules and standards. In the assimilationist strategy, the privileged groups implicitly define the standards according to which all will be measured. Because their privilege involves not recognizing these standards as culturally and experientially specific, the ideal of a common humanity in which all can participate without regard to race, gender, religion, or sexuality poses as neutral and universal. The real differences between oppressed groups and the dominant norm, however, tend to put them at a disadvantage in measuring up to these standards, and for that reason assimilationist policies perpetuate their disadvantage. Later in this chapter and in Chapter 7 I shall give examples of facially neutral standards that operate to disadvantage or exclude those already disadvantaged.

Second, the ideal of a universal humanity without social group differences allows privileged groups to ignore their own group specificity. Blindness to difference perpetuates cultural imperialism by allowing norms expressing the point of view and experience of privileged groups to appear neutral and universal. The assimilationist ideal presumes that there is a humanity in general, an unsituated group-neutral human capacity for self-making that left to itself would make individuality flower, thus guaranteeing that each individual will be different. As I argued in Chapter 4, because there is no such unsituated group-neutral point of view, the situation and experience of dominant groups tend to define the norms of such a humanity in general. Against such a supposedly neutral humanist ideal, only the oppressed groups come to be marked with particularity; they, and not the privileged groups, are marked, objectified as the Others.

Thus, third, this denigration of groups that deviate from an allegedly neutral standard often produces an internalized devaluation by members of those groups themselves. When there is an ideal of general human standards according to which everyone should be evaluated equally, then Puerto Ricans or Chinese Americans are ashamed of their accents or their parents, Black children despise the female-dominated kith and kin networks of their neighborhoods, and feminists seek to root out their tendency to cry, or to feel compassion for a frustrated stranger. The aspiration to assimilate helps produce the self-loathing and double consciousness characteristic of oppression. The goal of assimilation holds up to people a demand that they "fit," be like the mainstream, in behavior, values, and goals. At the same time, as long as group differences exist, group members will be marked as different – as Black, Jewish, gay – and thus as unable simply to fit. When participation is taken to imply assimilation the oppressed person is caught in an irresolvable dilemma: to participate means to accept and adopt an identity one is not, and to try to participate means to be reminded by oneself and others of the identity one is.

A more subtle analysis of the assimilationist ideal might distinguish between a conformist and a transformational ideal of assimilation. In the conformist ideal, status quo institutions and norms are assumed as given, and disadvantaged groups who differ from those norms are expected to conform to them. A transformational ideal of assimilation, on the other hand, recognizes that institutions as given express the interests and perspective of the dominant groups. Achieving assimilation therefore requires altering many institutions and practices in accordance with neutral rules that truly do not disadvantage or stigmatize any person, so that group membership really is irrelevant to how persons are treated. Wasserstrom's ideal fits a transformational assimilation, as does the group-neutral ideal advocated by some feminists (Taub and Williams, 1987). Unlike the conformist assimilationist, the transformational assimilationist may allow that group-specific policies, such as affirmative action, are necessary and appropriate means for transforming institutions to fit the assimilationist ideal. Whether conformist or transformational, however, the assimilationist ideal still denies that group difference can be positive and desirable; thus any form of the ideal of assimilation constructs group difference as a liability or disadvantage.

Under these circumstances, a politics that asserts the positivity of group difference is liberating and empowering. In the act of reclaiming the identity the dominant culture has taught them to despise (Cliff, 1980), and affirming it as an identity to celebrate, the oppressed remove double consciousness. I am just what they say I am – a Jewboy, a colored girl, a fag, a dyke, or a hag – and proud of it. No longer does one have the impossible project of trying to become something one is not under circumstances where the very trying reminds one of who one is. This politics asserts that oppressed groups have distinct cultures, experiences, and perspectives on social life with humanly positive meaning, some of which may even be superior to the culture and perspectives of mainstream society. The rejection and devaluation of one's culture and perspective should not be a condition of full participation in social life.

Asserting the value and specificity of the culture and attributes of oppressed groups, moreover, results in a relativizing of the dominant culture. When feminists assert the validity of feminine sensitivity and the positive value of nurturing behavior, when gays describe the prejudice of heterosexuals as homophobic and their own sexuality as positive and self-developing, when Blacks affirm a distinct Afro-American tradition, then the dominant culture is forced to discover itself for the first time as specific: as Anglo, European, Christian, masculine, straight. In a political struggle where oppressed groups insist on the positive value of their specific culture and experience, it becomes increasingly difficult for dominant groups to parade their norms as neutral and universal, and to construct the values and behavior of the oppressed as deviant, perverted, or inferior. By puncturing the universalist claim to unity that expels some groups and turns them into the Other, the assertion of positive group specificity introduces the possibility of understanding the relation between groups as merely difference, instead of exclusion, opposition, or dominance.

The politics of difference also promotes a notion of group solidarity against the individualism of liberal humanism. Liberal humanism treats each person as an individual, ignoring differences of race, sex, religion, and ethnicity. Each person should be evaluated only according to her or his individual efforts and achievements. With the institutionalization of formal equality some members of formerly excluded groups have indeed succeeded, by mainstream standards. Structural patterns of group privilege and oppression nevertheless remain. When political leaders of oppressed groups reject assimilation they are often affirming group solidarity. Where the dominant culture refuses to see anything but the achievement of autonomous individuals, the oppressed assert that we shall not separate from the people with whom we identify in order to "make it" in a white Anglo male world. The politics of difference insists on liberation of the whole group of Blacks, women, American Indians, and that this can be accomplished only through basic institutional changes. These changes must include group representation in policymaking and an elimination of the hierarchy of rewards that forces everyone to compete for scarce positions at the top.

Thus the assertion of a positive sense of group difference provides a standpoint from which to criticize prevailing institutions and norms. Black Americans find in their traditional communities, which refer to their members as "brother" and "sister," a sense of solidarity absent from the calculating individualism of white professional capitalist society. Feminists find in the traditional female values of nurturing a challenge to a militarist world-view, and lesbians find in their relationships a confrontation with the assumption of complementary gender roles in sexual relationships. From their experience of a culture tied to the land American Indians formulate a critique of the instrumental rationality of European culture that results in pollution and ecological destruction. Having revealed the specificity of the dominant norms which claim universality and neutrality, social movements of the oppressed are in a position to inquire how the dominant institutions must be changed so that they will no longer reproduce the patterns of privilege and oppression.

From the assertion of positive difference the self-organization of oppressed groups follows. Both liberal humanist and leftist political organizations and movements have found it difficult to accept this principle of group autonomy. In a humanist emancipatory politics, if a group is subject to injustice, then all those interested in a just society should unite to combat the powers that perpetuate that injustice. If many groups are subject to injustice, moreover, then they should unite to work for a just society. The politics of difference is certainly not against coalition, nor does it hold that, for example, whites should not work against racial injustice or men against sexist injustice. This politics of group assertion, however, takes as a basic principle that members of oppressed groups need separate organizations that exclude others, especially those from more privileged groups. Separate organization is probably necessary in order for these groups to discover and reinforce the positivity of their specific experience, to collapse and eliminate double consciousness. In discussions within autonomous organizations, group members can determine their specific needs and interests. Separation and self-organization risk creating pressures toward homogenization of the groups themselves, creating new privileges and exclusions, a problem I shall discuss in Chapter 8. But contemporary emancipatory social movements have found group autonomy an important vehicle for empowerment and the development of a group-specific voice and perspective.

Integration into the full life of the society should not have to imply assimilation to dominant norms and abandonment of group affiliation and culture (Edley, 1986; cf. McGary, 1983). If the only alternative to the oppressive exclusion of some groups defined as Other by dominant ideologies is the assertion that they are the same as everybody else, then they will continue to be excluded because they are not the same.

Some might object to the way I have drawn the distinction between an assimilationist ideal of liberation and a radical democratic pluralism. They might claim that I have not painted the ideal of a society that transcends group differences fairly, representing it as homogeneous and conformist. The free society envisaged by liberalism, they might say, is certainly pluralistic. In it persons can affiliate with whomever they choose; liberty encourages a proliferation of life styles, activities, and associations. While I have no quarrel with social diversity in this sense, this vision of liberal pluralism does not touch on the primary issues that give rise to the politics of difference. The vision of liberation as the transcendence of group difference seeks to abolish the public and political significance of group difference, while retaining and promoting both individual and group diversity in private, or nonpolitical, social contexts. In Chapter 4 I argued that this way of distinguishing public and private spheres, where the public represents universal citizenship and the private individual differences, tends to result in group exclusion from the public. Radical democratic pluralism acknowledges and affirms the public and political significance of social group differences as a means of ensuring the participation and inclusion of everyone in social and political institutions.

RECLAIMING THE MEANING OF DIFFERENCE

Many people inside and outside the movements I have discussed find the rejection of the liberal humanist ideal and the assertion of a positive sense of group difference both confusing and controversial. They fear that any admission by oppressed groups that they are different from the dominant groups risks justifying anew the subordination, special marking, and exclusion of those groups. Since calls for a return of women to the kitchen, Blacks to servant roles and separate schools, and disabled people to nursing homes are not absent from contemporary politics, the danger is real. It may be true that the assimilationist ideal that treats everyone the same and applies the same standards to all perpetuates disadvantage because real group differences remain that make it unfair to compare the unequals. But this is far preferable to a reestablishment of separate and unequal spheres for different groups justified on the basis of group difference.

Since those asserting group specificity certainly wish to affirm the liberal humanist principle that all persons are of equal moral worth, they appear to be faced with a dilemma. Analyzing W. E. B. Du Bois's arguments for cultural pluralism, Bernard Boxill poses the dilemma this way: "On the one hand, we must overcome segregation because it denies the idea of human brotherhood; on the other hand, to overcome segregation we must self-segregate and therefore also deny the idea of human brotherhood" (Boxill, 1984, p. 174). Martha Minow finds a dilemma of difference facing any who seek to promote justice for currently oppressed or disadvantaged groups. Formally neutral rules and policies that ignore group differences often perpetuate the disadvantage of those whose difference is defined as deviant; but focusing on difference risks recreating the stigma that difference has carried in the past (Minow, 1987, pp. 12–13; cf. Minow, 1985; 1990).

These dilemmas are genuine, and exhibit the risks of collective life, where the consequences of one's claims, actions, and policies may not turn out as one intended because others have understood them differently or turned them to different ends. Since ignoring group differences in public policy does not mean that people ignore them in everyday life and interaction, however, oppression continues even when law and policy declare that all are equal. Thus I think for many groups and in many circumstances it is more empowering to affirm and acknowledge in political life the group differences that already exist in social life. One is more likely to avoid the dilemma of difference in doing this if the meaning of difference itself becomes a terrain of political struggle. Social movements asserting the positivity of group difference have established this terrain, offering an emancipatory meaning of difference to replace the old exclusionary meaning.

The oppressive meaning of group difference defines it as absolute otherness, mutual exclusion, categorical opposition. This essentialist meaning of difference submits to the logic of identity. One group occupies the position of a norm, against which all others are measured. The attempt to reduce all persons to the unity of a common measure constructs as deviant those whose attributes differ from the group-specific attributes implicitly presumed in the norm. The drive to unify the particularity and multiplicity of practices, cultural symbols, and ways of relating in clear and distinct categories turns difference into exclusion.

Thus I explored in the previous two chapters how the appropriation of a universal subject position by socially privileged groups forces those they define as different outside the definition of full humanity and citizenship. The attempt to measure all against some universal standard generates a logic of difference as hierarchical dichotomy – masculine/feminine, civilized/ savage, and so on. The second term is defined negatively as a lack of the truly human qualities; at the same time it is defined as the complement to the valued term, the object correlating with its subject, that which brings it to completion, wholeness, and identity. By loving and affirming him, a woman serves as a mirror to a man, holding up his virtues for him to see (Irigaray, 1985). By carrying the white man's burden to tame and educate the savage peoples, the civilized will realize universal humanity. The exotic orientals are there to know and master, to be the completion of reason's progress in history, which seeks the unity of the world (Said, 1978). In every case the valued term achieves its value by its determinately negative relation to the Other.

In the objectifying ideologies of racism, sexism, anti-Semitism, and homophobia, only the oppressed and excluded groups are defined as different. Whereas the privileged groups are neutral and exhibit free and malleable subjectivity, the excluded groups are marked with an essence, imprisoned in a given set of possibilities. By virtue of the characteristics the group is alleged to have by nature, the ideologies allege that group members have specific dispositions that suit them for some activities and not others. Difference in these ideologies always means exclusionary opposition to a norm. There are rational men, and then there are women; there are civilized men, and then there are wild and savage peoples. The marking of difference always implies a good/bad opposition; it is always a devaluation, the naming of an inferiority in relation to a superior standard of humanity.

Difference here always means absolute otherness; the group marked as different has no common nature with the normal or neutral ones. The categorical opposition of groups essentializes them, repressing the differences within groups. In this way the definition of difference as exclusion and opposition actually denies difference. This essentializing categorization also denies difference in that its universalizing norms preclude recognizing and affirming a group's specificity in its own terms.

Essentializing difference expresses a fear of specificity, and a fear of making permeable the categorical border between oneself and the others. This fear, I argued in the previous chapter, is not merely intellectual, and does not derive only from the instrumental desire to defend privilege, though that may be a large element. It wells from the depths of the Western subject's sense of identity, especially, but not only, in the subjectivity of privileged groups. The fear may increase, moreover, as a clear essentialism of difference wanes, as belief in a specifically female, Black, or homosexual nature becomes less tenable.

The politics of difference confronts this fear, and aims for an understanding of group difference as indeed ambiguous, relational, shifting, without clear borders that keep people straight – as entailing neither amorphous unity nor pure individuality. By asserting a positive meaning for their own identity, oppressed groups seek to seize the power of naming difference itself, and explode the implicit definition of difference as deviance in relation to a norm, which freezes some groups into a self-enclosed nature. Difference now comes to mean not otherness, exclusive opposition, but specificity, variation, heterogeneity. Difference names relations of similarity and dissimilarity that can be reduced to neither coextensive identity nor nonoverlapping otherness.

The alternative to an essentializing, stigmatizing meaning of difference as opposition is an understanding of difference as specificity, variation. In this logic, as Martha Minow (1985; 1987; 1990) suggests, group differences should be conceived as relational rather than defined by substantive categories and attributes. A relational understanding of difference relativizes the previously universal position of privileged groups, which allows only the oppressed to be marked as different. When group difference appears as a function of comparison between groups, whites are just as specific as Blacks or Latinos, men just as specific as women, able-bodied people just as specific as disabled people. Difference thus emerges not as a description of the attributes of a group, but as a function of the relations between groups and the interaction of groups with institutions (cf. Littleton, 1987).

In this relational understanding, the meaning of difference also becomes contextualized (cf. Scott, 1988). Group differences will be more or less salient depending on the groups compared, the purposes of the comparison, and the point of view of the comparers. Such contextualized understandings of difference undermine essentialist assumptions. For example, in the context of athletics, health care, social service support, and so on, wheelchair-bound people are different from others, but they are not different in many other respects. Traditional treatment of the disabled entailed exclusion and segregation because the differences between the disabled and the able-bodied were conceptualized as extending to all or most capacities.

In general, then, a relational understanding of group difference rejects exclusion. Difference no longer implies that groups lie outside one another. To say that there are differences among groups does not imply that there are not overlapping experiences, or that two groups have nothing in common. The assumption that real differences in affinity, culture, or privilege imply oppositional categorization must be challenged. Different groups are always similar in some respects, and always potentially share some attributes, experiences, and goals.

Such a relational understanding of difference entails revising the meaning of group identity as well. In asserting the positive difference of their experience, culture, and social perspective, social movements of groups that have experienced cultural imperialism deny that they have a common identity, a set of fixed attributes that clearly mark who belongs and who doesn't. Rather, what makes a group a group is a social process of interaction and differentiation in which some people come to have a particular *affinity* (Haraway, 1985) for others. My "affinity group" in a given social situation comprises those people with whom I feel the most comfortable, who are more familiar. Affinity names the manner of sharing assumptions, affective bonding, and networking that recognizably differentiates groups from one another, but not according to some common nature. The salience of a particular person's group affinities may shift according to the social situation or according to changes in her or his life. Membership in a social group is a function not of satisfying some objective criteria, but of a subjective affirmation of affinity with that group, the affirmation of that affinity by other members of the group, and the attribution of membership in that group by persons identifying with other groups. Group identity is constructed from a flowing process in which individuals identify themselves and others in terms of groups, and thus group identity itself flows and shifts with changes in social process.

Groups experiencing cultural imperialism have found themselves objectified and marked with a devalued essence from the outside, by a dominant culture they are excluded from making. The assertion of a positive sense of group difference by these groups is emancipatory because it reclaims the definition of the group by the group, as a creation and construction, rather than a given essence. To be sure, it is difficult to articulate positive elements of group affinity without essentializing them, and these movements do not always succeed in doing so (cf. Sartre, 1948, p. 85; Epstein, 1987). But they are developing a language to describe their similar social situation and relations to one another, and their similar perceptions and perspectives on social life. These movements engage in the project of cultural revolution I recommended in the last chapter, insofar as they take culture as in part a matter of collective choice. While their ideas of women's culture, Afro-American culture, and American Indian culture rely on past cultural expressions, to a significant degree these movements have self-consciously constructed the culture that they claim defines the distinctiveness of their groups.

Contextualizing both the meaning of difference and identity thus allows the acknowledgment of difference within affinity groups. In our complex, plural society, every social group has group differences cutting across it, which are potential sources of wisdom, excitement, conflict, and oppression. Gay men, for example, may be Black, rich, homeless, or old, and these differences produce different identifications and potential conflicts among gay men, as well as affinities with some straight men.

REFERENCES

Alexander, David. 1987. "Gendered Job Traits and Women's Occupations." Ph.D. dissertation, Economics, University of Massachusetts.

Altman, Dennis. 1982. *The Homosexualization of American Society*. Boston: Beacon.

Bastian, Ann, Norm Fruchter, Marilyn Gittell, Colin Greer, and Kenneth Haskins. 1986. *Choosing Equality: The Case for Democratic Schooling*. Philadelphia: Temple University Press.

Bayes, Jane H. 1982. *Minority Politics and Ideologies in the United States*. Novato, Calif: Chandler and Sharp.

Beatty, Richard W. and James R. Beatty. 1981. "Some Problems with Contemporary Job Evaluation Systems." In Helen Remick, ed., *Comparable Worth and Wage Discrimination: Technical Possibilities and Political Realities.* Philadelphia: Temple University Press.

Beitz, Charles. 1979. *Political Theory and International Relations.* Princeton: Princeton University Press.

——. 1988. "Equal Opportunity in Political Representation." In Norman Bowie, ed., *Equal Opportunity.* Boulder: Westview.

Bell, Derek. 1987. *And We Are Not Saved: The Elusive Quest for Racial Justice.* New York: Basic.

Boxill, Bernard. 1984. *Blacks and Social Justice.* Totowa, N.J.: Rowman and Allanheld.

Bulkin, Elly, Minnie Bruce Pratt, and Barbara Smith. 1984. *Yours in Struggle: Three Feminist Perspectives on Anti-Semitism and Racism.* New York: Long Haul.

Canter, Norma V. 1987. "Testimony from Mexican American Legal Defense and Education Fund." *Congressional Digest* (March).

Carmichael, Stokley and Charles Hamilton. 1967. *Black Power.* New York and Random House.

Cliff, Michelle. 1980. *Reclaiming the Identity They Taught Me to Despise.* Watertown, Mass.: Persephone.

Collins, Sheila. 1986. *The Rainbow Challenge: The Jackson Campaign and the Future of U.S. Politics.* New York: Monthly Review Press.

Cornell, Stephen. 1988. *The Return of the Native: American Indian Political Resurgence.* New York: Oxford University Press.

Coward, Rosalind and John Ellis. 1977. *Language and Materialism.* London: Routledge and Kegan Paul.

Cruse, Harold. 1987. *Plural but Equal: Blacks and Minorities and America's Plural Society.* New York: Morrow.

Cunningham, Frank. 1987. *Democratic Theory and Socialism.* Cambridge: Cambridge University Press.

Deloria, Vine and Clifford Lytle. 1984. *The Nations Within.* New York: Pantheon.

Delphy, Christine. 1984. *Close to Home: A Materialist Analysis of Women's Oppression.* Amherst: University of Massachusetts Press.

D'Emilio, Joseph. 1983. *Sexual Politics, Sexual Communities.* Chicago: University of Chicago Press.

Edley, Christopher. 1986. "Affirmative Action and the Rights Rhetoric Trap." In Robert Fullinwider and Claudia Mills, eds. *The Moral Foundations of Civil Rights.* Totowa, N.J.: Rowman and Littlefield.

Epstein, Steven. 1987. "Gay Politics, Ethnic Identity: The Limits of Social Constructionism." *Socialist Review* 17 (May–August) 9–54.

Fischer, Claude. 1982. *To Dwell among Friends: Personal Networks in Town and City.* Chicago: University of Chicago Press.

Gutmann, Amy. 1980. *Liberal Equality.* Cambridge: Cambridge University Press.

——. 1985. "Communitarian Critics of Liberalism." *Philosophy and Public Affairs* 14 (Summer): 308–22.

Haraway, Donna. 1985. "Manifesto for Cyborgs." *Socialist Review* 80 (March/April): 65–107.

Irigaray, Luce. 1985. *Speculum of the Other Woman.* Ithaca: Cornell University Press.

Jaggar, Alison. 1983. *Feminist Politics and Human Nature.* Totowa, N.J.: Rowman and Allanheld.

Karst, Kenneth. 1986. "Paths to Belonging: The Constitution and Cultural Identity." *North Carolina Law Review* 64 (January): 303–77.

Kleven, Thomas. 1988. "Cultural Bias and the Issue of Bilingual Education." *Social Policy* 19 (Summer): 9–12.

Laclau, Ernesto and Chantal Mouffe. 1985. *Hegemony and Socialist Strategy.* London: Verso.

Lader, Laurence. 1979. *Power on the Left.* New York: Norton.

Littleton, Christine. 1987. "Reconstructing Sexual Equality." California Law Review 75 (July): 1279–1337.

Livingstone, John C. 1979. *Fair Game? Inequality and Affirmative Action.* San Francisco: Freeman.

Lloyd, Genevieve. 1984. *The Man of Reason: "Male" and "Female" in Western Philosophy.* Minneapolis: University of Minnesota Press.

Lofland, Lyn H. 1973. *A World of Strangers: Order and Action in Urban Public Space.* New York: Basic.

Logan, John R. and Harvey L. Molotch. 1987. *Urban Fortunes: The Political Economy of Place.* Berkeley and Los Angeles: University of California Press.

Lowi, Theodore. 1969. *The End of Liberalism.* New York: Norton.

Lyotard, Jean-François. 1984. *The Postmodern Condition.* Minneapolis: University of Minnesota Press.

McGary, Howard. 1983. "Racial Integration and Racial Separatism: Conceptual Clarifications." In Leonard Harris, ed., *Philosophy Born of Struggle.* Dubuque, Iowa: Hunt.

Miles, Angela. 1985. "Feminist Radicalism in the 1980's." *Canadian Journal of Political and Social Theory* 9:16–39.

Minow, Martha. 1985. "Learning to Live with the Dilemma of Difference: Bilingual and Special Education." *Law and Contemporary Problems* 48 (Spring): 157–211.

——. 1987. "Justice Engendered." *Harvard Law Review* 101 (November): 11–95.

——. 1990. *Making All the Difference*. Ithaca: Cornell University Press.

Nickel, James. 1988. "Equal Opportunity in a Pluralistic Society." In Ellen Frankel Paul, Fred D. Miller, Jeffrey Paul, and John Ahrens, eds., *Equal Opportunity*. Oxford: Blackwell.

Omi, Michael and Howard Winant. 1983. "By the Rivers of Babylon: Race in the United States, Part I and II." *Socialist Review* 71 (September–October): 31–66; 72 (November–December): 35–70.

——. 1986. *Racial Formation in the United States*. New York: Routledge and Kegan Paul.

Orr, Eleanor Wilson. 1987. *Twice as Less: Black English and the Performance of Black Students in Mathematics and Science*. New York: Norton.

Ortiz, Roxanne Dunbar. 1984. *Indians of the Americas*. New York: Praeger.

Ross, Jeffrey. 1980. Introduction to Jeffrey Ross and Ann Baker Cottrell, eds., *The Mobilization of Collective Identity*. Lanham, Md.: University Press of America.

Rothschild, Joseph. 1981. *Ethnopolitics*. New York: Columbia University Press.

Said, Edward. 1978. *Orientalism*. New York: Pantheon.

Sartre, Jean-Paul. 1948. *Anti-Semite and Jew*. New York: Schocken.

Sawicki, Jana. 1986. "Foucault and Feminism: Toward a Politics of Difference." *Hypatia: A Journal of Feminist Philosophy* 1 (Summer): 23–36.

Scales, Ann. 1981. "Towards a Feminist Jurisprudence." *Indian Law Journal* 56 (Spring): 375–444.

Schmitt, Eric. 1989. "As the Suburbs Speak More Spanish, English Becomes a Cause." *New York Times*, 26 February.

Scott, Joan. 1988. "Deconstructing Equality-versus-Difference: Or the Uses of Post-Structuralist Theory for Feminism." *Feminist Studies* 14 (Spring): 33–50.

Sears, David O. and Leonie Huddy. 1987. "Bilingual Education: Symbolic Meaning and Support among Non-Hispanics." Paper presented at the annual meeting of the American Political Science Association, Chicago, September.

Sunstein, Cass R. 1988. "Beyond the Republican Revival." *Yale Law Journal* 97 (July): 1539–90.

Taub, Nadine and Wendy Williams. 1985. "Will Equality Require More than Assimilation, Accommodation or Separation from the Existing Social Structure?" *Rutgers Law Review* 37 (Summer): 825–44.

Treiman, Donald J. and Heidi I. Hartman. 1981. *Women, Work and Wages*. Washington, D.C.: National Academy Press.

Vogel, Lisa. 1990. "Debating Difference: The Problem of Special Treatment of Pregnancy in the Workplace." *Feminist Studies*, in press.

Williams, Robert A. 1986. "The Algebra of Federal Indian Law: The Hard Trail of Decolonizing and Americanizing the White Man's Indian Jurisprudence." *Wisconsin Law Review*, pp. 219–99.

Williams, Wendy. 1983. "Equality's Riddle: Pregnancy and the Equal Treatment/Special Treatment Debate." *New York University Review of Law and Social Change* 13:325–80.

Wilson, William J. 1978. *The Declining Significance of Race*. Chicago: University of Chicago Press.

Wolgast, Elizabeth. 1980. *Equality and the Rights of Women*. Ithaca: Cornell University Press.

Young, Iris. 1979. "Self-Determination as a Principle of Justice." *Philosophical Forum* 11 (Fall): 172–82.

——. 1981. "Toward a Critical Theory of Justice." *Social Theory and Practice* 7 (Fall): 279–302.

——. 1983. "Justice and Hazardous Waste." In Michael Bradie, ed., *The Applied Turn in Contemporary Philosophy*. Bowling Green, Ohio: Applied Philosophy Program, Bowling Green State University.

——. 1985. "Humanism, Gynocentrism and Feminist Politics." *Women's Studies International Forum* 8:173–83.

——. 1987. "Impartiality and the Civic Public: Some Implications of Feminist Critiques of Moral and Political Theory." In Seyla Benhabib and Drucilla Cornell, eds., *Feminism as Critique*. Oxford/Minneapolis: Polity/University of Minnesota Press.

——. 1989. "Polity and Group Difference: A Critique of the Ideal of Universal Citizenship." *Ethics* 99 (January): 250–74.

'World Alienation and the Modern Age: The Deprivations of Obscurity'

from *Our Sense of the Real: Aesthetic Experience and Arendtian Politics* (1999)

Kimberley Curtis

Editors' Introduction

Young, in the previous text, addresses the conditions, and among them the injustices, of a multi-ethnic, multicultural society. Her idea of civic society is based on difference between individuals and between groups but especially the latter. Hannah Arendt's political philosophy places a greater emphasis on the individual citizen, and on derivation of ideas from Greek models of democracy, but this is not surprising in the context of the period in which she wrote. *The Human Condition*, where such ideas are rehearsed, was first published in 1955 (Chicago, University of Chicago Press), when the model of a liberal society based on humane values and a paradigm of inclusion was more or less unchallenged, and contrasted – not least in the Nürnberg trials, and Eichmann's trial in Jerusalem, which Arendt attended as an observer – with fascism and anti-semitism.

Arendt's key concern in the material critically discussed by Curtis (Assistant Professor of the Practice of Political Science and Women's Studies at Duke University) is the crippling state of oblivion of those denied access to the exposure of difference in the public sphere (or publicity). Arguing that natality, seen as a second birth in which a mature rather than a given self comes into being, is a process possible only amidst the perception of others, Arendt sees the public sphere as a lighter place than the private or domestic (or family) sphere. This may now seem a romanticisation of the public sphere, after analysis of its gendering in postmodern geography and other fields, but for Arendt it is more a continuation of Enlightenment rationality. Curtis discusses Arendt's ideas both in terms of their period and more recent concerns for display and the performative subject; but she retains the core idea, which is central to modernity of freedom through human plurality.

Curtis, in a later part of her text, draws out of Arendt's writing a sense of creative tension between subject and other, expressed as loss in modernity: 'The deepest sign of the modern age is loss of the condition of the world understood as both the subjective-in-between that intangibly forms between people who share a common life and the thing-quality or objective-in-between' (p. 77). She adds that for Arendt both senses carry the subject's duality of relation and separation. Although the emphasis is different, this could be read, as well as alongside Arendt's own work, alongside that, too, of Martin Buber in *I and Thou* (translated by Walter Kauffman, Edinburgh, T. & T. Clark, 1970). Also suggested is Arendt's *Men in Dark Times* (London, Jonathan Cape, [1955] 1970) in which she considers the public sphere through the work of Rosa Luxemburg, Karl Jasper, Walter Benjamin, and Bertold Brecht, among others.

Curtis refers to four of Arendt's books by acronyms, using the following editions:

BPF: *Between Past and Future*, 1978, Harmondsworth, Penguin
MDT: *Men in Dark Times*, 1955, New York, Harcourt Brace Jovanovich

HC: *The Human Condition*, 1958, Chicago, University of Chicago Press
OR: *On Revolution*, 1963, Harmondsworth, Penguin

> The poor man's conscience is clear; yet he is ashamed. His character is irreproachable; yet he is neglected and despised. He feels himself out of the sight of others, groping in the dark. Mankind takes no notice of him. He rambles and wanders unheeded. In the midst of a crowd, at church, in the market, at a play . . . he is in as much obscurity as he would be in a garret or a cellar. He is not disapproved, censured, or reproached; *he is only not seen*. This total inattention is to him mortifying, painful and cruel. He suffers a misery from this consideration, which is sharpened by the consciousness that others have no fellow-feeling with him in this distress.
>
> (John Adams, *The Works of John Adams*)

Writing in the first year of his term as vice president of the United States, John Adams gives voice to a kind of suffering he thought peculiar to the poor. Such suffering creates, he wrote, "as severe a pain as the gout or stone [and] produces despair and detestation of existence."[1] Arendt was drawn to Adams's insights, moved by "the feeling of injustice" he expressed. Once self-preservation has been assured, she wrote, "the real predicament" of the poor is that they must suffer "the insult of oblivion ' as she called it. "Darkness rather than want is the curse of poverty" (*OR*, 69). No one articulated more poignantly than John Adams what Arendt referred to as "the crippling consequences" of such darkness. And no political theorist has cared more about and wrote more passionately against these consequences than Hannah Arendt.

It is not, of course, that Arendt wrote against poverty or championed the cause of social injustice. Indeed, she seems somewhat tone-deaf to such concerns.[2] Arendt's political theorizing was fired by a feeling for a different kind of injustice – the insult of oblivion. This form of injustice is, I suggest, prior to social injustice in the sense that the degradation of obscurity is a primary precondition of our capacity to inflict and sustain the suffering involved in the many forms social and economic inequality take.[3] As I shall argue, the point is not only the uncertain moral claim that if we could but fully "see" others we could neither commit nor sustain social injustices against them. It is also that the condition of oblivion itself weakens those who suffer it in ethically troublesome ways. Thus, in what has been perceived as Arendt's insensitivity, even cold inhumanity, I find the essence of a particularly compelling humanism that owes much to her phenomenology and wells out of her concern with how to intensify our awareness of reality.

What, then, are the crippling consequences of this form of injustice? In the context of the American Revolution, Arendt wrote about the political predicament of the vast majority of the population who suffered not from want but from "continual toil and want of leisure." Here we can begin to trace Arendt's thought. The essential injustice suffered by this majority was that while, in varying degrees, they were represented and could choose their representatives, "these essentially negative safeguards by no means open the public realm to the many, nor can they arouse in them that 'passion for distinction' . . . which according to John Adams, 'next to self-preservation will forever be the spring of human actions'" (*OR*, 69).

Here Arendt hones in on the crucial triad: public space, arousal, and action. Excluded from the place where we appear to others and they to us, the play of arousal, the provocation between those who see and those who are seen, remains largely dormant. That aesthetic-existential urge to make our presence felt – the urge Arendt theorized as an active response to being perceived – is thwarted, the passion to excel unawakened. And with this comes suffering that is crippling.

It is tempting to interpret this crippling largely within the frame adopted by modern theorists of individual liberty who were influenced by the Romantic expressive tradition. John Stuart Mill, Alexis de Tocqueville, and Wilhelm von Humboldt, for example, argue that individuals denied the proper cultivation of their inward forces suffer distortions.[4] Here, however, the focus is on self-development, and although Arendt is not uninterested in this type of individual flourishing, her primary concern is with existential crippling. Those relegated to oblivion suffer a loss of feeling for their own existence, their own reality, as well as for the larger world and their relationship to it. They feel, as Adams put it, "despair and detestation of existence." It is this crippling of the urge to appear that makes the insult of oblivion prior to social injustice. In rendering the victim invisible, it disarms the impulse that is the spring of human action, creating a condition in which resistance to social injustice is unlikely.[5]

Although Arendt developed her understanding of the insult of oblivion, following Adams, in relation to the "curse" of poverty, it forms in her writing the central injustice suffered in the modern age more generally. Indeed, it is the normative standpoint from which she develops her critique of modern political thinking and modern social and political institutions and practices. Yet it is perhaps strange to call this an injustice endemic to the modern age, for if it is crippling, most of us remain dramatically unaware of its impact. "Modern sensibility," Arendt argues, "is not touched by obscurity" (*OR*, 70). Indeed, the odd thing is perhaps the lack of outrage, the absence of pain, despite Adams's passionate depiction. Thus we have the paradoxical argument for an elemental, widely suffered form of injustice that almost no one recognizes. I suggest that we understand the paradox in this way: that Adams is right, that the passion for distinction is a real existential need, but because it is intangible and requires institutional support, it is easily overwhelmed by substitutes. In the modern age, the allure of riches as the road to engaging the "passion for distinction" easily displaced the deeper path this passion can tread.

Arendt's political theory offers us an idiom with which to identify, even experience, perhaps, become more susceptible to this kind of suffering *as a phenomenon*. In this chapter I seek to elaborate this idiom through an examination of Arendt's conception of freedom, the public sphere, and humanism. I then turn to selected features of her critique of the modern age, highlighting her ontological concerns and approaching them with the seriousness I believe they merit. I discuss those modern conditions that widely extend the form of suffering with which she is concerned – conditions that dramatically attenuate the power of that mutual aesthetic provocation through which our sense of the real is born. I conclude by arguing for privileging a public sensibility that is attentive to the insult of oblivion but fruitfully mediates the tensions between the public passion for distinction and the suffering inflicted by diverse forms of social injustice.

THE ARENDTIAN IDIOM

What is freedom? It is, Arendt elegantly says, "the freedom to call something into being which did not exist before" (*BPF*, 151). This capacity, she argues, springs from our own beginning in the world, is, as it were, a response to it. Hence to be free is to begin, and our impulse to freedom "animates and inspires" all human activities (*BPF*, 169).

Yet our question is more specific: What is freedom as it relates to politics? Although born of the same responsive provocation to being in the world, freedom as it relates to politics is distinct. It is not the same as the inner freedom we may experience in solitude as our mental life's two-in-one plays with itself. Neither is it the freedom of the artist whose responsive creativity precedes the product and thus does not show itself to others, except as reified testimony to and promise of the human impulse to freedom. Politically relevant freedom, by contrast, is manifest only amid the sensuous aesthetic provocation of speaking and acting beings. It emerges only where people are together in this manner. It carries some of the exhilaration of free corporeal movement, untied and uncoerced by direct need. Indeed, to be free requires first that we be liberated from the toil upon us as nature's creatures. Won either by coercing others or through the invention of implements to replace our own labor, we become freed for a togetherness governed not by interest or need but by the impulse to freedom.[6]

Arendt argues that this worldly quality of freedom pertinent to politics is testified to in both Greek and Latin. In both languages there are two words for action. In Greek, *archein* means "to begin", "to lead," "to rule," and *prattein* means "to carry something through," "to achieve," "to finish." In Latin, a similar structure prevails: *agere* means "to set something in motion," and *gerere* means, originally, "to bear." Arendt finds here evidence for the "original interdependence" of action: that the one who begins "depends" on others to complete what was begun, and that those who carry forth the act to completion in turn "depend" on the beginner for their opportunity to act (*HC*, 189). By contrast, our single term, *action*, has lost the second cluster of meanings, so that for us to act is synonymous with leading and ruling. We think of action on the model – indeed, the ideal – of sovereignty, and with this change we lose the politically relevant relationship of action to freedom.

As fascinating as is this etymological path toward reviving an older understanding of action, the discrete nature of the terms and the different selves to which, in any given action, they refer should not obfuscate the phenomenon of freedom as Arendt theorizes it. Freedom emerges through the active and reciprocal intertwining of appearing beings. "The actor," she writes, "always moves *among* and *in relation* to other acting beings" (*HC*, 190; my italics). That is, we simultaneously act and are acted upon, and this "reversibility," in which our urge to make our presence felt conjoins simultaneously with our sense of the presence of others, is the birth of freedom. It is the birth of the fearsome miracle upsurge out of human plurality of the new beginning. Freedom as it is relevant to politics cannot be specified apart from an account of this very profound reciprocal provocation on the part of appearing beings, this "original interdependence."[7] Indeed, it is this reversibility that gives rise to the *deinon* attributes of human freedom: action's essential unpredictability, its boundlessness, and its irreversibility, all of which are attributes alone of this nonmodern sense of freedom. Perhaps, too, one of action's most quixotic but essential qualities is also born in this reversibility – the capacity to generate power and mobilize people freely on behalf of a common cause.

Politically relevant freedom is thus, as Arendt theorizes it, never something we *possess*. It is, rather, a mode of being whose specific virtuosity shows itself in how well we, as nonsovereign beings, act to answer the world's demanding, often urgent call. Those too burdened by private need or interest are poorly capable of such responsiveness.

Like all performing arts, then, the appearance of freedom in the sense of the capacity to begin something new in word and deed is ephemeral. Freedom *is* only in performance, only in acts. It has a fugitive, apparitional quality, and hardly seems real. In Chapter 2 I argued that what reality things in our appearing world might have depends on the witnessing presence of others. Each appearance is simultaneously a solicitation that others might give testimony to and receive the impact of its coming into the world. Our sense of reality in general and the reality of specific phenomena is called forth in the theater of display and witness. In a profound way, this is true of all phenomena. Yet some phenomena require a more specialized theater for their appearance; some require a "publicly organized space" if they are to have much reality for us (*BPF*, 154). Such a phenomenon is the freedom with which we are concerned.

Indeed, one of the most important purposes of the public sphere is to give this capacity some tangible reality. Only in the beckoning publicity of worldly spaces designed for its appearance can freedom come out of hiding and give us confirmation of its reality. This publicity of a distinctively public sphere – the fact that things there can "be seen and heard by everybody" (*HC*, 50) – is both one of the uniquely defining qualities of that sphere in contrast to other spheres of life and one of the most elemental reasons for Arendt's preoccupation with it. The publicity of the public realm intensifies our sense of the reality of that which appears in it. Arendt's argument here is not that our sense of the reality of our capacity for freedom, more generally understood, will be entirely extinguished without a public sphere to solicit and confirm its being into the world. The capacity will survive in underground forms in most times of severe political tyranny and closure, and in this form we may not be unaware of it as a phenomenon. But in such forms it will be irrelevant to the task of making the world we inhabit a human one, for, as I shall argue, this is, in Arendt's view, an essentially political capacity. And thus, the weight upon the existence of publicly organized spaces of appearance so that freedom might be tangible is enormous.

We may well need to add further to this burden, for Arendt argues that it is from this political or worldly experience of freedom that our other experiences of freedom are derived. All forms of inner freedom – of thought, of creative design, of will, she argues – are "derivative" (*BPF*, 146). They develop out of this prior performative experience of freedom. The argument that freedom is primordially a worldly phenomenon runs along two different but parallel tracks – the one phenomenal, the other historical.

First, Arendt argues, we "become aware of freedom or its opposite in our intercourse with others, not with the intercourse with ourselves" (*BPF*, 148). The elemental movement out into the world, together with the freedom to speak and act in that larger world of others, are the experiences through which we first grasp the phenomenon of freedom. The "original field" of freedom, Arendt argues, is "the realm of politics and human affairs in general" (*BPF*, 145). Phenomenologically, freedom is first and foremost a worldly engagement, and from such original engagements "derive" our experiences of freedom in other domains of human experience.

The second sense in which our experiences of inner freedom are derivative is historical. In the West, Arendt points out, the idea of freedom was not applied to inner life until late antiquity. Reacting to the severe loss of political freedom, the Stoics transferred it to the inner domain, where it was transfigured into an experience of total control over the self. There, without worldly impediment or interference, freedom had found a new home. The ideal of this new inner freedom was no longer performative virtuosity but sovereignty, and it required retreat, reverie, and solitude. This paved the way for early Christianity's "discovery" of the phenomenon of the will, and, with the powerful writings of Augustine, freedom as free will entered the history of philosophy and became (and has remained) both the prevailing understanding and practice of freedom in the West.[8]

These arguments concerning the twofold derivative character of inner freedom raise the interesting question of how much the vitality of freedom in this inner sense itself depends on being replenished by our experience of freedom in its "original field." If, phenomenologically speaking, we first know freedom in the world with others, does our freedom of will, of thought, and of creative design become desiccated if the world offers us fewer and less and less potent invitations "to enter, with other humans, through language, into the order and disorder of the world," as Adrienne Rich puts it?[9] When the worldly opportunities to experience freedom have receded, are these derivative experiences threatened as well?[10] If so, Arendt's understanding of freedom illuminates the depth of lethargy and despair, the "detestation of existence" suffered by groups entrenched in the darkness of generational poverty.[11]

Thus to summarize "the crippling consequences" of obscurity we find the following. Denied that movement in relation to others in a public sphere, denied the dense and pressing presence of speaking and acting beings, our own urge to appear remains unprovoked. Our potential to call forth something that had never been before, to "change every constellation" is crippled (*HC*, 190). And if this were not sufficient cause for despair, the denial of this political freedom over time may well undermine our capacity for inner freedom as well.

Still, to leave matters here would not do justice to the full ethical import of Arendt's concern with this kind of suffering, for the incessant "palaver" in the *polis*, that press of self-display and receptivity as it gives birth to the new, also humanizes. Thus to suffer the insult of oblivion is to be denied the full capacity for participating in humanizing the world. To unfold this argument requires further direct engagement with the enormous ontological importance with which Arendt charges this ephemeral phenomenon of freedom and the publicly constituted space for its appearance. For, as we shall see, implicit in Arendt's humanism – non-essentialist, qualified and tentative though it is[12] – is the contention that whether we are, in fact, to be *human* in any specific sense in our postmetaphysical appearing world will depend on our capacity to belong to and care for one another *in such a way as* to make our sense of the real fuller, deeper.

If our capacity to sense reality emerges in the contrast between a repetitious background and an eruptive novel ground, if it depends on the appearance of "the infinitely improbable," then the existence of an organized sphere where the capacity to begin can appear and be confirmed is of greatest ethical significance. In the bright light created by the testifying presence of all, the shock of the improbable can be collectively felt. Here I speak, much as I did in Chapter 2, of the capacity to register the elemental impact of existence that arises only in the face of the provoking, even incursion-like presence of others.

Such an elemental capacity to sense reality is of course possible without a public sphere, and this was one of the points of my account of Arendt's ontology of display. But the very essence of the public sphere is to arouse the impulse to freedom and to let it shine, and this it does, in contrast to other forms of social existence and private experience, as an *open* and *extended* domain of human plurality. As such, it offers a space in which the unrelated, the new, and uncertain events and developments can become relatable, a space in which those who share the public world can take stock, and meaning can be born.[13] Through the fragile, uncertain, often foreboding encounters between multiple, particular perspectives, our sense of a common world is won. When these encounters diminish, so, too, does our sense of the world as a specific, shared, and ongoing human project. Without public spaces to engender a world held in common, human existence is bewildering, and qualities of nonrelatedness and non-sensicalness prevail. And with this our world becomes dimly lit. In darkness we are inhabited by forces whose nature we hardly grasp.

So in these senses – that we better register and receive the shock of the improbable and actively constitute a sense of a commonly held world – the public sphere and freedom as Arendt theorizes them possess a particularly potent capacity to sustain and replenish our sense of reality. If a specifically human manner of living is to emerge, the existence of a public sphere in which freedom can appear is crucial. Now let me push further toward Arendt's humanism.

Arendt often referred to "the deeper significance" of the public-political realm. It was, she argued, also a "spiritual realm" in which can be seen not *what* you are (an inspired teacher, fat, from the middle class, a gifted mechanic, and so on) but *who* you are (*MDT*, 73). Implicit in all our relations but barely known to ourselves, who we are can appear fully only to others when we enter the public sphere (*MDT*, 73–74). It is a spirit-apparition and concerns a person's "living essence."[14]

We should not obscure the worldly nature of this spirit. Arendt seeks to name the essence of this *being-in-the-world*, and in so doing she is profoundly influenced by the phenomenological notion, so poetically formulated by Merleau-Ponty, that "the subject is a process of transcendence *towards the world*."[15] Who we are is born of our unique, ongoing, and difficult embrace of the world.[16] And it is in the public sphere in particular, Arendt argues, that this living essence can be humanized.

Those who enter the public sphere must be "speaking persons" (*MDT*, 79), and this places certain demands on them. Foremost among these are the courage and passion for the activities of dialogue: for the listening, the lingering over things, and the luring of them into existence that makes them worth talking about and responding to through action.[17] Through this commitment with which they abandon themselves to the public world – to its many strangers, its flux, its alien quality and often unwanted happenings, they can acquire the height of humanness, what Arendt calls "the valid personality" (*MDT*, 73). Not subjective, yet valid without being objective, this humanness wells out of an openness, gladness, even gratitude toward the world in all its density and recalcitrant plurality. When we say what we "deem truth," we humanize the world because what we see arises out of this profound conjoining with and "tragic pleasure" in the world. The world is the bearer, if you will, of our *human* particularity, and through our embrace there simultaneously occurs the upsurge of the world as a humanly meaningful place.

Thus without this venture out into a publicly organized space, the reality of what we do and who we are grows dim. Free action in the public sphere gives rise to a certain luminosity. And it is alone out of this deeper, spiritual sense of politics that our "sympathy," as Arendt put it, for human particularity can arise (*MDT*, 30). We must cultivate such sympathy to care for our world and one another in such a way as to make our sense of the real fuller, deeper. The insult of oblivion deprives us of the worldly relations through which this sympathy might dawn. And this is the same as being denied the opportunity to participate and live in a humanized, meaningful world. This, too, is one of the "crippling consequences" of this form of injustice. I turn now to Arendt's account of the conditions of the modern age that have made this kind of suffering so widespread.

NOTES

1 *The Works of John Adams, Second President of the United States: With a Life of the Author, Notes and Illustrations*, vol. 4, ed. Charles Francis Adams (Boston, 1851), 234.

2 For example, she writes in her controversial chapter "The Social Question" in *On Revolution*, "All rulership has its original and its most legitimate source . . . in the old and terrible truth that only violence and rule over others could make some free" (114). The way her relative lack of distress in passages such as this borders on inattentive complacency has rightly disturbed many of her readers.

3 Nancy Fraser offers a useful typology of such injustices, including *exploitation* (having the fruits of one's labor appropriated for the benefit of others), economic marginalization (being confined to undesirable or poorly paid work or being denied access to income-generating labor altogether), and *deprivation* (being denied an adequate material standard of living). *Justice Interruptus: Critical Reflections on the "Postsocialist" Condition* (New York, 1997), 13.

4 See, for example, J. S. Mill, *On Liberty* (New York, 1956), 67–90.

5 The insult of oblivion is analytically distinct from what Nancy Fraser calls "cultural and symbolic injustices" that fuel the contemporary politics of recognition and its identity-based claims. Although the insult of oblivion probably always receives symbolic and cultural articulation, keeping it analytically distinct allows us to avoid too complete an immersion in debates over cultural value and cultural relativism that can divert our attention from theorizing our profound human need to be in the presence of others. See Fraser's *Justice Interruptus*, 14.

The insult of oblivion is also different from non-recognition and the crippled self found in the works of Charles Taylor and Axel Honneth. Their vocabulary of integrity and respect is attentive to injuries to selfhood, whereas I am concerned, as was Arendt, with injuries to movement toward others. Again, these are not unrelated, but are importantly analytically distinct. See Charles Taylor, *Multiculturalism: Examining the Politics of Recognition*, ed. Amy Gutman (Princeton, N.J., 1994), and Axel Honneth, "Integrity and Disrespect: Principles of a Conception of Morality Based on a Theory of Recognition," *Political Theory* 20 (May 1992): 187–201.

6 These purges Arendt makes in order to isolate her phenomenon – politically relevant freedom – will strike and have struck the modern ear as perverse, and there exists a long and vigorous critique in this vein. See, for example, Pitkin, "Justice"; Elshtain, *Public Man/Private Woman*; Bakan, "Hannah Arendes Concepts"; Dietz, "Slow Boring of Hard Boards"; and Habermas, "Hannah Arendt's Communications Concept."

7 In Arendt's early works it is not difficult to read freedom as a capacity that arises sui generis. With the help of her late ontology, however, we understand this "original interdependence" at a deeper level. I amplify it further as I discuss the concept of dimensionality in the context of Arendt's work on judging in Chapter 4.

8 Arendt argues that Greek philosophy had no concept of freedom at all, as it was founded in opposition to a politics that, for the Greeks and Romans, was the (rightful) home of freedom. See *Between Past and Future*, 157.

9 *What Is Found There: Notebooks on Poetry and Politics* (New York, 1993), 6.

10 It may appear that the very experience of Christianity suggests otherwise, for precisely this inner freedom burned, Arendt herself argues, through the long centuries of the Christian world, during which politically relevant freedom was nowhere in evidence (*BPF*, 165). What makes these questions serious for us, however, is a singular distinction Arendt rightly draws between the Christian world and our own. Both were profoundly worldless times, but the Christian principle of charity alone created a kind of community in which worldless people felt a bond to one another. This "political principle" established not, to be sure, a public sphere but a kind of relatedness that gave rise to a "worldless world" in which initiating action of a kind could occur. A corollary to this "Christian political principle" we do not have (*HC*, 53). In this sense we seem more completely lacking in bonds of relatedness and are thus more vulnerable to losing, in our anomie, the derivative forms and experiences of freedom.

11 I follow Arendt in distinguishing poverty from abject misery. The suffering of the emmiserated is qualitatively different.

12 Here I am in full agreement with Jeffrey Isaac's arguments concerning the humanistic foundations of Arendt's political theorizing. See "Arendt, Camus, and Postmodern Politics," *Praxis International* 9 (1989): 63–68.

13 As I write, for example, the world, having been presented with Dolly the sheep, attempts, rather feebly thus far, to render meaningful and therewith register the reality of the practice of cloning.

14 Here Arendt uses a version of the phenomenological reduction that, in the effort to grasp the "living relations of experience," requires that we "break with our familiar acceptance of things," that we "suspend" our relationship to the whatness of the world. Merleau-Ponty calls this a suspension of our common sense or, alternatively, our natural attitude toward the world. The need for suspension in order to see this living essence, together with Arendt's fierce faith in the spiritual dimension of public life, perhaps helps explain the sense it made to her to exclude those too governed by interest or need from public life. See Merleau-Ponty, *Phenomenology*, xiii–xiv.

15 This notion was developed by Husserl and stands in contrast to Kant's transcendentalism in which the world is imminent in the subject. See Merleau-Ponty, *Phenomenology*, xiii–xiv; my italics.

16 The worldly nature of Arendt's conception of spirit is obscured in certain passages. Compare, for example, *The Human Condition* (179–81) with *Men in Dark Times* (71–75).

17 I am drawing heavily on Arendt's "Karl Jaspers: A *Laudatio*" (*BPF*, 71–80).

'Changing the Public Space: The Mothers of the Plaza de Mayo'

from D. Agrest, P. Conway, and L. K. Weisman (eds), *The Sex of Architecture* (1996)

Suzana Torre

Editors' Introduction

In the first, theoretical part of this essay, Suzana Torre references Hannah Arendt's concept of a space of public appearance, or, in more recent terminology, a symbolic realm of social representation. This space, Torre adds, is controlled by dominant political and economic interests, but is complex in that the forms of representation which take place there are not identical and overlap. In the case of the Argentinean mothers who, during the years of the military government following a coup in 1976, placed themselves in public space to ask, regularly, repeatedly, and at risk to their safety, where were their disappeared children, two meanings of the act collide. On one hand, as Torre points out, is the convention of the mother, whose assigned script in the dominant society is to support but remain in domesticity; on the other, the resistance, in an autocratic society, of making a public presence visible.

Plaza de Mayo is Argentina's equivalent of the Washington Mall, as Torre notes, though smaller. It is a space in which power is represented and in which people are constituted as an audience of power. It is also a hub in what Torre calls a 'Haussmannization' of Buenos Aires which took place from 1884. The acts of the Mothers who occupied the Plaza de Mayo raise issues of gender as well as the suppression of political difference, and extend a private sense of grievance into public space. Their refusal to go away or to constrain their grievance in national interests – a cause not exclusive to oppressive regimes – constitutes resistance, and is attended by physical danger in the conditions in which it took place. Against the symbolic architecture of the site, the Mothers adopted a personal architecture of white handkerchiefs as common identifier. Moving in pairs to avoid prosecution under a law preventing public demonstrations they circulated through the space, eventually attracting international press attention. The generals have since been prosecuted, though much remains undiscovered in the history of their regime. But what emerges, as the essay explains near the end, is that this kind of embodied architecture – of people moving in space and occupying it in a way which highlights the absence of audience – is a counter to the hegemonic architecture of most urban zones of government. It is a small way in which power is made to leak, but one which claims a lingering place in the imagination.

The role of women in the transformation of cities remains theoretically problematic. While women's leadership in organizations rebuilding communities and neighborhoods and their creation of new paradigms for monumentality are sometimes noted in the press, these interventions have yet to inform cultural discourse in the design disciplines or in the history and theory of art and architecture.

The largest body of current feminist scholarship on women in urban settings is concerned with the construction of bourgeois femininity in nineteenth-century European capitals. Within this framework, women are seen as extensions of the male gaze and as instruments of the emerging consumer society and its transformative powers at the dawn of modernity. In other words, they are described as passive agents rather than engaged subjects. When women have assumed transformative roles, feminist critics and biographers have seen them as exceptional individuals or female bohemians, publicly flaunting class and gender distinctions; in contrast, women in general, and working-class women in particular, are presented as unintentional agents of a collective social project, acting out assigned scripts. As a class, women share the problematic status of politically or culturally colonized populations. Both are seen as passively transformed by forced modernization rather than as appropriating modernity on their own and, through this appropriation, being able to change the world that is transforming them.

From this perspective it is difficult to see the current individual and collective struggle of women to transform urban environments as anything of cultural significance, or to reevaluate the enduring influence of traditional female enclaves originated in the premodern city. Many of these enclaves continue to serve their traditional functional and social roles, like the public washing basins in major Indian cities or the markets in African villages, while others have persisted as symbolic urban markings, like the forest of decorated steel poles that once held clotheslines in Glasgow's most central park. Some of these enclaves have even become a city's most important open space, like River Walk in San Antonio, Texas, where women once congregated to wash laundry and socialize.

A literature is now emerging, focused on the participation by marginalized populations in the transformation of postmodern cities and establishing the critical connection between power and spatiality, particularly within the disciplines of art and architectural history and architectural and urban design. To these contributions, which have revealed previously unmarked urban sites as well as the social consequences of repressive urban planning ideologies, should be added feminist analyses of women's traditional urban enclaves and of women's appropriations of public sites that symbolized their exclusion or restricted status. These appropriations, whether in the form of one of the largest mass demonstrations ever held on the Washington Mall (in favor of abortion rights) or in the display of intimacy in very public settings (such as the private offerings and mementos that complete Maya Lin's Vietnam Memorial and compose the monumental Names Quilt commemorating AIDS victims), continue to establish women's rights not merely to inhabit but also to transform the public realm of the city. It is in such situations that women have been most effective in constructing themselves as transformative subjects, altering society's perception of public space and inscribing their own stories into the urban palimpsest.

As in all instances where the topic of discussion is as complex as the transformative presence of women in the city – and particularly when this topic does not yet operate within an established theoretical framework – the main difficulty is to establish a point of entry. In the present essay I propose entering this territory through the examination of one dramatic case of a successful, enduring appropriation: the Mothers of the Plaza de Mayo in Argentina. This small but persistent band of women protesters first captured international attention in the mid-1970s with their sustained presence in the nation's principal "space of public appearance," as Hanna Arendt has called the symbolic realm of social representation, which is controlled by the dominant political or economic structures of society. This case illustrates the process that leads from the embodiment of traditional roles and assigned scripts as wives and mothers to the emergence of the active, transformative subject, in spite of – or perhaps because of – the threat or actuality of physical violence that acts of protest attract in autocratic societies. As we will see, this case is also emblematic of architecture's complicity with power in creating a symbolic system of representation, usually of power hierarchies. The hegemony of this system has been threatened ever since the invention of the printing press and is now claimed by electronic media and its virtual space of communication. Finally, the Mothers of Plaza de Mayo's appropriation of the public square as a stage for the enactment of their plea is a manifestation of *public space* as social production. Their redefinition of that space suggests that the public realm neither resides nor can be represented by buildings and spaces but rather is summoned into existence by social actions.

THE MOTHERS OF THE PLAZA DE MAYO

In March 1976, after a chaotic period following Juan Perón's death, a military junta wrested power from Perón's widow Isabel, in order (as the junta claimed) to restore order and peace to the country. The first measures toward achieving this goal were similar to those of General Pinochet in Chile three years earlier, and included the suspension of all civil rights, the dissolution of all political parties, and the placement of labor unions and universities under government control. It would take seven long, dark years for a democratically elected government to be restored to Argentina, which at last permitted an evaluation of the extent of open kidnappings, torture, and executions of civilians tolerated by the military. Because of the clandestine, unrecorded activities of the paramilitary groups charged with these deeds, and because many burial sites still remain undisclosed, agreement as to the exact number of "disappeared" may never be achieved, but estimates range from nine thousand to thirty thousand. Inquiries to the police about the fate of detainees went unanswered. Luis Puenzo's 1985 film, *The Official Story*, offers glimpses into the torture and degradation endured by thousands of men, women, and even babies, born in detention, some of whom were adopted by the torturers' families.

"Disappearances" were very effective in creating complicitous fear: many kidnappings were conducted in broad daylight, and the victims had not necessarily demonstrated open defiance of the military. In fact, later statistics show that almost half of the kidnappings involved witnesses, including children, relatives, and friends of those suspected of subversion. Given the effectiveness of arbitrary terror in imposing silence, it is astonishing that the public demands of less than a score of bereaved women who wanted to know what had happened to their children contributed so much to the military's fall from power. Their silent protest, opposed to the silence of the authorities, eventually had international resonance, prompting a harsh denunciation of the Argentinean military, which led, finally, to the demise of state terrorism and the election of a democratic government.

The actions of the "Mothers," as they came to be known, exemplified a kind of spatial and urban appropriation that originates in private acts that acquire public significance, thus questioning the boundaries of these two commonly opposed concepts. Gender issues, too, were not unimportant. The Mothers' appropriation of the plaza was nothing like a heroic final assault on a citadel. Instead, it succeeded because of its endurance over a protracted period, which could only happen because the Mothers were conspicuously ignored by the police, the public, and the national press. As older women they were no longer sexually desirable, and as working-class women they were of an inferior ilk. Nevertheless, their motherhood status demanded conventional respect. Communicating neither attraction nor threat, they were characterized by the government as "madwomen." The result of their public tenacity, which started with the body exposed to violence, eventually evolved into a powerful architecture of political resistance.

Plaza de Mayo is Argentina's symbolic equivalent of the Washington Mall. It is, however, a much smaller and very different kind of space: an urban square that evolved from the Spanish Plaza de Armas, a space that has stood for national unity since Creoles gathered there to demand independence from Spain in May of 1810. The national and international visibility of Plaza de Mayo as *the* space of public appearance for Argentineans is unchallenged. Originally, as mandated by the planning ordinances of the Law of the Indies, its sides were occupied by the colonial Cabildo, or city council, and the Catholic Cathedral. Today the most distinctive structure is the pink, neoclassical Casa Rosada, the seat of government.

Military exercises, executions, and public market commingled in the plaza until 1884, when Torcuato de Alvear, the aristocratic mayor, embarked on a Haussmannian remodeling of the center of Buenos Aires shortly after important civic structures – such as Congress and the Ministries of Finance and Social Welfare – had been completed: A major element of Alvear's plan was Avenida de Mayo, an east–west axis that put Congress and the Casa Rosada in full view of each other. Such a potent urban representation of the checks and balances of the modern, democratic state was achieved through selective demolition, including the removal of the plaza's market stalls and the shortening of the historic Cabildo's wings by half their original length. Currently, the plaza's immediate area includes several government offices, the financial district, and the city's most famous commercial street, Florida. This densely populated pedestrian thoroughfare links Avenida de Mayo to Plaza San Martin, another major urban square. A plastered masonry obelisk, the May Pyramid, erected on the square in 1811 to mark the first anniversary of the popular uprising for independence, was rebuilt as a taller, more ornate structure and placed on the axis between Congress and the Casa Rosada. In this new position, it became a metaphorical fulcrum in the balance of powers.

The now well-known image of a ring of women with heads clad in white kerchiefs circling the May Pyramid evolved from earlier spontaneous attempts at communication with government officials. At first, thirteen wives and mothers of the "disappeared" met one another at the Ministry of the Interior, having exhausted all sources of information about their missing children and husbands. There a small office had been opened to "process" cases brought by those who had filed writs of *habeas corpus*. One woman well in her sixties, Azucena Villaflor de Vicente, rallied the others: "It is not here that we ought to be," she said. "It's the Plaza de Mayo. And when there are enough of us, we'll go to the Casa Rosada and see the president about our children who are missing." At the time, popular demonstrations at the plaza, frequently convened by the unions as a show of support during Juan Perón's tenure, were strictly forbidden, and gatherings of more than two people were promptly dispersed by the ever-present security forces. The original group of thirteen women came to the plaza wearing white kerchiefs initially to identify themselves to one another. They agreed to return every Thursday at the end of the business day in order to call their presence to the attention of similarly aggrieved women. The Mothers moved about in pairs, switching companions so that they could exchange information while still observing the rule against demonstrations. Eventually they attracted the interest of the international press and human rights organizations, one of which provided an office where the women could congregate privately. Despite this incentive to abandon the plaza for a safer location, the Mothers sustained a symbolic presence in the form of a silent march encircling the May Pyramid. That form, so loaded with cultural and sexual associations, became the symbolic focus of what started as a literal response to the police's demand that the women "circulate."

The white kerchiefs were the first elements of a common architecture evolved from the body. They were adopted from the cloth diapers a few of the Mothers had worn on their heads in a pilgrimage to the Virgin of Luján's sanctuary. The diapers were those of their own missing children, whose names were embroidered on them, and formed a headgear that differentiated the Mothers from the multitude of other women in kerchiefs on that religious march. In later demonstrations the Mothers constructed full-size cardboard silhouettes representing their missing children and husbands, and shielded their bodies with the ghostly blanks of the "disappeared."

By 1982, the military had proven itself unable to govern the country or control runaway inflation of more than 1000 percent per year. The provision of basic services was frequently disrupted by the still powerful Peronista labor unions, and many local industries had gone bankrupt due to the comparative cheapness of imported goods under an economic policy that eliminated most import taxes. Then, in the same year, the military government embarked on an ultimately ruinous war with Great Britain over the sovereignty of the Falkland/Malvinas Islands. With the help of the United States satellite intelligence and far superior naval might, Great Britain won with few casualties, while Argentina lost thousands of ill-equipped and ill-trained soldiers. The military government, which had broadcast a fake victory on television using old movie reels rather than current film footage, was forced to step down in shame by the popular outcry that followed. Following the collapse of the military government, the Mothers were a prominent presence at the festivities in Plaza de Mayo, their kerchiefs joyously joined as bunting to create a city-sized tent over the celebrants. They have continued their circular march to this day, as a kind of living memorial and to promote their demands for full accountability and punishment for those responsible for the disappearance of their husbands and children.

After the election of a democratic government, the military leadership was prosecuted in civil rather than military court, resulting in jail sentences for a few generals and amnesty for other military personnel. Although the amnesty was forcefully contested by the Mothers and other organizations, the protest was seen by many as divisive. Nevertheless, the Mothers and a related organization of grandmothers pressed on with attempts to find records about disappearances and fought in the courts to recover their children and grandchildren. Then, early in 1995, more than a decade after the restoration of democratic government, a retired lieutenant publicly confessed to having dumped scores of drugged but still living people from a helicopter into the open ocean, and he invited other military men on similar assignments to come forth. The Mothers were present to demonstrate this time as well, but now the bunting had become a gigantic sheet that was waved overhead as an angry, agitated sea.

The Mothers were able to sustain control of an important urban space much as actors, dancers, or magicians control the stage by their ability to establish a presence that both opposes and activates the void represented by the audience. To paraphrase Henri Lefebvre, bodies produce space by introducing direction, rotation, orientation, occupation, and by organizing a *topos* through gestures, traces, and marks. The formal structure of these actions, their ability to refunctionalize existing urban spaces, and the visual power of the supporting props contribute to the creation of public space.

What is missing from the current debate about the demise of public space is an awareness of the loss of architecture's power to represent the *public*, as a living, acting, and self-determining community. Instead, the debate focuses almost exclusively on the physical space of public appearance, without regard for the social action that can make that environment come alive or change its meaning. The debate appears to be mired in regrets over the replacement of squares (for which Americans never had much use) with shopping malls, theme parks, and virtual space. But this focus on physical space – and this ideological potential to encompass the public appearance of all people, regardless of color, class, age, or sex – loses credibility when specific classes of people are denouncing their exclusion and asserting their presence and influence in public life. The claims of these excluded people underscore the roles of *access* and *appearance* in the production and representation of public space, regardless of how it is physically or virtually constituted. They also suggest that public space is produced through public discourse, and its representation is not the exclusive territory of architecture, but is the product of the inextricable relationship between social action and physical space.

'Re-imagining the Global: Relocation and Local Identities in Cairo'

from Ayşe Öncü and Petra Weyland (eds), *Space, Culture and Power* (1997)

Farha Ghannam

Editors' Introduction

Cairo, *Umm al-Dunya*, Mother of the World, is a mega-city with a population above ten million people. A significant number of these inhabitants live in informal housing, for instance in the old cemetery, or in peripheral areas. Some self-build developments are now being brought within a legal framework. A significant number of migrants to the city from rural areas live, too, in government housing schemes. The rate of growth is rapid, the city having had only three million inhabitants at the time of Egyptian independence in 1952. Carolyn Fluehr-Lobban, in *Modern Egypt and Its Heritage* (Pittsburgh, Carnegie Museum of Natural History, 1990), writes that Cairo's poor neighbourhoods are regionally diverse and overcrowded, but that a combination of 'poverty and crowded housing conditions has not resulted in the high rates of crime that have afflicted our crowded U.S. cities' (p. 23). She sees the reason as a strength of extended family relationships. This may be one factor, but, as Farha Ghannam shows in this chapter from *Space, Culture and Power*, there are others, not least among them the collectivity of Islam, and continuity in identification with place even after displacement.

Migration is one cause of displacement, but enforced movement within cities to make spaces for redevelopment is another. The site of Ghannam's research (while a doctoral candidate in anthropology at the University of Texas at Austin) is the neighbourhood of al-Zawiya al-Hamra. A low-income population hitherto from Bulaq, a city-centre district redeveloped as a high-income area, near the Hilton Hotel and World Trade Centre – New York and Barcelona are only two of the cities which have these institutions – was relocated to government-built housing here. These new homes were segregated from adjacent private housing, and the new population stigmatised much as immigrants from rural areas might be. As Ghannam notes, government officials were keen to remove the inhabitants of Bulaq when transnational companies showed an interest in developing the neighbourhood. The displaced population, however, rediscovered a collective identity in memories of the old place of habitation which they transposed onto the new. Ghannam concludes that active social agents mediate the impact of global trajectories, and that collectivities become increasingly important in such circumstances. This, though from a contrasting situation, sits well enough beside Young's call for a recognition of group difference in her chapter reprinted above. The kinds of collectivity are not the same, but the strength to avoid oblivion is, in both cases, located in them.

GLOBAL DISCOURSES AND LOCAL IDENTITIES

Since Sadat started his open-door policy, Cairo has witnessed the introduction of new forms of communication,[1] more emphasis on international tourism, increasing importance of consumer goods, and a growing flow of ideas related to civil society, democracy and political participation. Theoretical developments in anthropology and cultural studies have demonstrated that these global processes are not producing one dominant culture but present a set of discourses and practices that are juxtaposed in complex ways in local contexts (Hall 1991a; 1993; Massey 1994; Lash and Urry 1994). Thus, contrary to the old conceptualization of the world as becoming a 'global village', local differences and identities are not destroyed but are being reinforced in many cases by global forces and processes (Hall 1991b; 1993; Ray 1993; Massey 1994).

With the growing connectedness between different parts of the globe and with the circulation of global discourses and images facilitated by new systems of communication, the Other is becoming more identified with the self in complex ways. The connectedness and tension between the self and the Other is crucial to understanding how identities are constructed and shift over time and how the Other is simultaneously desired and dreaded. One example can be found in how people in Cairo desire the global (in this case identified as the West) because it is organized, clean, rich and 'democratic' and at the same time they distrust it because it is associated with 'moral corruption', drugs and violence. The focus on the connectedness and tension with the Other is therefore a necessary step to theorize the different ways that the global is reshaping local identities and redrawing their boundaries. This focus will enable us to conceptualize local identifications not as static but as always in the process of formation and constructed of multiple discourses and as composed 'in and through ambivalence and desire' (Hall 1991b: 49).

'Globalization', however, should not be reduced to 'Americanization' as some authors tend to do (Hall 1991a; Hannerz and Lofgren 1994). While the 'American conception of the world' (Hall 1991a: 28) may be hegemonic in various contexts, people experience the influence of other 'globals'. People in al-Zawiya al-Hamra not only experience the American culture that is transmitted to them in movies starring Arnold Schwarzenegger but they also experience the global through oil-producing countries where their children and male relatives work as well as through the mixture of people who visit and work in Cairo from different Arab countries. For example, women use oil for their hair that comes from India via their sons who work in Kuwait and collect their wedding trousseau from clothes, sheets and blankets brought from Kuwait; others visit husbands in Saudi Arabia; and many have accumulated electrical appliances from Libya where husbands and sons work. Despite the fact that many of the consumer goods are produced in the West, their meanings are given to them by their users who live in al-Zawiya al-Hamra. For many, consumer goods are investments that can be exchanged for cash when needed. Several families use their refrigerators to cool water during the summer but turn them into closets during the winter to store household appliances. The global is also introducing new forms of identification between the subjects of its processes. People, for example, enjoy watching television, especially some of the global sports events such as the soccer World Cup. Young men and women follow these games very closely; they know the names of the Brazilian, German and Italian players. While watching these games, different identities compete for priority: they shift from supporting African and Arab teams to cheering for third world teams when they play against European teams (Brazil against Germany, for example). People, thus, do not experience the global as a coherent set of discourses and processes that are transmitted from the West to the rest of the world but experience fragments and contradictory pieces that are filtered through other centres and that do not necessarily present a unified 'conception of the world'. Therefore, 'the global', as an analytical concept, should be expanded to include a mixture of images, discourses and goods that are brought to people through various channels such as state-controlled media, commercial video tapes, audio tapes that are distributed by Islamic activists, and consumer goods brought to al-Zawiya al-Hamra by migrants to Arab countries.

It is also important to remember that, as a theoretical concept, 'the local' should not be confused, as Massey (1994) and Lash and Urry (1994) argue, with the concrete, the empirical or the authentic or a spatially bounded entity (Urry 1995). I use 'local' to mean 'acts of positing within particular contexts' (Tsing 1993: 31). People attach multiple meanings to their localities that vary from one context to another. Despite the fact that geographical space is used as a point of reference for several local identities in Cairo, these different contexts share a set of social relationships and identities that include those who are like us (local people) and exclude people who are not like us (outsiders). Thus, when people identify the relocated group as 'those from Bulaq', they are trying to exclude them from another collective identity that includes people who have been living in al-Zawiya and identify primarily with it. The relocated people still refer to themselves as 'people of Bulaq' despite the fact that they have been living in al-Zawiya for fifteen years. Many also still identify with their old villages and towns that they left more than fifty years ago. In short, there is not one 'local' but there are various 'locals' that are juxtaposed in complex ways with multiple 'globals'.

OLD PLACES, NEW IDENTITIES

> Here in al-Zawiya, you do not find Pizza Hut and Kentucky Fried Chicken. Such places can never get any profit in areas like this. People are poor and the money they will pay for one meal in one of these restaurants will feed the whole family for a week if not more. (A male shop-owner who works in al-Zawiya but lives in another middle-class neighbourhood)

To understand the local identities that are in the process of formation in al-Zawiya al-Hamra, it is important to remember that state practices and discourses were based on what Foucault calls 'dividing practices' (Rabinow 1984: 8). The project started by separating and stigmatizing the targeted population as an expedient rationalization of policies that aim to modernize, normalize and reintegrate them within the larger community. Not only were the housing conditions attacked by state officials, but the people themselves were stigmatized and criticized. A 'scientific' social study conducted to determine the needs of the relocated group revealed, as stated by the Minister of Construction and New Communities, that the area of Bulaq in general and one of its neighbourhoods (al-Torgman) in particular have been shelters for *qiradatia* (street entertainers who perform with a baboon or monkey), female dancers, pedlars and drug dealers (*al-Ahram*, 27 December 1979: 3). The 'locals' were also represented as passive, unhealthy and isolated people who did not contribute to the construction of the mother-country and who had many social ills.[2] After resettlement, these publicized stereotypes fostered a general feeling of antagonism towards the newcomers. In addition to repeating the same words that were circulated in the media, residents of al-Zawiya added other stereotypes to describe this group such as *labat* (trouble-makers) and *shalaq* (insolent). Women, in particular, were singled out (as they were also singled out by the Minister who described them as dancers or *Ghawazi*); they were described as rude and vulgar, and were used in daily conversation as an analogy for bad manners.

These negative constructions of the relocated group are supported and perpetuated by the physical segregation of their housing (*masaakin*) from the rest of the community. Their public housing is clearly defined and separated from other projects and private houses (*ahali*). Public housing is characterized by a unified architectural design (the shape, the size of the buildings, as well as the colours of walls and windows), whereas private housing has more diversified patterns. This unity in design and shape sharply defines and differentiates public housing from private houses and makes it easier to maintain boundaries that separate the relocated from other groups. In short, neither the discourse of the state nor the shape and location of the housing project enhance the dialogical relationship between the relocated group and other groups in al-Zawiya. After fourteen or fifteen years of resettlement, the relocated group continues to be stigmatized and its interaction with the rest of the neighbourhood is restricted.

The identity of Bulaq

With their stigmatization in the state discourse and by the residents of al-Zawiya, and with the hostility that faced them, the relocated population rediscovered their common history and identification with the same geographical area. While people used to live in Bulaq and identify strongly with their villages of origin, after relocation Bulaq became an anchor for the group's sense of belonging and took precedence over other identifications. The attachment to the old place is not single or one-dimensional and Bulaq is remembered and related to differently by gender and age groups. These differences are beyond the scope of this chapter but it is sufficient here to say that Bulaq is of great significance for most of the group in reimagining their communal feelings. Currently, their public housing is called after one of Bulaq's neighbourhoods (*masaakin al-Torgman*) and people express their strong attachment to their old place in songs and daily conversations. Despite the fact that relocation reordered relationships within the group and destroyed a major part of their support system, the old neighbourhood still structures parts of the people's current interaction. They still refer to the people who used to live in Bulaq as '*min 'andina'* (from our place) which not only creates a common ground for identification but also indicates certain expectations and mutual obligations between the people in the current area of residence. At the same time, Bulaq is the point of reference for their identification with those who still live in parts of Bulaq[3] and those who moved to 'Ain Shams.

Through relocation, the group lost, among other things, a major part of its 'symbolic capital'. This is mainly manifested in two important aspects related to group members' identification with the old location. First, they used to live next to an upper-class neighbourhood, Zamalek. Young men and women, as emphasized by the people themselves and documented in a famous old movie (*A Bride from Bulaq*), could even claim that they were from Zamalek because only 'a bridge' separated (or connected) the two neighbourhoods. People also lost the pleasure and satisfaction associated with looking at the beautiful buildings and knowing that people of Zamalek – and much to their shame, as described by one informant – used to see Bulaq with its old and shabby houses.

Second, the group used to live in an 'authentic popular' or *baladi* area and perceives its relocation in al-Zawiya as moving down the social ladder. In Bulaq, the 'authentic popular' quarter, people used to live next to each other, separated only by narrow lanes that allowed close interaction and strong relationships. They remember the old place in the way people used to cooperate and 'eat together'. Their rootedness in the same place over a long period of time provided people with a strong support system, open social relationships, and a sense of security and trust. In contrast, al-Zawiya is a relatively new neighbourhood. It was mainly agricultural fields until the 1960s, when the area started to expand rapidly with the state construction of the first public housing project. This project housed families from different parts of Cairo who could not afford to live in more central locations. Immigrants (mostly Muslims) also came to al-Zawiya from different parts of the countryside and many live in private housing. The heterogeneity of its population is used by its residents, especially members of the relocated group, and people around them to indicate that al-Zawiya is not 'an authentic popular quarter'. Its people are 'selfish', 'sneaky' and 'untrustworthy'. It is seen as located between *baladi*[4] and *raqqi* (upper-class areas) which places it, as described by a male informant, in a tedious or annoying (*baaykh*) position. Al-Zawiya, thus, is geographically and socially marginal compared to Bulaq.

A key word in understanding the differences between what is seen as an 'authentic' neighbourhood such as Bulaq and 'less authentic' newer neighbourhoods such as al-Zawiya is *lama*. This word refers to the growing mixture and gathering of people from different backgrounds who live in the same locality. People from various quarters, villages and religions are coming to live in the same neighbourhood, hang out at the same coffee shop, visit the same market, and ride the same bus. These spaces are defined as *lamin* as compared to a more homogeneous or less *lama* places such as the village and the 'authentic popular' quarter based on long established relationships. Being rooted in a certain area, that is, localized in a particular place, allows the development of strong relationships between people. *Lama* is used to classify different localities and points to the difference between a neighbourhood where people know each other by name and face as opposed to more heterogeneous areas where people are strangers and not to be trusted. *Masaakin* is *lama* as opposed to *ahali* housing. Al-Zawiya is *lama* compared to Bulaq and Cairo is *lama* compared to the villages where the inhabitants originally came from.

RELOCATION AND RELIGIOUS IDENTITY

> Hegemony is not the disappearance or destruction of difference. It is the construction of a collective will through difference. It is the articulation of differences which do not disappear. (Hall 1991b: 58)

Despite the significance of Bulaq in how people reimagine their communal feelings, this identity does not facilitate the group's interaction with the rest of the people who live in al-Zawiya al-Hamra. Relocation rearranged local identities and added to the old identifications: people are now identified with a village (the place of origin), as locals of Bulaq (where they resided for generations), as occupiers of *masaakin* (which is stigmatized by dwellers in private housing) and as inhabitants of al-Zawiya al-Hamra (not known for its good reputation in Cairo). But above all, they are mainly Muslims. Religion, rather than nationalism, neighbourhood and the village of origin, became a powerful discourse in articulating and socially grounding the various identities of the different groups residing in al-Zawiya al-Hamra.[5] Only the religious identity promises to articulate these identifications without destroying them.[6] Displaced families, *ahali* and *masaakin* inhabitants, people of Bulaq and al-Zawiya, rural immigrants, *Fallahin* (peasants who come from villages in Lower Egypt) and *Sa'idis* (immigrants from Upper Egypt), who are largely pushed from their villages to Cairo in their search for work and a better life, as well as residents who moved from other areas of Cairo can all find commonality in religion that is expressed in practices such as a dress code and the decoration of houses and shops.

Islam brings people together on the basis of a common religion. Despite the fact that Muslims do not know each other on a personal basis, religion creates a 'safe' space (the mosque), a common ground where they are connected to each other, and a sense of trust and rootedness. This is clearly manifested in how the mosque, of all public spaces, is gaining importance in facilitating the interaction of various groups and the formation of a collective identity. To start with, the mosque's growing centrality in daily life is manifested in the many modern services that are provided to the people in it. Through charitable organizations (*jam'iyyat khayriyya*), the mosque provides socially required, services such as affordable education, health care and financial support to the poor. It is also the place where discourses circulate that prescribe and/or forbid daily practices. Above all, it is the most acceptable and safest social space where various groups can meet and interact.

To understand the importance of the mosque, we need to go back to the word *lama*. As previously mentioned, people tend to distrust areas and public spaces that are labelled as *lama* such as the market, the coffee shop and the bus. These spaces are seen as 'dangerous' and people are very careful when visiting or utilizing them. Compared to such spaces, the mosque, which is a historical space that is legitimated through its naturalized relationship with religion, is currently being actively articulated to frame the interaction between members of different groups as well as to empower emerging meanings, identities and relationships. Those who are labelled as trouble-makers and rude (people who come from Bulaq and live in *masaakin*) as well as the untrustworthy and selfish (people of al-Zawiya as described by people of Bulaq) can all meet in the mosque and collectively identify themselves as Muslims.

Thus, the power of the mosque is being currently reinforced through its promise of an equal and unified community out of a heterogeneous urban population. It is accessible to all Muslims and brings them in on equal terms. The unity of prayers and the importance of communal feelings is manifested in the unifying discourse and the similar movements that are performed simultaneously. The Imam leads the prayer and coordinates the movement of all the attendees through his pronounced signals that indicate when one should bend forward on the knees or stand up straight, and so on. Emphasis is placed upon standing in straight lines, very close to other attendees, in a way that leaves no empty spaces through which the devil could enter among the devout and divide their collectivity.[7]

The feelings that are associated with being part of a collectivity were cited by many, especially by women, as one of the main reasons for going to the mosque. As is the case with most of her neighbours, relocation shattered most of the support system that connected the fifty-five-year-old Umm Ahmed with friends and neighbours who were relocated to 'Ain Shams or to different parts of the new housing project in al-Zawiya al-Hamra. Although she used to perform her religious duties on a regular basis in Bulaq, Umm Ahmed's religiosity gained a different meaning in al-Zawiya al-Hamra. In addition to her adoption of the *khimar* (a head garment that covers the hair and the shoulders), which is seen as the 'real Islamic dress', Umm Ahmed began attending local mosques on a daily basis. She explained that she goes to mosques because the presence of other people strengthens her will and provides her with more energy than when praying alone. Currently, Umm Ahmed frequents five local mosques to perform four out of the five daily prayers. For Friday prayer, she usually selects a large mosque, located within the boundaries of the *masaakin* but that is also attended by some worshippers from the *ahali*. She also visits two small mosques that are identified with an Islamic group active in al-Zawiya al-Hamra. She attends these two mosques, which are located in the *ahali* area, to listen to weekly lessons and participate in Qur'ān recital sessions. Another mosque, which is located next to the vegetable market in the *ahali* area, is a convenient site for the midday prayer when Umm Ahmed is shopping for the family's daily food. For the evening prayer, she chooses a smaller mosque on the edge of the housing project that is attended by a mixture of worshippers from *ahali* and *masaakin* areas. She prefers this mosque, as she explains, because she meets 'wise' women who like to talk to her. Over the last five years, Umm Ahmed has formed strong relationships with other women from different parts of the neighbourhood, especially from the *ahali* area, who attend the same mosque. If one of them does not come to the evening prayer, she goes with other women to ask about their absent friend. At the same time, the mosque not only brings people together from the same neighbourhood but also encourages people to move from one part of the city to the other. Young men and women, for example, use the city bus to tour the city in their search for the 'truth'. They cross the boundaries of their localities to go to other neighbourhoods to attend certain mosques where popular sheikhs preach.

The mosque is also becoming more open to women in al-Zawiya al-Hamra. This is perceived by some Islamic activists as essential to counter other spaces that are open to women, such as universities, the workplace, cinemas and nightclubs. Women are identified by men as more vulnerable to the influence of global (defined here as American) discourses and practices. Women's actions, dress and access to public life are seen as threatening the harmony of the Islamic community and as the source of many social ills. Women have internalized these ideas and hold themselves, and not men, responsible for the safety of the morals of the community. As women repeatedly emphasize, men are weak creatures and cannot resist the seduction imposed on them by women who do not adopt Islamic dress. At the same time, women can be very active in the construction of the Islamic community. More voices have emphasized the positive aspects associated with opening the mosque to women who, as mothers, sisters and wives, can be active agents capable of altering their own practices as well as shaping the actions and values of other family members. Thus, to contain the destructive potential of women and promote their constructive power in the formation of the Islamic community, more attempts are made by Islamic activists to incorporate women within the mosque. Currently, women, especially those without jobs and small children, go to the mosque on a regular basis for prayer and to attend weekly lessons, while working women usually attend the Friday prayer. Women are also becoming more active in the mosque through their roles as teachers, students, workers and seekers of social, educational and medical services. In addition, more women help in taking care of the mosque and participate in mosque-related activities such as preparing food and distributing it to the needy.[8]

GLOBALIZATION AND RELIGIOUS IDENTITIES

It is important not to confuse my previous discussion of religious identity with 'fundamentalism', 'extremism' or 'militant Islam', which have been the centre of attention of several studies (Ibrahim 1982; Kepel 1993). Fundamentalism especially has been the focus of studies that aim to examine the relationship between globalization and religion (see, for example, Turner 1994; Beyer 1994). Such studies limit discussion of the ideology of the leaders of some radical Islamic groups and tend to present these movements as 'responses' or 'reactions' to the global. The role of ordinary people as active agents in negotiating religious and global discourses in their daily life and the formation of their local identities is largely neglected.

Despite the fact that communal feelings based on religion can be politicized and used as the basis to mobilize the working class (as happened in 1981 in clashes between Muslims and Christians in al-Zawiya al-Hamra), at the daily level religious identity brings people together as connected selves rather than separated and isolated others. It articulates the presence of the group at the neighbourhood level, integrates its members into the mosque and secures a space for them in Cairo. People do not want to relive the past, as some fundamentalists seem to desire, but try to live in the present with its complexity and contradictions. They hence struggle against efforts of some extremist groups who try to impose restrictions on how they appropriate certain aspects of modernity. Nuha, for example, is a twenty-three-year-old woman with a high school diploma who works in a factory outside the neighbourhood. She hears things on the radio, in the mosque and from her friends and then lets her heart and mind judge what she will follow. She expresses her religiosity in adopting the *khimar*. At the same time, she opposes many of the restrictions that extremists try to impose on people, such as forbidding men from wearing trousers and prohibiting eating with a spoon because, as some argue, the Prophet did not do these things. She believes that had these things existed when the Prophet was alive, he would have used them. So it is not a sin (*haram*) to eat with a spoon but, if one chooses to eat with the hands, one will get an extra reward.

There are moments when people directly reject the 'American conception of the world' (Hall 1991a: 28) with its homogenizing tendencies and use this conception to explain the conflict between the state and some Islamic groups. A young woman explained the conflict between the government and religious groups as follows:

> The problem is that the government has strong relationships with the United States which hates Islam and Muslims and is trying to spread its ideas and practices all over the world, especially wearing short clothes, the domination of science, and the destruction of religion. My cousin, who is a Sunni,[9] explained to me that Americans have many methods to achieve their purposes, especially through schools. They try to prove that science is better than religion by using the comparative method. They bring, for example, a candle and a light bulb and ask which is better. The first represents religion and the second represents science. Of course, one will choose the second. They also compare two pictures, one of a man wearing a *gallabiyya* [a long loose gown that the Prophet used to wear] with a beard and a rotten look, while the second picture is of a handsome man who is shaved and looks very clean and tidy. Of course, anyone will choose the second. The whole idea is for science to replace religion and dominate the universe. Islam is compatible with science because one can find all answers in it if examined closely. Science should serve religion.

The opposition to the 'global West', however, is not sufficient to explain the growing importance of the mosque and religious identity in al-Zawiya al-Hamra. There are complex local and national forces juxtaposed with the global to produce religious identity. State oppression, the daily frustrations in dealing with state bureaucracy, alienation, the fragmentation of the urban fabric, and the ability of Islamic groups to utilize various discursive strategies that mobilize people are as important as the economic frustration, the unfulfilled expectations and desires, and the need to have a voice in the global in understanding why religious identity is becoming more hegemonic.[10]

As manifested in the services that are being attached to the mosque in al-Zawiya al-Hamra, certain global discourses and consumer goods are negotiated and appropriated. For example, to avoid state censorship of discourses circulated in the mosque, Islamic activists use cassette tapes to distribute the religious discourse to a large segment of the urban population. Especially for illiterate men and women, tapes provide a powerful means of communication that brings popular preachers (that is, those who are believed to tell the 'truth') from the mosque into the home, the workplace, the taxi and the street. These can be replayed until their meanings become clear to the listener. Women can also pass them on to friends and relatives. On several occasions, women gathered to listen to such tapes and expressed strong emotional reactions to the descriptions of death, the horrible torture of the grave and the soothing visions of heaven.

Various Islamic groups, however, relate differently to the global in general and the West in particular. While there were some Islamic activists who attack the influence of the West on the dress code and practices of Muslims, a major part of the lessons that I attended in the mosque as well as tapes I heard in al-Zawiya al-Hamra did not attack the West but emphasized the horrible nature of torture that unbelievers would go through in the grave and in hell, and, in contrast, the rewards that are awaiting the believers in heaven. Women who do not adopt Islamic dress are singled out and detailed descriptions are presented of how they will be hung from their hair and breasts while huge snakes bite their bodies as they are grilled in hell. Women shivered, cried and prayed hard asking God to protect them from the horrible torture that is awaiting unbelievers. These graphic descriptions contextualize the critique of people's 'un-Islamic' practices and the 'medicine', as one sheikh said, that is provided to heal the ills of the current situation and to win the eternal heaven. The prescribed medicine is to go back to God, ask for his forgiveness, and live according to his commands.

Although the emphasis on the dress code can be seen as a rejection of the influence of the West, I would argue that gender distinctions are the centre of the restrictions applied to women's dress code. Another interesting example could be found in how people negotiate their definition of Islam and modernity. Their rejection of many of the ideas that are circulated by some religious extremists is clearly manifested by the struggle over some consumer goods such as colour televisions, VCRS and tape-recorders which are rapidly becoming signs of distinction. Many families participate in saving associations (*gam'iyyat*) to secure money to buy these goods which are also seen as investments that can be easily exchanged for needed cash. Just as Amal, most people dream of consumer goods and better living conditions that television brings to their homes without objective means to satisfy them. Many families try to solve Ramadan puzzles (these are usually presented daily by popular Egyptian performers) and collect the covers of tea-bags and chocolate bars to mail to the manufacturer in the hope that they may win a 'dish', a familiar English word that is used to refer to the satellite dishes that are spreading in upper-class neighbourhoods, a VCR, a washing machine or a gas stove. Among all the consumer goods that people use, Islamic groups centre their struggle against the television set. This struggle can be interpreted as 'rejection of modernity', but such an analysis fails to see how other aspects of modernity are being selectively incorporated in the struggle of these groups. They use the fax machine, the tape-recorder, the computer and many other modern facilities to achieve their aims. To understand the struggle over the television set, one should look how this medium is being used in people's daily lives.

Television is one of the most popular goods that people incorporate as one of the basic elements of their daily lives. Except for very few people with extreme religious beliefs, there is no housing unit in al-Zawiya al-Hamra without a television set. Each family, regardless of its income, owns a television set that is the centre of attention of all the family members. The television set is a powerful medium that conveys to them many experiences and values that can be described as global and brings the Other closer than ever to the self. The television set and the mosque are competing with each other to connect Muslims in different parts of the world. People of al-Zawiya al-Hamra are connected with other Muslims whom they have never met and who are not assumed to be identical duplicates of the self but are identified as the Other that is closely connected with the self. It is the force that binds people of this neighbourhood with Muslims who fight in Bosnia, Afghanistan and Chechnya. Young men, who are frustrated with the state's restriction on their participation in fighting with the Bosnian Muslims, circulate stories about God's help and support of the Bosnians. People talk about invisible soldiers (angels) and unidentified white planes that bomb the Serbs. Islam, thus, is becoming a force in localizing the global and globalizing the local. The distinction between Muslims who live in al-Zawiya al-Hamra and those who live in the rest of the world (such as the Bosnians) is blurred. On the other hand, television is blamed by Islamic groups for corrupting the people, silencing them, and distracting their attention from God as well as from what happens in their country and the rest of the world. With the total state control of this powerful medium, various Islamic groups do not have any option but to denounce its role in society and try to forbid it.

People are capable of articulating different discourses within their religious identity without seeing contradictions in being oriented to the global and attempting to enjoy what it offers, and being rooted in their religious and local identities. A twenty-year-old factory worker, who was born in Bulaq and was relocated with his family in 1980, dreams of having enough money to buy a villa in Switzerland for skiing during the winter, another villa in India where he will hire singers and dancers to perform for him as he has seen on video tapes, and of a palace in Saudi Arabia to facilitate his performing of pilgrimage every year. As Hall (1991a; 1993) emphasizes, with identity there are no guarantees. The openness and fluidity of identities and the multiple discourses that are competing to shape them make it hard to guarantee whether an identity is going to be inclusive or exclusive.[11]

NOTES

1 For example, one can now watch CNN in Cairo if supplied with cable services, and satellite dishes are added to the roofs of upper-class residential units.

2 See Mitchell (1988) for an analysis of such constructions under the British colonization of Egypt.

3 As previously mentioned, with the death of Sadat, the removal of the rest of Bulaq stopped.

4 *Baladi* is a complex concept that signifies a sense of authenticity and originality. It has been discussed in several studies. See, for example, Early (1993) and Messiri (1978).

5 Although people strongly identify themselves as Egyptians, the state definition of 'modern' Egyptians is exclusive. Different groups are not seen as contributing positively to the construction of their country and the state believes that its task is to cure them of their social ills and pathologies in order to turn them into 'good Egyptian citizens' (Ghannam 1993).

6 Let us not forget that Sadat tried to present himself as *al-Ra'is al-Mu'min* or 'The Believing President'.

7 In the women's section, which I had access to, the Friday prayer was coordinated by a woman who made sure that we were standing correctly and made room to squeeze in newcomers.
8 Despite the incorporation of women, the mosque is still a highly gendered space. It manifests and shapes the ways in which gender is constructed. Inside the mosque, women are spatially separated from men, their access to the mosque is conditioned by the absence of their menstrual period, and within its confines they are required to wear long and loose dress and to cover the hair and chest.
9 Al-Sunniyyin (singular Sunni) is used in daily life to refer to Islamic activists who are considered strict followers of the Prophet's traditions or Sunna.
10 It is important to remember that, although these processes take place at the neighbourhood level, they are part of the transformations that Cairo and Egypt in general have been experiencing in the last two decades.
11 The relationship between Muslims and Christians, for example, in al-Zawiya al-Hamra is very complicated and beyond the scope of this chapter. Religion also plays a central role among Christians who live in al-Zawiya. The church serves a similar role to that of the mosque in bringing Christians together. There is also a strong tension between the two religious groups that resulted in clashes between them in 1981.

REFERENCES

Abd El-Razaq, H. (1979) *Egypt During the 18th and 19th of January: A Political Documentary Study* (in Arabic). Dar al-Kalima, Beirut.

Abu-Lughod, J. (1971) *Cairo: 1001 Years of the City of Victorious*. Princeton University Press, Princeton, NJ.

Berman, M. (1988) *All That is Solid Melts into Air: The Experience of Modernity*. Penguin Books, New York.

Beyer, P. (1994) *Religion and Globalization*. Sage, London.

Bourdieu, P. (1979) *Algeria 1960*. Cambridge University Press, Cambridge.

—— (1989) *Distinction: A Social Critique of the Judgement of Taste*, trans. R. Nice. Routledge, London.

Early, E. A. (1993) *Baladi Women of Cairo: Playing with an Egg and a Stone*. Lynne Rienner, Boulder, CO and London.

Friedland, R. and Boden, D. (eds) (1994) *NowHere: Space, Time and Modernity*. University of California Press, Berkeley.

Ghannam, F (1993) 'Urban Planning and the "Imagined Community": Relocation and the Creation of Modern Subjects'. Paper presented at the workshop on 'Social Problems in Urban Planning of Modern Middle Eastern and North African Cities', The American University of Beirut, 21–24 September.

Goonatilake, S. (1995) 'The Self Wandering between Cultural Localization and Globalization', in J. Nederveen Pieterse and B. Parekh (eds), *The Decolonization of Imagination: Culture, Knowledge and Power*. Zed Books, London.

Hall, S. (1991a) 'The Local and the Global: Globalization and Ethnicity', in A. D. King (ed.), *Culture, Globalization and the World-System*. SUNY, Binghampton.

—— (1991b) 'Old and New Identities, Old and New Ethnicities', in A. D. King (ed.), *Culture, Globalization and the World-System*. SUNY, Binghampton.

—— 1993 'Culture, Community, Nation', *Cultural Studies* 7 (3), 349–63.

Hannerz, U. and Lofgren, O. (1994) 'The Nation in the Global Village', *Cultural Studies* 8 (2), 198–207.

Harvey, D. (1990) *The Condition of Postmodernity*. Basil Blackwell, Cambridge.

Ibrahim, S. E. (1982) 'Islamic Militancy as a Social Movement: The Case of Two Groups in Egypt', in A. E. Hillal Dessouki (ed.), *Islamic Resurgence in the Arab World*. Praeger Special Studies, Praeger, New York.

—— (1987) 'Cairo: A Sociological Profile', in S. Nasr and T. Hanf (eds), *Urban Crisis and Social Movements*. The Europo-Arab Social Research Group, Beirut.

Ikram, K. (1980) *Egypt: Economic Management in a Period of Transition*. Johns Hopkins University Press, Baltimore and London.

Kepel, G. (1993) *Muslim Extremism in Egypt: The Prophet and Pharaoh*. University of California Press, Berkeley.

Lash, S. and Urry, J. (1994) *Economics of Signs and Space*. Sage, London.

Massey, D. (1994) *Space, Place, and Gender*. University of Minnesota Press, Minneapolis.

Messiri, S. el, (1978) *Ibn al-Balad: A Concept of Egyptian Identity*. E. J. Brill, Leiden.

Mitchell, T (1988) *Colonising Egypt*. The American University in Cairo, Cairo.

Rabinow, P. (1984) *The Foucault Reader*. Pantheon, New York.

Rageh, A. Z. (1984) 'The Changing Pattern of Housing in Cairo', in *The Expanding Metropolis: Coping with the Urban Growth of Cairo*. The Agha Khan Award for Architecture.

Ray, L. J. (1993) *Rethinking Critical Theory: Emancipation in the Age of Global Social Movements*. Sage, London.

Rugh, A. (1979) *Coping with Poverty in a Cairo Community*. Cairo Papers on Social Science. The American University in Cairo, Cairo.

Sadat, A. el- (1978) *In Search of Identity*. Collins, London.

—— (1981) *The Basic Relationships of the Human Being: His Relationships with God, Himself Others, the Universe, and Objects* (in Arabic). General Agency for Information, Cairo.

Tsing, A. L. (1993) *In the Realm of the Diamond Queen: Marginality in an Out-of-the-way Place*. Princeton University Press, Princeton, NJ.

Turner, B. S. (1994) *Orientalism, Postmodernism and Globalism*. Routledge, London and New York.

Urry, J. (1995) *Consuming Places*. Routledge, London and New York.

The following images are from *Touch Sanitation* (1979–84), a project by Mierle Laderman Ukeles. Initially, Ukeles attached herself without funding to New York City's Sanitation Department. Walking the five boroughs in a systematic set of routes she personally shook the hand of each garbage collector. This might seem a nice gesture, or simply another piece of late 1970s conceptual art, but it has more point. The city produces an extraordinary quantity of trash which until recently was taken in barges to a land-fill site on Statten Island. It now goes further afield. At Fresh Kills its smell in summer drifted over housing areas and a mall (see discussion of odour in the Introduction to Part Eight). The collection of garbage is often carried out early or late in the day, and, crucially, by people whose visibility is circumscribed by the perception that their hands touch the filth which those who produce it prefer to consign to oblivion. By shaking hands and saying thank you, Ukeles ruptured that contract of invisibility – see the text by Kimberly Curtis above. Later, Ukeles was offered and now uses an office in the Sanitation Department near Wall Street, and is developing proposals for the reclamation of the land-fill site – used to deposit the debris from '9/11'. She says: 'Oh no, no, no, no. The City would never do that. They would never mingle human remains in a place where they put garbage; that would collapse a taboo in our whole culture. That crosses a line' (Mierle Ukeles, 'Leftovers/It's About Time for Fresh Kills', *Cabinet*, 6, spring 2002).

Plate 24 **Ukeles and New York City Sanitation truck** (Mierle Ukeles). (Courtesy of Ronald Feldman Fine Arts, New York.)

Plate 25 **Handshake with garbage collector in cab** (Mierle Ukeles). (Courtesy of Ronald Feldman Fine Arts, New York.)

Plate 26 **Handshake with garbage collector at land-fill site** (Mierle Ukeles). (Courtesy of Ronald Feldman Fine Arts, New York.)

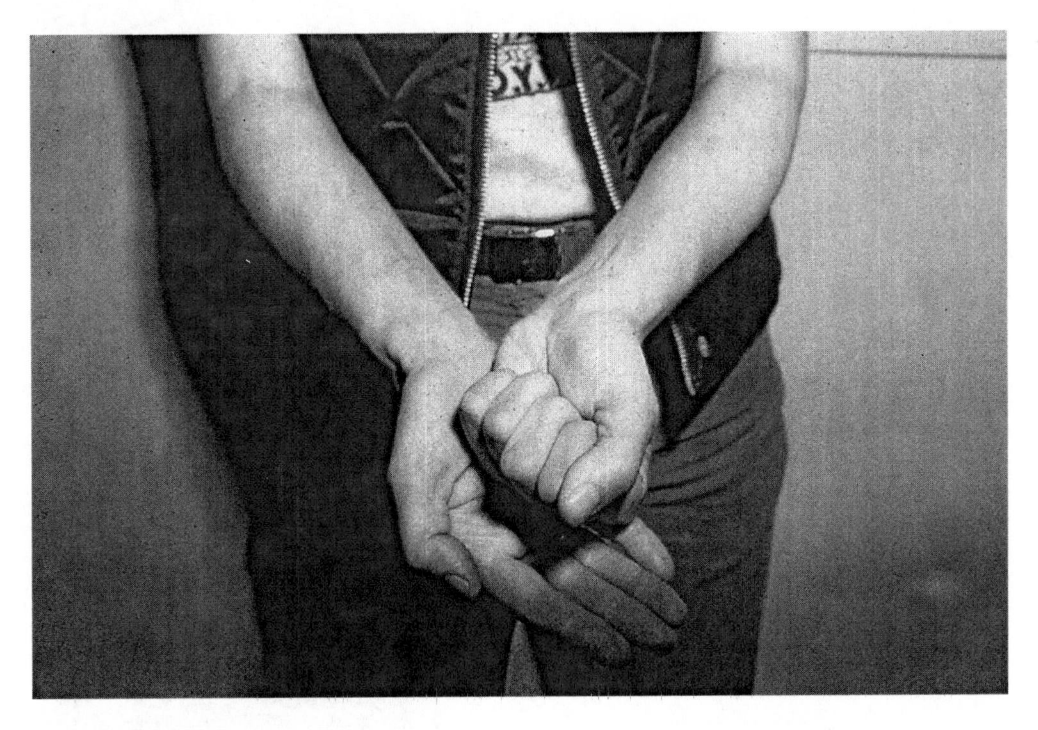

Plate 27 **The artist's hands** (Mierle Ukeles). (Courtesy of Ronald Feldman Fine Arts, New York.)

Plate 28 **Opening performance, 'Wiping Out the Bad Names', 9 September 1984** (Mierle Ukeles). (Courtesy of Ronald Feldman Fine Arts, New York.)

Plate 29 **Detail of performance, 'Wiping Out the Bad Names'** (Mierle Ukeles). (Courtesy of Ronald Feldman Fine Arts, New York.)

Plate 30 **Marine Transfer Station, 59th Street and Hudson River** (Mierle Ukeles). (Courtesy of Ronald Feldman Fine Arts, New York.)

Plate 31 **'Maintenance City/Sanman's Place'** (Mierle Ukeles). (Courtesy of Ronald Feldman Fine Arts, New York.)

PART EIGHT

Boundaries and Transgressions

INTRODUCTION TO PART EIGHT

The issues covered by the texts included in this section relate to those of Part Seven on the contestation of identities. The difference between the sections is that while Part Seven looks at how individuals and groups construct and contest their own identities, this part looks at the spaces which identities construct, and the border zones between them. Those borders need not be physical lines, but may be edges of mutual distrust, or the zones of suspicion felt by one group towards another. From such suspicions, some groups in a society are regarded as deviant or marginal, while others may voluntarily depart from conventions of sociation, economic activity or culture. The dominant society, too, now departs from norms in the move from metropolitan centres to dispersed kinds of settlement – the edge cities and exopolis discussed by Edward Soja in his text in Part Ten.

Of all borders those of the body's tactile, vulnerable surface of skin, which gives it a coherent presence but which contains many internal interfaces, are perhaps the most charged. But this border is also contested in a new world of cyber sensibilities. Architect William J. Mitchell sees an extension of sensory perceptions through new communication and information technologies as producing an ambivalence between the world perceived as being outside the body and that seen as within it: 'For cyborgs, then, the border between interiority and exteriority is destabilized. Distinctions between self and other are open to reconstruction. Difference becomes provisional' (William J. Mitchell, *City of Bits*, Cambridge, MA, MIT, 1996, p. 31). Mitchell sees this leading to a reformulation of the modern (or Cartesian) problem of mind–body as one of mind–network. This is speculative, but the point remains that constructions of identity are inseparable from attitudes to borders.

The texts by Ivan Illich and David Sibley below deal with spaces inhabited by bodies. But the spaces they portray are not spaces of commonality so much as of an enforcement of difference, of strategies for separation by a dominant element in society from bodies it regards as other and dangerous. At the simplest level, this could be seen as an elaboration of a basic premise of settlement, to separate out human from non-human habitats, turned back, as it were, on humanity. One function among others of early forms of human settlement may have been to keep wild animals away from humans and the farmed animals they used as food. Later, in the white conquest of north America as depicted in the art of the period, there is an identification of native Americans with wilderness, both regarded as worthless except for exploitation. Both were rolled back by industry, railways, and white notions of ownership until, after devastation, wilderness took on a new romantic signification, a sense of birthright for an entirely immigrant people. Perhaps, then, the finer spatial differentiations of the modern city, with their race, class and gender associations, do extend an early drawing of a line between safety and danger, the known and that outside the known. The two main difficulties in seeing social differentiations as having long historical roots in this way, however, is that, first, it is highly speculative since few reliable records exist; and second, more to the point, it can lead to a naturalising view as if what has always been must remain so despite its extension into new forms addressing new situations and expressing new feelings. While continuities and discontinuities are both present, it may be that to see the radical separations of bodies required in modern and postmodern societies as *distinct* from those of earlier European societies emphasises that these means of division are not only historically specific but also extraordinary in the lengths to which a dominant group will go to cast out those it rejects. There are historical precedents, for instance, in certain observances in organised religions such as the isolation of women during menstruation, but the scale of the operation in modern times means that division along social lines is a structuring mechanism of society.

What is also striking is the extent of concoction of evidence to support this. Illich writes in one of the extracts reprinted below from *H_2O and the Waters of Forgetfulness* of the notion of a miasma. This was a foul air which contaminated those who breathed it in with deadly disease. The architecture of the nineteenth-century hospital, with its long wards and large windows in place of the small rooms of the eighteenth-century clinic, follows in part from this fear and the need to circulate fresh air, though it is also a way to give those in charge of a ward a view of everyone in it at once. But before such buildings were designed, in the eighteenth century, the dead who had been buried for centuries in shallow graves in churchyards or under the naves of city churches began to be seen as a threat to the living. The smell of the poor was seen later in a similar way, consolidating class divisions. Readers of the report produced for the French Parliament in 1737 to which Illich refers in his text below, which accepts the notion of contagion by invisible miasma, may have believed this. Similarly, various explanations were accepted for the spread of cholera in the early nineteenth century before polluted drinking water supplies were identified as a cause, but the point in both cases is that what is believed at the time is defined less by evidence than by culture.

In the case of Europe during the period of the Enlightenment, the prevailing culture was one of purification, making the instrumental borders those between the pure and the defiled. In philosophy, Descartes purifies thought by rejecting the doubtful knowledges of the senses, travel, books, and memory (see the Introduction to Part Nine). In planning, the inscription of regular places purifies the city of disorder as represented by the accumulated, irregular spaces of the medieval city. In social organisation, from the eighteenth century onwards, systems are introduced to purge the visible society of elements regarded as disfiguring it, including the vagrant, the poor, the insane, the criminal, and the non-productive in whatever way. For all these categories suitable institutions and an appropriate architecture were invented, and for the most part remain in place in some form today. They have changed, so that contemporary schools and hospitals are seen as more welcoming environments than those of the nineteenth century, and prisons provide for rehabilitation as well as punishment – though they always did that by instilling discipline – but as institutional structures separating out social groups for training, repair, or re-education they are still a means to control social development.

A key commentator on the social process in which deviant elements are excluded from visibility and confined in specialist environments is Michel Foucault. In *Madness and Civilisation: A History of Insanity in the Age of Reason* (London, Tavistock, 1967), Foucault begins with a description of the empty places outside a European city's gates:

> At the end of the Middle ages, leprosy disappeared from the Western world. In the margins of the community, at the gates of cities, there stretched wastelands which sickness had ceased to haunt but had left sterile and long uninhabitable. For centuries, these reaches would belong to the non-human. From the fourteenth to the seventeenth centuries, they would wait, soliciting with strange incantations a new incarnation of disease, another grimace of terror, renewed rites of purification and exclusion (Michel Foucault, *Madness and Civilisation*, p. 3).

Foucault writes that the attitude to lepers was reconstituted in a similar form towards the vagrant, the criminal, and the insane. The incarceration of these at first largely undifferentiated groups began with the founding of the *Hôpital Général* in Paris in 1656. Foucault titles his chapter on this 'The Great Confinement'. The hospital (not the same as the clinic where physicians healed the sick and carried out medical experiments on the bodies of the poor) was a state within the state in which normal rights and sociation were suspended. The model spread throughout Europe. Meanwhile, those identified as destabilising the state or the property rights of its citizens were subjected to public execution by various means in different countries.

In time, as Foucault describes in *Discipline and Punish: The Birth of the Prison* (Harmondsworth, Penguin, 1991), prison replaces physical punishment or exile for most criminals, and replaces the public square as execution ground for others. The prison, for Foucault, is a system of self-discipline in isolation from others under the scrutiny of unseen guards – the panopticon which can be understood as both a concept and an architectural form which articulates it. Contemporary concerns about a culture of surveillance owe much to this formulation in which society's members, all of whom are assumed to have a capacity for deviance, become their own police force.

In cultural studies and the social sciences there is a tendency today, in recognition of this, to identify resistance in everyday acts which express deviance, from graffiti to giving false information to market researchers. Joel Kahn asserts that some such accounts are over-simplistic, seeing in them a 'bowdlerized amalgamation of the ideas of Gramsci and Foucault'. The writers of these accounts 'read representations of otherness as examples either of hegemony/knowledge-power or of resistance' (Joel Kahn, *Modernity and Exclusion*, London, Sage, 2001, p. 34). Kahn puts forward but does not unpack a bag of concepts: Gramsci's theories of hegemony in which a dominant interest in society maintains power through social institutions which normalise that power and seed its culture in all classes; Foucault's extension of the association of modern knowledge, as of the natural world, with power over it and other subjects; and a more recent concept of resistance at the level of ordinary living.

In terms of the latter, squatters' camps at national borders have been described as resistant settlements, which rather proves Kahn's point since those who live in them do so out of necessity not intention, are not resisting but surviving. But his scepticism should not mask a more developed understanding of the capacity of the subject, within ordinary living, to find a joy which is resistant of routine. This concept is found in the work of Henri Lefebvre, and in the term 'tactics' derived from Michel de Certeau (in *The Practice of Everyday Life*, an extract from which is reprinted in Part Six, with material from Lefebvre) and often used now to contrast with the more organised 'strategies' adopted by the dominant class in society. In *The Unknown City: Contesting Architecture and Social Space* (Cambridge, MA, MIT, 2001), the book's editors – Iain Borden, Joe Kerr, and Jane Rendell, with Alicia Pivaro – cite Foucault, Lefebvre and de Certeau, and define tactics as 'a more proactive response to the city: they are practices . . . that produce objects. These tactics may be words, images, or things; they may be theoretically and/or empirically based; and they may be romantically and/or pragmatically driven' (p. 13). The postmodern project of tactics, then, as an academic reading of the social practices of others than the ruling element in a society, is an attempt to recover or identify with the creative aspect of ordinary living. This is suppressed by social institutions, or perhaps it is more accurate to say social institutions lead subjects to repress it in themselves. Either way, the power of humour, love, and playfulness becomes an antidote to that of state authority and its mechanisms of purification.

David Sibley, in *Geographies of Exclusion* (London, Routledge, 1995), finds evidence of social purification in areas such as television commercials for washing powder. These, for instance, depict children as primitives – dressed as 'red Indians' – whose clothes are washed specially white by modern detergents. Whiteness is a loaded idea and sign for purity; an earlier instance of its use is the rejection by nineteenth-century German academics, for the most part, of Gottfried Semper's theory that Greek architecture was, like Greek sculpture, polychrome. The building thus becomes a support for the decoration applied to it. As Mark Wigley notes in 'Untitled: The Housing of Gender' (the final chapter in *Sexuality and Space*, edited by Beatriz Colomina, New York, Princeton Architectural Press, 1992, pp. 327–89), the real difficulty was not so much historical but that the purity of white marble in neoclassical aesthetics was undermined by Semper's insistence on decoration. Wigley writes that architecture ceased to be seen in Semper's analysis as a structure clothed, as it were, with ornament and became primarily ornamental: 'Building originates with the use of woven fabrics to define social space . . . The textiles are not simply placed within space to define a certain interiority. Rather, they are the production of space itself' (p. 367). Weaving, then, divides outside from inside, open space from the enclosed space of domestic life (before social life), in a foundational act not of structure or form, but of pattern-making and colouring, located in tents, or in rugs hung over poles. For Wigley, the clothing of the body follows from this clothing of space.

To return to Sibley, whiteness in contemporary society appears to have retained its charged status. Perhaps it is not only the cleanliness of washing using modern technologies which is being marketed, but also an older idea of a repression of deviance. The children, like original savages needing to be civilised, playing in the liminal space of a primitive encampment in the garden, dance into the modern kitchen with its appliances to be re-integrated in the dominant society through a change from dirty to clean clothes. This may seem harmless enough, the kind of narrative sophisticated, postmodern consumers would see right through, and a cultural-studies reading of it as a mythicisation of otherness may seem excessive. Yet Sibley's use of such material is convincing, and not all consumers of television and other kinds of advertising are sophisticated, postmodern subjects provided with a habit of irony and a vocabulary from semiotics. Sibley's reading is supported, for instance, by Richard Sennett's account of white suburbia as embodying a fear of others, in *The Uses of Disorder* (New York, Norton, 1970). Purification, then, takes the form of a demonisation of others who are seen as defiling a social model regarded as wholesome. Sennett charts this in terms of race, but it has a gendered dimension in women's exclusion from public spaces in the modern city except as objects of men's gazes; and in another kind of demarcation in the separation of transcendent, intellectual knowledges and practices from those in which mental and bodily energies remain intertwined in the domestic sphere, the latter assigned in modern societies to women (see Doreen Massey. 'Masculinity, Dualisms and High Technology', in *Body Space*, edited by Nancy Duncan, London, Routledge, 1996, pp. 109–26).

In the chapter from *Geographies of Exclusion* reprinted below, Sibley draws attention to the significance of borders and border zones, and the difficulties of liminal zones at the edges or margins of recognised (intellectual or geographical) territories. He explains the aversion felt towards liminality as, at root, a fear of pollution, translated from old religious to new secular forms, but found also in anthropological research. The chapter which follows this in *Geographies of Exclusion*, on 'Mapping the Pure and the Defiled', deals with the notion that the poor, or marginalised groups such as gypsies, are polluting.

Among Sibley's references is Alain Corbin's *The Foul and the Fragrant: Odour and the Social Imagination* (Basingstoke, Macmillan, 1996). Although Corbin limits his scope to French social and cultural history from the eighteenth century onwards he gives a more detailed and wide-ranging history of miasma than either Sibley or Illich, and his findings can be applied to other European societies in the same period. He summarises:

> Contemporary scientists [in the period 1750–1880], incomparable observers of odors, offered a fragmented, olfactory image of the town; they were obsessed with pestilential foci of epidemics. To escape this swamp of effluvia, the elite fled from social emanations and took refuge in fragrant meadows. There, the jonquil spoke to them of their 'I', inspired the poetry of the 'nevermore', and revealed the harmony between their being and the world (p. 230).

The traditional perfume of musk, made from the putrid intestines of musk deer and giving a smell associated with women's bodies, was replaced in the eighteenth-century court by more delicate scents made from flowers.

Along with decoration, an open availability of sexual pleasures appears threatening to the social order. This was one of the factors in the shift of bourgeois consumption in nineteenth-century Paris from the arcades, the gas-lit interiors of which with their upstairs rooms were associated with the sale of sexual services, to the safer, more regulated department stores. In several other European and north American cities, a fear of unlicensed sexual contact produced its containment in controlled vice districts. Some of these, in Vancouver and Barcelona for instances, were known as Chinatowns (in the latter, *barrio chino*). Kay Anderson writes about Vancouver's 'Little Corner of the Far East' in her chapter from *Body Space* reprinted in this section. The text is equally relevant to the previous section, but is included here because it addresses the spatiality of difference. Opium and gambling are added to sexual pleasures in a characterisation of Chinatown as mysterious, dangerous, and seductive. As Anderson writes, "it was the scenario of sexual liaisons between Chinese men and White women which was seen as the most threatening violation of all' (p. 204), while in an economic context Chinese labour was seen as docile and suitable to such tasks as railway construction. As Anderson demonstrates, monolithic categories such as race, class, and gender are inadequate to convey the range of associations and fears within the situation. Chinatown, too, had internal social differences.

One reason such differences are suppressed in official narratives which retain monolithic categories is because the existence, for example, of a capital-owning elite within the city's ethnic Chinese population was itself a threat to a mythicised white superiority of ownership, a border crossing. But perhaps a more pervasive problem is that of language itself, and the boundaries it sets to what can be said and thought. To give a simple example: the term sinister derives from a word in Latin meaning the 'left side', as *sinistra* means in Italian. But it gradually assumes, in English from the late middle ages, another meaning as dark, or deceitful, as in 'there is something sinister out there'. It can be argued that there is an association between the left side and constructions of femininity, so the linguistic model echoes a social dualism in which the masculine right is privileged. The language has a further trick in that to have equal use of both hands is to be ambidextrous, like having two right hands. There seems no way out of the trap – if, that is, linguistic transgressions render the sense of words no longer communicative. In poetry and some kinds of prose writing, in fictional dialogue perhaps, such transgressions are knowingly used to give new senses and nuances impossible in conventional usage. Yet language has a power which tends to be the more effective because it is so seldom questioned, as if the received use of words was in some way natural, or as if words had inherent, unalterable meanings, not identities as contested as those of the subjects who utter them. This situation has changed, however, through Saussure's insight that meanings (or signifieds) are mutable and contingent on conditions. But this is the problem faced by Luce Irigaray in her essay 'A Chance to Live', in *Thinking the Difference: For a Peaceful Revolution* (London, Athlone, 1994, pp. 1–36). A short extract is reprinted here, which looks at the gendering of language and the possibility of a revolution of public images as a step which might precede it. The extract is contextualised by Irigaray's reflection on what it means to live in a time of threatened mass destruction (after the breakdown of the Chernobyl nuclear power station) produced by a masculine, warlike form of social and economic organisation. As Irigaray points out, the dominant relationship in Western culture, which means largely Judeo-Christian culture, is the mother–son one. As a polemic, she proposes to remedy this social injustice by the display in public places of images of mother–daughter couples. The point, it would seem, is not to take this as a literal prescription for action but, first, as drawing attention to the absence of such images in western society; and, second, as stating the force of cultural narratives in shaping social attitudes.

The fifth text reprinted below is Dean MacCannell's essay on Celebration, Florida, 'New Urbanism and its Discontents'. This critiques a narrative in the form of the Disney town of Celebration, which conceals extreme regulation by offering consumers of housing many areas of choice. The pattern book – a revival of a nineteenth-century form in which designs for house facades and detailing were set out to enable builders to reproduce them – includes several house types of differing size, cost, and location. All the patterns are derived from conservation architecture, and the colour washes allowed are pastel shades in keeping with perceptions of nineteenth-century townscapes. But the web of regulation extends to the most minute details, such as the colour of curtains (white or beige only), kinds of shrubbery permitted in gardens, and frequency of garage sales. Furthermore, the town has a Disney executive in charge instead of an elected mayor, and, as MacCannell argues, a panopticon-like design of house interiors which leaves no space outside the gaze. Celebration serves, then, as a case of a culture of surveillance and control brought into postmodernity by using architecture's new eclecticism for what seem entirely repressive ends. This is driven by the suburban fear of difference of which Sennett writes, and its boundedness is relevant, too, to Sibley's argument on purification. So, the white picket fence which is a motif of Celebration, chosen to represent wholesome social relations, its colour as pure as snow (or the Parthenon in neo-classical iconography), stands, too, for a ruthless denial of otherness dressed up as nostalgia.

Celebration is a flagship of the 'new urbanism' increasingly liked by north American city authorities as an antidote to urban problems. Edward Soja comments on it briefly in *Postmetropolis* – see his text in Part Ten. Other accounts of Celebration are in essayist Bettina Drew's *Crossing the Expendable Landscape* (Saint Paul, Minnesota, Graywolf Press, 1998), and *Celebration, U.S.A.: Living in Disney's Brave New Town*, by journalists Douglas Frantz and Catherine Collins (New York, Henry Holt, 1999), who lived in Celebration for a year.

SUGGESTED FURTHER READING

Alain Corbin, *The Foul and the Fragrant: Odour and the Social Imagination*, Basingstoke, Macmillan, 1996; first published in English by Berg in 1986 as *The Foul and the Fragrant: The Sense of Smell and its Social Image in Modern France*, and originally by Editions Aubier Montaigne in 1982 as *Le miasme et la jonquille*

Tim Cresswell, *In Place, Out of Place: Geography, Ideology, and Transgression*, Minneapolis, University of Minnesota Press, 1996

Michel Foucault, *Madness and Civilisation: A History of Insanity in the Age of Reason*, London, Tavistock, 1967; first published as *Histoire de la folie* by Librairie Plon, 1961

Michel Foucault, *Discipline and Punish: The Birth of the Prison*, Harmondsworth, Penguin, 1991; first published as *Surveiller et punir: naissance de la prison* by Editions Gallimard, 1975

Joel Kahn, *Modernity and Exclusion*, London, Sage, 2001

Amerigo Marras, ed., *Eco-Tec: Architecture of the In-Between*, New York, Princeton Architectural Press, 1999

Jane Rendell, *The Pursuit of Pleasure: Gender, Space and Architecture in Regency London*, London, Continuum, 2002

'The Dirt of Cities, the Aura of Cities, the Smell of the Dead, Utopia of an Odorless City'

from *H_2O and the Waters of Forgetfulness* (1986)

Ivan Illich

Editors' Introduction

Ivan Illich was a radical anarchist whose work explored the archaeology and history of ideas. His project developed initially through four books published during the 1970s, *Deschooling Society* (1971), *Tools for Conviviality* (1973), *Energy and Equity* (1974), and *Medical Nemesis* (1976). Illich became concerned with providing a critique of the functioning of management and economics in society. These, he argued, are presumed to function to remedy problems of scarcity rather than to distribute welfare and resources. The thesis was also pursued in two collections *Toward a History of Needs* (1978) and *Shadow Work* (1981). The book from which this extract is taken, *H_2O and the Waters of Forgetfulness* (1985), examines the transformation in the idea of water. Initially imbued with mythic associations it came to mean nothing more than a liquid for the cleansing of cities. Like much of Illich's work it is polymathic in its approach, encompassing medicine, art, architecture and technology within the discussion.

Illich was born in Vienna in 1926 and lived and studied in south central Europe. He studied natural science, theology, philosophy, and history, obtaining a degree from the Georgian University, Rome and a Ph.D. in history from the University of Salzburg. During the 1950s he travelled to the United States where he worked as a priest amongst the Puerto-Rican community of New York. He later founded a centre for Cross-Cultural Studies in Puerto Rico and later in Cuernavaca, Mexico. Since the 1980s he has been Visiting Professor of Philosophy and of Science, Technology, and Society at Penn State University and has also taught at the University of Bremen. Ivan Illich died on 2 December 2002.

THE DIRT OF CITIES

When Aristotle drew up his rules for the siting of a city, he wanted the streets to be open to sunlight and to prevailing winds. Complaints that cities can become dirty places go back to antiquity. In Rome special magistrates sat under their umbrellas in a corner of the Forum to adjudicate complaints from pedestrians soiled by the contents of chamberpots. Throughout classical antiquity, beginning with the palace at Knossos (1500 B.C.), the dwellings of the wealthy occasionally had a special room for bodily relief. In Rome, wealthy households owned a special slave to empty the night-chairs. Most homes had no designated place for bodily relief. Like the sewers beneath the Athenian agora, the sewers beneath the imperial fora and pay seats in marble latrines were restricted to city areas covered with marble. In popular two-story dwellings, Roman ordinances required a hole at the bottom of the staircase. Otherwise, the street was assumed to be the proper place for such disposals. Medieval cities were cleaned by pigs. There survive dozens of ordinances which regulate the right of burghers to own them and feed them on public waste. In Spain and the Islamic areas, ravens, kites, and even vultures were protected as sacred scavengers. These customs did not change significantly during the baroque period. Only during the last years of Louis XIV's reign was an ordinance passed that made the removal of fecal materials from the corridors of the palace in Versailles a weekly procedure. Underneath the windows of the ministry of finance, pigs were slaughtered for decades, and their encrusted blood caked the palace walls. Tanneries were operated within the city, even though their smell in the valley of Ghinnom had become the symbol for hell (gehenna) in old Jerusalem. A survey carried out in Madrid in 1772 disclosed that the royal palace did not contain a single privy. These millennial city conditions prevailed in London when Harvey announced his discovery of the circulation of the blood.[1] Only after the great London fire of 1660 and after Harvey's death were "laystalls" set up on London street-crossings for the disposal of waste, and an honorary scavenger was appointed for each ward to supervise the rakers – men and women willing to pay for the privilege of sweeping the streets so that they could sell the refuse for a profit. In 1817 the powers of these scavengers and rakers were codified in the London Metropolitan Paving Act, which remained the statute until 1855. By this time the houses of the well-to-do in London usually contained one privy, from which the night-soil was removed several times each week. But for the larger part of London, the collection of night soil from the streets remained sporadic. In the late nineteenth century it was felt to interfere with rush hours. It was not until 1891 that the London County Council prescribe that privy cleaning had to be restricted in summer-time to the hours between 4 A.M. and 10 A.M. Quite obviously, throughout history cities have been smelly places.

THE AURA OF CITIES

Nevertheless, the perception of the city as a place that must be constantly washed is of recent origin. It appears at the time of the Enlightenment. The reason most often given for this constant toilette is not the visually offensive features of waste or the residues that make people slip on the street but bad odors and their dangers. The city is suddenly perceived as an evil-smelling space. For the first time in history, the utopia of the odorless city appears. This new aversion to a traditional characteristic of city space seems due much less to its more intensive saturation with odors than to a transformation in olfactory perception.

The history of sense perception is not entirely new. Linguists have dealt with the changing semantics of colors, art historians with the style in which different epochs see. But only recently have some historians begun to pay closer attention to the evolution of the sense of smell. It was Robert Mandrou who, in 1961, first insisted on the primacy of touch, hearing, and smell in premodern European cultures. Complex non-visual sense perceptions gave way only slowly to the enlightened predominance of the eye that we take for granted when we "describe" a person or place. When Ronsard or Rabelais touched the lips of their love, they claimed to derive their pleasure from taste and smell, which could only be hinted at. Even the eighteenth-century writer does not yet describe the loved body; at best the publisher inserts into the text an etching that illustrates the scene, an etching which, during the early part of the century, effectively hides whatever is individual, personal, "touching" in the scene the author describes. But while it is easy to follow historically the ability of poets and novelists to perceive and then paint the flesh and landscape in their uniqueness, it is much more difficult to make statements about the perception of odors in the past. To write well about this past perception of odors would be a supreme achievement for a historian because the odors leave no objective trace against which their perception can be measured. When the historian describes how the past has smelled he is dependent on his source to know what was there and how it was perceived. The case is the same whether he deals with odors perceived by lovers or those that help physicians recognize the state of the ill or those with which devils or saints fill the spaces within which they dwell.[2]

I still remember the traditional smell of cities. For two decades I spent much of my time in city slums between Rio de Janeiro and Lima, Karachi and Benares. It took me a long time to overcome my inbred revulsion to the odor of shit and stale urine which, with slight national variations, makes all unsewered industrial shantytowns smell alike. This smell is the characteristic for the early stage of industry; it is the stench of dwelling space that has begun to decay because it is threatened by imminent incorporation into the hygienic system of modern cities. It is distinct from the local atmosphere of a still vernacular town. A vernacular atmosphere is integral to dwelling space; according to traditional medicine, people waste away if they are sickened and repelled by the aura of a new place in which they are forced to live. Sensitivity to an aura and tolerance for it are requisites to enjoy being a guest. Many people today have lost the ability to imagine the geographic variety that once could be perceived through the nose. Because increasingly the whole world has come to smell alike: gasoline, detergents, plumbing, and junk foods coalesce into the catholic smog of our age. Where this smog mingles with the decay of vernacular atmosphere, as along the Rimac which carries Lima's sewage into the Pacific I learned to recognize the smell of development. It is there that I became sensitive to the difference between industrial pollution and the dense atmosphere of Paris between Louis XIV and Louis XVI. To describe it I shall draw heavily on Corbin.

THE SMELL OF THE DEAD

People then not only relieved themselves as a matter of course against the wall of any dwelling or church; the stench of shallow graves was evidence that the dead were present within its walls. This thick aura was taken so much for granted that it is rarely mentioned in contemporary sources. Universal olfactory nonchalance came to an end when a small number of citizens lost their tolerance for the smell of corpses. Since the Middle Ages, the corpses of clergy and benefactors had been entombed near the altar, and the procedures of opening and sealing these sarcophagi within the church had not changed over the centuries. Yet at the beginning of the eighteenth century, their miasma became objectionable. In 1737 the French parliament appointed a commission to study the danger that burial inside churches presented to public health. The presence of the dead was suddenly perceived as a physical danger to the living. Philosophical arguments were concocted to prove that the burial within churches was contrary to nature. An Abbé Charles Gabriel Porée, Fénelon's librarian, from Lyons argued in a book which went through several editions that, from a juridical point of view, the dead had a right to rest outside the walls. In his monumental history of attitudes toward death in the West since the Middle Ages, Philippe Ariès has shown that this new squeamishness in the presence of corpses was due to an equally new unwillingness to face death. Henceforth, the living refused to share their space with the dead. They demanded a special apartheid between live bodies and corpses at just the time when the innards of the live human body were beginning to be visualized as a machine whose elements were "prepared" for inspection on the dissecting table. Like the organs, the dead became more visible and less awesome; they also became increasingly more disgusting and physically dangerous for the living. Philosophical and juridical arguments calling for their exclusion from dwelling space went hand-in-hand with reported evidence of the deadly threat of their miasma. Corbin lists several instances of mass death among the members of a church congregation that occurred at the very moment when, during a funeral ceremony, miasma escaped from an opened grave. Burials within churches thereafter became rare – increasingly a privilege of bishops, heroes, and their like. The cemeteries were moved out of the cities. Though in 1760 the Cimetière des Innocents was still used for parties in the afternoon and for illicit love at night, it had been closed in 1780 by request of neighbors precisely because they objected to emanations from decomposing bodies. Yet even if the presence of the dead within the city was resented by rich and poor alike at the end of the ancien régime, it required almost two centuries to educate the lower classes to feel nausea from the odor of shit.

UTOPIA OF AN ODORLESS CITY

Both living and dead bodies have an aura. This aura takes up space and gives the body a presence beyond the confines of its skin. It mingles with the auras of other people; without losing its own personality, it blends into the atmosphere of a particular space. Odor is a trace that dwelling leaves on the environment. As fleeting as each person's aura might be, the atmosphere of a given space has its own kind of permanence, comparable to the building style characteristic of a neighborhood. This aura, when sensed by the nose, reveals the non-dimensional properties of a given space; just as the eyes perceive height and depth and the feet measure distance, the nose perceives the quality of an interior.[3]

During the eighteenth century it became intolerable to let the dead contribute their aura to the city. The dead were either excluded from the city or their bodies were encased in airtight monuments celebrating hygienic disposal, for which Père Lachaise became the symbol in Paris. In the process of their removal, the dead were also transmogrified into the "remains of people who have been," subjects for modern history – but no more of myth. Disallowing them shared space with the living, their "existence" became a mere fiction and their relics became disposable remains. In this process western society has become the first to do without its dead.

The nineteenth century created a much more difficult task for deodorants. After removing the dead, a major effort was undertaken to deodorize the living by divesting them of their aura. This effort to deodorize utopian city space should be seen as one aspect of the architectural effort to "clear" city space for the construction of a modern capital. It can be interpreted as the repression of smelly persons who unite their separate auras to create a smelly crowd of commonfolk. Their "common" aura must be dissolved to make space for a new city through which clearly delineated individuals can circulate with unlimited freedom. For the nose a city without aura is literally a "Nowhere," a *u-topia*.

The clearing of city space coincides with a new stage of the professionalization of architects. Their profession had formerly been in charge of building palaces, squares, fountains, city walls, and perhaps bridges or channels. They were now empowered to condemn dwelling space and transform it into garages for people. Observing the course of Peruvian settlement thirty years ago, John Turner has described what happens when dwelling *by* people is transformed into housing *for* people. Housing is changed from an activity into a commodity. This transformation requires making dwelling activities impossible, so that persons become domesticated docile residents within shelters which they rent or buy. Each now needs a street address with a house number (and, in some cases, an apartment number too). People have lost the aura that allowed their whereabouts to be sniffed out in the old days. When the idea of the new city, made up of residents, began to register in the minds of the leaders of the Enlightenment, everything that smacked of quality in space came to be objectionable. Space had to be stripped of its aura once aura had been identified with stench. Unlike the architect who constructed a palace to suit the aura of his wealthy patron, the new architect constructed shelter for a yet unidentified resident who was supposed to be without odor.

NOTES

1 For a general introduction to the history of sanitation, see Rawlinson and also Kennard. Gay is anecdotal and not documented. For London in late medieval times there are many facts on street cleaning and the technique of cesspool construction in Sabine 1934, 1937. For the hygienic conditions of Paris streets, Labande is full of details, Gaiffe dated and amusing.

2 Stench that kills not one but several persons on the spot is not new to the mid-eighteenth century. There are many previous reports of sinners killed on the spot by experiencing the devil's stench. What is new is the connection between the stench of decaying bodies and this physical effect. See Foizil and Ariès. All through the Middle Ages the sense of smell opened the gates of heaven and hell. Reports on the "odor of sanctity" perceived year after year by thousands of visitors to the grave of a saint are quite common. Deonna documents several hundred instances. Lohmeyer, Nestle, and Ziegler relate this experience to biblical texts. During the twelfth century the smell particular to the remains of saints was taken as evidence for the authenticity of such relics. There can hardly be any doubt about the widespread sharing of this experience. The perception of space and its characteristics by means of the sense of smell was taken for granted by the poets of the time. See Hahn ("Duftraum") and Ruberg, 89ff. A beautiful introduction to the "meaning" given smells is in Ohly. See also Ladendorf.

3 Giving off a smell is as much a part of a personality as casting a shadow, producing a mirror image, or leaving traces on the ground. In all of these "aura" becomes perceptible. People recognize one another by smelling out where they come from: "The Scots folks have an excellent nose to smell their countryfolk" (*OED*, 1756). One first relies on smell to discriminate among individuals: "What a man cannot smell out, a man may spy into" (*King Lear* 1, v, 23:1605). But "you can easily smell a rat," except that "where all stink, no one is smelled." The Latin proverb "mulier tum bene olet ubi nihil olet" quoted by Plautus has been variously translated: in 1529 as "A woman ever smelleth best, whan she smelleth of nothing" and in 1621 by Burton as "Then a woman smelleth best, when she hath no perfume at all." During this period "perfume" had changed its meaning. It had come into English as "odor", given off by incense or other burning substances and, by the time of Burton, had come to mean "scent." (See also Tilly, nos. S558 and R31, and F. Wilson, under "smell.")

During the second decade of the nineteenth century, the loss of "aura" becomes a major new motif in literature. It can be readily traced by following the influence of A. V. Chamisso's *Peter Schlemihl*, who sells his shadow to the devil in exchange for wealth. The loss or sale of one's "soul" was a well-known motif at the time, but by retelling the folktale and insisting on the loss of something visible and observable, Chamisso created a veritable school. In 1815 E. T. A. Hoffmann told the story of a young man whose mirror image was taken from him by a whore and an eerie physician. W. Hauff's hero of 1828 exchanges his heart for a stone counterfeit to save himself from bankruptcy. By the end of the century heroes have sold "sleep," "appetite," "name," "youth," and "memories" (for details, see Ludwig 1920 and 1921).

The shadow had always been part of the full personality (Bächtol 9, Nachtrag 126–42). Only when a Greek becomes luminous himself, in the presence of Zeus or when an Iranian becomes a saint, does he lose his shadow. According to Irish stories, if a person's shadow is pierced, he dies (Stith-Thompson D 2061.2.2.1). Among the Jews a ghost is recognizable by its lack of a shadow (ibid., G 302.4.4), just as it is said to leave no footprints (ibid., E421.2). The exchange in which the student of alchemy leaves his shadow to his master the devil as an honorarium is a motif that appears only in the eighteenth century. The shadow remains secondary in fairy tales and folk literature (Franz 1983). In fairy tales everybody is always everybody's shadow (24, 31). The idea of the shadow (or, for that matter, of the mirror image or act of memory) as a saleable commodity is a new and important motif that appears with possessive individualism in Chamisso. It fits into the period during which people's "smell," their aura and their "moral economy" (E. P. Thompson) were taken from them.

Ultimately the drugstore became the symbol of the industrialized aura; it is the supermarket of mass-produced glamour and scents for a deodorized population. People who obsessively scrub away their auras can pick and choose a better one there. Musil (v. 7, p. 895) has created a prophetic image: "Schlemil's guilt is his bourgeois nature, his refusal to admit his loss of his shadow, his incapability of creating genius from it."

'Border Crossings'

from *Geographies of Exclusion* (1995)

David Sibley

Editors' Introduction

David Sibley is a lecturer in geography at the University of Hull in the north of England. This extract is taken from his most well-known work *Geographies of Exclusion* (1995). In *Geographies of Exclusion*, Sibley brought together ideas that he had been exploring for some years on exclusion, transgression and boundary definition in western society, especially as a response to nomadic minorities. Such minorities are not confined to faraway places, as Sibley's research with gypsies in the United Kingdom has indicated.

In this sense there are broad comparisons with the ideas that Ivan Illich explored in the extract from *H_2O and the Waters of Forgetfulness* (1985). Sibley's work has been very influential within the geography and sociology communities in the United Kingdom and proved timely in that it prefigured a number of state controls of radical or simply 'other' groups in the United Kingdom including New Age Travellers, ravers, road protesters and asylum seekers introduced in the past few years. Sibley's critique thus seems even more relevant in the early twenty-first century than it did on its publication in 1995. In addition to this, David Sibley has written about the production of knowledge, the contributions of psychoanalysis and social anthropological theory to the spatial, and transgressive animals (especially feral cats) in the city.

The sense of border between self and other is echoed in both social and spatial boundaries. The boundary question, a traditional but very much undertheorized concern in human geography, is one that I will explore in this chapter from the point of view of groups and individuals who erect boundaries but also of those who suffer or whose lives are constrained as a result of their existence. Crossing boundaries, from a familiar space to an alien one which is under the control of somebody else, can provide anxious moments; in some circumstances it could be fatal, or it might be an exhilarating experience – the thrill of transgression. Not being able to cross boundaries is the common fate of many would-be migrants, refugees, or children in the home or at school. Boundaries in other circumstances provide security and comfort. I will start by examining some general characteristics of boundary zones and then describe some of the diverse ways in which boundaries are constructed, demolished and energized.

In a rather formalistic treatment of the boundary problem, Edmund Leach played with the idea of separating and combining categories, focusing particularly on the intersections of sets.[1] Like Leach, I will use Venn diagrams to introduce some of the boundary issues which are characteristic of social and spatial relations, issues which are central to questions of exclusion. The need to make sense of the world by categorizing things on the basis of crisp sets – A, not-A, and so on – is evident in most cultures, although I do not think it is a universal need, as Leach suggests. However, it is a good place to start because problems associated with this mode of categorization are readily identifiable.

Problems arise when the separation of things into unlike categories is unattainable. As Leach recognized,

'There is always some uncertainty about where the edge of Category A turns into the edge of Category not-A.'[2] For the individual or group socialized into believing that the separation of categories is necessary or desirable, the liminal zone is a source of anxiety. It is a zone of abjection, one which should be eliminated in order to reduce anxiety, but this is not always possible. Individuals lack the power to organize their world into crisp sets and so eliminate spaces of ambiguity.

To move from Leach's abstraction to the concrete, we might consider the case of the home. To the occupier, the home may represent a space clearly separated from the outside. Inside the home, the owner or tenant may feel that space is ordered according to his or her values.[3] However, problems can be created by entrances, breaches in the boundaries of the home. The entrance, the hallway or passage provides a link between the private and the public, but it constitutes an ambiguous zone where the private/public boundary is unclear and in need of definition and regulation in order to remove the anxiety of the occupier.[4] If you admit strangers to the house, are they confined to an entrance area or allowed to enter a living space? How do you cope with the Jehovah's Witnesses or with the person selling double-glazing? The response will depend on where the householder locates the boundary, but this may be variable, depending on how the outsider is perceived in relation to the occupier's conception of privacy.

A second example, child/adult illustrates a similarly contested boundary. The limits of the category 'child' vary between cultures and have changed considerably through history within western, capitalist societies. The boundary separating child and adult is a decidedly fuzzy one. Adolescence is an ambiguous zone within which the child/adult boundary can be variously located according to who is doing the categorizing.[5] Thus, adolescents are denied access to the adult world, but they attempt to distance themselves from the world of the child. At the same time, they retain some links with childhood. Adolescents may be threatening to adults because they transgress the adult/child boundary and appear discrepant in 'adult' spaces. While they may be chased off the equipment in the children's playground, they may also be thrown out of a public house for under-age drinking. These problems encountered by teenagers demonstrate that the act of drawing the line in the construction of discrete categories interrupts what is naturally continuous. It is by definition an arbitrary act and thus may be seen as unjust by those who suffer the consequences of the division.

BOUNDARY MAINTENANCE AND SOCIAL ORGANIZATION

In using these two examples, I am suggesting that liminality presents as many problems for highly developed capitalist societies as for the relatively simple agrarian and hunter-gatherer collectivities which have been the primary focus of anthropological research and where much of the theory of boundary dynamics has been developed. Dichotomies like traditional/modern or simple/complex do not seem to have much relevance to the questions of boundary drawing, inclusions and exclusions. Perhaps a meaningful distinction could be made between what Davis and Anderson term *high-density* and *low-density* social networks.[6] Albeit crude, this dichotomy does suggest varying attitudes to difference which might be attributed to the density of social interaction. Davis and Anderson suggest that in high-density networks 'most links are strong and one is likely to know and have direct ties to most people affected by the misbehavior [*sic*] of a member of one's network'. Conversely, they argue that in networks of low density, difference is less visible because there is less shared knowledge of individuals within the community. In preindustrial societies, people are enmeshed, involved in each other's lives through extended family, kin connections or clan membership coupled, in many cases, with simple physical propinquity. Gypsies and other semi-nomadic minorities demonstrate the characteristics of shared knowledge of members of the community and of physical nearness very clearly. An outsider in such a community is very exposed. However, similar forms of social organization are found within developed societies, both within traditional working-class and suburban neighbourhoods.

A division based on the density of social networks is fairly close to Durkheim's schema in which he distinguishes between societies exhibiting *mechanical* and *organic* solidarity.[7] Where social identity is based on mechanical solidarity, Durkheim argues that

> the society is dominated by the existence of a strongly formed set of sentiments and beliefs shared by all members of the community [so] it follows that there is little scope for differentiation between

individuals: each individual is a microcosm of the whole.

With shared beliefs, we can talk of a *conscience collective* which 'completely envelopes individual consciousness'.[8] By contrast, individualism is consistent with social solidarity in developed industrial societies because solidarity is organic, that is, deriving from contractual relationships which develop with an increasing division of labour. Durkheim assumed that these contracts were governed by norms which comprised the glue holding society together, but norms did not preclude individual difference. This dichotomy does considerable violence to reality and it was probably a fairly crude representation of varying forms of social organization when Durkheim was writing in the late nineteenth century. Like Davis and Anderson's view of social networks, it does provide some ideas about the way people might collectively react to difference, but we need knowledge of the social, political and geographical contexts of community responses to 'others' in order to say anything useful about conflicts based on difference. With the globalization of culture and the almost total penetration of capitalist forms of consumption, it certainly does not make sense to characterize societies in Durkheim's terms. Yet, there is something approaching a conscience collective in some middle-class North American suburbs[9] and on some local authority housing estates in Britain, manifest in reactions to the mentally disabled, Gypsies and Bangladeshis, for example. Some useful ideas on this kind of hostility to others comes from studies of small groups by social anthropologists. Here, the work of Mary Douglas and her critics is particularly illuminating.

POLLUTION, DISCREPANCY AND SMALL GROUP BOUNDARIES

At the social level, as at the individual level, an awareness of group boundaries can be expressed in the opposition between purity and defilement. In Mary Douglas's *Purity and Danger* and subsequent writing,[10] she developed this thesis, gathering support for her argument largely from fieldwork with tribal societies and ancient texts, particularly the Old Testament of the Bible as a record of Judaic ritual. Her key argument is that 'Uncleanness or dirt is that which must not be included if a pattern is to be maintained . . . in the primitive culture, the rule of patterning works with greater force and more total comprehensiveness [*sic*] . . . [than in a modern industrial society]'.[11] By patterning, Douglas means the imposition of a symbolic order 'whose keystone, boundaries, margins and internal lines are held in relation by rituals of separation'.[12] Separation is a part of the process of purification – it is the means by which defilement or pollution is avoided – but to separate presumes a categorization of things as pure or defiled. This can be illustrated by her analysis of a passage from Leviticus:

> The last kind of unclean animal is that which creeps, crawls or swarms upon the earth . . . Whether we call it teeming, trailing, creeping, crawling or swarming, it is an indeterminate form of movement. Since the main animal categories are defined by their typical movement, 'swarming', which is not a mode of propulsion proper to any particular element, cuts across the basic classification. Eels and worms inhabit water, though not as fish; reptiles go on dry land, though not as quadrupeds; some insects fly, though not as birds. There is no order in them.[13]

From such observations, Douglas proposes a rule for categorization in ancient Israel which forms the basis for a general rule, namely, that 'the underlying principles of cleanliness in animals is that they shall conform fully to their class. Those species are unclean which are imperfect members of their class or whose class itself confounds the general scheme of the world'.[14] Thus, it is those animals, people or things that are discrepant, that do not fit in a group's classification scheme, which are polluting. The evidence for this from records of the practices current in ancient Israel are quite convincing and Neusner has compiled a long list of polluting activities, conditions of the body, animals, and so on, which lends support to Douglas's thesis.[15] However, to generalize from one ancient culture, where there were strong rules of exclusion laid down by the rabbis who also wrote the texts that constitute the evidence, to all small groups and tribal societies is dangerous. Murray, in particular, questioned the empirical validity of her argument, suggesting that people were not really that concerned about defilement and happily mixed discrepant categories in their daily struggle for survival.[16] He asserted that: 'If there is any psychological reality to the "horror" purportedly inspired by such classification difficulties, it is confined to anthropologists intent on eliciting complete and exhaustive contrast sets.' According to Murray, the people who were supposedly engaged in these boundary rituals would be driven to anxiety 'by proscribing and even attempting to annihilate what is not readily classifiable from the world'. This criticism was hardly fair because Mary Douglas had herself recognized the limitations of her original thesis, long before Murray's assault.[17]

I think we can conclude that Douglas's argument about purification and defilement needs to be qualified in regard to time and place, but I would also argue that it has wider application than she recognized. In *gemeinschaft*-like groups, that is, closed, tightly knit communities with something approaching a conscience collective, it may be that adherence to the rules is more likely in times of crisis, when the identity of the community is threatened. However, my observations in English Gypsy communities suggest that poverty and family size are also factors affecting observance of pollution taboos. In order to ensure that things are not polluted (*mochadi* or *marime*), numerous separations are required, including among utensils which are used for washing food, clothes, and the body. In poor families with several small children, however, it is often impractical to comply with all these rules. There may be a shortage of water or not enough washing bowls and the mother may be too tired to meet the ritual requirements all the time. In affluent families, particularly with grown-up children, pollution taboos are much more likely to be observed. There is some support here for Murray's argument, but he overstates his case.

Boundary consciousness is also a characteristic of the mainstream in modern, western society, or, at least, it is in some kinds of locales and at certain times. The North American suburb has been represented as a particular kind of *gemeinschaft* within the swathe of individual anonymous worlds that are supposed to constitute the modern metropolis. The suburb was first described as an exclusionary, purified social space by Richard Sennett, in *The Uses of Disorder*,[18] the anatomy of the North American variety has been examined in some detail by Constance Perin;[19] and Mike Davis describes the enclosed communities, socially purified and defended fortress-like, against the supposed threat posed by the poor, which are an increasingly prominent feature of the geography of Los Angeles.[20] Affluent suburbs in Britain are similarly coming to resemble these closed communities where the discrepant is clearly identified and expelled, like a suburb in Bristol which hired a private security firm to patrol its leafy avenues and eject what the security officers described as 'hostiles', including young men in baseball caps.

In these suburbs, there is a concern with order, conformity and social homogeneity,[21] which are secured by strengthening the external boundary, but, as Davis recognizes, 'the greater the search for conformity, the greater the search for deviance; for without deviance, there is no self-consciousness of conformity and *vice-versa*'.[22] This process is seen by the members of the community as a virtuous one – it brings into being a morally superior condition to one where there is mixing because mixing (of social groups and of diverse activities in space) carries the threat of contamination and a challenge to hegemonic values. Thus, spatial boundaries are in part moral boundaries. Spatial separations symbolize a moral order as much in these closed suburban communities as in Douglas's tribal societies.

BOUNDARY ENFORCEMENT

Generalizations like Sennett's 'purified suburb' have to be qualified. While there is plenty of evidence that purified suburbs exist, with damaging consequences for the welfare of the rest of the population in metropolitan areas, not all suburbs are like this. Apart from an increase in racially mixed suburbs in the United States,[23] it has also been argued that (British) suburbs can provide a refuge for eccentrics.[24] A concern with privacy, minding your own business, is also characteristic suburban behaviour. However, communities which much of the time appear to be indifferent to others do occasionally turn against outsiders, particularly when antagonism is fuelled by moral panics. Moral panics heighten boundary consciousness but they are, by definition, episodic. Fears die down and people subsequently rub along with each other. Often, but not invariably, panics concern contested spaces, liminal zones which hostile communities are intent on eliminating by appropriating such spaces for themselves and excluding the offending 'other'.

In Stanley Cohen's classic *Folk Devils and Moral Panics*, he describes the phenomenon as

> a condition, episode, person or group of persona [which] emerges to become defined as a threat to societal values and interests: its nature is presented in a stylized and stereotypical fashion by the mass media . . . ways of coping are evolved

> or (more often) resorted to; the condition then disappears . . . or deteriorates and becomes more visible.[25]

One of the most remarkable features of moral panics is their recurrence in different guises with no obvious connection with economic crises or periods of social upheaval, as if societies frequently need to define their boundaries. As Cohen reflected on the Mods and Rockers episode in the 1960s,

> More moral panics will be generated and other, as yet nameless, folk devils will be created. This is not because such developments have an inexorable inner logic but because our society as presently structured will continue to generate problems for some of its members – like working-class adolescents – and then condemn whatever solutions these groups find.

The resonance of historical panics in modern crises is worth noting because it demonstrates the continuing need to define the contours of normality and to eliminate difference. This is as evident in what are claimed to be postmodern western societies, postmodern in the sense that they embrace difference,[26] as it was in Nazi Germany or England in the seventeenth century. I will describe one case from seventeenth-century England because of its remarkable similarity with modern panics, particularly in the way in which difference is represented by the popular media.

In the early modern period, before the industrial revolution in Europe and North America, Christian religion was an important source of conflict. The line between conformity and dissent, good and evil, light and darkness was of great concern for an established church attempting to isolate or eliminate religious minorities which appeared to threaten its theological hegemony. The Ranters were one such group who appeared threatening despite their very small numbers.[27] Their dissent from the established church was not unique but their views were publicized at a critical juncture in English history, just after the Revolution of 1649 and the establishment of the Commonwealth, when there was general political uncertainty. At this time, religion was politics and the collapse of religious authority was portrayed in sensationalist literature as 'a prelude to unbridled immorality and social chaos'.[28]

In this context, the Ranters represented a subversive and threatening group and their difference and deviance were amplified in a sustained press campaign which had little regard for the truth. Connections were suggested between religious dissent, atheism and immorality. Interestingly, as in witch crazes, Ranters were associated with inversions which threatened moral values, specifically, devil worship and promiscuity. While women generally had little scope for sexual relations outside marriage, Ranter women were portrayed as sexually unbounded: 'They were free to copulate with any man and did so enthusiastically and openly'. One broadsheet, 'Strange NEWES from the OLD-BAYLY', asserted in terms worthy of a modern tabloid that

> Some [Ranters] have confessed that they have had often meetings, whereat both men and women presented themselves stark naked one to the other, in a most beastly manner. And after satisfying their carnall and beastly lust, some have a sport called whipping the whore; others call for musick and fall to revelling and dancing.[29]

The broadsheet had a major role in developing the Ranter stereotype and in fostering a hysterical response to the minority, accentuating their deviance and so legitimating oppression. Modern analogies are cases of ethnic conflict where the media have an important role in demonizing the enemy, as in the use of the radio by the Hutu to generate opposition to the Tutsi in Ruanda, although cases like this, or the ethnic conflicts in the former Yugoslavia, are more enduring, not panics in Cohen's terms.

The Ranter panic is comparable to modern instances in the sense that the group is represented as a threat to core values. Core values in the seventeenth century were to a great extent religious values, whereas in modern secular societies, core values are embodied in the family, the home and the nation, and thus they have implications for 'deviant' youth, other sexualities and racial minorities. Although the folk devil comes in different guises, panics do not introduce a succession of new characters bearing no resemblance to each other. Rather, they are manifestations of deep antagonisms within society, for example, between adults and teenagers, blacks and whites, heterosexuals and homosexuals. The alterity personified in the folk devil is not any kind of difference but the kind of difference which has a long-standing association with oppression – racism, homophobia, and so on. The moral panic will be accompanied by demands for more control of the threatening minority, for the state to provide stronger defences for, say, white, heterosexual values. This may include a call for the stronger bounding of space to counter the perceived threat, as in attempts by the British government to exclude New Age Travellers from the countryside and secure this terrain for the middle classes.[30] An account of two recent sources of moral panics in North America and Britain – AIDS and mugging – demonstrates most of these points.

Media responses to AIDS went through a period in the 1980s when they were typically moralizing and exaggerated.[31] There was a moral panic, beginning in the early 1980s and lasting until the late 1980s, after which there was more restrained and informed reporting, at least, a more muted homophobia and racism when it was realized that AIDS was not a 'gay plague' or an 'African disease' but also affected the white, heterosexual population. The wilder threats of contagion could not be sustained as knowledge about the syndrome and advice on safer sexual behaviour were more widely disseminated. However, during the 1980s, there was a 'spiralling escalation of the perceived threat, leading to a taking up of absolutist positions and the manning of moral barricades'.[32] The moral barricades were manned on behalf of 'the family'. As Simon Watney comments, the press made its primary appeal to the family as the central site of consumption, thus of fundamental importance to the economy, and as the site of sexuality and child rearing – 'The family is positioned in newspaper discourse as [a] central term of professional journalistic know-how, establishing a fixed agenda of values, interests and concerns which are heavily moralized.' The family was threatened by alternative models presented by gays and by AIDS as a 'gay disease'.

One panic response in western societies was to advocate quarantine, the physical isolation of homosexuals with AIDS, a response which betrays ignorance of the epidemiology of the syndrome but is consistent with the idea of the threat of contagion, a metaphoric threat to the health of the family. A spokesperson for the right-wing Conservative Family Campaign in Britain advocated the removal of 'AIDS victims' to quarantine centres. A similar proposal in Australia indicated to one critic that the panic was not only 'a product of homophobia but was . . . tied to the belief that [Australians] can insulate themselves from the rest of the world through rigid immigration and quarantine laws'.[33] Family space, threatened by other sexualities, has its counterpart in national space, threatened by alien values. Homophobia will not go away while homosexuality is constructed as an 'other' which threatens the boundaries of the social self. AIDS, however, was the catalyst for a panic which temporarily reinforced boundaries. It brought homophobia into sharp relief.

Sexuality is a source of difference from which moral panics can emerge because it is fundamental to people's world-views and their relationships to others. Similarly, race, as it is culturally constructed, can be a source of social cleavage which can be magnified in an episodic fashion. Thus, a mugging episode in Britain in the 1970s emerged from the problem of racism, which is deeply rooted in the centres of former empires. The particular target of moral outrage in this case was young Afro-Caribbean males.

Two things came together in the reporting of street crime in the 1970s. The first was the notion of the British inner city as a *black* inner city, characterized by lawlessness and vice, so that inner city became a coded term for black deviance. The idea of a black inner city bore little relation to demographic or geographical reality, but the myth is more important than the reality. Thus, in a typical example of place labelling, Weaver claimed that the local press 'portrayed north central Birmingham as a violent, crime-ridden area, beset by problems rooted in the nature of its coloured [*sic*] residents rather than in the district's disadvantaged position in British urban space'.[34] Second, there was a rise in recorded street crime but a perceived rise, particularly in 1973, amounting to 'a national mugging scare',[35] and this fixed the idea of black youth as an inherently criminal minority and inner cities as inherently criminal localities. Susan Smith's quotations from Birmingham newspapers, like 'Society at limit of leniency' and 'Angry suburb', indicate a panic which was generated through the stereotyping of minority group and locality. This required a silence about policing, unemployment, the population composition of the district and comparative crime statistics for the city, which would have put a different complexion on the issue. The labels attached to the inner city during this panic strengthened the boundary separating the 'respectable white suburbs' and 'black inner areas' and decreased the likelihood of white people gaining knowledge of Afro-Caribbean and Asian communities through experience. A panic surrounding inner-city riots in the 1980s again confirmed the boundary, with material consequences for inner-city residents, such as the withdrawal of financial services. Different moral panics with slightly different scripts signalled the continuing presence of racism.

'Family', 'suburb' and 'society' all have the particular connotation of stability and order for the relatively affluent, and attachment to the system which depends for its continued success on the belief in core values is reinforced by the manufacture of folk devils, which are negative stereotypes of various 'others'. Moral panics articulate beliefs about belonging and not belonging, about the sanctity of territory and the fear of transgression. Since panics cannot be sustained for long, however, new ones have to be invented (but they always refer to an old script).

INVERSIONS AND REVERSALS

Moral panics bring boundaries into focus by accentuating the differences between the agitated guardians of mainstream values and excluded others. Occasionally, these social cleavages are marked by inversions – those who are usually on the outside occupy the centre and the dominant majority are cast in the role of spectators. Inversions can have a role in political protest in the sense that they expose power relations by reversing them and, in the process, raise consciousness of oppression. They energize boundaries by parodying established power relations.

In early modern Europe, inversions constituted a popular genre known as World Upside Down.[36] Broadsheets illustrating such a world were widely distributed among the illiterate, the themes being virtually unchanged for several hundred years. Illustrations showed, for example, the blind leading the sighted, sheep eating wolf, child punishing father, beggar giving alms to the rich. Their popularity with the oppressed could be accounted for by the fact that they fantasized about the existing order. Sometimes, reversals could serve as a symbol of actual revolt. In the Rebecca riots against the turnpike roads in Wales and the west of England, for example, men dressed as women and took Rebecca as a symbol of power against authority:

> And they blessed Rebecca and said unto her, 'Thou art our sister: be thou the mother of thousands of millions, and let thy seed possess the Gates of those who hate them.'[37]

This protest demonstrated the particular importance of 'women on top' as a symbolic reversal. It has a long history in Europe and Davis has suggested that

> males drew upon the sexual powers and energy of [unruly women] and their licence (which they had long ago assumed at carnivals and games) to promote fertility, to defend the community's interests and standards and tell the truth about unjust rule.[38]

The occasions when inversions assume a centre–periphery form, when the dominant society is relegated to the spatial margins and oppressed minorities command the centre, may represent a challenge to established power relations and, thus, be subject to the attentions of the state. There may be attempts to control or suppress such events because they harness the energies of groups which challenge mainstream values. As Stallybrass and White observe,

> There is no reason to suppose that capitalism should be . . . different from other societies in locating its most powerful symbolic repertoires at borders, margins and edges, rather than at the accepted centres of the social body.[39]

This is particularly the case with carnivalesque events which are licensed but have contested spatial and temporal bounds. For example, Caribbean carnivals in British cities have been grudgingly accepted by the state as legitimate celebrations of black culture in an avowedly pluralist society,[40] but, in the past, they have been heavily policed and contained. The appeal of the exotic for the white majority mixes uneasily with the images of black criminal stereotypes which have informed the responses of the control agencies. Similar conflicts occurred over the carnivalesque centre on festivals in rural England which attract New Age Travellers. With the Criminal Justice and Public Order Bill,[41] the British government is attempting to seriously limit or ban festivals which are seen as a threat to the cherished values of rural England. Inversions of this kind are thus important indicators of marginality. Responses to carnivalesque events demonstrate how the majority constructs the 'other'.

Other reversals may have less political currency although they can still be symbolically potent. One such case is the Gypsy pilgrimage to Saintes Maries de la Mer on the Camargue coast in the south of France. Gypsies have been relegated to the margins of French society for centuries, and being a Gypsy in the seventeenth and eighteenth centuries was under most regimes a crime warranting execution, mutilation, transportation or a life sentence in the galleys.[42] Since

1935, however, French and other European Gypsies have taken over the small town of Saintes Manes de la Mer on 24 and 25 May and again in October, for a ritual which inverts the practice of the established church in that the object of reverence is Sara, a black madonna. It was only in 1935 that French Gypsies gained ecclesiastical authority to venerate Sara, who has not been canonized by Rome. Although the pilgrimage has now been given a tourist gloss and the Gypsy veneration includes the other Saintes Maries, Salome and Jacobe, it is still a subversive event which expresses the collective but highly circumscribed power of European Gypsies and expresses the long history of racism to which they have been subject.

CONCLUSION

The propositions of object relations theory – the bounding of the self, the role of good and bad objects as stereotypical representations of others, as well as their representation as material things and places – can be projected onto the social plane. The construction of community and the bounding of social groups are a part of the same problem as the separation of self and other. Collective expressions of a fear of others, for example, call on images which constitute bad objects for the self and thus contribute to the definition of the self.

The symbolic construction of boundaries in small groups which have been studied by social anthropologists has its counterpart in the marking off of communities in developed western societies. Consciousness of purity and defilement and intolerance of difference secure some groups within the larger spaces of the modern metropolis. The outside is populated by a different kind of people who threaten disorder, so it is important to keep them at a distance. These fears, however, are fuelled by the exaggerated accounts of some sections of the media and the state who represent the claims of others for space, or simply for the right to dissent, as a threat to core values. Social and spatial boundaries in these circumstances become charged and energized. The defence of institutions like the family and spaces like the suburb becomes a more urgent undertaking during a moral panic. The oppressed, however, have their own strategies which challenge the domination of space by the majority, if only briefly and in prescribed locales. Ultimately, carnivalesque events confirm their subordination.

The problems that I have been discussing here concern, in part, territoriality, the defence of spaces and transgressions. Space is implicated in many cases of social exclusion, and in the next chapter I will try to identify in more detail the characteristics of exclusive social spaces and to relate these spaces to questions of power and social control.

NOTES

1 Edmund Leach, *Culture and Communication*, Cambridge University Press, Cambridge, 1976. Although Leach's penchant for A's and not-A's and his generally arid style is rather offputting and suggests an attempt to be scientific in his approach, he was sensitive to cultural difference. The formal treatment is only a means of exposing some key issues.

2 *Ibid.*, p. 35.

3 I am conscious of the fact that homes often contain families, the members of which may have conflicting interests. I discuss the question of power relations and exclusion in the home in Chapter 6 [not reproduced here].

4 This example comes from Roderic Lawrence, *Housing, Dwellings and Homes*, Wiley, Chichester, 1987.

5 The problematic nature of the category 'adolescent' is discussed by Allison James, 'Learning to belong: the boundaries of adolescence', in Anthony Cohen (ed.), *Symbolizing Boundaries*, Manchester University Press, Manchester, 1986, pp. 155–170.

6 Nanette Davis and Bo Anderson, *Social Control: The production of deviance in the modern state*, Irvington, N.Y., 1983.

7 Durkheim's concept of mechanical and organic solidarity is discussed in detail in Anthony Giddens, *Capitalism and Modern Social Theory*, Cambridge University Press, Cambridge, 1971, pp. 75–79.

8 *Ibid.*, pp. 75–76.

9 This is an assertion made by a number of writers, but, as I suggest later in this chapter, this kind of broad-brush portrayal of American suburbia needs qualifying. Apart from academic studies of purified suburban communities, there have been a few fictional ones, like the film *Edward Scissorhands*, which portrays the reaction of a conformist Californian suburban community to a gothic 'other', making a similar point to Richard Sennett in *The Uses of Disorder*, Penguin, Harmondsworth, 1970.

10 Mary Douglas's ideas on purity and defilement are further developed in *Natural Symbols*, Barrie and Jenkins, London, 1970.

11 Mary Douglas, *Purity and Danger*, Routledge and Kegan Paul, London, 1966.
12 *Ibid.*, p. 41.
13 *Ibid.*, p. 56.
14 *Ibid.*, p. 55.
15 Jacob Neusner, *The Idea of Purity in Ancient Judaism*, E. J. Brill, Leiden, 1973. Emphasizing the strength and pervasiveness of purification rituals, Neusner observes that:

The land is holy, therefore must be kept clean. It may be profaned by becoming unclean. The sources of uncleanness are varied and hardly cultic: certain animals, women after childbirth, skin ailments, mildew in the house, bodily discharges, especially the menses and seminal fluid, sexual misdeeds and the corpse.

16 S. Murray, 'Fuzzy sets and abominations', *Man*, 18, 1983, pp. 396–399.
17 In the preface to the first edition of *Natural Symbols*, Douglas argued that 'Each social environment sets limits to the possibilities of remoteness and nearness of other humans and limits the costs and rewards of group allegiance to and conformity to social categories.'
18 Sennett, op. cit. In this book, Sennett drew on the psychoanalytical theories of Erik Erikson in developing his thesis on purified identities and purified suburbs.
19 Constance Perin, *Belonging in America*, University of Wisconsin Press, Madison, 1988.
20 Mike Davis, *City of Quartz*, Verso Press, London, 1990.
21 These concerns emerged in a British study of people's ideal homes. Untouched wilderness in proximity to the home was severely frowned upon, presumably because it was not ordered or regulated nature. Distance from areas of racial tension was a predictable preference. *Designing and Selling Three and Four Bedroom Houses*, Research Associates Ltd, Stone, Staffordshire, 1988.
22 James Davis, *Fear, Myth and History: The Ranters and the historians*, Cambridge University Press, Cambridge, 1986.
23 This is documented by M. Baldasarre, *Trouble in Paradise: The suburban transformation of America*, Columbia University Press, New York, 1986.
24 Stanley Cohen and Laurie Taylor, *Escape Attempts*, Allen Lane, London, 1976.
25 Stanley Cohen, *Folk Devils and Moral Panics*, MacGibbon and Kee, London, 1972, p. 9.
26 Chris Philo writes about 'postmodern as object', which supposedly constitutes 'a complex (and seemingly quite chaotic) collision of *all manner of different objects* in the messy "collage" of contemporary people and places' (Paul Cloke, Chris Philo and David Sadler, *Approaching Human Geography*, Paul Chapman, London, 1991, p. 179). I am doubtful about this claim because I see little evidence of a loosening up of state controls. Both the state and capital impose order on societies through the regulation of space, as Mike Davis demonstrates graphically in the case of Los Angeles, or as the British government is attempting to do by restricting the activities of minorities in the countryside (see Chapter 6). The transgressions of folk devils are a symptom of this power. While there are some examples of unfettered expressions of difference, others are a product of exclusionary practices, or difference is recast as deviance in the process of exclusion.
27 Davis, op. cit.
28 *Ibid.*, pp. 104–105.
29 *Ibid.*, p. 106.
30 New Age Travellers are a diverse group of semi-nomadic people, primarily of urban origin and given a common identity by their rejection of mainstream aspirations and a search for autonomy in a rural setting. Richard Lowe and William Shaw, *Travellers: Voices of the New Age nomads*, Fourth Estate, London, 1993, is a useful ethnographic study.
31 In this account, I draw on Simon Watney, *Policing Desire: Pornography, AIDS and the media*, Methuen, London, 1987.
32 *Ibid.*, p. 40.
33 *Ibid.*, pp. 40–41.
34 Cited by Susan Smith, 'Crime and the structure of social relations', *Transactions, Institute of British Geographers*, NS, 9 (4), 1984, 427–442.
35 *Ibid.*
36 David Kunzle, 'World Upside Down: the iconography of a European broadsheet type', in Barbara Babcock (ed.), *The Reversible World: Symbolic inversion in art and society*, Cornell University Press, Ithaca, 1978, pp. 39–94.
37 I am grateful to Philip Jones of the School of Geography, Hull University, for this information.
38 Natalie Davis, 'Women on top: symbolic sexual inversion and political disorder in early modern Europe', in Babcock, op. cit., pp. 147–190.
39 Peter Stallybrass and Allon White, *The Politics and Poetics of Transgression*, Methuen, London, 1986, p. 20.
40 For an analysis, see Peter Jackson, 'Street life: the politics of Carnival', *Environment and Planning D: Society and Space*, 6, 1988, 213–227.
41 I discuss this legislation in relation to 'others' in the English countryside in Chapter 6.
42 Jean-Pierre Liègeois, *Gypsies: An illustrated history*, Al Saqi Books, London, 1986.

'Engendering Race Research: Unsettling the Self–Other Dichotomy'

from Nancy Duncan (ed.), *Body-Space* (1996)

Kay Anderson

Editors' Introduction

Kay Anderson is a cultural geographer based at the University of Durham whose interests include the spatialities of race, an area in which she is widely published, and which is explored in this extract. Originally from Australia she obtained a degree from the University of Adelaide and a Ph.D. from the Department of Geography, University of British Columbia in Canada. She has explored her research interests primarily through three types of contested, negotiated spaces of racial and cultural identity, Chinatowns, zoos and Aboriginal communities in cities. Some of her notable publications include *Vancouver's Chinatown: Racial Discourse in Canada, 1875–1980* (1999), the co-edited volume *Inventing Places: Studies in Cultural Geography* (1992) and contributions to journals such as *Transactions of the Institute of British Geographies*, on the cultural construction of nature at Adelaide Zoo, *Progress in Human Geography*, exploring critical geographies of domesticity and *Environment and Planning D: Society and Space*, on racialised constructions of humanity and animality.

INTRODUCTION

At a moment in feminist theorizing when scholars are grappling with ethnocentric presumptions of a 'generic woman' implicit within 'imperial feminism' (Amos and Parmar 1984), it is timely to note the paucity of attempts to unsettle the epistemology of separation implicit in much race research. The fictionalized collectivities of 'Black', 'White', 'European', 'Asian' and so on – the stock in trade of the field called 'race relations' – are often the corollaries of a dichotomized us/them framework that (unwittingly) obscures the subjectivities of identities internal to those categories. Such a framework also tends to overwrite the interconnections of privileged race positions with other sources of identity and power. Whereas the critique of Western feminism by Black, post-colonial and lesbian writers has challenged feminist consensus (Butler 1990; Collins 1991; hooks 1981, 1991; Larbalestier 1991: Singleton 1989), much race research – including work by anti-colonialists such as Said (1978) and Clifford (1988) – has worked with modernist presumptions of an ordered (racialized) reality whose subject positionings are, for the most part, fixed and undifferentiated (cf. Anthias and Yuval-Davis 1992; Donald and Rattansi 1992).

In this chapter I seek to problematize the polarity of race identities upon which rests the cohering argument of my earlier work *Vancouver's Chinatown* (1991). I aim to undertake such an auto-critique by feeding into the Chinatown story the discursive fields and social positionings of gender and sexuality, a task I undertake not for its own sake, but rather to sharpen the critical analysis of the many valences of social power. By extension, the chapter critiques other work in race relations that implicitly or explicitly disengages race identities from other historically situated oppressions such as those surrounding gender, class and sexuality. Without discrediting work that specifies the contribution that race-based oppression makes to structures of inequality, the chapter seeks to foreground the *multiplicity and mobility* of subject positionings, including those of race and gender.

Such a style of analysis maybe particularly revealing, because while racism has long structured socio-spatial relations in British settler nations such as Canada and Australia, it has been woven into a range of power-differentiated regimes out of which colonial relations have been organized into the present. Certainly the stories that emerge from a re-examination of select moments in the history of White/Chinese relations in Vancouver, Canada, reveal a more 'complex dominator identity' (Plumwood 1993) than a unified White oppressor. The projects of colonialism, themselves manifestly variable from place to place, relied on the imaginative and practical leadership – less of Whites *per se* (as if there exists such an abstract, uncontradictory 'self'; see Bhabha 1990a) than – of a specific 'master subject' who was White, adult, male, heterosexual and bourgeois (Rose 1993). To erase these refinements to the dominator perspective is to risk invoking a falsely tidy dichotomy of relations between *a* racialized 'us' versus *a* racialized 'them', when in reality the social processes constituting social relations were complex and differentiated. The oppressions through which 'colonialism's cultures' (Thomas 1994) were elaborated in Canada's western province of British Columbia had myriad and overlapping sources in the structures of capitalism, patriarchy and cultural domination by race and sexuality. And while pronouncements about the intersections of diverse idioms of oppression are now commonplace in theorizing in the social sciences – with many efforts at formulating multi-dimensional models of class, race and gender oppression (see e.g. Bottomley and de Lepervanche 1984; Bottomley *et al.* 1991; Jennett and Randal 1987) – empirical demonstration of the ways in which oppressions interacted and became mutually confirming is not extensive (although see e.g. Bear 1994; Pettman 1992; Ware 1992). Within the space constraints of a chapter, this is the challenge of the first section of what follows.

Racialized and gendered discourse did not always, however, operate in a fully efficacious complicity in late nineteenth and early twentieth-century British Columbia. To re-tell the story of Vancouver's Chinatown by emphasizing the power of an (albeit more differentiated, that is, European male) centre is to risk reifying further the master-perspective of the dominator identity. Indeed at certain moments in the history of Vancouver's Chinatown, alliances between White women and Chinese men trouble the falsely consensual understanding of domination that arises from one-dimensional race analyses. Thus, if we reorient the story of *Vancouver's Chinatown* around the subjectivity of White women, at least in select moments, we shall see the spaces where are upset images of the monolithic societal racism upon which neat race narratives depend. By foregrounding such spaces, as I do in the second part of this chapter, we begin to 'denaturalize' (Kobayashi and Peake, 1994) the racialized and gendered marking of subjects that has been so central to strategies of domination (see also Jackson 1994).

A parallel oversimplification in the governing logic of 'European hegemony' is the construct of *a* homogeneous racialized category pitted beneath one coherent oppressor. Yet the gender- and class-differentiated experiences and statuses within the category 'Chinese' also defy the essentialized configurations of binary (self/Other) models. By briefly drawing on published sources documenting the uneven experiences of Chinese men and women in early Vancouver, the second part of the chapter also unsettles notions of a stably positioned, internally unified and uniformly oppressed victim. Taken together, the examples support post-structuralist critiques of the 'centred subject' (see e.g. Donaldson 1992; Nicholson 1990) by highlighting the contradictory, multi-dimensional and strategic quality of identities. The examples also go some way to demonstrating how the relationships between dominant and minoritized groups are crossed not only by diverse discursive fields (Lowe 1991), but also by multiple *positionings* that are not reducible to the binary division of 'us' and 'them'.

The potential infinity of the fractual patterns of social relations might appear to paralyse the quest for explanation conceived around a single, controlling point of determination. If we dissect categories too far, we risk losing sight of the structuring threads of power that cohere in empirically specifiable ways. That case has been persuasively argued by 'post-postmodernists' such as Walby (1992). Certainly the argument in what follows should not be taken as a refutation of the force, persistence and profound material effect of that pernicious ideological and material regime that is racism. On the contrary, to specify the intersections of axes of oppression, as this chapter undertakes to do, is not necessarily to disperse or disable the critique of power. It is rather to offer glimpses of two things. First is revealed the often mutually constitutive boundary-making practices out of which colonialism's cultures were constructed in the late nineteenth- and early twentieth-century setting of western Canada. That racism often drew on gendered meanings and positionings in these processes is not to deny racism's strength; rather it is to appreciate the wider discursive network in which racism was inserted. Second is exposed the possibility of rupture of racialized regimes (and the readings they support) when alternative speaking perspectives to that of the elite master-subject are positioned at the centre of analysis. If racialized alliances are at times crosscut by gendered and classed struggles, such as we shall see occurred in Vancouver in the 1930s, and we resist assimilating such struggles into an epistemic regime of race domination, then we glimpse fresh (Chinatown) stories and alternative political alliances and possibilities.

This chapter now turns to a brief résumé of the work which I then propose to critically revisit in the light of these introductory comments.

VANCOUVER'S CHINATOWN (1991): A BRIEF RÉSUMÉ

My intention in writing *Vancouver's Chinatown* was to reconceptualize an enclave which had long been theorized in the social sciences as a colony of the East in the West, an 'ethnic' neighbourhood whose residents and streetscapes existed in natural connection to their Oriental difference and Chineseness. In contrast to that model, I developed an anti-essentialist conceptualization of Chinatown as a construct of Western imagining and practice.

Using the example of Vancouver, Canada, where people of Chinese origin settled in a few blocks of that city's East End from the late 1870s, I traced through time the discursive practices that shaped the definition and management of that district. This, from the time of negative stereotyping between the late 1880s and 1930 when Chinatown was classified by White Vancouver society as a vice district; through the period of the 1930s and 1940s when Chinatown's classification grew more complex and contradictory; that is, when the vice classification came to coexist with the district's first formal tourist definition as Vancouver's 'Little Corner of the Far East'. During the 1950s and early 1960s – the post-war era of modernist urban planning – Chinatown was targeted by Canadian federal and civic administrations as a 'slum' and came close to being completely destroyed by urban renewal and freeway plans. Come the 1970s, Chinatown became classified as an ethnic and heritage district – valued by White Canadians precisely *for* its Chineseness and refurbished in a radically new kind of targeting as an Oriental district with funds from all three levels of Canadian government. In that project, they were joined by Chinatown retailers who, in the rush for the spoils of multiculturalism, manipulated to their own advantage the racialized representations of identity and place that delivered them to White Vancouver society.

Lying beneath the phases of neighbourhood definition, I argued, was the continuity of a racialization process that is the book's structuring narrative, Over the hundred-year period, Chinatown was constructed – both ideologically and materially – out of manifestly variable guises of race classification on the part of those armed with the conceptual and instrumental power to define and regulate the area. The role of the three levels of Canadian state in sponsoring and enforcing that power is highlighted throughout. Chinatown was not just the object of biased depiction and 'prejudice', then, as liberal theses had argued (Ward 1978), but – following Foucault (1979) – of a particular cultural politics of discursive production that enabled one (European) set of truths to acquire the status of truth and normalcy. That this operation entailed a will to dominate ('hegemony') had already been persuasively argued for a different scale and context by Said (1978). Thus, Chinatown, like that mythical region of Western imagining called the Orient, was recurrently White Vancouver's Other, I argued, a place through which a dominant group forged its own cultural understanding of its identity, boundaries, status and privilege. Chinatown was a site through which were articulated diverse narratives of race, health, vice, civility, blight, heritage and ethnic pluralism. Unlike other critiques of orientalism, however, notably Said and the more recent work by Lowe (1991), my interest lay not only in the discursive struggles surrounding identity and Othering strategies, but also the social production of the district, its changing material form and fortunes.

NATION-BUILDING IN COLONIAL CONTEXT: DISPLACING THE NARRATIVE DEVICE OF UNITARY RACE POSITIONINGS

In late nineteenth-century British Columbia, notions of in-group and out-group drew on a complex network of raced and gendered discourses. Later in this section of the chapter, we shall see the interaction of these meanings at the scale of 'place', with specific reference to the discursive construction of Chinatown in early Vancouver. But the processes were equally evident at the scales of 'nation' and 'province'. The making of 'Canada' in its symbolic dimension entailed representational practices that were deeply saturated with race and gender concepts, and by highlighting their collusion, we further refine the identities and subjectivities out of which a dominant imagined geography of nation grew (see also Bhabba 1990a).

That Canadian officials of the late nineteenth century were seeking to create a White Canada was abundantly clear in the languages and debates recorded in the government texts that were the primary data sources for *Vancouver's Chinatown* (Anderson 1991). Federal legislation in the form of a head tax – passed in 1885 to contain Chinese immigration – thus sought to limit family settlement, while permitting a controlled amount of Chinese labour and capital. Bound up with the impulse to contain Chinese numerically was also a desire to prevent what was called 'miscegenation' or 'mixture of races'. To that end, the ultimate target of legislation appears to have been less the 'Chinese' as a collective racialized category, than the more narrow category of Chinese women. Moralities of race *and* gender fed an interactive discursive network. In the words of the Royal Commissioner for Chinese Immigration, John Chapleau, when arguing the case to the House of Commons for a head tax in 1885: 'If they came with their women they would come to settle and what with immigration and their extraordinary fecundity, would soon overrun the country' (Canada 1885b: 98). Canada's Prime Minister of the time, John MacDonald, held the same opinion: 'If wives are allowed, not a single immigrant would come without a wife, and the immorality existing to a very great extent along the Pacific Coast would be greatly aggravated' (Canada 1887: 643). Not just idle polemic, such views shaped policies that imposed constraints on Chinese family life in Canada and China well into the twentieth century (see Li 1988: Ch. 4). (By 1938, the Vancouver press reported that the 'ultimate solution' to the Chinese problem had been found in the severe sex imbalance in the local Chinese population which was constraining its ability to replenish itself, see *Province* 28 February 1938.)

Embedded in the projections of officials such as Chapleau and MacDonald was a particular construction of Chinese women, as wives and ipso facto reproductive beings who threatened the demographic strength and integrity of White Canada. Chinese women were never seen as single immigrants with the potential for waged (or unwaged) labour, yet we know that such women undertook a range of jobs within the enclave economy of Vancouver's Chinatown in the late nineteenth and early twentieth centuries (see Adilman 1984). And if Chinese women were seen by Canadian officials as anything other than 'fecund', they were cast as prostitutes – a still more ominous identity according to John Chapleau, in 1885, because 'they bring with them a most virulent form of syphilis and in a special way corrupt little boys' (Canada 1885b: xii). Such pronouncements filtered into media texts across the country to feed the image of a disease-bearing race with whom sexual liaisons would be ill-advised.

A sharpened gender awareness of the discursive processes at work in Canadian nation-building brings into view not just Chinese women immigrants, however, caught between While cultural superiority and male power. Running through the official constructions of Chinese immigration – and especially during the lead up to the Chinese Exclusion Act of 1923 (see Anderson 1991: 132–41) – was a racialized *and* gendered aesthetic that interconnected the spaces of nation and body. 'Canada' appears scripted in official texts as a pure space, one that if impregnated by the flow of alien material would become contaminated and offer up inferior 'stock'. In Figure 12.1 [not reproduced here], for example, taken from the Vancouver press during the anti-Chinese riot of 1907, we see how the iconic body of White woman was grafted onto the space of nation (and province), The rhetorical device within this system of representation was to symbolically construct the nation as passive and pure by affording it the attributes of an Anglo female body. Like other symbolic figurations of nation such as Miss Britannia and Lady Liberty (see Pateman 1988; Yuval-Davis 1991), Miss BC ismade to emblematically stand as an essence under peril of violation. Thus invested with agency is the wilful guardian of nation, Prime Minister Wilfrid Laurier, whose (masculinized) charge and call to action becomes no less than the heroic rescue of the imperilled British Columbia.

The 'captivity narrative' (Schaffer 1991) at work here – of defenceless, feminized space whose boundaries require protection from 'the commingling of blood' (Canada 1922: 1518, 1522, 1524, 1529) – structured the discursive terrain of province and nation-building. It created and appropriated the bounded notions of 'British Columbia' and (writ large) of 'Canada', drawing on specific codes of race, femininity and masculinity within Canadian culture. These were the constitutive cultural and political resources whose interaction we need to glimpse in order to appreciate how the 'ideological work' (Poovey 1989) of nation-building actually got done. That those codes were powerful and persistent is doubtless also the case, as witness the 'race hygiene' debates of the 1930s (see Anderson 1991: Ch. 4) when there was very explicit concern in Canadian policy circles about the 'mongrelization' of White purity by Chinese 'penetration' (see also Stepan 1991). It follows that we need to resist any reductive impulses to distil a governing logic of binary (race or gender) relations to more fully understand the irreducibly *broad* social projects entailed in the construction of colonialism's cultures.

SEX, VICE AND CHINATOWN: RACE, GENDER AND PLACE

Vancouver's texted place called 'Chinatown' was crafted out of a repertoire of images whose racialized *and* gendered content enhanced their cultural appeal and political effect. In this section I would like to elaborate some of the gender-silences in my earlier race analysis with reference to the field of knowledge surrounding Chinatown in the period between approximately 1880 to 1930, During that time period, it is evident that the concept of 'Chinatown' harnessed racialized images to its service that were already deeply gendered (see Ware 1992) and which drew on more than a more narrowly conceived Orientalist field of knowledge. By revisiting a series of illustrations from the Vancouver press earlier this century, I hope to expand the interpretive grid I cast on such historical materials and illuminate the discursive network within which operated race, gender, and sexuality languages and practices.

I previously argued that, of all the things that might have been said by early White Vancouver residents about Chinatown, only those aspects that fitted the racial categorization became filtered into the neighbourhood construct. In the early twentieth-century cartoons from Vancouver's *Saturday Sunset*, we see the nature of the material out of which Chinatown was ideologically constructed. Local knowledge drew upon the presumed proclivities for opium, gambling, sexual exploitation and overcrowding of that abstracted figure 'John Chinaman' to produce interlocking registrations of vice, mystery, danger and disease. For those seeking to render the place and its people eternally alien, a label into which could be assimilated all the things that Europeans sought to deny or repress in themselves, served a persistently useful function. Certainly the label became more than a package of (derogatory) meanings, words and texts. As I demonstrated in my earlier work, the Chinatown concept triggered and justified harassment campaigns for many decades as part of the state's management of ethnic pluralism.

One can go further, however, than arguing that European conceptions of 'a Chinese race' was the primary modality governing law enforcement practices in Chinatown, There seems also to have been constructed around Chinatown a deeply gendered 'moral panic' (Cohen 1972) that served to legitimize not just White Canadian intervention but a historically specific form of masculinity and moral guardianship. In the 1908 illustration 'Vancouver Must Keep this Team', civic guardians Chief of Police Chamberlain and Magistrate Alexander – icons of the law enforcement arms of the state – are clearing their heroic path through Vice-town. It is a site where 'difference' is being scripted as 'danger' to culturally dominant norms: a place that the 'axe man', Deputy Chief of Police McLennan, sought tirelessly to 'tame' (*Province* 16 July 1913). And that the norms these men felt moved to police were themselves crafted out of gendered (and heterosexist) material is plainly evident in the 1907 cartoon 'The Unanswerable Argument'. The ideal of Canadian, suburban, civilized, family life is here set up in opposition to, and at risk from, the pathologized modes of living on Carrall Street.

Although the competition presented to the White working class by cheap Chinese labour was a persistent theme within early Vancouver's colonial discourse, it was the scenario of sexual liaisons between Chinese men and White women which was seen as the most threatening violation of all. Indeed nothing served to congeal stereotypical knowledge about Chinatown more securely than the emblematic activity of John

Chinaman's predation. So while in certain instances, Chinese could be cast as 'a feminine race', to use the words of Royal Commissioner Dr Justice Gray in 1885 – 'docile' and well suited to the menial labour of railway construction (Canada 1885b: 69) – they could, to serve other purposes, be masculinized and construed as energetic pursuers of White women (see also Back 1994). The contradictions within the languages that constructed Chinese as alien were complex and do themselves point to the irreducible diversity of classed, raced, gendered and sexualized resources upon which colonial discourses drew.

There were other double standards. Whereas sexual relations between Chinese women and White men were rarely discussed (except to fuel alarm about the transmission of syphilis), the possibility of sexual relations between Chinese men and White women was deeply troubling (if also, perhaps, titillating) and supplied much discursive material for Chinatown's image-making. The 'moral blight' that was Chinatown certainly set a pressing agenda for Christian missions in early Vancouver. In the 1908 cartoon 'The Foreign Mission Field in Vancouver', Chinatown's opium dens are constructed as the natural habitat of the lascivious John. Inside them, White women – passive at the hands of the inscrutable Oriental – are induced to commit 'amoralities', in the words of many a civic official. As occurred in other 'inter-racial' settings, the scenario is made to stand as the most profound of violations, and it worked in a few ways: first by contacting the generalized fear, beginning at the point of immigration, that 'Canada' faced a threat from close contact with outsiders. Proximity of 'races' within the private sphere could thus also be construed as 'perilous' by image-makers such as Attorney-General Mason and the press, both of which played wickedly on the notoriety of 'Vice-town' in 1924 when a Chinese domestic ('China boy') allegedly murdered his employer, Janet Smith, of the high-income district of Shaughnessy (see *Province* 13, 28 Nov., 5 Dec. 1924; also Lee 1990: 65–9). This was one of a few occasions when alleged murders of White, wealthy women by Chinese servants was used to transform difference into 'danger' and to justify the enforcement of boundaries between Chinese men and White women (see also the controversy surrounding the death of Mrs Millard of the West End in 1914 in Anderson 1991: 116).

Second, the fantasies and anxieties surrounding sexual relations between Chinese men and White women fed into other cultural discourses. Within such discursive 'fellowships' (Foucault 1972) circulated languages of: presumptive heterosexuality (according to which White women were the exclusive preserve of White men); racialized manhood (such that the 'bestial negro' and 'wily Oriental' could be rendered lustful primitives); White femininity as an innocent and vulnerable essence; and of women as Othered objects, servicers of male bodily needs and desires. Like their insatiable pursuers, White women were also closer to nature than the rational, controlled, White male. Small wonder, then, that sex between Chinese men and White women became inscribed as the ultimate moral and political transgression, Not only did it compromise racial boundaries, it threatened White, male property. Safeguarding the virtue of White women thus became dignified as a prerogative, and was very often the pretext that law enforcers used in targeting Chinatown. In so doing, White men exercised not just their sense of race and gender supremacy, but also their power of definition over the criteria of normalcy in sexual conduct.

The sexual politics at work in civic missions to Chinatown trouble the binary frame of race analyses that assume an essential opposition of interests between 'Whites' and 'Chinese'. For one thing, the representations and practices surrounding Chinatown originated in social relations that included gender and sexual orientation, suggesting there was nothing unitary in the position of racial privilege (or, as I shall later argue, of racial subordination). White women's inferior positioning relative to White men, together with the privileging of heterosexist masculinity in Canadian culture, informed the very moralities that grew up around the 'race' question. Indeed they compromise any narrative characterizations of social relations that might collapse gender, sexuality and class arrangements into a larger governing conception of 'race' domination. Nor was it a case of a simple layering of race with gender meanings (as might be implied by linear and mechanistic 'additive models' of oppression: see Sacks 1989; Spelman 1988). The point underlined by the substantive discussion is that racist knowledges had gender and moral codings relating to family, sexuality, marriage and residence embedded within them, just as discourses surrounding gender, sex, citizenship and family life relied on race meanings for their cultural integrity. It follows that race identities cannot be decontextualized and separated off analytically or politically from the constitution of other identities and axes of power. Each division is practised in the rhetorical and interactive context of others. The representational practices surrounding Chinatown thus bring into view the insinuation through each other of the multiple hierarchies that underwrote early Vancouver society.

CHINATOWN REORIENTED: ALTERNATIVE SUBJECT POSITIONINGS AND STORIES

The contribution of racialized and gendered discourses to ethnic relations in early Vancouver might seem to have been so decisive as to support readings of colonialism as a pervasively efficacious venture. If the making of a *British* British Columbia was a relatively influential project, however, it was not a unitary one evolving from a singular source of ethnic superiority. We have seen that its sources were multiple and differentiated. It is also the case that, for all of colonialism's power at certain times and in particular places, the management of ethnic difference was no neat process of imposition. Rather it entailed struggle, and was often fragmented and frustrated by debate, contradiction and resistance by those it subordinated.

Analysis of colonialism's operation can also be constructed from a range of vantage points. Indeed the perspective of the elite White master-subject is possibly only legitimized by accounts, such as appear in the first part of this chapter, of the simultaneity of race, gender and heterosexist oppressions in colonialism's extension to British Columbia. Thus while it is important to continue to Illuminate the differentiated sources and forms of power under colonialism, it is also helpful to puncture the binding grip of a (master-) story of the inter-ethnic encounter – a grim tale, that is, of inexhaustibly coherent control on the part of a privileged Anglo group, In what follows, I seek in a most preliminary way, to diversify the Chinatown problematic by opening up the story-field to alternative subject and speaking positionings.

Although there can be little doubt that White Vancouver women were often complicit in the practices that marginalized Chinese in that city's early history, there were moments when White women broke ranks with White men and formed alliances that undercut the stable fixings of racialized boundaries. A more nomadic style of story-telling to the linear mode of *Vancouver's Chinatown* (1991) – one that weaves narrative threads through scattered moments – illuminates such apparent ruptures in the Orientalist logic of Occident versus Orient. To that end, vignettes that I earlier framed in race terms (see Anderson 1990; 1991: 116–20, 158–64), can be recast to disrupt logics of race complementarity and to highlight the possibility of political alliances that cut across racialized identities.

In 1920, the Vancouver city council decided to impose a hefty $100 licensing fee on vegetable peddling, a trade almost wholly dominated by Chinese in early Vancouver. The move on the part of council was undertaken out of support for the powerful Retail Merchant's Association which had grown concerned about the inroads being made into its business by the itinerant vendors, The pedlars – disinclined to accept the fee – decided to enlist the support of their clients as well as the Chinese ambassador. Interestingly, on this occasion, some 5,000 Vancouver women were more interested in avoiding long shopping trips to the city market than endorsing the vendetta of the Retail Merchant's Association. The women signed the petition in support of their Chinese produce suppliers and against the White male retailers who, in race readings of social relations, are cast as their compulsory partners.

The incident wasn't the only occasion when there were alliances *across* racialized boundaries that are written out by a logic of binary opposition. There were other moments of vulnerability in dominant discourses surrounding Chinatown. Fifteen years later, in 1935, Chief Constable W. Foster found cause to implement British Columbia's Act for the Protection of Women and Girls, specifically in Vancouver's Chinatown. That Act had been implemented back in 1919 out of fear for women's 'moral safety' in Chinese restaurants throughout British Columbia. In Vancouver, Foster argued that contact between Chinese men and White women was being set up inside restaurant booths and that after working hours, women would go to Chinese quarters where 'they were induced to prostitute themselves and immorality would take place' (cited in Anderson 1991; Ch. 5). This was no trivial matter for the retired colonel, and between 1935 and 1937, Foster and his 'moral reform squad' comprising Mayor M. Miller, License Inspector H. Urquart and City Prosecutor O. Orr set about banning White waitresses from Chinatown cafes by cancelling the licenses of businesses employing them.

The vendetta against Chinatown cafes met with angry resistance from Chinese owners of the restaurants, including the powerful and wealthy president of the Chinese Benevolent Association, Charlie Ting.

Perhaps the most revealing challenge, however, came from the women themselves, who in 1937 marched to City Hall to protest their dismissal by, in the words of one woman, 'the self-appointed directors of the morals of women in Chinatown'. The women were quite prepared to articulate the specificity of their experience as workers and defend their right to choose their employers and place of work. Certainly the women defied the image of the passive object of desire that we have seen was so useful in dignifying earlier male missions in Chinatown. One waitress, Kay Martin, told the press she would 'much prefer working for a Chinese employer than for other nationalities'. Another stated that 'if a girl is inclined to go wrong she can do it just as readily uptown as she can down here'. Another noted 'our bosses are honourable men who know that we must live'.

If it was the adversity of living conditions during the Depression that brought about the womens' defence of their employers, the action nonetheless upsets readings of relentless race polarization. Such interpretations risk flattening the experiences of White women as subjects, deducing them from the (putatively) immobile and deterministic position of race power. Yet as we have seen, 'Whites' were not always and necessarily fated to dominate. The apparent coherence of racism gives way before such evidence, which, while necessarily brief here, highlights the mutable configurations, crosscutting constituencies and contingent authorities out of which social relations are made.

Similarly unsettling of dualistic race readings are the distinctions of class, gender, ethnicity, generation, language and so on that pluralized Chinatown as deeply as they did White Vancouver. If the likes of Charlie Ting became, for a time, an ally of White women in Vancouver, those of his class may have been less 'honourable' in the eyes of the Chinese rank and file workers of the enclave economy in Vancouver's Chinatown. Many scholars have identified a socio-economic pyramid in the district, at the apex of which stood a tiny minority of men of capital who were some of the wealthiest individuals in early Vancouver (Wickberg *et al.* 1982; Yee 1988). The liability implied by the racial category 'Chinese' may well have been the asset of certain merchants who in Chinatown had a vulnerable and captive labour force at their disposal. This bloc of workers, unprotected by White unions, often laboured under punitive contracts for their Chinese bosses. There were also many unpaid workers, including women, who worked long hours sewing buttonholes and doing much of the handwork for Chinese tailors (see Adilman 1984). The experience of those workers was shaped by their subordinate status in an array of dualities, not least of gender. The women prefigured today's sweatshop workers whose notoriously exploited labour in other North American Chinatowns, such as New York City, tells of ongoing class and gender antagonisms which have only recently prompted agitation for reform on the part of Chinese women workers (Kwong, forthcoming).

Such gender-differentiated relations within the racialized category are erased by characterizations of the universally subjugated 'Chinese'. Not only do they suggest different racisms for different groups of 'Chinese' (see Satzewich 1989), they also lead us to consider the possibility that other oppressions – quite apart from the relation that places Whites in a deterministically antagonistic relationship to Chinese – might have been as decisive in shaping their everyday experience. Not all the realities and aspirations of the lives of Chinatown's residents would have been exhausted by the fact of racial subordination, as the growing number of fictionalized accounts by Chinese Canadian writers of life in Vancouver are beginning to reveal (Chong 1994; Lee 1990).

CONCLUSION

This chapter has sought to confront the tensions raised for analysis by the intersection of axes of socio-spatial inequality. I have sought to undo the privileging of *racialized* positionings – European versus Chinese – by foregrounding the gendered meanings and practices that at times reinforced, and at other times, disrupted those categorizations and relations. In the poststructuralist spirit of a more 'distrustful analysis' (Bottomley 1991: 108) that eschews the search for unitary subjects and singular explanatory frameworks, I have attempted to decentre the authorial paradigm of 'European hegemony' charted in my own *Vancouver's Chinatown* (1991). Without discrediting the case for racism's force, malleability and resilience in Canadian culture, or indulging a naive postmodern embrace of endlessly infinite identities, I have here tried to demonstrate that the cultural field is created and fractured by a range of social relations and subjectivities whose mappings invite what Pratt (1993) has called a 'restless storytelling'. By emphasizing the different centres of cultural authority surrounding race, gender and sexuality, and the invariant political alliances surrounding those idioms of identity and power, the chapter has sought to disrupt modernist notions of undifferentiated subjects, root causes and fixed trajectories. It follows that the Chinatown story-field might effectively be further opened up to re-tellings from the vantage points of the district's residents, themselves multiply and fluidly positioned in relation to each other and the wider society.

REFERENCES

Adilman, T. (1984) 'A preliminary sketch of Chinese women and work in British Columbia 1858–1950', in B. Latham, R. Latham and R. Pazdro (eds) *Not Just Pin Money*, pp. 53–78, Victoria: Camusun College.

Amos, V. and Parmar, P. (1984) 'Challenging imperial feminisms', *Feminist Review* 32.

Anderson, K. (1990) 'Chinatown re-oriented: a critical analysis of recent redevelopment schemes in a Sydney and Melbourne enclave', *Australian Geographical Studies* 28(2): 137–54.

—— (1991) *Vancouver's Chinatown: Racial Discourse in Canada, 1875–1980*, Montreal: McGill-Queens University Press.

Anthias, F. and Yuval-Davis, N. (1992) *Race, Nation, Gender, Colour and Class and the Anti-racist Struggle*, London: Routledge.

Back, L. (1994) 'The "white negro" revisited: race and masculinities in South London', in A. Cornwall and N. Lindisfarne (eds) *Dislocating Masculinities*, pp. 172–83, London: Routledge.

Bear, L. (1994) 'Miscegenations of modernity: constructing European respectability and race in the Indian railway colony, 1857–1931', *Women's History Review* 3(4): 531–48.

Bhabha, H. (1990a) 'The other question: difference, discrimination and the discourse of colonialism', in R. Ferguson, M. Gever, T. Minh-ha and C. West (eds) *Out There: Marginalisation and Contemporary Cultures*, pp. 71–88, New York: New Museum of Contemporary Art and Massachusetts Institute of Technology.

—— (ed.) (1990b) *Nation and Narration*, London: Routledge.

Bottomley, G. (1991) 'Representing the "second generation": subjects, objects and ways of knowing', in G. Bottomley, M. de Lepervanche and J. Martin (eds) *Intersexions: Gender/Class/Culture/Ethnicity*, pp. 92–110. Sydney: Allen & Unwin.

Bottomley, G., and de Lepervanche, M. (1984) *Ethnicity, Class, and Gender in Australia*, Sydney: Allen & Unwin.

Bottomley, G., de Lepervanche, M. and Martin, J. (eds) (1991) *Intersexions: Gender/class/Culture/Ethnicity*, Sydney: Allen & Unwin.

Butler, J. (1990) *Gender Trouble: Feminism and the Subversion of Identity*, New York: Routledge.

Canada (1885a) *Debates of the House of Commons*.

—— (1885b) *Sessional Papers*, No. 54a, Royal Commission on Chinese Immigration.

—— (1887) *Debates of the House of Commons*, 31 May.

—— (1922) *Debates of the House of Commons*, 8 May.

Chong, D. (1994) *The Concubine's Children: Portrait of a Family Divided*, Harmondsworth: Penguin.

Clifford, J. (1988) *The Predicament of Culture*, Cambridge, MA: Harvard University Press.

Cohen, S. (1972) *Folk Devils and Moral Panics*, London: MacGibbon & Kee.

Collins, P. (1991) 'Learning from the outsider within: the sociological significance of black feminist thought', In J. Hartman and B. Messer-Davidow (eds) *(En)gendering Knowledge: Feminists in the Academe*, pp. 40–65, Knoxville: University of Tennessee Press.

Donald, J. and Rattansi, A. (eds) (1992) *'Race', culture and difference*, Milton Keynes and London: The Open University/Sage.

Donaldson, L. (1992) *Decolonizing Feminisms: Race, Gender and Empire-building*, Chapel Hill and London: The University of North Carolina Press.

Foucault, M. (1972) *The Archaeology of Knowledge*, transl. by A.M. Sheridan Smith. New York: Harper & Row.

—— (1979) *Discipline and Punish*. New York: Viking.

hooks, b. (1981) *Ain't I a Woman? Black Women and Politics*, London: South End Press.

—— (1991) *Yearning: Race, Gender and Cultural Politics*, London: Turnaround.

Jackson, P. (1994) 'Black male: advertising and the cultural politics of masculinity', *Gender, Place and Culture* 1(1): 49–60.

Jennett, C. and Randal, S. (eds) (1987) *Three Worlds of Inequality: Race, Class and Gender*, South Melbourne: Macmillan.

Kobayashi, A. and Peake, L. (1994) 'Unnatural discourse: "race" and gender in geography', *Gender, Place and Culture* 1(2): 225–43.

Kwong, P. (forthcoming) 'Back to basics: politics of organizing Chinese women garment workers', *Social Policy*.

Larbalestier, J. (1991) 'Through their own eyes: an interpretation of Aboriginal women's writing', in G. Bottomley, M. de Lepervanche and J. Martin (eds) *Intersexions: Gender/Class/Culture/Ethnicity*, pp. 75–91, Sydney: Allen & Unwin.

Lee, S. (1990) *Disappearing Moon Cafe*, Vancouver: Douglas & McIntyre.

Li, P. (1988) *The Chinese in Canada*, Toronto: Oxford University Press.

Lowe, J. (1991) *Critical Terrains: French and British Orientalisms*, Ithaca and London: Cornell University Press.

Nicholson, L. (ed.) (1990) *Feminism/Postmodernism*, New York: Routledge.

Pateman, C. (1988) *The Sexual Contract*, Cambridge: Polity Press.

Pettman, J. (1992) *Living in the Margins: Racism, Sexism and Feminism in Australia*, Sydney: Allen & Unwin.

Plumwood, V. (1993) *Feminism and the Mastery of Nature*, London: Routledge.

Poovey, M. *Uneven Developments: the Ideological Work of Gender in Modern Victorian England*, Chicago: Chicago University Press, 1989.

Pratt, G. (1993) 'Reflections on feminist empirics', *Antipode* 25(1): 51–63.

Rose, G. (1993) *Feminism and Geography: The Limits of Geographical Knowledge*, Cambridge: Polity Press.

Sacks, K. (1989) 'Towards a unified theory of class, race and gender', *American Ethnologist* 16(3): 534–50.

Said, E. (1978) *Orientalism*, New York: Random House.

Satzewich, V. (1989) 'Racisms: the reactions to Chinese migrants in Canada at the turn of the century', *International Sociology* 4(3): 311–27.

Schaffer, K. (1995) 'The Elisa Fraser story and constructions of gender, race and class in Australian culture', *Hecate* 17(1): 136–49.

Singleton, C. (1989) 'Race and gender in feminist theory', *Sage* 6(1): 12–17.

Spelman, E. (1978) *Inessential Women: Problems of Exclusion in Feminist Thought*, Boston: Beacon Press, 1988.

Stepan, N. (1991) *The Hour of Eugenics: Race, Gender and Nation in Latin America*, Ithaca and London: Cornell University Press.

Thomas, N. (1994) *Colonialism's Culture: Anthropology, Travel and Government*, Melbourne: Melbourne University Press.

Walby, S. (1992) 'Post-post modernism? Theorizing social complexity', in M. Barrett and A. Phillips (eds) *Destabilizing Theory: Contemporary Feminist Debates*, pp. 31–52 Cambridge: Polity Press.

Ware, V. (1992) *Beyond the Pale: White Women, Racism and History*, London: Verso.

Ward, P. (1978) *White Canada Forever: Popular Attitudes and Public Policy Toward Orientals in British Columbia*, Montreal: McGill-Queens University Press.

Wickberg, E., Con, H., Johnson, G. and Willmott, W. (1982) *From China to Canada: A History of the Chinese Communities in Canada*, Toronto: McClelland & Stewart.

Yee, P. (1988) *Saltwater City*, Vancouver: Douglas & McIntyre.

Yuval-Davis, N. (1991) 'The citizenship debate: women, the state and ethnic processes', *Feminist Review* 39: 58–68.

'A Chance to Live'

from *Thinking the Difference: For a Peaceful Revolution* (1994)

Luce Irigaray

Editors' Introduction

Luce Irigaray was a feminist activist and academic, born in Belgium in the early 1930s. A close associate of Jacques Lacan, she became a psychoanalyst and a fierce critic of the phallocentric economy, including that of higher education. After an early academic career that included a Master's degree from the University of Louvain, teaching in a Belgium high school and a Master's degree in psychology from the University of Paris and two doctorate theses, she taught at the University of Vincennes until the cessation of her employment in the early 1970s. Since then she has written widely on feminist issues and been an important figure in the European women's movement. Her work has consistently explored the linguistic and psychoanalytical aspects of gender.

Her major publications include *Speculum: Of the Other Woman* (1985), *An Ethics of Sexual Difference* (1993), *Sexes and Genealogies* (1993). *Thinking the Difference: For a Peaceful Revolution* (1994), from which the extract included here is taken, explores the way in which women have been failed by institutions and cultural forms. Irigaray's book is distinctive, however, in that, rather than simply offering a critique, she proposes changes that she argues would rectify the failures that women have had to endure.

My first examples concern the mythological, religious and symbolic foundations of our contemporary cultural and social order.

- In all public, civil and religious arenas, it is always the *man's* father or mother that counts.
- In what are called matrilineal societies, the power at issue is often associated with *male* filiation on the mother's side: it is the mother's brother who is responsible and valued socially, thus the *son*, not the daughter, and this son ruptures the cultural relationship between mother and daughter.
- According to anthropologists, the taboo on mother–son or sister–brother incest is the basis of our sociocultural order.
- Father–son and mother–son relationships dominate our religious models. While the father–son relationship is supposed to be closer to perfection, to Christians the mother–son couple is the couple that incarnates God; it is represented at almost all religious sites, and mentioned in all Christian services.
- According to Freud, the mother–son relationship is the perfect model of desire, and love between a woman and a man is possible only if the woman has become the mother of a son and she transfers to her husband what she feels for her boy-child.

All this partakes of the same sociocultural models. But very few mythologists have explained the origins, the qualities and functions, the occasions and causes of disappearance of the great mother–daughter couples of mythology: Demeter–Kōrē, Clytemnestra–Iphigeneia and Jocasta–Antigone, to name only the best-known Greek figures.

To anyone who cares about social justice today, I suggest putting up posters in all public places with beautiful pictures representing the mother–daughter couple – the couple that illustrates a very special relationship to nature and culture. Such representations are missing from all civil and religious sites. This is a cultural injustice that is easy to remedy. There will be no wars, no dead, no wounded. This can be done before any reform of language, which will be a much longer process. This cultural restitution will begin to redress women's individual and collective loss of identity. It will cure them of some ills, including distress, but also rivalry, and destructive aggressiveness. It will help them move from the private sphere to the public, from the family to the society in which they live. The mother–daughter couple is always erased, even in places where a mother–daughter couple is honoured. Thus, the phenomenon of Lourdes – an event that attracts millions of pilgrims and tourists, involves many public gatherings and makes a great deal of money – concerns the relationship of a daughter to a so-called divine mother. Usually, however, the mother is represented without her daughter, most notably in churches, but also on street corners, and it is men who organize the worship, thus intervening in the relationship. But maybe this event commemorates the mother–daughter couple that was so important in the time before our patriarchal culture. Maybe – who knows? – it is a sign of things to come. In any case, it leaves no one indifferent.

We must not forget that in the time of women's law, the divine and the human were not separate. That means that religion was not a distinct domain. What was human was divine and became divine. Moreover, the divine was always related to nature. 'Supernatural' mother–daughter encounters took place in nature. Reintroduced into established religions, they are rarely interpreted from this none the less highly traditional perspective of women's religion. Why not? Subjected to the patriarchal churches for centuries, women have become sick of religion without giving any thought to their own divine traditions. Patriarchy has separated the human from the divine, but has also deprived women of their own gods or divinity(ies). Before patriarchy, women and men were potentially divine beings, which may mean social beings. In most traditions, all social organizations are chiefly religious. Religion gives the group cohesion. In a patriarchal regime, religion is expressed through rites of *sacrifice* or *atonement*. In women's history, religion is entangled with cultivation of the earth, of the body, of life, of peace. Religion became the opium of the people because it took over as the religion of mankind alone. It is actually a facet of the organization of society. But the deification, here and now, of sexual bodies is a different story: this exists in societies where women are not excluded from the organization of the culture. In India, for example, and at the beginning of our Greek culture – for to some extent this era still exists in India – sexuality was cultural, sacred. It was also an important source of energy for men and women. Patriarchy stripped women of divinity, taking it over in the places where men are amongst themselves, and often suspecting women's religion of devilry.

But few scholars or theologians have given any thought to the relationships of mother–daughter couples to fertility and respect for nature. After a certain era, women who were close to nature were called witches, practitioners of magic, whereas at the beginning of our History the mother–daughter couple was a positive representation of the site of worship of the body and the natural elements. Magic, holocausts and sacrificial or propitiatory rites did not come into play until after this relationship with nature was broken off – the only universal that could be both immediate and mediated with neither obscurity nor occultism.

Male religion masks an appropriation that interrupts the relationship with the natural universe and perverts its simplicity. It represents a social universe organized by men, but this organization is based on a sacrifice: the sacrifice of nature and the sexual body, particularly woman's. It imposes a spirituality cut off from its natural roots and environment and therefore cannot fulfil humanity. Spiritualization, socialization and cultivation must start with what is. Patriarchal regimes do not do so, because they seek to obliterate the means that they use to take control: (a) a wresting of power from the domain of the other sex and (b) excessive privilege for the family over the sexual couple.

Putting up images – photographs, paintings, sculptures, etc., not advertisements – of mother–daughter couples in all public places today would show respect for the social order. The social order is not made up of mothers and *sons*, as patriarchal culture represents it, with its own virginal ideals that it often assimilates to money, with its reproductive issues, with its incestuous games, and with its reduction of love to natural fertility, the release of social entropy, and so on.

Women's inability to organize themselves and agree on what they want makes some people smile and discourages others. But how could they unite when they have no representation, no example, of such an alliance? There was not always such a deficiency. At one time mother and daughter formed a natural and social model. The mother–daughter couple was the guardian of the fertility of nature in general, and of the relationship with the divine. In that era, food consisted of the fruits of the earth. The mother–daughter couple safeguarded human food and the site of oracular speech. This couple preserved the memory of the past, and thus the daughter respected her mother, her ancestry. This couple was also concerned with the present: the earth produced food in peace and quiet. It was possible to foresee the future thanks to women's relationship with the divine, with oracular speech.

And were men harmed by this organizational structure? No. In respecting life, love and nature this way, neither sex was destroyed by the other. The two sexes loved each other without the institution of marriage, with no obligation to bear children – which never put an end to procreation – and with no censorship of sex or the body.

That is probably what monotheistic religions are telling us in the myth of earthly paradise, a myth that corresponds to centuries of History now called Prehistory, the primitive era, etc. The people of these so-called archaic times were perhaps more cultured than we are now. Traces of their artistry still remain: temples, sculptures and paintings, but also myths and tragedies, especially as expressions of the transition to the so-called historic era. The closest that this era came to our own time was the start of the Golden Age of Greece.

The beginnings of patriarchal power as we know it – which means the power of the man as the legal head of the family, tribe, people, state and so on – coincided with the separation of women from each other and especially the separation of daughters from their mothers. The mother–daughter relationship – the most fertile from the point of view of preserving life in peace – was destroyed to establish an order tied to private property, to the handing down of property within the male line of descent, to the institution of monogamous marriage so that property, including children, belong to this line of descent, and to the establishment of men-only social organizations for the same purpose.

'New Urbanism and its Discontents'

from Joan Copjec and Michael Sorkin (eds), *Giving Ground: The Politics of Propinquity* (1999)

Dean MacCannell

Editors' Introduction

Dean MacCannell has employed the techniques of semiotics and anthropology to understand the twentieth century phenomena of mass tourism. His writings in this field have established him as a major figure in the study of tourism. His long-standing interests in the cultures, communities and contradictions of mass, global tourism have prefigured and influenced recent debates around consumer cultures, authenticity and identity by many, from sociology, geography and cultural studies, who have explored the contours of postmodern society. His texts on tourism and the tourist have become seminal readings on the subject. MacCannell's particularly erudite critiques have exposed the 'cannibalism' of mass tourism in which cultures and artefacts are commodified for the tourist gaze and purse. In this chapter MacCannell broadens his focus while remaining concerned with the contradictions between consumerism and social justice. He asks the question: 'How is it possible to create middle-class neighbourhoods supposedly founded on principles of ecological sustainability, civic involvement and so on, and at the same time neutralize in advance any convergence of people's interests which might frustrate the corporate definition of citizen as pure consumer?'

Dean MacCannell obtained a Ph.D. in development sociology at Cornell University in 1968 and is currently Professor of Applied Behavioural Sciences and Sociology at the University of California at Davis. Amongst his publications the most widely known and influential are: *The Tourist: A New Theory of the Leisure Class* (1976), *Empty Meeting Grounds: The Tourist Papers* (1992).

> The Other that we experience through [religion] is omnivalent. It is precisely what is called, in Christianity, the neighbor. It is a way to nullify extimacy; it grounds what is common, what conforms, conformity. It belongs fundamentally, as universal, to this conformity.
>
> (Jacques-Alain Miller, *Extimité*)

Big Capital is currently in the process of pushing the United States' middle class closer together in high-density suburban developments and urban infill designed to resemble nineteenth-century towns. This movement is called the 'New Urbanism' or sometimes 'neotraditionalism'. The ideas for this kind of community planning have been around for a long time on the intellectual fringes in the ecological movement. Only in the last ten years has it captured mainstream attention.

There is more happening here than just a move uptown of a part of Environmentalism. For a start, an entire new 'class solidarity' is being marketed along with the physical amenities of 'New Urban' developments: 'shared values', a 'renewed sense of community', 'neighborliness', 'co-operation', 'closeness', and 'harmony'.[1] The first phase of the experiment has already been bought by eager new home owners at Seaside in Florida, Harbor Town in Memphis, Haymount and Carlyle in Virginia, Battery Park City in New York, Southport near Sacramento, and The Crossings in Mountain View, California. The boldest expressions of New Urbanism are Disney Development's town of Celebration in Florida and Steven Spielberg's Dreamworks town of Playa Vista in Los Angeles.

The New Urbanism can be seen – and this is one of the ways it sees itself – as a straightforward antidote to the kind of suburbs built after World War II. In the 1950s, low-density development effectively thwarted local control and discouraged lasting local intimacies. Major corporations freely transferred their executives to distant offices knowing they would find a home of equivalent value in a neighborhood where the social life would be at least as superficial as the one they just left. A certain level of intra-community hostility was supported as families were encouraged to build backyard bomb-shelters and equip them with guns to ward off neighbors who had failed to prepare themselves for attack.

This is perhaps the least significant way in which, for the last fifty years, the 'politics of propinquity' has been driven by nuclear strategic considerations. Proletarian and subproletarian ethnic minorities have been jammed together with a few die-hard liberal professionals at ground zero. The conservative, white middle class has been distributed around the countryside in low-density suburbs.[2] This was horrendously uneconomical, requiring massive investment in highways and other infrastructure duplication, overdrafts of fossil fuel, the destruction of America's smallest farms, and so on.

But cost was no object as the United States reconfigured itself as an enormous defensive weapon, a nuclear military-demographic masterpiece, a society that could mask its racism as a certain casual bravado in the face of nuclear threat. Every time the Soviets built a new type of bomb, the United States built five new white suburbs and found new negative terms for inner-city existence: 'welfare cheats', 'gang related', 'drug infested', 'psychotic homeless', 'drive-by shooting'. If this is permissible as 'social consciousness', its unconscious would be a desire for a nuclear hit on the city. An attack by our enemy would solve all our worst problems. Of course, no sensible person could actually believe such a thing, but it was official United States policy.

Now we are entering an era wherein the 'politics of propinquity' is supposedly no longer warped by strategic concerns. Today, economics is the driving force. In the absence of nuclear threat, it is both more economical and ecologically correct to push the white middle class together. It always would have been except for the 'nuclear thing.' Now we hear enthusiasm for 'urban in-fill' and the New Urbanism from quarters where such ideas would have been unthinkable only ten years ago – banks, developers, and mainstream planners.[3]

The deep pathos of the New Urbanism is that its proponents do not see their plans as symptomatic of nuclear trauma. The town of Celebration's copyrighted logo is pure kitsch: 'a little girl with a ponytail riding a bicycle past a picket fence under a spreading oak tree as her little dog chases along behind.'[4] The entire ensemble is symptomatic of an unavowed desire to rewind the life of the people from the present back to 1945 and replay it as if it had not been lived under threat of nuclear annihilation. The Celebration logo reproduces the opening scene of the infamous 1950s civil defense film 'Duck and Cover': a boy happily riding his bike past picket fences in Anytown, US is hit by a nuclear blast. The phrase 'a sense of' – as in 'a sense of security', 'a sense of community', 'a sense of family values', 'a sense of involvement', 'a sense of mutual interdependence' – forcefully reminds us of the impossibility of living 'as if' the last fifty years could be erased from collective memory. Yet this impossible desire is precisely the aim of neo-traditionalism.

THE AHWAHNEE PRINCIPLES

Two years after the Berlin Wall was taken down, just as the American people began to believe that they might not be vaporized in a nuclear holocaust, a group of architects, planners, community activists and lawyers got together at Ahwahnee Lodge in Yosemite and laid out a conceptual framework for the New Urbanism. Set forth in fifteen short, easy-to-understand points, the *Ahwahnee Principles* contain new language suggesting respect for the natural environment, but otherwise they are quite similar to the *Thirteen Points of Traditional Neighborhood Development* set forth earlier by Andres Duany and Elizabeth Platte-Zyberg, both of whom were at the Ahwahnee meeting.

The *Principles* are implicitly critical of the low-density suburbs developed after World War II. The New Urban, or neo-traditional neighborhood should have a discernible center. Most of its dwellings should be a short walk from the center. It should have a variety of housing, ranging from rental apartments to substantial single-family dwellings. Shops within walking distance (at the center or at the edge) should provide for the residents' weekly needs. The elementary school should be within walking distance for most children. There should be small playgrounds no more than 700 feet from every house. The houses should be set close to tree-lined streets and the streets should be narrow, slowing vehicular traffic and emphasizing the importance of pedestrian traffic and the pedestrian experience. Parking is located in the backs of the homes, accessed by alleys. The transportation corridors for motor vehicles, bicycles and pedestrians should be integrated, lighted, and otherwise designed in such a way as to encourage walking and bicycle use and discourage high-speed traffic. The natural terrain, drainage and vegetation should be preserved wherever possible. Community systems should be designed in such a way as to conserve water and energy and to minimize waste. Decision-making should be local and democratic. The single signifier of 'neo-tradition' is a generous, covered front porch.

Setting aside the sentimentality, there are several socially and environmentally sound design concepts contained in the *Principles*. Seaside is apparently one of the most successful planned communities in America. Duany and Platte-Zyberg have gone beyond the earlier position papers, suggesting that residents be allowed to build secondary structures in their backyards to be used as rental property or workshops. As often happens with a good idea, the heart of the matter seems to have gotten lost in its corporate interpretations. The only development to implement the secondary structures idea maintains a separate deed to the backyard shop, causing homebuyers to have to buy two pieces of property. For the most part, the history of the implementation of the *Ahwahnee Principles* has been a history of perversion.

Real estate developers instantly embraced the New Urbanism. Norman Blankman, writing in *Real Estate Finance Journal* remarks:

> Housing policy must be fundamentally altered. The single most important thing that should be done to bring affordable homing within reach for millions of people is to change zoning codes to permit more compact development The first steps have been taken by a nationwide movement to reform US urbanism. Its principles are applied to a project proposed for Suffolk County, New York, which is an epitome of suburban sprawl. Calthorpe Associates prepared a master plan embodying neo-traditional principles. The plan provided on 840 acres the essential elements that a developer proposed on 2150 acres. The plan . . . establishes a strong sense of community.[5]

In every statement made on behalf of New Urbanism by developers and builders, expressions of concern for 'community' thinly veil broad hints about new ways to make profits.

COMMUNITY SOLIDARITY AND THE NEW URBANISM

John Gardner, writing in the 'Inaugural Issue' of *The Celebration Journal* comments, 'The forces of disintegration have gained steadily and will prevail unless individuals see themselves as having a positive duty to nurture their community and continuously reweave the social fabric.' But what is meant here by community'? Gardner is not specific about the process of 'reweaving', and what exactly are those 'forces of disintegration'?

Emile Durkheim argued that pushing people together, or increasing population density and the social complexity of the community causes what he called 'greater moral density'. He writes:

> [T]here occurs a drawing together of individuals who were separated from one another . . . Hence movements take place between the parts of the social mass which up to then had no reciprocal effect upon one another. [Social relationships] consequently become more numerous, since they push out beyond their original boundaries on all sides.[6]

Durkheim goes on to say, in effect, the more complex the better, because a differentiated population is not in competition for the same resources. As the citizens 'perform different services, they can perform them in harmony'.[7] Thus, we might look to the New Urban community for creative coalition building on a local level, and improvements in participatory democracy, local self-definition and autonomy.

Historically, middle-class solidarity at the local level very often takes the form of obstinate resistance to the free play of large corporate interests: environmental activism, anti-development initiatives, consumer boycotts, food safety movements, subscription farming, local money, and so on. This is not the kind of citizen action the developers of New Urban towns have in mind when they invoke the various concepts of local solidarity. It is another, even opposing, kind of solidarity, a solidarity of consumer and corporate interests.

> Celebration will make its critical mark – for it breathes to life an intangible heritage, and that heritage is one of hope: a hope that large corporations can and will work with existing communities and local governments to accomplish great things, beneficial to both corporate life and to local entities.[8]

The corporate interpretation of the New Urbanism marks either the end of the twentieth century or the beginning of the twenty-first. Already there is talk that Disney will retreat from the field of town building, having discovered that they cannot 'storyboard' real life drama.[9] If Celebration is to be the last Disney town and not the first of many, as was initially claimed, perhaps it is because Michael Eisner, and not some neoleftist, was quick to see one of those pesky social contradictions that refuse to go away even after the fall of the Soviet Socialist Republic. How is it possible to create new middle-class neighborhoods supposedly founded on principles of ecological sustainability, civic involvement and so on, and at the same time neutralize in advance any convergence of people's interests which might frustrate the corporate definition of the citizen as pure consumer? The problems would go away if, and only if, the needs and desires of the citizen were precisely coincident with the goods and services produced by the corporation. But so far this is only a corporate utopian ideal.

DISNEY DEMOCRACY: CAUSE FOR CELEBRATION?

In its currently unanalyzed state, the New Urbanism is fully shaped by the nuclear unconscious. Its politics are built around a nostalgic submission to the kind of absolute authority that replaced civic life during the nuclear age. From the very beginning, even before the first house was built, Celebration was mobilized against its citizens, against any possibility of meaningful citizen involvement. In all the official materials there is a compulsive emphasis on surface detail, and a corresponding vagueness when it comes to decision-making, governance, and so on.

> Celebration will surely be one of the most visible and potentially influential designs as we enter the next century, and it presents a subtle and complex reading of our future that many will find a welcome antidote . . . Celebration [is] new but tradition inspired . . . The architects decided to design some buildings to resemble large houses that had later been converted to apartments . . . Color, too was the subject of long discussion . . . Residential neighborhoods will have variety but a . . . neutral palate . . . The collection of favorite house types appropriate to Celebration led to a choice of six designated styles: Classical, Victorian, Colonial Revival, Coastal (raised cottage), Mediterranean and French (the last two giving spice to the mix). (pp. 37 and 41)

Alongside these remarks *The Celebration Journal* contains an article on the one-room school: 'older students taught younger students – there was a sense of mutual interdependence – today those same conditions can be recreated' (p. 19). Another article extols the virtues of picket fences 'just tall enough to keep in the chickens – and to keep small boys out of the petunias – but not so tall as to form an impediment to adult discourse' (p. 54). And yet another article explains the philosophy behind Celebration Health, the community HMO, a 'hospital without walls' that will reduce health care costs by accessing 'the gift economy' the 'untapped resource of thousands of volunteers' (p. 65).

Clearly what is being sold is not a two-hour celluloid fantasy but an entire fantasy life. And, by now, it is pretty widely known that not everyone who came to Celebration is having a good time. Some have expressed concern that Celebration is over-designed:

> [E]very last visual detail my eyes had taken in during my two-hour walk, from the precise ratio of lawn to perennials in the front yards to the scrollwork on the Victorian porches to the exact relationship of column, capital and entablature on the facades of

every Colonial Revival, had been stipulated . . . I knew all that, yet now I felt it, too, and how it felt was packaged, less than real, somewhat more like a theme park than a town.'[10]

And perhaps also like a graveyard. The shadow of death and its denial is a central theme which gives tension to many Disney products.[11]

Some residents are not happy that the town manager is a Disney executive and that conflict, even conflict over the Disney rules, is relegated to 'focus groups' run by professional facilitators hired by Disney. It seems that everyone tries to live comfortably within these rules, which require, among other things, white or beige window covering, no pick-up truck parking in the streets, only one garage sale per household in any given twelve-month period, and political expression limited to a single sign, no more than eighteen by twenty-four inches, posted no more than forty-five days prior to the election. Master planner and board member Robert A.M. Stern explains the rules:

> In a free-wheeling Capitalist society you need controls – you can't have community without them. It's right there in Tocqueville: in the absence of an aristocratic hierarchy, you need firm rules to maintain decorum. I'm convinced these controls are actually liberating to people. It makes them feel their investment is safe. Regimentation can release you.[12]

Stern's philosophical fatuity aside, he probably won't get much argument from the residents of Celebration, it is not as though he told them, '*Arbeit Macht Frei.*' These are, after all, surface matters, not much more intrusive than the required architectural details in the Disney pattern book.

But the 'one-room' schoolhouse is something else. At Celebration, it is not exactly 'one room', but several rooms, each modeled on the one-room school, that is, with grades one through six taught together. The planners were apparently not mindful that one-room schools in America were historically understood by those who attended them, and who taught in them, to be inferior to graded classrooms; that they were replaced by graded schools as soon as population density increased to the point that the community could afford education differentiated by age; that whatever virtues they may have had were dictated by necessity. The 'one-room school' concept is not based so much on a consideration of sound educational policy as on a desire for yet another nostalgic signifier of 'tradition'. It also, incidentally, prepares future citizens to live in arbitrarily configured, artificial social environments.

Ironically, this and some other features of the Celebration school (its insistence on using the discredited 'whole language' approach to reading) have deeply divided the residents of Celebration, giving us our first glimpse of postmodern politics at the neighborhood level. Instead of the usual political divisions between Democrats and Republicans, conservatives and liberals, and so on, Celebration is divided into 'Positives' and 'Negatives'. The Corporation was quick to label residents who complained about the school 'the Negatives' and the Negatives in turn labeled the Corporate organized supporters of the school the 'Positives', or sometimes the 'Pixie Dust Parade'. And how does the postmodern community deal with the Negatives? It encourages them to leave. Home buyers in Celebration must sign a contract promising not to profit from the sale of their homes within a specified time. The Corporation has informed vocal Negatives that it will not hold them to that contract if they take their profits and go quietly, that is, if they sign an agreement 'promising never to reveal their reasons for leaving Celebration'.[13]

On the side of the 'Positives' there is ample evidence that they could not be happier with the paternal set-up. A 'Positive' homeowner is reported to have responded to a question about the corporate form of government, 'Come on! Disney gives me a sense of security. They will insure a quality product and keep home values up.' Another defines Disney democracy: 'It is definitely a democracy because we can go to town hall and express our feelings. It's a very responsive government.'[14] A third remarks, 'If it was anyone other than Disney, we would never have done this. We just feel that they represent first class all the way. Anything they do is quality.'[15]

The panoptic house

A glance at the three house plans schematized in the Realty pamphlet, 'Welcome to Celebration', reveals one design element that is featured in all plans, independent of the price level of the house, namely: from the front door through the house and into the backyard, nothing obstructs one's line of sight. There are almost no architectural options that would allow one to move from room to room without being seen by someone presenting him or herself at the door. This has been remarked upon by architects as if it were some kind of design error.

> The [indoor] layouts make for great expanses of awkward space and such dubious innovations as downstairs master bedrooms off the dining areas of two-story homes. Robert A.M. Stern, the New York architect who is one of Celebration's master planners, described the interiors . . . as 'horrendous'.[16]

Rymer labels this a 'dissonance of surface and depth' between the inside and the outside of Celebration homes, but closer examination reveals a unifying logic: the panoptic style of house answers to a nostalgia for central authority that penetrates the most intimate details of life. It is designed to replace the unconscious. Perhaps because it does not require changes in local zoning laws, the panoptic house precedes the creation of entire 'new urban' towns. In the last five years the panoptic house has spread through virtually every new middle-class and upper-middle-class tract in America.

The panopticon domestic dwelling eliminates the possibility of discovering anything that might disconfirm the hypothesis of deep spiritual harmony. The entry hall has become the entire house. When one enters this space just beyond the front door, one has a sense of expansiveness. To the left is the 'living area'; just ahead, and up a step, is the 'dining area'; to the right, a large stairway leads up to a balcony that traverses the scene like a theater stage. The doors into the bedrooms are clearly visible just beyond the stage-balcony railing. Toward the back of the dining area, beneath the balcony, through expansive archways are the kitchen to the left and family room to the right, and still beyond through 'French' doors is the backyard. Never mind that each one of the functional spaces is smaller than its counterpart in the box-full-of-boxes suburban tract home built five years ago; it looks huge.

What is more important than the visual lie is the absence of any sacred center, any place of privacy, any place where some kind of local craziness might thrive unexamined by everyone who happens to come up to the front door. The hearth is not a hearth, but an opening in the living room wall covered with glass and surrounded by a thin veneer of natural material – slate, marble or limestone. There can be no sacred center, no place for being, because it is *all* center. This is a new kind of domestic space, without shadows, with nothing to be opened up by story or memory; it is a successful totalitarian attempt to remove habitability.[17]

THE THRESHOLD OF DIFFERENCE

The interpretation of the New Urban community and the panoptic house at Celebration and elsewhere constitutes a suppression, even an erasure, of human difference except as a (very) few demographic categories recognized by corporate community planners. This erasure takes the form of enforced, deep homogeneity, not just on matters of aesthetics, but on what constitutes proper authority, and even on the details of life beyond the domestic threshold. The panoptic house leaves no room for questioning the existence of homogeneous neighborhood standards, including those extending to the most intimate details of domestic life.

Neighborhood and subject before the new urban convergence

Neighborhoods as such are experienced from the outside. That is their essence. You can live next door to people for years, never cross their threshold or know their name, and still recognize them as being 'from the neighborhood'. If self-understanding is nuanced by imagined responses to neighborhoods and strangers who live in them, it is shaped by external signs, by documentary evidence. Viewed from their streets, each suburb, *barrio*, bourgeois urban enclave, tenement district; each place in the United States where people live side by side, easily lends itself to a totalizing experience of its visible elements, its surfaces.

There are exceptions – Houston's Third Ward, and beach and river-front communities – but overall, neighborhoods in the United States are supposed to exude socio-aesthetic homogeneity. Postmodern theory's assertions of depthlessness is in this respect suspiciously similar to official doctrine: that is, to representations of neighborhood advanced by banks, insurance companies and realtors, and to the enforcement of zoning ordinances, building codes, and laws governing public conduct. There has always been a moral structure to neighborhood representation, and this moral structure is very often linked to the definition of self.

When an individual enters a neighborhood, it is an occasion for a broad-based appraisal. Is the neighborhood safe or dangerous? Is its economic standing below, the same as, or above that of the visitor? Are the locals trusting, or suspicious and menacing? Is it too quiet or too noisy; too clean or too messy? Is the housing aesthetically appealing? Is the visitor comfortable around the people encountered there? Are there too few shops and services or too many? Even if this interrogation is conducted just outside the range of conscious thought, it is not about the neighborhood so much as it is about the visitor him- or herself. The New Urbanism rounds off the edges of all these questions in advance.

If the visitor can be said to have a self, it is composed in some measure of the answers she or he gives to these kinds of questions; answers that may be shaped as much by the desires and prior experiences of the visitor – and by the policies of banks, zoning departments, and so on – as by actual characteristics of the neighborhood. Thus both the self and the neighborhood first appear in the realm of the imaginary. A similar unconscious dialogue may occur when an individual is introduced to a new country or person, but *neighborhood* is among the most ubiquitous, accessible, and fine-grained progenitors of self. Micro-responses to neighborhood symbols and signs, to the range of public behaviors encountered there and to human difference, map being.

The apparent drive of the New Urbanism is to forge a vapid unity of self and place, unconstrained by history, seemingly unconstrained by what was once called 'the human'. At Celebration, the undoing in advance of any edge that might be capable of producing a human contour is accomplished by the creation of a 'backstory' for the town. Peter Rummel, president of Disney Design and Development, explained to Russ Rymer, 'One of the things we do particularly in imagineering, is we often create a story, a backstory. You write a whole mythology about something, and it helps you to stay true to your design of a show or a ride or whatever you are doing.' The mission of the Celebration Foundation is to make the town 'feel like it has a tradition, even though it doesn't.'[18]

Michel de Certeau wrote that all urban space is built on broken pieces of the city's past, ultimately on the extermination of forests, of aboriginal inhabitants, of 'hidden places where legends live'.[19] He called the city a 'suspended symbolic order'. But it is clear that he did not think it possible for the urban to suppress its violent origins absolutely. He said 'haunted places are the only ones people can live in – and this inverts the schema of the *panopticon*.'[20] What de Certeau intended to suggest by this is that in a very real way the things and events buried in our neighborhoods, under our basements, are looking back at us. He refers to the relationship between the neighborhood and that which it suppresses as a kind of silent partnership – as 'stories in reserve'.

Let there be no mistake on this point: Celebration and every other New Urban community has its traditions, its 'stories in reserve', just like every other inhabited place on the face of the earth, but its 'stories in reserve' bear no resemblance to the 'backstory' provided by the real estate development company. The hidden places where legends live are the corporate back offices, and the all-but-forgotten heroes are Michael Eisner, Peter Rummel, Robert Stern and others.

The difference between the founding heroes of the New Urbanism and the other forgotten ancestors is attributable to the fact that the developers have gone to great lengths to produce a fiction of unity, one that blocks in advance the emergence of any human difference. There may always have been a collective desire for such styling clichés and affectations which would affirm an underlying sense of local neighborhood homogeneity. But until the emergence of the New Urbanism this desire had been attenuated by one of the most powerful extant social norms, that against unannounced and uninvited visits beyond the doorway or front hall of a home. Neighborhoods may have been marked by apparent unity, but the households were equally marked by extreme circumspection on the matter of identity, subjectivity and local practices. These latter were not neighborhood matters but domestic, and the uniqueness of this sphere has been protected by custom and law.[21] Such laws are a clear example of what is meant by the 'paternal metaphor': a symbolic function that allows us to replace the alpha male whose responsibility it was to defend 'his' women and children from any kind of intrusion.

Before the New Urbanism, neighborhoods, especially middle-class neighborhoods, strained to represent deep structural equivalences among local life forms (and here I am purposefully including plant life, animals, insects, and germs and viruses, because neighborhood moral integrity eventually references all of these). But there was an additional social contract to leave these forms unexamined except under staged conditions – an invitation – or extreme conditions – a subpoena, a search warrant. On the one hand there was the assumption that people living next to one another ought to be deeply like one another, and on the other hand there was an agreement not to test the assumption of propinquitous moral homogeneity.

Cops

Some measure of the actual diversity of neighborhoods, and of domestic scenes, in the United States is available on the television show, *Cops*. The neighborhoods on *Cops* are a New Urbanist's nightmare – the evil sibling of 'neo-traditionalism'.[22] Watching *Cops* one learns that calls to domestic disturbances often take the patrol officers into neighborhoods where tall weeds grow out of cracks in the streets and sidewalks; that there are mobile home parks with planned landscaping and temporary skirting around the base of the homes between the floor-level and the ground; that there are other parks with no plants, no skirts and the wheels still on the trailer, as if the entire neighborhood is ready for a quick getaway.

Even if the neighborhood has a look of drab anonymity, it has a distinctive 'look' which, for better or worse, is recognizable to residents, cops and outsiders. This overall appearance is grounded in a maze of local norms and local meanings of pride, shame and limit, all of which is made manifest in the forms and condition of the housing, the public spaces and the objects found there. Are the lawn chairs and barbecues broken, rusted and fallen over, or are they pristine? Are the garden hoses neatly coiled or snaking across sidewalks? Are the street gutters filled with fast-food wrappers? Off to the side of the action on *Cops* we see details of local practices which have unexpectedly become a part of the scene of the action. Here, someone has converted an abandoned right-of-way to a garden; there, the right-of-way has become a hangout for kids. The look and feel of the place is manifest in the moveable stuff – pulsations of human presence, accumulations of vehicles, the periodic movement to the curb and back (or not) of garbage cans.

It would seem that very little work would be required to restore the fictional moral unity of the neighborhood. But often, calling the police has the opposite of the intended effect. Bill Nichols, who has made an important contribution to the literature on 'reality TV', has commented specifically on *Cops*.[23] Nichols suggests that *Cops* patrols 'the boundary of normalcy', that 'we are there on the street with the cops', and that 'we share their point of view and subjectivity'. What is intriguing about *Cops* is that this 'boundary of normalcy' is not always, or even usually, set up around a part of the community that is openly taboo – a brothel, a crack house, the scene of a murder. The boundary most often crossed on *Cops* is usually the interior of a neighborhood residence or the sanctity of a backyard, garage or tool shed. What *Cops* does nightly is to break the rule against sudden, unannounced penetration into deep domestic space. It is repeated exposure of domestic back regions that distinguishes *Cops* from other 'reality TV' shows.

After Lacan, Jacques-Alain Miller has suggested that racism and other forms of hatred for one's neighbors is based on a theft of enjoyment:

> We may well think that racism exists because our Islamic neighbor is too noisy when he has parties. However, what is really at stake is that he takes his *jouissance* in a way different from ours. Thus the Other's proximity exacerbates racism: as soon as there is closeness, there is a confrontation of incompatible modes of *jouissance*. For it is a simple matter to love one's neighbor when he is distant, but it is a different matter in proximity.[24]

One could argue from this position that the norm against uninvited visiting is designed to block this kind of jealousy, to prevent neighbors from witnessing one another's enjoyment, thereby permitting them to maintain the fiction of deep subjective homogeneity. The facts of intrusion suggest otherwise. Certainly differences in the ways of pleasure are revealed, but more often what one finds are differences in everyday, abject pain. The common denominator is *difference*: human difference which exceeds fictional imagination; difference that exceeds scientific finding now that sociology and anthropology have abandoned their mandate to describe us to ourselves. What is exposed on *Cops* goes beyond any representational conventions found in regular TV and motion picture neighborhood and domestic depictions.

Sometimes it is hot pursuit: a kid caught with a weapon or dope makes a break for it and is followed by the police and the cameraman into a neighbor's house, through the living room, the hall, the back bedroom, out the window, across the yard, over the fence. Often it is a domestic disturbance. In one episode, a seven-year-old boy refuses to leave his grandparents' home to return to the care of his mother and her boyfriend. The police find the mother and her boyfriend cowering in the street, complaining that the grandfather brandished a gun at them. The discussion over custody is brought into the grandparents' home, first in the living room, then in a back bedroom. Every square inch of the living room wall from floor to ceiling is covered with shallow shelves crowded with ceramic animals. In the bedroom there is a picture on the dresser that appears to be the daughter when she was younger. In the picture, she is wearing a white dress with red ruffle trim. The curtains on the bedroom windows are white with red ruffle trim outline. The feeling of unavowed madness is palpable. Certainly any sense of deep *petit bourgeois* homogeneity is rendered unsustainable.

A *Cops* segment shot in Kansas City begins with officers Mark Horkheimer and Tony White going to a house to serve a warrant. They knock at the door.

TW: 'POOlice!' (Long pause.) 'POOlice!'

Officer Horkheimer goes around the side of the house. White pushes on the front door and it comes open. He sticks his head in. The camera follows the line of sight inside the house; a scene of total disarray with furniture, clothing, utensils, packing boxes, in heaps and jumbles. In the middle of the mess, staring straight into the camera is a large, hostile pig. The pig breaks wind.

TW: 'Oh, that doesn't sound good.'

The officers call 'Animal Control' and begin interviewing the neighbors. They find a pretty girl about ten years old in the alley carrying a baby in her arms.

MH: 'Who's seen the pig before?' Has she seen the pig?
Girl: 'Yeah, I have.'
MH: 'What's goin' on with the pig?'
Girl: 'I don't know.'
MH: 'Does it ever come out and run in the street, or . . .'
Girl (interrupting): 'Yeah, it comes over here. My oldest brother, he was afraid, and he goes running home.'
MH: 'He goes and comes as he wants?'
Girl: 'Uh huh. There was a dog, too, with the pig. Sometimes they go out together.'
MH: 'So the dog and the pig run together?'
Girl: 'Uh huh. Sometimes they go out and sometimes they don't. But the dog ran away.'

Horkheimer rejoins White to compare notes

MH: 'Apparently the pig has been coming and going as it pleases and terrorizing the neighborhood with its buddy the dog. But the dog left the pig – like a divorce.'

This segment illustrates, among other things, the elasticity of the human capacity to normalize neighbors, no matter how different. There were a few smiles and puns, but the only discourse readily available treats the pig as an occasionally annoying neighbor, in a domestic relationship (in this case with a dog), a couple that 'comes and goes' and has their differences, as neighbors are wont to do. All of this would seem to substantiate Miller's thesis, except that everyone extended a kind of courtesy toward the pig – much more than occurs when human neighbors are upset with one another. The pig is the figure of excess, perhaps even excess enjoyment, routinely engaging in at least one of the seven deadly sins, and this particular pig lived up to its symbolic role. But everyone – cops and other neighborhood residents – was fully prepared to accept the pig as a kind of eccentric neighbor, not as someone (or something) to be despised for living a life of excess, for having a home just like theirs even though it was 'only' a pig.

Not every house or apartment in *Cops* is 'the same' as the one beside it, but they are ineffably equivalent in the moral totalization of the 'neighborhoods'. This moral homogeneity appears to be based on contractual and quasi-contractual agreements. This is not just a matter of demographics, of lending policies, zoning laws, residential inspection practices, and the like. And it is certainly not a matter of centralized rules governing the color of curtains, or parking for pick-up trucks. Local tolerance of disorder and noise levels, the appearance of front yards, behavior of children and dogs, even pigs, of what are taken to be 'appropriate' ways of appearing to be different, of the local limits of difference; all of this must be worked out between neighbors as a living agreement. It is precisely the human capacity to arrive at this kind of agreement with one's neighbors which has been denied by the New Urbanism.

When agreements break down, someone calls the

cops. We know from *Cops* that these calls are often for banal delicts and infractions: a neighbor's dog poops in the yard; kids play on cars instead of in the playground; and so on. Across the United States there are countless variations on these agreements and much is made of these variations by realtors, residents and visitors. As a system of socio-cultural differentiation it has enormous integrity. The differences *between* neighborhoods in the United States are perhaps as great as have existed in any other time or place. But intro-neighborhood variation easily falls within Freud's 'narcissism of small differences'. The shirtless skinhead with jail tattoos on every visible part of his body does not stand out in the crowd that gathers for his arrest in his own neighborhood.

Not all the results have been exemplary, but the enormous variation of neighborhoods in America and local agreements on which this variation is based are the laboratories in which new cultural arrangements are being created by the people. The New Urbanism is opposed to naturally occurring cultural variation. One might go so far as to suggest that this is its reason for being.

THE MUTATION OF 'COMMUNITY'

What remains to be shown is the opposition between society and culture that opens the way for the construction of postmodern/New Urban/neo-traditional places. There are two kinds of 'significant' social relations or relations of human adjacency: *statistical* and *symbolic*. These two are not mutually exclusive, occupying at the same time the space or gap between people.

Statistical significance rests on the determination of frequencies of the occurrence of a characteristic or quality in the population. The determination of frequency requires the use of number, and number is based on the assumption of individuation. In this framework, a person, originally and in the first place, is not a part of a group. A person is an individual. All 'groups' are artificial, that is, statistical aggregates. Postmodern neighborhoods are aggregates based on specified ranges of household income, ethnicity and other selection criteria.

One cannot speak of the 'statistical significance' of the relationship between x and y unless x and y are conceived to be separate series. Statistics does not specify any direct relation between x and y, if any exists only a co-relation. Any real relationship that might hold between conservative political beliefs and the rules requiring 'earth tone' exterior home colors in some Orange County neighborhoods (for example) cannot be specified statistically. What can be specified is the probable incidence of political conservatism (x) occurring by chance, compared to the probable incidence of earth tone exterior home colors (y) occurring by chance. When x and y co-vary and certain conditions of measurement and sampling are met that make it possible to assert that the observed incidence of x and y occurring *together* by chance is fewer than five times in 100, or one time in 100, their co-variation is 'statistically significant'.

In short, any causal connection that may hold *between* x and y is not relevant to statistical significance. Thus, *statistically*, we have neighbors who are like ourselves in several socio-economic particular – skin color, income level, life stage, family size, and so on – without ever needing to *relate* to them beyond the polite exchange of clichéd platitudes. And so long as nothing upsets the balance of life in postmodern neighborhoods, the people living in them can pretend that the *statistically* significant relationships between them are also actually *socially* significant. They are held together by the bonds of consumerism. This could be called 'Yuppie Solidarity'.

The substitution of a statistical for a social relation is an administrative ideal, and as such it is the basis for the organization of increasingly large spheres of human life on both smaller and larger scales than neighborhood composition. For example, the content and timing of jokes in television situation comedies is based on a statistical analysis of the demographic characteristics of the population that is known to be flipping channels at the moment the joke is planned – the joke is written to hook and hold a significant proportion of channel surfers.[25] Every neighborhood that has been created as a part of a postmodern or neo-traditional 'housing development' is based on similar statistical analyses of the 'target market'. Let there be no mistake on this – the people who live in such developments do not necessarily see themselves as living in a bad joke. The people of '92707' live in housing tracts which pioneered theme of statistical models developed in the first place for electronic media audience analysis. They have adopted for themselves the slogan, 'Another Day In Paradise'.

Just because the statistical relation implies no actual connection between people does not mean that the people will refuse to value it. The statistical relation, perhaps because it contains nothing of the human, is jealously guarded by those who live it. Some of the sharpest conflicts in the postmodern world are over the presumed right to maintain a strict numerical definition of identity and the person: for example, the recent insistence by University of California Regents that 'minority groups' should not be given any statistical advantage in admission policies. A dominant pop philosophy of our current epoch summarizes its own position in the phrase 'looking out for number one'.

The *symbolic relation* specifies a human tie that is symbolically mediated, classically by language and contractual obligation. Both Freud and Durkheim provided accounts of symbolically mediated group formation wherein the function of other persons for an individual was to serve as model, object, helper or opponent. This should not be taken to suggest that symbolic relations are good (while statistical relations are bad). The symbolic tie is transparent to good and evil. It differs from the statistical relation in that *number* is not a basis for meaning or value. Two people can be connected via spoken language by thousands of symbols; and millions of people can be united by a single symbol.

There is, however, a point at which a fantasy form of the symbolic relation and the statistical relation converge. This convergence is precisely the principle governing relations with one's neighbors in New Urban developments, which continue to run smoothly according to corporate definitions of clan solidarity, that is to say for as long as any real solidarity is effectively undermined. According to Freud (reading McDougall's *Group Mind*)[26] individuals in a group may have a common interest in a symbolic object, a similar emotional bias toward that object, and a degree of reciprocal influence – a contagion of emotions – which pleasurably carries them toward a single goal. Often it is a goal which none of the members of the group holds individually. This formation radically simplifies the range of possible outcomes for communication and joint co-operative activities that is inherent in the symbolic relation *and* in the diverse interests of the members of the group.

Freud points out that in groups that have *leaders*, the leader becomes the symbolic object of identification for the members of the group. Each individual narcissistically views the leader as the missing piece of his or her own ego. So long as they believe the leader fills in for their lack, they can love the leader and the group exactly as they would love themselves if only they were perfect and whole. So long as the leader appears to love all of them equally, every member of the group, even complete strangers, can believe themselves connected on a basic level, united by the same psychic lack which the group and its leader seems to fill. According to Freud, a 'primary group' (he also calls them 'artificial groups') is a number of individuals who have put the same object in the place of their ego ideals and have therefore identified themselves with one another in their egos. Thus, group solidarity based on the fantasy symbol that promises to complete the ego is just the flip side of envy. 'I want what you possess' has been 'democratized' to read, 'Everyone must be the same and have the same'.[27] This is also the postmodern/neo-traditional/New Urban ideal.

Postmodern neighborhoods may have administrators, even elected administrators, but they do not have leaders. An absence of leadership is one of their most distinctive features. We have seen that in Celebration, potential leaders are run out of town. So, on first examination they would seem to fall outside Freud's critique of artificial group formation. Or do they? I would suggest that the so-called neo-traditional, New Urban neighborhood *itself*, its demographic definition, aesthetics and paternal administration, replaces 'the leader' as the external fantasy object of ego-identification.

Let's go back to a symbolic neighborhood, one that has not been permitted to degenerate into a human aggregate, one that is not yet susceptible to psychic administration, the kind of neighborhood that appears on *Cops*. Within the symbolic-as-language, other people help, hinder, serve as positive and negative models, perhaps as 'pawns in the game of life'. Even a pawn in a language game retains more human potential than the highest ranking 'number' in the statistical or pseudo-symbolic relation that communicates only via career and consumer codes. Symbolic communication involves efforts to express oneself verbally and in other ways, calculated and naive moves to conceal expression, and strategic efforts to uncover information which others may wittingly or unwittingly hide. Culture guarantees *human* intelligence by not offering any guarantees on human interaction.

This kind of 'full communication', in Goffman's sense of the term, all but disappears in postmodernity. The New Urban development comes with guarantees. And the scope of the 'guarantee' now extends well beyond the boundaries of the New Urban community. Creative activities once requiring intense communication and risk for their accomplishment (for example, starting a successful small company, establishing democratic processes in a small community, or running a grass-roots neighborhood organization) are now done according to formulae. Success and funding depend on manuals, advanced degrees in non-fields (for example, 'Arts Administration'), how-to seminars, and hyper-specialized consultants for every defined stage in the process. Courtship and seduction, once highly language-dependent activities, can now be accomplished by simply comparing resumés or otherwise providing proof that one has reached an acceptable level on an appropriate career track. If there is no courtship beyond a mutual disclosure of professional accomplishments and material possessions, biological reproduction can occur without there having been a sexual or other human relation. IBM inventor, Tom Zimmerman, is testing prototype wet wire data transfer devices so that when two people ('wet wires') stand close to each other in an elevator (let's say) or brush by each other in the street, or shake hands, they will exchange biographical or other facts stored on a credit-card sized chip in their wallets. 'Business cards', 'marital status', and so forth can be exchanged, to be downloaded later, just by rubbing elbows. Lacan's apocalyptic-sounding aphorism has in fact become a stated goal of postmodern medicine and reproduction in families headed by two professional adults.

Durkheim, and the sociology that is based on his work, did not believe that anything like this could ever happen. He thought that hyper-specialization in a complex division of social labor would ensure that human beings would always have to come together and negotiate their differences. He further believed that an evolving symbolic and legal order would serve as the singular medium for the negotiation of human difference. Durkheim understood that no individual is self-sufficient and that the symbolic order exists precisely to balance and mediate competing human needs under conditions of absolute interdependence. More than this, Durkheim also believed that modern complex societies would exhibit strong *moral integration* because everyone would recognise their mutual dependence, at least at the level of a collective conscience.

This is perhaps the nicest theoretical picture that could have been devised for us. Unfortunately, the model of social complexity Durkheim provided is flawed. What he did not understand is an all too human tendency to abrogate, be blind to, abridge, and short-circuit interdependence, to ignore the rules of restraint that make life together possible.

In fact, the social circuit is, itself, designed so as to make possible a human blindness to the symbolic imperatives. In the simplest model of social complexity, person A and person C need not be related directly, but can be related through a third party, usually 'God'. If B is not a god, but merely a bureaucrat, he is able to keep the A:B relationship completely separate from the B:C relationship. In fact, his success as a bureaucrat is dependent on his ability to compartmentalize his relationships. A and C may be necessary in the division of labor, but their efficiency as specialists is predicated on their not having other than an abstract, instrumental relationship. If the boss (B's) supervision of C is affected by the fact that he worried about a complaint that A has threatened to file against him, that is, if A is *actually* influencing C, the system of differentiated roles and functional integration breaks down.

In actual situations governed by the principle of task specialization and functional integration, there is a necessary double effacement of both 'self' and 'society'. Functionally specialized and integrated workplaces, homes, neighborhoods, and so on, are of necessity composed of the absence of human relationships. Of course, human relationships may subversively occur there, but that is not what such settings are about. Thus, the secular division of labor in society was set up in the first place so that it might evolve into a kind of postmodern pretense of community. If it was a test, humankind is failing it.

It is sad to return to Durkheim's community, composed of every imaginable human type, with the realization that no one ever really *belonged* or had a place there, and that no one ever really *had* to enter into a relationship with anyone else, even those upon whom they were completely dependent. Hopefully, every postmodern individual will have a few human relationships made up on the spot for no apparent reason, even if structurally they can get along without them, that is, continue to misrecognize their mutual dependency. Hopefully some postmodernites will dream of making a difference instead of just *being* a difference. But these dreams can very quickly turn into bourgeois fictions of 'place', 'roots', or 'fame' that are routinely substituted for the missing links that might actually connect one person to others. The various postmodern fictional versions of the social relation masquerade as social life. This simply could not occur in a universe that was designed according to theoretical principles laid down by Durkheim, or even by Marx, to the extent that Marx believed the 'superstructure' to occupy an important supporting role.

A fiction of deep subjective unity at the neighborhood level cannot be based on possessions, whether these are material, mental or biological. Whenever assignment to a group is made on the basis of possessions, the group is designed to fall apart. Credentials are challenged, someone may claim to have more of the requisite possession or a better version of it. As soon as possession is the criterion, there is the possibility of a '*plus de jouir*' and the neighborhood devolves into hostility, as Jacques-Alain Miller suggests. Deep unity at the group level can only be based on shared *lack*. The basis for any spiritually homogeneous group is some *external* object that everyone desires and no one can have. Until recently, it was *security* in the face of nuclear threat that everyone lacked.

There are some historical groups for which it has been possible to specify the external missing object: security, the phallus, a healthy ego, 'agency.' In the postmodern world it cannot be specified – it is a phantom object that simply signifies lack, one's own lack and the same lack in the other. In the 1950s, lack was signified by the various guarantees of privacy. Uninvited visitors were not permitted beyond the threshold because they might see what the family did not have. So long as no unannounced visitor entered the private space of another, everyone could collectively maintain the fiction that they all were equally deprived of something. If any members of the group actually possessed the object rather than simply holding it as an ideal, the group would fall into jealous fighting over it. The norm against surprise visits permitted every family and other domestic unit co-operatively to uphold the fiction that no one has 'It', or possesses 'The Thing'. The norms of privacy also permitted every family to congratulate itself, if it wished, for secretly having It so long as they kept It sufficiently hidden to prevent anyone else from becoming jealous. Since no one knew what It was, any family would congratulate itself in this way, so long as no one would surprise them in their conceit.

In the 1990s lack is signified differently, by adherence to a small number of 'approved' surface details and by opening up domestic space for all to see. The entire home is visible from the front door and there is nothing to see. Everything is as in a stage-set. The entire design of the neighborhood is built around a fantasy guarantee of shared lack. The New Urban development precludes any possibility of unwanted intrusions into the back regions of the community – and the mind – by the simple expedient of eliminating any such space from the community.

The human prospect here is ultimately more horrifying than jealous racism, and could even produce a nostalgia for a time when a theft of enjoyment or a '*plus de jouir*' even in the 'other' was possible. The New Urban neighborhood realizes the impossible ideal of a solidary neighborhood in which *everyone* imagines themself to be *uniquely* in possession of the universal object of desire. In this way, they are, in fact, deeply spiritually homogeneous. They are collectively guilty because there is a gap or separation between the ego and in ideal. But they have all filled in the space of their guilt with the same fantasy of lack transformed into the appearance of a generosity of pure space painted white, a pure absence of being.

NOTES

1 All invoked in the 'Inaugural Issue' of *The Celebration Journal* (Celebration, Fla: Walt Disney, n.d.)
2 Some historic evidence for this relationship between the bomb and the suburb is examined in my early paper, 'Baltimore in the Morning After,' *Diacritics* Summer 1984.
3 See, e.g., the 1996 Bank of America Position Paper opposing 'leap frog' development and low-density suburbs in favor of urban infill.
4 As reported by Russ Rymer in *Harper's Magazine*, October 1966, p. 67.
5 Norman Blankman, *Real Estate Finance Journal*, vol. 9, no. 4, 1984, pp. 70–74.
6 Emile Durkheim, *The Division of Labor in Society*, pp. 200–201.
7 Ibid., p. 110.
8 'Editorial' in *The Celebration Journal*. (All further references to this journal will be from the editorial and will be made in the text.)
9 For a good discussion of the environment of corporate decision-making, see Jon Lewis, 'Disney after Disney', in Eric Smoodin, ed, *Disney Discourse*, New York, Routledge 1994, pp. 87–105.
10 Michael Pollan, 'Downtown Building is No Mickey Mouse Operation,' *The New York Times Magazine*, Dec. 14, 1997, p. 62. The same point was made earlier by Michael Sorkin in his 'Introduction', *Variations on a Theme Park*, New York: Noonday 1992.
11 For a fuller treatment, see my *Empty Meeting Grounds: the*

tourist papers, New York, Routledge, 1992, pp. 74ff. Diane Ghirardo observes, along these same lines, that in his original plan for EPCOT (which was supposed to be a living community with real inhabitants, not just a theme park), Walt Disney stipulated that no older people would be allowed to reside there.

12 Reported in Pollan, 'Downtown Building is No Mickey Mouse Operation' p. 80.
13 Ibid., p. 76.
14 Ibid., pp. 80 and 81.
15 Rymer, in *Harper's Magazine*, p. 75.
16 Ibid., p. 70.
17 Brian Block has pointed out to me in conversation that this is the obverse of the experience reported by Michel de Certeau in 'Walking in the City', *The Practice of Everyday Life*, Berkeley: University of California Press, 1984, pp. 91ff.
18 Rymer, p. 18.
19 De Certeau, 'Walking in the City' p. 106.
20 Ibid., p.108 (my emphasis).
21 At the individual level, and the level of the domestic establishment, these laws are designed to guarantee and protect privacy. The function of the same laws at the social level is to protect human difference. Structurally, it is very difficult to separate legal guarantees of individual privacy from guarantees of social difference, though the right is always trying to make this separation.
22 *Cops* is shot, edited, and presented in a 'real TV' format. According to the baritone male voice-over, the camera crew rides on patrol with 'the men and women of law enforcement' in cities and rural areas of every region of the country. At the beginning of each episode, a small block letter title in the lower left corner of the screen names the jurisdiction, e.g., North Boston or Lee County, Florida. The name of the officer in the car appears in the title space as he or she speaks about joining the force or about why he or she likes police work. Whatever the other messages of *Cops* may be, it is valuable, for the purposes of this paper, because it provides a unique, spontaneous glimpse of neighborhoods across the United States.
23 Bill Nichols, *Blurred Boundaries: Questions of Meaning in Contemporary Culture*, Bloomington and Indianapolis: Indiana University Press, 1994, p. 44.
24 Jacques-Alain Miller, 'Extimité', *Lacanian Theory of Discourse: Subject, Structure and Society*, ed. Mark Bracher, New York and London: Routledge, 1994, pp. 79–80.
25 Nick Browne has described this and similar strategies in 'The Political Economy of the Television (Super)Text', in *Television USA*, ed. Nick Browne, New York: Harwood Academic Publishers, 1994.
26 Sigmund Freud, *Group Psychology and the Analysis of the Ego, The Standard Edition of the Complete Psychological Works of Sigmund Freud*, London: Hogarth Press, 1955, vol. 18.
27 This transformation has been analysed by Juliet MacCannell in 'The Postcolonial Unconscious, or the White Man's Thing', *Journal of the Association for the Psychoanalysis of Culture and Society*, vol. 1, no. 1, 1996, especially pp. 30–31. She attributes the discovery of the structural relation of democracy to envy not to Freud, but to Rousseau.

PART NINE

Utopias and Dystopias

INTRODUCTION TO PART NINE

The concept of utopia – a place which is not anywhere, or no-place – is essentially literary and begins in accounts framed as traveller's tales of far-off lands, such as Thomas More's *Utopia* or Tomasso Campanella's *City of the Sun*. The literary location of utopia allows the writer to imagine any kind of settlement, though usually it takes the form of an ideal city, or an island on which there is an ideal city. Utopia can take as well any form of social organisation, though this tends to follow a pattern of hierarchy. In distinction from the make-do world of reality, it allows a direct relation between the two, so that the built and social architectures resemble each other. Neither is cause to the other's effect but both are twin manifestations of a principle of order. But because utopia is nowhere in particular but, like a day-dream (a more consciously directed fantasy than a night-dream) occupies a space which is psychological, the imagined city or state is separated from present reality by an unbridgeable chasm. All that can be done to go there is to formulate it as a concept, or a story. Utopia is thus like the aesthetic dimension, separated from the rest of life, a realm of fantasy which offers a harmony absent in ordinary experiences. It can compensate for the tribulations of those experiences, the alienation of toil, the immiseration of the working class, the deadness of routines, and so forth, but not actively change the conditions of their production (see Herbert Marcuse's essay on affirmative culture in Part Two).

Utopia, despite the precedent of Plato's *Republic*, is also decidedly modern. The term modern is used here in the general historical sense, not the art-historical, to mean the period of modern thought from the sixteenth century until the beginning of postmodernism in, say, the 1970s. The modern is now encapsulated in history, but asking how it differs from the postmodern foregrounds its characteristics: the modern deals in certainties; it makes new; and it orders. These overlap, in that the confidence lent by a knowledge of certainties leads to the wish to build a new world; this in turn, because it is built on a blank space, a *tabula rasa* like the white sheet of paper on which it might be drawn (or the blank computer screen), offers scope to design a totality, to construct a universe amidst a chaos (chaos being the zone over the edge of the map, where ships fall off the sea into nothing).

To an extent, this extends the form of the citadel as a place of safety to a space in which seasonal change, or the vicissitudes of conflict and division, and the dangers of defilement by dirt (matter out of place), are absent. Hence the Enlightenment planned city embodies principles of spatial order which are translated into materiality in regular and clean streets, facades and open spaces set out according to a preconceived plan which is itself a model of geometric order, and social boundaries. Later, the spaces are lined by trees in a reintroduction of the natural to the cultural, but initially it is the proportions and the regularity of the city which make it *appear*.

The idea that the city appears, implying a contrary that the landscape does not but is just drably there, is not specifically modern. Indra Kagis McEwen uses a philological method to reveal its presence in Greek thought. In *Socrates' Ancestor* (Cambridge, MA, MIT, 1993), she argues that practices including ceremonial dancing, weaving, and ship-building are the basis of a spatiality which brings the city into being. The city, she writes 'was not a vessel with a fixed form, but, like the appearing surface of a woven cloth – of all the traces of material culture one of the most perishable – had continually to be mended or made to reappear'. It *appeared* in the harmony of an orthogonal street plan, though this was not used universally but mainly in

Greek colonial cities and in Piraeus. She continues: 'if we think of the city in terms of weaving, as I believe the early Greeks did, the intention made manifest in orthogonal street layouts becomes quite precise', and explains that weaving consists of placing threads at right-angles to each other. She adds: "*Harmonia*, close fitting, can be a feature of the tightly woven cloth only: a textile with a loose weave is not, so to speak, "harmonious". It does not, properly speaking, appear at all' (all quotations p. 83). McEwen sees a perversion of this order, which takes place through human interaction with land and material, in the Roman concept of a standardised spatial plan. This sense of an ordering of space may be one aspect of the planned city of the Enlightenment, but the latter also has an optimism which is new. It is as if in making new the troubles of existence can be shed, the dirty slate wiped clean.

Optimism is arguably at the heart, too, of Descartes's use of an architectural metaphor in his *Discourse on the method of rightly conducting one's reason and seeking the truth in the sciences, and in addition the Optics, the Meteorology and the Geometry, which are essays in this Method* of 1637 (to use the abbreviated title he gave the work). The *Discourse* can be seen in this way as a utopian text. Descartes writes:

> I was at one time in Germany, attracted thither by the wars which are not yet ended . . . when winter brought me to a halt in quarters where, with no society to distract me, and no cares or passions to disturb me, I spent the day in a stove-heated room . . . Among these [thoughts], one of the first that came to my mind was that there is often less perfection in what has been put together bit by bit, and by different masters, than in the work of a single hand. Thus we see how a building, the construction of which has been undertaken by a single architect, is usually superior in beauty and regularity to those that many have tried to restore . . . So, too, those old places which, beginning as villages, have developed in the course of time into great towns, are generally so ill-proportioned in comparison with those an engineer can design at will in an orderly fashion (*Descartes: Discourse on Method*, translated by A. Wollaston, Harmondsworth, Penguin, 1960, pp. 44–5).

His image of the engineer making regular places by design – the original text says '*compassées*' and '*à sa fantaisie*' – sums up the modern project, as represented by the towns of Nancy (1588) and Charleville (1605) which Descartes may have had in mind in writing this.

Descartes's fantasy is (in a sense different in French from English) the imagination's freedom to reason without constraint; the engineer uses the compasses to draw lines in space, making a diagram of a place which does not yet exist (see Claudia Brodsky Lacour, *Lines of Thought: Discourse, Architectonics, and the Origin of Modern Philosophy*, Durham, NC, Duke University Press, 1996). Richard Sennett, in *Flesh and Stone* (London, Faber and Faber, 1995, pp. 261–70), sees the Enlightenment city not as a plan, or cosmos without a deity, but as a site of circulation based on discovery of the circulation of the blood in bodies, a circulation the stopping of which is fatal. This shares with Descartes's text the sense of reason moving at will, but its emphasis is utilitarian. Descartes's rationality is not, however, a reaction to fear of a blocked circulation but a response to doubt. It is doubt which leads him to distrust bodily sensations, memories, the learning of books, and the knowledge of the world gained through travel, and to seek certainties in the self-contained systems of geometry and mathematics.

An antithesis of such certainty, and of utopia, is Daniel Defoe's description of a fictional island in *Robinson Crusoe* (1719). As a traveller's tale it shares its form with that of More's utopia, but the island it describes is not a society but a wild land. In a way this prefigures the idea of a noble savage, located in places untouched (or corrupted) by industrial civilisation, developed by Jean-Jacques Rousseau; in another it is a brutal world of necessity. Crusoe is self-sufficient, but his efforts are to civilise the island by making plantations, becoming happier as he maps the island in his mind.

Ambiguities of hope and fear, the undefiled and the controlled, as well as of doubt and certainty, permeate the literature of utopia. The plan of Campanella's ideal city can be interpreted either as an optimistic roaming of intellect creating a cosmos, or as a retreat to geometry. Both the City of the Sun and Amaurotum in Utopia (one of fifty-four identical cities on More's fabled island) are contained within thick walls with high towers, on a central hill. Campanella describes the defences of his imagined city as concentric rings dissected by four streets with gates, set out according to the points of the compass, and moves quickly to their fortified state: 'Furthermore, it is so built that if the first circle were stormed, it would of necessity entail a double amount of energy to storm the second' (*City of the Sun*, translated by T. W. Halliday, London, George Routledge and Sons, 1887, p. 218). The city's walls are decorated with representations of *all* knowledge, freely available to all citizens and used as a teaching aid for the young, but as if there will be nothing unforeseen and nothing to add later or to change through new understandings. Indeed, the difficulty in most utopian systems is this inability to admit change, which reflects their separation from a world of contingency (see discussion of postmodern identities in the Introduction to Part Seven). It might also be noted that Campanella was in prison when he wrote *City of the Sun*. Yet there is a sense, too, of that free imagining which makes regular places (according to the compass as well as the compasses) where nothing is but where consequently anything can be. Perhaps the problem which utopias express is a tension between imagination and a fear of what imagination might produce if unleashed.

François Choay, in *The Rule and the Model* (Cambridge, MA, MIT, 1997), writes of More:

> In the economy of More's project, the spatial model thus seems to be a response to a problematic of identity which arose at a precise moment in European history. More has discovered that a society can transform itself, can construct itself to be *other* than the tradition which has determined it. He opts for this transformation . . . But at the same time as he steels himself against the intoxication caused by his freedom, he annuls its solvent effect. He safeguards himself against the dispersion and disappearance of the individual community to which he belongs (Choay, p. 154).

More's project remains a speculation not to be realised, hence appropriate to literature. The same could be said of countless utopias produced since. These, however, differ from some nineteenth-century utopian writing, particularly in France, where new social models with complementary spatial or architectural prescriptions *are* intended as plans for a new society – as in Charles Fourier's phalansteries. Fourier describes a society of Harmony where work is libidinal, in communities of 1,600 people living in specially designed groups of buildings, with workshops, social spaces, and sleeping areas, surrounded by farmland. These are to spread throughout France, not at first to replace cities, from which he expects large numbers of visitors to the phalansteries, but as models of the new within the old (see Jonathan Beecher and Richard Bienvenu, *The Utopian Vision of Charles Fourier*, Columbia, MO, University of Missouri Press, 1983, pp. 233–70).

In a more localised way, but realised as a pilot venture, the Charterville Allotments near Minster Lovell in Oxfordshire, England, represent an attempt to build an alternative society in rural England in the 1840s. The cottages are still there, though most have been modified. When built they were one-storey buildings adequate for a family, with two to three acres of land attached on which to grow crops and keep animals, thus to be self-sufficient in food. A ballot was held to select the occupants, but many of the allottees had no experience of working on the land and the soil was full of large stones. The experiment failed and the Chartist Land Company with it. Other experiments in the form of model villages such as Port Sunlight in Merseyside, or Bourneville outside Birmingham – intended to provide decent housing for working-class people employed in the adjacent factories – have lasted better due to their integration in a system of production. The difficulty with such experiments in England is that they tend to be anti-urban. That is, they are either situated in the countryside, or mimic the spatiality of a rural village. Port Sunlight, despite being next to a soap factory with a belching chimney, has half-timbered houses, a village green, and stream, all in a pastiche of Merry England. The Garden City movement of Ebenezer Howard is the most developed of such tendencies (see Peter Hall and Colin Ward, *Sociable Cities: The Legacy of Ebenezer Howard*, Chichester, Wiley, 1998). This effort to transplant the city with its multiple deprivations to an abundant countryside which lacked only employment prospects was naive, but more importantly it was nostalgic, seeing the city as the problem and the solution as release from it to the green fields of an imagined countryside.

Some of these issues are touched on by Judith Shklar in her essay reprinted below. She cites nineteenth-century utopians such as Fourier and Saint-Simon, also Marx, and begins with Karl Mannheim's division of thought into utopian and ideological categories – the former striving for change on the part of those below, and the latter smothering that change by those above seeking to retain power. She cites Hannah Arendt as seeing in Plato an understanding that the model, a concept to which literary utopias such as More's and Campanella's correspond, is enjoyed for its own sake and not as a plan to be realised. This is a briefer form of Choay's argument noted above and implies that utopias are (contrary to Fourier's plans) aesthetic entities.

Shklar sees a nostalgia for the perceived if unreal unity of the Greek city-state – the *polis* – in socialism: 'The ideal of its unity, of its homogeneous order, coloured all their visions of the future. Certainly Marx was no stranger to the nostalgia for that cohesive city or its medieval counterpart' (from 'Melancholy to Nostalgia', in Frank E. Manuel, ed., *Utopias and Utopian Thought*, London, Souvenir Press, 1973, pp. 101–15, reprinted here on pp. 404–12). In this respect, discussion of utopia is close to that of culture in the mid-nineteenth century (for Matthew Arnold, for instance) as an antidote to the industrial revolution – see the Introduction to Part Two. Shklar, writing in the early 1970s, begins by asking why there are no utopias today, and ends by saying that utopian thought is disabled by its continued attachment to classicism, and that the concept is, in certain key respects, now irrelevant.

An opposite view is held and elaborated at great length in *The Principle of Hope* (Cambridge, MA, MIT, [1959] 1986, 3 vols) by Ernst Bloch. Rather than include an extract from a book which is eclectic and rhetorical, and even less given to extraction than Benjamin's writing (see reference in the General Introduction), a text about Bloch by Vincent Geoghegan is included in this Part. In a discussion between Bloch and Adorno on the contradictions of utopian longing (*Widersprüche der utopischen Sehnsucht*) in 1964, Adorno opens the conversation by saying 'If I may be allowed to say something first, even though I may not be the correct person to begin, since my friend Ernst Bloch is the one mainly responsible for restoring honor to the word "utopia"' (in *The Utopian Function of Art and Literature*, translated by J. Zipes and F. Mecklenburg, Cambridge, MA, MIT, 1988, p. 1); and then argues that the subject-matter of most utopian dreams has been technologically realised while their content of happiness has been forgotten. Herbert Marcuse approached the same problem in his *Essay on Liberation* (Harmondsworth, Penguin, 1969). Geoghegan gives a summary of Bloch's ideas contextualised by brief details of his life – accepting a Chair of Philosophy at Leipzig in the German Democratic Republic in 1949, for instance – and concludes that Bloch's achievement was to show that utopianism is not confined to high culture but is present as much in day-dreams and in popular fiction. This might be one way to deconstruct utopian nostalgia. But perhaps of more interest is Bloch's theory that glimpses of hope, which recur in all societies and social conditions and are given form in art, music, and literature, act to educate the mind, which begins with fantasies of a better world but shapes these.

Modernism, in planning and architecture if less in visual art, was seen by many of its adherents as a utopian project, an effort to engineer a new society. Leonie Sandercock describes its death and possible replacement by a postmodern praxis. While the Modernist model was in the end seen to be either a rationalisation or an aestheticisation of space, and efforts to transpose it to the actualities of building social housing were problematic, Sandercock's proposal for a planning praxis is seen very much as rooted in experience and applicable to the complex situations in which planners find themselves in a multi-ethnic environment. Voices from the margins make themselves heard (see reference to Sandercock in the Introduction to Part Seven), and because they make *themselves* heard instead of being allowed to be heard, they are heard in a different way from the conventional voices of the rational planning model in which a professional elite make supposedly a-political decisions on city growth. Again, there is a parallel with feminist refusals of assimilation to men's cultural structures discussed in the Introduction to Part Seven. This process produces different knowledges rather than more of the same kind of knowledge: 'They argue that theory is always embodied. They celebrate the value of experiential and other alternative ways of knowing, learning, discovering, including traditional ethnic or culturally specific modes: from talk to story telling, the blues to rap, poetry and song' ('The Death of Modernist Planning: Radical Praxis for a Postmodern Age', p. 181). The means to gain such knowledges for the planner include spending a lot of time 'hanging out' with dwellers, much as a contemporary ethnographer might seek to become semi-integrated as a participant observer of a group within society.

The rational planning model which postmodern praxis replaces is not dead, however. Rosalyn Deutsche describes its residual operation in a functional attitude to space in policies for gentrification. The extract reprinted below from Deutsche's long essay 'Uneven Development: Public Art in New York City', first published in October in 1988, is the beginning of the essay. It can be read in conjunction with another essay by Deutsche, 'Alternative Space' (in *If You Lived Here: The City in Art, Theory, and Social Activism*, edited by B. Wallis, Seattle, Bay Press, 1991, pp. 45–66). Deutsche cites Mayor Koch's pronouncement that homeless people, some of them perhaps evicted as a consequence of gentrification, cannot use Grand Central Station because they are not there to take trains. Deutsche, obviously, argues for a multi-valent use (and concept) of the city, and notes that the Situationists saw this kind of objectification as a key way in which capitalism shields itself from challenges. In 'Alternative Space', Deutsche writes in a similar vein that 'Art history purports to simply discover, rather than to construct, the objects it studies – art, the city, society' (p. 46). In both cases concealment of the *production* of space shields its representation from contradictions.

That shield may be distantly related to the walls of the city in More's and Campanella's accounts, as bastions against contingency. The notion of an orderly city, without the mess of social dis-harmony and the dirt of displacement, is in its way and while being a massively impoverished interpretation of rationality, a utopian idea. At root it is a yearning for a city free from conflict. But this is found only in dreams, or infancy.

This section pairs utopia with dystopia in its heading, but has avoided the several well-known texts which dwell on urban disintegration and decay – Mike Davis' 'Fortress Los Angeles', for instance, in *City of Quartz* (London, Verso, 1990). This decision was taken because, first, the disaster scenario is not only well-rehearsed but also addictive (see Malcolm Miles, 'Strange Days', in *Urban Futures*, edited by Malcolm Miles and Tim Hall, London, Routledge, 2003, pp. 44–59); and, second, because that scenario has little to offer in the way of deeper understandings of urban change.

The final inclusion, by one of the Editors, returns to critique, and the work of the Frankfurt School. This text began as a record of a walk through Rotterdam one lunchtime during a break in an academic conference. It is not a critique of dystopia, nor as such of utopia, but argues that there is a fundamental difficulty in the concept of radical social transformation which disables its realisation. This might be called the utopian content, and is put in one form succinctly by Murray Bookchin: 'But what is most disquieting about Marx's vision of social change is the extent to which it denies the power of speculative thought to envision a new society long before the old one becomes intolerable or is bereft of any room for development' (*Urbanization Without Cities: The Rise and Decline of Citizenship*, Montreal, Black Rose, 1992, p. 187). The problem is put in another form when Marcuse admits, in discussion after a lecture on the end of utopia in Berlin in 1967, that the new, to happen, needs an abolition of the institutions of the old, consciousness of which is located in the new. In other words, tomorrow never comes because it is not today.

SUGGESTED FURTHER READING

Ernst Bloch, *The Utopian Function of Art and Literature*, edited by J. Zipes and F. Mecklenburg, Cambridge, MA, MIT, 1988

Murray Bookchin, *Urbanization Without Cities: The Rise and Decline of Citizenship*, Montreal, Black Rose, 1992

Martin Buber, *Paths in Utopia*, Syracuse, Syracuse University Press, 1996

Peter Hall and Colin Ward, *Sociable Cities: The legacy of Ebenezer Howard*, Chichester, Wiley, 1998

Jonathan Hughes and Simon Sadler, eds, *Non-Plan: Essays on Freedom, Participation and Change in Modern Architecture and Urbanism*, Oxford, Architectural Press, 2000

Andrew Light, ed., *Social Ecology after Bookchin*, New York, Guilford Press, 1998

Theodore Roszak, *Where the Waste Land Ends: Politics and Transcendence in Postindustrial Society*, London, Faber and Faber, 1973

Brian Wallis, ed., *If You Lived Here: The City in Art, Theory, and Social Activism*, Seattle, Bay Press, 1991

'The Political Theory of Utopia: From Melancholia to Nostalgia'

from Frank E. Manuel (ed.), *Utopias and Utopian Thought* (1973)

Judith Shklar

Editors' Introduction

The book from which this text is taken also includes essays by urbanist Lewis Mumford ('Utopia, the City and the Machine'), literary critic Northrop Frye ('Varieties of Literary Utopias'), and theologian Paul Tillich ('Critique and Justification of Utopia'), among several others. Several of the texts first appeared in *Daedalus*, the Journal of the American Academy of Sciences, in 1965. Shklar's was selected for inclusion here as giving a broad critical view of the concept of utopia. It is in a section headed 'Utopia is Dead', with an account of socialism and utopia by Adam Ulam. This should not be interpreted as a view on the part of the Editors of this volume that utopian thought, or socialism, for that matter, has ceased to have validity; but it reflects a need, now as in the 1960s and 1970s, to reconsider the idea of utopia and ask why it has so constantly failed to be realised. Ulam, in his essay, concludes that in many cases of emerging nations in the post-colonial period, socialism was fused with nationalism, and unable to resist modification in face of consumerism; and that, despite advances in the sciences which promise new kinds of better worlds, social organisation has come to be seen in terms of a new sobriety. Shklar, however, takes a less historical and more theoretical approach, looking analytically at the concept of utopia rather than at its manifestations and revisions. Her conclusion is that because utopia is based on a false notion of a European past – Arcadianism – it is not relevant other than as a psychological problem of nostalgia. In terms of political philosophy, she sees it as 'a comment upon an intellectual situation' which does not lead to solutions.

But, as Shklar implies throughout her essay, the status of utopian thought as comment rather than plan does not render it uninteresting. While a common criticism of utopias is that they are future states unable to accept change without denting their brittle sense of perfection, Shklar's emphasis is on their *backward-looking* derivation from a past which, outside interpretation and imagination, may be simply absent. Utopia, then, constitutes 'an anguished recollection of antiquity' (p. 105), in which Greek *polis* and Roman republic are drawn together. Hence virtue is projected back in time, in effect stating the present as a time in which it lacks. Psychologically, this is an inversion of a present but unacknowledged absence displaced onto a distant past, so distant, it could be added, that it can bear almost any imprint. Shklar adds to this argument that the utopia of classicism in Europe was an intellectual concern, which reflects the isolation from certain other realities of the class in society by and for whom it was produced and consumed.

What does the plaintive question, "why are there no utopias today?" mean? Does it merely express the nostalgia of those who were young and socialist in the thirties? Is it just that they resent the lack of sympathy among younger people? Do some of the latter, perhaps, long to re-experience the alleged political excitements of the romanticized thirties, but find that they cannot do so? For it is pre-eminently a question about states of mind and intellectual attitudes, not about social movements. And it says something about the historical obtuseness of those who ask, "why no good radicals?" that they do not usually consider the obvious concomitant of their question, "why no Nazism, no fascism, no imperialism and no bourbonism?" If the absence of utopian feeling mattered only to that relatively small number of intellectuals who are distressed by their inability to dream as they once did, then it might concern a social psychologist, but it would scarcely interest the historian.

There is, however, more to the question than the temporary malaise of a few relics of the inter-war period. The questions "after socialism, what?" and "can we go on without utopias?" were already being asked before 1930, specifically by Karl Mannheim. Here, an entire theory of history and of the historical function of utopian thought was involved.[1] Mannheim's now celebrated proposition was that all the political thought of the past could be divided into two classes, the utopian and the ideological. The former was the "orientation" of those aspiring classes that aimed at the complete or partial overthrow of the social structure prevailing at the time. Ideology, on the other hand, was the typical outlook of the dominant classes, intent upon preserving the established order. It is, of course, more than questionable whether the vast variety of Europe's intellectual past can be squeezed into this Manichean strait-jacket. And, in fact, it was a perfectly deliberate falsification of history on Mannheim's part. As he blandly admitted, the historian's concern with actual differences, contrasts, and nuances was a mere nuisance to one who sought to uncover the "real," though hidden, patterns beneath the actual men and events of the past. As seen through the spectacles of the "sociology of knowledge," history had to show successive waves of revolutionary fervor as the chief constant feature of European intellectual and social life. This meant, among other things, that so marginal a figure as the "chiliastic" Thomas Müntzer had to be pushed to the very front ranks of intellectual luminaries. He was the first in a series that included such first-class thinkers as Condorcet and Marx. Karl Kautsky had indeed allowed Sir Thomas More to share with Müntzer the honor of being the first socialist, but he saw More rather as a unique intellectual prophet of the socialist future than as a mere class manifestation.[2] Mannheim, however, rejected Sir Thomas More summarily as a figure of no sociological significance in the "real" history of utopian thought. This entirely Marxian view of the past as dominated by incidents of revolutionary conduct and its reflected thought is of considerable importance, because it is what makes the contemporary absence of such zeal appear so entirely new, unique, and catastrophic and thus gives the question, "why no utopias?" its tense historical urgency. Certainly it had that effect upon Mannheim. If "art, culture and philosophy are nothing but the expression of the central utopia of the age, as shaped by contemporary social and political forces," then indeed the disappearance of utopia might well mean the end of civilization. And since Mannheim assumed that the classless society was at hand, and that no challenging, utopia-inspiring classes would again appear, the new "matter-of-factness" seemed threatening and ominous indeed. The disappearance of "reality transcending doctrines" brings about "a static state of affairs in which man himself becomes no more than a thing" and in relinquishing utopia men lose the will to shape history and so the ability to understand it. What, above all, is to become of the heirs of Müntzer, Condorcet, and Marx, of the intellectual élite who, until now, have been the producers of utopias? Mannheim's response, natural under the circumstances, was to provide a blueprint of a future society to be run by an intellectual élite trained in the sociology of knowledge, capable of both transforming and controlling history in the interests of freedom, democracy, and rationality.[3]

Since the social role and ideas of the intellectuals are the central concern of the sociology of knowledge, and since Mannheim, unlike Marx, seemed to believe that this élite was of supreme importance in shaping the pattern of history, it is not at all surprising that their notions of utopia should have differed so much. While Mannheim accepted Marxian ideas about ideology, utopia was for him the intellectuals' vehicle of self-expression and it was they, not the voiceless classes, who ultimately shaped the ages. Yet Marx's and Engels' views on utopian thought were historically far more sound in at least one respect. The classical utopia, the critical utopia inspired by universal, rational morality and ideals of justice, the Spartan and ascetic utopia was already dead after the French Revolution. Doomed to impracticability, since the material conditions necessary for its realization had not been prevalent, the classical utopia could be admired even though it had lost its intellectual function with the rise of "scientific socialism." This judgment, according to Marx and Engels, was also applicable to such socialist precursors as Owen, Fourier, and Saint-Simon. Their successors, and indeed all non-Marxist socialists, were, however, utopians in a very different sense. For these rivals *were* in a position to understand the true course and future of bourgeois society and to act accordingly. *They* had the benefit of Marx's theories of surplus value and of dialectical materialism. They could recognize both his "scientific" truths and the necessity for revolutionary activity. Instead they produced "duodecimo editions of the New Jerusalem," preached the brotherhood of man to the bourgeoisie, and ignored the Eleventh Thesis against Feuerbach.[4] Here, "utopian" clearly becomes a mere term of opprobrium for un-Marxian, "unscientific" socialists. One wishes that Marx and Engels might have chosen another epithet. Certainly many useless verbal wrangles over the "true" meaning of the adjective "utopian" might have been avoided. What remains relevant in their views, however, is the serious importance they attached to the classical utopia and their recognition that it was a thing of the past because socialism had replaced it in their own age. To this, one must add that it was not only Marxian socialism, but all forms of socialism and, indeed, all the social belief systems (especially Social Darwinism) which prevailed in the nineteenth century, that joined in this task. Moreover, all of these, in spite of Mannheim's ideosyncratic vocabulary, are now called ideologies. In short, it was ideology that undid utopia after the French Revolution.

To understand why the classical utopia declined, not yesterday, but almost two hundred years ago, demands a more detailed analysis of its character than either Marx, Engels, or Mannheim offered. It also requires a return to that historical way of looking at the past which they despised, because it does not try to uncover real patterns, nor to establish laws. Instead of concentrating on paradigmatic, even if obscure, figures which fit a preconceived scheme, it looks at the acknowledged masters of utopian literature: at Sir Thomas More and his successors. Paradoxically, this utopia is a form of political literature that cannot possibly be fitted into either one of Mannheim's categories, for it is in no sense either revolutionary and future directed or designed to support the ruling classes. All the utopian writers who followed More's model were critical in two ways. In one way or another all were critical of some specific social institutions of their own time and place. But far more importantly, utopia was a way of rejecting that notion of "original sin" which regarded natural human virtue and reason as feeble and fatally impaired faculties. Whatever else the classical utopias might say or fail to say, all were attacks on the radical theory of original sin. Utopia is always a picture and a measure of the moral heights man could attain using only his natural powers, "purely by the natural light." As one writer put it, utopia is meant "to confound those who, calling themselves Christians, live worse than animals, although they are specially favored with grace, while pagans, relying on the light of nature manifest more virtue than the Reformed Church claims to uphold."[5] No one doubts the intensity of Sir Thomas More's Christian faith, but the fact remains that his Utopians are not Christians, "define virtue as living according to nature,' pursue joy and pleasure and are all the better for it – which is, of course, the main point.[6]

The utopian rejection of original sin was, however, in no sense a declaration of historical hopefulness – quite the contrary. Utopia was, as Sir Thomas More put it, something "I wish rather than expect to see followed." It is a vision not of the probable but of the "not-impossible." It was not concerned with the historically likely at all. Utopia is nowhere, not only geographically, but historically as well. It exists neither in the past nor in the future. Indeed, its esthetic and intellectual tension arises precisely from the melancholy contrast between what might be and what will be. And all utopian writers heightened this tension by describing in minute detail the institutions and daily lives of the citizens of utopia while their realization is scarcely mentioned. "Utopus" simply appears one day and creates utopia. This is very much in keeping with the Platonic metaphysics which inspired More and his imitators as late as Fénelon. For them, utopia was a model, an ideal pattern that invited contemplation and judgment but did not entail any other activity. It is a perfection that the mind's eye recognizes as true and which is described as such, and so serves as a standard of moral judgment. As Miss Arendt has said, "in (Platonic contemplation) the beholding of the model, which . . . no longer is to guide any doing, is prolonged and enjoyed for its own sake."[8] As such it is an expression of the craftsman's desire for perfection and permanence. That is why utopia, the moralist's artifact, is of necessity a changeless harmonious whole, in which a shared recognition of truth unites all the citizens. Truth is single and only error is multiple. In utopia, there cannot, by definition, be any room for eccentricity. It is also profoundly radical, as Plato was; for all historical actuality is here brought to judgment before the bar of trans-historical values and is found utterly wanting.

If history can be said to play any part at all in the classical utopia, it does so only in the form of an anguished recollection of antiquity, of the polis and of the Roman Republic of virtuous memory. This is a marked feature also of utopias not indebted to Platonic metaphysics, even those of libertine inspiration. The institutional arrangements of Plato's *Laws*, Plutarch's *Lycurgus*, and Roman history also served as powerful inspirations to the utopian imagination. Thus, to the melancholy contrast between the possible and the probable was added the sad confrontation between a crude and dissolute Europe and the virtue and unity of classical antiquity. It is this, far more than the prevalence of Platonic guardians, perfected and effective education, and rationalist asceticism, ubiquitous as these are, that marks utopia as an intellectualist fantasy. Until relatively recently nothing separated the educated classes from all others more definitely than the possession of laboriously gained classical learning. It might be more correct to say classicism possessed them. They identified themselves more deeply and genuinely with the dead of Athens and Rome than with their own despised and uncouth contemporaries. And inasmuch as utopia was built on classical lines it expressed the values and concerns of the intellectuals. It was to them, not to unlettered lords or peasants, that it was addressed. As such, it was the work of a socially isolated sensibility, again not a hope-inspiring condition. But it survived even the literary and scientific victory of the Moderns over the Ancients. Nothing seemed to shake the long-absorbed sense of the moral and political superiority of classical man. That is why wistful Spartan utopias were still being written in the second half of the eighteenth century.

Of course, the political utopia, with its rational city-planning, eugenics, education, and institutions, is by no means the only vision of a perfect life. The golden age of popular imagination has always been known, its main joy being food – and lots of it – without any work. Its refined poetic counterpart, the age of innocence, in which men are good without conscious virtue, has an equally long history. The state of innocence can exist, moreover, side by side with a philosophic utopia and illuminate the significance of the latter. Plato's Age of Kronos and Fénelon's Boetica, in which wisdom is spontaneous, are set beside rational, Spartan-style models.[9] The state of innocence is what moral reason must consciously recreate to give form and coherence to what all men can feel and imagine as a part, however remote, of their natural endowment. Both utopias, in different ways, try to represent a timeless "ought" that never "is."

Among the utopias that do not owe anything to classical antiquity at least one deserves mention here: the utopia of pure condemnation. Of this genre Swift is the unchallenged master, with Diderot as a worthy heir. The king of Brobdingnag, the city of giants, of supermen, that is, notes, after he hears Gulliver's account of European civilization, that its natives must be "the most pernicious Race of little odious Vermin that Nature ever suffered to crawl upon the Surface of the Earth." A comparison of his utopian supra-human kingdom with those of Europe could yield no other conclusion. Gulliver, then, tastes the delights of a non-human society of horses, an experience which leaves him, like his author, with an insurmountable loathing for his fellow-Yahoos. Here, utopia serves only to condemn not merely Europe, whether ancient or modern, but the human louse as such. To the extent that Diderot's account of Tahiti slaps only at European civilization, it can be said to be more gentle. However, after observing the superiority of primitive life, his European travelers return home wiser in recognizing the horrors of their religion, customs, and institutions, but in no way capable or hopeful of doing anything about them. The aim, as in Swift, is to expose absurdity and squalor simply for the sake of bringing them into full view.

This all too brief review of classical utopia should suffice to show how little "activism" or revolutionary optimism or future-directed hope there is in this literature. It is neither ideology nor utopia in Mannheim's sense, but then neither is most of the great critical political literature before the end of the eighteenth century. Machiavelli, Bodin, Hobbes, Rousseau: were they "reactionary ideologues" because they were not "revolutionary utopists"? Significantly, it is only during the course of the English Civil War that action-minded utopists appear. However, even the two most notable among them, Harrington and Hartlib, were concerned with constitutional and educational reform, respectively, rather than with full-scale utopias. Nevertheless, for once imminent realization was envisaged. As for poor Winstanley and his little band of Diggers, they have merely been forced to play the English Müntzer in Marxian historiography in search of precursors and paradigms. These were all voices in the wilderness, part of a unique revolutionary situation. It is only as partial exceptions to the rule that they are really illuminating. They seem only to show how revolutionary the general course of utopian thought and political thinking was before the age inaugurated by the French Revolution. The end of utopian literature did not mark the end of hope; on the contrary, it coincided with the birth of historical optimism.

Utopia was not the only casualty of the revolution in political thought. Plutarchian great-man historiography and purely critical political philosophy were never the same again either. Nor was it solely a matter of the new theory of historical progress. As Condorcet, one of the first and most astute of its authors, observed, the real novelties of the future were democracy and science, and they demanded entirely new ways of looking at politics. If a democratic society was to understand itself it would need a new history: "the history of man," of all the inconspicuous and voiceless little people who constitute humanity and who have now replaced the star actors on the historical stage. The various historical systems of the nineteenth century, with their "laws" – whether progressive, evolutionist, dialectical, positivist, or not – were all, in spite of their endless deficiencies, efforts to cope with this new history. To write the history of the inarticulate majority, of those groups in society which do not stand out and therefore must be discovered, was a task so new and so difficult that it is scarcely surprising that it should not have succeeded. After all, contemporary sociology is, in a wiser and sadder mood, still plodding laboriously to accomplish it. As for science, Condorcet recognized not only that technology, that is, accumulated and applied knowledge, would transform material and social life, but also that science was not just an acquisition, but an entirely new outlook. With its openmindedness and experimentalism it had to replace older modes of thought which were incompatible with it. As such it was not only the vehicle of progress, but also the sole way in which the new society-in-change could be understood and guided. Scientific thought was inherently a call to action. The new world, as Condorcet saw it, would be so unlike the old that .its experiences could be grasped, expressed, and formed only by those who adopted the openness of scientific attitudes.[10] It was this that his systematizing successors, the victims of classical habits of thought, did not appreciate in the slightest. Whatever one may think of Condorcet's own historiography with its simple challenge-and-response ladder of improvement, he, at least, had the great merit of understanding why classical history and critical political theory had to be replaced by more democratic, dynamic, and activist social ideas.

Given the revolutionary changes of nineteenth-century Europe, the preoccupations of the classical utopists were no longer relevant. Original sin and the critical model were not vital interests. Marx, in spite of his protestations, was not the only one to take over the critical functions of the old utopists and to expand them into a relentlessly future-directed activism. All his rivals were just as intent upon action as he was. It was merely that some of them thought, as Saint-Simon had, that philosophers would exercise the most significant and effective authority by ruling over public opinion, rather than by participating directly in political action. Thus "the philosopher places himself at the summit of thought. From there he views the world as it has been and as it must become. *He is not only an observer. He is an actor.* [Italics added.] He is an actor of the first rank in the moral world because it is his opinions on what the world must become which regulate human society."[11] This bit of intellectualist megalomania could be illustrated by endless quotations not only from socialist sources, but also from liberal writings and from the distressed conservative deprecators of both.

The activism of the age was, moreover, not a random one. The future was all plotted out, and beckoning. "The Golden Age lies before us and not behind us, and is not far away."[12] The inevitable had only to be hastened on. Certainly there was no point in contemplating the classical past. However, the impact of the polis was not quite gone. Especially in socialist thought, even if not so openly as in Jacobin rhetoric, the ancient republic was still an inspiration. The ideal of its unity, of its homogeneous order, colored all their visions of the future. Certainly Marx was no stranger to the nostalgia for that cohesive city or its medieval communal counterpart. It was the liberal Benjamin Constant who noted that efforts to impose the political values of classical antiquity upon the totally dissimilar modern world could lead only to forms of despotism, which, far from being classical, would be entirely new.[13] John Stuart Mill, following him, found this notion, "that all perfection consists in unity," to be precisely the most repellent aspect of Comte's philosophy.[14] Indeed, the engineered community, whose perfect order springs not from a rational perception of truth, but from a pursuit of social unity as a material necessity, provides neither ancient nor modern liberty. The imagery is revealing. Cabet delighted in the vision of factory workers who displayed "so much order and discipline that they looked like an army."[15] Bellamy's "industrial army" speaks for itself.

The form of such works as Cabet's *Voyage to Icaria* and Bellamy's *Looking Backward* should not lead one to think that these pictures of perfected societies are in any sense utopias. Precisely because they affect the external format of the classical utopia they demonstrate most effectively the enormous differences between the old and new ways of thought. The nineteenth-century imaginary society is not "nowhere" historically. It is a future society. And, it too is a summons to action. The purpose of Cabet's expedition to set up Icaria in America was not simply to establish a small island of perfection; it was to be a nucleus from which a world of Icarias would eventually spring. No sooner had Bellamy's work appeared than Bellamy societies, often (and not surprisingly) sponsored by retired army officers, appeared to promote his ideas. Theodore Hertzka's *Freeland* led to similar organized efforts, as its author had hoped it would. He declared frankly that the imaginary society was merely a device to popularize social ideas which he regarded as practical and scientifically sound. This, in itself, suffices to account for the literary feebleness of virtually all nineteenth-century quasi-utopias. There was nothing in them that could not have been better presented in a political manifesto or in a systematic treatise. They were all vulgarizations and were devised solely to reach the largest possible audience. The form of the classical utopia was inseparable from its content. Both were part of a single conception. The social aspirations of the nineteenth century found their literary form in the realistic novel, not in the crude and unstylish fiction of social theorists turned amateur romancers. Even the "utopias" based on scientific, rather than social, predictions were either childish or tedious. Either their fancies displayed no insight into the real potentialities of technology or, if they were well-informed, they were rendered obsolete by the actual developments of technology. Not even the last and most talented of latter-day contrivers of imaginary societies, H. G. Wells, could save the genre. He at least saw that his utopia had nothing in common with the classical works of that name. Now the perfect model is in the future, that is, it has a time and a place and is, indeed, already immanent in the present. It must be world-wide, devoted to science, to progress, to change, and it must allow for individuality. Only the intellectual ruling class of "samurai" is left to remind one of the classical past.[16] As a liberal socialist, Wells was, no doubt, especially aware of the need to root out the remnants of that illiberal, self-absorbed, and closed social order which classicism had left as its least worthy gift to the democratic imagination. However, the novel of the happy future did not prosper in Wells' or in any other hands. For it was simply superfluous. Its message could be presented in many more suitable ways. Certainly it was in no way a continuation of the classical utopian tradition.

It has of late been suggested that the radicalism of the last century was a form of "messianism," of "millennialism," or of a transplanted eschatological consciousness. Psychologically this may be quite true in the sense that for many of the people who participated in radical movements social ideologies fed religious longings that traditional religions could not satisfy. These people may even have been responding to the same urges as the members of the medieval revolutionary millennialist sects. In this sense one may well regard radical ideology as a surrogate for unconventional religiosity. It should, however, not be forgotten that millennialism always involves an element of eternal salvation.[17] And this was entirely absent in the message of even those social prophets who called for new religions as a means of bringing social discipline to Europe. For them it was only a matter of social policy, not of supraterrestrial truth. Marxism and social Darwinism, moreover, did not even involve this degree of "new" religiosity. Whatever they did for the fanaticized consciousness that eagerly responded to them, the intellectual structure of radical doctrines was not a prophetic heresy either in form or in intent. It represented an entirely new chapter in European thought. As Condorcet had clearly seen, it was a matter of new responses to a new social world; and the aspirations, methods of argument, and categories of thought of these historical systems were correspondingly unique, however primordial the human yearnings that they could satisfy might be. One ought not to forget the rational element, the effort of intellectual understanding that is perfectly evident in the writings of Saint-Simon, Marx, Comte, and all the rest. The various political revolutions after 1789 gave more than a semblance of reality to a vision of social history as a perpetual combat between the forces of progress and of conservatism. Conservative and liberal social observers no less than socialists took that view of historical dynamics. Even John Stuart Mill, who recognized that the categories of order and progress were inadequate concepts for a deeper analysis of politics, could see the past as a sequence of struggles between freedom and repression.[18] The theory of class war was by no means the only one that, in a projection of nineteenth-century experience into the past, saw the history of Europe as a series of duels. Some saw it as progress, some as doom, but all perceived the same pattern. Looking back, of course, the century before the First World War appears infinitely more complicated than that, and so do the eras preceding it. However, if it is quite understandable why one should be sensitive to pluralistic social complexities today, it is also not difficult to see why dualistic patterns tended to dominate the historical imagination of the nineteenth century. Nor is it totally irrational that the experience of these rapid changes, so unlike those of the past, should lead men to entertain great expectations of the future. Neither the view of history as a dualistic combat of impersonal social forces nor the confident belief in a better future which would at last bring rest to mankind was a "millennial" fancy, nor was either really akin to the chiliastic religious visions that inspired that apocalyptic sects. If they were not utopias, neither were they New Jerusalem. The desire to stress similarities, to find continuities everywhere, is not always helpful, especially in the history of ideas, where the drawing of distinctions is apt to lead one more nearly to the truth.

The reason that ideology has been represented so often as a type of religiosity is, of course, a response to the terrifying fervor expressed by the members of modern mass-movements. It is the emotional element in Nazism, communism, and other revolutionary movements all over the world that is so reminiscent of many of the old popular heresies. The dynamism of mass parties, however, is really at stake here, not the actual systems of ideas which were produced in such great quantity by nineteenth-century Europe. Ideology, however, when it refers to those systems of ideas which were capable of replacing all the inherited forms of social thought, utopia among them, was clearly more significant intellectually than the brutish "isms" that animate both the leaders and the led of these movements. The latter should not be confused with either ideology, or utopia, or even with the religious extravagances of other ages. Nor is the question, "why no utopias?" really concerned with the organization of mass-parties. Indeed, even in Mannheim's theory, utopia and ideology refer to highly developed modes of thought and not to quasi-instinctive mental reactions. If, unlike Mannheim, one does not identify utopia with the charted mission of the intellectual class, one can recognize that "the end of utopia" involves not sociological, but philosophical, issues. It is the concern of political theory – of the high culture of social thought. What is really at stake is the realization that the disintegration of nineteenth-century ideology has not made it possible simply to return to classical-critical theory, of which utopia was a part. The post-ideological state of mind is not a classical one, any more than an ex-Christian is a pagan. On the contrary, the end of the great ideological systems may well also mark the exhaustion of the last echo of classicism in political theory, even if, occasionally, a nostalgic appreciation for the integral classicism of the more distant past can still be heard. The occasional contemporary efforts to construct pictures of perfect communities illustrate the point. They are compromises between the old utopia and the newer historical consciousness. Thus, for example, Martin Buber and Paul Goodman argue only for the historical non-impossibility of their plans, not for their inevitability. *Kvuzas*, or perfectly planned cities, are feasible, and certainly their admirers hope for their realization.[19] These are still calls to action, but modest ones. Their scope, moreover, is limited, and their very essence is a revival of that dream of the polis, of the "authentic" small community that truly absorbs and directs the lives of its inhabitants. These relatively mild and moderate proposals, and the more general concern with *real* community life, do show, among other things, the lingering power of classical values. Here the longing for utopia and nostalgia for antiquity are inseparable. And indeed the question, "why is there no utopia?" expresses not only an urge to return to antiquity, but also, and far more importantly, a sense of frustration at our inability to think as creatively as the ancients apparently did.

Classicism, in one form or another, was, as we have seen, an integral part not only of utopia, but also of most political thinking. Hobbes and Bentham in their firm rejection of the conventional classical model were intellectually far more radical than the later ideologists. For in spite of occasional liberal protests, socialist doctrines were by no means the only ones that contrived to perpetuate classical notions throughout the nineteenth century. Long after Platonic metaphysics and the critical-contemplative mode of thought had been abandoned, classical imagery and values retained their hold on the political imagination, and classical methods of description and argument continued to hold the expression of political ideas in a social context in which classicism had ceased to be relevant. In this respect, all the ideologies served to retard political thinking. Their decline now has left political theory without any clear orientation and so with a sense of uneasiness. It is not that political theory is dead, as has often been claimed, but that so much of it consists of an incantation of clichés which seem to have no relation to social experiences whose character is more sensed than expressed. Could it be that classicism, not only as a set of political values and memories, but as a legacy of words, conceptions, and images, acts as a chain upon our imagination? Is it not, perhaps, that language, mental habits, and categories of thought organically related to a social world completely unlike our own are entirely unsuitable for expressing our experiences? May this not be the cause of our inability to articulate what we feel and see, and to bring order into what we know? Certainly a vocabulary and notions dependent upon Greek and Latin can no longer be adequate to discuss our social life-situation. Nor will the continual addition of implausible neologisms composed of more Greek and Latin words help, for they do not affect the structure of thinking. The malaise induced by this state of affairs is responsible for much of the ill-tempered and ill-informed hostility of many humanists toward the natural sciences which do not share these inherited difficulties. It also accounts for many ill-considered efforts to "imitate" science by the metaphorical or analogical use of words drawn from biology or physics. Nor is analytical philosophy of much use, for it does not address itself to concerns which are more nearly felt than spoken and which involve not so much what can be said as the difficulty of saying anything at all. To be sure, nostalgia is the least adequate response of all to these discomforts. And that is just what the question, "why is there no utopia?" does express in this context.

With these considerations, the question, "why is there no utopia today?" has, hopefully, been reduced to its proper proportions, which are not very great. To the extent that it depends on an erroneous and dated view of the European past, it is simply irrelevant. As a psychological problem its interest is great, but of a clinical nature. Lastly, it is only one item in the far more complex range of questions that concern the possibilities of contemporary political philosophy. Here, however, it does at least have genuine significance, even though it does not ask for an answer. For it is more a comment upon an intellectual situation than a real query. That is why a journey, however quick, through the utopian and ideological past seemed a fitting response, since it might show what the question implies, even if it does not offer any solutions.

NOTES

1 Karl Mannheim, *Ideology and Utopia*, tr. by Louis Wirth and Edward Shils (New York: Harvest Books, n.d.), pp. 193, 195–197, 205, 222, 255–257, 263, et passim.
2 Karl Kautsky, *Thomas More and his Utopia*, tr. by H. J. Stenning (New York: Russell, 1959), pp. 1–3. 171.
3 Karl Mannheim, *Freedom, Power and Democratic Planning* (New York: Oxford University Press, 1950).
4 Lewis S. Feuer, ed., *Marx and Engels: Basic Writings* (New York: Anchor Books, 1959), pp. 37–39, 70, 81, 90, 245.
5 Gabriel de Foigny, *Terra Australis Incognita*, in Glenn Negley and J. Max Patrick, *The Quest for Utopia. An Anthology of Imaginary Societies* (New York: Schuman, 1952), p. 402.
6 Sir Thomas More, *Utopia*, tr. and ed. by H. V. S. Ogden (New York: Appleton-Century-Crafts, 1949), pp. 48–49.
7 *Ibid.*, p. 83.
8 Hannah, Arendt, *The Human Condition* (Chicago: University of Chicago Press, 1958), p. 303.
9 H. C. Baldry, *Ancient Utopias* (Southampton: University of Southampton, 1956); Fénelon, *The Adventures of Telemachus*, in Negley and Patrick, *op. cit.*, pp. 424–437.
10 A.-N. de Condorcet, *Sketch for the Historical Picture of the Progress of the Human Mind*, tr. by June Barraclough (London: Weidenfeld and Nicolson, 1955), pp. 163–164, 168–170, 184–202.
11 Quoted in Frank E. Manuel, *The New World of Henri de Saint-Simon* (Cambridge: Harvard University Press, 1956), p. 151.
12 Edward Bellamy, quoted in Negley and Patrick, *op. cit*,., p. 80.
13 *Ouevres* (Paris: Bibliothèque de la Pléiade, 1957), pp. 1044–1058.
14 J. S. Mill, *The Positive Philosophy of Auguste Comte* (New York: H. Holt and Co., 1873), p. 128.
15 Etienne Cabet, *Voyage to Icaria*, in M. L. Berneri, *Journey Through Utopia* (London: Routledge & Kegan Paul, 1950), p. 234.

16 H. G. Wells, *A Modern Utopia*, in Negley and Patrick, *op. cit.*, pp. 228–250.

17 On this important point, see Sylvia L. Thrupp, "Introduction," in *Millennial Dreams in Action, Comparative Studies in Society and History* (The Hague: Mouton, 1962), p. 11.

18 J. S. Mill, *Representative Government*, in *Utilitarianism* (London: Everyman's Library, 1944), pp. 186–192, in contrast to the dialectical argument of *On Liberty*.

19 Martin Buber, *Paths in Utopia*, tr. by R. F. C. Hull (London: Routledge & Kegan Paul, 1949), pp. 127–148; Paul Goodman, *Utopian Essays and Critical Proposals* (New York: Vintage Books, 1964), pp. 3–22, 110–118; and Paul and Percival Goodman, *Communitas* (New York: Vintage Books, 1960), passim.

‘Ernst Bloch and the Ubiquity of Utopia’

from *Utopianism and Marxism* (1987)

Vincent Geoghegan

Editors’ Introduction

While the previous text is drawn from a wide-ranging collection of critiques, this is from a more specific enquiry into utopian socialism and anarchism in Europe since the nineteenth century. Other chapters cover Marx and Engels, Stalinism, and Herbert Marcuse’s synthesis of Marxism and psychoanalysis, as well as an opening chapter on French utopian thinkers such as Saint-Simon and Fourier. These could be read in conjunction with Fourier’s own writing in *The Utopian Vision of Charles Fourier: Selected Texts on Work, Love, and Passionate Attraction*, translated and edited by J. Beecher and R. Bienvenu (Columbia, MO, University of Missouri Press, 1983). Geoghegan is also author of a standard critical biography of Bloch: *Ernst Bloch* (London, Routledge, 1996), and is Reader in Politics at Queen’s University, Belfast.

Geoghegan writes, on the first page of the biography, that Bloch ‘is an exhilarating thinker’ who ignores traditional scholarly divisions of labour to roam over wide intellectual and cultural fields, but also that his ‘vast scheme will strike many as positively antediluvian’ (p. 3). He also notes that the three volumes of *The Principle of Hope* (published in German in 1959 while Bloch was Professor of Philosophy at the University of Leipzig in the German Democratic Republic) are intimidating: ‘Bloch’s writing is forbidding; structural complexity and formal eclecticism are combined with a writing style studded with opaque metaphor’ (p. 2). The reader may be pleased, then, that an account of Bloch rather than his own text is reprinted here. In it Geoghegan brings out the theme of hope’s ubiquity, which runs through and structures Bloch’s work. While Adorno maintained an elitism of culture, seeing mass culture only as mass deception, Bloch saw in popular novels (including the wild-west stories of Karl May) a presence of utopian thought on which the reader’s imagination could seize. Such glimpses of a world other and better than that around the reader served, in Bloch’s theory, to educate hope, to bring it nearer a stage of realisation of its object, which is freedom. Bloch retains an orthodox Marxist view that freedom is the objectively given end of history, and tries – it would seem forlornly but that can be debated – to establish hope as having the scientific status of equivalence to a Freudian drive. Much of his book is taken up with this, and much more with an almost endless range of cases of the utopian image in past cultural production. Yet, even if Bloch was wrong, it could still be that what he saw as a ubiquitous and ever-present hope remains valid. As Geoghegan summarises, Bloch’s place in a history of utopian thought is that he extends it from the elitist, intellectual concept (in which Shklar locates it) to popular culture. Hence the creative arts, rather than being the playthings of urban developers and property speculators, might be the paths to revolution.

THE PRINCIPLE OF HOPE

For Bloch, the enemies of hope are confusion, anxiety, fear, renunciation, passivity, failure and nothingness. Fascism was their apotheosis. But since all individuals daydream, they also hope. It is necessary to strip this dreaming of self-delusion and escapism, to enrich and expand it and to base it in the actual movement of society. Hope, in other words, must be both educated and objectively grounded; an insight drawn from Marx's great discovery: 'the subjective and objective hope-contents of the world'.[1]

The Principle of Hope is an encyclopaedic account of dreams of a better existence; from the most simple to the most complex; from idle daydreams to sophisticated images of perfection. It develops a positive sense of the category 'utopian', denuded of unworldliness and abstraction, as forward dreaming and anticipation. All the time, however, the link between past, present and future is stressed – concern with what one might be is the royal road to what one has been, and what one is: 'we need the most powerful telescope, that of polished utopian consciousness, in order to penetrate precisely the nearest nearness'.[2] This whole project is examined under five headings, each referring to a distinct form of hope:

1 'Little Daydreams': all those flights of fancy and reveries that occupy people throughout their day.
2 'Anticipatory Consciousness': the very basis of hope; the roots and purpose of dreaming in the individual.
3 'Wishful Images in the Mirror': the expression of hope in such forms as display, fairy tale, travel, film and the theatre.
4 'Outlines of a Better World': planned or outlined utopias – medical, social, technological, architectural and geographical utopias, plus the 'wishful landscapes' of painting and literature.
5 'Wishful Images of the Fulfilled Moment': the most powerful conceptions of authentic humanity.

'Little daydreams'

This delightful little section examines, with great sensitivity and acuity, a range of everyday hopes and fantasies. Bloch vividly recaptures the dreams of childhood: the secure hiding places, voyages to exotic lands, far-away castles, unlimited power; also the adolescent fantasies of love where the 'street or town in which the loved one lives turns to gold, turns into a party',[3] or the dream of returning home in triumph to the once unfeeling, but now awestruck, parents. Ever mindful of the experience of fascism, Bloch notes how these early yearnings were often captured, how the 'often invoked streak of blue in the bourgeois sky became . . . a streak of blood'.[4] With maturity comes the wishful rewriting of history, where the wrong turn is righted and the missed opportunity achieved, and related to this is the dream of revenge. With great personal bitterness, Bloch evokes the murderous, anti-semitic fantasies of the petit-bourgeois in the Weimar Republic and of their cynical manipulation by the bourgeoisie. He also details the various compensatory sexual fantasies of individuals: 'a dream forest of randy eyes and spread legs',[5] and the visions of financial success and domestic comfort. The inevitable limitations of bourgeois dreaming are emphasized – and most graphically exemplified in the figure of the jaded and bored rich man who has had the misfortune (to use Shaw's phrase) to get his heart's desire. By contrast, what Bloch terms the 'non-bourgeois dreamer', looks beyond the existing range of options to the socialist vision of true equality, freedom and community. Inevitably, these yearnings are 'considerably less distinct than those which need only reach into the existing window-display'.[6] They are, however, of a much higher status, and represent the way forward. This leads on to Bloch's touching and bittersweet account of the dreams of old age, where he contrasts the unnecessary hardship of the old under capitalism with the vision of wisdom and rest, evening and house, of 'authentic life in old age'.[7] Throughout this section on little daydreams one is struck by powerful images and evocative phrases: of how, for the young visitor to the big city, 'the houses, the squares, the stages seem bathed in a utopian light',[8] or of the brutality, malice and repulsiveness of petit-bourgeois dreams 'as pervasive as the smell of urine',[9] or, again of this latter class: 'it is also quite happy to put its clenched fist back into its pocket when crime is no longer allowed a free night on the town by those at the top'.[10]

'Anticipatory consciousness'

In this section Bloch goes back a stage and seeks to establish the basis of human dreaming, human aspiration: the basis, in other words, of hope. He distinguishes a whole series of interlinked tendencies within the individual – urging, striving, longing, searching, driving, craving, wishing and wanting – all of which propel us beyond ourselves. But what is behind these? He rejects the various Freudian explanations of motivation: they are saturated with bourgeois assumptions; they are oriented to the past ('there is nothing new in the Freudian unconscious');[11] they are obsessed with the libido ('it emphasises solely spicey drives';)[12] they disembody human impulses and consequently ignore basic socio-economic factors, and fail to grasp the historical mutability of human drives. Freud's one-time disciples come in for particular condemnation; Adler's 'will to power' is dismissed as an apology for capitalism, whilst Jung. that 'fascistically frothing psychoanalyst',[13] is accused of a racist and irrational primevalism. A much better candidate for a basic drive, Bloch argues, is hunger, 'the drive that is always left out of psychoanalytic theory',[14] and, as regards 'complexes', he suggests 'the one which Franziska Reventlov so unmedically called the money complex'.[15] Both rest on the only real basic drive – self-preservation – though even this is experienced differently in different environments. Self-preservation, however, turns into self-extension, as basic appetites are satisfied and give way to ever more sophisticated forms; ultimately, 'out of economically enlightened hunger comes today the decision to abolish all conditions in which man is an oppressed and long-lost being'.[16] Dreaming is an integral part of this process. Bloch is at pains to counter Freud's minimizing of the differences between day-time and night-time dreaming. Although wish-fulfilment occurs in nocturnal dreams, it is in an essentially regressive, repressive and highly distorted form. Daydreams, by contrast, combine clarity, open-endedness and future orientation. However, even the dreams of the night contain material which can be transformed into a utopian form in waking consciousness. The crucial element in all of this is what he terms the 'Not-Yet-Conscious'. This is a pre-conscious faculty in individuals, from which all novel material is generated: it is 'the psychological birthplace of the New'.[17] The New, however, does not come out of the blue, nor is it pure subjective creation; rather, it is drawn from the objective possibilities of the developing real world: 'inspiration . . . emerges . . . from the meeting of subject and object, from the meeting of its tendency with the objective tendency of the time, and is the flash with which this concordance begins'.[18] Only in Marxism is there this combination of hope and concreteness. This involves a combination of the 'warm stream' and the 'cold stream' of Marxism, where coldness is the rigorous scientific aspect and warmness its libertarian intent; this is expressed elsewhere as the unity of sobriety and enthusiasm. Again, there is much more in this section than this very bare summary suggests. Throughout, Bloch branches off into all manner of fascinating discussions – from art to folklore, history to religion; philosophy to psychology – in which he deploys a truly awesome erudition (as well, let it be said, as a deal of pomposity and wilful obscurity). This whole section, of nearly three hundred pages, is the theoretical core of Bloch's project. The underlying ideas are attractive. The concept of the 'Not-Yet-Conscious' avoids much of the insulting reductionism present in Freudian psychology. Its image of the individual is not that of the battered and screwed-up end-product of obscure childhood traumas, but of a person endowed with much greater independence and capacity for creative self-development. Instead of brooding on the hidden – and usually base – roots of people's desires and wishes, it focuses on the desirability of the goals and the beneficial function of the dreams. It also lends itself much more readily to an overall Marxist framework than do attempts to harness Freud to this end. On the other hand, sceptics might reply that this is due to the highly general and abstract nature of the concept 'Not-Yet-Conscious': its reliance on the author's intuition and on cultural authorities – the fact, in short, that it hasn't deigned to soil its hands with the clinical procedures of the Freudians.

'Wishful images in the mirror'

This is an exciting, original and important section. Bloch's achievement is to have uncovered the utopianism in (the often despised) mass or popular culture. He is fully aware of the exploitative nature of this culture, but equally of its link with wish-fulfilment (again, these are not always healthy wishes). As he notes of fashion and display:

> people cannot make of themselves what has not already previously begun with them. Equally, in terms of pretty wrappings, gestures and things, they are attracted outside only by what has already existed for a long time in their own wishes, even if

> only vaguely, and what is therefore quite willingly seduced. Lipstick, make-up, borrowed plumes help the dream of themselves, as it were, out of the cave. Then they go and pose, pep up the little bit that is really there or falsify it. But not as if it were possible for someone to make themselves completely false; at least their wishing is genuine.[19]

The travelling fair and circus also, amidst their tackiness and exploitation of the not-normal, are said to contain 'a bit of frontier land . . . with preserved meanings, with curiously utopian meanings, conserved in brutal show, in vulgar enigmaticness'.[20] Bloch shows the influence of his beloved Karl May in his assertion of the utopian content of adventure tales – a genre which is a 'castle in the air *par excellence*, but one in good air and, as far as it can be true at all of mere wishing-work: the castle in the air is right'.[21] Our author casts his net far and wide for vehicles of wishing such as, for example, travelling, stamp collecting, gardening, delight in wild weather. Dance 'paces out the wish for more beautifully moved being';[23] mime points to another region, as does film. Here again, the double-edged nature of the phenomenon is stressed: Jitterbug and Boogie-Woogie are 'imbecility gone wild'[23] and Hollywood is condemned as 'Dream-factory in the rotten . . . sense'.[24] But rotten dreams are not the ultimate enemies: those enemies are, rather, pessimism and nihilism – the absence of dreams:

> artificially conditioned optimism . . . is nevertheless not so stupid that it does not believe in anything at all . . . For this reason there is more possible pleasure in the idea of a converted Nazi than from all cynics and nihilists . . . Thus pessimism is paralysis *per se*, whereas even the most rotten optimism can still be the stupefaction from which there is an awakening.[25]

In the 1930s, Bloch had analysed the paradoxical nature of fascism – its mixture of the progressive and the reactionary. This was due to the fact that 'not all people exist in the same Now'.[26] Many sections of the population carried within themselves consciousness from earlier times because they were not fully integrated into contemporary society. Fascism managed to harness these older currents to its own chariot. Marxists, on the other hand, failed to see the progressive dimension in elements of this earlier consciousness; failed to see that these elements were produced by inadequacies in earlier societies. These elements are the

> still subversive and utopian contents in the relations of people to people and nature, which are not past because they were never quite attained . . . These contents are, as it were, the gold bearing gravel in the course of previous labour processes[27]

Socialists cannot simply dismiss the whole phenomenon, they must attempt to integrate this valid dimension into their politics. As Bloch concluded at the time, fascism will continue 'as long as the revolution does not occupy and rebaptize the living Yesterday'.[28] Socialism, the ultimate goal, can therefore draw upon myriad sources deep in every individual. This is Bloch's great service in this section: he points to a transmission belt between the small-scale, the mundane – and even the seemingly reactionary – and the grand *telos* of communism. Where one might fault him is in his rather arbitrary distinctions between authentic and inauthentic: where, for example, adventure stories are placed on a higher footing than 'syrupy stories' in glossy magazines. This was due partly to an inevitable personal quirkiness, plus the broader influences of the milieu in which he grew up. There does also appear to be in these distinctions a strong dose of the anti-Americanism found in many of the 'emigration' generation of Weimar Germany. There is some loss of sensitivity as a result. This, however, is only a minor caveat; the overall perspective is truly impressive.

'Outlines of a better world'

In the nearly five hundred pages which make up this section, Bloch assembles the different conscious attempts to depict a better world – the more usual meaning of the word 'utopia'. There are the various medical utopias, deeply rooted in perennially human concerns, with their abolition of disease and pain. However, even the perennial is rooted in a particular historical context – 'utopias have their timetable',[29] Bloch insists, and they cannot be understood outside their time. We are then treated to an encyclopaedic account of historical utopias. Along with the usual Plato we get Solon, Diogenes and Aristippus. The Bible is seen as a treasure-house of utopian imagery; Moses is credited with the creation of a liberation God: 'the God he imagines is . . . no masters' God . . . Yahweh begins as a threat to the Pharaoh: the volcanic God of Sinai becomes Moses' god of liberation, of flight from slavery'.[30] Jesus is interpreted as the harbinger of a new world: 'the eschatological sermon has precedence for Jesus over the moral one and determines it.'.[31] Augustine is included, as is the fascinating medieval heretic Joachim of Fiore, with his dream of the Third Kingdom. And so on, through More and Campanella, Rousseau and Fichte, Owen and Fourier, Cabet and Saint-Simon, past Stirner, Proudhon and Bakunin, on to Weitling. He includes a rather odd section of women's utopias, which many modern feminists would find patronizing and sexist, as they would many of his references to women throughout the work. He claimed that the women s movement 'is at once outmoded, replaced and postponed',[32] in that capitalism is more than willing to extend its worthless equality and class struggle has primacy over sexual struggle. It is, however, only postponed in that women have a utopian dimension to contribute to future socialist society, a contribution defined in terms of the 'special qualities' of women. This gives way to a discussion of Zionist visions ('Zionism flows out into socialism, or it does not flow out at all'),[33] then the utopian works of Bellamy, Morris, Carlyle and George. The account is seemingly endless: technological utopias, architectural utopias from 'Dreams on the Pompeian wall' to Le Corbusier, geographical utopias ('Eldorado and Eden'), wishful landscapes in painting, opera and literature ('Pieter Brueghel painted his Land of Cockaigne exactly as the poor folk always dreamed it to be')[34] and so on and so on. The erudition and colossal scale are quite breathtaking. It is an absolute gold mine for those interested in utopias. Bloch's purpose, however, is not antiquarian – rather it is both to demonstrate the historical ubiquity of this type of dreaming forward and to argue for a synthesis of dreaming, stripped of illusion, with a Marxism stripped of positivism and empiricism, where 'everything inflamed in the forward dream is thereby removed as is everything mouldy in sobriety'.[35] This is the concept of 'concrete utopia'.

'Wishful images of the fulfilled moment'

In this concluding section, Bloch presents what he considers to be the most sublime images of existence, the ones which throughout history have possessed an aura of profound otherness. These are the golden seams of human dreaming. They also provide a window on the deepest beliefs and values of Bloch himself. Historically, these images have often appeared in contradiction to one another: contradictions which will develop into dialectical syntheses. Thus there will be a life combining the old opposed ideals of danger and happiness, in which courage and adventure prevent enervation and boredom, and felicity prevents brutality, insecurity and emptiness; there is the new tactical ideal of 'neither non-violent hesitation nor cunning abstractness of violence, but violence concretely mediated'.[36] The same is said of the other dualisms – body and soul, action and contemplation, solitude and friendship, individual and collective. Two important areas of focus in this section are music and religion. The notion of art as a bridge to an order outside and beyond the given has a long history; so too has an appreciation of the subversive role of art. The exalted status of poetry in classical Greece is well known, as is Plato's desire to control it in his ideal society. In modern times, Schiller developed the concept of art as a vehicle for a normative ideal – a standard from which to criticize current arrangements and a goal for political change. For Bloch, music, like all phenomena, has an ideological dimension, rooted in its time: 'it extends from the form of the performance right to the characteristic style of the tonal material and its composition, to the expression, the meaning of the content. Handel's oratorios in their festive pride reflect rising imperialist England'.[37] This does not exhaust its content, for as Bloch argues elsewhere:

> a 'significant' work does not perish with the passing of time, it belongs ideologically, not creatively, to the age in which it is socially rooted. The permanence and greatness of major works of art consists precisely in their operation through a fulness of pre-semblance and of realms of utopian significance.[38]

In fact, 'no art has so much surplus over the respective time and ideology in which it exists'[39] as music. The complex qualities of music have made it a particularly rich vehicle for the expression of utopian content and historically it has expressed the most sublime longings of humanity: 'thus music as a whole stands at the frontiers of mankind'.[40] Bloch attempts the (as he would himself admit) impossible task of articulating some of these images of liberation. In the case of religion we are dealing with a phenomenon which has been looked down on in the Marxist tradition. It is sometimes forgotten by Marxists that Marx's remark that 'religion ... is the opium of the people' occurs in a passage where the critical element of religion is also referred to: religion, for Marx, is also 'the sigh of the oppressed creature';[41] it is 'the expression of real distress and the protest against real distress'.[42] That this was forgotten in the Marxist tradition is no doubt due to Marx's own opinion that religion's critical moment had passed – that it had nothing further to contribute to a revolutionary politics. Bloch, on the other hand, develops an impressive analysis of the critical and anticipatory elements in the world's religions and argues for the continuing relevance of religion in Marxism. The religious impulse, stripped of its illusory aspects, is thus profoundly revolutionary. This involves:

> the elimination of God himself in order that precisely religious mindfulness, with hope in totality, should have open space before it and no ghostly throne of hypostatis. All of which means nothing less than just this paradox: the religious kingdom-intention as such involves atheism, at least properly understood atheism.[43]

or, as he pithily put it later: 'Only an atheist can be a good Christian, only a Christian can be a good atheist'.[44]

The book throughout displays Bloch's pro-Soviet Marxism-Leninism. Thus we are informed that 'The Soviet Union faces no question of women's rights any more, because it has solved the question of workers' rights'[45] (and he doesn't mean by abolishing them!) and that the Soviet Union is in the forefront of progress across the board. This was a long-standing theme in Bloch's work. He defended the Moscow Show Trials in 'A Jubilee for Renegades' (1937). Of the critics, 'the renegades', he wrote:

> Although many of them have loved the beginning of the Russian Revolution, during the last two years they have lost their enthusiasm. They cannot get over the fact that this 20-year-old bolshevist child must rid itself of so many enemies, and that it discards them so ruthlessly.[46]

After the war, Bloch accepted a university post in East Germany at Leipzig, and for a number of years appears to have found nothing particularly objectionable in the Marxist-Leninist concept of the communist party. In a lecture to an East German audience he couched his utopian perspectives in the language of Marxist-Leninist planning:

> adventure is in the vanguard of the dialectical-material process, together with a plethora of real problems which evoke courage in order to survive the venture, as well as penetratingly concrete reason – needed in order to perceive the tendency. This wisdom, the always keen and well-thought-out wisdom of Lenin, watches over the path to the classless society. Out of this non-schematic approach new intermediate analyses of situations, always more concrete and expanded two-year, five year plans of theory and practice are always arising.[47]

The party would thus appear to be the ultimate directional force, the guardian of analysis and utopia. On such puny legs, therefore, did Bloch rest his great edifice. In the wake, however, of party harassment and revelations of the nature of Soviet Stalinism, Bloch became disillusioned and in 1961 took up permanent residence in West Germany. In his new home he argued that Marxism's own shortcomings had enabled the authoritarian centralism of Stalinism to triumph. Neglect of the natural law tradition had resulted in concern with human happiness not being matched by concern for human dignity:

> There are men [*sic*] who toil and are burdened, those are the exploited. But in addition there are also men who are degraded and offended ... The factor of vexation and degradation urgently deserves a name and a concept. Stalinism was able to impose itself without resistance because this term was hardly heard in Marxism after 1917.[48]

Another lacuna, present in Marx's own work, was a 'comparatively weak emphasis on personal freedoms'[49] which assisted the anti-democratic nature of Stalinism. Bloch cites Rosa Luxemburg's slogan 'No socialism without democracy' as an example of a healthier trend.

This then is Bloch's great masterpiece. His achievement was to see that utopianism is not confined to intellectuals and their various blueprints of a better life. He saw that, in countless ways, individuals are expressing unfulfilled dreams and aspirations – that in song and dance, plants and plaster, church and theatre, utopia waits.

NOTES

1 E. Bloch, *The Principle of Hope*, Oxford: Basil Blackwell 1986, three volumes, 7.
2 *Ibid.*, 12.
3 *Ibid.*, 26.
4 *Ibid.*, 29.
5 *Ibid.*, 32.
6 *Ibid.*, 35.
7 *Ibid.*, 41.
8 *Ibid.*, 28.
9 *Ibid.*, 31.
10 *Ibid.*, 31.
11 *Ibid.*, 56.
12 *Ibid.*, 64.
13 *Ibid.*, 59.
14 *Ibid.*, 64.
15 *Ibid.*, 66.
16 *Ibid.*, 76.
17 *Ibid.*, 116.
18 *Ibid.*, 125.
19 *Ibid.*, 339.
20 *Ibid.*, 366.
21 *Ibid.*, 369.
22 *Ibid.*, 394.
23 *Ibid.*, 394.
24 *Ibid.*, 409.
25 *Ibid.*, 445-6.
26 E. Bloch, 'Nonsynchronism and the obligation to its dialectics', *New German Critique*, 11, 1977, 22.
27 *Ibid.*, 38.
28 *Ibid.*, 27.
29 Bloch, *The Principle of Hope*, 479.
30 *Ibid.*, 496.
31 *Ibid.*, 500.
32 *Ibid.*, 595.
33 *Ibid.*, 611.
34 *Ibid.*, 813.
35 *Ibid.*, 622.
36 *Ibid.*, 947.
37 *Ibid.*, 1063.
38 M. Solomon (ed.), *Marxism and Art*, Hassocks: Harvester 1979, 584.
39 Bloch, *The Principle of Hope*, 1063.
40 *Ibid.*, 1103.
41 K. Marx and F. Engels, *On Religion*, London: Lawrence & Wishart 1958, 42.
42 *Ibid.*, 42.
43 Bloch, *The Principle of Hope*, 1199.
44 E. Bloch, *Atheism in Christianity*, New York: Herder & Herder 1972, frontispiece.
45 Bloch, *The Principle of Hope*, 595.
46 E. Bloch, 'A jubilee for renegades', *New German Critique*, 4, 1975, 18.
47 E. Bloch, *On Karl Marx*, New York: Herder & Herder 1971, 139.
48 M. Landmann, 'Talking with Ernst Bloch: Korcula, 1968', *Telos*, 25, 1975, 171.
49 Bloch, *On Karl Marx*, 169.

'Uneven Development: Public Art in New York City'

from *Evictions: Art and Spatial Politics* (1996)

Rosalyn Deutsche

Editors' Introduction

The version of this essay from which a short extract is given here was revised for the collection *Evictions*, but largely follows the text first published in *October* (no. 47) in 1980, and in *Out of Site: A Social Criticism of Architecture*, edited by Dianne Ghirardo (Seattle, Bay Press, 1991, pp. 157–219). A further essay by Deutsche is 'Alternative Space', in *If You Lived Here: The City in Art, Theory, and Social Activism*, edited by Martha Rosler and Brian Wallis (Seattle, Bay Press, 1991, pp. 45–66). These texts were influential during the 1990s in moving criticism of art in public places from an aestheticist to a socially critical position. Much of their content can be applied to and concerns questions in urban theory and planning. In 'Alternative Space', for instance, Deutsche argues that the ways in which cities have been conventionally depicted in art, and discussed by art historians, is as merely reflecting or recording social trends rather than constructing them. Among the theoretical material on which she draws is the work of Henri Lefebvre, Jürgen Habermas and Manuel Castells.

Only the opening of 'Uneven Development' is reprinted here, as indication of some rather dystopian conditions prevailing in New York in the late 1970s – the period of SoHo's gentrification. The essay is long and needs careful reading, but this extract is included in part because it cites New York's Mayor Koch as representative of the rational planning model in a state of decay. While homeless (that is, in many cases evicted) people used public space in the city as living space, having none other, Koch sought to clear them on the grounds of a mono-functional approach to urban space, along the lines of train stations as only for people taking trains. Deutsche sees Koch's attitude as serving a dominant interest, portrayed as a coherent majority and requiring a coherent, single interpretation of spatiality. This she contrasts with a view from social critic Raymond Ledrut, that space is socially produced through interaction and the co-presence of different practices in the same space. Lefebvre makes a similar case in *The Production of Space*, though perhaps with less emphasis on contestation and more on the overlay of different meanings in space (see the extract from his writing in Part Six). This text could be read, too, with Iris Marion Young's in Part Seven.

The true issue is not to make beautiful cities or well-managed cities, it is to make a work of life. The rest is a by-product. But, making a work of life is the privilege of historical action. How and through what struggles, in the course of what class action and what political battle could urban historical action be reborn? This is the question toward which we are inevitably carried by our inquiry into the meaning of the city.

(Raymond Ledrut, "Speech and the Silence of the City")

BEAUTY AND UTILITY: WEAPONS OF REDEVELOPMENT

By the late 1980s it had become clear to most observers that the visibility of masses of homeless people interferes with positive images of New York, constituting a crisis in the official representation of the city. Dominant responses to the crisis took two principal, often complementary forms: they treated homelessness as an individual social problem isolated from urban politics or, as Peter Marcuse contends, tried "to neutralize the outrage homelessness produces in those who see it."[1] Because substantial efforts to deal with homelessness itself would have required at least a partial renunciation of its immediate causes – the commodification of housing, existing employment patterns, the social service policies of today's austerity state – those committed to preserving the status quo tried, instead, through strategies of isolation and neutralization, to cope with the legitimation problems that homelessness raises.

Exemplary of the "social problem" approach is a widely circulated report issued in June 1987 by the Commission on the Year 2000. Obedient to its governmental mandate to forecast New York City's future, the panel described New York as "ascendant," verifying this image by pointing to the city's "revitalized" economy and neighborhoods. Conspicuous poverty and patent stagnation in other neighborhoods nonetheless compelled the commission to remark on the unequal character of this rise: "We see that the benefits of prosperity have passed over hundreds of thousands of New Yorkers."[2] But the group's recommendations – prescribing the same pro-business and privatizing policies that are largely responsible for homelessness in the first place – failed to translate this manifest imbalance into a recognition that uneven economic and geographical development is a structural, rather than incidental, feature of New York's present expansion. The panelists' own expansive picture required, then, a certain contraction of their field of vision. Within its borders, social inequities appear as random disparities and disappear as linked phenomena. An optical illusion fragments the urban condition as "growth" – believed to occur in different locations at varying paces of cumulative development but ultimately to unfold its advantages to all – emerges as a remedy for urban decay, obscuring a more integrated economic reality that is also inscribed across the city's surface. For in the advanced capitalist city, growth, far from a uniform process, is driven by the hierarchical differentiation of social groups and territories. Residential components of prosperity – gentrification and luxury housing – are not distinct from, but in fact depend upon, residential facets of poverty – disinvestment, eviction, displacement, homelessness. Together, they form only one aspect of the city's comprehensive redevelopment, itself part of more extensive social, economic, and spatial changes, all marked by uneven development. Consequently, redevelopment proceeds not as an all-embracing benefit but according to social *relations* of ascendancy, that is, of domination. Consensus-oriented statements such as *New York Ascendant* disavow these relations, impressionistically offering proof of growth side by side with proof of decline; both acquire the appearance of discrete entities. But today there is no document of New York's ascendancy that is not at the same time a document of homelessness. Municipal reports, landmark buildings. and what we call public spaces are marked by this ambiguity.

Faced with the instability pervading New York's urban images, the second major response to homelessness – the neutralization of its effects on viewers – attempts to restore a surface calm that belies underlying contradictions. To legitimate the city, this response delegitimates the homeless. In the spring of 1988 Mayor Ed Koch demonstrated the neutralizing approach while speaking, fittingly, before a group of image makers, the American Institute of Architects (AIA), convened in New York to discuss (even more appropriately) "Art in Architecture." Answering a question about Grand Central Terminal – landmark building and public place – Koch, too, emphasized the dual significance of New York's urban spaces by directing his listeners' attention to the presence of the homeless people who now reside in the city's train stations:

> These homeless people, you can tell who they are. They're sitting on the floor, occasionally defecating, urinating, talking to themselves many, not all, but many – or panhandling. We thought it would be reasonable for the authorities to say, "You can't stay here unless you're here for transportation." Reasonable, rational people would come to that conclusion, right? Not the Court of Appeals.[3]

The mayor was denigrating the state court's reversal of an antiloitering law under which police would have been empowered to remove the homeless from transportation centers. Even had police action succeeded in evicting the homeless, it is doubtful that it could have subdued the fundamental social forces threatening the station's appearance as an enduring symbol of New York's beauty and efficiency. Deprived of repressive powers, however, Koch could protect the space only by ideological means, proclaiming its transparency, in the eyes of reasonable people, to an objective function – transportation.

To assert in the language of common sense that an urban space refers equivocally to intrinsic uses is to claim that the city itself speaks. Such a statement makes it seem that individual locations within the city and the spatial organization of the city as a whole contain an inherent meaning determined by the imperative to fulfill needs presupposed to be natural, simply practical. Instrumental function is the only meaning signified by the built environment. This essentialist view systematically obstructs – and this is actually its principal function – the perception that the organization and shaping of the city as well as the attribution of meaning to spaces are social processes. Spatial forms are social structures. Seen through the lens of function, spatial order appears instead to be controlled by natural, mechanical, or organic laws. It is recognized as social only in the sense that it meets the purportedly unified needs of aggregated individuals. Space, severed from its social production, is thus fetishized as a physical entity and undergoes, through inversion, a transformation. Represented as an independent object, it appears to exercise control over the very people who produce and use it. The impression of objectivity is real to the extent that the city is alienated from the social life of its inhabitants. The functionalization of the city, which presents space as politically neutral, merely utilitarian, is then filled with politics. For the notion that the city speaks for itself conceals the identity of those who speak through the city.

This effacement has two interrelated functions. In the service of those groups whose interests dominate decisions about the organization of space, it holds that the exigencies of human social life provide a single meaning that necessitates proper uses of the city – proper places for its residents. The prevailing goals of the existing spatial structure are regarded as, by definition, beneficial to all. The ideology of function obscures the conflictual manner in which cities are actually defined and used, repudiating the very existence of groups who counter dominant uses of space. As the urban critic Raymond Ledrut observes, "The city is not an object produced by a group in order to be bought or even used by others. *The city is an environment formed by the interaction and the integration of different practices.* It is maybe in this way that the city is truly the city."[4]

NOTES

1 Peter Marcuse, "Neutralizing Homelessness," *Socialist Review* 18, no. 1 (January/March 1988): 83. Marcuse's premise – that the sight of homeless people is shocking to viewers but that this initial shock is subsequently counteracted or neutralized by ideological portrayals – implies that viewers responses to the presence of homeless people in New York today are direct and natural. It thus fails to acknowledge that current experience of beggars and "vagrants" by housed city residents is always mediated by existing representations, including the naming of such people as "the homeless" in the first place. The form and iconography of such representations not only produce complex, even contradictory, meanings about the homeless – the object of the representation – but also, in setting up the homeless as an image, construct positions in social relations. It is important to question these relationships as well as the content of representations of the homeless. Despite Marcuse's ingenuous approach to the issue of representation, his description of official attempts to neutralize the effects of homelessness and his effective efforts to contest these neutralizations are extremely useful. This is especially true now, as, encouraged by the final years of the Koch administration, the media seem determined to depict homeless people as predators, to encourage New Yorkers to refuse donations to street beggars, and to create the impression that adequate city services exist to serve the needs of the poor and homeless.

2 *New York Ascendant: The Report of the Commission on the Year 2000* (New York: Harper and Row, 1988), 167.

3 David W. Dunlap, "Koch, the 'Entertainer,' Gets Mixed Review," *New York Times*, May 19, 1988, B4.

4 Raymond Ledrut, "Speech and the Silence of the City," in M. Gottdeiner and Alexandros Ph. Lagopoulos, ed., *The City and the Sign: An Introduction to Urban Semiotics* (New York: Columbia University Press, 1986), 122.

'The Death of Radical Planning: Radical Praxis for a Postmodern Age'

from Mike Douglass and John Friedman (eds), *Cities for Citizens* (1998)

Leonie Sandercock

Editors' Introduction

Leonie Sandercock established herself as a leading voice in radical planning through her book *Towards Cosmopolis* (Chichester, J. Wiley & Sons, 1998). At the time of researching and writing it, Sandercock was Professor of Landscape, Environment, and Planning at RMIT (she is now in Vancouver), seeing at first hand how a city coped with immigration from other countries at the same time as a not always happy effort to come to terms with the presence of an indigenous population. Several of the ideas examined in that book are given in a briefer way in the chapter reprinted here from *Cities for Citizens*. Her point of departure is the Los Angeles riot of 29 April1992, in which parts of the city burned. But rather than launch into an apocalyptic or nightmare narrative of approaching urban destruction, Sandercock moves to a critical history of planning as the means by which decisions about the kinds of city and space people occupy as citizens are made. The general direction of this history is from the rational planning model of the Chicago sociologists of the 1920s and 1930s – see the text by Ernest W. Burgess reprinted in Part One as exemplar of this – through advocacy planning and other more progressive models, to a radical planning model in which the professional planner becomes a collaborator with active groups at the local level. This entails listening to voices outside conventional planning circles, and spending a considerable part of the time working with mobilised community groups.

In some ways, this likens the planner to a participant observer in contemporary ethnographic research, with the difference that the planner is not obliged to spend six months living with the mobilised community being assisted in order to write a thesis about them. The planner's role is more radical, in being an active force for change working through others who, given the necessary information, language, and confidence, can empower themselves. In this respect, radical planning goes considerably beyond advocacy planning in its revision of what planners do. It is not unlike the model of self-empowerment proposed by Paolo Freire in his seminal *Pedagogy of the Oppressed* (Harmondsworth, Penguin, 1972). What has this to do with utopia? Apart from explaining the limitations of the modern project as interpreted in rational planning, it offers a way in which a new society can be imagined, not as an aesthetic entity, but in ways conducive to its realisation at a local level. It may not work, nor solve global problems, but as Sandercock concludes "the very notion of certainty has been abandoned in favour of a more complex, historically informed and situation-specific reading of competing claims" ('The Death of Radical Planning – Radical Praxis for a Postmodern Age', p. 184). A further text on multi-ethnic issues in city development which it would be helpful to read with this is Jane M. Jacobs' *Edge of Empire: Postcolonialism and the City* (London, Routledge, 1996).

I am angry
It is all right to be angry
It is unfortunate what people do when they are
frustrated and angry.

The fact of the matter is,
whether we like it or not,
riot
is the voice of the unheard.

(Congresswoman Maxine Waters, quoted in Anna Deveare Smith 1993)

THE BACKDROP: MAPPING POSTMODERN SPACE

Twilight, Los Angeles, 29 April 1992. The city is burning. As the smoke and glow from the fires begins to rise over the city, millions of horrified citizens huddle in front of TV sets which transmit the images of rioting and looting. At first the media portray these events as black–white rage riots, a return of the ghetto insurrections of the mid-1960s, triggered by the "not guilty" verdict against four white cops accused of beating the African American motorist Rodney King. But a closer look, in the blazing days that follow, shows that these riots, this looting, have a new character. Few of the businesses destroyed were white owned. About half belonged to Koreans, another third to Latinos (Mexican Americans and Cubans). Many of the looters were black, but there were also Latinos and whites in the space, which at times appeared carnivalesque, a joyous celebration of the breakdown of law and order. In fact, the week of riots was more inferno than carnival, an inferno in which 52 people were killed, a third of them Latinos; 16291 people were arrested, a third of them black, another third Latino: 500 fires were set and 4000 businesses destroyed. About 1200 of those arrested were illegal aliens, roughly 40% of whom were turned over to the Immigration and Naturalization Service for immediate deportation (Oliver, Johnson and Farrell 1993:118). An article in New York's *Voice* described these turbulents days as "the first multicultural riots" (Kwong 1992). Subsequent analyses refer to the nation's first "multiethnic rebellion" (Oliver, Johnson and Farrell 1993:130).

Los Angeles has become a favored and perhaps overused image of the postmodern urban condition (see Soja 1989; Davis 1990; Jameson 1984), a worst-case scenario of racial and ethnic conflict, social polarization and residential segregation. It is a city marked by enormous gulfs: between the corporate elite of the "network society" (Castells 1996) and the informal sector workers who service them; between rich, well-guarded suburbs and decaying and crime-plagued inner cities; between citizens and non-citizens; between a dominant culture and minority cultures. This gulf may be described as the divide between inclusion and exclusion, and its contours are not unique to Los Angeles. Seyla Benhabib (1995) notes that every Western nation state in Europe today "faces the challenge of multiculturalism and multinationalism." In countries like Australia, Canada, New Zealand and the United States, she writes, multinationalism – "the presence of more than one 'we' community within the same political constitution" – is the norm.

> Where "we" are today globally is a situation in which every "we" discovers that it is in part a "they": that the lines between "us" and "them" are continuously redefined through the global realities of immigration, travel, communication, the world economy, and ecological disasters.
>
> (Benhabib 1995:244)

The cities and regions of the end of the twentieth century are multiethnic, multiracial, *multiple*. The cities and regions of the next century will be more so. This chapter argues that the economic and demographic restructurings which cities and regions across the globe have been experiencing since the 1970s – during the transformation of a modernist landscape into a postmodern one – require a parallel theoretical restructuring in our thinking about planning in this postmodern age.

Most commentators on this restructuring of cities and regions have focused on its economic dimensions (Friedmann and Wolff 1982; Friedmann 1986; Sassen 1991; Harvey 1989; Jameson 1991), some on the socio-spatial consequences (Soja1989; 1996). But the relationship between socio-cultural and socio-spatial reshapings, between the crisis of Fordist modernization and the crisis of modernism itself, has been less well studied. I want to argue that there have been and are three major socio-cultural factors at work in the reshapings of cities and regions, each of which has contributed to the emergence of a "new cultural politics of difference" (West 1990), undermining the modernist paradigm on which planning practices have been constructed. I want to forge a productive encounter between the spaces of contemporary cities and regions, the recent theorizations of feminism, postmodernism and postcolonialism the spaces of an insurgent urbanism (the rise of civil society), and the practice of planning.

THE SOCIO-CULTURAL RESHAPING OF CITIES AND REGIONS

The age of migration

> The last decade of the twentieth century and the first of the twenty-first century will be the age of migration.
>
> (Castles and Miller 1993:3)

International migration is not an invention of the late twentieth century, nor even of modernity in its twin guises of capitalism and colonialism. Migrations have always been part of human history. But there has been a growth in the volume and significance of migration since 1945, and particularly since the mid-1980s. An ensemble of factors has contributed so this: growing inequalities in wealth between North and South impel people to move in search of better ways of living; political, ecological and demographic pressures force some people to seek refuge beyond their homeland; ethnic and religious struggles like those in Palestine and Yugoslavia lead to mass exodus; and the creation of new free-trade areas causes movements of labor. In some developing countries, emigration is one aspect of the social crisis which accompanies integration into the world market and modernization (Castles and Miller 1993:3–4). There have been and are many geographical expressions of these forces: from periphery to center; from former colony to heartland of empire; from South to North; from countryside to big city . . . The result is postmodern cities and regions of extraordinary cultural diversity and the attendant problems of living together in one society for ethnic groups with diverse cultures and social practices. Migrations can change economic, demographic and social structures, and the associated cultural diversity can call into question longstanding notions of citizenship and national identity. Influxes of migrants lead eventually to the spatial restructuring of cities and regions, in which sometimes the very presence of new ethnic groups is destabilizing of the existing social order. In this new "ethnoscape" (Appadurai 1990:7) ambivalent new communities are thrust together with anxiously nostalgic old ones, and xenophobic fears can quickly turn into a territorially based racist politics as the new mix of cultures projects itself on to the urban landscape (Jacobs 1996; Cross and Keith 1993).

Twenty years ago I lived in an inner-city neighborhood of Melbourne, Australia, which had undergone a postwar transition from its Anglo-Celtic mid-nineteenth century working-class origins to a significant Greek influx. The high street in the 1970s was dominated by good cheap Greek restaurants and coffee shops where Greek men seemed to play cards all day. Revisiting this neighborhood in 1995, walking down that same street, I encountered a class of schoolchildren, 30 or so 10-year-olds, marching in more or less orderly pairs behind their teacher. Twenty-five of these kids were Vietnamese, and the remaining few were the once more familiar blond Anglo-Celtic stock. The Greek restaurants and coffee shops had gone, replaced by Vietnamese grocery stores, restaurants and other small businesses. A couple of blocks away, off the high street, amid the high-rise public housing towers that some of us had fought unsuccessfully against in the 1960s (not because they were public housing but because they were high rise), what had previously been wasteland around these towers had been converted into dense vegetable gardens by Vietnamese residents.

We all know such stories, have had such experiences. Some of us find them exhilarating, others threatening. When residents with different histories, cultures and needs appear in "our" cities, their presence disrupts the normative categories of social life and urban space. The urban experiences of new immigrants, their struggles to redefine the conditions of belonging to "their" new society, are reshaping cities the world over, but particularly the so-called world titles (Friedmann and Wolff 1982; Friedmann 1995; Knox and Taylor 1995) or global cities (Sassen 1995) of the advanced capitalist economies. As new and more complex kinds of ethnic diversity come to dominate cities, the very notion of shared community, of "the public interest," becomes increasingly exhausted. These struggles over belonging take the form of struggles over citizenship, in its broadest sense, of rights to and in the polis. James Holston (1995) has called the sites of these struggles "spaces of insurgent citizenship."

> Citizenship changes as new members emerge to advance their claims, expanding in realm, and as new forms of segregation and violence counter these advances, eroding it. The sites of insurgent

> citizenship are found at the intersection of these processes of expansion and erosion. (Holston 1995:44)

These sites vary with time and place. Today they include ethnic neighborhoods, suburban migrant labor camps, sweatshops, places of worship and the zones of the new racism (which does not exhaust the list). The multicultural (multiethnic, multiracial) city is continually creating these sites of struggle. They are part of the *landscape of postmodernity*, which is a landscape of/marked by difference. Negotiating these spaces, claiming them, making them safe, imprinting new identities on them, is a central socio-cultural and political dynamic of cities and regions, in which planners have a pivotal role. Planners' historic role has been above all to control the production and use of space. In this state-designated role, they have acted as spatial police, regulators of bodies in space, deciding who can do what and be where, and even when. They have closed public parks at night so that the homeless can't sleep in them, created ordinances to prevent street vendors, blocked permits for mosques, determined what kinds of housing renovations are permissible, and invented no end of both blunt and subtle ways of keeping certain bodies (*marked* bodies, marked by color, by race, by gender, by sexual preference and by physical ability) out of the sight and out of the way and out of the neighborhoods of certain other bodies. These discriminations and repressions are the *noir* history of planning (see Sandercock 1995b). They mark the profession as reactionary, as complicit with the dominant culture. If our future role is to be less reactionary, then we have to come to terms with the new and complex ethnic and cultural diversity that is coming to characterize the multicultural city, to develop a multicultural literacy more attuned to cultural diversity, and to redefine and reposition planning according to these new understandings.

THE AGE OF POSTCOLONIALISM AND THE RISE OF INDIGENOUS PEOPLES: THE POSTCOLONIAL POLITICS OF URBAN AND REGIONAL SPACE

In New World settlements the world over, in the era of colonialism, settlers usually occupied space at the expense of existing inhabitants, who were referred to as native and regarded as "primitive". While the details of colonial occupations vary, the process of city building and the clearing of regions for farming and other extractive industries required an ordering of urban and regional space by a whole range of spatial technologies of power, such as the laws of private property, the practices of surveying, naming, mapping, and the procedures of urban and regional planning. The effects of these various sorts of legal and/or violent arrangements and appropriations were the effective dispossession and exclusion of indigenous peoples. The desire to establish settler colonies depended on the "will of erasure" or, when that failed, the "systematic containment" of the original inhabitants (Jacobs 1996:105). In the United States and Canada, "treaties" were struck with Indian nations, who were then forced on to reservations. In Australia, "this erasure was inaugurated by the notion of *terra nullius*, land unoccupied, which became the foundational fantasy of the Australian colonies" (Jacobs 1996:105). While spatial segregation of indigenous peoples was the almost universal intent of colonizers, that intent was only ever imperfectly realized, quickly giving way to more disorderly and permeable spatial arrangements in which individuals or groups found their way into cities and settled areas, occupying what came to be the unseen or unincorporated parts of the city (McWilliams 1990), or the fringes of urban areas or country towns.

Since the 1970s there has been a global movement on the part of indigenous peoples to reverse these foundational injustices and dispossessions. At the heart of this movement are land claims that are profoundly or potentially destabilizing of established practices of land-use planning, land management and laws of private property – all of which constitute part of the core of planning practice. While most of these land claims have been lodged and contested in relatively remote regions, where they may clash with mining and farming interests, there are also urban implications and sitings of these postcolonial struggles. Jacobs (1996) details a case in Perth (a city of almost one million people, capital of its province, on the west coast of Australia) in the late 1980s, in which a former brewery came to be seen by the then state government as a heritage site and as a perfect opportunity for recycling for use as a tourist/leisure centre featuring restaurants, retail outlets and galleries. This vision was confounded by Aboriginal claims that the site was the home of the Waugal serpent; that is, an Aboriginal site of significance which must be protected. For almost a year Aboriginal people from the Perth region occupied the site and residents and urban authorities were confronted with the unexpected presence of the Aboriginal sacred in the city.

Studies of the postmodern city such as those by Ed Soja (1989) and David Harvey (1989) speak exclusively of a secular city. "This erasure of the sacred speaks precisely of the modernist underpinnings of 'western' urbanism as well as the recalcitrant modernism of these accounts of the 'post' modern urban" (Jacobs 1996:131). In the contemporary city, as Kong (1993) has shown for multireligious Singapore, sacredness is planned for, not to protect is from the threat of the secular but because the secular, the accumulative drive of the urban, needs to be protected from the irrreverence – the unproductivity – of the sacred. The Aboriginal sacred is deeply antagonistic to urban modernity's need so keep the sacred apart from the secular and to regulate is as if it were just another land use.

While Aboriginal protests failed on this occasion, and development proceeded, subsequent events on the national stage in the 1990s – what has come to be known as the Mabo decision and the subsequent Native Title Act of 1993 – have even more profoundly unbound existing laws governing the ownership and use of land, "dismantling the established spatial architecture of existing land rights provisions . . . which comfortably placed a spiritualised, 'tribalised', lands-rights deserving, Aboriginality well away from the urban centres" (Jacobs 1996:112). I return to these developments at the end of this chapter, as an example of what a postmodern regional planning practice might be. In the claims of indigenous peoples for the return of or access to their lands, planners are confronted with values incommensurable with modernist planning and the modernization project which it serves, a planning which privileges "development" and in which exchange value usually triumphs over use value. If the voices and desires of indigenous peoples are to be respected, acknowledged and honored, the foundations of the modernist planning paradigm itself must be abandoned and replaced.

The age of feminism: gender, sexuality and space

Beginning in the 1970s a new wave of feminist writing and activism began to dismantle/deconstruct the city as it had come to be understood in modernist thought. The spatial order of the modern industrial city came to be seen as a profoundly patriarchal spatial order; that is an arrangement of space in which the domination of men over women was written into the architecture, urban design and form of the city. Cities, built and planned by men, for men, confined women to the suburbs, to the home, to the private sphere, and then, having segregated them, doubly disadvantaged them by not recognizing that their needs in the city were different from those of men, based as they were primarily around home, neighborhood and caring for children and the elderly. From the routes of public transport to the location of key educational, cultural and health facilities in downtown urban centers, cities and their planning processes could be seen as excluding women from participation. Not surprisingly then, urban social movements advocating women's needs in the city – needs for more and better public transport, for childcare, for community facilities, for safety, for a right to occupy public space, day and night, and so on – have flourished since the 1970s. This critical onslaught, combined with social activism and subsequent demands to engender planning practices, at first seemed containable within a modernist paradigm. Feminist planners working within the planning system addressed themselves both to raising the consciousness of male colleagues and to the identifiable demands of the feminist urban social movements described above. We swapped "war stories" and gave each other support at conferences, creches and in coffee shops. We made some progress, and shared many disappointments and frustrations.

But then, as of the early 1980s, something profoundly destabilizing happened. The "we" of feminist urban analysis was challenged by "Other Women" who argued that the "we" had never included "them." These were the voices of women of color, of lesbians and of the physically challenged, who claimed that the voices who had hitherto represented "women" were really only representing the white, middle-class, able-bodied, heterosexual and nuclear family oriented, and metropolitan-based women who actually already lived relatively privileged lives, and whose very privilege was usually built on the "backs" of women of color/Third World women who cleaned their houses, did their shopping and sometimes cooking for them, and looked after their children.

These "voices from the borderlands," the voices of women who are in one way or another on the margins (from whom we will hear more in the penultimate section of this chapter), are the third major challenge to the modernist planning paradigm. Each of these socio-cultural changes is intricately interwoven, as cause and effect, with the crisis of modernization itself, the crisis of the Fordist economy, The voices from the borderlands also inhabit and embody the new cultural politics of difference, complicating that politics with their intersectionalities of race, class, ethnicity, gender and sexual preference formations of "difference". Together they suggest that social justice in postmodern cities and regions is inseparable from a respect for and an engagement with the politics of identity and difference. How might planning praxis – that is a theoretically informed practice – engage with all of the transformations described above, from the globalization of migration to the rise of indigenous peoples to the challenges of gender and sexuality, in all of their spatially inscribed everydayness?

THE STAGES OF PLANNING PRAXIS

I want to trace and sketch six theories of good planning practice that have emerged since the 1940s and argue that only a revised version of the sixth, a theory of radical practice or empowerment, is capable of addressing the issues of social transformation inherent in our concern with diversity and equity and sustainability. Embracing this model amounts to embracing a new professional identity: deconstructing the pillars of modernist planning wisdom and reconstructing them as pontoons rather than pillars – as something more fluid and adaptable to the space and pace of change in postmodern cities and regions; and recognizing that a counter-hegemonic planning practice cannot be neutral with respect to race and gender, and must be sensitive to the multiple forms of oppression, domination and exclusion that exist in today's cities rather than privileging any one form of counter-analysis – such as class analysis.

Within the modernist paradigm, there have been a succession of competing theories, each claiming the intellectual and/or moral highground at different times over the past 50 years. What follows is a brief sketch of six competing theories or traditions of good planning practice, arranged in the form of a chronological narrative. The interesting point is that as each new theory emerges, it seeks to redefine precisely what it is that planners "do," not so much in terms of substantive fields but in terms of approach, process and allegiance. Each contains subtle epistemological shifts, but the sixth model marks a significant epistemological break with the Enlightenment tradition in which modernist planning has been embedded.

The rational comprehensive model

For two full decades after the Second World War this model, shaped by and exported from the University of Chicago planning program, dominated the field. With its origins in Enlightenment epistemology, a belief in the possibility of greater rationality in public policy decision making has informed this mode of theorizing ever since Herbert Simon first proposed his synoptic model of decision making in 1945. Theorists from Simon (bounded rationality) to Lindblom (incremental decision making, mutual partisan adjustment) to Etzioni (mixed scanning) have shared a faith in *instrumental rationality* (Simon 1976; Lindblom 1959; Etzioni 1978). For them it was a given that technology and social science could make the world work better, and that planning could be an important tool for social progress. Planners, in this model, are handmaidens to power, "speaking truth to power" as Wildavsky's (1979) much-quoted phrase put it. Planners are also part of an ambitiously *comprehensive* public policy process, attempting to coordinate more and more specialized and narrowly defined activities. Here was planning at its most *heroic*, confident in its capacity to discern and implement the public interest.

In this model the planner was indisputably the "knower," relying strictly on "his" professional expertise and objectivity to do what was best for "the public". The notion of "the public", never critically examined, implied an undifferentiated, homogeneous group in which differences of class, or race, or gender, were not considered relevant "input." This is a model which assumes a benign state, and a state whose structure is neutral with respect so questions of gender and sexual preference, race and ethnicity, rather than a state which may be, for example, patriarchal, homophobic, racist and allied to capital.

Within two decades this model was to face challenges from within as well as from without: from the dilemmas inherent in instrumental rationality, as well as from community critiques of this form of top-down planning as anti-democratic. Yet despite apparently definitive critiques and ongoing community opposition, this model continues to win adherents and to spawn new theorists from Andreas Faludi (1973, 1986) to Franco Archibugi (1992a, 1992b) and Ernest Alexander (1992). Why did this model become the dominant paradigm in planning? By offering rules (of decision making) by which planners could proceed and through which they could consider alternative procedures and consequences, it seemed to offer professional legitimacy and so make planning "scientific." And is dovetailed neatly with the economists' paradigm of rational economic man and rational resource allocation models. But why does the paradigm persist, in the face of mounting critiques?

One explanation could be that we continue to teach it in planning schools. By emphasizing rational/objective analysis through such courses as quantitative methods, modelling, use of computers and so on, we create expectations that favor explicit preestablished goals which can be met by planning procedures that favor such methods. A whole planning culture has been built around privileging such methods. And for the *longue durée* of the modernization project, the model of applied rationality was the perfect handmaiden. But that hegemonic project began to be challenged in the 1960s by those groups which had been excluded from its domains and its fruits, or who were concerned about the environmental consequences of the global drive to modernization.

The advocacy planning model

The first challenge to the rational comprehensive model was the concept of advocacy planning that emerged in the mid-1960s in the United States. Significantly, there had been major riots in American cities in 1964–5, and the Civil Rights movement by then had had a decade of momentum, which created a climate in which dissenting opinion might be heard. A new approach coalesced around an article written by Paul Davidoff and published in the *Journal of the American Institute of Planners* in 1965, "Advocacy and Pluralism in Planning." Concerned that the rational model of planning was obsessed with means, he warned us that the question of ends remained. He stressed the role of politics in planning. The public interest, as he saw it, was not a matter of science but of politics, and he urged planners to participate in the political arena. He called for many plans, rather than one master plan, and for full discussion of the values and interests represented by different plans. He brought the question of who gets what – the distributional question which the rational model had so carefully avoided – to the foreground.

The idea of advocacy planning was that those who had previously been unrepresented would now be represented by advocacy planners, who would go to poor neighborhoods, find out what those folks wanted, and bring that back to the table in the planning office and city hall. With his lawyer's faith in due process and enlightened plural democracy, Davidoff had outlined a model which, although seemingly at odds with its predecessor, would in fact serve to perfect both the rational model and pluralist democracy as a result of the advocate informing the public of all the social costs and benefits and formulating alternatives which would be incorporated into a better master plan. His approach found an immediate following among left-wing liberal intellectuals, mostly white middle-class professionals, who soon headed off into poor neighborhoods and black communities to offer their advocacy skills.

Advocacy planners' experiences in Harlem and Boston were sobering. ARCH, the Architects Renewal Committee in Harlem, quickly found that what the poor lacked most was not the technical skills that the advocates were offering, but the power to control action. Eventually, ARCH dropped the advocacy label altogether and decided to help provide the means by which the community could represent itself. In other words, it turned away from advocacy to an empowerment model (Heskin 1980).

Similarly in Boston, UPA (Urban Planning Aid), a group of predominantly white professionals (including such prominent names as Lisa Peattie, Chester Hartman and Robert Goodman), wanted to assist the poor by taking their ideas and translating them into the technical language of plans, to make them forceful in the policy arena. Peattie later described their work as "the manipulator model" (Peattie 1968). Professionals set the agenda, conceptualized the problem and defined the terms in which a solution to the problem would be sought. She noted that the issues raised were likely to be those with which the professionals were most comfortable, rather than those which were highest on the community's list. Robert Goodman provided an even more devastating critique of this work in his *After the Planners* (1972), describing planning professionals as agents of social control, as the "soft cops of the system. He argued that taking the poor off the streets and encouraging their participation in planning was not empowering them but robbing them of their power.

Clearly, advocacy planning represented a significant expansion of the definition of what it is that planners do. Under this model some planners would now explicitly think about and represent "the poor" in the planning process – without, however, actually giving them a voice in that process. Instead, advocates became the ventriloquists for poor communities. *Advocacy planning expanded the role of professionals and left the structure of power intact, confident in the workings of plural democracy.*

From this experience with advocacy, different planners drew different conclusions. Some, like Norman Krumholz and Pierre Clavel, were deeply inspired and saw the possibility of perfecting the advocacy concept by planners allying themselves with progressive politicians and doing "equity planning" (Krumholz 1994; Clavel 1994). Others drew lessons that focused more on process, and either became advocates for citizen participation or began to rethink the role of planners as experts, and so formulate new ideas of transactive planning, mutual learning and social learning (Friedmann 1973). A third group drew the more radical lessons (of UPA and ARCH), and moved towards an empowerment model (Heskin 1980; Leavitt 1994; Peattie 1968, 1987; and, much later, Friedmann 1992). But before I tell those stories, there is another model that emerged in the mid-1970s, in response to new theoretical analyses of the structural relationship between planning and capitalist society. It is to that Marxist political economy critique of planning that I now turn.

The radical political economy model

Just as debates about participation, mutual learning and empowerment began to gain an audience in the planning profession in the early 1970s, an entirely new narrative and analysis erupted on the periphery of the planning world. With the publication in 1973 of geographer David Harvey's *Social Justice and the City*, and the English translation of urban sociologist Manuel Castells' *The Urban Question* in 1976, the story of planning began to be rewritten. In the Marxist story, planning was no longer the hero but more like the divine fool, naive in its faith in its own emancipatory potential, ignorant of the real relations of power which it was serving and in which it was deeply and inextricably implicated. The works of Marxist urban scholars in university departments of geography, sociology and urban studies enjoyed a decade or so "in the sun," as a powerful critique of mainstream planning, focusing on planning as a function of the capitalist state.

Manuel Castells produced one of the first (and much imitated) case studies of the role of urban planning in the development of the growth pole of Dunkirk (Castells 1978). He identified three functions of planning: as an instrument of rationalization and legitimation; as an instrument of negotiation and mediation of the differing demands of the various fractions of capital; and as a regulator of the pressures and protest of the dominated classes. Whether the focus of these radical political economists is on production or on consumption, on the state's role in capital accumulation or its role in providing collective goods and thereby maintaining legitimacy, the conclusion is the same with respect to the function of urban planning. Far from being the progressive practice that the profession claims for itself, in the Marxist narrative planning can always and only be in the service of capital.

The emergence of this new literature presented a challenge to planning schools and planners, exacerbating already existing divisions between theorists and practitioners. While some of the more theoretically oriented planning faculty tried to bring this new work into the planning field, more practice-oriented folks denied its relevance. The latter reaction is understandable in the sense that Marxist analyses denied planners a role in social transformation and that this kind of theorizing seemed to have a paralyzing effect on policy debates. The lasting value of this model is at the level of critique rather than action. Marxist or radical political economy theory locates planning as an inherently political activity within a capitalist state which is itself part of a world capitalist system. We can no longer ignore this structural reality in our analyses of planning practices and policies. Further, Marxist critique has demystified the idea that planning operates in the public interest, making it very clear that class interests are always the driving force.

But there is also a problem with this last insight, for in insisting on the primacy of class interests in their counter-analysis, Marxists have ignored (or tried to subsume into their class analysis) other forms of oppression, domination and exploitation, such as those based on gender, race, ethnicity and sexual preference. However, the ultimate weakness of this model has been its inability to provide a new definition of what it is that planners can do. Generic or half-hearted answers such as "The planner can become the revealer of contradictions, and by this an agent of social innovation" (Castells 1978:88) have not proved sufficient to inspire a new generation of radical planners. For that, we must turn back to those who have developed the legacy of advocacy planning in the three directions outlined in the previous section.

The equity planning model

While the advocacy movement of the 1960s began outside the aegis of city halls, one group of planners who see themselves as inheritors of the advocacy tradition have developed the tradition in the direction of making alliances with and working for progressive politicians. The two most prominent practitioners and definers of this new model of equity planning are Norman Krumholz and the late Robert Mier, the former in his career as chief planner for the city of Cleveland for a decade is the 1970s, the latter in his work as head of economic development planning during the regime of Harold Washington as mayor of Chicago in the 1980s.

Krumholz defines equity planners as those who consciously seek to redistribute power, resources or participation away from local elites and toward poor and working-class city residents. In his book with John Forester, *Making Equity Planning Work* (1990), Krumholz tells his "war stories" from Cleveland as a sort of inspirational tale of how good can be done, and precisely what it is that equity planners do, within given structural constraints. Accepting that planning is the handmaiden of politics, equity planners choose the politicians for whom they want to work. There are costs involved, in the sense of a willingness to be reasonably mobile, not expecting to stay in one job forever, but only for as long as the planner has the support of a progressive regime. Given such a regime, Krumholz argues that planning within the local state can be both meaningful and ethically defensible. The state isn't a monolith but rather a terrain of political struggle, and planners with the interests of the poor and unrepresented in mind can do good and constrain evil.

Interestingly, this model retains a belief in the planners' expertise and doesn't say much about drawing on local knowledge. The planner is still the center of the story, the key actor. But what the planner does is now defined much more broadly than in the rational comprehensive model. Krumholz stresses the importance of *talk*: at local meetings, at county and state testimonies, speeches to the profession, op-ed pieces, interviews with reporters, speech writing for mayors and councilors, and engaging in dialogue with other city agencies. The planner is a *communicator*, a tireless propagandist, and good communication skills are critical. The planner is also a gatherer of information and analysis, a problem formulator. By reformulating a problem, planners have some power to shape debates, to direct public attention to issues which planners see as important. Equity planning is still engaged in speaking truth to power, still engaged in a top-down politics, only now it is a consciously politicized practice, and its allegiances are consciously directed to those who have been excluded. It is an attempt at a *top-down inclusionary politics*, in which the poor, the marginalized, are still not part of the action, and do not feature as active agents in the narrative or theory of "making equity planning work."

The social learning and communicative action models

Paul Davidoff believed in opening up the political process, overtly espousing competition among plans. The first generation of advocacy planners under his influence took their technical skills into poor communities, intending to offer assistance so that alternative plans could be created which took into account the needs of such communities. Some advocates learned a different lesson from this, a lesson about local knowledge and about the political skills that exist within poor communities. Reflections on this lesson by a number of planning theorists and practitioners have led to the emergence of models of social learning and communicative action.

In *Retracking America* (1973) John Friedmann wrote of a crisis of knowing, which was reflected in the emerging conflict between expert/processed knowledge and personal/experiential knowledge. He described the growing polarity between so-called experts and their "clients," a polarity exacerbated by the inaccessible language in which professionals usually formulate problems, and he argued that neither side ever has all the answers. The obvious solution was to bring the two together to engage in a process of *mutual learning*, to develop a personal relationship between expert and client through the adoption of what he called a *transactive style of planning*. Friedmann characterized transactive planning as the life of dialogue, emphasizing human worth and reciprocity in contrast to the traditionally arrogant and aloof stance of the professional. This involved acceptance of the authenticity of the other person; a fusion of thinking, moral judgment, feeling and empathy; a recognition of the importance of the non-verbal as well as the verbal; and an acceptance of and willingness to work with and through conflict. What is radical about this approach is its epistemological shift away from the monopoly on expertise and insight by professionals to an acknowledgment of the value of local, or experiential, knowledge. It is also a shift away from a static conception of knowledge (as in "body" of knowledge) to a more dynamic concept and metaphor of learning.

Beginning with the same fundamental observation that planning is, above all, an interactive, communicative activity, another group of scholars coalesced in the 1980s around the study of planning as a communicative practice. Inspired principally by the work of John Forester, who calls his theory *critical planning*, basing it on the Habermasian concept of communicative action, the work of this group has moved from the instrumental rationality of the earlier model to an emphasis on communicative rationality. They rely more on qualitative, interpretive inquiry than on logical deductive analysis, and they seek to understand the unique and the contextual, rather than to make general propositions about a mythical, abstract planner (Innes 1995:184). Forester's detailed observational study of planners at work, *Planning in the Face of Power* (1989), stands as the closest thing to a bible for this group. For Forester, planning is primarily a form of critical listening to the words of others, and observing their non-verbal behaviour. It is a mode of intervention that is based on speech acts, on listening and questioning, and learning how, through dialogue, to "shape attention." Forester is interested in what story is being told in any planning situation, because these stories embody and enact the play of power, the selective focusing of attention, the presumption of "us and them," the creation of reputations, and the shaping of expectations of what is and is not possible as well as the production of politically rational strategies of action. What planners say "involves power and strategy as much as is involves 'words"' (Forester 1991:23).

If equity planners can be said to be trying to perfect planning as an Enlightenment project by representing the interests of the poor and the marginalized in city halls, then communicative action theorists might be said to be trying to perfect the Enlightenment's democratic project by removing the barriers to communication, by creating a model of open discourse, by removing distortions. The emphasis is less on what planners know and more on how they use and distribute their knowledge; less on their ability to solve problems and more on opening up debate about them. In this model, planning is about talk, argument, shaping attention. But the primary actor and source of attention is still the formally educated planner working through the state.

While this is certainly a more inclusive theory of planning practice than its predecessors, it does have some serious weaknesses from a counter-hegemonic perspective. It does not attempt to address the issue of empowerment raised by the third group of critics of the early advocacy model, except in terms of speech acts. It acknowledges, but then brackets, the problem of structural inequalities. And it treats citizenship as an unproblematic concept which is gender and race neutral, following the Habermasian and Rawlsian use of universal categories, and in the process suppressing the crucial questions of difference and marginality and their relationship to social justice. To find a forum for discussing these last issues, we need to turn to those who have been trying to elaborate a theory of radical planning, or empowerment.

The radical planning model

Among the relatively small community of scholars who have sought to outline a radical or emancipatory practice in the pass two decades, some are direct descendants of advocacy planning (Heskin 1980, 1991; Peattie 1987, 1994; Leavitt 1994); others have arrived via a feminist critique (Leavitt 1994; Hayden 1980); and still others through engagement in the civil rights movement and ongoing struggles around racism (King 1981; Leavitt and Saegert 1990); or contemporary debates around multiple forms of oppression and exploitation (Starr and Lee 1992; Hooper 1992; Sandercock 1995a); or working on problems of poverty and exclusion in an international development context (Friedmann 1992).

Radical practices emerge from experience with and a critique of existing unequal relations and distributions of power, opportunity and resources. The goal of these practices is to work for structural transformation of these systemic inequalities and, in the process, to empower those who have been systematically disempowered. Obviously the focus of radical practice will depend on the focus of the critique. The dominant radical critique of urban inequalities from the 1960s through to the 1980s was that of class analysis, particularly after the rise of Marxist urban political economy and urban sociology in the mid-1970s. But those white professionals who actually went into poor communities were faced with more complex situations, intersections of racism, poverty, sexism, homophobia, anti-immigrant sentiments, to name the most obvious. It began to occur to some planners that "the poor" and "the oppressed" were not a homogeneous "mass" but rather spoke with many voices. By the 1980s, feminist activists within planning were developing their own spatial as well as social and political analyses of gender inequalities; people of color (and some white allies) were drawing attention to racist practices within and effects of planning; and gay and lesbian activists were documenting a history of oppressive spatial and social practices affecting their lives in cities. Questions about social justice and the city have correspondingly been expanded from the earlier, class-driven formulations of Marxists like Harvey and (the early) Castells to include "what would a non-sexist city be like?" (Hayden 1980) and, by extension, what would a non-racist and non-homophobic city be like? But the toughest question of all is what can planners do about any of these inequalities? Radical planners have given various answers, in their theory and practice. Most of these answers are related to community organization, urban social movements and issues of empowerment, rather than to working through the state. Weaving itself through each of these answers is an ongoing angst about the relationship between professional identity and radical practice.

Some of today's radical planners were involved in the advocacy movement of the late 1960s, which served to clarify for them the insoluble dilemmas of working as a planner in the bureaucracy. Allan Heskin (1980) and Jacqueline Leavitt (1994) drew similar conclusions from that experience. In Leavitt's words, "on the one hand, advocacy planning couldn't fight city hall, on the other hand it didn't deal well with conflicting interests in the community" (Leavitt 1994:119). The, to them, obvious conclusion was that in order to make a difference in the lives of the poor, the excluded, the marginalized, an empowerment approach was required, and that such an approach could only be practiced outside the bureaucracy. Leavitt describes what the does as entering a community, gaining trust, allocating time, listening, arguing and letting others speak. The primary requirement of this kind of community-based practice is allocating enormous amounts of time to "hanging out" with the mobilized community.

Heskin and Leavitt's work has revolved around housing struggles in multiethnic, multiracial, poor communities. Other radical practices have been organized around plant closures and worker buyouts; women mobilizing to get a whole range of women's needs in the city addressed (Women Plan Toronto); the establishment of credit unions in previously red-lined neighborhoods; community gardens, childcare, bicycle paths; opposing the siting of environmental hazards like toxic waste incinerators in poor and minority communities . . . Taken separately, none of these struggles may seem at all system-threatening (a bicycle path?), but taken together they do constitute a challenge, because they have the potential for making people less dependent on global capital, increasing their social power and experiencing their own political power, albeit at a local level. But it is precisely through action at the local level that people begin to get some handle on how to make a difference to their own lives and concerns as well as those of fellow citizens – concerns about jobs, housing, schools, health.

Working in and with such mobilized communities, planners' roles are not the heroic ones described in the rational model. Rather, in working for social transformation in community-based organizations, planners acknowledge that theory and practice become everyone's concern and that responsibilities for both are multiple and overlapping. Planners bring to radical practice general and specific/substantive skills: everything from skills of analysis and synthesis to grantsmanship, communication and the managing of group processes, as well as specific knowledge of labor markets or environmental law or transportation modeling or housing regulations. But they also recognize the value of the contextual and experiential knowledge that those in the front line of local action – the mobilized community – bring to the issue at hand. And they are open to learning through action, through experience. Above all, radical practice depends for its effectiveness on interpersonal relations of trust (Friedmann 1987:402; Leavitt 1994), and a social learning approach based on a "radical openness" (hooks 1990:148).

Radical practice, then, does not lie on a logical continuum with rational planning. It is not about participation in projects by the state. More often than not, radical planners will find themselves in opposition to either state or corporate economy, or both. This implies an epistemological break with past ways of thinking and doing (Friedmann 1987:191), of what it means to be a planner, and what it is that planners do. *It requires nothing less than a new professional identity.* How might that identity be described? There are two somewhat conflicting portraits in the current literature. Radical practitioners like Heskin and Leavitt state very clearly that the allegiance of the activist planner is to the community with whom she or he is working. There is a "crossing-over" implied here, in which the professionally educated planner sheds her or his professional status/identity and chooses, instead of loyalty to professional codes, loyalty to the poor and the oppressed. For Heskin this is a clear choice about class allegiance. A planner cannot cling to her professional class status and hope to be helpful to "the community," To contribute to community empowerment, she must not see the community as a client, but instead see herself as the ally of that community, helping people to clarify their goals, enabling them to achieve collective self-determination.

Here is a dramatic shift from the other five models of planning, in which the planner is still the key actor, the driving force. In these other models, the professional planner, by definition, works through the state, even if, as an equity planner, for example, her goal is to achieve some kind of redistribution of resources on behalf of the poor. It is the *on behalf of* that is the problem for the radical planner. In Heskin's description of radical planning, it is the community that initiates, and the planner who enables, assists, but never imposes her solutions and only offers advice when asked. Similarly in Leavitt's work, she immerses herself in the community, hangs out with them, helps them with research and preparation of documents, advises on how to deal with bureaucracies, but never does these things *for* the community, always *with* them. The identity of the radical planner in these works is that of a person who has, essentially, gone AWOL from the profession, has crossed over "to the other side", to work in opposition to the state and corporate economy. This does not mean that community-based planners have nothing to do with the state. There is a clear acknowledgement in Heskin's and in Leavitt's work of the need to think strategically about the state, to make alliance with those planners who do work within state agencies and who might be regarded as friendly to the cause. And the knowledge that these activist planners possess of the workings of the state is invaluable to the communities with whom they are working. But there is a clear line being drawn in the sand. Choose the community and you are choosing professional/class death. Choose to work for the state and you retain your professional identity, but don't delude yourself about whose interests you are serving.

In the radical planning model elaborated by John Friedmann there is a different take on this notion of professional identity. Friedmann insists that a radical planner has to maintain a notion of critical distance. She or he does not, ultimately, cross over. "Radical planners must not become absorbed into the everyday struggles of radical practice . . . as mediators, they stand neither apart from nor above nor within such a practice" (Friedmann 1987:392). While not denying that radical planners must be committed to the group's practice and to the global project of emancipation, he nevertheless posits "an optimum critical distance between planners and the front line of action" (Friedmann 1987:404). Friedmann's radical planner is then a tightrope walker, trying to maintain some autonomy *vis-à-vis* the radical group or community. But why is this autonomy/critical distance necessary? Why not just cross over?

Friedmann talks about the relevant actors in the struggle for a new society (the global project of emancipation) as being (1) individual households who have opted for the alternative; (2) organized social groups based in local communities; and (3) larger, more inclusive movements not bounded by territorial limits. Heskin and Leavitt, on the other hand, talk almost exclusively about the "struggle for community" (Heskin 1991) or "community-based planning" (Leavitt 1994). For Heskin and Leavitt the state is and can only be an adversary. For Friedmann, any social advances achieved through a radical planning that by-passes the state "will quickly reach material limits. To go beyond these limits, appropriate actions by the state are essential" (Friedmann 1987:407). Clearly, the state has been the missing ingredient so far in this discussion of radical planning. And while it may well be a contradiction in terms to think of the state engaging in radical planning, it may be equally misleading to think that radical planning can do without the state.

Lyotard has argued that there is no such thing as Reason, only reasons (Lyotard, in Van Riejen and

Veerman 1988). The same might be said about Community – there is no such thing as Community, only communities. In the writings of most radical planners, "the community" has been reified and romanticized. If the state is the enemy (the implicit argument seems so be), then the mobilized community can do no wrong. But we are all familiar with specific communities (straight, white, Christian, mentally or physically able . . .) who try to use planning to exclude specific other communities (gay, black, Jewish, Muslim, mentally or physically challenged . . .). *What rights should communities as collectivities have* vis-à-vis *individual rights on the one hand, and the rights of the larger society on the other*? This is a very difficult question in political theory – and it is a question which planners cannot avoid, especially in the contemporary context of the rise of mobilized groups/communities in civil society, asserting and demanding respect and space for their "difference." I will turn next and finally to this contemporary challenge of "difference."

But before doing so, I want to draw attention to the repressive potential of mobilized communities just as, in their past analyses, radical planners have emphasized the repressive practices of the state. We need to remember that, in the conflict over legal segregation in the southern United States during the 1960s, the Federal Government eventually intervened in local affairs and acted against local authorities, in a clear case of the transformative power of the state. The lesson of this paradox is:

> that planning needs to engage not only the development of insurgent forms of the social but also the resources of the state to define, and occasionally impose, a more encompassing conception of right than is sometimes possible to find at the local level . . . Above all, planning needs to encourage a complementary antagonism between these two engagements. (Holston 1995:49)

It is this antagonistic and yes also dialectical relationship between the state and the mobilized community that radical planners have yet to address. The first step is to get beyond the notion of the state as always and only the adversary. Instead, we need to think about the complementary as well as antagonistic relationship between state and civil society and of the possibility of social transformation as a result of the impact on the state of mobilized groups within civil society. If modernist politics could be defined as the bi-polar struggle between capital and labor, in which the state was allied to capital, then a postmodern urban politics is perhaps best understood as a multiplicity of struggles around multiple axes of oppression, in which the role of the state is not a given (not simply "the executive committee of the bourgeoisie"), but is dependent on the relative strength of the social mobilizations and their specific context in space and time.

In other words, there is an unresolved, and unresolvable, tension between the transformative *and* repressive powers of state-directed planning practices, *and* their mirror image, the transformative *and also* repressive potential of the local, the grassroots, the insurgent. In order to work with this tension, perhaps Friedmann's concept of critical distance assumes a new importance. And in maintaining a productive tension between state-driven planning and the insurgent practices of mobilized communities, radical planners do need a different kind of professional practice, different in both objective and method. This difference amounts to a reconceptualization of the field and of the notion of professional identity. Rather than the "crossing-over" outlined in Heskin's and Leavitt's work, the appropriate image may be that of crossing back and forth, of blurring boundaries, of deconstructing ("community", "the state") and reconstructing new possibilities.

In terms of methods, an *epistemology of social learning and of multiplicity* is the theory of knowledge underlying radical practice. This means that action is primary, but that we need to develop new ways of knowing and being as well as new ways of acting. One possible source of guidance for such new ways of knowing and being and acting comes from the voices of women and people of color, postmodern and postcolonial voices resonant with experiences of marginality, exploitation and domination.

LISTENING TO THE VOICES OF DIFFERENCE

These "voices from the borderlands" (Sandercock 1995a) belong to people who dwell in cultures of displacement and transplantation, to cultures with a long history of oppression, to people who have been marginalized for hundreds of years, but who are now insurgent, and who are turning their marginality into a creative space for theorizing and "a space of insurgent citizenship" (Holston 1995). They challenge dominant notions of theory and practice, of epistemology and ontology. They challenge us to acknowledge and respect diversity or difference in our theory and practice. If we listen to these feminist, postmodern and postcolonial voices, then we need to revise our radical model of planning, from its epistemological base to its theory of practice, to incorporate (but not uncritically) the concept of "difference".

The writings of bell hooks, Gloria Anzaldua, Cornel Wear, Guillermo Gómez-Peña, Himani Bannerji, and many more, are exploring the borderlands, the margins – searching for a new consciousness or, as Anzaldua says, "a new story, to explain the world and our participation in it, a new value system with images and symbols that connect us to each other and to the planet" (Anzaldua 1987:81). Their range of concerns includes identity and difference; more inclusive ways of thinking about justice; how knowledges are invented and institutionalized; new spaces of opposition, resistance and consciousness; the importance of story telling; multiple voices; and the search for new ways of being alongside new and old ways of knowing. Their central theme is the meaning and importance of difference. They are speaking from experience about the violence, physical and spiritual, done to those whom the dominant white racist and homophobic culture labels as different. But they are also (re)defining and celebrating their cultural difference and demanding that it be acknowledged and respected. These are the voices not only of immigrants but of indigenous peoples, African Americans, gays and lesbians. We need to hear their stories because they are telling us something about the death of the modernist city with its drive/will to homogeneity, and of the need to create a different kind of planning, one which eases rather than resists the transition to genuinely multicultural (that is multiracial and multiethnic) cities, now and into the twenty-first century, a more democratic planning for a heterogeneous public and an expanded notion of social citizenship. There are lessons for a radical planning practice in listening to these voices of "difference" – lessons to do with new ways of being, knowing and acting.

As a way of being, these writers describe the state of living on/in the borderlands, living in between, on the margins. They accept the necessity of living with uncertainty, without universals, without the panaceas of revolution or progress. But they do not live without hope or meaning. They embrace uncertainty as a potential space of radical openness which nourishes the vision of a more experimental culture, a more tolerant and multifocal one. And they reject the notion of a static identity in search of a safe place, a homogeneous community, a clearly defined border between "us" and "them", a space that is "yours only."

Beyond these existential issues (of being), each of these writers is concerned with *ways of knowing*. How does knowledge get produced, by whom, for what purposes and with what effects? How is theory defined? Who is included and excluded from theoretical discourses? They argue that there are different ways of "doing theory" and that the voice of Western abstract logic and reason is not the only possible theoretical voice. They argue that theory is always embodied. They celebrate the value of experiential and other alternative ways of knowing, learning, discovering, including traditional ethnic or culturally specific modes: from talk to story telling, the blues to rap, poetry and song; and visual representations, from cartoons to murals, paintings and quilts. This is what I call an epistemology of multiplicity. This suggests a whole different planning practice in which communicative skills, openness, empathy and sensitivity are crucial; in which we respect class, gender and ethnic differences in ways of knowing, and actively try to understand and practice those ways in order to foster a more inclusive and democratic planning.

These voices from the borderlands are also concerned with *new ways of acting*, with an emancipatory politics. They challenge us to acknowledge and respect difference without being paralyzed by it. And while they have not lost sight of certain key principles of the Enlightenment – like justice and equality – they have made transparent the exclusivity involved in the historical application of those principles. They are calling for a more inclusive moral vision, shaped by consciousness of the borderlands/margins, informed about racism and sexism xenophobia and homophobia, homelessness and poverty. They offer a theory of social transformation based on an oppositional world view (West 1993) and powerful critique of dominant culture (hooks 1990; Anzaldua 1987; Gómez-Peña 1993); rooted in communities of resistance (hooks and West 1991); informed by principles of justice and equality, caring, tenderness and love (hooks 1990; Jordan 1992; West 1993). And every one of them advocates the practice of coalition politics – of forging coalitions that force us to move out of that safe place with our own people, to build bridges, to move beyond the "womb stage" of identity politics, insisting on the interdependence of our struggles, "white black straight queer male female" (Anzaldua and Moraga 1983:iv). This last is particularly important if the current attention to difference is not to move in the direction of dividing us (as some radicals fear), rather than fostering a broad emancipatory project. *Is there a contradiction between the values of social justice in the city and those of respecting difference? Or does the very achievement of broader social justice require a politics of difference?*

These voices from the margins are reaching out and connecting with others, speaking for and across social groups or collectivities, offering a vision beyond the old class politics and the new identity politics, with their eyes on a more just, equitable, humane and sustainable society. *Their common theme is that difference must be incorporated into the quest for social justice in the multiracial, multiethnic city of the present and future.* How do planners respond to the complex challenges of difference? How do we allow for, build difference into, our practice?

There are many ways of answering. There are principles and there are stories. The principles include the idea of planning for a heterogeneous public, rather than for the (modernist) "public interest;" the empowerment of specific (hitherto excluded) social groups; representation for those groups in decision making; participation without assimilation; developing a multicultural literacy, not just literally – in terms of languages – but in terms of other ways of knowing, an *epistemology of multiplicity*; and planning with and by mobilized communities as well as state-directed planning. It means thinking beyond tolerance. Difference is not just to be tolerated but valorized, given value by the dominant culture. Difference, in the multicultural city, speaks to us with a collective voice, with the voice of social groups. Their claim is to be allowed to be different within an inclusive society, within a society of citizens – with the right to make claims on the political community and to participate in it. Difference then means not just different interests, nor just a reincarnation of the old familiar pluralism, but a different way of being in the world. *This involves the need, and the right, to give expression to difference in the public sphere.* This means planning needs to examine its practices in so far as they are spatially policing groups who are "different" – historically, women, gays, people of color and certain immigrant groups (see Sandercock 1997).

RADICAL PLANNING IN A POSTMODERN AGE: AN UNFOLDING STORY

Finally, and always, there are stories. I began this chapter by evoking Los Angeles as an (all too familiar) avatar of a postmodern city/region, a place where the real and potential conflicts inherent in multiracial, multiethnic coexistence are all too often resolved by resort to violence, and where the history of planning has notoriously been the history of the land development industry (see Davis 1990). I close by crossing the Pacific to a remote part of Australia, to the west coast of Cape York Peninsula in the far north of Australia, an area exoticized by the movie *Crocodile Dundee*, but home to the Wik people, the traditional Aboriginal occupants of land that is now under pastoral lease to white folks. Here a postmodern planning story is unfolding.

Over the past three decades the Wik peoples have campaigned vigorously, and occasionally successfully, against the mining industry and against successive Queensland (i.e. provincial) governments. During the 1970s the Wik community often clashed with the government of the day by resisting its assimilationist policies. The local population centre, Aurukun, had been established in 1904 as a Presbyterian mission after the region was declared an aboriginal reservation. Since the 1970s, through the out-station movement, many of the Wik peoples shifted from Aurukun back to their tribal homelands. Unlike some Aboriginal communities in northern Australia, the Wik has been steadfastly opposed to mining and mining interests, and has been an incubator for a strong land-rights movement.

In the mid-1970s John Koowarta, a Wik man, clashed with the (ultraconservative) provincial government over land rights. In 1976 that government stopped Koowarta from purchasing a grazing lease on his traditional land, presumably because Aboriginal people were somehow unfit to run a cattle station, despite a long tradition of working in the Cape York pastoral industry. With the backing of the church, Koowarta fought an anti-discrimination case all the way to the High Court where, in 1982, he secured a finding that the Racial Discrimination Act was valid, and that the Federal Government could compel the provincial government to comply with it. The provincial government responded by immediately turning the cattle station into a national park, foiling Koowarta's effort to regain his land before his death in 1991. But, a decade after his High Court victory, and 12 months after his death, the High Court made its historic Mabo ruling, which relied heavily on the findings in Koowarta's case (Kennedy 1996).

Mabo: the very word evokes a sea-change in black–white relations in Australia, but has in turn produced a divided response from the national community. In 1992 the High Court of Australia handed down a decision that found in favor of Eddie Mabo's claim that his people's entitlement to the Murray Islands in Torres Strait had not been extinguished by the settlement of Australia by the British. In the making of this decision the court found that the common law in Australia recognizes a form of native title which is also applicable to the mainland. This decision effectively unbound the previous limits of existing land-rights provisions for Aboriginal peoples by displacing the fallacy of *terra nullius*, of land unoccupied prior to British settlement in 1788. It opened the way for the passage and implementation of the Native Title Act of 1993. The Mabo decision implied that all lands, including urban lands, were once legitimately native lands and potentially open to claim. It heralded the possibility of a more equitably and uniformly applied land rights provision, including the possibility of meaningful urban land rights" (Jacobs 1996:112).

Mabo was a radical gesture, sweeping aside orthodox doctrine concerning land laws in the name of justice, of historical truth and of difference. But it also left many questions unanswered and created a climate of uncertainty around two specific issues. First, Aboriginal people do not yet know how much and what kind of proof they have to come up with to persuade a court that they have native title. Second, holders of pastoral leases are unsure whether the existence of a pastoral lease automatically extinguishes native title rights. This second question was put to the High Court by the Wik people when they registered a claim for native title in 1993. The judgment which has just been handed down by that court is a fascinatingly complex one which necessitates a postmodern planning response.

In December 1996 the High Court decided that no pastoral lease, anywhere in Australia, automatically extinguishes native title. Instead, the court deemed that where proposed Aboriginal uses seem to be in harmony with an existing pastoral lease, the two can coexist, and where there appears to be conflict between the two, then the outcome must be decided by a stocktaking of rights in each situation as it arises – the rights given the leaseholders by a myriad of state (provincial) laws on the one hand, and those given to blacks by their ancient customs on the other. Again, there is an apparently unanswerable question – how can native title, a concept just declared in Australia, be reconciled with land laws which have been developing for 150–200 years in denial of any such concept? The answer is with difficulty and disagreement, with conceptual paradox and practical uncertainty, says the judgment in the Wik case. And that is a negation of the certainty which graziers and miners had hoped to achieve from this case (Lane 1996). The Wik people must now establish their own native title rights through the National Native Title Tribunal, as will other Aboriginal communities who want to make land claims under the Native Title Act. But where the disputed land is under pastoral lease, negotiation and mediation are now the name of the game.

In this radically altered planning framework, the very notion of certainty has been abandoned in favor of a more complex, historically informed and situation-specific reading of competing claims. For this new framework to be successful requires a very different mindset and set of skills to those associated with modernist planning. Land and resource management under Native Title must begin with an acknowledgment of historical injustice, and an understanding and positive valuation of difference, a respect for and honoring of a different way of life and of a desire to preserve that difference. This presupposes a literacy about another culture; an ability to communicate with people whose life ways and daily rhythms are very different and do not include familiarity with bureaucratic procedures of decision making and conflict resolution; and a genuine openness in negotiation and mediation processes. It requires face-to-face mutual learning, compassion, empathy and patience. It may require acknowledgment of incommensurable life ways and new definitions of social justice based on that acknowledgment. It is a breathtaking decision that the High Court has taken, reflecting a faith in the possibility of some kind of national maturity that has yet to be proven. It is a leap into the postmodern age of planning, with its potential to empower indigenous people in their quest for the reterritorialization of regional Australia and for self-management. Educating planners to work in these circumstances is a very different task to educating them as implementors of statutory planning codes and suggests that socio-cultural or multiracial literacy must become a foundation of a life in planning in the postmodern age.

REFERENCES

Alexander E (1992) *Approaches to Planning*, 2nd ed., Philadelphia, Gordon and Breach

Anzaldua G (1987) *Borderlands / La Frontera*, San Francisco, Aunt Lute Books

Archiburgo F (1992a) 'Towards a New Discipline of Planning', working paper, Rome, Planning Studies Centre

Archiburgo F (1992b) 'The Resetting of Planning Studies', reprint A92.7, Rome, Planning Studies Centre

Benhabib S (1995) 'Cultural Complexity, Moral Interdependence, and the Global Dialogical Community', in Nussbaum M and Glover J, eds *Women, Culture and Development*, Oxford, Clarendon Press, pp235–55

Castells M (1978) *City, Class and Power*, London, Macmillan

Castles S and Miller K J (1993) *The Age of Migration: International Population Movements in the Modern World*, New York, Guilford Press

Cross M and Keith M, eds (1993) *Racism, The City, and the State*, London, Routledge

Davis M (1990) *City of Quartz*, London, Verso

Etzione A (1978) *The Active Society: A Theory of Social and Political Processes*, New York, Free Press

Faludi A (1973) *Planning Theory*, Oxford, Pergamon Press

Faludi A (1986) 'Procedural Rationality and Ethical Theory', *Planning Theory in Practice* vol. 1, Turin

Forester J (1991) 'On Critical Practice: The Politics of Storytelling and the Priority of Practical Judgement', Clarkson Lecture, Buffalo, SUNY

Friedman J (1986) 'The World City Hypothesi' *Development and Change*, vol. 17, 1, pp69–84

Friedman J (1992) *Empowerment: The Politics of Alternative Development*, Oxford, Blackwell

Friedman J and Wolff G (1982) 'World City Formation', *International Journal of Urban and Regional Research* vol. 6, 2, pp309–44

Harvey D (1989) *The Condition of Postmodernity*, Oxford, Blackwell

Hayden D (1980) 'What Would a Non-Sexist City be Like?', in Stimpson C *et al*, eds, *Women and the American City*, Chicago, University of Chicago Press, pp167–84

Heskin A D (1991) *The Struggle for Community*, Boulder, Westview Press

Holston J (1995) 'Space of Insurgent Citizenship', in Sandercock L, ed. *Making the Invisible Visible*, Milan, Angeli, pp35–52

hooks b (1990) *Yearning: Race, Gender and Cultural Politics*, Boston, South End Press

Hooper B (1992) 'Split at the Root', *Frontiers* vol 13, 1, pp45–80

Innes J (1996) 'Planning Through Consensus Building', *Journal of the American Planning Association*, vol. 62, 4, pp460–72

Jacobs J M (1996) *Edge of Empire*, London, Routledge

Jameson F (1984) 'Postmodernism, or, The Cultural Logic of Late Capitalism, *New Left Review*, vol 146, pp53–92

Kennedy F (1996) 'Title Fight' *The Weekend Australian*, December 28–29

Knox P and Taylor P, eds (1995) *World Cities in a World System*, Cambridge, Cambridge University Press

Kong L (1993) 'Ideological Hegemony and the Political Symbolism of Religious Buildings in Singapore', *Environment and Planning D: Society and Space*, vol. 11, 1, pp23–46

Kwong P (1992) 'The first multicultural riots' *Voice*, June vol 9

Leavitt J (1994) 'Planning in the Age of Rebellion', *Planning Theory*, vol. 10/11, pp111–30

Leavitt J and Saegert S (1990) *From Abandonment to Hope: Community Households in Harlem*, New York, Columbia University Press

Lindblom C (1959) 'The Science of Muddling Through' *Public Administration Review*, vol 19, 2, pp79–99

McWilliams C [1948] (1990) *North from Mexico*, New York, Praeger

Oliver M, Johnson J H, and Farrell W C Jr (1993) 'Anatomy of a Rebellion: A Political Economic Analysis', in Gooding-Williams R, ed. (1993) *Reading Rodney King: Reading Urban Uprising*, London, Routledge, pp117–41

Peatie L (1987) *Planning: Rethinking Ciudad Guyana*, Ann Arbor, University of Michigan Press

Sandercock L (1995) 'Voices from the Borderlands: A Meditation on a Metaphor', *Journal of Planning Education and Research*, vol 14, 2, pp77–88

Sassen S (1991) *The Global City: New York, London, Tokyo*, Princeton, Princeton University Press

Sassen S (1995) 'On concentration and centrality in the global city', in Knox P and Taylor P, eds, *World Cities in a World System*, Cambridge, Cambridge University Press, pp63–78

Simon H [1945] (1976) *Administrative Behaviour*, New York, Free Press

Soja E (1989) *Postmodern Geographies*, London, Verso

Starr A and Lee C (1992) 'Indigenous Planning: Paradigm and Curriculum for Relevance to Communities of Color', unpublished paper, Urban Studies and Planning department, MIT

West C (1990) *Race Matters*, New York, Random House

'Café-Extra: Culture, Representation and the Everyday'

from Nick Stanley and Ian Cole (eds), *Beyond the Museum: Art, Institutions, People* (2000)

Malcolm Miles

Editors' Introduction

One of the starting points of this essay was an effort to find a way of writing about theoretical questions in utopianism and critical cultural theory, without lapsing into a completely abstract way of thinking; hence the location of the text in Rotterdam, a city in transition. Rotterdam was European City of Culture in 2002, though that is not really relevant; it has, however, a distinct cultural quarter of museums and small galleries, with many cafés and bars. A detailed (if now out-dated) examination of its cultural policy is given by Maarten A. Hajer in 'Rotterdam: re-designing the public domain', in *Cultural Policy and Urban Regeneration: The West European Experience*, edited by Franco Bianchini and Michael Parkinson (London, Routledge, 1993). Another starting point, more tangential to the argument, was a reading of Richard Sennett's *Conscience of the Eye* (New York, Norton, 1990), in which Sennett, writing as narrator, describes his walk through central Manhattan from Greenwich Village north over 14th Street to a mid-town restaurant, and a feeling of disappointment that he remains a detached observer who in the end does not say whether or not he goes into the restaurant, let alone what he eats and drinks. This may be outside his scope, or none of the reader's business, and have no intellectual weight, yet the lack of materiality in the description remained nagging. Hence interludes in this essay of noticing the material conditions in which a particular problem was reconsidered.

The theoretical problem is that faced and not resolved by Herbert Marcuse at the Free University in Berlin in 1967 (see Marcuse's *Five Lectures*, Harmondsworth, Penguin, 1970). The essay below cites his response to a question as to how radical change can happen when its need becomes felt only in the new consciousness it produces. Broadly, the essay seeks to address the problem (not to solve it) by shifting its ground from time to space. The text can be read alongside Doreen Massey's remarks on Ernesto Laclau – who sees time as the dimension of change and space as static – in *Space, Place and Gender* (Cambridge, Polity, 1994, pp. 250–69). Marcuse's title was 'The End of Utopia'. But what he meant was the end of the dream and inception of a reality. The next section in this book seeks to put some flesh on those bones.

INTRODUCTION

This essay argues that the utopian content of art is no longer viable amidst a cultural fragmentation and fusion which changes the shape but retains the conservatism of art's institutions, but that it is evident in a different form in the creativity of everyday life. Within this scenario, the essay addresses a problem raised by Herbert Marcuse following a lecture titled "The End of Utopia" (Marcuse, 1970: 62–82) at the

Free University, Berlin in 1967 – that the realisation of a new society requires the prior abolition of the institutions of the old, whilst the need for that abolition is realised only once the new has dawned. This is as if to say, simplistically, we could have an omelette if the eggs will break themselves. The advantage of the everyday is that, because it is already there, it does not need to usher in its own dawn. This moves the argument from the abolition of (art's) institutions to their supersession, as the duality of centre-margin gives way to a pluralism which, whilst entirely contingent, is not value-free.

EXPERIENCE AND DISTANCE

The everyday has always been there, providing material for narratives of the city; a conventional form of these in the nineteenth and twentieth centuries is that of the distanced observation of the stroller (*flâneur*) – from Baudelaire to Walter Benjamin and Richard Sennett. But if Baudelaire is like an ancestor figure in urban literature, there are significant differences between Benjamin's account of a year in Moscow, in *Moscow Diary* (trans. Sieburth, 1986), and Sennett's representation of New York in *Conscience of the Eye* (1990). Benjamin's narrative engages the reader in the details of daily life – his walks, the cafés he enters, what he eats, what he sees in the streetmarket, the colours of women's shawls. Benjamin writes: "The inventory of the streets is inexhaustible." (January 1st, 1927; p58). Floating through the diary's pages, also, is the presence, or absence, of Latvian theatre director Asja Lacis, object of Benjamin's emotional life at the time. For Benjamin, the diary represents an effort to convey the experience of his visit as it was, recalled but unmediated by prior conceptualisation: "My presentation will be devoid of all theory . . . I hope to succeed in allowing the creatural to speak for itself . . . I want to write a description of Moscow at the present moment in which 'all factuality is already theory' and which would thereby refrain from any deductive abstraction . . ." (Benjamin, 1986: 6).

Sennett's writing differs in two respects: firstly, the book presents a hypothesis of urban society, in which experience illustrates theory; the reader is in the position of listener, as Sennett imparts his interpretation of the world. Secondly, though Sennett describes details of the street, there is a distancing, a sublimation represented in the third-person voice (as in this commentary), and in the visuality of his text; he writes with the intriguing fluency of a novelist, but his narrative of a walk from Greenwich Village to a mid-town French restaurant is framed by his critical position, and, perhaps crucially, he never tells us what he eats, or even if he reaches the restaurant. Sennett laments the lack of engagement in the city:

> New York should be the ideal city of exposure . . . a city of difference par excellence . . . collecting its population from all over the world . . . By walking in New York one is immersed in . . . differences . . . but precisely because the scenes are disengaged they seem unlikely to offer themselves as significant encounters . . . The leather fetishist and the spice merchant are protected by disengagement . . . social classes, who mix but do not socialise. (Sennett, 1990: 128)

But he is himself disengaged, enjoying a masculine freedom to ponder the shop-window displays of everyday things in the lives of strangers, a stroller for whom the city's edges are "not too disturbing, if I keep moving" (Sennett, 1990:124). If Sennett offers a panorama, Benjamin offers glimpses at street-car level, though these, too, are framed by the format of the diary.

For Benjamin and Sennett, however, the separate categories of art and everyday life are retained. Benjamin seeks out Russian toys in shops and museums, but does not call these art. They offer instead the authenticity associated with peasant cultures, unmediated by tests and requiring no interpretation – the decorative as implicit identity. Art, in contrast for Benjamin, is mediated by the rituals of its institutional display which produce its aura. For Sennett, in *The Uses of Disorder* (1970), graffiti plays the part of the raw against the city's cold abstraction, as, in *Conscience of the Eye*, the arrays of the spice and leather shops denote actuality, though his narrative becomes an aestheticisation of the city.

But the world moves on, and an easy differentiation between art and everyday life is no longer viable when advertising derives its vocabulary from art, when consumer durables are designed like sculptures, when sculpture resembles freezer cabinets, and vacuum cleaners are exhibited in art galleries. The boundaries are dissolved in an all-encompassing visual culture, in which the level of sophistication is often higher in advertising than in art, as is the level of financial support. Appearances, however, can be deceptive, and the fusion of categories does not mean a democratisation of art's meta-consciousness, a spreading of its aura into the street.

To take a specific experience: walking through Rotterdam in 1998, in Witte de Withstraat, I saw a café sign with blocks of red, yellow and blue, referencing de Stijl in a city which has at least three de Stijl houses (in a row near the Architect Institute); but the rectangles of primary colour are juxtaposed with a cut-out beer bottle and a Coca-Cola sign. Is the bottle pop-art and the cola sign sub-Warhol? The question is self-defeating – it doesn't matter now, everything is culture, and everything is co-opted to consumption. Across the road was a billboard, placed neatly in the vertical space of the wall so that its edges were aligned with the sign for a gallery on the ground floor, as if intended; Triumph bras and panties (not by Barbara Kruger) signify (or do not) the XX multiple gallery, though the windows reveal the usual abstractions. Various stock interpretations of this alignment can be rehearsed: galleries market art like billboards market underwear; workers in the underwear factory and poorly paid gallery assistants are both exploited; art and bras are used in the selling of lifestyles, though art is more expensive. And, post-John Berger and Laura Mulvey, we know that women's bodies are objectified by the masculine gaze, in the billboard as in a nearby exhibition of Picasso's paintings of women – most of whom are dissected, and several crying. Then there is the erasure of context signified by the billboard's white background, like the white walls of the value-free space of the art gallery. None of this explains why the woman wearing black is at the end, neatly above the gallery sign's XX. But what emerges from the experience is the dissociation of signs from evident interpretations, the impossibility of meaning – not now an aestheticisation of the city but its anaesthetisation (Leach, 1999). Walking back, after eating lunch in a café opposite – I am obliged by my remark above to say that this consisted of grilled goat's cheese with peppers, salad, and warm bread, and a glass of wheat-beer – I saw another sign on a traffic-island, near the Willem de Kooning Academy, mixed up with lights and direction signs, which read 'PSYCHOTIC'.

ORDER

Categories are convenient, dividing up what is otherwise continuous, allowing observations of fragmentation or fusion; and they are historically specific, even in their dissolution. In Rotterdam, art is reproduced as a street sign, advertising seems to signify art (which, post-Duchamp and Hirst, is often accidental in effect), and the whole place is crazy. But perhaps a more interesting approach than the interpretations advanced above (though these remain valid) is to ask what the signs convey, The red, yellow and blue rectangles are removed from the context of Modernism, as a café sign, but they do not indicate food either – the café does not sell red, yellow and blue sandwiches, and the little biscuits that come with the coffee are round. What is indicated is the commodity of culture itself, which signifies affluence. In a street which denotes an emerging cultural quarter, a zone of galleries and museums of contemporary art attracts up-market tourism, and purchases the city's place on an international culture map. Leaving aside a critique of cultural regeneration, the fusion which is taking place is between art, as a category of objects with a meta-level of meaning, and information, as a category of the undifferentiated norm. Within critical theory there are two approaches to mass culture. Both are underpinned by a history in which art is, in the eighteenth century, a category of taste underpinned by the unified subject-citizen of liberalism. Yves Michaud writes; "Art gained its autonomy *as art* in the middle of the eighteenth century . . . not for aesthetic but rather for social reasons: as it became an object to be savoured and discussed by a public of art lovers" (Michaud, 1998: 42). In the bourgeois period, according to Marcuse in his essay on the affirmative character of culture (in *Negations*), this becomes a realm for displacement of hope for a world which is better, a land of dreams in which present ills are appeased not by the future salvation of religion but in the absent-present of the aesthetic dimension (Marcuse, 1968a: 97–8). Building on this, Marcuse's position in *The Aesthetic Dimension* (1978) draws close to Adorno's in *Aesthetic Theory* (1997). Art's distancing from society now offers a safe house in a hostile world. In face of a grim reality, the present phase of which is globalisation (Bauman, 1998), art's autonomy is a transcendent space in which the perceptions of the dominant society can be resisted.

This gives rise to a new problem: how are the insights gained from distance translated into intervention? But the question is already flawed, in that intervention supposes a place from which to intervene, reproducing the duality of art and society. For Adorno, preservation of this duality, and a refusal of closure of the argument between art's aesthetic and social dimensions in any resolution, is the only possibility for criticality. He sees what he calls the culture industry as a means only of mass deception, and employs this term, instead of mass culture, because the culture purveyed to the mass public is not produced by that public but imposed on it (Adorno, 1991; 85). The same critique could be applied to what are now called the cultural industries, seen by city planners as a magic answer to urban regeneration.

Benjamin (1973) takes a different view, seeing in film a democratisation of culture – the movie as mass dreaming and liberation from the given in narratives of what might be rather than what is – though he writes that Russian films suffer from bad acting and lack of resources, and that it is Russian avant-garde theatre which is the more revolutionary medium. Either way, Benjamin sees glimpses of resistance in mass culture; and, in his essay "The Work of Art in the Age of Mechanical Reproduction", in *illuminations*, images of everyday life reproduced photographically rupture the surfaces of the dominant culture. Here, for Benjamin, is a wholly accessible realm of imaginative futures, freed from the enforcement of art's aura in its ritual observance in the museum. And though Adorno may in the end be right, as the market subsumes all resistant forms to itself, and the mass promotion of lifestyles turns demands for liberation into demands for designer jeans and sun-glasses, that process spawns its own subversion in the fake.

TRANSGRESSION AND SUBLIMATION

The limitation of the everyday is its lack of transcendence. But if it cannot transcend, can it transgress? For Peter Fuller (1980), influenced by Marcuse (1978) in his shift from Berger to Ruskin, transcendence is art's unique claim, and is found, as he sees it, in the universal subject-matter of the mother-child image – which is usually the mother-son, as Irigaray (1994: 8–14) has noted. The biological base for art – all humans have been infants wanting the breast, which is both good and bad, offered and withheld – lent Fuller a rational explanation for its enduring attraction regardless of the circumstances of its productions. This is one use of psychoanalytic knowledge (from Millner, Klein and Winnicott), but differs from Marcuse's (from Freud), for whom art offers, in its autonomy from the oppression of social life in an unfree society, a meta-level of reflection and critique, a space for transgression beyond the constraints of the reality principle in the ego. This, metaphorically at least, opens the path to an erotic society. Marcuse begins this fusion of Marx and Freud in *Eros and Civilisation* (1955), but it remains the framework of his late work, and takes on a specific vitality in *An Essay on Liberation* (1969), written amidst the student revolts of 1968 in Paris, Berlin and the USA.

But does the idea of a transcendent art hold now? The major difficulty is not so much that art is a historically-specific and socially constructed concept, as that it is defined, more specifically, by what artists do within the institutional and market structures which support them. Art's currency depends on critical interpretation within a structure which today turns art into a branch of the entertainment industry. It tends to conservatism as much in the deceptive authenticity of Tracy Emin's messed-up bed (which is a knowing manipulation of the gallery system) as in dreams of Arcadia, or modern art's appropriations of primitivism. These are all cases of the pseudo-uncooked, promising direct experience but delivering its sublimation. They do not transgress, but affirm a world of privilege, interpretation, alienation and disempowerment.

Another argument, leaving aside art, might see history as immanent rather than transcendent, its promise of liberation here and now. This is the millenarian vision of Joachim of Fiore, who proclaimed, c.1200, the Third Kingdom in which hierarchies, privileges, and the individual ownership of property were abolished, not in heaven but on this Earth (Bloch, 1986; 509–15) – the meek shall inherit the Earth today, not at the landlord's convenience. This is not literary utopianism, which encapsulates the better life on the page or is a distant island which does not exist. And, in 1968, it is another millenarian vision which informs Marcuse, when he argues that art, in its most abstract forms, ruptures the surfaces of the dominant society by refusing that society's normalised codes of perception. In his paper to the *Dialectics of Liberation* Congress at the Roundhouse in 1967, Marcuse speaks of society as a work of art – the most radical possibility for change, involving liberation from performance of the reality principle. In the same year, on an album titled *Strange Days*, The Doors sang, "We want the world and we want it now." Society as a work of art, however, is not an aestheticisation of society, but its libidinisation: and for Marcuse, the libidinal society is possible when the economic problem of scarcity is solved by new technologies of production. This allows a shift from labour as alienating toil, to an eroticisation of work and social relations. Rather than leading to "a society of sex maniacs" (Marcuse, 1955: 201), the relaxation of the reality principle leads to spreading of sexuality to all aspects of life:

> The pleasure principle reveals its own dialectic. The erotic aim of sustaining the entire body as subject-object of pleasure calls for the continual refinement of the organism, the intensification of its receptivity, the growth of its sensuousness. The aim generates its own projects of realisation: the abolition of toil, the amelioration of the environment, the conquest of disease and decay the creation of luxury. (Marcuse, 1956: 212).

Luxury here denotes Baudelaire's *Invitation to the Voyage*, subject-matter of various paintings by Matisse, including *Luxe calm et volupté* (1905), and this vision of a relational society – which is close to Fourier's concept of a libidinal society – constitutes society as a work of art. At the *Dialectics of Liberation* Congress, Marcuse says that the qualities of a socialist society seem utopian:

> this is precisely the form in which these radical features must appear if they are really to be a definite negation of the established society: if socialism is indeed the rupture of history, the radical break, the leap into the realm of freedom – a total rupture. (Marcuse, 1968b: 177).

But how does this transgression in which consciousness assumes the autonomy characteristic of art take place? How does the aim produce its own realisation? Marcuse sees discontent in the mass public, and increasing repression in capital's manipulation of life (Marcuse, 1968b: 187), but in answer to a question after his Berlin lecture accepts he does not know:

> You have defined what is unfortunately the greatest difficulty in the matter. Your objection is that, for new, revolutionary needs to develop, the mechanisms that reproduce the old needs must be abolished. In order for the mechanisms to be abolished, there must first be a need to abolish them. That is the circle in which we are placed, and I do not know how to get out of it. (Marcuse, 1970: 80).

Marcuse sees a convergence of the political and the psychological which gives rise to a new structure of psyche, but is to see this to be there? Perhaps it was in the alternative cultures of 1967. But as art infused with the genres of visual culture, its criticality is more difficult to see.

APORIA

The problem today is no longer art's isolation. Art is big business, embellishes urban development with a new generation of monuments, is utilised in lifestyle consumption, makes news and produces stars, and is explained in gallery education and outreach programmes. In efforts to widen access to art, or to engage new publics by re-siting art from the gallery to the street and the school playground, there is seldom a sense of the playful. And the move from permanent commissions to temporary projects in public art which took place during the 1990s was a product of the desire of public art's new managers to be like gallery curators, assembling mixed exhibitions in the rain. Art is part of everyday reality, but of its dominant aspect.

In this context artists continue, as they began in the 1890s with the *Salon des Indépendents*, to assert autonomy by appropriating the means of art's production and dissemination – today in the warehouse studio and alternative exhibition venue. A recent variation on this is the use by Laura Godfrey-Issacs of her south-London house as a gallery. But this remains within the limits of the art world, an adaptation rather than overthrow of its institutional structures.

Is this unavoidable? Does art carry within its practice its own institutionalisation? Michaud takes a pessimistic view, perceiving "the rehearsal of a strange comedy" in which "democracy has exhausted utopia and we are left . . . with a legitimation and motivation deficit with regard to the social system." He continues, after saying that art offers only a mirage: "One can imagine the importance of being attentive to other more modest and less lofty forms of legitimation and motivation. The crisis of art raises the issue of new concepts which have to be formed in order to think through radical democracy" (Michaud, 1999: 156). This restates in a gloomy way Marcuse's dilemma – but, still, how does it happen?

Adorno, not known for his optimism, argues that efforts to grant art a social function are doomed: "It is uncertain whether art is still possible: whether, with its complete emancipation, it did not sever its own preconditions" (Adorno, 1997: 1); its aesthetic and social aspects are mutually disabling, its only option something on the edge of darkness and silence: "Art must turn against itself, in opposition to its own concept, and thus becomes uncertain of itself" (Adorno, 1997: 2) or: "Radical art today is synonymous with dark art; its primary colour is black . . . what is written, painted and composed is also impoverished; the most advanced arts push this impoverishment to the brink of silence" (Adorno, 1997: 39–40). But, Adorno also says:

> Art therefore falls into an unsolvable aporia. The demand for complete responsibility on the part of artworks increases the burden of their guilt; therefore this demand is to be set in counterpoint with the antithetical demand for irresponsibility. The latter is reminiscent of an element of play, without which there is no more possibility of art than of theory. (Adorno, 1997: 39)

Perhaps here, in play, is hope: play, as Marcuse sees it, as alternative to performance of the reality principle. Marcuse (1955) writes of an anticipatory memory of freedom as a state of psyche prior to the dominance of the reality principle, and though this mythicises a projection, he sees the drives which for Freud play of their own accord in the *Es* (id) as historically specific, open to change – in a world free from toil, the drives shift towards a libidinal model, not through conscious intention, but in the unconscious. In this respect, the new sensibility does appear to produce itself, but does so amidst specific conditions, and in these is a possibility for intervention.

EXITS

Marcuse is very close to solving his own dilemma. But he seeks a solution, in keeping with the classical model of emancipation (Laclau, 1996) which involves the ending of one state in order that the next might begin. The difficulties here include that the new is conceived within, and to an extent will be a reaction against (conditioned by) the old; and that if the new has apocalyptic states, descending from the heavens like New Jerusalem, or rising from the sea like New Atlantis, remains unattainable – all that descends from the sky is ultra-violet light through the hole in the ozone layer, and what arises from the ocean is toxic algae. Marcuse's problem is set in time. A possibility, following the renewal of interest in space following translation of Lefebvre's work and its galvanisation of disciplines such as geography and sociology in the anglophone world, is to set the problem instead in space: in two parallel spaces of, to borrow Lefebvre's categories, conceptual and lived space. In the dominant culture, the concept is privileged. But, as Lefebvre is at pains to indicate, the experiential does not go away and is evident in traces of occupation which are seen by the dominant culture as margins to its centres, but are not cancelled, are the ground of an alternative sensibility.

Conceptualisation, even of an oppositional culture, sublimates its radical content as interpretation; the alternative is recognition, rather than construction, of a non-repressive sensibility which is present in moments of liberation within the routines of everyday life. To end with an ordinary example: the men's lavatory on the waterfront at Hull – the attendant, seeing its potential as a small conservatory, has turned the interior into his version of Kew Gardens; plants such as papyrus, fern and geranium occupy every available space, amidst gleaming white china fittings and polished copper pipes. The site has become a local landmark, to which people take their visitors even when they do not need to use it, and it stands as a sign of that creativity which extends also to allotments, the decoration of houses, and the ingenuity of so-called outsider-art. Perhaps the quality Benjamin found in Russian toys is of a similar kind – a quality sometimes perceived (problematically) as primitivism and sublimated in the idealising forms of art by Kandinsky and Brancusi.

All of which is very well; but, the question now becomes, how is the consciousness which is not new but extant, yet marginalised, to be the ground for a new social organisation? The grand solutions are as useless as the grand narratives which postmodernism abolished in an unmediated relativism. And just as fascism in the 1930s appropriated the semblances of the left in its banners and parades, turning carnival into its grotesque parody as the torchlight procession and the architecture or searchlights, so, in the United Kingdom in 2000, the Countryside Alliance and the thugs who run the oil industry (and their petit-bourgeois hired hands) appropriate direct action. But if there are no certainties, only contingencies, then attention, as Michaud argues, no longer lofty in its objects, turns to the analysis of conditions, and the possibility to chip away, widening the cracks which appear in the dominant society, and must appear because that society is built on a contradiction, going back to 1789, between its promise of universal liberty and its prevention of liberty in the mechanisms of exchange and representation.

Opposite the XX multiple gallery, at Witte de With Centre for Contemporary Art, on a side wall, is a poster left over from an exhibition several years ago. A woman sits at her desk; beside her is the text "Melly Shum hates her job". To work in the cracks, then, like frost, like water.

REFERENCES

T.W. Adorno, *The Culture Industry*, London, Routledge, 1991.

—— *Aesthetic Theory*, London, Athlone Press, 1997.

Z. Bauman, *Globalization: the Human Consequences*, Cambridge, Polity, 1998.

W. Benjamin, *Moscow Diary*, ed. G Smith, trans. R Sieburth, Cambridge (MA), Harvard, 1986.

—— *Illuminations*, London, Collins, 1973.

E. Bloch, *The Principle of Hope*, Cambridge (MA), MIT, 1986.

D. Cooper (ed.), *The Dialectics of Liberation*, Harmondsworth, Penguin, 1968.

P. Fuller, *Art and Psychoanalysis*, London, Readers and Writers Cooperative, 1980.

L. Irigaray, *Thinking the Difference*, London, Athlone Press, 1994.

E. Laclau, *Emancipation(s)*, London, Verso, 1996.

N. Leach, *The Anaesthetics of Architecture*, Cambridge (MA), MIT, 1999.

H. Lefebvre, *The Production of Space*, Oxford, Blackwell, 1991.

H. Marcuse, *Eros and Civilization* Boston, Beacon Press, 1995.

—— *Negations*, Harmondsworth, Penguin, 1968a.

—— "Liberation from the Affluent Society", in Cooper, 1968, pp 175–192

—— *An Essay on Liberation*, Harmondsworth, Penguin, 1969.

—— *Five Lectures*, Harmondsworth, Penguin, 1970.

K. Marx, "Theses on Feuerbach", in (ed.) F. Engels, *Ludwig Feuerbach*, (n.d.) London, Lawrence & Wishart pp73–5, 1845.

Y. Michaud, "The End of the Utopia in Art" in Schaeffer, 1999, pp131–156.

M. Miles, *Art, Space & the City*, London, Routledge, 1997.

J.-M. Schaeffer, (ed.), *Think Art: Theory and Practice in the Art of Today*, Rotterdam, Witte de With Centre for Contemporary Art, 1999.

R. Sennett, *The Uses of Disorder*, New York, Norton, 1970.

—— *The Conscience of the Eye*, New York, Norton, 1990

This series of images documents Hermann Prigann's *Ring der Erinnerung (Circle of Remembrance)*, a project first conceived in a drawing in 1985 and begun in practical terms in 1989 when it seemed likely the two Germanies – Federal and Democratic – would unite. The site is a remote area of forest on high ground in the Harz Mountains, between Braunlage and Blankenburg. The images show the surviving border tower left on the site to remind visitors, of whom there is a steady flow, of the border's past existence, then the making of the work, and its appearance in 2002. It cannot be shown finished because part of the work is a slow encroachment by succession growth on what has been made, the natural recolonisation of a circular earth bank (built on an armature of tree trunks which will rot) by wild fruit bushes, wild roses, and wild grasses and flowers. In this case, the re-growth is poignant for two reasons: the site was previously badly affected by acid rain; and because the border itself which dissects the ring, some of the concrete fence posts still there in a line through it, is gradually being dissolved away. The re-growth, however, follows a shape made by human intervention and will never be as if that intervention had not been made.

The space inside the ring is uncanny. All around the dark, tall forest is silent while in the distance a deer runs through a clearing. There are few birds flitting through the dappled light.

The following text was written by Prigann at the time of the work's construction (reprinted by permission of the artist):

> The *Ring of Remembrance* is to be seen in the context of the artistic programme 'Metamorphic objects – sculptural locations', on which Prigann has been working since . . . 1980. For him, it is an irrefutable recognition 'that we must develop and evolve for our time an aesthetic awareness of the BEAUTY in nature. Without this awareness, cultural and ecological problems can never be solved. This BEAUTY is defined in a landscape environment which shows signs everywhere of destruction, as something that transforms this destruction into locations of artistic and ecological change and . . . into a new experience of nature and the landscape.
>
> The ring is conceived as a large circular precinct, an embankment constructed of withered trees from the immediate vicinity. The interior space will be 70 metres diameter, the embankment varying in height between 4 and 5 metres. Four passageways, facing the points of the compass, will be marked out on the ground by great stone slabs. A fifth stone slab will define the centre of the precinct. The stone slabs represent the five concepts: EARTH–FAUNA–FLORA–AIR–WATER. The first stage in the development of the *Ring of Remembrance* will be concluded with the completion of the wall. The wooden wall, as a metaphoric object, will fall into decay during the course of the years, it will rot and the blackberry bushes and dog roses planted all around it will overrun everything with time.
>
> On the former border strip, a work of art will thus come into being which will awaken intricate associations. On the one hand it will stand as a pointer to the ecological problems of our times; on the other hand it will symbolize the union of the two German nations in an aesthetic configuration.
>
> The circular precinct will link three different themes: The wooded landscape of the Harz region portrays considerable signs of being a dying forest. Therefore, the material from which the ring wall of the precinct is to be built will be dead wood from this region. The image of the fallen forest is thus contained within the ring wall. The dead wood of the ring wall will rot in the course of time and return to the earth. The planting of blackberries and other climbing perennials will lead to the dead wood becoming overgrown. Thus, from a dual viewpoint, the metamorphic process will be reconstructed at this ring all: fall and rise.

Plate 32 **Watchtower on the ex-German Democratic Republic side of the border** (Photo: M. Miles). (Courtesy of H. Prigann.)

Plate 33 **Drawing for the project, 1990** (Courtesy of H. Prigann.)

Plate 34 **Site before construction of the Ring, 1992** (Courtesy of H. Prigann.)

Plate 35 **Construction of the Ring, 1992** (Courtesy of H. Prigann).

Plate 36 **Raising withered trees, 1992** (Courtesy of H. Prigann).

Plate 37 **Installation of stones at the entrance to the site, 1992** (Courtesy of H. Prigann).

Plate 38 **The Ring in 2001** (Photo: M. Miles). (Courtesy of H. Prigann).

Plate 39 **Entrance stone, 2001** (Photo: M. Miles). (Courtesy of H. Prigann).

PART TEN

Possible Futures

INTRODUCTION TO PART TEN

The first literary utopias were distant in space. More recent cases tend to be projected in time, to either near or distant futures. They may derive their forms from imagined classical or medieval pasts, or be critiques by oblique means of the present state of affairs, but the beautiful society is a beautiful tomorrow. In this it is, as said in the Introduction to the previous section, like the aesthetic dimension which displaces the imagination of a world which might or ought to be to a compensatory rather than an active realm. But if utopias are beautiful dreams and dystopias are nightmares, this will translate into action only with difficulty. Change then remains as distant as the images of it are distanced from reality. And yet, in a world which has the technological capacity to end scarcity at the level of basic needs (for food, shelter, and clothing, for instances), and possibly far more than that, there is a widening division between those who possess the basics of life and those who lack them, or are deprived of them by systems of exchange and distribution. As Susan George writes in *How the Other Half Dies* (Harmondsworth, Penguin, [1976] revised 1986): 'We are no longer living in the seventeenth century when Europe suffered shortages on an average of every three years and famine every ten. Today's world has all the physical resources and technical skills necessary to feed the present population . . . the problem is not a technical one . . . Hunger is not a scourge but a scandal' (p. 23). The theme of this final section of the book is, then, the possibility of a better tomorrow; and one which is not distanced to a never-never land but is being made through human action today. That is, the theme is how alternatives to a dominant, in many ways destructive, development trajectory are beginning to be stated through new practices and alternative critical and theoretical positions in keeping with them. Although the problem of sustainability touches most aspects of life, the book's concern is with cities and the cultures which shape and take shape in them, and the texts selected address issues in that context – either directly dealing with, say, architecture, or more generally with social and cultural process such as the planting of abandoned lots in New York's lower east side – a kind of squatting in the form of inner-city agriculture and horticulture called avant-gardening in the text by Bill Weinberg reprinted below. For the most part, then, the section is optimistic, following a similar avoidance of melodramatic and apocalyptic material in the previous section. That inevitably means taking a position on what constitutes a desirable future, interpreted here as what constitutes a sustainable future.

Sustainability has been interpreted in many ways, including those convenient to the global economy. In some cases, such as the report of the Bruntland Commission of the World Commission on Environment and Development, (formed by the United Nations in 1984), the emphasis is on economic impacts on the environment. For this book, however, it is seen more broadly as having cultural and social as well as economic and environmental dimensions.

At its most conservative, the concept of sustainability means maintaining economic development on the western, technologically advanced model for the perpetual future, without destroying the planet. The implications of that are, on one hand, an opportunity to continue to export the apparatus of an economic system to the entire world, in which culture, as in fast food and the movies, plays a key role; on the other, a responsibility to ensure that resources, particularly of energy and raw materials, are not depleted at a rate faster than that at which they can be replaced. In terms of the burning of fossil fuels, where geological production takes millions of year, it obviously means finding alternatives, as major oil companies are beginning to do to a small extent. Jennifer Elliott, in *An Introduction to Sustainable Development* (London, Routledge, second edition, 1999), writes that sustainability raises issues of what each generation passes on to the next in terms of the assets of human ingenuity as well as natural capital (the planet's resources); and of how limits to growth are set, whether through supposedly neutral technical judgements, or in a more holistic way. There is a parallel here with the difference between the modern rational planning model and its contemporary, postmodern and radical alternatives described by Leonie Sandercock in the previous section. Elliott goes on to suggest that sustainability requires appropriate political, economic, technological, and governmental systems, as well as respect for the ecological systems which are the base for all development.

The part can be seen then as, if in a tangential way, a critique of urban development in late capitalism. There is, however, a balance of material in the section. The first text reprinted in it is an excerpt from Edward Soja's *Postmetropolis* in which he considers new urban forms in a world restructured by globalisation at the beginning of the twenty-first century. This is a view from the affluent world, since much of Soja's work is informed as well as contextualised by his location in Los Angeles. But Soja begins this book with a chapter based on archaeological evidence from the archaic Anatolian city of Çatal Hüyük which demonstrates that, contrary to much romantic fantasy, the city is a primary model of settlement, not an amalgamation of villages. The villages are satellite collections of dwellings linked to animal husbandry, into which nomadic hunter-gatherers were drawn. Soja sees the city with its density of population, as the driving force of cultural development: 'By reducing the friction of distance in everyday life while increasing population densities, human interaction and sociality were creatively intensified' (p. 46). His description of new, dispersed modes of settlement seems to go against this grain of several millennia. Of Orange County outside Los Angeles, for example, Soja says 'It is almost as if a new category of city is being invented, one that does not fit any conventional definitions' (p. 237). Two contexts for this come to mind: the adoption of new technologies of communication which make physical space obsolete for other than social purposes; and the growth of mega-cities (with populations above ten million) which may push those creative ingenuities Soja sees in Çatal Hüyük to breaking point.

Some of the mega-cities, such as New York, Los Angeles, and Tokyo, are in the affluent world but most are in the non-affluent world, previously and patronisingly called the third world and now more often the south, including São Paulo, Mexico City, Lagos, Cairo, Shanghai, Mumbai, Calcutta, Jakarta, and Manila (some of which are in fact north of the equator). The move to a new suburbanism in either dispersed cities or, on a smaller scale, the Disney town of Celebration, Florida (critiqued by Dean MacCannell in Part Eight) is a flight from cities gone wild. In a contrary direction, the rapid and extensive growth of informal or squatter settlements on the peripheries of large cities is a flight from rural poverty and the impact of the industrialisation and now globalisation of agriculture. These are not confined to far-away places: such settlements exist in Lisbon, within the affluent world, while the United States–Mexico border, despite intensive security and policing, is a site of constant illegal migration with squatters' camps on the Mexican side. In some cases, in Egypt, Turkey, and Brazil, for examples, informal settlements of reasonable duration are being legalised, provided with basic energy and water sources, and brought into the city's official patterns of regulation. What transpires, from the pioneering work of John Turner in the 1970s and from contemporary experiences in new non-affluent settlements, is that the poor exert and apply precisely that ingenuity of which Soja wrote in terms of archaic city-states. In some ways, informal settlements are the most stable kind available. In others, of course, they are sites of massive deprivation which would be seen as criminal in a world in which social justice was the prevailing principle (rather than economic monopoly). But perhaps something can be learned here.

That is the content of Jeremy Seabrook's chapter from *Pioneers of Change: Experiments in Creating a Humane Society* (1993) reprinted below. The chapter follows presentation of a Right Livelihood Award to John Turner in 1988 (by the Right Livelihood Foundation). Among other recipients of the Award was Egyptian architect Hassan Fathy (see *Architecture for the Poor*, Chicago, Chicago University Press, 1973). Seabrook sees the poverty of urban migrants, and the better chance of economic growth and health when settlers have control over their own immediate environments and can build shelters in flexible ways. Turner gained much of his experience in Peru and Mexico, finding in research that government housing schemes tended to produce worse poverty, health, and sociation than self-build development. Turner's writing can be read in conjunction with Seabrook's text, in *Housing by People* (London, Marion Boyars, 1976) and in his (with R. Fichter) *Freedom to Build: Dweller Control of the Housing Process* (London, Collier-Macmillan, 1972). More recent work on informal settlements and their status is found in *Illegal Cities: Law and Urban Change in Developing Countries*, edited by Edésio Fernandes and Ann Varley (London, Zed Books, 1998). The idea that such settlements are models for a revision of urban policy in the affluent world may require some effort of persuasion before being widely adopted, particularly by urban and state authorities and, even more, by major companies whose vested interests would be severely damaged by a serious pursuit of sustainability. Yet in the field of development studies research shows that people in local situations in the non-affluent world often have highly appropriate answers to the environmental problems they face.

Faraway, in what are called developing countries (again, a patronising term because it implies there is only one trajectory of development to which all countries aspire), local knowledges are seen to be appropriate and insightful. In *Liberation Ecology* (London, Routledge, 1996), Richard Peet and Michael Watts write of a link between approaches to practical environmental problems – which may be the local effects of global changes – and a resurgence in the idea of civil society:

> It is perhaps unsurprising that the enhanced emphasis within current development discourse on consolidating and promoting civil society has often drawn from the various strains of populism, in other words ideas about the power of what the World Bank called 'ordinary people'. Populism here implies not only a broadly-specified development strategy – that is to say, the promotion of small-scale, owner-operated, anti-urban programs which stand against the ravages of industrial capitalism . . . but also a particular form of politics, collective popular will and an 'ordinary' subject (p. 26).

Although they see this more direct democracy as in part anti-urban, it seems compatible with urban social life as expressed in the phenomenon of avant-gardening described by Bill Weinberg in the text reprinted below. Growing plants has a history within urban society, from the gardens of factory workers and miners which were a source of fresh food in the nineteenth century, to the allotments provided by British local authorities in the twentieth, and, in this case, to a new status as resistance to a process of gentrification which leaves lots empty while residual populations are drawn elsewhere or removed.

But what differentiates avant-gardening from a more utilitarian history of vegetable production – apart from that the gardens are often filled with flowers, as meaningful urban decoration – is its evidence of self-organisation on the parts of poor and marginalised groups within the dominant society. This is, if seen in the context of Soja's model of the dense, archaic city in which close sociation leads to inventiveness and creativity, evidence in favour of cities as high-density settlements. This is one aspect of the argument put forward by Graham Haughton in 'Environmental Justice and the Sustainable City', a chapter from *Sustainable Cities* (edited by David Satterthwaite, London, Earthscan, 1999) reprinted in this section. His text covers several areas of discussion and, with Soja's, provides a broad context within which more specific texts such as Seabrook's and Weinberg's can be understood. Haughton goes through the various aspects of a sustainable city (with a stronger urban focus than Elliott, cited above), and argues that two kinds of change are now required: better political, legal, and economic systems of regulation (this, it can be recalled, in a period of globalisation which rests on decreasing levels of regulation); and a new sense of environmental responsibility on the part of those who determine urban and environmental futures. Although he does not say it in so many words, this means a new culture of urban development. What he does say is that 'Fundamentally, sustainable development is about altering behavioural patterns not just directly in relation to the environment, but about changing the broader systems which shape human behaviour' (p. 65).

Sustainability may require new approaches to building and designing a city. The last text in this section is from Anne Pendleton-Jullian's account of the Open City at Ritoque in Chile. This represents an alternative form of building, and a re-visioning of architecture via surrealist poetry. A city built on sand within a stone's throw of the Pacific Ocean, from a distance it looks like a dispersed informal settlement, but is linked to the dominant city in as much as its inhabitants' children go to school there, and food is brought from supermarkets there. Perhaps they go to the movies there as well. Is this a city of tomorrow built today? There are probably too many singularities for that to be the case, as there are with two other experimental cities, Paolo Soleri's Arcosanti in the Arizona desert, and Auroville, a meditation city in south India. But Ritoque may have more to offer than, say, Le Corbusier's fantasy planning schemes requiring the wholesale demolition of urban centres, or his concrete utopia at Chandigarh – now surrounded by non-Modernist housing and unplanned shelters in which the majority of workers and the poor live outside the administrative district designed by Le Corbusier. If, as Iain Borden argues in 'What is Radical Architecture?' (in *Urban Futures*, see suggested further reading below), it is time to go beyond despair at the failure of the Modernist avant-garde by working knowingly within the dominant society 'not wholly oppositionally but ironically and irritatingly' (p. 118), then this will produce a diversity of localised solutions to the problems of urban development.

SUGGESTED FURTHER READING

David Drakakis-Smith, *Third World Cities*, second edition, London, Routledge, 2000

Jennifer A. Elliott, *An Introduction to Sustainable Development*, second edition, London, Routledge, 1999

Edésio Fernandes and Ann Varley, eds, *Illegal Cities: Law and Urban Change in Developing Countries*, London, Zed Books, 1998

Alan Gilbert and Joseph Gugler, eds, *Cities, Poverty and Development: Urbanization in the Third World*, Oxford, Oxford University Press, 1992

Peter Hall, *Cities of Tomorrow: An Intellectual History of Urban Planning and Design in the Twentieth Century*, Oxford, Blackwell, 1988

Malcolm Miles and Tim Hall, eds, *Urban Futures: Critical Essays on Shaping the City*, London, Routledge, 2003

Michael Peet and Richard Watts, eds, *Liberation Ecologies: Environment, Development, Social Movements*, London, Routledge, 1996

Paolo Soleri, *The Urban Ideal*, edited by J. Strohmeier, Berkeley, Berkeley Hills Books, 2001

'Exopolis: The Restructuring of Urban Form'

from *Postmetropolis* (2000)

Edward Soja

Editors' Introduction

Edward Soja is known for his two previous books, *Postmodern Geographies: The Reassertion of Space in Critical Social Theory* (London, Verso, 1989), and *Thirdspace: Journeys to Los Angeles and Other Real-Imagined Places* (Oxford, Blackwell, 1996), and is a seminal voice in geography's re-invention of its field in the 1980s and early 1990s, with David Harvey and Doreen Massey. Part of his contribution is to have introduced the work of Henri Lefebvre, though he and Massey find different emphases in it. For both, it draws attention to space as a vehicle for social processes, but while Soja interprets it in *Thirdspace* largely as a model of thinking, Massey tends to place more emphasis on experiential and personal experiences. In *Space, Place and Gender* (Cambridge, Polity, 1994), Massey takes Soja to task for retaining a bird's eye perspective, looking on Los Angeles from a high tower in the old-fashioned masculine way. However, Soja remains an important commentator, and not only on Los Angeles. In *Postmetropolis*, he cites a wide range of other reading, making the book almost a reader in its own right. But he critiques rather than reproduces the texts he cites, pointing out, among other things, the limitations of Mike Davis's disaster scenario in *City of Quartz* (London, Verso, 1990). In the chapter from which material is reprinted here, he discusses the rise of mega-cities (of more than ten million inhabitants), new urban forms which have been spawned by new technologies of communication, and a more general encroachment of cities on their hinterlands – the new exopolis.

Soja begins by mentioning the division between discourses of globalisation and economic restructuring, and those more empirically based on spatial form and sociation. He sees the division as artificial, and, in the extract reprinted, introduces a question as to what is the underlying causes of new processes of urbanisation. Next he writes about mega-cities, and the development of outer city spaces, which challenge the modern idea of a city as being centred on a business or administrative district. He looks at post-suburbia, edge cities, and urban nostalgia in the further extract reprinted below. The remaining parts of his chapter cover new urbanism – critiqued by Dean MacCannell in his text in Part Eight (which could be read in conjunction with this text) – and a view specifically of Los Angeles. Near the end he states:

> The restructuring of urban form, in combination with mass movements against increasing taxes and for smaller government, is creating a new round of fiscal crises in postsuburbia, leading an increasing number of the constellation of local governments that comprise the postmetropolis to the edge of bankruptcy, and beyond . . . But the crises that are affecting the off-the-edge cities are even deeper and more difficult to address (p. 261).

This may be Soja's own disaster scenario, but in context of the idea that cities rather than villages are *the* primary form of human settlement.

Soja's position can be compared with that of writers who see cyberspace as constituting another suburbia, such as Norman M. Klein in *The History of Forgetting: Los Angeles and the Erasure of Memory* (London, Verso, 1997). Klein writes of 'exurban extension', stating 'cyberspace is the next suburb' (p. 298). He suggests architect William T. Mitchell's *City of Bits: Space, Place and the Infobahn* (Cambridge, MA, MIT, 1995) as the best guide to it.

METROPOLIS TRANSFORMED

That there have been pronounced changes in the spatial organization of the modern metropolis over the past thirty years, and that these changes are inducing significant modifications in the "urban condition" and the ways we interpret it is the provocative premise of the third discourse. Relatively little attention is given within the discourse to the causes of this deep and broad reorganization of cityspace, for such causality is either implicitly or explicitly imputed to the restructuring processes discussed in the preceding two chapters. What is emphasized here are the *geographical outcomes* of the new urbanization processes and their concrete effects on everyday life, the planning and design of the built environment, and the uneven patterning of intra-urban economic growth and development.

The impact of globalization and economic restructuring has generated an extraordinary array of new terms and concepts to describe the reconfigured spatial specificities of the postmetropolis, triggering increasingly heated debates about how best to capture the most important features of contemporary postmetropolitan geographies. More than in any of the other discourses, the debates on the restructuring of urban form have become embroiled in a naming game, with a multiplicity of metaphorical terms competing to capture the essence of what is new and different about cities today. I will represent the discourse here through a series of these different yet related nominal encapsulations . . .

Outer cities, postsuburbia, and the end of the Metropolis Era

In 1976, a small monograph appeared with the title *The Outer City: Geographical Consequences of the Urbanization of the Suburbs.*[1] Written by Peter O. Muller, it consolidated an ongoing debate in the USA on the changing geography of urbanism and introduced several terms that continue to shape the discourse on postmetropolitan cityspace.[2] Although still steeped in the canons of traditional urban geographical analysis, this work was one of the first to demonstrate clearly that something very un-suburban was happening to American suburbia. By this time, suburbia had been recognized by academic writers and the popular media as a social and cultural milieu quite distinct from popular and academic notions of "the city." No longer just a commuting zone for the urban agglomeration, suburbanism had become its own way of life with its own spatial specificities, most revolving in one way or another around the automobile and the detached owner-occupied home and household. Suburbia was described by its leading historians as the product of the search for "bourgeois utopias" (Fishman, 1987) on the "crabgrass frontier" (Jackson, 1985), the new heartland of American culture and ideology. But as these historians and geographers noticed, suburbia was being significantly transformed in the second half of twentieth century in the development of a seemingly new form, perhaps most simply called the *Outer City*, arising from a process involving the *urbanization of the suburbs*, with both concepts literally and figuratively rife with oxymoronic connotations. If suburbs were becoming urbs, where are we then when we venture outside the city?

The urbanization of suburbia and the growth of Outer Cities has generated its own tracks of reconceptualization, not just of the erstwhile suburban milieu but of the modern metropolis as a whole. In recent years, *postsuburbia* has emerged as one of the catchall terms, with Orange County, the heartless center of postsuburban California (Kling et al., 1991), as its most representative case. There are other descriptive metaphors: "the metropolis inverted," the "city turned inside-out," "peripheral urbanization," and, in a more comprehensive sense, the term postmetropolis itself. What all these descriptions share, implicitly or explicitly, is the notion that *the era of the modern metropolis has ended.* I hasten to say that this does not mean that the modern metropolis has disappeared, only that its social, cultural, political, and economic dominance as a distinctive organizational form of human habitat is no longer what it once was; and that a new urban form and habitat is emerging, not as a total replacement but as the leading edge of contemporary urban development.

In our earlier discussion on the evolution of urban form in North America the period following the First World War and extending into the 1970s was described in conjunction with the rise of Fordism and the effects of Keynesian state management on mass production, mass consumption, and urban development. It can also be seen retrospectively as the Era of the Modern Metropolis, a period in which the metropolitan region, with its distinctively dualized configuration of a monocentric urban world surrounded by a sprawling suburban periphery, consolidated as the dominant and defining habitat and source of local identity for the majority of the national population. Earlier roots and routes of the modern metropolitan region can be traced back to the urban restructuring that occurred in the last three decades of the nineteenth century, when major cities began to spawn satellite industrial centers and, closer by, "streetcar suburbs," reshaping what had been the earlier and simpler form of the more compact industrial capitalist cityspace.[3] But it was only in the 1920s, with the onset of what would later be described as Fordist mass production and mass consumption, that the regional metropolis began to take on its most representative form, marked by a distinct and cosmopolitan urban world concentrated in the core or central city, where the most important economic, political, and cultural activities (along with the positive and negative synekisms) were most densely packed; and by a more extensive and culturally homogeneous, administratively fragmented, and relatively disarticulated "middle-class" suburban world, drawing selectively on the attractions of both the central city and the more open spaces of the countryside, and increasingly dependent on the automobile to allow both city and countryside to be at least potentially accessible.

In the traditional discourse, the regional morphology of cityspace was seen most broadly as a product of the continuous interplay of centrifugal and centripetal forces emanating from a dominant and generative "central city." The center, as has almost always been the case for cities, was the translucent vortex of urban life, the defining node for concentric, radial, and other patternings of urban behavior and land use; for the stimulating but also often frustrating densities of urbanism as a way of life; and for the accretion of residential communities into an expanding cosmopolitan urban realm, defined by the official boundaries of what was generally recognized as *the* City. The taken-for-granted center, signified in such terms as "downtown" (curiously, never "downcity") or the CBD (central business district) was the focal point for concurrent processes of clustering and dispersal, for the simultaneous and systematic creation of urban and suburban life.

Fordism simultaneously accentuated centrality, with the concentration of financial, governmental, and corporate headquarters in and around the downtown core; and it accelerated decentralization, primarily through the suburbanization of the burgeoning middle class, manufacturing jobs, and the sprawling infrastructure of mass consumption that was required to maintain a suburban mode of life. The literature on the development of the modern metropolis, not surprisingly, has focused primarily on suburban decentralization or dispersal, for in effect centralization was presumed to be given. Many metropolitan downtowns did not experience significant growth during this period and only a few developed the large and often downtown-adjacent heavy industrial zones of Fordist mass production. But every major metropolitan region experienced significant suburbanization, as growth by annexation slowed down and the formal boundaries of the central city became relatively stabilized.[4]

In the now classic works of urban historians, suburbia was viewed primarily as a product of voluntary residential decentralization, initially of a wealthy elite, but soon followed, closer to the city center, by working-class inner suburbs and further out by primarily white middle-class "pioneers" pushing ever outward the suburban "frontier," following the grand American tradition of civilizing frontier settlement. The search for better housing, backed by improved public transportation facilities and promoted by eager real estate developers, was seen as the major driving force behind suburbanization, and the end result was a sprawling dormitory landscape of detached and privately owned homes, a culturally homogeneous and "consumerist" suburbia where most jobs (and those proverbially satanic mills) remained outside the local milieu. Carried along with these histories was the old dichotomy of city and countryside, now reconstituted in the modern metropolis around the division between urban and suburban landscapes or worlds, each with their distinctive "ways of life."

That these contrasting worlds were inherently shaped by class, race, and gender relations spawned an early round of what today is called critical urban studies. Feminist urban scholars, for example, saw the construction of the metropolitan region as not only male-dominated in conception and implementation, but also as intensifying patriarchal power. In particular, it was argued that women as "housewives" were "trapped" in suburbia, subservient to the dominant male "breadwinner" and subsumed into unpaid household labor, with all its ostensibly labor-saving appliances. Although given less attention, the urban core was also seen as a masculinist space, a built environment designed to control, often through violence, women's access to the primary sites of male power. Parallel and occasionally connected arguments were made about how the spatial organization of the modern metropolis, and particularly its division into urban and suburban worlds, was shaped by discriminatory practices based on class, race, and ethnicity, producing in the normal workings of the dualized metropolitan region two distinct systems for producing and reproducing social inequalities.

This simplified structure of the modern metropolis continues to dominate the urban imaginary of scholars, the media, and most popular discourse. It is becoming increasingly clear, however, not only that the metropolitan region today no longer fits the older model as much as it once did, but also that, when viewed from a contemporary perspective, significant modifications may have to be made in conventional historical interpretations of the Metropolis Era itself. In *Magnetic Los Angeles* (1997), for example, Greg Hise draws upon the new geopolitical economy to re-explore the historical relation between Fordism and the modern metropolis. He begins by challenging the very distinction between "city" urbanization and "countryside" suburbanization, arguing for a recombinant alternative that views urban and suburban development as a process of dispersed nucleation or *citybuilding* right from the start. Moreover, it is not housing and residential choice that drives this process so much as the decentralization of industrial production and employment. This formative budding-off process is accompanied by the opportunistic actions of "community builders," private developers as well as public entrepreneurs, to attract residents and infrastructural investment to the nascent urban nucleations. The development of these multiple nuclei can be described as demand-driven, but the demand was not as much from households as for labor.

Hise also argues that these citybuilding processes were most often highly planned and plotted, and that what has been called uncontrolled suburban sprawl was actually, for the most part, carefully organized and often fairly well designed and planned urban development. The target population for these Fordist "edge cities," as Hise describes them, was not primarily a homogeneous white elite seeking exclusive enclaves but the working class seeking better jobs. There were many areas where speculative building arrived well before the local availability of attractive jobs, creating true "dormitory suburbs" occupied almost entirely by the white and predominantly white-collar middle class and frequently sustained by racially restrictive covenants and other regulations on entry. But where the employment nucleations did form and grow, labor demand was such that most of these restrictions were lifted or bypassed, producing more racially mixed populations than were usually assumed to exist in classic suburbia.

There are thus significant continuities between the Metropolis and Postmetropolis eras, just as there are between Fordism and postfordism, modernity and postmodernity. But again the discursive question revolves around whether a certain tipping point or threshold has been reached where the interpretive power of studying the "intensified" new forms and functions outweighs a revisioning of the continuities that link the present to the past. Emphasizing the new while recognizing the persistence of long-established geohistorical trends, it can be argued that during the past thirty years the growth of Outer Cities has both decentered and recentered the metropolitan landscape, breaking down and reconstituting the prevailing monocentric urbanism that once anchored all centrifugal and centripetal forces around a singular gravitational node. Deindustrialization has emptied out many of the largest urban-industrial zones and nucleations of Fordism, while postfordist reindustrialization has concentrated high-technology industries in new industrial spaces far from the old downtowns. These "greenfield" sites, the urban equivalent of the Newly Industrialized Countries (NICs) in the global economy, are not just satellites but have become distinctive cities and gravitational nodes in their own right. The most successful have spawned and sustained the mall-centered hives of consumerism that are the popular hallmarks of the postfordist, postmetropolitan Outer City.

Although the decentralization of industrial production and employment began in the last half of the nineteenth century, it was only in the last third of the twentieth century that the regional balance of industrialization in many postmetropolitan areas was reversed, with the majority of production and jobs located in the outer rings rather than in the inner cities of the conurbation. As much as anything else, it is this *role reversal in the geography of industrial urbanism* that has led observers such as Sudjic to claim that the new urban form marks the moment when the industrial city "finally shook off the last traces of its nineteenth century self."

In a process that can no longer be simply described as sprawl, post-metropolitan cityspace has been both stretched out and pinned down to cover a much larger regional scale than ever before. It reaches out and connects to a network of interdependency that is now global in scope, a hierarchical hinterland that blurs the discreteness of both the City and the older metropolitan region, and dilutes the degree to which cityspace represents the "culmination of local and territorial cultures" (Chambers, 1990: 53). If sprawling suburbia is no longer what it used to be, the same is true for the urban core. In a strange contrapuntal movement, the densest urban cores in places like New York are becoming much less dense, while the low-rise almost suburban-looking cores in places like Los Angeles are reaching urban densities equal to Manhattan. What once could be described as mass regional suburbanization has now turned into *mass regional urbanization*, with virtually everything traditionally associated with "the city" now increasingly evident almost everywhere in the postmetropolis. In the Era of the Postmetropolis, it becomes increasingly difficult to "escape from the city," for the urban condition and urbanism as a way of life are becoming virtually ubiquitous.[5] And in the wake of these changes, the ways in which the metropolitan region is patterned by class, race, and gender relations have become more complex and opaque.

Edge cities and the optimistic envisioning of postmetropolitan geographies

> Americans are creating the biggest change in a hundred years in how we build cities. Every single American city that *is* growing, is growing in the fashion of Los Angeles, with multiple urban cores. These new hearths of our civilization – in which the majority of metropolitan Americans now work and around which we live – look not at all like our old downtowns. Buildings rarely rise shoulder to shoulder . . . Instead, their broad, low outlines dot the landscape like mushrooms, separated by greensward and parking lots.
>
> (Joel Garreau, *Edge City* (1991): 3)

No other book has captured the nominal imagery of the postmetropolis quite like Joel Garreau's *Edge City: Life on the New Frontier* (1991).[6] A senior writer for the *Washington Post* and author of *The Nine Nations of North America* (1981), a book that attempted optimistically to transfigure the regional makeup of the continent in much the same way that *Edge City* approaches contemporary urbanism, Garreau has become the Pied Piper of Postsuburbia, a guru to a mass national audience of businessmen, academics, and just plain folks trying to understand what has been happening to the cities of North America in the late twentieth century. Garreau's view of the restructuring of urban form emphasizes not just the increasingly polycentric nature of postmetropolitan cityspace but revolves specifically around its most visible landmarks, the shopping mall and office-centered developments he calls, and nimbly trademarks as, Edge Cities.

According to Garreau's defining criteria, the Edge City (1) has 5 million square feet or more of leasable office space – the workplace of the Information Age; (2) has 600,000 square feet or more of leasable retail space – the equivalent of a fair-sized mall; (3) has more jobs than bedrooms; (4) is perceived by the population as one place; and (5) was nothing like "city" as recently as thirty years ago (1991: 6–7). Southern California has the largest total number of existing (16) and "emerging" (8) Edge Cities, closely followed by Washington DC (16 and 7) and New York (17 and 4). According to the *Edge City News*, a newsletter published by the Edge City Group Inc. and subtitled "Tools for the New Frontier," there are nearly 200 Edge Cities already built in the USA, more than four times the number of comparably sized "old downtowns." They now contain two-thirds of America's office space, a huge jump from the approximately 25 percent that was there in 1970. The figures for the number of jobs located in Edge Cities against those for the old downtowns also show a dramatic increase.

Garreau takes a few other rough cuts at categorizing different types of Edge Cities. "Uptowns" are built on the commercial renaissance of older, established nodes such as Pasadena CA or Stamford CT, and take on a wide diversity of forms and flavors befitting their more complex historical cityspaces. "Boomers" are "the classic kind of Edge City," centered on a mall at the intersection of freeways and etched into place in three different shapes: the Strip, the Node, and the Pig in the Python (multinodal strips). The "Greenfield" version is "increasingly the state of the art, in response to the perceived chaos of the Boomer." They are figuratively built "at the intersection of several thousand acres of farmland and one developer's monumental ego." Irvine, in California's Orange County, and the more recently Disneyed world near Orlando, in Florida's Orange County, are two prime examples.

For Garreau and many of his interlocutors, the Edge City becomes a frontier outpost of epochal proportions. Here are just a few of his discursive tropes: "in-between triumphant," "a vigorous world of pioneers and immigrants," "the third wave of our lives pushing into new frontiers," "places to make one's fame and fortune," "anchored by some of the most luxurious shopping in the world," "the crucible of America's urban future," "the forge of the fabled American way of life well into the twenty-first century," a release "from the shackles of the nineteenth-century city," "another Garden . . . the best of both worlds," "the philosophical ground on which we are building our Information Age society," "the most purposeful attempt Americans have made since the days of the Founding Fathers to try to create something like a new Eden," "the result of Americans striving once again for a new, restorative synthesis," "the search for Utopia at the center of the American Dream." I think you get the idea.

Garreau fundamentally ignores the explanatory arguments of the industrial urbanism discourse and rarely speaks of globalization except to intone connections to the New Information Age.[7] Instead, he invents his own causal nexus for the emergence of Edge Cities, beginning, promisingly enough, with what he calls "the empowerment of women." What was once suburban entrapment for women is inverted by Garreau into Edge City liberation.

> Edge Cities doubtless would not exist the way they do were it not for one of the truly great employment and demographic shifts in American history: the empowerment of women . . . It is no coincidence that Edge Cities began to flourish nation-wide in the 1970s, simultaneous with the rise of women's liberation . . . they were located near the best-educated, most conscientious, most stable workers – underemployed females living in middle-class communities on the fringes of the old urban areas. (1991: 111–12)

In a long chapter on Atlanta, Garreau also enlists another "revolutionary" development: the emergence of a new black middle class. Here again, his fixation on selected success stories blinds his vision of the downside of the post-metropolitan transition.

> The rise of Edge Cities contained a nightmare possibility for America: that because so many jobs were moving out to the fringe, frequently into what had been lily-white suburbs, an entire race would be left behind, trapped, in the inner city, jobless, beyond reach of the means of creating wealth. Such fears, however, have not been confirmed, despite the plight of the black underclass. A black suburban middle class is booming, statistics show. And it is emerging at the same time and in the same places as Edge Cities. (1991: 144)

He uses these statistics to argue that "the rise of Edge Cities is primarily a function of class – not race" (1991:152). "Karl Marx was right," he writes, "issues of class are what control" (1991: 165), a progressive-sounding but inadequately explored causal argument.

Los Angeles looms large in Garreau's *Edge City*, and vice versa. Garreau has become well connected to a local network of journalists and others who are similarly spin-doctoring optimistic pictures of paradigmatic Los Angeles and life on its "new frontier" in local and national newspapers. One particularly influential contact has been Christopher B. Leinberger, author (with Charles Lockwood) of "How Business is Reshaping America" (*Atlantic*, October 1986), an article which perhaps more than any other drew the popular attention of the East Coast to the emergence of Los Angeles as the leading edge of American urban trends. Leinberger's notion of "urban villages" is closely akin to Garreau's Edge Cities. Both are filled with allusions to the original Garden City concept of Ebenezer Howard (and also to an even earlier Edenic garden), where one can obtain the best of both worlds, city and countryside wedded together by the electronic possibilities of the new Information Age and a radically optimistic vision of the coming together of gender, race, and class divisions.

Another important local contact is Joel Kotkin, a former colleague at the *Washington Post* and currently a contributing editor to the Opinion section of the *Los Angeles Times*, senior fellow at the Center for the New West in Ontario and the Pepperdine University Institute for Public Policy in Malibu. Kotkin and his colleague David Friedman have become the leading entrepreneurial spin-doctors of the new Los Angeles. Unlike Garreau, they pay particular attention to the restructured postfordist geopolitical economy and the local and global literature on industrial urbanism in their rosy localized reconstitutions of the Edge City idea.[8] Often capturing the attention of local decision-makers involved in citybuilding (mayors, architects, developers) as well as the popular urban imaginary in Los Angeles, Kotkin in particular has launched a crusade against all those who speak of a downside to the contemporary globalized multicultural postfordist metropolis. Included here as special targets are the critical urban planners, geographers, and sociologists of UCLA who featured so prominently in the previous two chapters. They are seen by Kotkin as the leading purveyors of the "declinism," "unmitigated negativity," and "loss of confidence" allegedly plaguing and biasing the contemporary image wars over the region's future. These occasionally Red-baiting attacks feed into and on the confused public realm, desperate for good news and easy solutions to the current concatenation of crises facing the edgy postmetropolis of Los Angeles.

Like Ronald Reagan's hyperreal interpretation of the causes of stagflation in the 1980s ("it's all in the mind of the beholder"), the new-wave boosters locate the problems of the 1990s in Los Angeles (and by paradigmatic implication, everywhere else in urban America) in the popular state of mind, in downside thinking, in the wrong attitudes. Factuality is subordinated to image-making and appropriate faith in the economic system, critics are admonished as destructive doomsayers, and the future is hinged on getting the right and proper spin. They have become the latest entry in a long line of public–private interfacers attempting to boost a new Los Angeles by sublimating its darker geohistory.[9] Similar booster-entrepreneur journalists, often backed by affiliations with local universities, now exist in almost every urban region, making a living by promoting a soothing vision of exaggerated optimism over the contemporary restructuring of urban form.

NOTES

1. Peter O. Muller, *The Outer City: Geographical Consequences of the Urbanization of the Suburbs*, Resource Paper 5: 7S–2, Washington, DC: Association of American Geographers, 1976. For a summary of the British debates on the same subject, see John Herington, *The Outer City*, London: Harper and Row, 1984.
2. At the time, the term that prevailed in describing this changing urban geography was *counter-urbanization*, coined by the University of Chicago geographer, Brian J. L. Berry. See Berry, "The Counter-Urbanization Process: Urban America since 1970," in Berry ed., *Urbanization and Counter-Urbanization*, Urban Affairs Annual Review 11, Beverly Hills: Sage Publications, 1976.
3. Herington (1984) begins his first chapter, "The City Beyond the City," with a prescient quote from H. G. Wells, writing in 1902: "The country will take upon itself many of the qualities of the city."
4. The slowdown in central city annexation has been given relatively little attention in interpreting the rise of the modern metropolis. In many ways, however, it both defined and accentuated the growth of suburbia. Had the central city continued to grow by absorbing the budding urban centers on its fringes, the officially suburban realm would have been much smaller and the statistical differences between "urban" and "suburban" population growth much less pronounced.
5. For some cinematic references, see the recent films *Escape from New York* and *Escape from Los Angeles*. Many other references can be made to the wave of illustrative films on the postmetropolis and the contemporary urban condition but I will generally leave such textual enrichment to the reader.
6. *Edge City: Life on the New Frontier*, New York: Doubleday. See also Garreau's *The Nine Nations of North America*, Boston: Houghton Mifflin, 1981.
7. Industry in *Edge City* is equated to commerce and office space. Garreau hardly speaks of manufacturing at all, and when he does (briefly, in the chapter on New Jersey) he claims that "Industrial and warehouse space does not create anything urbane. No dense centers ever evolve" (1991: 31). Garreau's failure to see the important and continuing links between urbanization and industrialization (see chapter 6 and the discussion of Greg Hise's work in the preceding section) significantly weakens his conceptualization of Edge Cities.
8. See, for example, the glossy report by Kotkin and Friedman, *The Next Act: Southern California's New Economy*, Ontario, CA: Center for the New West, 1994.

The report concludes with the following admonition: "But upon one thing all else depends: the restoration of public confidence and faith in the region's long term prospects . . . Southern Californians must realize – and communicate to the world – that this region still possesses a remarkable economic, cultural and creative dynamism unmatched by any major urban region in North America" (1994: 24).

9 For a very perceptive history of this tradition of spin-doctoring time and place in Los Angeles, see Norman M. Klein, *The History of Forgetting: Los Angeles and Erasure of Memory*, London and New York: Verso, 1997.

REFERENCES

Iain Chambers, 1990, *Border Dialogues: Journeys in Postmodernity*, London, Routledge

Robert Fishman, 1997, *Bourgeois Utopias: The Rise and Fall of Suburbia*, New York, Basic Books

Joel Garreau, 1991, *Edge Cities: Life on the New Frontier*, New York, Doubleday

Greg Hise, 1997, *Magnetic Los Angeles: Planning the Twentieth-Century Metropolis*, Baltimore, Johns Hopkins University

Kenneth Jackson, 1985, *Cities and the Wealth of Nations: Principles of Economic Life*, Oxford, Oxford University Press

Rob Kling, Spenser Olin, Mark Poster, eds, 1991, *Postsuburban California: The Transformation of Orange County since World War II*, Berkeley, University of California Press

'Environmental Justice and the Sustainable City'

from David Satterthwaite (ed.), *Sustainable Cities* (1999)

Graham Haughton

Editors' Introduction

The book which contains this essay includes sections on sustainability as a concept, its link to urbanism, sectors of development (such as health, transport, and industry), city-level initiatives, and a regional and global context. The book takes a balanced view of sustainability, including voices from more and less radical positions. Graham Haughton, who leads the Centre for Urban Development and Environmental Management (CUDEM) at Leeds Metropolitan University, writes on the idea of environmental justice as applied in terms of urban sustainability. This, in effect, transposes the modern concept of social justice, and the rights of citizens, to the environmental debate. The chapter is reprinted in full apart from the author's acknowledgements and one chart. Further reading on the topic is given in the list of references.

Haughton begins by linking social and environmental justices, and setting both in the local–global context. He then gives a series of principles for sustainable development derived from such concerns, emphasising the need to view the problem broadly, in keeping with the increasing breadth and complexity of urban footprints of trade, resources, and waste. He notes that high-density settlements may, whatever their internal difficulties, be less unsustainable (or environmentally destructive) than dispersed cities of the kind described by Soja in the previous text. He draws a preliminary conclusion that if cities are to be or become sustainable then change is needed in economic, legal and regulatory processes; and in the understanding of a responsibility for environmental consequences on the part of the producers of environmental problems: 'New systems are needed which not only require those who cause environmental problems to share in their remediation, but which also move beyond this to requiring changes in behaviour patterns to reduce or halt environmentally degrading activities' (p. 65). The remaining sections of the text reconsider urban development with a view to attaining those changes, for instance by adopting new models such as the self-reliant city and the fair shares city. Haughton is helpfully non-prescriptive, but argues for social equity as the most likely condition of sustainability. The worst-case scenario for Haughton is myopia, though he accepts that planners lack the necessary range of tools to achieve sustainability at present.

JUSTICE, EQUITY AND SUSTAINABILITY

Recent years have seen a re-emergent interest in issues surrounding social justice and environmental justice, with the two increasingly seen as interlinked (Friedmann 1989; Harvey 1992; Hofrichter 1993; Smith 1994; Hay 1994; see Harvey 1973, and Berry and Steiker 1974 for earlier discussions). This resurgent interest in examining justice issues has been accompanied by a more general reawakened interest in normative theoretical approaches within planning and geography, with detailed examination of values (rights, ethics, quality of life) being reinserted with renewed confidence into recent work within these disciplines (Beatley 1994; Bourne 1996; Smith 1997; Sayer and Storper 1997). The need to re-evaluate the ethical underpinnings of policy and analysis has been given additional impetus by the emergence of sustainable development debates – Lipietz (1996 p 223), for instance, argues that the greatest achievement of the 1992 Rio Earth Summit and related conferences may well be the widespread popular and political acknowledgement of the need for 'new rights and obligations to be incorporated within social norms', involving 'the recognition, at first moral, of new rights, new bearers of rights and of new objects of rights'. In this view, the discourse of sustainable development has enlarged consideration of rights through its explicit attention to the rights of future generations, of present-day socially marginalized groups to a 'good' environment, and also to the need to consider 'Other' (non-human) dimensions of the natural world as having rights to continued existence, as recognized in biodiversity treaties.

Reflecting theoretical debates over local-global dimensions of economic restructuring, social justice and environmental justice debates involved equity issues at a range of scales, from the local to the global, and also the broader economic, social and political systems which foster and perpetuate inequalities between different social groups and different areas. In environmental terms this requires looking both at systems which generate environmentally degrading activities and also at differential access to environmental 'goods' and environmental 'bads,' notably as expressed in differential impacts on different social groups, sectors and geographical areas. A particularly powerful illustration is provided by environmental racism debates which have highlighted how cases of environmental dumping, such as the concentration of toxic waste incineration plants found in poor areas, have exercised disproportionate adverse impacts on areas with large concentrations of people of colour in many instances (Bullard 1990; 1993). Environmental justice and social justice are seen as intrinsically connected in such analyses, with both treated in this article in this larger conception of addressing both the underlying systemic causes of injustice and the more traditional distributive justice concerns of seeking to redress inequalities of outcome. This is important since some commentators have begun to question whether a narrow 'equity' concern with distributional aims (eg decisions on who comes to be most polluted and where) may have the unintended perverse effect of overshadowing more broadly constituted 'justice' concerns with addressing underlying structural issues – that is, engaging with systems to reduce or prevent pollution rather than distribute it more equitably (Young 1990; Pulido 1994; Heiman 1996; Lake 1996).

This article sets out a small number of interlinked principles for sustainable development which derive from these concerns, and then moves on to assess how four different sets of approaches to creating a sustainable city measure up against them. One of the basic premises of the analysis here is that a sustainable city cannot be achieved purely in internal terms: a sustainable city is essentially one which contributes effectively to the global aims of sustainable development, where sustainable development is seen as much as a process as an end-product. With the emergence of ever-thickening and extending patterns of global economic trading, and increasingly global exchanges of environmental resources and waste streams, it is futile and indeed virtually meaningless to attempt to create a 'sustainable city' in isolation from its broader hinterland area. Moreover, it is necessary to see that cities make other contributions to global well-being which make a purely local focus on their environmental impacts potentially unhelpful. Or, to put it another way, a densely developed, highly populated city could well be deemed 'unsustainable' if looked at solely at the local level, in terms of its dependence on the appropriated environmental assets of other regions; alternatively, at a regional or even global scale, this form of city may be preferable to sprawling, low density, low-level developments, consuming considerable agricultural land and requiring considerable energy for transportation between dispersed activities. In a global sense then, a high-density city which overcrowds and displaces the already transformed 'natural' environment at the local level may be preferable to lower density forms of urban development where at the local level 'nature' survives rather better. In part, how one makes a judgement on such issues depends on whether a weak or strong definition of sustainability is adopted – that is, the extent to which it is seen as acceptable to replace natural capital with human capital, and in particular the approach taken to preserving critical natural stocks. In the four approaches to sustainable urban development outlined below, it becomes clear that each embodies a different perspective in this respect, although the issues are not teased out in full here – the emphasis is rather on equity and justice issues. In urban management terms, these tensions in defining what the sustainable city is, or might be, are important in guiding attention to the need to look at the underlying philosophical bases of local actions in support of sustainable development, whilst also directing attention to look globally as well as locally when assessing the overall impacts of policies in support of sustainable urban development.

It is valuable to begin by elaborating on what sustainable development actually means and on some of its tensions, Ultimately sustainable development means the long-term survival of the planet and its processes of dynamic evolution, including the wide range of species which currently live on it, not least humankind, For humans, it specifically requires achieving a position which allows us to live in harmony with the rest of the planet, so that we neither destroy ourselves nor the systems which support other life-forms. The essential threat to sustainable development is that the human species is attempting to live beyond the capacity of the Earth to sustain both humans and other species, most notably as we destroy the natural balance of critical natural protective systems, from depletion of the ozone layer to the creation of the greenhouse effect. Moving towards sustainable development requires achieving economic and social systems which encourage environmental stewardship of resources for the long-term, acknowledging the interdependency of social justice, economic well-being and environmental stewardship. The social dimension is critical since the unjust society is unlikely to be sustainable in environmental or economic terms in the long-run, since the social tensions which are created undermine the need for recognizing reciprocal rights and obligations, leading in all manner of ways to environmental degradation and ultimately to political breakdown.

The tensions between economic development and environmental stewardship are a central feature of the sustainable development debate, with controversy surrounding the Brundtland Commission's (WCED 1987) declared view that economic development is essential in order to meet the social goals of sustainable development. The critics of this view hold that economic growth within the current dominant market-driven capitalist mode is largely responsible for environmental degradation, therefore we need to question whether continued economic expansion of this type is either desirable or acceptable (Seabrook 1990; O'Connor 1993, 1994). It is precisely because the Brundtland Commission's growth-compatible vision of sustainable development version is more politically palatable that its definition gained so much political support, whilst more radical views have remained a marginalized part of the sustainable development discourse.

This brings the analysis back to the varying possible interpretations of sustainable development: these have been described as running along a spectrum from 'very weak' or 'light green' versions to 'very strong' or 'deep green' interpretations (Turner, Pearce and Bateman 1994). Proponents of the weak version are held to have a largely anthropocentric worldview, which sees considerable scope for technological solutions to environmental problems, and in particular for the substitutability of natural capital with human capital – for instance, replacing fossil fuel-derived energy with new technologies for creating nuclear energy or capturing tidal energy. Versions of 'environmentalism' which support 'sustainable profits' or 'sustainable accumulation' by private enterprises are generally regarded as 'light green' in that they tinker at the edges with the existing system of accumulation, rather than seek radical transformations in favour of preserving natural assets. By contrast, proponents of deep-green sustainability views tend to emphasize that market-led systems of capitalist growth reductively consign nature to a role of mere market 'inputs' or 'outputs', limited only by the capacity of the market to make profits out of these natural assets, in the process inevitably leading to environmental degradation. The strong version of sustainability holds a more nature-centred worldview, which seeks to prevent destruction of natural assets beyond their regenerative capacities by reducing overall consumption levels and to avoid unnecessary high risks associated with some untested quick-fix technological solutions. In particular, the strong sustainability perspective argues against wholesale 'substitutability' of natural with human assets, involving a more widely constructed definition of critical natural stocks – that is, those natural assets which cannot be used beyond their natural regenerative capacities without major damage to the integrity of ecosystems.

In order to move towards sustainable development it is essential to address the way in which our current political, economic and social systems allow widespread cost-transference to take place, where many of the negative environmental and related impacts of the activity of a person, company or even region, are in effect displaced elsewhere (Haughton 1998). At the urban level resource demands can exercise major impacts on other areas – for instance, in the valleys where reservoirs are built to supply cities, whilst urban pollution of water can have negative impacts on downstream users and natural aquatic ecosystems. We have currently evolved sophisticated systems for hiding our responsibilities for the deleterious effects of our behaviour patterns, whilst embarking on a series of risky technology-driven projects without full consideration and knowledge of their impacts (from the introduction of CFCs to the development of nuclear energy). Indeed, it can even be argued that much of the recent growth of Western capitalism can be traced to the ability of corporations (and governments) to externalize more and more of their social and environmental costs in pursuit of cost-cutting competitive gains, even at the same time as achieving efficiencies in energy and raw material usage (O'Connor 1991; O'Connor 1993). Such externalities can effectively divorce people, businesses and governments from responsibility for their actions, fostering irresponsible behaviour patterns. Extending rights and obligations to ensure that externalities are brought into the decision-making frame, whether through the market mechanism (via the pricing mechanism, eg adopting 'green' taxes), legal sanction or other means, is essential to moving away from current patterns of widespread profligacy in resource usage and unthinking disposal of wastes.

At two levels, then, we need to change our ways. Firstly, we need improved political, economic, regulatory and legal systems, allied to enhanced information and educational systems, which bring home to individuals and groups the way in which their activities exercise direct and indirect impacts beyond the local scale and over a very long time horizon. Secondly, and related, we need to devise systems which ensure that those responsible for making environmental demands assume the main responsibility for the consequences of their activity – they should not expect other people, other species or other places to absorb the associated costs of environmental and social breakdown. New systems are needed which not only require those who cause environmental problems to share in their remediation, but which also move beyond this to requiring changes in behaviour patterns to reduce or halt environmentally degrading activities. Sustainable development, then, is about recognizing and accepting our responsibilities not just for the places where we live, but more widely for the environment at a global scale. In order to do this we need to look beyond the environment itself, to the broader economic, social and political systems within which human decisions are made. Fundamentally, sustainable development is about altering behaviour patterns not just directly in relation to the environment, but about changing the broader systems which shape human behaviour.

FIVE CENTRAL EQUITY PRINCIPLES FOR SUSTAINABLE DEVELOPMENT

In trying to establish what it is that makes a concern for sustainable development different from the existing concerns of environmental planning, it is helpful to highlight five interconnected 'equity' principles which move to centre place in any discussion of sustainable development, representing the essential environmental justice dimension of the concept. I would argue strongly that if these equity conditions are not addressed singly and collectively, then inevitably the ability to move towards sustainable development will be critically undermined. This said, each of these 'equity' principles in themselves represent contestable goals since no clear definitive state of final 'achievement' recognizable by all is ever likely to occur – it is the process of moving towards them, of changing human practices in the spirit of them, which is important, not some elusive readily quantifiable end-goal. This initial analysis provides the beginnings of a normative framework for environmental justice against which it is possible to evaluate different approaches to fostering the sustainable city, a task which is undertaken in subsequent sections.

Firstly, there is the principle of *intergenerational equity*, or the principle of futurity as it is sometimes known. This is perhaps the most widely acknowledged ingredient of sustainable development, drawing from the Brundtland definition of sustainable development as being 'development which meets the needs of the present without compromising the ability of future generations to meet their own needs' (WCED 1987). A second principle is also alluded to in this statement, and more fully elsewhere in the Brundtland report, that of *intragenerational equitiy* or, more generally, contemporary social equity or social justice. As argued earlier, since equity and justice can be argued to have different emphases for some commentators, the emphasis here is on the wider conception of social justice – that is, seeking to address the underlying causes of social injustice, not simply dealing with redistributive measures. Whilst not without some ambiguity, it is possible to introduce these two principles here just briefly since they have been so fully developed elsewhere in the literature (see, for instance, WCED 1987; Haughton and Hunter 1994).

The third key principle for sustainable development is that of *geographical equity*, or transfrontier responsibility. Transfrontier responsibility requires that local policies should be geared to resolving global as well as local environmental problems. All too often, decision-makers adopt a parochial concern for protecting localized corporate or environmental interests whilst effectively ignoring external impacts of their decisions (Beatley 1991). Of particular concern is that too often, external impacts which affect areas outside the particular jurisdictional domain of the host polluter can be ignored if the polluter feels no responsibility for the recipient area and is beyond formal systems of legal sanction, such as liability to pay compensation. In a variant of the old saw 'out of sight, out of mind', activities are effectively encouraged which degrade distant areas, creating major environmental externalities – that is, uncompensated costs which are passed on to someone or somewhere else, and the further afield they are, the more administratively and politically separate their legal system, the easier it is to perpetrate such transfers of costs.

Geographical equity concerns are apparent from the neighbourhood level to the global, from issues of environmental dumping and concern over environmental racism to transfrontier acid deposition and degradation of the Amazon rainforests. As consumers, when we have information about the products of tropical rainforests, whether they are taken from a sustainably managed resource or a short-term commercial pillaging of an area, we can all take part in this process. The policy need is to go beyond current systems to ensure that environmental information is more widely available and that systems are in place to prevent trade in resources from non-sustainable sources, In addressing these concerns, it is essential to ensure that political or jurisdictional boundaries are not to be used to shield individuals, companies and governments from the negative impacts of their activities, as argued in the first section of this paper.

This in turn relates to a fourth principle, that of *procedural equity*. This principle holds that regulatory and participatory systems should be devised and applied to ensure that all people are treated openly and fairly. In its narrowest interpretation, this concern with procedural equity is applied solely within a particular legal jurisdiction, which can create problems in an increasingly globalized economy and in an era when environmental impacts are increasingly large-scale in impact, paying no regard to political boundaries. Pulido (1994), for instance, charts the way in which south-western US firms have sought to evade local environmental regulations by threatening to move across the border to Mexico, effectively transferring the problem rather than solving it. In the present broader definition of procedural equity, which links closely to geographical equity, a central concern is that political boundaries should not be used to allow polluters to be immune from prosecution by affected people in other jurisdictions. Those affected by pollution in other countries, for instance, should have the same rights to legal standing to defend themselves against polluters as those in the host country would (Haughton and Hunter 1994). Critical to making this form of equity operational is a right of equal access to information, and beyond this a more general right of access to information for all interested parties on activities which exert deleterious environmental impacts, locally and globally. Added to this is a concern over which decision-making processes procedural equity might cover, given that to be involved solely over how the environmental burdens of contemporary society are equitably distributed would be a highly truncated definition. Procedural equity requires an extended definition which encompasses engagement with 'the gamut of prior decisions affecting the production of costs and benefits to be distributed' (Lake 1996, pp 164–166).

The concern with procedural equity here also covers what is sometimes referred to as the principle of participation. In general I have tended to be cautious about adopting this as a separate principle, reflecting my deep concern that some forms of participation can undermine rather than support democratic processes of engagement – for instance, middle-class community groups lobbying to prevent homes for the mentally ill opening in their neighbourhoods, or tenacious mavericks who seek to usurp community participation channels for narrow sectarian interests or even to pursue personal grudges and vendettas. Alternatively, it is clear that participation is central to achieving effective and sustainable processes of regeneration, owned and mobilized by the general public as well as state authorities.

Reflecting these concerns, it is possible to argue that

procedural equity is about much more than legalistic and bureaucratic procedures for establishing and enforcing obligations and rights. In addition, it needs to embrace wider processes of public engagement, where multiple democratic and participative forms and channels are brought into play to foster participation and engagement with processes of change. People need an appropriate framework of democratic political processes, and with this responsibilities, ranging from the local, urban, regional and national scales, to multinational decision-making bodies. This concern suggests that all people should have access at different points into public decision-making processes (in particular at the junction of public and private decision-making, for instance over corporations siting hazardous waste facilities). This requires a balancing of democratic and participative methods of engagement with decision-making, rather than a displacement of necessary democratic responsibilities by other bodies.

The earlier discussion on different interpretations of sustainable development contrasted those who hold what might be termed largely anthropocentric views with those with more nature-centred values. Following on from this, there is a view that more attention needs to be paid to what might be termed *interspecies equity* – that is, placing the survival of other species on an equal basis to the survival of humans. This is not to suggest the moral equivalence of humans with other life-forms, rather to highlight the critical importance of preserving ecosystem integrity and maintaining biodiversity. Other species have intrinsic rights, though these are not necessarily the same as those for humans. In a sense, then, the argument is that nature has certain rights, whilst humans also have obligations, to nature and to each other, to ensure that individual animal species and indeed whole ecosystems are not degraded to the point of non-sustainability. It is in this latter sense, then, that I incorporate a concern with interspecies equity here, to reflect a broader concern with environmental stewardship.

FOUR APPROACHES TO ACHIEVING SUSTAINABLE URBAN DEVELOPMENT

The literature about promoting sustainable urban development in Western nations can be broadly categorized into four approaches. These approaches, or models, are far from being mutually exclusive, and indeed it is possible to see that whilst at times the advocates of the different approaches appear to be in 'ideological' conflict with each other, in reality there is a fairly strong shared base of common assumptions and common policy directions. This said, each approach has its own distinctive traits, its own ways of rationalizing its policy stances and its own sets of priorities for both how a city should be laid out and managed. In this sense each model in fact embodies distinctive assumptions of what creates non-sustainable behaviour patterns in cities and, more broadly, from this, how policies in support of sustainable urban development should be framed.

Self-reliant cities

The self-reliant city approach is centred on attempts to reduce the negative external impacts of a city beyond its own bioregion, seeking to: reduce overall resource consumption; use local resources where possible; develop renewable resource-based consumption habits, always in a sustainable fashion; minimize waste streams; and deal with pollution *in situ* rather than 'exporting' it to other regions (Morris 1982, 1990). Whilst coming in many guises, perhaps the most distinctive variants of the self-reliant city are those propounded by West Coast bioregionalists in the USA. The bioregion is usually seen as a central construct, replacing artificial political boundaries with natural boundaries, based typically on river catchment areas, geological features or distinctive ecosystem types, although it is readily conceded that precise boundaries are usually difficult to define (Register 1987; Andruss et al 1990). Urbanization in a bioregional context is usually argued as best being smaller in scale and more decentralized, whilst calling for greater efforts to design with nature – that is, designing cities in ways which bring 'nature' into the city, from open spaces to urban forests and roof gardens.

The politics of the bioregion are similarly usually envisaged as more decentralized and openly participative than conventional politics (Bookchin 1974, 1980, 1992; Berg 1990), bringing a direct concern with the 'equity of engagement' issues discussed earlier. Callenbach's (1975) novel *Ecotopia*, for instance, talks of decisions being arrived at only by debate and consensus, with no formal voting. Social equity issues are a dominant theme, encouraging people to realize their full potential and accept the need to act responsibly to others and towards the natural environment more generally. In this version of the self-reliant city, the intention is to build ecocities which blend into their natural environment and which enhance 'life, beauty and equity' (Register 1987, p 13); quite explicitly, this critique holds that cities 'built for maximum profit . . . or to confer maximum wealth on all citizens equally' cannot emerge as ecocities (ibid). The distinctive approach to interspecies equity becomes one of respecting the need to preserve external habitats and also to encourage a more complex urban ecology, where cities are designed to 're-establish or permit natural life forms to co-exist with the city by giving them sizable slices of their natural habitat around the city (greenbelt instead of suburbia), and within it (in parks, along restored creeks and shorelines)' (Register 1987, p 18). A preservationist stance to critical natural resources is dominant, together with a commitment to encouraging biodiversity at all spatial scales, based on a belief that natural ecosystems tend towards complexity, characterized by increasing numbers of life-forms complexly interrelated. This is carried forward into a widely held belief that cities need to emulate this complexity by fostering environmental, social and economic diversity, avoiding social and economic 'monocultures' as much as environmental ones.

Given its considerable political agenda for fundamental institutional transformation in the quest for sustainability, the self-reliant city is the most radical of the approaches outlined here. The main problem with this approach is the danger of taking regional autarky to unacceptable levels. Wallner et al (1996, p 1770) capture something of this in their discussion of islands of sustainability, arguing that cities need to balance the need to build internally and connect externally, since areas which are wholly self-reliant may survive but 'do not make any contribution to the evolution of the whole economic system towards sustainability', neither learning from nor sharing with other regions. Alternatively, the great strength of the self-reliant approach is its explicit concern with equity issues. In particular, relative to the other models of sustainable urban development, what the self-reliant city approach adds is a clear emphasis on interspecies equity, procedural equity and also social equity, bringing these to the forefront of both problem diagnosis and the processes of devising policies for the sustainable city.

Redesigning cities

This is perhaps the dominant approach adopted by most Western planners and architects. In essence the environmental problems of cities are seen to be linked intrinsically to poor design of the urban fabric, in particular 20th-century additions predicated on the assumption of cheap and readily available fossil fuels for homes, work and transport. Of special concern are the problems associated with the rise of the motor vehicle, from the spread of low-density residential sprawl to the need to build a substantial infrastructure, including road systems and parking lots. From this perspective, a central feature of moving towards sustainable development has to be the redesigning of the city, including the very layout of the city at the regional scale – should we opt for concentrated decentralization, eg in new towns, corridors of urban expansion, or continue to sprawl (Breheny and Rookwood 1993; Haughton and Hunter 1994)? At more local scales, there are important issues of building design, promoting higher residential densities, and attempts to foster greater mixed land use, the latter widely advocated in order to minimize the need to undertake long journeys from home to work, school, shops and leisure facilities. Using the land resource more effectively through reshaping the city, it is argued, can lead to substantial energy savings (Owens 1986; see Breheny 1995 for a critique), whilst more localized efforts to improve building insulation and to aligning buildings to capture more natural sunlight can also bring significant energy savings.

The redesigning the city movement can be seen as essentially concerned with an approach which focuses less on bringing nature back into the city itself and more on creating a city on human-terms – that is, one which is socially and economically vibrant and viable, creating and celebrating distinctive built environments with their own cultural assets and aesthetics. With the emphasis on higher residential density developments in particular, the intention is less to create a fabric which embraces 'nature' within the city, rather to create a vital urban centre in its own terms, which exercises less impact on external areas – for instance, by reducing rural land take for expansion and more generally by reducing energy consumption. Natural capital is assumed to be substitutable with human capital, within certain limits, so that preserving an ancient cathedral may be more valuable than creating a new wildlife habitat; this said, this is substitutability within limits, with critical natural stocks preserved as far as possible and technologies generally being re-orientated to work *with* rather than *against* nature. In terms of equity considerations, most are indirectly present – that is to say, implicit rather than explicit. However, the emphasis on sorting out the problems of the city from within does mean that external impacts are not unpacked in any detail – it is just assumed that reducing land take and energy usage will have desirable impacts elsewhere. Little attention is paid to what the impacts might be of the remaining imported resources flows and exported waste streams of the city.

Externally dependent cities

Conventional economists within what Rees (1995) refers to as the 'expansionist' paradigm, tend to argue that the best way to address environmental concerns is to pursue the current dominant Western path of high economic growth which, it is said, will provide the wealth to address social inequalities and come up with solutions to environmental problems. This is in marked contrast to the 'steady state' approaches of ecological economics which emphasize the incerconnectedness of economy and ecology within a context of finite possibilities offered up by the ecosphere, in terms of resource availability and ability to absorb wastes. Ecological economics tends to argue that the environment should be seen as providing both opportunities and limits to economic growth, so that growth needs is shaped to ensure that it does not exceed local and global environmental carrying capacities (Ekins and Max-Neef 1992).

The externally dependent city essentially follows the conventional or 'neoclassical' view that environmental problems can be addressed effectively through improving the workings of the free market within existing capitalist systems. Typically, light-green approaches emphasize the power of economic growth to generate problem-solving technologies, as opposed to problem-creating technologies, with the implicit assumption of considerable substitutability between human and natural capital stocks, For example, Simon (1981) challenges the view that loss of agricultural land to an out-of-town shopping mall is necessarily a bad thing, since the market is sending clear signals that the land is worth more commercially developed for retailing than it is for selling crops, reflecting overall consumer preferences and utilities. He also argues that although the loss of land maybe associated with considerable 'externality disutilities', there are also evident externality benefits to those who use the mall or those with adjacent land which rises in value.

In the free-market model, it is generally assumed that critical natural stocks can be adequately protected by the market itself supported by a minimalist regulatory system, setting minimum standards in respect of preservation. Following in the footsteps of Coase (1960), advocates of the free-market approach generally envisage expanding the areas of human activity subject to 'market disciplines' by pricing externalities and extending property rights, aiming to use market disciplines, backed by legal sanction, to change human behaviour patterns in support of resource conservation and reduced pollution. For instance, extending property rights to cover ambient air quality could discourage factories from emitting pollutants if they were subject to paying compensation for any environmental deterioration experienced by other property owners, provided that this made it economically cheaper to purchase equipment to reduce pollution at source than to pay compensation (LeGrand et al 1992).

For the advocate of using the market system to address environmental problems, the essential problems are usually seen as those of inadequate pricing, underdeveloped systems of property ownership, poorly constructed markets, overregulation, poor regulation and no regulation. These can all lead to market failures, which in turn lead to environmental problems. Central to this critique are market externalities – that is to say, aspects of behaviour not adequately captured by conventional market pricing signals and not readily amenable to formal regulatory intervention. So, excessive petroleum usage in cars in this view is connected directly to the fact that certain impacts of this usage are not captured in the pricing mechanism. These externalities might include the costs of pollution in terms of, for instance, impacts on asthmatics, contributions to the greenhouse effect, and the costs of road deaths and accidents. If such costs were calculated and added to the price of petrol, this ought to lead people to prefer more petrol-efficient cars, sending signals in turn to car manufacturers to produce less energy-profligate vehicles. Considerable intellectual energy is currently being expended by economists to find ways of shadow pricing such externalities and bringing them into appropriate pricing systems, within the inevitable political constraints – for instance, voter resistance to governments which push up petroleum taxes rapidly (Pearce, Markandya and Barbier 1989). It should be noted that it is not just conventional economists who see merit in this approach – some radical commentators also argue for altering pricing systems to incorporate externalities more effectively – the main differences perhaps relate to the relative emphasis on the role of state regulatory systems (Jacobs 1991). Alternatively, there is a strong radical critique which holds that pricing reform is essentially ephemeral, since capitalist market systems are inherently incapable of moving towards sustainable development – in this view, more far-reaching transformations of the economy and its regulation are required (O'Connor 1993).

In a similar vein, from the perspective of many conventional economists pursuing a market-centred approach, the environmental problems of the city are fundamentally ones of market failure or government failure. In the externally dependent city, trading around the world is seen as essentially unproblematic as long as market externalities can be identified, assessed and brought into the market mechanism by some means. Urban environmental problems are seen to be related not so much to cities themselves, rather to more general market failures. Cities just happen to be major generators of market externalities by virtue of their size, which leads to concentrations of environmental problems in them (Button and Pearce 1989). The free market view is most evident in the work of the World Bank in trying to address the environmental problems of Third World cities. Here, the environmental problems of cities are reduced to issues of improving the city overall: '[t]he challenge of urban policy is, in abstract terms, how to maximise the agglomeration economies and their positive externalities while minimising the diseconomies and their negative externalities' (World Bank 1991, p 53; see also Button and Pearce 1989).

Following from this problem identification, the World Bank's (1991) proposed solutions flow rapidly in terms of measures to improve market efficiency; for instance, introduce land reforms, including land registers, and use the market to provide incentives to alter behaviour patterns. In particular it is argued that pricing policies may need to be changed since '[b]y pricing resources and services at cost, excessive resource use can be discouraged and costly investments postponed . . . especially in countries with seriously distorted prices, improved pricing policies can be an incentive for more efficient resource use and reduced air and water pollution' (ibid. p 74). Whilst equity considerations are not absent from such analyses (see Pearce 1992), they are sometimes in practice reduced to secondary elements for policy concern, where it is assumed that appropriate market adjustment would in any case begin to address inequities.

Fair shares cities

The final approach to sustainable urban development is one which I term FairShares cities, which sets out to ensure that environmental assets are traded on a fair basis, with a particular view to ensuring that exchange does not take place in ways which degrade donor environments, economies and societies. To achieve this, it is important to ensure that adequate compensation is provided for the transfer of environmental assets. Similarly, waste streams which effectively 'appropriate' the environmental health of other areas need to be regulated so that they do not impact adversely on recipient area ecosystems, economies and societies, and, to the extent that they do, adequate compensatory mechanisms should be established. Given the emphasis on reducing use and pollution streams, many elements of the self-reliant city, redesigning the city and externally dependent city policy directions are present. In the Fair Shares model, however, the additional dimension is to bring about institutional transformations which directly link the actions of those responsible for degrading environments, within the city and beyond, to the means of repairing or compensating for this damage. As a precondition for trading in environmental externalities it is essential to take into account the carrying capacity and tolerance levels of host and recipient environments. In this model, critical natural stocks are preserved, whilst there is a conditional form of substitutability in other respects, where environmental exchanges are subject to increased concern about ensuring that adequate compensation is made for any damaging environmental impacts. In overall policy terms, then, it is important in the first instance to seek to minimize adverse impacts in aggregate, and to ensure that access to environmental assets is equitably distributed, and in the second instance to ensure that adequate compensatory mechanisms are in place to compensate for the transfer of environmental externalities between individuals, groups and geographical areas.

The two dimensions of change are very much interconnected. If negative externalities can be identified and attributed to their source, requiring full compensation under the principles of geographical and procedural equity in particular, this will lead to some changes in behaviour patterns. Such sentiments are easy to express, yet they remain surrounded by ambiguities which make it difficult to begin to see how they can be converted from broad principle into operational practice. Realistically, there will always be some trading of environmental goods and bads. For instance, a city may well 'export' a small amount of air pollution to a neighbouring underdeveloped region which still has sufficient natural assimilative capacity to absorb and neutralize the pollution. In this case the environment of the recipient area remains largely undamaged and it might be possible to devise compensation mechanisms for the usage of this spare capacity, which could help to develop the area concerned. These need not be financial – they might also include preferential rights to market access of polluter countries or changes to migration rights for people in the recipient areas (White and Whitney 1992). The danger remains, however, of socially unacceptable impacts arising, as demonstrated by regular protests by those opposing exports of nuclear or other toxic wastes. What is deemed an 'underpolluted' area by one person, may well be someone else's preferred pristine environment. Clouding the issue still futher, there will also be instances of tension between environmental and social equity – for instance, where a poor nation pollutes a richer one, making the notion of 'transfer' compensation payments more politically difficult to negotiate. In addition, there will be many instances of ambiguity over where a pollutant is sourced from and the relative impacts of different sources – for instance, among West European nations causing acid deposition problems in Scandinavia.

If we accept that some exchanges of environmental value are both inevitable and potentially beneficial, the policy imperative becomes to identify and in some sense measure these exchanges, and then to devise systems to provide adequate compensation for any adverse impacts. These might be simply changes to pricing systems, but as the distribution of resulting income tends to be geographically and socially unconnected to where impacts are felt (contributions to the general tax base or feeding the profits of large multinationals, for instance), then it is also likely to require some form of linked system for reparations. White and Whitney (1992), for instance, take the view that it should be possible to devise systems of reparations which link areas benefiting from an exchange of environmental value to those degraded by it.

It needs to be emphasized that there are enormous practical and conceptual difficulties in isolating and gauging the net impacts of these flows. The word 'net' here is important, since cities will inevitably generate a series of positive impacts on parts of their hinterland areas, providing much needed investment and jobs, for instance (Jacobs 1984). Reducing the calculus to environmental inputs and outputs is immensely problematic in this sense. Even if restricted solely to environmental considerations alone, there remain practical difficulties of measuring the multiplicity of tangible environmental flows involved and also their interactions – for instance, accidental combinations of air pollutants from different sources (Haughton 1997). There are problems, too, in identifying how revenues would be raised, and also in respect of precisely who should pay for, and who should receive, any reparations.

Given the problems of identifying externalities and trade-offs plus workable compensatory mechanisms at anything other than a large scale, the 'fair shares' approach is one which is perhaps easier to operate at the level of the nation state rather than at the individual city, except where strong city-states exist, such as Hong Kong or Singapore. This said, it is possible to see ways in which cities can begin to work out their 'ecological footprint' as one first step towards increasing awareness of hinterland impacts and developing the policies to address them. Even small neighbourhood areas within cities can set up trading relationships with hinterland areas on more favourable terms or engage in targeted remedial action. An example of this is the Halifax EcoCity proposal in Adelaide, which envisages buying and restoring rural land in its hinterland area as part of a 'remedial-compensatory' approach (Downton 1997).

In terms of equity considerations, the 'fair shares' approach potentially seeks to address all dimensions given that it is constructed here as building on the best of each of the other models. It is strongest in addressing geographical and procedural equity issues, whilst the attention to local and global carrying capacities signals a strong concern for interspecies equity. Where the model falls down perhaps is in its preoccupation with institutional transformation which mirrors the similarly problematic technical and design solutions of the redesigning the city approach and the overemphasis on economic tools in the externally dependent city model.

EQUITY PRINCIPLES AND MODELS OF SUSTAINABLE URBAN DEVELOPMENT

It is difficult to provide a definite set of judgements on the relative merits of each of the four approaches to sustainable development outlined here, not least because in practice there is often considerable overlap between the policies adopted under each approach – the shifts towards higher residential densities and mixed land uses advocated under the 'redesigning the city' approach, for instance, are also clearly central to the self-reliant cities approach. As in this example, what differentiates the approaches is sometimes as much what is not present as what is, in terms of both intellectual baggage and practical policy approaches. The approach here is to reflect on some of the equity criteria outlined earlier to begin the process of evaluating how the four approaches match up to the equity principles underpinning notions of sustainable development. There are problems here, too, since in truth to be meaningful the principles need to be seen not separately but in combination, given that they are all interrelated to some degree. It also needs to be admitted from the start that these equity principles represent just one way in which the various approaches could be evaluated, and undoubtedly others would prefer to highlight various dimensions of economic efficiency, political accountability or social viability, for instance.

Despite the necessarily crude nature of trying to judge each model in a reductive fashion which reduces complex trade-offs to one person's (ie my) judgement, it is a valuable exercise to begin to see how each of the four models of sustainable urban development embodies different sets of concerns. Figure 4.1 [not reproduced here] provides a summarized version of how each model of sustainable urban development might be seen to fare when judged against the main principles for sustainable development outlined earlier. The intention here is not to *prove* or *assert* any one model as better than another; rather, it is to **highlight** the key areas of concern in relation to fundamental principles for sustainable development which I see as being linked to each model. In a sense, it is for other people to use their own judgement to undertake similar exercises since I do not claim this as in any way being a definitive view – even for myself, there are tensions and contradictions which I recognize still need to be explored. For all its problems, this initial analysis does begin to highlight some of the possible weak points and the strengths of each approach to sustainable urban development.

Having briefly alluded to some of the key equity issues associated with each model in the previous section. the present section reworks this by providing a brief overview of each principle in turn, reflecting on the relative merits of each model. In terms of intergenerational equity, I tend to give the proponents of each approach the benefit of the doubt and assume that it is an overriding concern for all of them, although this appears to me to be largely implicit in the case of the externally dependent city model and more explicit in the other models. It should be said that, in practice, whilst intergenerational equity may be very much a concern for those who are seeking to integrate sustainable development considerations into the free-market approach, such work represents just a small part of the totality of work in this vein, much of which continues seemingly oblivious to the challenges raised by sustainable development.

The most problematic area in respect of social equity concerns extreme versions of the free-market model which rely heavily on the market mechanism rather than state regulation to achieve their distributive goals. This said, it is clear that transforming markets to reduce externalities and associated cost-transferring activities is likely to be a central ingredient in shifting towards processes of sustainable development within existing capitalist systems. As such, simply rejecting the market modification approaches would be short-sighted; alternatively, expecting the market approach to bring about social equity goals without major 'directive' transformations and regulatory intervention would be unrealistic. Whilst both self-reliant and fair shares approaches invariably put social equity considerations to the fore, this is not always true of 'redesigning the city' approaches, which sometimes tend towards tinkering with technocratic systems of doing things for people or making people do things differently (which is also one reason why this model achieves slightly less than top-grading in respect of procedural equity).

In terms of geographical equity, I have a concern that extreme versions of the self-reliant city approaches could have detrimental impacts on regions which formerly relied on income gained from trade with them, so for this reason I rate it as having potentially (but not necessarily) perverse implications in respect of this principle. Similarly, there is a residual concern that areas which become too introspective and look to solutions from some romanticized low technology past will miss out on some of the benefits of emergent technologies – an isolated community stuck in a time-warp of 50 years ago would be stuck with a lot of energy-inefficient technologies. for instance. Whilst externally dependent city approaches have the potential to create policies which address spatial inequities, as a general rule the tools of the free-market economist tended to limit consideration of geography to the 'friction of distance effects' associated with location decisions, whilst underplaying the way in which space is used in the creation of certain externalities.

My analysis of procedural equity in each of the models follows that for geographical equity, with the exception that implicitly at least self-reliant cities are more coherently committed to setting in place systems which reduce or compensate for adverse external impacts. whilst being very clearly committed to improving internal procedural issues, in particular participation mechanisms.

With respect of interspecies equity, as noted earlier, judgements on potential impacts of urban development approaches may well vary according to which spatial scale is used in studying the city-region and the personal judgements about whether it is better to encourage natural areas, agriculture and wildlife habitats closer to the city centre (self-reliant city) or to protect hinterland environments by building more compact cities with only limited public open space. One example of such tensions in terms of redesigning the city approaches concerns attempts to build higher density, more compact cities, which may well lead to the loss to residential development of valuable brownfield open space sites, for instance, which once hosted considerable wildlife activity. Alternatively, such policies may well forestall or even prevent residential encroachment on greenfleld sites at the edge of the city. It very much reflects my personal beliefs that I see more problems with 'concreting' over the city than with approaches which encourage the extension of green areas within the city; for this reason, the redesigning the city and the externally dependent city models both appear as potentially problematic since in some versions both undervalue the role of encouraging greenspace in the city either by relegating it in importance or, in the case of the free-market approach, relying on commercial land values to reflect broader aesthetic values.

The question of which approach overall is 'best' is evidently more tricky than it seems, made more problematic in that so many of the features of the different models are in practice shared. The approach which appears to emerge most favourably is the Fair Shares city, although by imbuing it with most of the positive qualities of the other models, I have perhaps begun to give it a degree of concreteness and desirability that it does not really yet have. In practice, my view is that each of the models is in a sense dominated by particular professions, often linked into specific policy and academic disciplinary areas (eg ecologists, planners or economists).

In conclusion, the main value of highlighting the differences in approach is to draw attention to the parallel possibilities which exist in addressing the urban contribution to sustainable development, to try to ensure that policy myopia does not set in. This seems particularly important to planners: we do not have all the tools necessary to move towards sustainable development, just some of them. We can see aspects of the Tinbergen principle at work here – for a policy to achieve its goals policy-makers require at least as many tools as there are goals. Too often policies have been framed which are too ambitious relative to the resources and instruments available to implement them. This needs to be recognized in our training of planners too, as we seek to integrate ecological and economic perspectives more thoroughly into our curricula. So I am not against any of these approaches which are, after all, rather artificially constructed here to highlight the different prevalent schools of thought; rather, I would argue that all bring particular insights of value to the goal of moving towards the sustainable city and sustainable development more generally. The challenge is to draw on their respective concerns and strengths towards a more integrated policy approach for the future.

REFERENCES

Andruss, V, Plant, C, Plant, J and Wright, E, eds 1990. *Home! A bioregional reader* Philadelphia: New Society Publishers

Beatley, T 1991. A set of ethical principles to guide land use policy. *Land Use Policy*, 8 (1), January: 3–8.

Beatley, T 1994. *Ethical Land Use: principles of policy and planning*, Baltimore: John Hopkins University Press
Berg, P 1990. Growing a life-place politics. In *Home! A bioregional reader* eds Andruss: V *et al* 137–44. Philadelphia: New Society Publishers
Berry, D and Steiker, G 1974. The concept of justice in regional planning: justice as fairness. *Journal of the American Institute of Planners* 6 (4): 414–21
Bookchin, M 1974. *The limits of the City*. New York: Harper Colophon
Bookchin, M 1980. *Towards an Ecological Society*, Montreal, Canada: Black Rose Books
Bookchin, M 1992. *From Urbanization to Cities: towards a new politics of citizenship*, London: Cassell
Bourne, L S 1996. Normative urban geographies: recent trends, competing visions, and new cultures of regulation. *The Canadian Geographer*, 40 (1), 2–16
Breheny, M 1995. The compact city and transport energy consumption *Transactions of the Institute of British Geographers*, NS 20 (1): 81–101
Breheny, M and Rookwood, R 1993. Planning the sustainable city region. *In Planning for a Sustainable Environment*, ed, A Blowers, 150–89, London: Earthscan
Bullard, R 1990. *Dumping in Dixie: race, class and environmental equality*. Boulder, Colorado: Westview Press
Bullard, R. ed 1993. *Confronting Environmental Racism: voices from the grassroots*. Boston, Mass: South End Press
Button, K and Pearce, D 1989. Improving the urban environment: how to adjust national and local government policy for sustainable urban growth. *Progress in Planning* 32 (3): 135–184
Callenbach, E 1975. *Ecotopia*. Berkeley, CA: Banyan Tree Books
Coase, R H 1960. The problem of social cost. *Journal of Law and Economics*, 3: 1–44
Downton, P 1997. Ecological community development. *Town and Country Planning*, 66 (1), January: 27–29
Ekins, P and Max-Neef, M, eds 1992. *Real-life Economics: understanding wealth creation*, London: Routledge
Friedmann, J 1989. Planning, politics and the environment. *Journal of the American Planning Association*, Summer: 334–341
Harvey, D 1973. *Social Justice and the City*, London: Edward Arnold
Harvey, D 1992. Social Justice, postmodernism and the city. *International Journal of Urban and Regional Research*. 16 (4): 588–601
Haughton, G 1997. Developing sustainable urban development models. *Cities* 14 (4):189–195
Haughton, G 1998. Geographical equity and regional resource management: water management in southern California *Environment and Planning B*, 25 (2): 279–98
Haughton, G and Hunter, C 1994. *Sustainable Cities*. London: Jessica Kingsley Publishers
Hay, A 1995. Concepts of equity, fairness and justice in geographical studies. *Transactions of the Institute of British Geographers* NS 20 (4): 500–508
Heiman, M 1996. Race, waste and class: new perspectives on environmental justice. *Antipode* 28 (2): 111–121
Hofrichter, R, ed 1993. *Toxic Struggles: the theory and practice of environmental justice*. Philadelphia: New Society Publishers
Jacobs, J 1984. *Cities and the Wealth of Nations*. Harmondsworth, UK: Penguin
Jacobs, M 1991. *The Green Economy: environment, sustainable development and the politics of the future*. London: Pluto Press
Lake, R W 1996. Volunteers, NIMBYs, and environmental justice: dilemmas of democratic practice. *Antipode*, 28 (2): 160–174
LeGrand, J, Propper, C and Robinson, R 1992. *The Economics of Social Problems*. Third edition, Basingstoke, UK: Macmillan
Lipietz, A 1996. Geography, ecology, democracy. *Antipode*, 28 (3): 219–228
Morris, D 1982. *Self-Reliant Cities: energy and the transformation of urban America*. San Francisco: Sierra Club Books
Morris, D 1990. The ecological city as a self-reliant city. In *Green Cities*, ed D Gordon, pp 21–35. Montreal, Canada: Black Rose Books
O'Connor, J 1991. Socialism and ecology, *Capitalism, Socialism, Nature*, 2 (3): 1–12
O'Connor, M 1993. On the misadventures of capitalist nature. *Capitalism, Socialism, Nature*. 4 (3): 7–40
Owens, S 1986. *Energy, Planning, and Urban Form*. London: Pion
Pearce, D 1992. Economics, equity and sustainable development in *Real-life Economics: understanding wealth creation* eds P Ekins and M Max-Neef, pp 69–76, London: Routledge
Pearce, D, Markandya, A and Barbier, E B 1989. *Blueprint for a Green Economy*, London: Earthscan

Pulido, L 1994. Restructuring and the contraction and expansion of environmental rights in the United States. *Environment and Planning A* 26 (6): 915–936

Ravetz, J 1994. Manchester 2020 – a Sustainable city Region Project. *Town and Country Planning* 63 (3): 181–185

Rees, W 1995. Achieving sustainability: reform or transformation? *Journal of Planning Literature* 9 (4): 343–361

Register, R 1987. *Ecocity Berkeley: building cities for a healthy future*, Berkeley, CA: North Atlantic Books

Sayer, A and Storper, M 1997. Guest editorial essay. Ethics unbound: for a normative turn in social theory. *Environment and Planning D* 14 (4): 1–17

Seabrook, J 1990. *The Myth of the Market.* Bideford, Devon, UK: Green Books

Simon, J L 1981. *The Ultimate Resource.* Oxford, UK: Martin Roberston

Smith, D M 1994. *Geography and Social Justice.* Oxford: Blackwell

Smith, D M 1997. Back to the good life: towards an enlarged conception of social justice. *Environment and Planning D* 15: 19–35

Turner, R K, Pearce, D, and Bateman, I 1994. *Environmental Economics.* Hemel Hempstead, UK: Harvester Wheatsheaf

Wallner, H P, Narodoslawsky, M, and Moser, F 1996. Islands of sustainability: a bottom-up approach towards sustainable development. *Environment and Planning A* 29 (10): 1763–1778

WCED (World Commission on Environment and Development) 1987. *Our Common Future*, Oxford, UK: Oxford University Press

White, R and Whitney, J 1992. Cities and environment: an overview. In *Sustainable Cities: urbanization and the environment in international perspective*, eds R Stren, R White, and D Whitney, pp 8–52. Boulder: Westview Press

World Bank 1991. *Urban Policy and Economic Development: an agenda for the 1990s.* World Bank Policy paper, Washingion DC: World Bank

Young, I M 1990. *Justice and the Politics of Difference.* Princeton, NJ: Princeton University Press

'The Urban Poor: An Invisible Resource'

from *Pioneers for Change* (1993)

Jeremy Seabrook

Editors' Introduction

Jeremy Seabrook has written extensively on the conditions of urban life in the non-affluent world. Other books by him, mainly based on direct experience, include *Victims of Development: Resistance and Alternatives* (London, Verso, 1993) and *In the Cities of the South: Scenes from a Developing World* (London, Verso, 1996). Part of his text reprinted here draws on the former (pp. 43ff), but in a different context: that of the Right Livelihood Awards presented annually by the Right Livelihood Foundation. This was established in 1980 by Jakob von Uexkull, who initially approached the Nobel Prize Committee with a view to contributing funds to an environmental prize. The approach was not successful, and von Uexkull raised the initial endowment for the new Foundation by selling his stamp collection to a museum in Saudi Arabia. It has since grown through individual donations.

Seabrook's focus in this chapter is the wealth of knowledges possessed by the urban poor, which remain largely unrecognised by planners and city authorities. John Turner's work in central and south America drew attention to the viability of informal settlements and the – in many ways – better prospects they offered than government housing schemes. While the latter tended to be inflexible as well as poorly designed, built, and sited, self-build housing displayed ingenuity, flexibility, and suitability to purpose. A comparison can be made with Hassan Fathy's efforts to reintroduce the skills of mud-brick building and design in Nubia – a scheme which produced the village of New Gourna, which remains a viable settlement today, but was frustrated in the late 1940s by Egyptian government bureaucracy, and compromised by an agenda of population clearance from older sites in the archaeological zone – and with the growing revival of alternative building technologies today, using compacted earth, old tyres, and rubble as basic materials. Turner's (and Seabrook's) point, however, is that the affluent world can learn from the experiences of the non-affluent: 'the poor may be seen as offering both hope and instruction; not through their poverty, but through their energetic striving for a secure and modest sufficiency' (p. 81). Seabrook expands his theme through reference to other award winners, Ela Bhatt, whose work advanced women's industrial organisation, and Winefreda Geonzon, a lawyer from the Philippines striving for legal rights for the poor. These sub-sections may seem less directly relevant to urban development, but if Haughton's argument for a *social* as well as technical and economic understanding of sustainability is accepted then they are equally relevant. The whole chapter is reprinted here on that basis.

In the faces of the migrants to the cities of the South can be read an epic story of rural dislocation, environmental ruin and extreme social injustice. Within the next two decades, the urban population of the world is expected to exceed that of the countryside. Although there is now evidence that some earlier UN predictions of bloated city populations will not be realised, and despite the greater fluidity that exists between town and country than is sometimes understood by the idea of 'urbanisation', there will still be growing pressure, both on unwieldy urban agglomerations and on the exploited rural areas that must feed them.

At ten o'clock in the morning, the dim gothic interior of São Paulo's Catedral dé Se is full. The people are not praying, but sleeping, a ragged humanity, slumped in the hard pews and covered with coarse blankets; a frieze of rough statuary. Here, at least, they are safe from the predators who roam the Praça de Se outside. A cathedral of impoverished dreamers, cold and silent, a *Totenschiff* with its drifting cargo of refugees from economic and developmental violence.

Urbanisation in the South today is, in many respects, unlike the experience of the early industrial era in Britain. Then a severe discipline, required by the factory system, was imposed upon a raw rural sensibility. Now, the same process is accompanied by an orgiastic consumerism, an absence of controls, which means that resistance to social disruption and exploitation is harder to organise.

The life of migrants is penetrated by an iconography of privatised dreams, which also calls into existence strange new divisions of labour. Economic development to the poor woman from Mato Grosso means paid employment, cleaning the floor in the porno movie house, where the men masturbate in mechanistic accompaniment to images of brutal sex. The vendors on the viaduct of Sta Ifigenia stand guard over the offal of industrial society, as though entrusted with its greatest treasures – T-shirts emblazoned with obscenities, value-added, nutrition-subtracted foodstuffs, magazines with titles like *Girls Who Dig Girls* and *Anal Sex*, tarot readings, barbie dolls and war toys.

The square outside the cathedral is part market-place, part fairground. It has the aspect of a rural meeting place. Gypsy women from the North in lime-green or scarlet flounces, read the hands of credulous passers-by; sellers of traditional medicines, barks, herbs and roots, offer relief from syphilis, ulcers, cancer and worms; a man attracts a crowd by breaking glass bottles with his bare feet. The ornamental fountain serves as bathing place for the homeless, and the faded washing of the poor is laid out on the green-painted benches. At night, in the Rua Marquesa de Itu, the young transvestites, gaudy as tropical birds, cluster around the purring Opels and Cadillacs, and climb in behind the smoked-glass windows.

There are a million unemployed in São Paulo. Last year, there were more than 2,500 murders. New death squads have appeared, eliminating the young small-time crooks, children really, who prey on poor neighbourhoods. These, and feuds between rival drug gangs, leave a bloody cargo of human wreckage for the municipal ambulances to clear from the streets. In one parish after a recent weekend, there were 43 funerals – more than half of the dead had been murdered; the others, young children, were the victims of that other violence: malnutrition and curable disease. The city is overcrowded, scarred by alcoholism, drugs, crime, the breakdown of families. In addition, São Paulo, like the other growing cities of the South, spread circles of desolation around themselves, as they draw ever more resources and people from the spoiled hinterland.

And yet, people always say that this is better than where they have come from. Their lives, scarred by migration and upheaval, are also suffused with hope, creativity and formidable energy.

According to Jorge Hardoy and David Satterthwaite at the International Institute for Environment and Development in London, there are now 1.3 billion people in urban centres in the Third World; at least 600 million of them in life- and health-threatening homes and neighbourhoods. It is common for between 30 and 60 per cent of the population of big cities to live in illegal settlements or tenements.

City governments still regularly clear squatters from pavements, from private and public land in cities as diverse as Bombay, Seoul and São Paulo. But many now recognise that those who migrate to the urban centres have taken a step that cannot easily be reversed. It is acknowledged that even the flimsiest, most makeshift shelters of newcomers will, over time, be transformed into decent neighbourhoods and communities, provided that people are given security of tenure of the land, and the opportunity to build for themselves. The slums may appear to the tidy-minded to be unsightly and random collections of people who blight the city landscape and drain its resources. In fact, the people create work for themselves in the urban economy, performing valuable services – domestic work, laundry, selling fruit, vegetables and snacks, offering transport and other amenities to the very people who are the most vociferous about their encroachments.

If governments are now coming to understand that the people themselves are the most effective architects, planners and builders, especially the women, to whom much of the labour of creating a home falls, this is due in considerable measure to the pioneering work of the late Hassan Fathy, Egyptian architect and author in 1973 of *Architecture for the Poor* (University of Chicago Press, and Right Livelihood Award winner in 1980). This work has been amplified and carried forward by John Turner.[1]

Turner insists that neither the state nor the market can deliver adequate housing to the majority of the city poor. The market requires too great a capital investment, which is not available for such uncertain returns, and the state not only rarely has the resources, but also takes away the freedom of the people to construct for themselves. Nowhere are these shortcomings more apparent than in the *cortiços* (tenements) of São Paulo, where, in one infamous apartment block, more than 10,000 people occupy apartments that have been illegally sub-divided. All over the South, 'low cost housing has been appropriated by the only people who can afford it, the middle class.'

In Manila, in the Philippines, only about 15 per cent of the people can pay for typical market-provided housing. The Freedom to Build Corporation in Manila is directly inspired by John Turner. The Horacio de la Costa project at Novaliches, on the periphery of the city, provides householders with loans of up to 80,000 pesos at 6 per cent interest. For this, they receive a basic unit of grey breeze-block construction on plots of 55 square metres. Each structure is simply the most rudimentary shell. As people's income rises, and their work prospects improve, they add to, embellish and extend the simple building, according to their tastes and means. The back wall can be easily opened up, and there is room for a second storey to be added. Some people have already installed new doors and windows, even though the development at Novaliches is not yet finished; others have built a terrace, created a garden, erected a grille and railings. Most of the people here had come from insanitary, insecure slums. It is a place of great vitality, tangible hope.

Even in areas where people first settle in the cities, or where invasions of land are organised, the transformation in the space of a few years can be dramatic. 'They should not be called invasions,' said the organisers of one community in São Paulo, 'because that makes it sound like criminal activity. We call them occupations.' John Turner says:

> That the mass of the urban poor are able to seek and find improvement through home ownership (or de facto possession), when they are still very poor by modern standards, is certainly the main reason for their optimism. If they were trapped in the inner cities, like so many of the North American poor, they too would be burning instead of building.

When the inhabitants of an area have security, can control the major decisions, can make their own contributions to design, construction and management of housing, both the process of participation and the environment which this produces, stimulate social and individual well-being. When they have no control over and no responsibility for key decisions in their own housing, dwelling environments may well become a barrier to personal fulfilment and a burden on the economy. A 'good' environment is not necessarily of a high material standard: a cheap shelter may release money for the family's more urgent priorities, especially when the children are young. The functional relationship between inhabitant and habitat is what counts; location close to work may well be more important than high standards of amenity.

Many of John Turner's insights have been reached through an imaginative understanding of the social and psychological journey of the urban dweller. Turner's experience was gained in Peru, at a time when half the city of Lima did not even appear on any map, and 50 per cent of the adult population had been born in the provinces.

> When people move to the city, traditional sources of security – the family, subsistence farming – are undermined, and must be substituted by the job market; and home ownership, no matter how basic, offers an alternative form of economic and social security.
>
> Successful urban planning depends upon the alignment of government action with the priorities of the forces of popular settlement. Official housing policies have often 'telescoped' the organic growth of cities as embodied in people's own efforts at self-improvement. The imposition of modern minimum standards on popular urban housing is an assault upon the traditional function of housing as a source of social and economic security and mobility.

John Turner is an advocate of the maxim that, 'appropriate technology is technology that people can appropriate.' People have been ground between the millstones of state control and market forces: from unworkability in Eastern Europe to unsustainability in the West. In Eastern Europe, the state undermined all popular initiatives; in the West, the community base is so eroded that it hardly exists. In the South, the initiative is still there, although constantly under attack; but at least there, people can do things for themselves, seize land, create spaces where there is scope for their own self-determination and control.

More recently, Turner has sought to bring back to the North some of the insights gained in the South. He seeks to reintegrate the functions of neighbourhood, so that living, working and leisure are less dispersed and fragmented. 'In the most highly maldeveloped societies – to which the rest aspire – most people live alone or with one other.' (In Britain, 25 per cent of households consist of a single person; a proportion that is expected to rise to one in three early in the next century.)

> The great majority commute to employments which are themselves increasingly remote from the production of goods and services. Typically, work is eliminated from neighbourhoods in urban-industrial societies, either by restrictive zoning or by the elimination of local manufacturing or trading networks. Divorced from home and work life, creative, recreative and cultural activities are restricted to leisure time. Industrialised, institutionalised and professionalised patterns of settlement and building types waste space and land, time and life, energy and materials – our irreducible inventory of irreplaceable resources. Most surviving neighbourhoods that work and struggle to maintain and regenerate themselves, their own economies and cultures, are materially poor. Hope for the future lies in the fact that so many of the poor in poor countries manage to do so much with so little. While the rich do so little with so much, there can be no future.

Not for the first time, the poor may be seen as offering both hope and instruction; not through their poverty, but through their energetic strivings for a secure and modest sufficiency. It is a lesson which the rich are bound to resist; for just as the shining imagery of the West is projected across the world, detached from the human, social and environmental costs it involves, so, in exchange, what comes back to the West are the scenes of desolation, backwardness and poverty of the South, shorn of any of the humanising influences of community and custom, especially of the work of women, which makes it more bearable than those bare images alone could possibly suggest. This highly tendentious two-way flow of partial information serves as a constant reminder to the privileged of the fate that awaits them, should they be so foolish as to jeopardise their good fortune by questioning too closely the wisdom of the existing arrangements. The noses of the poor press constantly against the windows of the TV screens of the West. It is extraordinary how *images* can serve to block perception; and the creativity and hopefulness of the poor remain invisible to the troubled, though largely unresponsive rich.

Even before they seek shelter, the priority of the city poor is work. For 20 years, Ela Bhatt's SEWA[2] has been a focus for the poorest women to organise themselves. It has strengthened and succoured groups of workers previously believed to be beyond the scope of trade union, or indeed, any other social organisation.

The women are mostly self-employed. This means that they have created work for themselves at the lowest level of reward. Great profits have been made, mainly by men, out of their 'invisible' labour. They have been described as 'marginal', 'peripheral', part of the 'informal' economy. They are *chindi*-workers (quilters), vegetable vendors, home-based workers, *bidi*-makers, block-printers, headloaders, sellers of second-hand clothes, makers of *papad* and *agarbatti*, cane or handicraft workers, recyclers of paper, metal, rags or plastic. If they have remained invisible to the eyes of economists and planners, their presence has been all too clear to middlemen and wholesalers, who have taken advantage of their isolation to pay them at pitifully low rates. One of the greatest inhibitions upon their ability to upgrade their work and skills has been the non-availability of credit, which has led to dependency on money-lenders. The formation of the SEWA bank has transformed the lives of many women. The level of default is lower than in commercial banks.

SEWA found that many traditional skills had been lost or degraded through 'development'. About 20,000 women in Ahmedabad had become dependent on paper-picking for a living, which requires no skill and brings poor rewards. SEWA is concerned to consolidate and enhance skills, and to facilitate the marketing of products through co-operatives. Applying skills collectively has achieved what could never have been accomplished by individuals alone. A new and flexible form of trade unionism has emerged, which empowers, makes connections and raises consciousness through *doing*.

SEWA identified three levels of exploitation: the immediate exploiter – the cruel policeman, the rapacious employer, the vicious contractor. Supporting these is a second level of injustice – government agencies and legal structures: the Labour Department, for instance, corrupted by employers, helps these to avoid labour laws. The municipality criminalises poor vendors: there is no space for vendors, therefore they are illegal, no matter how vital their function in provisioning the city. Finally, there is injustice at the highest level, institutionalised in laws and policies, and buttressed by international networks of wealth and power.

The philosophy of SEWA is that members will work outwards, from a perception of the local constraints on their lives, and come to understand little by little, the complexity of the wider oppressive forces, nationally and globally. Many women in the city know that they have been driven there by technological change, evicted from settled patterns of living by 'development.' For instance, a vendor borrows 50 rupees a day to buy vegetables, and pays back 55 rupees in the evening. This woman may have been forced to leave land that had been degraded by river-effluents or salination: memories of her former agricultural function live on in her vegetable dealing. A farm labourer, in the season when there was no work, used to weave cloth. She has been displaced by the availability of cheaper mill-produced cloth, and has become a rag picker, in a ghostly gesture to her spoilt purpose. Bamboo workers find they can no longer buy bamboo, because it has all been sold to paper-mills. The forest-dweller who harvested grass, seeds and honey, is now banned from the forests. Many other craft skills of seasonal agricultural workers have been lost because of dwindling markets.

While a proportion of men migrate to Ahmedabad and find work in the formal economy, women often remain in the village; and this is the reason for SEWA's expansion into rural areas in the past five years. In some places, up to 80 per cent of villagers are in debt. Women who used traditional techniques to make day-to-day utility items have been ousted by mass-produced industrial products. The disemployment of weavers, potters, cobblers, has led SEWA to think about the survival of traditional crafts. This can be achieved only by upgrading skills, gaining direct access to raw materials, making and selling goods through bulk orders or directly to consumers. In India, 'labour saving' means 'human wasting'. A form of development which creates wealth by mechanising production can only add to an already vast reservoir of unemployed and underemployed; and it is to counter the effects of this that SEWA has extended its activities.

SEWA has also initiated a wide range of socially supportive services for women – health care, child care, water and sanitation programmes, housing projects, protection against violence and sexual harassment, legal aid. Exploitative employers have been pursued in the courts, and there have been campaigns to regularise child labour (its abolition would damage families in which children often contribute essential income), and to rehabilitate the victims of communal violence. The great majority of the members of SEWA are Untouchables, or from other Backward Castes, Muslims and tribal women. Their common plight has proves a formidable solvent of communal and caste differences; the more so since the last two years have seen an upsurge in communal conflict all over India, following the rise of the fundamentalist Bharatiya Janata Party.

The self-employed sector comprises 90 per cent of manufacture in India. Although SEWA was originally a dissident offspring of the Textile Labour Union, it has become a unique and seminal force, serving as model and inspiration to those seeking to organise the excluded. In 1987, the National Commission of Self-Employed Women was set up, and the report it produced was accepted by the government. In 1989, the V.P. Singh administration included Ela Bhatt in the Planning Commission. In spite of this, the linking of the struggles of poor workers, especially women, to the wider framework of trade unionism, remains to be accomplished. There is still much work for SEWA to do; it remains a dynamic, living movement.

Although martial law was lifted in 1981, and the Marcoses deposed after the display of Corazon Aquino's people power in 1986, there has been no improvement in the plight of the poorest. Quite the contrary. The persistence of civil conflict provides a cloak for official and semi-official violence against the poor and against those fighting for social justice. Indeed, after Aquino's Total War policy against the insurgents from 1987, there was a worsening of human rights violations in the country. There are now over one million internal refugees in the Philippines, displaced as a consequence of military action or 'pacification' programmes.

The poor are criminalised. The jails of Cebu City are full: people accused principally of 'economic crimes', driven to despair by loss of livelihood, unemployment, social violence. For many of them, justice remains a travesty. The police regularly raid the poor areas, round up young men at random, and accuse them of unsolved crimes. It is easy for the poor to be thrown into prison and forgotten, children as well as adults. In the city jail, I saw ten-year-old twins, who had been accused of theft. Minors who can be bailed to their parents are allowed home, but those with no responsible adult to care for them just stay in jail: the deprived suffer, and the most deprived suffer most.

Cebu is a spreading ramshackle city, and has the air of an unfinished place, with buildings in rough concrete and breeze blocks, rusty metal roofs, pot-holed roads, festoons of cables across the streets. It seems the buildings are there simply to bear the logos of transnational companies – Mitsubishi, Sanyo, Sharp, Coca-Cola, Marlboro, Nabisco. The banks are windowless fortresses guarded by armed security workers. The shops are full of imported junk – Pee Wee crunchies, squid rings, corn- and potato-based snacks, Mutant Teenage Ninja Turtles merchandise, electronic keyboards, kitsch holy pictures, plastic flowers, fast food, Washington Red apples and California oranges (in a country which has some of the most succulent fruits in the world, mangosteens, custard apples, rambutans, mangoes), powdered hair-dye, soft toys, cameras, junk medicines. The church of Santo Nino in the downtown areas was the first Christian church in the Philippines. Destroyed by fire, the effigy of the saint miraculously survived the flames. Outside the church, along the stone weathered to a spectral grey-white, a long line of human misery presses itself against the walls, immobile as carved buttresses – crippled children, malformed adults, the blind, sick and defeated, hands outstretched in supplication. Inside, the scalloped altar piece, with barley-sugar columns and effigies of saints, chandeliers and plaster mouldings, delicate fretted woodwork, dwarf and diminish the people, from whom the low hum of prayer is stirred by the humid draught from straw fans. In Cebu, the contemporary cargo-cult of consumerism exists side by side with the ossified religion of the former conquerors. On Mactan Island is the statue of Lapu-Lapu, who killed Magellan in 1521; but Magellan's cross remains near the pier, commemorating the spot where King Humabon of Cebu, with his queen, daughters and 800 subjects were baptised.

Duljo is a wretched slum area reclaimed from the sea, which has more than its share of residents in prison. The houses are of bamboo nailed to batons, with metal roofs rusting in the rain; narrow mirey alleys, with pigs on a leash in tiny yards. This is the area close to the docks, warehouses for grain, timber and sugar; fish market, metal workshops, battery-rechargers, recyclers of rubber tyres which are made into floats for fishing, cabinet makers, carvers of holy statues, shellcraft workers; there are horse-drawn *caleças* and cycle rickshaws for hire. The unemployed young men stand idly in the rain. Danilo was in prison for eight years for a murder he didn't commit, and is still serving his sentence, although on parole. He and his wife work selling food they cook on a wayside stall. George, too, was in prison for murder, and even though the complainant has withdrawn the accusation against him, the case has still to be heard. He works whenever there is a ship to be unloaded, and is paid 50 centavos for each sack of grain he lifts. Each sack weighs 50 kilos; he weighs 49. The roads are crowded with women selling fish, firewood, charcoal or coconut husks for fuel; there are cycle-and radio-repair shops, men at sewing machines, flower sellers, lighter-repair stalls, vendors of lottery tickets. Some people are roasting a pig on a spit for a wedding; it rotates over a fire protected from the rain by a length of polythene. The Flores de Mayo festival is being celebrated in a little church: girls in pink and yellow, with silvered cardboard wings, carry bunches of crimson tea roses. Some boys sit under the canopy of a cycle rickshaw to shelter from the rain; they grin, as though unaware of their fate as tomorrow's sweated labour or jail fodder.

The city jail is a concrete structure with round watchtower, high walls surmounted by barbed wire. On the edge of Cebu City, it occupies the site of the former airport. The bare cells with cement walls are a series of iron cages, a forest of bars. Each was designed for 35 men, but currently contains 70 or 80. There are buckets for slopping out, and juveniles are not separated from adults. There are people here accused of murder ten years ago, who have never come to trial. Others have seen the judges in their case promoted, die or emigrate; or their papers have been lost, which has plunged them into official non-existence, another form of living death. Over the cells there is a huge effigy of Christ crucified.

Among the prisoners was Julius Gacayan, who is 20. He was arrested when he was 16 for stealing a motorbike with two friends. They both blamed him, and he was taken into custody. On his way to court, he slipped away and ran into the hills, where his relatives have a farm. The police could not pursue him, because the area was controlled by the New People's Army. Because of this, he was assumed to be a supporter of the insurgency. Later, he was picked up by vigilantes of the Citizens' Army. One of their members had been killed, and they accused Julius of the murder.

As a result, Julius's father has had to leave Cebu for Manila The forces of the Citizens' Army threatened his life, and have declared the whole family a menace to the community. This means that the family is now ostracised by their neighbours. Their hut was burned down. The mother comes to jail to visit her son. She touches his face with her fingertips; it is as though she is saying goodbye to a condemned man. Julius buries his face in his hands. He says he knew nothing about politics. They have never been subversives. All he did was commit an act of youthful foolishness.

Paolo, who is 46, is accused of rape. He was originally charged with non-payment of a debt. Released, he was apprehended the following day and accused of 'acts of lasciviousness'. Not told who the complainant was, he later learned that she was a member of the family to whom the debt was owed. His case was filed away in the archive because of administrative error and he has not come to court. That was in 1984. His family has stopped visiting him.

The provincial jail, in contrast to the city jail in the centre of Cebu, is more humanely run. At least there, juveniles are separated from adults. There are about 30 young people in a cell apart from the main body of the prison. Many come from Consolacion, a settlement some 15 kilometres away, where many squatters were relocated about ten years ago. Ruela Patigaon is 20. He has been in prison three years, accused of homicide. The lawyer dealing with his case has died and has not been replaced. Arnel Rosalita is a slight boy, whose 15th birthday it is. The others sing 'Happy Birthday to You' in English. He is a second-year high school student, who stabbed to death a 19-year-old. It was, he says, in self-defence; when attacked by the older boy, he seized a knife from a nearby food stall and struck out with it. A small 15-year-old is accused of rape and murder The boy is undersized, incapable of rape, but he was with an older man who ran away. As accomplice, he must bear the whole charge. He has no father, and lives with his grandmother in the mountains. Three boys, Rodolfo Blanco, 15, Marcillo Granada, 13, and Jerry Caneta, nine, are accused of robbery. They have been here four months. When a Free LAVA worker telephoned the court to find out when the case was coming to court, nothing was known. The case had been filed and forgotten. Roy Sotto is 20. His mother left the family to find work in Manila as a maidservant. Roy has been accused of drug pushing, but says that the police planted ten sticks of marijuana on him.

Free LAVA has a threefold approach to the work: crime prevention, free legal advice and assistance and rehabilitation of offenders, including employment loans. There is also a programme of recommending prisoners for release, pardon and parole. Many of the workers at Free LAVA have themselves been wrongfully imprisoned, including Nonoy, who spent eight years in jail accused of murder. Among the young people in the provincial jail, LAVA has started a scouting troop, run by Tony Auditor, himself a former prisoner, and as part of the prevention programme, scout troops operate in the community. Winefreda Geonzon believed that constructive community work is the most helpful way of preventing young people from coming to the attention of the police. At the summer scout camp in the hills overlooking Cebu, many of the mothers came to see their children assemble to receive their colours and awards. Their pride in their sons and their affection for Winefreda were very moving.

Later, I went with her to visit the homes of some of the families she has worked with. Jut behind the Capitol, the US-style government building, with dome and classical façade, runs the Guadalupe River, which divides the North of Cebu from the South. On the steep banks of the river bed, the people have built their homes. After months of drought, the river has been reduced to a muddy trickle. There is a wide expanse of dry bed, where young people have set up a baseball post. People sit out in the cool of early evening. Palm trees grow out of the riverbank, and creepers and shrubs have sprouted, now that the drought has just broken. The houses are of rattan, bamboo and nipa, cardboard and wood. The waste water trickles out of pipes, and spills in a spread of moss and algae onto the rip-rap of concrete and stone walls that have been raised up to prevent flooding. Those who have built up the bank beneath their house will be safe when the heavy rain comes; others will see their houses swept away, as occurs almost every year. Teofila says her house was carried away downstream by last year's floodwaters. The neighbours helped her and her husband to drag it back, and now they have made breeze-block foundations to strengthen it. He drives a jeepney, for which he pays the owner 225 pesos a day. He starts work at six in the morning, and finishes late at night, for a profit of only 60–70 pesos a day, well below the minimum wage. Many of the men here work as security guards at banks, hotels or private companies, for a salary of around 3,000 pesos a month. The women wash for 150 pesos a month from each family they wash for. They do their washing in the muddy waters of the Guadalupe. They dig a pit in the sand, close to where the river flows; and the water is filtered through the sand, so that when it reaches the pit, it is perfectly clean. The washing dries in the strong disinfecting sunlight, and smells clean and sweet. It is unlikely, says Teofila, that the owners of the washing ever think about the places where their linen is cleaned. Drinking water has to be purchased from the owner of a brick house on top of the river bank, five centavos for four litres. Some women work on shellcraft, which is a major industry in Cebu. Formerly, they worked in a factor, where they acquired the skills, and then they set up on their own account, with a loan from Free LAVA. They buy shells, an electric drill, and wicker baskets which they adorn with festoons of coloured shells; these are sold to a stall-holder for 44 pesos; he sells them for about 100 pesos. Others go for export to Japan through a middleman, who visits the area every few weeks to collect the finished work. The women say that all they ask of life is to be left alone to improve their lives and the prospects for the next generation. Modest enough ambitions, but scarcely realisable when their children's role is to fulfil that of criminal or hoodlum, in order to justify the vast expenditure that goes on police, paramilitary groups, vigilantes, the armed forces and corrupt politicians.

Winefreda Geonzon[3] was born in the hills North of Cebu. During the Japanese occupation, when she was 15 months old, her father was out in the fields, and he met with a group of Japanese soldiers. As he turned to avoid them, they shot him. Winefreda's mother was left with seven children, the youngest only three months old. They had a small piece of land, but it was rocky ground on the side of a hill, impossible to cultivate since they had no money for seed. The children worked in other people's fields at harvest time, while their mother stitched for a garment maker. Winefreda would go to the seashore, catch fish for the family, and gather shells for sale to shellcraft workers. She worked on farms for payment in corn, worked in the corn-mills, separating grain from chaff. Later, she became a teacher, which continued for nine years, and then a stenographer, while she studied law. In response to the human rights abuses of the Marcoses, she felt she was called to work with the victimised and unrepresented. Her faith and courage protected her; and she continued to work until a few days before her death from liver cancer in July 1990.

In spite of the squalors of city life, few people wish to return to their home-place. One of the most degraded urban settings in the world must be the Dantesque landscape of Manila's 'Smoky Mountain', a vast garbage heap which has been colonised by around 5,000 families, in defiance of the sulphurous fumes and smoke still rising from its smouldering core. That those living here should express their satisfaction with it gives some idea of the violence and poverty which they have fled. To see the huts constructed on levelled terraces in the flank of Smoky Mountain in the drenching summer rain, when the slurry and mud effaces the margins of the road, and even the rusty jagged metal of the go-downs and huts seem to melt in rust-coloured liquid, is, at first sight, an appalling spectacle. A generation of children has already been born here, in a junk yard culture of throwaway goods. But the people cheerfully explain that they are waiting for the waste carts to arrive with the daily cargo of Manila's garbage. If they work through the night on the detritus, recycling metal, paper, plastic, wood, rags, they will earn up to 150 pesos, which is more than the minimum wage. In the city, an income is the first necessity. And in due course, when they have levelled the land, planted trees and shrubs around the houses, the land will become valuable, even desirable. The people have a vision of the future that is not accessible to those who can see only the present horrors.

NOTES

1 John F.C. Turner. Right Livelihood Award Winner 1988. John Turner has spent over 40 years working with the practice and developing the theory and tools for self-managed home and neighbourhood building in Peru, the United States and the United Kingdom. He worked in Peru from 1957 to 1965, mainly on the advocacy and design of community action and self-help programmes in villages and urban squatter settlements. He later lectured at the Joint Centre for Urban Studies of the Massachusetts Institute of Technology, and at the Development Planning Unit, University College London. Since 1983, he has devoted himself to his non-profit consultancy, which he

established in 1978. His publications have had a great deal of influence on housing policies worldwide. *Uncontrolled Urban Settlement: Problems and Policies* was published in 1966, *Freedom to Build, Dweller Control of the Housing Process* by Macmillan in 1972; *Housing by People, Towards Autonomy in Building Environments* by Marion Boyars in 1976. From 1983 to 1986, Turner was co-ordinator of Habitat International Coalition's Habitat (HIC) Project for the United Nations International Year of Shelter for the Homeless, 1987. The HIC Project carried out a global survey of recent and current local initiatives for home and neighbourhood improvement. From over 200 Third World cases identified, 20 were selected for in-depth documentation, and were published as *A Third World Case Book*, ed. Bertha Turner, (BCB, April 1988).

2 Ela Bhatt, Self-Employed Women's Association. Right Livelihood Award Winner 1984. Ela Bhatt is a former lawyer and social worker, who in 1968 was the chief of the women's section of the Textile Labour Association of Ahmedabad. In this position, she became directly aware of the conditions suffered by poor self-employed women, in the city, and elsewhere in South Asia. In order to address the poverty and lack of control over their working conditions, Ela Bhatt set up the Self-Employed Women's Association in 1972. Within three years, SEWA had 7,000 members, and was registered as a trade union. By 1988, there were 100,000 members in six Indian states. Through organisation and solidarity, SEWA members have new negotiating power with their employers. They established their own bank in 1975. This now has over 22,000 accounts, and has rescued thousands of women from money-lenders, pawnbrokers and middlemen on whom their labour was dependent. The average repayment rate on loans is 94 per cent (see Kalim Rose, *Where Women are Leaders: The Self-Employed Women's Association*, 1992, Sage Publications India; and Zed Books, London, 1992) Ela Bhatt has taken the struggle for justice and recognition on behalf of self-employed women into the national and international arenas. SEWA is campaigning for a convention on Home-based Workers' Recognition and Protection for the United Nations' International Labour Office.

3 Winefreda Geonzon, Free Legal Assistance Volunteers Association. Right Livelihood Award Winner 1984. Winefreda Geonzon, who died in 1990, was a Filipina lawyer, who in 1978 became the Legal Aid Director of the Integrated Bar of the Philippines in Cebu, which brought her into contact with the many injustices and abuses of the legal system which occurred during the martial law years of the Marcos regime. People, including young children, were jailed without charges or trial; imprisoned beyond their term; tortured and brutalised; or simply forgotten in prison. In response, Geonzon set up the Free Legal Assistance Volunteers Association (Free LAVA), as a free legal aid office for victims of violations of human rights, poor prisoners who could not afford to hire a lawyer, and people whose cases had implications for social justice. As its reputation grew, Free LAVA involved growing numbers of lawyers, students and community groups in its work. A Documentation and Research Group gathered legal evidence for abused or wrongly imprisoned people, a Legal Services Group undertook their representation, a Civic Assistance Team sought to administer to their basic needs in prison and provide for their rehabilitation. By 1987, 26 community groups were involved in Free LAVA's work.

'¡Viva Loisaida Libre!'

from Peter Mabourn Wilson and Bill Weinberg (eds), *Avant-Gardening* (1999)

Bill Weinberg

Editors' Introduction

This text comes from a series of small, inexpensively produced books on contemporary urban and cultural issues. Its subject-matter is the spread of squatter gardens in New York's lower east side ('loisaida') during the 1980s. The area concerned is known as Alphabet Village, from street names such as Avenue A and so forth, and is multi-ethnic. Through the period which saw SoHo – on which Sharon Zukin writes in her much-cited *Loft Living: Culture and Capital in Urban Change* (New Brunswick, Rutgers University, [1982] second edition 1989), and Neil Smith in *The New Urban Frontier: Gentrification and the Revanchist City* (London, Routledge, 1996) – become a fashionable and high-rent arts neighbourhood, other parts of Manhattan also saw property developers move into undervalued areas of real estate in order to clear and then re-invent them. One of these was the lower east side. Displacement has continued at an uneven pace, but its impact has been to peripheralise the poor. Sarah Ferguson writes in another text in the same book that the Chico Mendez Mural Garden was subject to an eviction notice in 1997 to make space for Del Este Village, a complex of apartments selling at $103,000 upwards. Shortly before but unannounced to the tenants, the plot was granted to the New York City Housing Partnership, which saw the garden as a vacant lot. Ferguson says: 'The oversight is sadly typical of the Giuliani Administration, for whom gardens are "interim sites", space savers for future development' (from 'The Death of Little Puerto Rico', in *Avant-Gardening*, p. 60).

Weinberg's aim is to arrest and reverse the culture of city governance and property speculation which discards the rights and creativity of tenants and the poor in favour of the interests of a new inner-city bourgeoisie. Against a monolithic and largely mono-cultural administration he sees a loose, multi-ethnic coalition seeking to identify common interests and causes. His text reads like a manifesto, and foresees a self-organised autonomy in Loisaida. If this is fanciful, the specifics are practical: job creation, pedestrianisation, and a moratorium on new commercial development. Rather as Charles Fourier (see Introduction to Part Nine) predicted, flocks of city-dwelling visitors to his model communities, the phalansteries, so Weinberg writes that tourists will flock to the Lower East Side Autonomous Zone. Interestingly, Weinberg sees Loisaida being in part at least able to grow its own food. This may be the most valuable proposition he makes. If effected it would change the urban footprint through reduced transportation, and reconstruct the modern relation of city as centre to countryside as the margin which feeds it. Weinberg's text has threads of anarchism running fairly visibly through its texture, and is far-reaching and at times utopian, but it is nonetheless a serious manifesto for a very different kind of city zone produced in local democracy and partial self-sufficiency.

What can be done to arrest and reverse the real estate industry's enclosure of Manhattan's Lower East Side? The demolition of gardens to make way for yuppie condos, the displacement of low-rent tenants, the eviction of squatter buildings, the squeezing out of mom-and-pop bodegas and private bookstores by corporate chains, the redesign and semi-privatization of East River Park, the erasing of New York City's most diverse and tolerant neighborhood's dreams, memory and identity – what, as someone once asked, is to be done?

THE LOWER EAST SIDE AUTONOMOUS ZONE

We have a vision. We have a vision that one day in the near future, the residents of the Lower East Side will start to meet and talk with each other in our tenements, on our blocks, in our gardens; Puerto Ricans and Dominicans and Central Americans, Poles and Ukrainians and Slovaks, Bengalis and Chinese and Korean, Blacks and Jews and Italians, punkers and hipsters and homeboys, artists and activists and squatters; and decide to find our common interest in reclaiming our neighborhood from the occupying forces of speculators, developers, landlords, organized crime, police and automobiles.

Community Board 3, a toothless tool of party patronage with neither accountability nor binding power, will be challenged by a Popular Community Board, which will emerge from the tenement and block meetings, drawing its legitimacy from the genuine loyalty of active and aware Lower East Side residents.

The Popular Community Board will declare itself a body with binding power, rather than a mere "advisory" rubber stamp for policies created by the City Hall/Real Estate nexus. Representatives on the board will be bound to vote according to the wishes of the residents they represent, following policies hashed out in the neighborhood meetings. All representatives will be recallable by popular vote at any time.

Effectively seceding from the increasingly ungovernable entity known as New York City, the Lower East Side Autonomous Zone (LESAZ) will be declared – popularly known as *Loisaida Libre*. At a minimum, its boundaries would stretch from 14th Street on the north to Pike Slip on the south, and from the East River to Bowery/Fourth Avenue.

Among the first declarations of the LESAZ will be the banning of absentee landlords. All buildings not owned by neighborhood residents will be expropriated without compensation and turned over to the tenants to be run cooperatively. NYC's own Buildings Department would, of course, be considered an "absentee landlord," and all city-owned buildings will be similarly expropriated and turned over to low-income or homeless neighborhood residents for immediate housing.

A moratorium will be declared on new businesses whose prices and aesthetics offend the traditional working-class character of the neighborhood. Funds for rehabilitating crumbling tenements will be made available from heavy taxation of those yuppie boutiques and eateries which choose to remain in the newly autonomous 'hood. Such establishments would include the Gap on St. Marks Place and any other corporate chains or franchises. As for the McDonalds on First Avenue, they are simply too noxious. They will have to leave.

There will be no taxes for tenants, employees or the unemployed! But the Lower East Side Autonomous Zone will impose such heavy taxes on Avenue A's yuppie eateries that NOBODY in our territory will know want!

Jobs would be created by the tenement rehab program. After lead, asbestos and other toxins are removed from the buildings, after floors are replaced and stairs repaired, then plumbing systems will be completely revamped. Graywater recycling systems will be installed, sending wastewater from the sink into gardens. The effort to convert our tenements into decent and ecologically sound housing will create gainful employment for neighborhood residents for years to come.

Automobiles will be barred from the LESAZ, with barricades erected across the intersections leading into Loisaida Libre. Our neighborhood will become an island of fresh air as our streets are liberated from the *auto*-cracy of toxinspewing death machines. The bicycle will become the predominant mode of transportation. Workshops such as The Hub on East Third Street (currently threatened with eviction) will organize a network of bicycle collectives throughout the LESAZ, providing bicycle repair, rental and even production for LESAZ residents at reasonable rates. LESAZ-designed human-powered taxis and even buses will provide transportation within the LESAZ for residents whose personal health does not permit them to cycle themselves. The cottage industry (or tenement industry) will produce creative human-powered vehicles for export to surrounding neighborhoods and municipalities.

Tourists will doubtless flock to the LESAZ to witness the new society under construction, or simply to escape the noise and pollution of the rest of the city, much as New Yorkers now head to Central Park for an afternoon of relative sanity. Guided tours for outsiders will both help spread the ideas of the alternative Lower East Side and provide an additional source of neighborhood revenue.

The Tompkins Square bandshell will be rebuilt.

As City Hall strikes back, responding to the Lower East Side's secession with the withdrawal of city services, neighborhood residents will exercise creativity to invent their own alternative institutions. Many who remember the Tompkins Square riots will no doubt be happy to see "New York's Finest" depart. The police will be replaced with neighborhood watch groups and rotating block patrols of local residents. Guns and uniforms will become unnecessary as responsibility for keeping peace in the neighborhood is decentralized and democratized. Popular self-defense classes organized by block committees, and a restoration of the neighborhood's sense of community, will evaporate the climate of fear and alienation in which violent crime thrives. The young, strong and healthy will take responsibility for protecting the backs of the elderly, infirm and disabled. Burglars, thieves, muggers, rapists and coke and heroin dealers will have to face a citizenry which directly responds to them, rather than a citizenry which cowers behind locked doors waiting for the police to arrive. Apprehended miscreants will be photographed before being escorted to the neighborhood's borders – or thrown into the East River.

As the city withdraws sanitation services, a program of total recycling will be instated. The Lower East Side Ecology Center on 7th Street, which has already turned a waste disposal site into an extremely pleasant garden, will become a model to be emulated throughout the LESAZ. Vegetable waste will become compost for community gardens. Products with excess packaging will be banned from importation into the LESAZ, and plastics will be generally discouraged. Thriving tenement industry will turn all manner of metal, wood, glass and cardboard "garbage" into useful products of every variety, from vehicles to furniture to tablewares to art. When markets within the LESAZ have been saturated, these products can be exported. Eventually, none of the neighborhood's waste will be going to the NYC Sanitation Department's bloated Staten Island landfill. The LESAZ may even be paid to import metal "waste" from the rest of the city, and then transform it into useful products to be resold back to the city. However, entry into the LESAZ of any hazardous substances will be scrupulously banned.

To facilitate this thriving trade, docks will be rebuilt and revitalized on the East River waterfront, and a fleet of merchant barges will ply the city's waterways. Inter-neighborhood and inter-municipality trade will restore a localized, labor-intensive, ecologically-sound and small-scale industrial economy to the metropolitan area, creating a real alternative to the Capital-intensive and unviable model of real estate and high finance now in place. The waterfront revitalization project, which will be vital to this effort, could be undertaken jointly with the Williamsburg Autonomous Zone directly across the river.

Partial or eventually complete autonomy from the Con Edison grid could be achieved through the construction of turbines which generate power by exploiting tides in the East River.

Loisaida Libre could become at least partially self-sufficient in food as the neighborhood's vacant lots are turned over to community gardens, the soil nourished with rich compost created from organic "garbage" (coffeegrounds, eggshells, fruit and vegetable scraps, etc.), which adjoining neighborhoods will pay us to accept. Rooftop and raised-bed gardening will prevent contamination of crops with toxins left over from destroyed tenements.

In addition to growing fruits, vegetables and corn, Loisaida gardens will also grow herbs for use in the neighborhood's network of Free Clinics, which will also be funded by heavy taxation on the outside economic interests that continue to operate within the LESAZ. Taking the lead from the Lower East Side Needle Exchange and the alternative treatment advocates within ACT UP, the Free Clinics will emphasize prevention, self-healthcare and informational programs on hygiene and nutrition, minimizing reliance on antibiotics and pharmaceuticals.

However, much of our food will be imported from local organic growers in Hudson Valley and New Jersey farm communities, building on the system already used by the Union Square Green Market and the Good Food Coop on Fourth Street. Eventually these farm communities could become markets for Loisaida tenement industry products in exchange for organic produce. Revitalization of the Lower East Side waterfront will also help stimulate this trade, thereby creating a symbiotic relationship with the expansion of organic agriculture in nearby rural areas. As our country cousins work to restore local watersheds, native species and self-sufficient farm communities, the alternative economy will start to spread throughout the bioregion defined by the Hudson River Valley.

A network of inter-dependent Autonomous Zones can begin to form an economic and political counter-structure to the existing grid of suburbanizing municipalities which surround the city. This Hudson Bioregional Confederation of Autonomous Zones, usually refered to simply as the Hudson Bioregion, will act as a solidarity network. For instance, if New York City authorities threaten to withhold water from Loisaida Libre or other autonomous zones serviced by the aqueduct system, this move could be effectively barred by our bioregional allies in the Catskill Autonomous Zones, who will be working to maintain the city's mountain reservoirs by preserving the watersheds that feed them through reforestation efforts and monitoring runoff from farms, cesspools and lawns. As the resource base becomes increasingly threatened by ecological mismanagement, those with the skills, vision and wherewithal to implement the sustainable alternatives will be in a position of power to negotiate with – and perhaps eventually supersede in importance – the established authorities of municipal bureaucracy and private power.

In generations to come, entities such as New York City will become irrelevant as Autonomous Zone networks become the new arbiters of political power. Similarly, New York State and New Jersey will be superseded by the Hudson, Mohawk, Delaware, Finger Lakes, St. Lawrence and Seneca Bioregions, roughly conforming to watersheds and sharing the mountain divides of the Ramapo, Catskills and Adirondacks. The United States of America will cease to exist, replaced by a decentralized North American Bioregional Federation. The Shasta Bioregion of Northern California and Cascadia Bioregion of the Pacific Northwest have already been mapped and declared. Those interested in identifying and preserving native flora and fauna and reclaiming a sense of place in the more developed and densely-inhabited East Coast are spreading the new idea of social organization which will eventually replace both the nation-state and the multinational corporation.

As the inevitable breakdown of nations, already well underway in Eurasia, begins to spread to the West, it will be up to us to find the models and unifying principles which can create *positive* alternatives to the old structures and avoid nightmares such as those now playing themselves out in the Balkans, the Caucasus and Central Africa. We can already see the potential for local nightmares in such situations as the Black-Hasidic rivalry in Brooklyn's Crown Heights, where desires for local autonomy have become linked to notions of racial purity and chauvinism on both sides of the conflict. The long and proud progressive tradition of the Lower East Side, however, points to a better possibility: that of linking desires for local autonomy to notions of cultural diversity, social justice and ecological sustainability.

As New York continues the process of political fragmentation which we can see already, Loisaida Libre is likely to become a refuge for progressive exiles from areas of the city where local reactionary forces have taken the reins – such as South Queens and parts of Staten Island. Loisaida Libre will be a haven for free thought and cultural experimentation.

AGAINST THE NEW MUNICIPAL ORDER

The Lower East Side Autonomous Zone's legality and legitimacy will be based on the unassailable premise that the city government's redevelopment of the neighborhood is illegal and in violation of the City Charter. Since the courts have provided no redress for these grievances, the Lower East Side has no alternative but to seize control from below through mechanisms of direct democracy.

In southern Mexico, the Maya Indian support base of the rebel Zapatista movement have declared their outlying hamlets and farming communities to be Autonomous Zones and "New Municipalities," which govern from below in defiance of the fraudulently-elected, brutal and corrupt official governments of the municipal seats. Similarly, Loisaida Libre will become a "free municipality" in opposition to the illegitimate and illegal regime in Gracie Mansion. The Loisaida Autonomous Zone would stand in opposition to the reigning New Municipal Order in which the notion of genuine citizenship has been betrayed as surely as in the New World Order which it mirrors.

The clearest illegality has been the closure and "renovation" of Tompkins Square Park. The year-and-a-half closure of the park from 1990, and the demolition of the historic bandshell which had been a showcase for neighborhood talent for over a generation, were unilaterally decreed by Mayor David Dinkins, in violation of the City Charter, which calls for City Council and Community Board approval in such decisions. A legal challenge to the plan on this basis failed to halt demolition.

Plans now underway for East River Park are apparently even worse, with the abandoned amphitheater there slated to become a privately-owned yuppie coffee shop – which most of the residents who now use the park would be unable to afford to patronize. If efforts to halt this irresponsible plan through legal mechanisms fail, the case for a Lower East Side Autonomous Zone would become stronger.

This issue cuts to the heart of whether we contemporary urbanites live in a democratic culture. Parks are supposed to be neighborhood meeting places, public communities and, when necessary, forums for free speech and protest available to anyone, regardless of income or social status. Private administration and heavy policing is antithetical to this vision. Tompkins Square may be cleaner and prettier now than before the renovation, but nothing can ever compensate for the destruction of the bandshell. The Tent City shantytown which established itself in the old Tompkins Square may have seemed unsightly, but the whole point was to say *no*, we *cannot* continue to ignore and shunt aside the homeless.

In Tompkins Square, for the last century and a half, renovations have been scheduled to defuse working class movements and unrest in the neighborhood – after the Civil War anti-draft riots, after bread riots in the 1870s depression, and most recently following the August 6, 1988 riot and subsequent Tent City and related protests. In the evolution of the city's parks we can follow the decline of urban citizenship: from the utopian populist vision of Frederick Law Olmsted to the standardized machine politics of Robert Moses to the yuppie New Enclosure of semi-private groups like the Grand Central Partnership, a power nexus of midtown business (which briefly hired its own goon squad to roust and assault the homeless).

Such groups boast of parks reclaimed from drug dealers and anti-social scum. This demonization of the petty criminals at the bottom of the social hierarchy masks, as always, a slide into a police state – increasingly, this time, a private-sector police state where those disenfranchised from the New Municipal Order of Real Estate/Finance/Big Media are squeezed out of one of the few physical and social spaces still open to them. Bryant Park behind the New York Public Library on 42nd Street, policed and maintained by the Grand Central Partnership, is a study in post-Orwellian private-sector totalitarianism. Appropriation of public space by Business Improvement Districts and private foundations is the face of the New Enclosure.

Even the city's security forces are not immune from the New Enclosure. The privatization of the New York Police Department is already underway, with a new precinct slated for the Wall Street area funded entirely by local BIDs, and officers offered for hire as rent-a-cops at the high-class functions of the city's elite.

This agenda has, of course, dramatically advanced with the election of Mayor Rudolph Giuliani. Mayor Giuliani's agenda for the city is spelled out in the position papers of the Manhattan Institute for Policy Research, a right-wing think tank on municipal policy. The 1992 Giuliani campaign's top economic advisor Stephen Kagann is a top contributor to the Manhattan Institute's "prestigious" *City Journal*, which was apparently a "hot book" at Giuliani for Mayor HQ in 1992. Articles in the *City Journal* proposed that New York sell off its public hospitals and transportation lines, and dramatically cut spending on social services. True to form, Giuliani has turned vast areas of the city over to Walt Disney and even proposed privatizing the Brooklyn Bridge.

The Manhattan Institute also published *The Dream & The Nightmare*, a book by *Fortune* magazine writer Myron Magnet which argues that "pathological poverty came into being not in spite of, but because of, the cultural shift that began in the '60s." In one gem of right-wing victim-bashing rhetoric, Magnet writes that "cockeyed ideas of economic victimization got mixed up with an appropriate horror at racism to produce the belief that the state had to compensate the poor, especially poor blacks, for their plight. That compensation turned out to be welfare, which has become a machine for perpetuating the underclass by . . . undermining values and supporting the most blighted family structures."

Not surprisingly, bigshot investment banker Felix Rohatyn stepped down as chairman of the city's Municipal Assistance Corporation just as Giuliani was coming in. Rohatyn's MAC had long had *de facto* veto power over the city budget, reining in spending on public services in the name of "fiscal responsibility." Although they weren't called such, these MAC-enforced cuts were essentially "austerity measures" of the same type that the World Bank and IMF impose on India and Brazil. Rohatyn's departure from the MAC, which he had headed since its founding in the fiscal crisis of 1975, indicates that the city's financial establishment – the so-called "permanent government" – finally had a man they can completely trust in City Hall.

The only role for low-income neighborhoods in the New Municipal Order is as a toxic dump site. Giuliani (like the outgoing Dinkins, after he flip-flopped on the issue), supported construction of the Brooklyn Navy Yard trash incinerator, which would spew deadly poisons into the air of the Latino and Jewish immigrant neighborhood of Williamsburg (and was being underwritten by Lazard-Freres, Felix Rohatyn's investment firm). Due to aggressive neighborhood organizing efforts and environmentalist opposition, it now appears that incinerator will not be built. Instead, however, Williamsburg appears slated for the equal and opposite nightmare of rampant gentrification, as the yuppie wave from the "East Village" rolls across the river to Brooklyn, just as it had rolled in from the West Village a generation earlier.

In the New Municipal Order, the city only owes police protection to the affluent – not services to the poor. Only those who cannot afford private hospitals would go to the nightmarish HHC hospitals. Giuliani's agenda is to follow the national trend towards privatization of city services – further squeezing out access for citizens not among the yuppie elite. Giuliani's envisioned New York is strictly a "global" city – a city for high finance, high art, big media, international diplomacy and trade. It is not a city of neighborhoods and working people. The labor is cheaper in Mexico and Guatemala, and with the trade barriers going down there are no obstacles to exploiting that labor. Cities like New York and LA – where Mayor Richard Riordan was elected on a similar agenda – are now merely centers for elite global management and electronic paperwork. New York's working class has outlived its usefulness. It only remains to be driven out of town – or, when that is not an option, behind bars – by a wave of privatization and repression which will make life untenable for all but the monied management class.

Robert Fitch provides an excellent analysis of the cycles of Gotham politics in his book *The Assassination of New York*. The corrupt, entrenched Democratic machine generally holds sway, until a fed-up populace votes in a Liberal-Republican "fusion" candidate once every few decades – last time around, as the New Deal Republican strongman Fiorella LaGuardia; this time around in the law-and-order get-tough-on-(smalltime)-crime guise of the incumbent schmuck. Giuliani was elected on an openly reactionary platform, playing to white racists, while the Machine pols rely, to greater or lesser degrees, on the "minority vote" (remember, whites are the *real* minority in this town) that they control through their system of patronage. Eventually the Machine recoups its losses in the voting booths and the cycle begins all over again.

The Giuliani regime represents a distinct 1990s version of the Fusion phenomenon. Beginning in the 1950s and climaxing in the 1970s, New York was subject to "white flight" as the middle class fled to the suburbs. Then, in the 1950s, suburban offspring of the generation which had fled started returning to the city to reap the gains of the booming stock market – the yuppie class. Giuliani is the candidate of this new generation. He grew up in Long Island's Nassau County, and has returned to govern a city which he manifestly sees as threatening and alien. He represents the suburbs' revenge on the city; the paradoxical suburban colonization of the decayed urban center. This is especially obvious in the turning over of 42nd Street and other vast pieces of Manhattan Island to the Walt Disney company.

The naked racism which lay at the root of the "white flight" syndrome is no less evident in the suburban recolonization regime. Without hyperbole or exaggeration, Giuliani can be characterized as a proto-fascist. The holster-sniffing cop-glorification and contempt for democracy is there. The drunken and openly racist police orgy at City Hall during the campaign against Dinkins recalled Hitler's 1923 Beerhall Putsch, while echoes of Hitler's later designs can be heard in the forcible internment of thousands of Haitian refugees in a *de facto* concentration camp at Camp Krome, Fla, that Giuliani headed up when he was number-two man in the Reagan Justice Department. When the refugees sued, Giuliani made a little junket to Port-au-Prince where he got schmoozed up by Baby Doc Duvalier, and came back giving his torture regime an entirely predictable clean bill of health. All those human rights violations – nuthin' to 'em! Them boat people just want a free ride to the US of A.

Giuliani now rules New York City by decree, his executive mandating the destruction of tenements with no notice to the residents, privatizing community gardens to real estate developers, opening the parks to billboard advertisers – all without regard for legal requisites or public oversight.

In short, whatever pathetic facade of neighborhood

democracy existed under the Democrat machine is being rapidly dismantled. The era of looking to bureaucrats for solutions is definitively over. Citizens will only be able to make their power felt from below. The most dramatic exponent of this kind of resistance is witnessed in the militant human rights movement sparked by the police slaying of East African immigrant Amadou Diallo, with nearly 2,000 arrested in March 1999. Ultimately, how much of the New Municipal Order agenda Giuliani will be able to get away with will be decided in the city's streets, schools, tenements and workplaces.

FOR THE INDEPENDENT CITY-STATE OF NEW YORK!

Another basis for declaration of the Lower East Side Autonomous Zone is New York City's *failure* to consider secession from New York State! The June 1997 battle in Albany over the rent control laws provided ample evidence that such a move is long overdue. With upstate and suburban legislators demanding the rent laws be sunset, New York City leaders had a responsibility to at least threaten secession. Not a one rose to the occasion.

In the saturation of feel-good free-market landlord propaganda it is forgotten that the rent laws were passed for a *reason*. It is absolutely obscene that decontrol is even an issue, given that homelessness has been at crisis proportions in this city for at least fifteen years. Remember back in the '80s, when "the homeless" was a fashionable liberal issue. Now we're all supposed to be suffering from "compassion fatigue," and the legions of street-dwelling refugees from rent-gauging and "planned shrinkage" have returned to their conveniently invisible status. But we tenants are going to have some heavy karma to pay for this insensitivity and betrayal of human solidarity.

Look at how much ground we've lost. "If you can't afford to live in New York City – move!" is the message, succinctly summed up by our former Mayor Ed Koch. We say: Fuck that! This is *our* city! Let Disney and Starbucks decamp for California and Seattle! Like those damn Hollywood crews that are setting up shop all over the sidewalks of the East Village, treating the surly locals with undisguised contempt. What an insult!

There were, and remain – for the state rent laws will come up for renewal again in 2003 and the city laws in 2000 – two surefire tactics to oppose any erosion of tenant protections:

1 Credible threat of a city-wide universal rent strike if the tenant protection laws are at all eroded in Albany. Start organizing your tenant strike committee now!
2 A movement for the Greater Metropolitan Area to secede from Upstate. If New York State is going to gut New York City's rent regulations, New York City should secede from New York State.

We city-dwellers are not the constituents to whom conservative upstate legislators like State Senate Majority Leader Joe Bruno (who led the battle against rent control) must answer in the voting booth. Therefore our opinions mean nothing to them. However, the landlords who oppress us are free to lobby Joe Bruno and his ilk – and make donations to their campaigns (with *our* rent money, no less!). This is inherently unjust.

It is criminal that they got away with rent deregulation in Boston, where in 1994 Massachusetts voters decided in a referendum to do away with all rent protections statewide. Since then, Bostonians have been paying the price. The Boston Herald reported on Dec. 26 1996, "Boston Mayor Thomas M. Menino yesterday blasted 'greedy' landlords who have begin issuing 'astronomical' rent increases to elderly tenants about to lose their rent control protections. 'We're hearing about $1,000 [a month] rent increases for elderly couples. . . .' said Menino"

Now, why should someone in Amherst or Great Barrington get to vote on the lives of Bostonians, we'd like to know? Similarly, why should Joe Bruno get to vote on the lives of all us Jews, Negroes, faggots and welfare-cheats down here in the Big Apple?

A new housing study, commissioned by the Rent Stabilization Association (RSA), an Orwellianly-named landlord group which seeks to do AWAY with rent stabilization, finds that if rent regulations were abolished, rents in stabilized apartment buildings would rise by up to 51% in some neighborhoods.

The assumptions and methodology of the study are being questioned by tenant activists because the purpose was ostensibly to *allay* fears of even greater rent increases. So the rent increases listed by the study are probably entirely too optimistic from the tenant's point of view (or pessimistic from the landlord's point of view). The rent increases by neighborhood would be:

- 51% on the Upper West Side
- 30% on the Upper East Side

- 30% in Greenwich Village
- 21% in Chelsea
- 16% in Stuyvesant Town
- 10% in the Lower East Side/Chinatown
- 5% in Harlem, Morningside Heights, and Washington Heights/Inwood

Outside of Manhattan, the report predicts that rents would rise 8%, with increases ranging from 0% in Staten Island to 19% in Queens neighborhoods. Even these increases would mean the difference between precarious month-to-month housing and eviction for countless families. Households which are now one paycheck away from eviction would be pushed over the edge.

If we can't afford to live in New York, move? Where are we supposed to go? All those upstaters who support rent deregulation because they think we are all whining minorities looking for a hand-out certainly wouldn't want us moving into *their* backyards! A few letters making this point from NYC residents with obviously "ethnic" last names to the newspapers of Albany, Kingston and Rochester might be appropriate. Don't forget Rensellaer, Bruno's home turf.

Joe Bruno showed his true colors in the 1994 budget battle in Albany, when he attacked those he sees as truly responsible for breaking the state's budget: "It is the lowest-income people. It's the blacks the Hispanics, and I only say that because look at the numbers . . . 90 percent of those people support him [meaning Assembly Speaker Sheldon Silver of the Lower East Side] . . . Why? Because they are people that got their hands out. They are the ones fighting for welfare."

The damned impertinence and arrogance (not to mention undisguised racism, anti-Semitism and homophobia) of these upstate politicians! They demonize us to their constituents as a bunch of whiny parasites! Meanwhile, who is raising the lion's share of the tax revenues for Albany's budget? It sure as hell ain't Buffalo! What interest do we have in sticking with New York State any longer? Let's see how long they can get along without us.

The rent struggle could provide the spark for the secession movement. Let the City Council pass new, *more* restrictive legislation if the state rent regulation laws are ever allowed to sunset. Let Albany try to fight it in the courts if they so choose. If Albany wins in the courts, we secede!

From the USA, if need be.

This is the only realistic position. However, since our elected leaders are too bound by the confines of mundane conventionality and fealty to the Permanent Government to utter even a peep of such notions, we have no alternative other than go beyond the media-hyped spectacle of American "representative" democracy, to a kind of democracy which is grassroots and community-based and radically participatory, and secede on a neighborhood-by-neighborhood basis – with the Lower East Side in the vanguard. Federative relations with progressive upstate communities can then be established on an equalitarian basis.

Secessionism and local autonomy need not remain the domain of the redneck right and cowboy militia types in Texas and Montana. Our Republican and radical-right enemies are giving anarchy a bad name; we have to reclaim our turf. Traditional leftist notions of class struggle without any sensitivity to questions of local autonomy have brought such nightmares as Stalinism. Contemporary Republican and radical-right bluster about local political autonomy without any opposition to the vast class disparities and centralizations of economic power will lead to the equal and opposite nightmare of direct Corporate Rule. The so-called "decentralization" of those who oppose welfare and rent control laws is just a pseudo-populist sugar coating on the final dismantling of the last remnants of the New Deal order, a weaker federal state in the interests of greater centralized power for global capital. In other words: Meet the new boss.

Don't be fooled again! On towards the Manhattan People's Soviet!

'Architecture Co-generated with Poetry, because the Word is Inaugural, it Conveys, it Gives Birth'

from *The Road That is Not a Road and the Open City at Ritoque, Chile* (1996)

Ann M. Pendleton-Jullian

Editors' Introduction

The Open City Amereida at Ritoque, near Valparaiso, Chile, was conceived and constructed by the Faculty of Architecture of the Catholic University of Valparaiso during the 1970s and 1980s. It occupies a strip of dune-land next to the ocean, divided by a highway and a small freight railway. Because the structures are built on sand, they must be accommodating of its shifting nature, and are constructed using light, mainly recyclable materials without the aid of heavy plant (which could not traverse the dunes). The land was purchased by members of the faculty, and the idea of a city re-visioned there through a collaboration between designers, engineers, and poets – among them Surrealist poet Godofredo Iommi and architect Alberto Cruz (appointed to head the Architecture School in 1952). It is a collection of buildings without a plan, each site being founded by a poetic act such as the co-production of a poem. Decisions are made collectively, though the founders would seem to have a certain presence in them. Architect Ann M. Pendleton-Jullian, who teaches at MIT, describes the Open City's structures as made of timber and iron, roofed with metal, plywood and plastic – not unlike those of informal settlements. She notes that some are occupied by young teachers or groups of students, and that maintenance is constant due to the fragility and lightness of the structures.

The extract reprinted here is headed in Spanish '*Arquitectura co-generada con la poesia, porque la palabra es inaugural, lleva, da a luz*' (translation in heading above) and deals with the poetic preoccupation which informed the city's generation. A parallel is drawn between changes to language in modern poetry and architecture's need for a new means of expression and relation to site. The original publication includes photographs of several structures, including the *hospederias* which act as residential sites but with a function of hospitality as well as domestic house. Pendleton-Jullian writes 'the hospederias are open to all those who may come, receiving food and/or lodging in exchange for a sharing of experiences and ideas' (p. 29). This is the city's openness. But what of its relation to the big city nearby? Without the University it would not have existed, and members of the Open City use the nearby city for schooling, shopping, and so forth, in the usual way. The Open City seems to be a kind of urban experimental laboratory, provoking new architectural thinking and forms through its immersion in modern poetry while co-existing with the dominant society. Pendleton-Jullian writes later: 'It specifically employs modern theoretical and conceptual tools: analysis and critique filtered by phenomenology . . . These tools and methodologies belong implicitly and explicitly to an ideology that is not only the *foundation* for the production of physical space but also the *way* to producing physical space' (p. 131). This might be, almost, another way to say praxis.

The central poetic preoccupation of the group, which was considered its "unconventional" approach, was the relationship of poetry to architecture, sculpture, and painting. It was to be a direct relationship, without mediating elements, and one which stated, and required, the bond of action to word. It was not poetry as a bias or sentiment but rather poetry as a way of acting and of doing creatively. The poetry around which the group united was that of the modern French poets – the *poètes maudits* – and of the surrealists: Baudelaire, Mallarmé, Rimbaud, Verlaine, Lautremont, Breton. This poetry involved a passionate quest in which poetry, no longer a commodity, transformed itself into poetic activity that aimed at recuperating the mind's original powers. The poet was an alchemist who employed the imagination to transform reality both mentally and physically, who embraced the mystery and adventure of creative activity as reality was opened up to a different reading, a different understanding, a different reality, ignited by the power of words to embrace multiplicity and plurality within the unifying body of poetic language.

The appearance of the poètes maudits can be attributed to a confluence of two key conditions: first, the deterioration of the social status of poetry and poets as wealthy patrons of the nineteenth century disappeared in the wake of the growth of the European bourgeoisie, who were neither entertained nor enlightened by poetry. Poetry lost in value as a commodity, as a means of employment. Concurrently, there was the birth of modernity and a heritage of a vision of the modern world formed by the German romantics Hölderlin and Novalis in which divine systems were no longer sufficient or relevant to existential meaning within modern reality. Together, these provided the ground for the liberation of the poetic word from privileged claims and allowed it, in fact required it, to return to the source of language: to man. Man begins the return to himself for meaning and the responsibility that that implies. The "antagonism between the modern spirit and poetry begins as an agreement. With the same decision of philosophical thought, poetry tries to ground the poetic word on man himself. The poet does not see in his images the revelation of a secret power. . . . Poetic writing is the revelation of himself that man makes to himself."[1]

For this revelation to form, it was imperative that the traditional subject matter supplied by previous poetries be dismissed along with the emotionalism attached to such subject matter. The effusive description of elements of the natural world and the tedious articulation of feelings or thoughts supplied by these poetries were no longer relevant as they obscured the perception of reality. In the modern French poetry, reality and "real things" are not described but are consciously put aside so that transparent contact with the profundity of reality may be discovered. It is a self-conscious program that begins from an intuitive suspicion that our relationship with reality is attached not only to surface perceptions but, even more important, to an enigmatic mental dimension that underlies physical reality in which all things are linked by "correspondences" between them. Not through similarities but through Charles Baudelaire's idea of the almost mystical familiarity and intimacy between things that are never the same, that can never be the same. They are connected while retaining a distance that can never be collapsed. Access to this unknown region is achieved by the imagination of the poet through its receptivity to the nonconcrete, through its use of language as a "forest of symbols" where mysterious and interconnected signs transport the mind and every sense.

Described as being "the original faculty of all human perception" by the English poet Coleridge, the imagination is capable of operating through reason, or of transcending reason, to perceive, select, judge, edit, transform. With the modern French poets, the making of poetic images combined with a necessary altering of language to discover as well as convey that which transcended reason. Words were not used to describe but to create images attached to the rhythm of language. For this reason, every new poem was to be a total recasting of its author's means of expression, disdaining all preestablished conventions and any innovations made prior to it in ways of thinking or ways of saying. Poetry was about discovery. It became adventurous. The fantastic and marvelous were stalked as the new poetry argued the reattachment of life to art: life consisting of the day to day, minute to minute, the mundane; and art being conceived through the word, specifically, because it is the element through which one engages both the world one acts within and the activity of the mind. It is a tool of perception and a tool of contemplation. Therefore, through the word, the poet is empowered to unite the processes of interpretation and transformation. "So we manage to have a synthetic attitude combining the need to transform the world radically and to interpret it as completely as possible."[2]

To achieve this interpretation and transformation, it was necessary to alter language radically. Its very matter was operated on as words were forced to "lay bare their hidden life and reveal the mysterious trade they indulge in, independent of their meaning."[3] It was asked to return to its origin, to discard the conventional relationship language had developed with meaning through arbitrary mental selection processes based on prescribed and learned value judgments. The new language was highly transparent and fluid, capable of joining sensation and thought by moving between the two with such facility as to totally obliterate the boundary between interior and exterior. It was to be, as Arthur Rimbaud described: "of the soul, for the soul, summarizing all, scents, sounds, colors, of thought hooking onto thought and pulling." The poem itself takes on an autonomous, ritualized quality as product of the imagination, and it becomes keyed to the participation of the reader as an experience in itself. Whereas poetry, the product, had previously been the ambition when it was marketable, it now becomes the by-product of poetic activity.

Language engaged in poetic activity created poems in which the poet engaged in a mental communication with the world exterior to himself as well as an interior world. The poet becomes the medium between exterior and interior, and the poem becomes the site for the communication. The site that the poet creates contains both the object and the subject, the world and the poet, within its borders. By introducing the subject into the site of the poem, in a role more significant than that of narrator, modern poetry proclaims its commitment to reengaging us in a relationship with our world unmediated by divine systems or the false promise of reason's omnipotence. Subjectivity over reason's objectivity was part of the need to understand, or recreate, this relationship for the modern poet and, by extension, for modern society. Poetry was not personally motivated but aspired to reach into the *other* dimension of all human existence where we are positioned at the center of a "forest of symbols" that we initiated when we created language. Modern poetry embraced both the despotic ego and the id common to all men. It worked off of the assumption that the poet, as an individual in a certain cultural and historical context, is representative of the species of mankind in the same context and that the mental operation that consisted of going from being to essence on an individual level could discover value on a more general level. But, it is the poet who has the capacity to interpret and extend this value to restore to the species its integrity. "It is from poets, in spite of everything over the centuries, that it is possible to receive and permitted to expect the impulses that may succeed in restoring man to the heart of the universe, extracting him for a second from his debilitating adventure and reminding him that he is, for every pain and every joy exterior to himself, an indefinitely perfectible place of resolution and resonance.[4]

The absolute conviction in the power of poetry and poetic activity to interpret and transform life generated a consequent attitude toward poetry making, both as a craft and as a *sublime* activity. Mallarmé claimed that "Poetry is the expression, through human language restored to its essential rhythm, of the mysterious meaning of existence: it thus grants authenticity to our time on earth and constitutes the unique spiritual task." Included in the spiritual task were the poets and others who had the capacity to engage the poem with the imagination of the poet. No longer a commodity for wealthy patrons, religions or secular, and elevated out of the popular base it had lodged in during romanticism as well, poetry became an autonomous and even conflictive endeavor. It did not serve to entertain or enlighten but instead had the capacity to change life where other methods of revolution had failed. It demanded that act be consequent with word. And, moreover, that poetry initiate action: "Poetry will no longer set its rhythm with the action, it *will be there in advance* [of the action]."[5]

For the modern French poets, the world's real torment lay in the human condition, much more so than in the social condition. Therefore, the action that was demanded was to be radical and extreme, but it was to be primarily transcendental in nature.[6] Poetry as revolution "shows a double face: it is the most revolutionary of revolutions and, simultaneously, the most conservative of revelations, because it consists only in reestablishing the original word."[7] From the German romantics Hölderlin and Novalis and onward, for all the European poets, the greatest promise of modernity was the spiritual liberation of humanity. Freed from the prescriptive powers of lords and kings, religion and its gods, the individual was now in possession of an authentic and unmediated relationship with his or her own life and death – this was the promise. The difficulty then becomes the responsibility of that freedom and a reordering of existential meaning in the modern era in which we have become distanced from ourselves and from others. Through the reestablishment of the original word and the reordering of language and its relationship to life, in a revolutionary manner, it was believed that the distance could be removed by revealing the alikeness and correspondences between things in a world no longer fractured or organized by systems of power.

Reestablishing the original word required a rediscovery of language, which meant that the process of making poetry, along with its reception, became more intuitive. A more subliminal use of language was initiated, and poetic form found a more consequent relationship with its subject matter. Rhetoric was relaxed if not completely abandoned as the poem engaged in potential as a vehicle for the spontaneous and discontinuous activity of the unconscious mind. Formally, one finds strong imagery elaborated by expressive sound patterns and nonrhyming rhythms that generate a feeling of spatial three-dimensionality. Free verse was initiated to strengthen the verbal transparencies and simultaneities as a complete shift from conscious activity to the discovery initiated by unconscious activity occurred.

Incited by the work of Freud, it became an assumption of the surrealist movement especially that the enigmatic dimension underlying reality was received through the unconscious mind. More than just receiving or reading this dimension, however, the motivation of the surrealist poets, as André Breton described it, was to find and fix the "point of the mind at which life and death, the real and the imagined, past and future, the communicable and the incommunicable, high and low, cease to be perceived as contradictions"[8] and to establish a conducting wire, *un fil conducteur*, between these two worlds too long disassociated. To achieve this, the constraints that weighed on supervised thought – the mind's subjection to immediate sensory perceptions that falsify the course of ideation, and the critical spirit imposed on language – had to be removed along with all obstacles created by mental logic, morality, and taste. Breton conceived of the subverting of the conscious mind and its closed rationalism, which is dominated by solicitations from the external world as well as by individual preoccupations and sentimentality, to release the imagination by making a direct link between poem and unconscious mind through psychic dictation. In the "First Surrealist Manifesto" he defines surrealism as psychic automatism: "psychic automatism in its pure state, by which one proposes to express – verbally, by means of the written word, or in any other manner – the actual functioning of thought. Dictated by thought, in the absence of any control exercised by reason, exempt from any aesthetic or moral concern."[9] Psychic dictation transcends the logic and reasoning processes of the conscious mind by exorcising intellectual activity and literary convention from the process. It employs language not merely as a means but allows it to exist as its own entity, relying on word play, verbal association, free tonal and audio association, original poetic analogies, chance associations, and so on. It exalts language as the material of the *fil conducteur*.

The surrealist program went beyond the poetic activity of the modern European poets in several aspects. First, poetry and the poem itself became even less important as the poetic experience broke the boundaries of all previous conventions related to this activity. For the surrealists, it was truly the experience and not the poem that took on value. Attached to this experience was the ambition to completely transform life in all its facets into poetry. Poetic activity engulfed all aspects and moments of life. It was an active transformation as well as an active revelation. Additionally, both the modern French movement and surrealism relied on subjectivity, but it was the surrealist group that inserted the subject so forcefully into the object that any vestige of objective reality was fractured beyond recognition. What is more, the imagination, which for the modern poets was a privileged tool belonging to the poet, became for the surrealists something that was "owned" and employed collectively. Therefore, ultimately, the subject disappeared as well. The concept of the poet as the prestigious interpreter and medium suffered terribly at the hands of the surrealists, who believed that all men, being equal, have poetic capacity. Poetic capacity is the active foundation for inspiration and "the socialization of inspiration leads to the disappearance of poetic works, dissolved into life. Surrealism does not propose the creation of poems as much as the transformation of men into living poems."[10] The surrealist poets approached society and in historical context in a direct and open way, challenging both by their activity and their inactivity. There was no escape from life through the poem. Poetry was life and, more specifically, an attitude toward life in which inspiration was politic.

The surrealists, specifically the surrealist poets, used many different methods to spark the transmission along the "conducting wire" connecting the unknown regions of thought to the conscious mind. All of them relied on the self-conscious creation of an external context from which a spontaneous response or set of responses could erupt disassociated from conventional causality. Performances or "acts" were patterned after cabaret shows and entitled with sensational descriptions to entice a large audience participation. Dreamlike trance states were induced but were later abandoned when they became too dangerous physically. Excursions or wanderings were planned with gratuitous or nonexistent goals:

> *We all agreed at the time that great adventure was within our reach. "Leave everything. . . . Take to the highways": that was the theme of my exhortations in those days. . . . But what highways could we take? Physical highways? Not likely. Spiritual ones? Hard to imagine. Nonetheless, it occurred to us that we might combine these two types of roads. Out of this came a four-man stroll. . . . We started out from Blois, a town that we had picked at random on the map. It was agreed that we would head off haphazardly on foot, conversing all the while, and that our only planned detours would be for eating and sleeping. In actual practice, the project turned out to be quite peculiar, even fraught with danger. The trip, which was scheduled to last for about ten days, but which we finally cut short, immediately took an initiatory turn. The absence of any goal soon removed us from reality, gave rise beneath our feet to increasingly numerous and disturbing phantoms. . . . All things considered, the exploration was hardly disappointing, no matter how narrow its range, because it probed the boundaries between waking life and dream life.*[11]

NOTES

1 Octavio Paz, The Bow and the Lyre, trans. Ruth L. C. Simms (Austin, Texas: University of Texas Press, 1973), 215.

2 'Ainsi parvenons-nous à concevoir une attitude synthétique dans laquelle se trouvent conciliés le besoin de transformer radicalement le monde et celui de l'interpréter le plus complètement possible." André Breton, les vases communicants (Paris: Gallimard, 1955), 148. Translated into English by Mary Ann Caws and Geoffrey T. Harris under the title Communicating Vessels (Lincoln, Nebraska: University of Nebraska Press, 1990), 127.

3 André Breton, trans. Mark Polizzotti, Conversations: The Autobiography of Surrealism (New York: Paragon House, 1993), 83.

4 "C'est des poètes, malgré tout, dans la suite des siècles, qu'il est possible de recevoir et permis d'attendre les impulsions susceptibles de replacer l'homme au coeur de l'univers, de l'abstraire une seconde de son aventure dissolvante, de lui rappeler qu'il est pour toute douleur et toute joie extérieures à lui un lieu indéfiniment perfectible de résolution et d'echo." Breton, les vases communicants, 169–170; English translation: Caws and Harris, 146.

5 "la Poésie ne rythmera plus l'action; elle sera en avant." Arthur Rimbaud, "Lettre à Paul Demeny, 15 mai 1871 (Lettre du Voyant)," in Claude Edmonde Magny's Arthur Rimbaud Poètes d'Aujourd'-hui 12 (France: Pierre Seghers, Éditeur, 1956), 71.

6 Although the surrealists, because of their insistence on the reattachment of life to art and because of the revolutionary nature of their art, embraced the ideas of social revolution as delineated by Marxism, their dependency on pure subjective necessity over ideology placed them in a contentious contradictory position that eventually caused them to abandon the Communist group.

7 Paz, The Bow and the Lyre, 222.

8 André Breton, transl. Richard Seaver and Helen R. Lane, Manifestos of Surrealism (Ann Arbor: The University of Michigan Press, 1972), 123.

9 *Ibid.*, 26

10 Paz, The Bow and the Lyre, 222.

11 Breton, Conversations, 59–60.

ILLUSTRATION CREDITS

Plates 1–10: Images of Barcelona, all photographs by M. Miles, 2002.
1. Claes Oldenburg and Coosje van Bruggen, *Book-matches*, 1991, Parc de la Vall d'Hebron; 2. Museum of Contemporary Art, Barcelona (MACBA); 3. Old and new apartment blocks in el Raval: the absence of balconies in the new facades; 4. World Trade Centre, Port Vel; 5. Hotel des Arts and casino, 1992 waterfront development; 6. 'Beach Club'; 7. Parc de Poblenou, designed by X. Vendrell and M. Ruisànchez; 8. Preserved nineteenth-century brick chimney seen from mall development site; 9. The new urban landscape; 10. Beyond the construction site: the three chimneys of the power station at 'red Besòs'.

Plates 11–15: *Capital Arcade* by John Goto (originals in colour), reproduced by permission of the artist.
11. Welcome to Capital Arcade; 12. Next; 13. Mothercare; 14. B & Q; 15. McDonald's

Plates 16–23: *The Pursuit of Happiness* and other works by Marjetica Potrč, 2001, reproduced by permission of the artist and Max Protetch Gallery, New York.
16. Bear Falling (ink-jet print); 17. Coyote (ink-jet print); 18. Pool Bear (ink-jet print); 19. Racoon (ink-jet print); 20. Urban Bear (ink-jet print); 21. Emergency Drinking Water (installation); 22. Survival Kit (installation); 23. Pepper Spray for Bears (installation)

Plates 24–31: *Touch Sanitation*, 1979–84, by Mierle Laderman Ukeles, reproduced courtesy of Ronald Feldman Fine Arts, New York.
24. Ukeles and New York City Sanitation truck; 25. Handshake with garbage collector in cab; 26. Handshake with garbage collector at land-fill site; 27. The artist's hands; 28. Opening performance, 'Wiping Out the Bad Names', 9 September 1984 (multimedia installation); 29. Detail of performance, 'Wiping Out the Bad Names'; 30. Marine Transfer Station, 59th Street and Hudson River; 31. 'Maintenance City/Sanman's Place' installation at Ronald Feldman Fine Arts and the Marine Transfer Station, 59th Street and Hudson River.

Plates 32–39: *Ring der Erinnerung*, Herman Prigann, photographs by the artist and M. Miles, reproduced by permission of the artist.
32. Watchtower on the ex-German Democratic Republic side of the border; 33. Drawing for the project, 1990 (photo H. Prigann); 34. Site before construction of the Ring, 1992 (photo H. Prigann); 35. Construction of the Ring, 1992 (photo H. Prigann); 36. Raising withered trees, 1992 (photo H. Prigann); 37. Installation of stones at the entrance to the site, 1992 (photo H. Prigann); 38. The Ring in 2001; 39. Entrance stone, 2001

COPYRIGHT INFORMATION

Every effort has been made to contact copyright holders for their permission to reprint selections in this book. The publishers would be very grateful to hear from any copyright holder who is not here acknowledged and will undertake to rectify any errors or omissions in future editions of this book. Following is copyright information.

PART ONE

SIMMEL Georg Simmel, 'The Metropolis and Mental Life' [1903], from Kurt H. Wolff, ed. and trans., *The Sociology of Georg Simmel*, New York, 1964, pp. 409–24. Reprinted by permission of The Free Press, a division of Simon & Schuster Adult Publishing Group.

BURGESS Ernest W. Burgess, 'The Growth of a City: An Introduction to a Research Project', from Robert E. Park, Ernest W. Burgess and Roderick D. McKenzie, eds, *The City*, Chicago, 1925. Reprinted by permission of University of Chicago Press.

MUMFORD Lewis Mumford, 'What is a City?', from *Architectural Record* LXXXII, November 1937. Copyright © McGraw-Hill Inc., all rights reserved. Reproduced by permission of the publisher.

KRACAUER Siegfried Kracauer, 'The Hotel Lobby', from *The Mass Ornament: Weimar Essays*, trans. and ed. and with an introduction by Thomas Y. Levin, Cambridge, MA, Harvard University Press, 1995, pp. 179–85. Reprinted by permission of the publisher.

WILSON Elizabeth Wilson, 'World Cities', from *The Sphinx in the City: Urban Life, the Control of Disorder and Women*, Berkeley, University of California Press, 1991, pp. 121–34. Reprinted by permission of the Regents of the University of California and the University of California Press.

PART TWO

WILLIAMS Raymond Williams, 'Metropolitan Perceptions and the Emergence of Modernism', from *The Politics of Modernism*, London, Verso, 1999, pp. 37–49. Reprinted by permission of the publisher.

MARCUSE Herbert Marcuse, 'The Affirmative Character of Culture' [1937], from *Negations: Essays in Critical Theory*. © 1968 by Herbert Marcuse, translation from the German, © 1968 by Beacon Press. Reprinted by permission of Beacon Press, Boston.

DUNCAN Carol Duncan, 'The Art Museum as Ritual', from *Civilizing Rituals: Art Museums*, London, Routledge, 1995. Reprinted by permission of the publisher.

DEBORD Guy Debord, 'Separation Perfected', from *Society of the Spectacle*, 1983 edition, Detroit, Black and Red, non-copyright.

TOMLINSON John Tomlinson, 'Globalised Culture: The Triumph of the West?', from *Culture and Global Change*, ed. T. Skelton and T. Allen, London, Routledge, 1999, pp. 22–9. Reprinted by permission of the publisher.

PART THREE

BOURDIEU Pierre Bourdieu, 'The Production of Belief: Contribution to an Economy of Symbolic Goods', from *The Field of Cultural Production*, Oxford, Blackwell, 1993, pp. 74–112, extract pp. 82–93. Reprinted by permission of the publisher.

HALL Tim Hall, 'Opening Up Public Art's Spaces: Art, Regeneration and Audience', from *Cultures and Settlements*, ed. M. Miles and N. Kirkham, Bristol, Intellect Books, 2003. Reprinted by agreement with the author and Intellect Books.

ROSLER Martha Rosler, 'Fragments of a Metropolitan View'. © 1998, from *If You Lived Here . . .: the City in Art, Theory, and Social Activism*, ed. Martha Rosler and Brian Wallis, 1991, pp. 15–43 (extract pp. 31–40). Reprinted by permission of the New Press, New York (800) 233–4830.

CRAWFORD Margaret Crawford, 'The World in a Shopping Mall', from *Variations on a Theme Park*, Michael Sorkin, New York, Hill & Wang, 1992. Essay copyright © 1992 by Margaret Crawford. Compilation copyright © 1992 by Michael Sorkin. Reprinted by permission of Hill & Wang, a division of Farrar, Strauss and Giroux LLC.

FAINSTEIN Susan Fainstein, 'Justice, Politics and the Creation of Urban Space', from *The Urbanisation of Injustice*, ed. Andy Merrifield and Erik Swyngedouw, London, Lawrence & Wishart, 1996, pp. 18–44, reprinted by permission of the publisher.

PART FOUR

ADORNO Theodor W. Adorno, 'The Culture Industry Reconsidered', from *The Culture Industry*, ed. J. Bernstein, London, Routledge, 1991, pp. 85–92. Reprinted by permission of the publisher.

URRY John Urry, 'Reinterpreting Local Culture', from *Consuming Places*, London, Routledge, 1995, pp. 152–62. Reprinted by permission of the publisher.

DODD Dianne Dodd, 'Barcelona: The Making of a Cultural City', from *Planning Cultural Tourism in Europe*, ed. D. Dodd and A. van Hemel, Amsterdam, Boekmanstichting, 1999. Reprinted by permission of Boekmanstichting, study centre for arts, culture and related policy, Amsterdam, Netherlands.

HICKMAN Mary J. Hickman, 'Difference, Boundaries, Community: The Irish in Britain', from *Recoveries and Reclamations*, ed. Judith Rugg and Daniel Hinchcliffe, Bristol, Intellect Books, 2002, pp. 131–9. Reprinted by permission of the author and Intellect Books.

PHILLIPS Patricia Phillips, 'Out of Order: The Public Art Machine', from *Artforum*, December 1988, pp. 92–6, © Artforum. Reprinted by permission of the author and *Artforum*.

PART FIVE

SHELLER and URRY Mimi Sheller and John Urry, 'The City and the Car', from *International Journal of Urban and Regional Research*, 24, 4, 2000, pp. 737–57. Reprinted by permission of the editors and Blackwell Publishing Ltd.

GRAHAM and MARVIN Stephen Graham and Simon Marvin, 'Telecommunications and the City', from *Telecommunications and the City: Electronic Spaces, Urban Places* London, Routledge, 1995, pp. 209–13. Reprinted by permission of the publisher.

SCHELLING Vivian Schelling, 'The People's Radio of Vila Nossa Senhora Aparecida: Alternative Communication and Cultures of Resistance in Brazil', from *Culture and Global Change*, ed. Tracey Skelton and Tim Allen, London, Routledge, 1999, pp. 167–79. Reprinted by permission of the publisher.

TERRANOVA Tiziana Terranova, 'Posthuman Unbounded: Artificial Evolution and High-tech Subcultures', from *FutureNatural*, ed. George Robertson, Melinda Mash, Lisa Tickner, Jon Bird, Barry Curtis, and Tim Putnam, London, Routledge, 1996, pp. 165–80. Reprinted by permission of the publisher.

CRITICAL ART ENSEMBLE Critical Art Ensemble, 'Electronic Civil Disobedience', from *Electronic Disobedience and other Unpopular Ideas*, New York, Autonomedia, 1996, pp. 7–34. Reprinted by permission of Autonomedia.

PART SIX

LEFEBVRE Henri Lefebvre, part XVII of 'Plan of the Present Work', from *The Production of Space*, [1974] trans. Donald Nicholson-Lord, Oxford, Blackwell, 1991, pp. 38–46, translation © Donald Nicholson-Lord 1991. Reprinted by permission of the publisher.

DE CERTEAU Michel de Certeau, 'Railway Navigation and Incarceration', from *The Practice of Everyday Life*, trans. Steven Rendell, Berkeley, University of California Press, 1984, pp. 111–14, © The Regents of the University of California 1984. Reprinted by permissions of the Regents of the University of California and the publisher.

LIPPARD Lucy Lippard, 'Home in the Weeds', from *The Lure of the Local: Senses of Place in a Multicentred Society*, New York, The New Press, 1997, pp. 216–24. Copyright © 1998 *The Lure of the Local: Senses of Place in a Multicentred Society* by Lucy Lippard. Reprinted by permission of The New Press (800) 233–4830.

BETANCOUR and HASDELL Ana Betancour and Peter Hasdell, 'Tango: A Choreography of Urban Displacement', from *White Papers, Black Marks*, ed. Lesley N. N. Lokko, London, Athlone Press, an imprint of Continuum, 2000, pp. 148–75 (extract 148–9, 151–9, 162–7). Reprinted by permission of Continuum.

BORDEN Iain Borden, 'A Performative Critique of the City: The Urban Practice of Skateboarding', first published in *Everything*, 2, 4, March 1999, © Iain Borden. Reprinted by permission of the author.

PART SEVEN

MASSEY Doreen Massey, 'Space, Place and Gender', from *Space, Place and Gender*, Oxford, Blackwell, 1994, pp. 185–90. Reprinted by permission of the publisher.

YOUNG Iris Marion Young, 'Social Movements and the Politics of Difference', Princeton, Princeton University Press, 1990, pp. 156–90 (extract 158–72). Reprinted by permission of Princeton University Press.

CURTIS Kimberley Curtis, 'World Alienation and the Modern Age: The Deprivations of Obscurity', from *Our Sense of the Real: Aesthetic Experience and Arendtian Politics*, Ithaca, Cornell University Press, 1992, pp. 67–92 (extract pp. 67–75), Copyright © Cornell University Press. Used by permission of the publisher.

TORRE Suzana Torre, 'Changing the Public Space: The Mothers of Plaza de Mayo', from *The Sex of Architecture*, edited by D. Agrest, P. Conway, and L. K. Weisman, New York, Henry N. Abrams Inc., 1996, pp. 241–50. Reprinted by permission of the publisher.

GHANNAM Farha Ghannam, 'Re-Imagining the Global: Relocation and Local Identities in Cairo', from *Space, Culture and Power*, edited by Ayşe Öncü and Petra Weyland, London, Zed Books, 1997, pp. 119–39. Reprinted by permission of Zed Books.

PART EIGHT

ILLICH Ivan Illich, 'The Dirt of Cities, The Aura of Cities, The Smell of the Dead, Utopia of an Odorless City', from *H_2O and the Waters of Forgetfulness*, London, Marion Boyars, 1986, pp. 47–58. Reprinted by permission of Marion Boyars Publishers Ltd.

SIBLEY David Sibley, 'Border Crossings', from *Geographies of Exclusion*, London, Routledge, 1995, pp. 32–48. Reprinted by permission of the publisher.

ANDERSON Kay Anderson, 'Engendering Race Research: Unsettling the Self–Other Dichotomy', in *Body-Space*, ed. Nancy Duncan, London, Routledge, 1996. Reprinted by permission of the publisher.

IRIGARAY Luce Irigaray, part 1 of 'A Chance to Live', from *Thinking the Difference: For a Peaceful Revolution*, London, Athlone Press, an imprint of Continuum, 1994, pp. 8–14. Reprinted by permission of Continuum.

MACCANNELL Dean MacCannell, 'New Urbanism and its Discontents', from *Giving Ground: The Politics of Propinquity*, ed. Joan Copjec and Michael Sorkin, London, Verso, 1999. Reprinted by permission of Verso.

PART NINE

SHKLAR Judith Shklar, 'The Political Theory of Utopia: From Melancholia to Nostalgia', from *Utopias and Utopian Thought*, ed. Frank E. Manuel, London, Souvenir Press, 1973, pp. 101–15. Reprinted by permission of the publisher.

GEOGHEGAN Vincent Geoghegan, 'Ernst Bloch and the Ubiquity of Utopia', from *Utopianism and Marxism*, London, Methuen, 1987, pp. 87–97, rights owned by Routledge. Reprinted by permission of Routledge.

DEUTSCHE Rosalyn Deutsche, 'Uneven Development: Public Art in New York City', from *Evictions: Art and Spatial Politics*, Cambridge, MA, MIT, 1996, pp. 49–107 (extract pp. 49–52) [previously published in *October* 47, winter 1980, and in *Out of Site: A Social Criticism of Architecture*, ed. Dianne Ghirardo, Seattle, Bay Press, 1991, pp. 157–219; the text is in a revised version in the 1996 MIT edition]. © Massachusetts Institute of Technology. Reprinted by permission of the author and MIT Press.

SANDERCOCK Leonie Sandercock, 'The Death of Radical Planning: Radical Praxis for a Postmodern Age', from *Cities for Citizens*, ed. Mike Douglass and John Friedman, Chichester, Wiley, 1998, pp. 163–84. Reprinted by permission of J. Wiley & Sons Ltd.

MILES Malcolm Miles, 'Café-Extra: Culture, Representation and the Everyday', from *Beyond the Museum: Art, Institutions, People*, ed. Nick Stanley and Ian Cole, Oxford, Museum of Modern Art, 2001, pp. 30–7; © the author. Reprinted by permission of the author and with agreement of the Museum of Modern Art, Oxford.

PART TEN

SOJA Edward Soja, 'Exopolis: The Restructuring of Urban Form', from *Postmetropolis*, Oxford, Blackwell, 2000, pp. 233–63 (extracts pp. 234–5 and 238–46). Reprinted by permission of the publisher.

HAUGHTON Graham Haughton, 'Environmental Justice and the Sustainable City', from *Sustainable Cities*, ed. David Satterthwaite, London, Earthscan, 1999, pp. 62–79. Reprinted by permission of Kogan Page.

SEABROOK Jeremy Seabrook, 'The Urban Poor: An Invisible Resource', from *Pioneers of Change, Experiments in Creating a Humane Society*, London, Zed Books, 1993, pp. 77–90. Reprinted by permission of Zed Books.

WEINBERG Bill Weinberg, '¡Viva Loisaida Libre!', from *Avant Gardening*, ed. Peter Lamborn Wilson and Bill Weinberg, New York, Autonomedia, pp. 38–56. Reprinted by permission of Autonomedia.

PENDLETON-JULLIAN Ann M. Pendleton-Jullian, part 3, '*Arquitectura co-generada con la poesía, porque la palabra es inaugural, lleva, da a luz*' ('Architecture Co-generated with Poetry, because the Word is Inaugural, it Conveys, it Gives Birth'), from *The Road That is Not a Road and the Open City at Ritoque, Chile*, Cambridge, MA, MIT, 1996, pp. 23–45, © Massachusetts Institute of Technology. Reprinted by permission of MIT Press.

Index